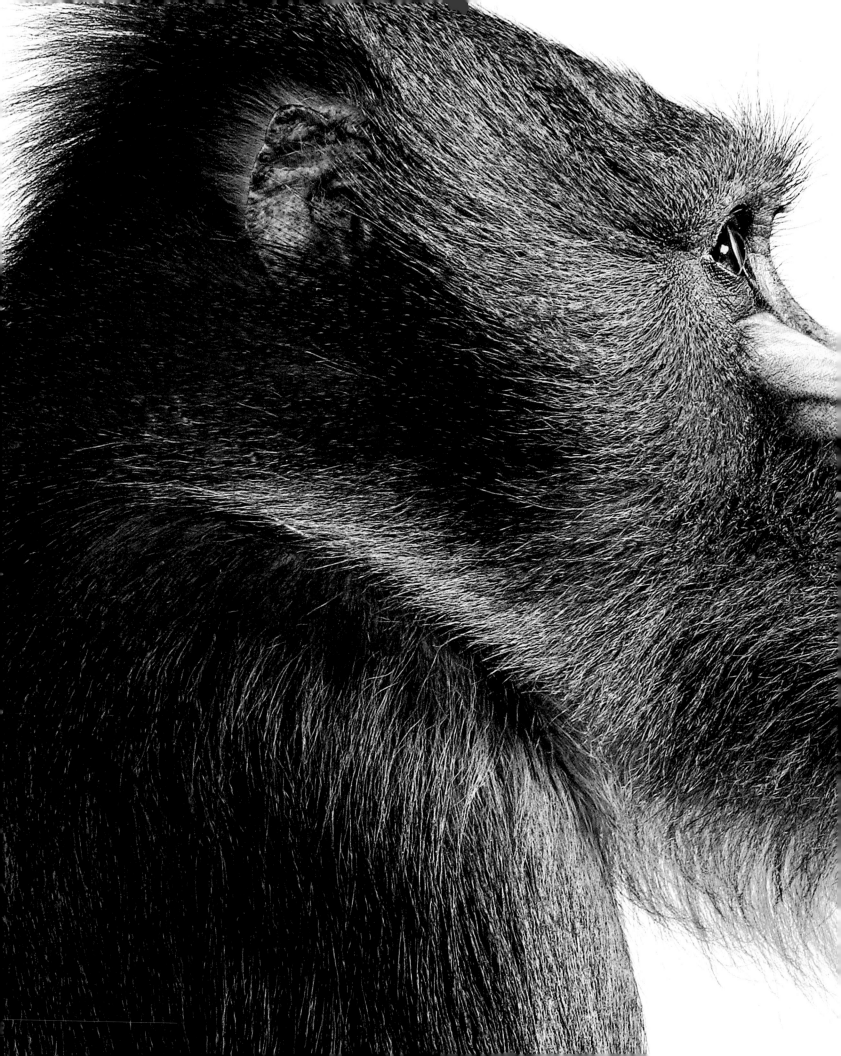

NATURAL

THE ULTIMATE VISUAL GUIDE TO EVERYTHING ON EARTH

HISTORY

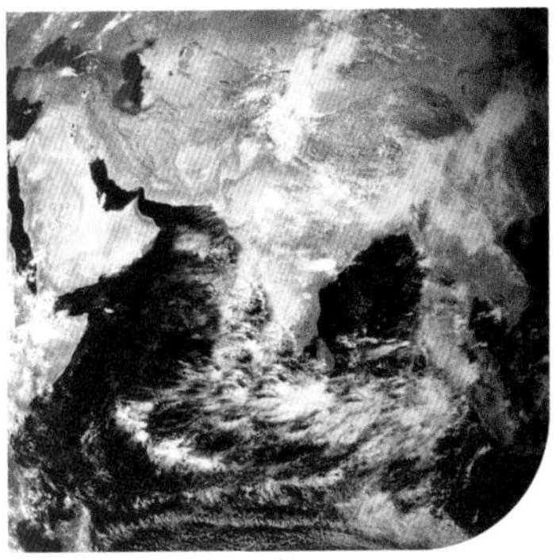

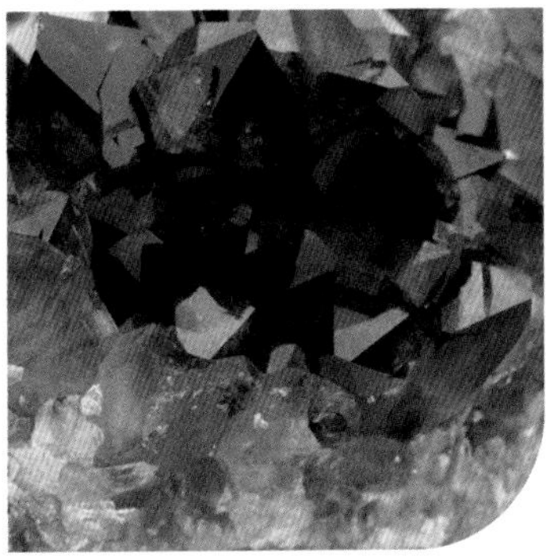

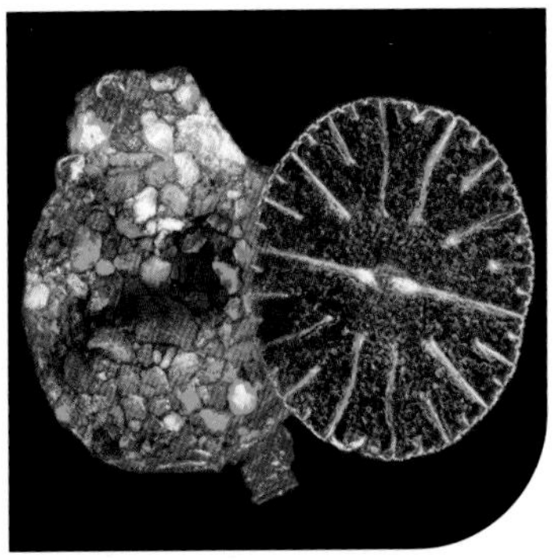

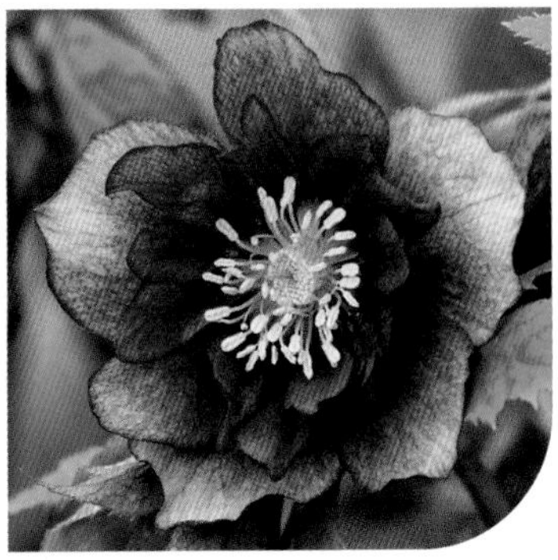

LIVING EARTH

MINERALS, ROCKS, AND FOSSILS

MICROSCOPIC LIFE

PLANTS

CONTENTS

DK | Penguin Random House

SECOND EDITION

SENIOR EDITOR Gill Pitts
SENIOR ART EDITOR Ina Stradins
US EDITOR Karyn Gerhard
MANAGING EDITOR Angeles Gavira Guerrero
MANAGING ART EDITOR Michael Duffy
ART DIRECTOR Karen Self
ASSOCIATE PUBLISHING DIRECTOR Liz Wheeler
PUBLISHING DIRECTOR Jonathan Metcalf
DESIGN DIRECTOR Phil Ormerod
SPECIAL PHOTOGRAPHY Gary Ombler
MEDIA ARCHIVE Romaine Werblow
PRODUCTION EDITOR Gillian Reid
SENIOR PRODUCTION CONTROLLER Meskerem Berhane

DK INDIA

PROJECT EDITOR Tina Jindal
SENIOR ART EDITOR Mahua Mandal
EDITOR Nidhilekha Mathur
ART EDITOR Karan Chaudhary
PROJECT PICTURE RESEARCHER Deepak Negi
PICTURE RESEARCH MANAGER Taiyaba Khatoon
SENIOR MANAGING EDITOR Rohan Sinha
MANAGING ART EDITOR Sudakshina Basu

SENIOR DTP DESIGNER Jagtar Singh
DTP DESIGNER Jaypal Chauhan
PRE-PRODUCTION MANAGER Balwant Singh
PRODUCTION MANAGER Pankaj Sharma
JACKET DESIGNER Tanya Mehrotra
DK INDIA EDITORIAL HEAD Glenda Fernandes
DK INDIA DESIGN HEAD Malavika Talukder

FIRST EDITION

SENIOR PROJECT EDITOR Kathryn Hennessy
PROJECT EDITOR Victoria Wiggins
SENIOR ART EDITORS Gadi Farfour, Helen Spencer
EDITORS Becky Alexander, Ann Baggaley, Kim Dennis-Bryan, Ferdie McDonald, Elizabeth Munsey, Peter Preston, Cressida Tuson, Anne Yelland
US EDITORS Jill Hamilton, Christine Heilman, Jane Perlmutter
DESIGNERS Paul Drislane, Nicola Erdpresser, Phil Fitzgerald, Anna Hall, Richard Horsford, Stephen Knowlden, Dean Morris, Amy Orsborne, Steve Woosnam-Savage
SPECIAL PHOTOGRAPHY Gary Ombler
PICTURE RESEARCH Neil Fletcher, Peter Cross, Julia Harris-Voss, Sarah Hopper, Liz Moore, Rebecca Sodergren, Jo Walton, Debra Weatherley, and Suzanne Williams
DK PICTURE LIBRARY Claire Bowers
DATABASE Peter Cook, David Roberts
PRODUCTION EDITOR Tony Phipps

SENIOR PRODUCTION CONTROLLER Inderjit Bhullar
MANAGING EDITOR Camilla Hallinan
MANAGING ART EDITOR Karen Self
ART DIRECTOR Phil Ormerod
ASSOCIATE PUBLISHER Liz Wheeler
REFERENCE PUBLISHER Jonathan Metcalf

DK INDIA

MANAGING EDITOR Rohan Sinha
ART DIRECTOR Shefali Upadhyay
PROJECT MANAGER Malavika Talukder
PROJECT EDITOR Kingshuk Ghoshal
PROJECT ART EDITOR Mitun Banerjee
EDITORS Alka Ranjan, Samira Sood, Garima Sharma
ART EDITORS Ivy Roy, Mahua Mandal, Neerja Rawat
PRODUCTION MANAGER Pankaj Sharma
DTP COORDINATOR Sunil Sharma
SENIOR DTP DESIGNERS Dheeraj Arora, Jagtar Singh, Pushpak Tyagi

This American Edition, 2021
First American Edition, 2010
Published in the United States by DK Publishing
1450 Broadway, Suite 801, New York, NY 10018

Copyright © 2010, 2021 Dorling Kindersley Limited
DK, a Division of Penguin Random House LLC
21 22 23 24 25 10 9 8 7 6 5 4 3 2 1
001–315218–Sept/2021

All rights reserved.
Without limiting the rights under the copyright reserved above, no part of this publication may be reproduced, stored in or introduced into a retrieval system, or transmitted, in any form, or by any means (electronic, mechanical, photocopying, recording, or otherwise), without the prior written permission of the copyright owner.
Published in Great Britain by Dorling Kindersley Limited

A catalog record for this book is available from the Library of Congress.
ISBN 978-0-7440-3501-8

DK books are available at special discounts when purchased in bulk for sales promotions, premiums, fund-raising, or educational use. For details, contact: DK Publishing Special Markets, 1450 Broadway, Suite 801, New York, NY 10018
SpecialSales@dk.com

Printed and bound in China

www.dk.com

FSC www.fsc.org MIX Paper from responsible sources FSC™ C018179

This book was made with Forest Stewardship Council ™ certified paper—one small step in DK's commitment to a sustainable future. For more information go to www.dk.com/our-green-pledge

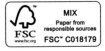

FUNGI

ANIMALS

Smithsonian

Established in 1846, the Smithsonian Institution—the world's largest museum and research complex—includes 19 museums and galleries and the National Zoological Park. The total number of objects, works of art, and specimens in the Smithsonian's collections is estimated at 155.5 million, the bulk of which is contained in the National Museum of Natural History, which holds more than 126 million specimens and objects. The Smithsonian is a renowned research center, dedicated to public education, national service, and scholarship in the arts, sciences, and history.

SMITHSONIAN CONSULTANTS

Dr. Stephen Cairns, Dr. Allen Collins, Dana M. De Roche, Dr. Carla Dove, Leslie Hale, Dr. M. G. (Jerry) Harasewych, Gary Hevel, Dr. Rafael Lemaitre, Dr. Chris Meyer, Dr. Jon Norenburg, Dr. David L. Pawson, Paul Pohwat, Dr. Jeffrey E. Post, Dr. Klaus Rutzler, Dr. Hans-Dieter Sues, Dr. Michael Vecchione, Dr. Warren Wagner, Dr. Jeffrey T. Williams, Dr. Don E. Wilson, Dr. George Zug.

ADDITIONAL CONSULTANTS

Dr. Matthew D. Kane, Dr. James D. Lawrey, Dr. Diana Lipscomb, Dr. Robert Lücking, Dr. Thorsten Lumbsch, Andrew M. Minnis, Dr. Ashleigh Smythe, Dr. William B. Whitman.

SMITHSONIAN ENTERPRISES

PRODUCT DEVELOPMENT MANAGER
Kealy Gordon

DIRECTOR, LICENSED PUBLISHING
Jill Corcoran

DMM, ECOM AND D-TO-C
Janet Archer

PRESIDENT, SMITHSONIAN ENTERPRISES
Carol LeBlanc

CONSULTANT EDITOR

David Burnie is a former winner of the Aventis Prize for Science Books, and the editor of DK's highly successful *Animal*. He has written or contributed to more than 100 books and is a fellow of the Zoological Society of London.

CONTRIBUTORS

Richard Beatty, Dr. Amy-Jane Beer, Dr. Charles Deeming, Dr. Kim Dennis-Bryan, Dr. Frances Dipper, Dr. Chris Gibson, Derek Harvey, Professor Tim Halliday, Rob Hume, Geoffrey Kibby, Dr. Richard Kirby, Joel Levy, Chris Mattison, Felicity Maxwell, Dr. George C. McGavin, Dr. Pat Morris, Dr. Douglas Palmer, Dr. Katie Parsons, Chris Pellant, Helen Pellant, Michael Scott, Carol Usher, Professor Mark Viney, Dr. David J. Ward, Dr. Elizabeth Wood.

FOREWORD

When I was a boy, I would visit the local public library and spend as long as I was allowed looking at science and reference books. The books that especially enthralled me were the forerunners of this book. Packed with color diagrams, images of exotic species, and faraway places, coupled with informative text, they were the siren call for a lifetime of studying and teaching. The natural world around me was then an unknown entity and I wanted to learn about all of it. It is a very human desire to learn, but nobody knows which bit of biology they will be most attracted to and I went on to pick the largest and most diverse animal group of all. If I were to live a lifetime of lifetimes, I would never know everything there is to know about insects.

No book could ever be a comprehensive study of Earth's extant life forms, which incidentally comprise only one percent of those that have ever lived. For that you would need a library of hefty tomes. What you have in your hands is much more serviceable. It is a clear roadmap to guide you on a journey that will take you from the steaming forests of the tropics to the freezing cold of the polar regions, and from mountain peaks to the depths of the ocean. It is an abridged exploration of the products of a couple of billion years of evolution that will show you some of the several million ways of living on Earth.

It is relatively easy to write screeds of detailed information on any topic you care to name, but this approach is not going to be of use to anyone other than experts and it is unlikely to encourage anyone else to study the living world. No, the real skill and the very purpose of this book is to distil a vast body of natural history knowledge amassed by countless researchers to provide an authoritative, yet accessible, account.

There may come a time in the distant future when we are able to venture elsewhere in our galaxy and beyond. Will we find extraterrestrial life on a far distant rocky planet and if so, will it be simple, single-celled organisms or complex creatures to rival humans or humpbacked whales? To imagine that in the vastness of the Cosmos, our inconsequential planet is the single, solitary place where life has evolved—an unbelievable fluke of stellar serendipity—seems vanishingly improbable. What is certain is that we inhabit a planet of staggering complexity and breathtaking beauty, and one that is constantly changing. What we must realize, and quickly, is that we have no dominion over it but are simply part of an immense community of interdependent species.

Newcomers to natural history will find themselves engrossed and even specialists, steeped in the minutiae of one tiny part of the tree of life, will value the majestically broad sweep of this meticulously researched volume. If I could have my life all over again, I'd start by reading this.

DR. GEORGE McGAVIN

Honorary Research Associate,
Oxford University Museum of Natural History
Senior Principal Research Fellow, Imperial College London

ABOUT THIS BOOK

The Natural History Book begins with a general introduction to life on Earth: the geological foundations of life, the evolution of life forms, and how organisms are classified. The next five chapters form an extensive, accessible catalog of specimens—from minerals to mammals—interspersed with fact-filled introductions to each group and in-depth feature profiles.

for easy reference, visual contents panels list the subgroups within each section, and the page number where each subgroup can be found

SECTION INTRODUCTION >
Each chapter is divided into sections representing major taxonomic groupings. The section introduction highlights the characteristics and behaviors that define the group, and discusses their evolution over time.

on each introduction, classification boxes display the current taxonomic hierarchy—the level of the group under discussion is highlighted

PHYLUM	CHORDATA
CLASS	REPTILIA
ORDERS	4
FAMILIES	92
SPECIES	11,050

debate boxes tackle scientific controversies and taxonomic discussions arising from new discoveries

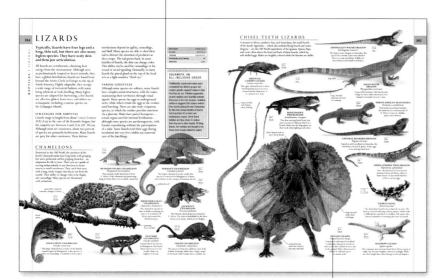

∧ GROUP INTRODUCTION
Within each section—for example, reptiles—lower-ranking taxonomic groups, such as lizards, are explored. Key features are described, including their distribution, habitat, physical characteristics, life cycle, behavior, and reproductive habits.

species-specific information accompanies each image

a male of the species ♂   ♀ *a female of the species*

SPECIES CATALOG >
Pictorial galleries profile some 5,000 species, showing the distinctive visual features of each one. Closely related species are placed together for useful comparison, while essential data highlights unique and interesting aspects of each organism.

*data sets display key
details at a glance,
such as size, habitat,
distribution, and diet*

SIZE 4½–9½ ft (1.4–2.9 m)
HABITAT Forests, swamps, scrub thickets,
savanna, and rocky landscapes
DISTRIBUTION India to China, Siberia, the Malay
peninsula, and Sumatra
DIET Mainly hoofed animals, such as deer and
pigs; may also catch smaller mammals and birds

⌄ FEATURE PROFILE
Zooming in on single specimens,
feature profiles use close-up
photographs to provide in-depth
portraits of some of the world's
most spectacular species.

*each feature includes a side
profile of the animal, plant, or fungus*

TIGER
Panthera tigris

463

ST VINCENT PARROT
Amazona guildingii
F: Psittacidae

*species' common names are
highlighted in bold, scientific
names are in italics; in some
cases the family (F) name
is given below*

12 in
30 cm

*size boxes give the most
appropriate measurement
for the organism they
accompany (see panel, right)*

MEASUREMENTS

The approximate sizes of the organisms in this book are given in
data sets and size boxes. Below is a list of the dimensions used:

MICROSCOPIC LIFE
Length

PLANTS
Maximum height above ground, except for:
Height above water rushes
Spread aquatic plants

FUNGI
Width (of widest part), except for:
Height stinkhorn, dog stinkhorn

INVERTEBRATES
Adult body length, except for:
Height sponges, feather stars, common hydra, feather
hydroid, pink-hearted hydroid, fire "coral" hydroid, tall
anemones, tall corals
Diameter snailfur, blue button, eight-ribbed hydromedusa,
cup hydromedusa, freshwater jellyfish
Diameter, excluding spines echinoderms
Diameter of medusa jellyfish
Wingspan butterflies and moths
Length of colony bryozoans
Length of shell mollusks, shelled gastropods
Spread of tentacles octopuses

FISHES, AMPHIBIANS, AND REPTILES
Adult body length from head to tail

BIRDS
Adult body length from bill to tail

MAMMALS
Adult body length, excluding tail, except for:
Height to shoulder elephants, apes, even-toed ungulates,
odd-toed ungulates

PLANT ICONS

The basic shape of all trees, shrubs, and woody plants is
described using one of the following symbols. Herbaceous
perennials which die back each winter are not given symbols.

TREES	SHRUBS
Broadly columnar	Bushy, mound-forming
Broadly conical	Bushy, suckering
Large weeping	Compact, bushy
Small weeping	Erect, treelike
Multi-stemmed tree	Loose, open
Narrowly columnar, flame-shaped	Open, spreading
Narrowly columnar	Rounded, bushy
Narrowly conical	Spreading, prostrate
Rounded, broadly columnar	Upright
Rounded, broadly spreading	Upright, arching
Single-stemmed palm	Upright, vigorous, bushy
Multi-stemmed palm, cycad or similar	Sprawling, climbing

ABBREVIATIONS

SP.: species (used where species name is unknown)
MYA: million years ago
H: hardness of a mineral, measured on the Moh's scale
SG: specific gravity—a mineral's density is measured by
comparing its weight to that of an equal volume of water

LIVING EARTH

Our blue planet, spinning in the vastness of space, is the only proven home of living things. Over nearly 4 billion years, life has evolved from the simplest of beginnings. Most species have become extinct, but life itself has flourished, endlessly diversified, and recovered from repeated extinctions. The result is an extraordinary variety of living things, which scientists continue to study as they piece together the story of life on Earth.

A LIVING PLANET

Earth is uniquely equipped to support a wide diversity of life, both on land and in the seas. Without heat and light from the sun, plentiful supplies of water, the protection provided by the atmosphere, and chemicals from the rocks that form the basis of Earth's ecosystems, life would perish.

DYNAMIC EARTH

Within our solar system, Earth seems to be uniquely placed to support abundant life. The third planet from the sun, Earth is neither too close to nor too far from the sun's heat. It therefore retains an outer atmosphere of oxygen and other gases, and a hydrosphere of plentiful surface water. Together, these form a protective, insulating layer that enables life to flourish. In contrast, the other planets in the solar system are either too hot or too cold, and devoid of the levels of water and oxygen required to support detectable life.

Earth has a layered structure, with an extremely hot, solid metallic core at its center, surrounded by an outer molten layer. This, in turn, is surrounded by a thick and hot silicate mantle, which rises to a thin, cool, and brittle outer crust. The mantle is constantly churned by heat rising from the core, which generates new ocean floor rocks and breaks the crust into large "plates." Over geological time, some oceans expand, carrying continents with them. Other oceans become smaller as old, cool crust descends into the mantle. Continents collide to form mountain ranges. Life has had to adapt to the constant change in the Earth's environment.

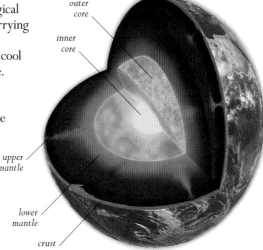

outer core

inner core

upper mantle

lower mantle

crust

EARTH'S STRUCTURE >
The liquid mantle is constantly stirred by heat rising from the core. This moves the plates of the outer crust, causing earthquakes and volcanic eruptions on the surface.

SUN AND MOON

The sun and the moon both have a direct impact on life on Earth. Without the sun's energy, in the form of heat and light, there would be no life. Solar energy heats Earth's atmosphere, the oceans, and the land, producing our varied climate. Because Earth rotates at an angle while it orbits the sun, the sun's radiant energy is unevenly distributed over its surface. This results in daily, seasonal, and annual variations in light, heat, and living conditions for plants and animals. Even at the equator, there are marked temperature changes between day and night. The orbit of Earth's satellite moon and its gravitational pull raise tides in Earth's oceans and seas. Tidal cycles are especially influential on coastal life, which has to adapt to changing conditions.

∧ SOLAR FLARES
The sun's energy is dramatically released from the surface in periodic explosions, which heat its atmosphere to form solar flares of hot ionized gas.

WATER AND LIFE
Life depends upon water, which forms more than 50 percent of all living tissues. Most rainfall is driven by evaporation from the oceans, which hold 97 percent of Earth's surface water, and a vital network of rivers flow in all but the world's hottest, coldest, and driest places.

FRAGILE ATMOSPHERE

Earth's atmosphere is 75 miles (120 km) thick. It is made up of several layers, each with its own temperature and gas composition. Its density decreases with height, until it becomes the sparse, outermost layer, called the ionosphere. The ozone layer, in the lower atmosphere, plays a vital role in protecting life, because it absorbs harmful radiation such as ultraviolet light, which damages living cells. Before the ozone layer formed, life was confined to the seas, whose waters offered some protection against ultraviolet light.

The majority of water vapor and weather activity is restricted to the lowest 10 miles (16 km) of atmosphere, known as the troposphere. Earth's surface water and gaseous atmosphere interact to recycle water from the surface into the atmosphere and—through clouds, rain, and snow— redistribute it over the land and sea. From the land, water flows back into the sea, although large quantities are held back in lakes, ice, and under the ground.

∧ BLUE PLANET
About two-thirds of Earth's surface is covered with water, which supports the abundance and diversity of life.

∧ LAYERS OF ATMOSPHERE
Earth is surrounded by a thin, layered atmosphere, composed of water vapor and various gases, which trap solar energy and heat the surface.

ATMOSPHERIC GASES >
Nitrogen and oxygen make up over 99 percent of Earth's atmosphere, along with small but important volumes of water vapor, carbon dioxide, and several other gases.

other gases, including carbon dioxide, methane, and ozone
argon 0.9%
nitrogen 78%
oxygen 21%

VARIED ROCKS

There are about 500 different kinds of rocks on Earth, made up of varying combinations of thousands of naturally occurring minerals. All rocks have a specific composition and properties, and can be divided into three main categories: igneous rocks were originally molten; sedimentary rocks are deposited on the Earth's surface; and metamorphic rocks result from the alteration of existing rocks within Earth's crust. These different types of rocks are exposed on the surface by a mixture of uplift, driven by Earth's moving crust, and surface processes such as weathering and erosion. Erosion also modifies the rocks to produce multiple kinds of landforms, soils, and sediments. These are the inorganic elements on which life depends.

IGNEOUS ROCKS
The cooling and solidification of molten rock produces crystalline igneous rocks. Their composition and texture vary—rapid cooling produces fine grained rocks, and slow cooling produces coarse grained rocks.

BASALT

METAMORPHIC ROCKS
The application of heat and pressure to existing rocks deep within Earth's crust can change their form and mineral composition, resulting in metamorphic rocks such as slate, schist, and marble.

BIOTITE SCHIST

SEDIMENTARY ROCKS
Layers of sediment and the remains of plants and animals are deposited by wind and water. Over time, older and deeper buried sediments undergo compression and chemical change which transforms them into rock.

SANDSTONE

ACTIVE EARTH

Earth's surface is constantly changing, thanks to dynamic geological processes driven by its internal heat energy. The plates of Earth's brittle outer crust are always in motion, altering the shape of oceans and continents as they do so.

PLATE TECTONICS

Over geological time, Earth's surface—and the distribution and size of the continents and oceans—has constantly changed, driven by the process of plate tectonics. The cool and brittle rocks of Earth's outermost crust are broken into a number of semirigid slabs known as tectonic plates. There are seven major continent-sized plates and about a dozen smaller ones. Over time, crustal plates have jostled against one another. Hot magma rising from the mantle erupts from rifts in the crust

to form new ocean floor crust. This spreads away on either side of the rift to form divergent plate boundaries, mainly beneath the oceans. And, since Earth itself cannot expand, the creation of new oceanic crust requires that the crust is shortened elsewhere by the same amount. This reduction occurs at convergent boundaries where either one plate overrides the other—a process known as subduction—or the plate margins are compressed and buckled to form mountains.

∧ SAN ANDREAS FAULT
Stretching some 810 miles (1,300 km) through California, this dramatic fault is the product of a transform boundary between the Pacific and North American plates, which slide against one another.

ridge appears at site of new plate formation

plates moving apart

DIVERGENT BOUNDARY
As plates are pulled apart they stretch and break, forming faulted rifts and volcanically active ridges.

∨ PLATE BOUNDARIES
This map shows the main plates that form the jigsaw surface of Earth's crust. A study of earthquake locations worldwide established where the plate boundaries lie.

MAP KEY
- ▬ Convergent boundary
- ▬ Divergent boundary
- ⋯ Deep-sea trench
- ▬ Transform boundary

the thinner, denser plate is forced down into Earth's crust

plates moving together

SUBDUCTION ZONE
Where two plates converge and one plate is thicker, it pushes the thinner plate beneath it in a process called subduction.

plates moving past each other

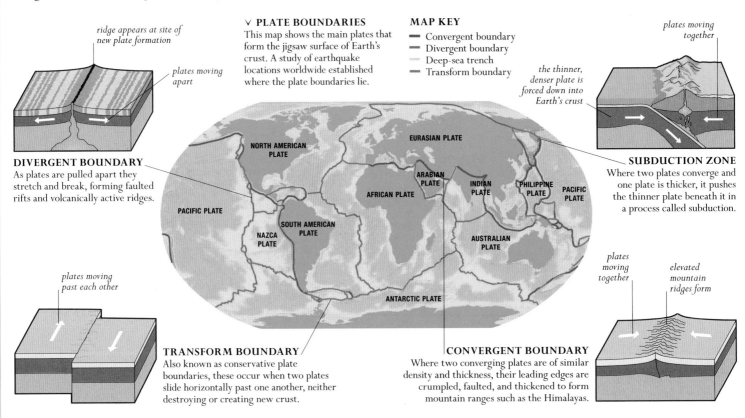

NORTH AMERICAN PLATE

EURASIAN PLATE

ARABIAN PLATE

AFRICAN PLATE

INDIAN PLATE

PHILIPPINE PLATE

PACIFIC PLATE

PACIFIC PLATE

NAZCA PLATE

SOUTH AMERICAN PLATE

AUSTRALIAN PLATE

ANTARCTIC PLATE

plates moving together

elevated mountain ridges form

TRANSFORM BOUNDARY
Also known as conservative plate boundaries, these occur when two plates slide horizontally past one another, neither destroying or creating new crust.

CONVERGENT BOUNDARY
Where two converging plates are of similar density and thickness, their leading edges are crumpled, faulted, and thickened to form mountain ranges such as the Himalayas.

< FOLD MOUNTAINS
The intense pressure caused by converging plate margins folds, faults, and thickens the rocks of the crust. Crustal thickening elevates mountain ranges.

ACTIVE VOLCANOES >
Most volcanoes form at plate margins. Deep down, rocks melt to produce hot magma which rises and erupts at the surface. Even dormant volcanoes may one day erupt as plates shift beneath them.

MOUNTAINS AND VOLCANOES

One of the major factors controlling the movement and distribution of life on Earth is its varied topography—its surface features—including the towering mountains and volcanoes on land and under the sea. On land, mountains not only hinder the movement of wildlife but alter weather, climate, and local plant life, which in turn impacts upon animal life. Active volcanoes also affect their surroundings when they erupt—initially by destroying life, but also, in the longer term, when the weathering and erosion of erupted lava and ash provide new mineral nutrients that fertilize the area. Submarine mountains and volcanic eruptions affect the movement of water, marine life, and the fertility of ocean waters.

WORN AWAY
Weathering by wind and water can result in the significant erosion of rocks and the sculpting of land surfaces, as seen here in the dramatic landscape of Bryce Canyon, Utah.

WEATHERING AND EROSION

Many rocks are formed below Earth's surface, and when they are exposed by uplift from pressure within Earth's crust or by retreating seas or rivers, they react in many ways with the atmosphere, water, and living organisms. The physical and chemical processes resulting from the interaction of rocks and minerals with the atmosphere are known as weathering. The processes by which rock material is loosened, dissolved, and transported away is known as erosion. The combination of weathering and erosion wears down Earth's rocky surfaces, layer by layer. Exposed rocks on mountain tops and on the exterior of buildings, for example, are subject to chemical weathering by acid rain and to physical weathering by temperature change and the splintering freeze-thaw effect of ice formation. Bare rock surfaces may also be physically eroded by sand particles carried by the wind. The combined effect of weathering and erosion dissolves some rocks, and reduces others to fragments. As rock debris is broken down and transported by wind, water, and ice, the sediment created becomes increasingly available to life forms, providing important mineral nutrients and a new surface for anchorage and growth.

< LANDSLIDE, RIO DE JANEIRO
Even with a well-developed plant cover, the impact of high rainfall on an area of steep slopes has the potential to cause landscape-altering and even life-threatening occurrences such as avalanches and landslides.

SOIL FORMATION

The production of a soil requires the initial weathering and erosion of the parent rock, which is broken down into small, mineral-bearing particles known as regolith. The addition of humus—organic matter formed from the remains of plants and animals—forms the basis for a soil. The soil in turn becomes the bed in which more life flourishes.

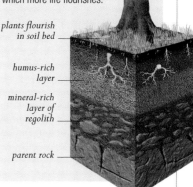

plants flourish in soil bed

humus-rich layer

mineral-rich layer of regolith

parent rock

CHANGING CLIMATES

The features of our seasons—for example, hot, dry summers and cold, icy winters—make up the climate of each region. Earth's climate has always changed at different rates, places, and times, and this variation in conditions has considerable and continuing impact upon the evolution of life.

ARCTIC FOX
IN SUMMER

ARCTIC FOX
IN WINTER

WHAT IS CLIMATE?

Climate is a region's average weather over a long period of time, produced by all the atmospheric conditions such as temperature, rainfall, wind strength, and pressure. The climate of any given place is also partly controlled by other factors such as human input, height above sea level, local topography, proximity to seas and oceans—with their prevailing winds and water currents—and, most importantly, its latitudinal position between the equator and pole. Latitude controls the amount of solar radiation received

in different regions of the world. For instance, there is a major difference between climates of polar regions, which receive the least light and heat, and of tropical regions around the equator, which receive the most.

CHANGING CONDITIONS

Global climates are generally classified according to the average temperature and the amount of rainfall each area receives, and their combined effect on plant growth. For instance, equatorial regions at present are hot and wet because oceans dominate there, whereas deserts are dry and polar regions are cold. However, this has not always been the case. Factors controlling weather conditions over geological time have affected the climate of the planet, from ice ages through to global warming.

< CHANGING TREE LINE
As the air temperature declines with increasing altitude, plant life changes. Broad-leaved trees are replaced by conifers, then shrubs.

∧ SEASONAL ADAPTATION
Annual climate change can bring acutely different living conditions from one season to the next. Animals and plants have various means of adapting to these changes; for example, the Arctic fox grows a thick coat in winter, which it sheds in summer.

LIFE IN THE DESERT
Plants such as cacti have adapted to life in semiarid environments by slowing growth, reducing the leaves to spines (to stop water loss from transpiration), and evolving special tissues that retain water.

< EVIDENCE OF CLIMATE CHANGE
Studies of polar ice core samples have revealed
details of past climate change. The chemical analysis
of trapped gas bubbles helps provide an approximate
measure of the temperature of the atmosphere at
the time the ice was formed.

**< ICE CORE
SAMPLE**
This is a close-up of
an ice core sample
recovered from Lake
Bonney, Antarctica,
which has a permanent
ice cover. It shows
trapped bubbles of air
and sediment particles
from the lake bed.

CLIMATE CYCLES

There is clear evidence from rocks and fossils
that Earth's climate has changed significantly over
time and that this has affected the evolution and
distribution of life, and led to the extinction of
many species. There are a number of causes of this
natural climate change, including volcanic activity—
which pollutes the atmosphere with ejected gas and
dust—and changes in ocean currents, which move
heat around the globe. Change is also driven by the
orbital and rotational cycles of the planet, which
affect the amount of solar radiation that reaches
Earth's surface. This, in turn, influences the planet's
temperature and climate, and triggers Earth's cycles
of ice ages and greenhouse periods.

CHANGING GEOGRAPHY
Over time, continents have been displaced as
oceans have expanded and contracted due to plate
tectonics. As continents moved from one hemisphere
to another they passed through different climatic
zones and, at times, formed supercontinents. The
size of these huge landmasses affected regional
climates. In addition, the changing shape of ocean
basins has altered water circulation, which in
turn changes the temperature and humidity of
the atmosphere above, and so affects climate.

GREENHOUSE AND ICE AGES
Long-term changes in climate are divided into cold
periods—ice ages, when there were long-lasting
ice sheets at the poles—and warmer, greenhouse
periods with largely ice-free poles. Warm phases
are linked to the release of greenhouse gases, such
as carbon dioxide, by plants into the atmosphere.
These trapped heat in Earth's atmosphere and
produced huge, shallow seas, arid zones, and lush
forests that, in the time of the dinosaurs, provided
them with plentiful food. Ice ages, lasting millions
of years, can be traced from the impact that
glaciation has left on the landscape. Fossils show
how rapid climate change associated with ice ages
had a dramatic impact upon life on a global scale.

Plant growth depends upon an
exchange of gases between the
atmosphere and the plant tissue,
through special gaps between the
leaves' cells called stomata. The
stomata open and close to take in
carbon dioxide for photosynthesis
and allow waste water and oxygen to
escape. In general, plants adapt to high
levels of atmospheric carbon dioxide,
associated with warm "greenhouse"
climates, by evolving high densities of
stomata on the leaf surface. Fossil
evidence of changes in the stomatal
density of certain plants tracks changes
in atmospheric carbon dioxide over time.

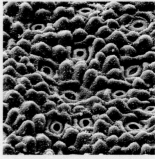

EUCALYPTUS STOMATA

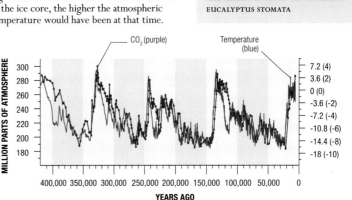

∧ DEVONIAN CORAL REEF
Earth's changing climate is spectacularly illustrated by this limestone outcrop in
The Kimberleys, W. Australia. In the Devonian Period (400 million years ago),
the area was under water and the cliff was a barrier reef.

**∨ CARBON DIOXIDE
AND TEMPERATURE**
Gas bubbles trapped in polar ice cores indicate
the fluctuating temperature of the planet—the
higher the amount of carbon dioxide detected
in the ice core, the higher the atmospheric
temperature would have been at that time.

HABITATS FOR LIFE

Earth's uniquely varied habitats enable it to support an abundantly rich diversity of animal and plant life—from the extreme depths of the ocean floors to the highest mountains, and from arid deserts and grasslands to the warm, wet tropics.

KEY

Polar region	Coniferous forest
Desert	Mountains
Grassland	Coral reef
Tropical forest	Rivers and wetland
Temperate forest	Oceans

Each life-form has its preferred habitat—the one to which it has adapted over thousands or even millions of years. However, Earth's varied environments allow many different kinds of animals and plants to live in the same habitat—a phenomenon known as biodiversity. Over geological time, life has had to adapt to environmental change, especially rapid climate change, with both extinctions and the evolution of life-forms to colonize new habitats. The presence of these pioneering organisms, in turn, produced changes in the environments they colonized—by forming soils, for instance—and these modifications encouraged further colonization by new life forms.

Differences in habitats are produced by many factors, ranging from an area's height above sea level, its distance from the equator, and its topography (or physical shape). Some areas of the globe are biodiversity "hotspots," rich in animal and plant life—most notably tropical reefs and rainforests—while other areas with more extreme conditions support only a few, though often richly populated, species.

LEVELS OF LIFE

Organisms rarely exist entirely on their own, even in the remotest places on Earth. The interactions between wildlife create different levels—from the individual organism to the overall ecosystem which it inhabits and shares with others.

INDIVIDUAL
A single, usually independent and habitat-restricted member of a population.

POPULATION
A group of individuals of the same species that occupy the same area and interbreed.

COMMUNITY
A naturally occurring collection of plant and animal populations living within the same area.

ECOSYSTEM
A biological community and its physical surroundings, which support one another.

MAP OF BIOMES
A biome unites ecosystems that have developed under similar climatic and soil conditions in different parts of the world. Biomes are defined by a variety of factors such as plant types, climate, geology, and topography.

GRASSLAND
The evolution of grass plants some 20 million years ago, and the colonization by grazing mammals, transformed Earth's landscapes. Temperate grasslands are generally treeless and have extremely fertile soils. Savanna grasslands, as shown here, are more like open woodland, featuring scatterings of trees and scrubs.

BISON

DESERT

Extreme lack of rain and soils for sustained plant growth creates deserts, which at present account for about a third of Earth's landscapes, although this proportion is increasing. The largest desert is the African Sahara.

RATTLESNAKE

TROPICAL FOREST

The richest wildlife habitats on land are found in the forests of the tropics —Earth's hottest areas, which lie by the equator. Their numerous ecosystems are important yet increasingly vulnerable biodiversity hotspots.

STRAWBERRY POISON-DART FROG

TEMPERATE FOREST

Temperate environments lie between the tropics and the polar regions. The influence of both tropical and polar air masses promotes vast forests with considerable biodiversity. However, clearing by humans has greatly reduced their extent.

RED DEER

CONIFEROUS FOREST

Coniferous forest trees, such as redwoods, spruce, and fir, belong to an ancient plant group and are the world's toughest trees. Evergreen with small leaves, they thrive where few other trees can, in cold areas and in mountain ranges.

BROWN BEAR

MOUNTAINS

Reaching as high as 5½ miles (9 km) above sea level, Earth's mountains are home to many different environments. A single mountain can rise from temperate woodland up to arctic conditions as the climate changes with altitude.

PEREGRINE FALCON

RIVERS AND WETLAND

A wide range of animal and plant life thrives in Earth's rivers and lakes. Landscapes saturated with water, either permanently or seasonally, form distinctive wetlands, where areas of open water mix with dense vegetation.

DRAGONFLY

CORAL REEF

Coral reefs develop from the skeletons of marine organisms in shallow, sunlit tropical waters. Supporting an immense variety of life, they are the rainforests of the aquatic world.

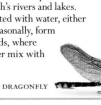

YELLOW TANG FISH

ARCTIC OCEAN

ASIA

PACIFIC OCEAN

INDIAN OCEAN

AUSTRALIA

SOUTHERN OCEAN

ANTARCTICA

POLAR REGIONS

The Arctic and Antarctic experience extreme seasons, with 24-hour daylight in summer and perpetual darkness in winter. Dominated by snow and ice, and with vast, dry areas of polar desert, both are highly vulnerable to climate change.

ROCKHOPPER PENGUIN

OCEANS

Life in oceans is found at all levels, from the sunlit surface to the deepest depths. Covering two-thirds of the planet, oceans form the largest continuous habitat on Earth and support very varied life forms— from microscopic plankton to the largest living mammal, the blue whale.

LOBSTER

∧ SCARRED LANDSCAPE
The growth of industry has required the exploitation of raw materials. Their extraction, such as at this copper mine, has changed global landscapes forever.

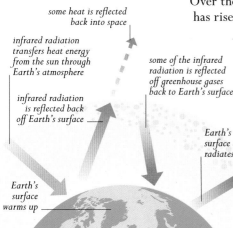

∧ ATMOSPHERIC POLLUTION
Slash-and-burn forest clearing for agriculture not only releases atmospheric pollutants but also diminishes the capture of carbon dioxide by the plant life.

∨ GREENHOUSE EFFECT
An excess of greenhouse gases in the atmosphere creates a shield which prevents some of the sun's energy from radiating back into space.

some heat is reflected back into space

infrared radiation transfers heat energy from the sun through Earth's atmosphere

infrared radiation is reflected back off Earth's surface

some of the infrared radiation is reflected off greenhouse gases back to Earth's surface

Earth's surface also radiates heat

some heat is reflected back into space

Earth's surface warms up

heat absorbed by greenhouse gases radiates back to Earth's surface

HUMAN IMPACT

The rapid growth of the human population has had an enormous impact upon Earth's natural environments, affecting the climate and countless species of plant and animal life. Some of these changes are irreversible.

ENVIRONMENTAL CHANGE

Earth has a long history of climate change with dominant warm "greenhouse" climates, widespread forestation, and no polar ice punctuated by glacial "ice ages." Global warming is known to be linked to high levels of greenhouse gases in the atmosphere, such as carbon dioxide and methane, which trap incoming solar energy and raise the temperature of the oceans, land, and air. In the past, natural increases in atmospheric carbon dioxide have been balanced by the development of forests on land and lime-rich sediments in the seas, which eventually became coal and limestone and stored excess carbon dioxide effectively. Since the industrial revolution of the 19th century, human activity has released huge amounts of carbon dioxide and other greenhouse gases back into the atmosphere through the mining and burning of fossil fuels, the clearing of forests, and the raising of cattle.

THE OCEANS

The health of Earth's oceans is vital to all life. Marine life depends upon the circulation of oceanic waters that contain enough oxygen and nutrients to support the food chain from plankton and shellfish to all the other animals that depend upon them for food. Fossil records show that when the oceanic environment deteriorated in the past, it led to extinctions of life. Today, overfishing and pollution, especially with plastics, by humans are affecting the marine environment.

THE ATMOSPHERE

For thousands of years human activity has affected the atmosphere. Initially, it was restricted to the release of pollutants from domestic fires and forest clearance. In Roman times metal production released the first industrial pollutants into the atmosphere, leaving traces that can be seen in polar ice cores. Over the last 200 years, pollution by gases and particulates has risen steeply. This has produced acid rain and smog, and greenhouse gases that are linked to global warming and depletion of the ozone layer, which helps screen out harmful ultraviolet radiation.

THE LAND

Since settlements and agriculture became widespread 8,000 years ago, humans have had a growing impact upon Earth's landscapes. Worldwide population growth, land settlement, and demand for food has left few untouched landscapes. Greater awareness of the impact of human activity on the environment is now leading to efforts to conserve natural habitats.

∨ POLAR ICE SHELF COLLAPSE
Rising temperatures are leading to the collapse of the polar ice shelves. The release of such large volumes of water is raising sea levels, which in turn threatens coastlines.

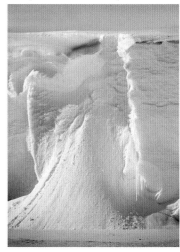

EXTINCTION

The inability of many organisms to adapt to environmental change has led to a huge turnover of species throughout geological time. In fact, the vast majority of nature's species are now extinct. Only the fittest organisms survive, usually through gradual adaptation, but sometimes by the sudden elimination of competitors. For example, when a huge asteroid hit Earth 66 million years ago, it set in motion a chain of events that led to the demise of many life-forms, including the dinosaurs on land and the ammonites in the seas. But it also opened the way for the rapid evolution of mammals, including humans. More recently, the arrival of modern humans in different regions around the world has contributed to the extinction of particular species, such as the woolly mammoth in Europe and Asia (see far right). Today, as the human population expands, our actions are placing a growing number of species, such as the tiger, in imminent danger of extinction.

∧ DECLINE OF THE CRESTED IBIS

Once widespread in Asia, hunting and habitat loss has restricted the crested ibis to a small wild population in China. Captive breeding has allowed it to be reintroduced to Japan.

< PÈRE DAVID'S DEER

Extinct in the wild, this east Asian deer has survived only in captive herds bred in England since 1900. Père David's deer reintroduced to China in the 1980s now form a population some 700 strong.

SCIENCE
MAMMOTHS NO MORE

Woolly mammoths were elephants that were adapted to the cold. They migrated across ice-age Europe and Asia in vast herds. Archaeological evidence such as cave paintings shows that they were actively hunted by humans about 30,000 years ago, which may have contributed to the extinction of most woolly mammoths by 11,000 years ago.

CAVE PAINTING, PECH MERLE, FRANCE

ORIGINS OF LIFE

The fossil record shows that life first appeared on Earth at least 3.7 billion years ago, and that all complex life has evolved from these first simple forms. Today, diverse life-forms range from single-celled organisms to mammals with a complex anatomy.

WHAT IS LIFE?

Several features distinguish living organisms from inanimate, inorganic matter, although the existence of viruses blurs the boundary between the living and nonliving. These features include an ability to take in and expend energy, to grow and change, to reproduce, to adapt to its environment, and—in more complex living organisms—to communicate.

The cell is the fundamental unit of life, capable of replicating itself and carrying out all living processes. Even the smallest independent organisms are made up of at least one cell, and almost every cell of every living organism has its own set of molecular instructions. Within each cell, the threadlike chromosomes carry hereditary information in the form of genes,

∧ **PHOTOSYNTHESIS**
Plants use the pigment chlorophyll to capture light energy and convert water and carbon dioxide into sugars and oxygen. This benefits other life-forms that feed on the plants and breathe in oxygen.

which are responsible for the particular characteristics of an organism. The set of instructions in a gene are mainly recorded in the form of a molecule called chromosomal deoxyribonucleic acid (DNA). The DNA of an organism carries information from one generation to another, allowing certain characteristics to be passed on from parent to offspring.

∧ **VITAL ENERGY**
To sustain itself, life must obtain energy from the environment. Energy flows mostly from photosynthesis in plants—but also occasionally from chemosynthesis in bacteria—to the animals that feed upon them and are, in turn, eaten by other animals.

DIVISIONS OF LIFE

The huge array of life on Earth is divided into three domains or superkingdoms—Archaea, Bacteria, and Eukaryota—which encompass all life-forms from plants and fungi to animals. Archaea and bacteria are primitive, single-celled organisms, and were probably the earliest form of life on Earth. The more advanced eukaryotes are distinguished from archaea and bacteria by having a cell nucleus, which contains the cell's genetic material, DNA. Eukaryotes vary enormously in shape and size, ranging from single-celled organisms to complex, multicelled plants and animals.

< **GROWTH**
A capacity to grow and repair is one of the key defining features of life. All organisms, from simple fungi to mammals, grow mostly by cell division but also increase in cell size.

EARLY LIFE

The very earliest forms of life evolved in the seas, and evidence of this comes from two main sources: living primitive organisms and the fossil record. The most primitive life-forms today are archaea and bacteria, which can survive even in extreme temperatures and acidic conditions. Such microorganisms may be similar to those that first evolved in the extreme environments of early Earth.

Fossil evidence for early life is contentious. Chemical evidence of carbon derived from living organisms has been found in rocks from Western Greenland and dates back some 3.7 billion years. The most convincing of ancient records of life are layered, mound-shaped stromatolites (see right).

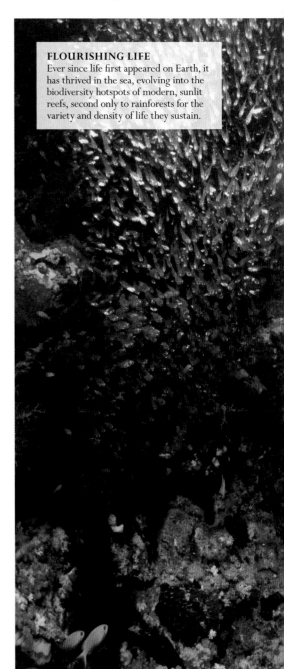

FLOURISHING LIFE
Ever since life first appeared on Earth, it has thrived in the sea, evolving into the biodiversity hotspots of modern, sunlit reefs, second only to rainforests for the variety and density of life they sustain.

From the first appearance of simple life, it took at least 2.7 billion years before complex life-forms appeared. The fossil of a microscopic and multicelled red algae called *Bangiomorpha* provides the first evidence of the existence of specialized cells. These cells evolved for sexual reproduction, and also for the development of a holdfast to attach the algae to the sea floor.

< STROMATOLITES
For billions of years these layered structures have built up in shallow tropical seas by alternating layers of sediment and sheets of microorganisms, including cyanobacteria (blue-green algae).

About 650 million years ago, fossils of various frond- and disk-shaped, soft-bodied, and mostly sessile organisms appeared. These marine animals were known as the Ediacaran Fauna. By the beginning of the Cambrian Period, 545 million years ago, numerous multicelled marine organisms had evolved, including burrowing worms and a variety of small, shelled mollusks, whose bodies had muscle tissues and organs such as gills for respiration. About 510 million years ago, the first vertebrates appeared, with an internal skeletal support for the body. By late Devonian times, about 380 million years ago, vertebrates had begun to emerge from the seas onto land.

∧ BURGESS SHALE
The fossils of the Burgess Shale in Canada show that marine life diversified rapidly in Cambrian times, from sponges and arthropods to vertebrates.

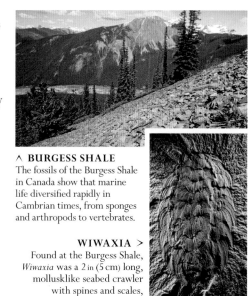

WIWAXIA >
Found at the Burgess Shale, *Wiwaxia* was a 2 in (5 cm) long, mollusklike seabed crawler with spines and scales, and a soft underside.

EVOLUTION AND DIVERSITY

Until the 19th century, when a number of theories were proposed, it was a matter of speculation as to how such remarkably diverse life-forms had developed on Earth. Today, the theory of evolution and diversification, alongside geological evidence for changes in the distribution of continents, give a fascinating insight into the ever-changing life on our planet.

CHANGE OVER TIME

All living things can change and adapt to their surroundings. Changes that are passed from generation to generation are hard to see, but over time they can alter the way a species looks or behaves. This process, known as evolution, is not gradual but has seen significant extinctions and bursts of growth over time.

The study of fossils to unravel the history of life was in its early stages in Charles Darwin's day, and since then a vast amount of information supporting the theory of evolution has emerged. We now know that life evolved in the oceans at least 3.7 billion years ago, and that it was from these early simple life-forms that all current life on Earth evolved—including plants, fungi, and animals.

As life-forms became more complex and moved from sea to land, the first forests and land-living animals evolved. The Mesozoic Era, which began 252 million years ago, with its successions of evolving plants and animals, produced the dominant dinosaur reptiles and their bird descendants. A mass extinction event 66 million years ago ushered in the Cenozoic Era. The reptiles were largely replaced by mammals both in the seas and on land, where flowering plants and their pollinating insects also became abundant and diverse.

< GIANT SALAMANDER
This extremely rare fossil skeleton of an *Andrias* (giant salamander) was mistaken for a human victim of the biblical Flood until French anatomist Georges Cuvier identified it as an amphibian in 1812.

EVIDENCE OF EVOLUTION
Comparison of the anatomy of vertebrate limb bones from various species show that, despite different appearances and functions, they derive from the same basic developmental plan and the same genes.

FROG
The frog's leg, arm, and finger bones are modified for swimming. Large muscles enable it to jump powerfully—essential for catching prey and escaping from predators.

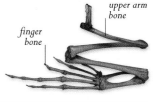
upper arm bone
finger bone

OWL
The wing of a bird is powered by flight muscles attached to the upper arm and bones of the wrist, with greatly modified and extended fingers.

upper arm bone
finger bone
lower arm bone

CHIMPANZEE
The arm of a chimpanzee is anatomically very similar to our own, but has slightly different proportions, with elongated fingers and a short thumb.

finger bone
upper arm bone
lower arm bone
wrist bone

DOLPHIN
The arm bones of whales and dolphins form a flipper—with shortened, flattened, and strengthened arm bones and greatly lengthened second and third fingers.

shoulder blade
finger bone
upper arm bone
lower arm bone

LAMARCK LEADS THE WAY

The 18th-century French biologist Jean-Baptiste Lamarck developed the first comprehensive evolutionary theory that higher forms of life had "evolved" from simpler organisms. Based on his extensive research of invertebrates in particular, Lamarck argued that necessary characteristics could be acquired during an organism's lifetime—through the desire to gain food, shelter, and mates—and those not needed could be lost, with the resulting transformation passed on to its offspring. Although modern genetics disproves this theory of "soft inheritance," Lamarck's concepts were an important starting point, and were developed further by the Scottish anatomist Robert Grant, who tutored Charles Darwin in Edinburgh. Darwin himself did not entirely rule out a Lamarckian mechanism, which he thought might be supplementary to natural selection.

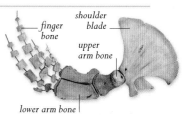
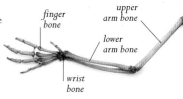

< ^ INNER STRIVING
Lamarck believed that evolution worked through a process of "inner striving." Thus the giraffe developed a long neck to reach the leaves of trees, and the heron grew long legs in order to wade.

∧ GALAPAGOS FINCHES
On his voyages, Darwin collected
many different specimens of Galapagos
finches which he believed may have
descended from a common ancestor.

*butterfly
wing*

DARWIN AND WALLACE

In the mid-19th century, British naturalists
Charles Darwin and Alfred Russel Wallace
independently came up with the theory of evolution
by natural selection. They both had experience of
fieldwork in the tropics—an environment boasting
high biodiversity, competition for resources, and
marked differences in the organisms living in
separate areas. They both wondered how and why
such natural phenomena had arisen. On his travels
Wallace collected specimens for study and sale, and
it was in the Malay Archipelago that he formulated

his explanations for the geographical distribution
of organisms—biogeography—and realized the
role of natural selection in evolution. Meanwhile,
Darwin's five-year voyage on HMS *Beagle* around
the southern hemisphere as a gentleman naturalist
provided him with much material to formulate his
own theory of evolution. In 1858, Wallace and
Darwin produced a joint publication on natural
selection, and the following year Darwin expanded
the theory to produce his famous and influential
book, *On the Origin of Species*.

∧ COLLECTION BOX
Both Darwin and Wallace were fascinated
by insect diversity, especially as found in
the tropics, and were avid collectors.

Cynognathus
*therapsid fossils
from the
Triassic Period*

AFRICA

INDIA

Lystrosaurus
*therapsid fossils
from the Triassic*

< BIOGEOGRAPHY
The distribution of certain
reptile, therapsid, and plant
fossils across the southern
continents shows that they were
once joined as a supercontinent,
called Gondwana.

SOUTH AMERICA

Mesosaurus
*reptile fossils from
the Permian Period*

AUSTRALIA

Glossopteris
*plant fossils from
the Permian Period*

ANTARCTICA

∧ EARLY BIRD
The discovery of the fossil *Archaeopteryx*
in 1861 revealed characteristics that
provided an evolutionary link between
two major groups: reptiles and birds.

EVOLUTION IN PROGRESS

Darwin and Wallace proposed the theory of natural selection, but it was the discovery of the gene that gave scientists the mechanism by which selection takes place. Understanding the gene has since become the key to understanding evolution.

NATURAL SELECTION

A key evolutionary mechanism, natural selection favors the survival of the fittest. In other words, individuals that possess characteristics best adapted to their current environment have a better chance of surviving to reproduce and pass on those favorable traits to another generation. Natural genetic variation within populations produces differences such as size, shape, and color, and some of these may help promote survival. For instance, a particular body coloration may provide a camouflage that better hides the animal from predators than other colors would. If this leads to the survival of that animal and to its reproduction, that same improved coloration will be passed on to some of its offspring. As an environment

∧ INDIVIDUAL VARIATION
Litters of domestic cats commonly include individuals with varied coloration, especially where the parents' colors differ.

changes over time, a different coloration might be more beneficial, so natural selection will ensure another change. A split into two populations may be imposed by a physical barrier, with each new population becoming adapted to slightly different conditions. Eventually, this may lead to one species becoming two, in a process known as speciation.

< ∨ SEXUAL DIMORPHISM
There are often marked differences between males and females of the same species. Male frigate birds have inflatable throat pouches for attracting the females.

SCIENCE

CREATIONISM: BELIEF VERSUS SCIENCE

Most of the world's religions provide a theory of creation, which gives an explanation for the formation of Earth and life. Many of these creation stories originated long before scientific data and theories were available to offer alternative explanations and understanding. Some believers in the Western Judeo-Christian tradition maintain that the complexity of "design" in many organisms implies that there must have been a "designer"—God—behind their creation, and for this reason they dispute the theory of evolution.

GENES AND INHERITANCE

Particular traits pass from parents to offspring through the transmission of genetic material. Genes preserve, encoded in their DNA, all the information necessary for the replication of a cell's structure and its maintenance. Genes are therefore the basic units of heredity. Individual chromosomes—the threadlike part of a cell—hold thousands of genes on long strands of DNA. During sexual reproduction, the fusion of sperm and egg cells produces two complete sets of gene-bearing chromosomes: one copy from the father and one from the mother.

a forest fire wipes out large populations of butterflies

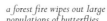

∨ GENES AND CHANCE
Sometimes individuals are eliminated at random, with the result that their genes are not passed on to the next generation.

by chance, the survivors are mostly yellow butterflies

only the genes of the survivors are passed on

next generation has fewer purple butterflies

a chance event leads to total loss of purple butterflies

ADAPTING TO EXTREMES
The lesser flamingo has evolved to occupy a very specialized niche: to feed on algae in extremely alkaline African lakes. In the absence of competitors, it can thrive there in huge numbers.

ISLAND EVOLUTION

Isolated islands provide natural laboratories for unusually fast evolution, with intense competition for limited resources leading to rapid speciation. In 1835, Darwin's visit to the Galapagos Islands allowed him to collect many bird specimens, particularly finches. He noted slight variations between the specimens from island to island. He also heard about the differences between the giant tortoises on separate islands, and subsequent visits to other Pacific islands made him wonder about the possibility of new species evolving from a common ancestor. The ornithologist John Gould was able to identify Darwin's finches as a new group of 12 separate species, rather than just varieties of the same species.

This persuaded Darwin that species could change under certain conditions such as island isolation. Island wildlife continues to be an important research area for modern evolutionary biologists.

< FLIGHTLESS BIRDS
Many species of flightless birds, such as kiwis, evolved on the islands of New Zealand, which lacked serious predators until humans arrived.

ARTIFICIAL SELECTION

Over millennia, humans have domesticated many different kinds of animals and plants, from dogs and cattle to fruit trees and cereal crops. Before the discovery of the gene, this was achieved simply by selectively interbreeding organisms bearing the desired characteristics—such as the ability to run fast or produce more succulent fruits—over many generations, until the selected traits became dominant. Today, biotechnology achieves the same result much faster by directly manipulating genes, both to enhance beneficial traits and to remove problematic ones.

∧ GENETIC MODIFICATION
Alteration to the genetic makeup of organisms can remove undesired characteristics and add more useful ones, such as resistance to disease.

∧ CLONING
Genetically identical individuals can be created by the transfer of a nucleus—with its genetic information—from an adult cell into a host egg cell.

CLASSIFICATION

Global diversity is estimated to range from 10 million species to billions. Nearly 2 million have been described, with many more added each year. All species are named and classified using a system devised over 250 years ago.

For centuries people have studied the natural world. Initially they were limited to what they could find locally and to reports from travelers, as it was impossible to preserve and send specimens any distance. Later, as travel became easier, explorers were paid to collect plants and animals and ship's artists would draw them. By the early 1600s, natural history collections in Europe were substantial, and many specimens had been described, but there was no formal arrangement that made these specimens or accounts of them easily accessible.

The aim of the early taxonomists, or scientists who describe and classify species, was simple—to organize living things so that they reflected God's plan of creation. Between 1660 and 1713, John Ray published works on plants, insects, birds, fishes, and mammals, forming his groups on the basis of morphological (structural) similarities.

The science of morphology, along with other criteria such as behavior and modern genetics, forms the basis of classification today. In 1758, the tenth edition of *Systema Naturae* was published. It was written by the Swedish botanist Carolus Linnaeus. He and his friend Peter Artedi had decided to divide the natural world between them and classify everything in it, fitting the 7,300 described species into the same hierarchical framework. Although Artedi died before his book was finished, Linnaeus completed the work and published it along with his own.

LATIN NAMES

All living things now have a unique Latin name—such as *Panthera leo* for the lion—which is made up of the genus name, starting with a capital letter, and a descriptive species name. Linnaeus devised this binomial method for identifying different organisms to replace the arbitrary descriptions that existed previously. The new method put an end to confusion caused by the same name being given to several species, or to single species being know by several names.

Within a species, distinct subspecies can sometimes be recognized in different locations. In the 1800s, Elliot Coues and Walter Rothschild adopted a trinomial Latin system to accommodate them. This convention for naming species and subspecies is still used today.

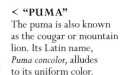

"DOG ROSE" >
Known variously as the dog rose, wild briar, witches' briar, dogberry, and the hip tree, this plant has only one Latin name, *Rosa canina*, which identifies it for everyone everywhere.

< "PUMA"
The puma is also known as the cougar or mountain lion. Its Latin name, *Puma concolor*, alludes to its uniform color.

TRADITIONAL CLASSIFICATION

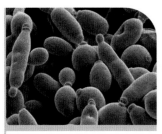

DOMAIN
Eukaryota
The latest taxonomic unit to be created is the domain. It is based on whether organisms possess cells with a nucleus (Eukaryota: protists, plants, fungi, and animals) or not (Archaea and Bacteria).

KINGDOM
Animalia
In recent years, the traditional kingdoms of plants and animals have been further subdivided. Kingdom Animalia now includes only multicellular organisms that must eat other species in order to survive.

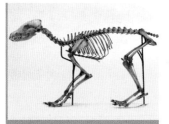

PHYLUM
Chordata
A phylum (known as a division in plants) is a major subdivision of a kingdom and is made up of one or more classes that share certain features. Members of the phylum Chordata have a notochord—the precursor of the backbone.

CLASS
Mammalia
Introduced by Carl Linnaeus, a class contains one or more orders. Class Mammalia includes only animals that are warm-blooded, have fur, a single jaw bone, and that suckle their young on milk.

> CONTRIBUTING TO CLASSIFICATION
Over the years many scientists have attempted to organize the natural world, combining earlier ideas with new research, culminating in the traditional system of classification shown above and the use of bi- and trinomial Latin names. Some scientists were particularly influential, making significant contributions to taxonomy.

ANIMALS AND PLANTS

Aristotle was the first person to classify living things, and he introduced the term *genos* (meaning race, stock, kin)—genus in Latin. He separated animals into those with blood and those without, not realizing blood need not be red. This division is very close to the modern classification of vertebrates and invertebrates.

ARISTOTLE, 384–322 BCE

ORDER FROM CHAOS

John Ray classified organisms based on their overall morphology rather than just a part of it. In doing so he could establish relationships between species more easily and organize them into groups more effectively. He also divided the flowering plants into two major groups of orders—monocotyledons and dicotyledons.

JOHN RAY, 1627–1705

150
173
179
174
185
152
170
175
58

CLASSIFYING INSECTS
About 20,000 species of butterfly have now been described. In order to identify them or recognize new species, museums hold vast reference collections, like this one, for comparative purposes.

ORDER
Carnivora
Orders are the next level down in Linnaeus's hierarchy, and contain one or more families. The Carnivora have modified cheek teeth (carnassials) and large canine teeth specialized for biting and shearing.

FAMILY
Canidae
A subdivision of an order, a family is made up of genera and the species within them. The family Canidae has 35 living species, all with nonretractile claws and two fused wrist bones. All but one species have long, bushy tails.

GENUS
Vulpes
A term first used by Aristotle in Ancient Greece, a genus identifies subdivisions of a family. *Vulpes* is a genus within the family Canidae. All foxes have upright, triangular ears and a long, narrow, pointed snout.

SPECIES
Vulpes vulpes
The basic unit of taxonomy, species are populations of similar animals that breed only with one another. *Vulpes vulpes*, known for its bright red fur, breeds only with other red foxes.

ANIMAL, VEGETABLE, OR MINERAL

Linnaeus divided the natural world into three kingdoms—animals, plants, and minerals. He then devised a hierarchical system of classification based on class, order, family, genus, and species, and established the convention of giving species binomial Latin names.

CARL LINNAEUS, 1707–1778

A NEW KINGDOM

Historically organisms were classed as being either animals or plants, but in 1866 Ernst Haeckel argued that microscopic organisms formed a separate group, which he called Protoctista (now Protista). There were now three kingdoms of life: Animalia, Plantae, and Protista.

ERNST HAECKEL, 1834–1919

INTRODUCING THE ARCHAEA

Recognized by Carl Woese and George Fox in 1977, the Archaea are microscopic organisms that live in very extreme environments. Initially grouped with the Bacteria, their DNA turned out to be so unique that a new three-domain taxonomic system was introduced.

CARL WOESE, BORN 1928

ANIMAL GENEALOGY

In the 1950s, a revolutionary new way of classifying organisms was proposed. Called phylogenetics, it allowed taxonomists to investigate evolutionary relationships between species by placing them in hierarchical groups called clades.

Phylogenetics, also known as cladistics, is based on the work of entomologist Willi Hennig (1913–1976). He assumed that organisms with the same morphological characters must be more closely related to each other than to those that lacked them. Therefore, these organisms must also share the same evolutionary history and have a more recent ancestor in common. Like the traditional taxonomy of Linnaeus, this method of classifying organisms is hierarchical, but due to the volume of data involved, computers are used to generate the family trees, known as cladograms.

For a morphological feature to be useful in cladistic analysis, it must have altered in some way from the so-called "primitive" ancestral condition to a "derived" one. For example, the legs and paws of most carnivores are considered primitive in the cladogram on the opposite page, compared to the derived condition of flippers in the seals, fur seals and sea lions, and walruses. This derived character, called a synapomorphy, is useful because it is shared between at least two taxonomic groups, suggesting

they are more closely related to each other than to groups without flippers. Characters that are unique to one group are useful for recognition purposes but say nothing about relationships. So cladistic analysis is based entirely on the identification of synapomorphic characters.

UNDERSTANDING ANCESTRY

The more derived characters organisms have in common, the closer their relationship is assumed to be. For instance, brothers and sisters look more alike than other children—they may have the same eyes, the same chin, and so on. This is because in terms of common ancestry they have the same parents but are only distantly related to other people.

Cladistics is now most commonly based on genetics—except in the study of fossils—and has exposed some unexpected shared ancestry. For example, a genetic cladogram surprisingly revealed that the whale's closest living relative is the land-dwelling hippopotamus—a relationship that Linnaeus would certainly not have expected.

∨ **CLOSE RELATIONS**
The giraffes and kudus are both even-toed ungulates, so are more closely related to each other than to the odd-toed zebras. As fur-bearing mammals, all three are more closely related to each other than to the feathered birds around them.

LOOKING AT CLADOGRAMS

To do a cladistic analysis, the different groups of organisms are scored on a set of characters that are either primitive or derived. Their distribution is not always as straightforward as is shown in the diagrams below. Often the resulting cladogram can be constructed in a number of different ways and taxonomists have to choose between them. To do this they adopt the principle of parsimony—they choose the cladogram involving the least number of steps or character transformations to explain the observed relationships between the groups.

< MILKY DIET
All mammals possess mammary glands. Unique to the class Mammalia, this feature is synapomorphic at this taxonomic level. To find additional familial relationships within the class, characters that are synapomorphic at familial level are used.

CHARACTER	CANID	BEAR	SEAL	FUR SEAL & SEA LION	WALRUS
Feeding young on milk	1	1	1	1	1
Short tail	0	1	1	1	1
Forelimbs modified into flippers	0	0	1	1	1
Very flexible spine	0	0	1	1	1
Hind limbs turn forward under body	0	0	0	1	1
Presence of tusks	0	0	0	0	1

< CHARACTER SET
Most modern cladograms are based on genetics using DNA codes. The codes that were used to generate the cladogram below have been replaced with more familiar morphological descriptions in this character set. One character, feeding young on milk, is shared by all the groups shown; some characters are found in only some of the groups; and one character, the presence of tusks, is unique to the walrus.

∨ CLADOGRAM

In this cladogram the canids are considered the most primitive group, known as the outgroup, and the walrus is the most derived. All characters on the cladogram are shared by the groups to the right of each number—for example, a short tail is shared by the bears, seals, fur seals and sea lions, and the walrus.

KEY

0	primitive character
1	derived character

| CANIDS | BEARS | SEALS | FUR SEALS AND SEA LIONS | WALRUS |

the outgroup does not share character 1

two characters differentiate the seals, fur seals and sea lions, and walruses from the bears

only the fur seals and sea lions, and the walruses share character 4

walrus is the most derived genus in this cladogram

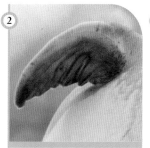

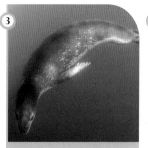

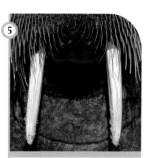

SHORT TAIL
Bears, seals, fur seals and sea lions, and the walrus all have short tails (character 1). Canids, however, display the primitive condition and have long, bushy tails. Character 1 is therefore a synapomorphic character shared by all the carnivore families shown except the canids.

FLIPPERS
Among carnivores, the modified limbs of seals, fur seals and sea lions, and the walrus are unique. Character 2 (forelimbs modified into flippers) is therefore synapomorphic at this level, suggesting these three groups are more closely related to one another than to the bears.

FLEXIBLE SPINE
Character 3 (flexible spine) operates at the same level as character 2, further supporting the relationship indicated by possession of flippers. The more synapomorphic characters that appear at a particular level, the more convincing the suggested relationship becomes.

SUPPORTING LIMBS
In both the fur seals and sea lions and the walrus, the pelvic girdle can be rotated to allow the hind limbs to aid locomotion on land. This suggests they share a more recent common ancestor with each other than they do with the seals, which are far less mobile when out of water.

TUSKS
This is a unique character (autapomorphy) of the walrus that reveals nothing about its relationship to other mammal families. The same is true of the character "feed young on milk"—present in all the groups, it provides no clues as to how they interrelate, so it is not plotted on the cladogram.

TREE OF LIFE

Using a branching tree as a way of showing the diversity of life was first suggested by the German naturalist Peter Pallas in 1766. Since then many such trees have been constructed. Initially treelike, complete with bark and leaves, they later became more diagrammatic, and took account of evolutionary theories. Modern, computer-generated trees of life present many different ideas of how living organisms are related.

DARWIN'S FIRST TREE

The first person to produce a tree of life that reflected the concept of evolution was Charles Darwin. He sketched the first of his ten evolutionary trees of life in 1837. It was a simple branching diagram, which he developed further before publishing it in the *Origin of Species* in 1859. The lettered branches show how he thought his theory might work—the more branch points separating an organism from its ancestor (numbered 1), the more different the organism will be. In 1879, Ernst Haeckel took the idea further, with a tree that showed animals evolving from single-celled organisms.

Today, DNA and protein analyses as well as morphology are used to construct evolutionary trees and establish the genetic relationships between organisms. Vast data sets require computers to generate the trees, which are continually refined as new species and information are discovered.

Traditionally, trees of life emphasised vertebrate groups and relationships as those of the microscopic archaea and bacteria, and protists (those eukaryotes not classified as plants, animals, or fungi) were more problematic. However, modern genomic studies have revealed an extraordinary diversification and complexity of relationships within the bacteria and within the early, or basal, eukaryote groups.

MASS EXTINCTIONS

Mapping all life forms that have ever existed on a tree is difficult as, over time, more than 95 percent of all species have become extinct. A mass extinction occurs when a large number of species dies off at the same time. This has happened five times in the past. The best-known extinction, which wiped out the dinosaurs, occurred at the end of the Cretaceous Period; it is thought to have been caused by a meteor impact combined with volcanic activity. As habitats are rapidly destroyed by human activity, it is likely that there will be another extinction event in the future.

EXTINCTION TIMELINE

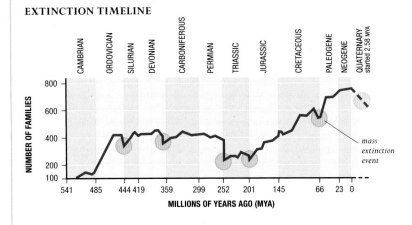

mass extinction event

READING THE TREE

This diagram shows how life evolved from simple organisms, such as the Archaea—which appeared about 3.4 billion years ago—to complex life forms, such as animals, which appeared about 650 million years ago. It also shows the diversity of the vertebrates (see pp.34–35), which have a disproportionate representation here. The circles indicate points where two or more groups of organisms have branched from a common ancestor at about the same time. Only extant species are shown, so extinct groups, such as the reptilian dinosaurs, are missing.

LIFE BEGINS

ARCHAEA

BACTERIA

STRUCTURE OF LIFE

All forms of life belong to one of three domains: Archaea, Bacteria, and Eukaryota. Archaea and bacteria are unicellular and lack a cell nucleus—they used to be classified together as Prokaryota. Eukaryotic organisms tend to be multicellular and each cell contains a nucleus, within which DNA is stored. This table shows the three domains and the four eukaryote kingdoms. Despite appearances, the Archaea and Bacteria are the largest groups—although only about 20,000 species have been described, estimates exceed 4 million species. Among eukaryotes, the groups that make up the protists and invertebrates are far more numerous in terms of species than vertebrate groups.

ARCHAEA	EUKARYOTES
BACTERIA	**PROTISTS**
	PLANTS
	MOSSES
	LIVERWORTS
	HORNWORTS
	LYCOPHYTES
	FERNS AND RELATIVES
	CYCADS, GINKGOS, AND GNETOPHYTES
	CONIFERS
	FLOWERING PLANTS
	FUNGI
	MUSHROOMS
	SAC FUNGI
	LICHENS
	ANIMALS
	INVERTEBRATES
	CHORDATES

CYANOBACTERIA

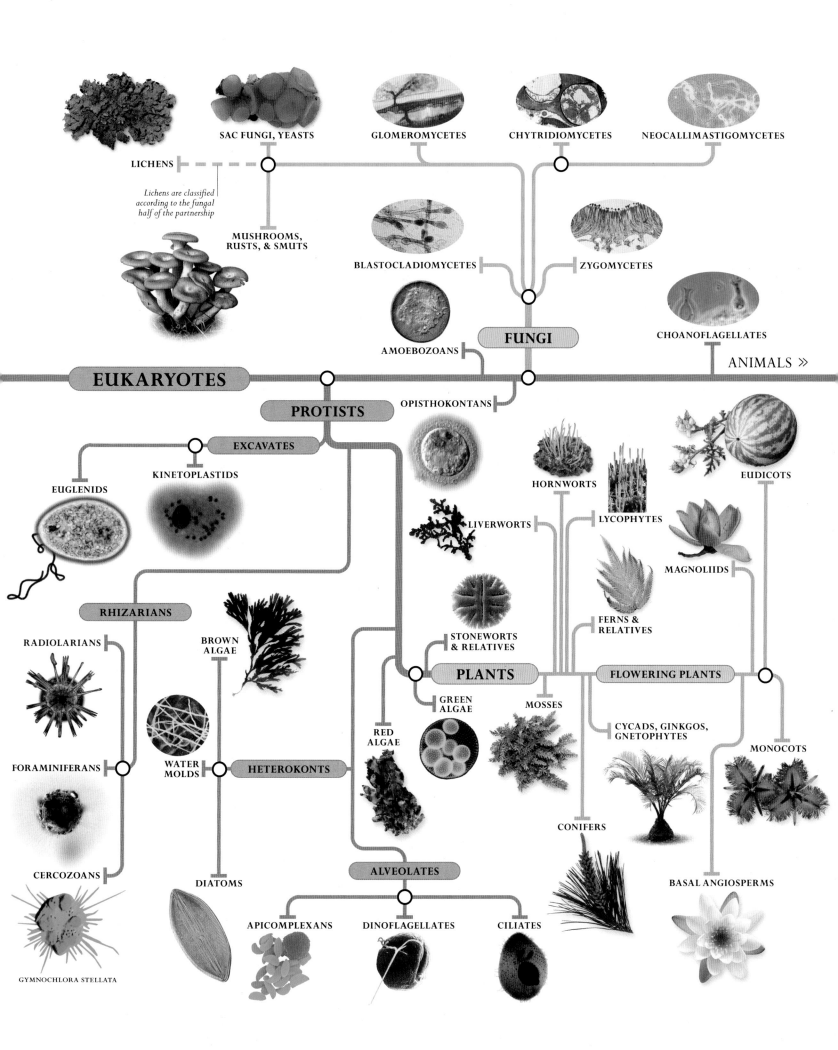

LICHENS

SAC FUNGI, YEASTS

GLOMEROMYCETES

CHYTRIDIOMYCETES

NEOCALLIMASTIGOMYCETES

Lichens are classified according to the fungal half of the partnership

MUSHROOMS, RUSTS, & SMUTS

BLASTOCLADIOMYCETES

ZYGOMYCETES

AMOEBOZOANS

FUNGI

CHOANOFLAGELLATES

ANIMALS »

EUKARYOTES

PROTISTS

OPISTHOKONTANS

EXCAVATES

KINETOPLASTIDS

HORNWORTS

LYCOPHYTES

EUDICOTS

EUGLENIDS

LIVERWORTS

MAGNOLIIDS

RHIZARIANS

BROWN ALGAE

RADIOLARIANS

FERNS & RELATIVES

STONEWORTS & RELATIVES

PLANTS

FLOWERING PLANTS

GREEN ALGAE

MOSSES

FORAMINIFERANS

RED ALGAE

WATER MOLDS

HETEROKONTS

CYCADS, GINKGOS, GNETOPHYTES

MONOCOTS

CONIFERS

CERCOZOANS

DIATOMS

ALVEOLATES

BASAL ANGIOSPERMS

GYMNOCHLORA STELLATA

APICOMPLEXANS

DINOFLAGELLATES

CILIATES

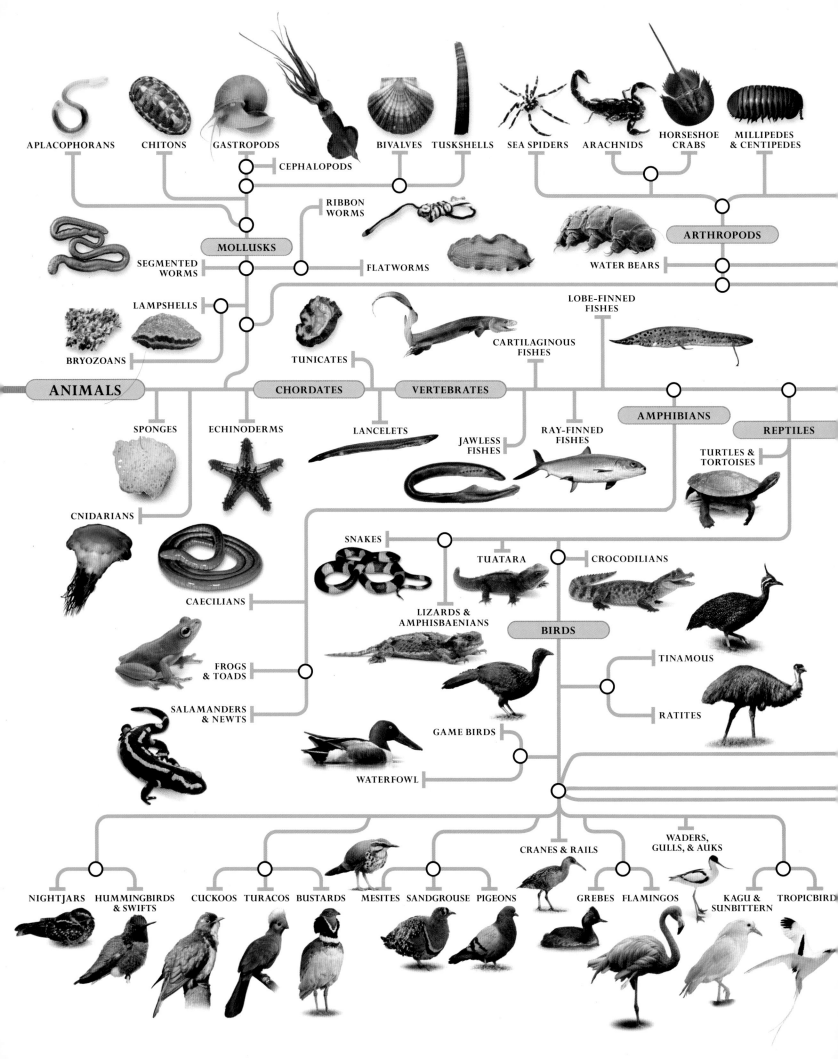

APLACOPHORANS

CHITONS

GASTROPODS

CEPHALOPODS

BIVALVES

TUSKSHELLS

SEA SPIDERS

ARACHNIDS

HORSESHOE CRABS

MILLIPEDES & CENTIPEDES

RIBBON WORMS

ARTHROPODS

MOLLUSKS

SEGMENTED WORMS

FLATWORMS

WATER BEARS

LOBE-FINNED FISHES

LAMPSHELLS

CARTILAGINOUS FISHES

BRYOZOANS

TUNICATES

ANIMALS

CHORDATES

VERTEBRATES

AMPHIBIANS

REPTILES

SPONGES

ECHINODERMS

LANCELETS

JAWLESS FISHES

RAY-FINNED FISHES

TURTLES & TORTOISES

CNIDARIANS

SNAKES

TUATARA

CROCODILIANS

CAECILIANS

LIZARDS & AMPHISBAENIANS

BIRDS

TINAMOUS

FROGS & TOADS

RATITES

SALAMANDERS & NEWTS

GAME BIRDS

WATERFOWL

CRANES & RAILS

WADERS, GULLS, & AUKS

NIGHTJARS

HUMMINGBIRDS & SWIFTS

CUCKOOS

TURACOS

BUSTARDS

MESITES

SANDGROUSE

PIGEONS

GREBES

FLAMINGOS

KAGU & SUNBITTERN

TROPICBIRD

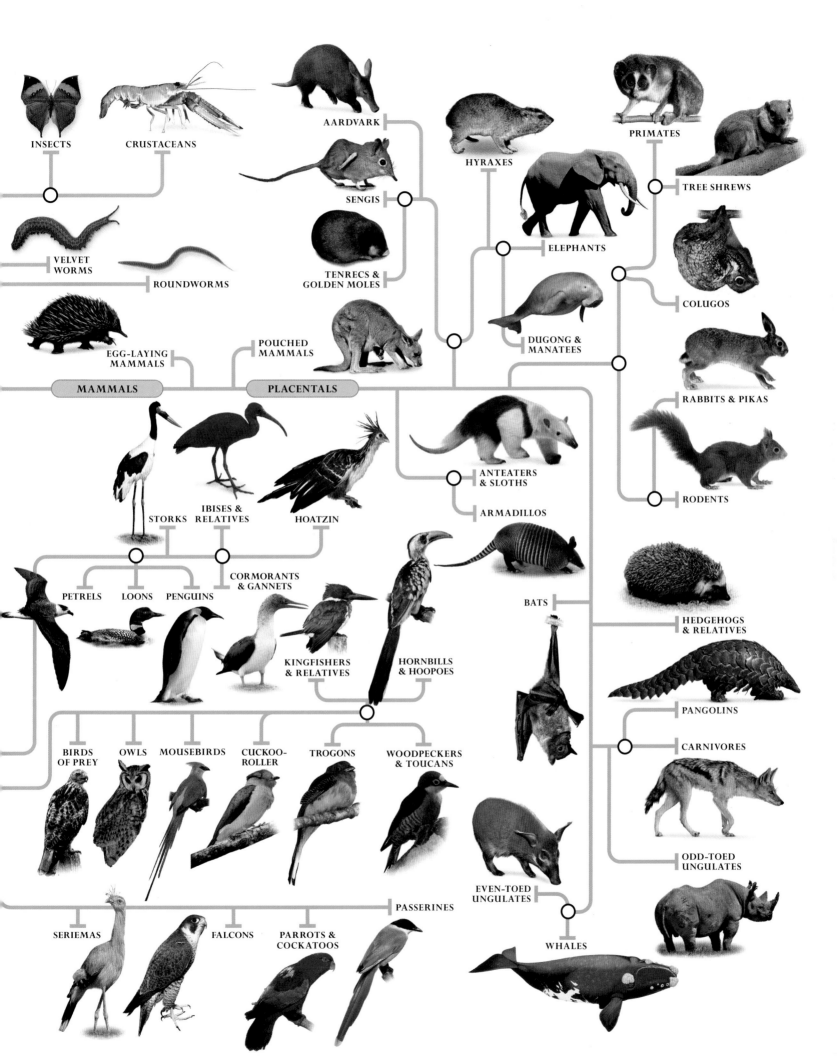

INSECTS

CRUSTACEANS

VELVET WORMS

ROUNDWORMS

EGG-LAYING MAMMALS

MAMMALS

AARDVARK

SENGIS

TENRECS & GOLDEN MOLES

POUCHED MAMMALS

PLACENTALS

HYRAXES

ELEPHANTS

DUGONG & MANATEES

PRIMATES

TREE SHREWS

COLUGOS

RABBITS & PIKAS

RODENTS

STORKS

IBISES & RELATIVES

HOATZIN

PETRELS

LOONS

PENGUINS

CORMORANTS & GANNETS

KINGFISHERS & RELATIVES

HORNBILLS & HOOPOES

ANTEATERS & SLOTHS

ARMADILLOS

BATS

HEDGEHOGS & RELATIVES

PANGOLINS

CARNIVORES

ODD-TOED UNGULATES

BIRDS OF PREY

OWLS

MOUSEBIRDS

CUCKOO-ROLLER

TROGONS

WOODPECKERS & TOUCANS

PASSERINES

EVEN-TOED UNGULATES

SERIEMAS

FALCONS

PARROTS & COCKATOOS

WHALES

MINERA
ROCKS,
FOSSILS

LS,
&

Life on Earth is shaped by the rocks that lie beneath our feet. Made of different combinations of minerals, they have a far-reaching influence on the landscape, on vegetation, and on the soil. Preserved within these rocks, fossils form a highly detailed record of life in the distant past, showing the path that evolution has followed over hundreds of millions of years.

The building blocks of rocks, minerals typically have a crystalline structure. Several thousand exist in Earth's crust, but fewer than 50 minerals are common and widespread.

Classified according to how they formed, rocks are constantly broken down and reformed. The oldest rocks date back 3.8 billion years, when Earth's crust first solidified.

Most fossils preserve hard remains, such as teeth and bones. They can also include traces such as footprints, and carbon impressions in rock, which reveal the past existence of living things.

MINERALS

Minerals are the materials from which rocks are made. More than 4,000 different minerals exist, each one with its own chemical composition, found naturally on Earth. Many minerals are hard and crystalline, and, while some are extremely abundant, others—including diamond—are very rare, and highly prized.

Copper often appears in a dendritic form, branching like a tree. It is a very important economic mineral.

Malachite can be botryoidal, with small, rounded shapes, or massive, with no definite shape.

Crocoite, which is lead chromate, frequently occurs as slender, elongated prismatic crystals.

Minerals are of great economic significance. They provide us with countless useful materials, from metals to industrial catalysts, and they also include objects of great intrinsic beauty, particularly when cut and polished as gems. But on an even broader scale, minerals are essential for life itself. In soil and in water, soluble minerals release a steady stream of chemical nutrients that plants and other organisms need to grow. Without them, the world's ecosystems would no longer operate.

Minerals are classified according to their chemistry. A few minerals, such as gold, silver, and sulfur, can exist in a native state, meaning they contain a pure chemical element, and nothing else. All other minerals are chemical compounds. Quartz, for example, contains two elements—silicon and oxygen—which are bound together very tightly, giving quartz its exceptional hardness and strength. Under the Strunz classification system, quartz (silicon dioxide) is classified as an oxide, as are chalcedony and opal, but under the Dana system, used in this book, they are included with the silicates. The largest group of minerals, the silicates make up about 75 percent of Earth's crust. Other common mineral groups include sulfides, oxides, carbonates, arsenates, and halides.

IDENTIFYING MINERALS

With experience, many minerals can be identified by their appearance alone. Important clues include their color, their luster—the way light is reflected from their surfaces—and above all, their habit, and/or crystal form. Crystals are grouped into six systems according to their symmetry (see panel, below). Minerals do not always occur as crystals but can take different forms: they can be dendritic (branching) or botryoidal (like a bunch of grapes), for example.

Minerals also differ in density, or specific gravity (SG)—measured by comparing the weight of a mineral to that of an equal volume of water—and in their hardness (H). On the 10-point Mohs scale, which measures hardness, talc is rated 1, whereas diamond, the hardest mineral, is rated 10. A fingernail (hardness 2½), copper coin (3½), and steel knife blade (5½) all make useful benchmarks for testing hardness. Surprisingly, size is not a useful clue. Gypsum crystals, for example, are usually less than ³⁄₈ in (1 cm) long, but the largest specimens ever discovered are the size of a two-story house.

VOLCANIC MINERALS >
The ground at Dallol in Ethiopia's Danakil Desert is pockmarked by volcanic vents and covered in sulfur, a native element.

MINERALS OR NOT?

Minerals are traditionally considered to be inorganic. Although some organic materials are used as gemstones and for other decorative purposes, and some of these have a similar chemical composition to minerals, they are not true minerals. Amber is the hardened resin from trees; most amber is of Cretaceous, Palaeogene, and Neogene age and can contain fossilized insects. Jet, a soft black rock, is a type of impure coal, and became highly fashionable for jewelry in Victorian times. Shell, pearl, and coral (the use of which is not sustainable) are rich in calcite and are used decoratively.

CRYSTAL SYSTEMS

Crystals in the **cubic** system are relatively common and easily recognized. Such crystals have three equal axes at right angles; shapes include the cube and the eight-faced octahedron.

Hexagonal and trigonal systems are very similar to each other, with four axes of symmetry. Their crystals are often six-sided prisms with pyramidal tops (left).

Crystals in the **tetragonal** system have three axes of symmetry, all at right angles. In a tall prism (left), the vertical axis is the longest. Squat prisms are also common.

Crystals in the **monoclinic** system have three unequal axes of symmetry, only two at right angles. Tabular (flattened, left) and prismatic habits are common.

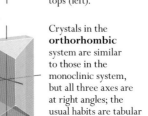

Crystals in the **orthorhombic** system are similar to those in the monoclinic system, but all three axes are at right angles; the usual habits are tabular and prismatic (left).

Crystals in the **triclinic** system have a low degree of symmetry, as the three axes are unequal in length, with no right angles between them. Prismatic habits are common.

NATIVE ELEMENTS

Of the many natural elements, only about 20 are found in the native state; that is, uncombined with other elements. They are divided into three groups. The metals rarely form as distinct crystals, tend to have a high specific gravity, and are relatively soft. Semimetals, such as antimony and arsenic, commonly occur as rounded masses. Nonmetals, including sulfur and carbon, often form as crystals.

ANTIMONY
Trigonal
H 3–3½ • SG 6.69
This rare semimetal occurs in hydrothermal veins, often with arsenic and silver minerals. The silvery gray masses are coated white when oxidized.

GRAPHITE
Hexagonal • H 1–2 • SG 2.09–2.23
A form of pure carbon common in metamorphic rocks, graphite is dark, soft, and greasy, and makes ideal pencil leads.

NATIVE COPPER

COPPER
Cubic • H 2½–3 • SG 8.94
Native copper occurs mainly as irregular masses or branching or wirelike forms. It is most notably associated with basaltic lava. A good conductor, it is widely used in the electrical industry.

COPPER ON GOETHITE GROUNDMASS

dendritic habit

distinct, isolated diamond crystal

DIAMOND
Cubic • H 10 • SG 3.51
The hardest of all minerals, diamond, a valuable form of carbon, occurs in igneous rocks called kimberlites, which originate in deep volcanic pipes.

rock groundmass

resinous luster

ARSENIC
Trigonal • H 3½ • SG 5.72–5.73
Highly poisonous arsenic usually forms as pale gray, rounded masses in hydrothermal veins. If heated, it smells of garlic.

SULFUR
Orthorhombic • H 1½–2½ • SG 2.07
Native sulfur forms striking yellow crystals and powdery crusts around volcanic vents. It is mined for use in sulfuric acid, dyes, insecticides, and fertilizers.

PLATINUM NUGGET

PLATINUM
Cubic • H 4–4½ • SG 21.44
A rare metal, native platinum forms as scales, grains, and nuggets in igneous rocks and is found in alluvial sands. Its high melting point means it is useful in industry, for example in aircraft spark plugs.

uneven surface

PLATINUM

GOLD IN QUARTZ

GOLD NUGGET

GOLD
Cubic • H 2½–3 • SG 19.3
Prized for its color and malleability, gold forms in hydrothermal veins and often weathers out to be found as nuggets in river sands.

IRON
Cubic • H 4½ • SG 7.3–7.87
Most native iron occurs at Earth's core, and near the surface it readily combines with other elements.

BISMUTH
Trigonal
H 2–2½ • SG 9.7–9.83
Native bismuth is relatively rare. Hardly ever found as distinct crystals, it more often has a granular or branching form.

mercury globules in rock cavities

MERCURY
Trigonal
H Liquid • SG 14.38
This is the only metal that is liquid at normal temperatures. In liquid form, mercury appears as silvery globules.

SILVER
Cubic • H 2½–3 • SG 10.5
Widely distributed but rarely found in abundance, native silver occurs mainly as twisted wires, scales, and branching masses.

SULFIDES

The sulfides are a large group of minerals, in which sulfur is combined with one or more metals. Many sulfides have a high specific gravity and a metallic luster. They often form as excellent crystals. Sulfides occur in many geological situations, but frequently in hydrothermal veins. The group includes the majority of the economically important metal ore minerals.

COBALTITE
Orthorhombic • H 5½ • SG 6.33
Cobaltite is an uncommon sulfide of arsenic and cobalt. It is an important cobalt ore in, for example, Sweden and Norway.

GALENA
Cubic • H 2½ • SG 7.58
Lead sulfide is one of the more abundant and widely distributed sulfide minerals. It is mined extensively as lead ore.

GREENOCKITE
Hexagonal
H 3–3½ • SG 4.82
Named after Lord Greenock, on whose Scottish property it was discovered in 1840, this rare cadmium sulfide can be yellow, red, or orange.

USUAL, CRYSTALLINE FORM OF SPHALERITE

MASSIVE SPHALERITE

SPHALERITE
Cubic • H 3½–4 • SG 3.9–4.1
A sulfide of zinc with variable iron content, sphalerite is the most heavily mined zinc-containing ore.

long, curved crystals resemble blades or swords

indistinguishable crystals form large masses

CINNABAR
Trigonal
H 2–2½ • SG 8–8.2
Red-colored mercury sulfide has been the main source of mercury through the centuries. It occurs around hot springs and volcanic vents.

STIBNITE
Orthorhombic • H 2 • SG 4.63–4.66
A sulfide of antimony, this dark gray mineral is the main ore of antimony. Large deposits occur in China, Japan, and the W. US.

BORNITE CRYSTALS

MASSIVE BORNITE

BORNITE
Orthorhombic • H 3 • SG 5.08
This copper iron sulfide is a coppery red color, tarnishing to shimmering purple and blue. It is an important copper ore.

COMMON, MASSIVE FORM OF CHALCOPYRITE

CHALCOPYRITE
Tetragonal
H 3½–4 • SG 4.35
A sulfide of copper and iron, chalcopyrite has a deep brassy yellow color. It has significant value as a copper ore.

CHALCOPYRITE CRYSTALS

ACANTHITE
Monoclinic
H 2–2½ • SG 7.22
A sulfide occurring as dark, metallic, sometimes spiky crystals, acanthite is the main ore of silver.

ORPIMENT
Monoclinic
H 1½–2 • SG 3.49
Named from the Latin for "golden paint," this arsenic sulfide occurs as foliated, columnar masses around hot springs.

REALGAR
Monoclinic • H 1½–2 • SG 3.56
A bright orange-red sulfide of arsenic, realgar was historically used as a pigment.

GLAUCODOT
Orthorhombic • H 5 • SG 6.05
This sulfide of cobalt, iron, and arsenic occurs as silver-white pyramidal crystals and brittle masses.

MOLYBDENITE
Hexagonal
H 1–1½ • SG 4.62–4.73
This molybdenum sulfide is lead-gray in color. It has a greasy feel, which is due to weak bonds within a layered atomic structure.

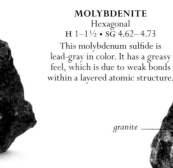

granite

thin hexagonal crystals in layers

MARCASITE
Orthorhombic
H 6–6½ • SG 4.89
A sulfide of iron that is lighter and more brittle than pyrite, marcasite frequently occurs as cockscombs and spear-shaped "twins."

indigo-blue covellite

COVELLITE
Hexagonal • H 1½–2 • SG 4.68
Although covellite is rare, it is a widespread copper sulfide. Its shining indigo-blue color makes it attractive to mineral collectors.

slender prismatic crystals

HAUERITE
Cubic • H 4 • SG 3.46
Hauerite is a very rare sulfide of manganese. The brown octahedral crystals can form when certain minerals are altered in the caps of salt domes.

ARSENOPYRITE
Monoclinic
H 5½–6 • SG 6.07
Silvery-colored arsenopyrite is a sulfide of arsenic and iron. With almost 50 percent arsenic content, it is a principal ore of arsenic which is poisonous to humans.

STANNITE
Tetragonal • H 4 • SG 4.3–4.5
Stannite is a sulfide of tin, copper, and iron mined for the tin. Its name comes from the Latin for tin, *stannum*.

PENTLANDITE
Cubic
H 3½–4 • SG 4.6–5
This nickel and iron sulfide is found in basic igneous rocks. It is an important source of nickel.

PYRITE
Cubic
H 6–6½ • SG 4.8–5
Nicknamed "fool's gold" because of its light goldish color, this iron sulfide is the most common of all sulfide minerals.

calcite groundmass

MILLERITE
Trigonal
H 3–3½ • SG 5.3–5.5
This sulfide of nickel occurs in limestones and ultramafic rocks. It is sought after as a nickel ore.

PYRRHOTITE
Monoclinic • H 3½–4½ • SG 4.58–4.65
An iron sulfide with variable iron content, pyrrhotite has a magnetism that increases as the iron content decreases.

CHALCOCITE
Monoclinic • H 2½–3 • SG 5.5–5.8
Dark gray to black in color, this copper sulfide has been mined for centuries. It is one of the most profitable copper ores.

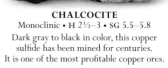

BISMUTHINITE
Orthorhombic • H 2 • SG 6.78

This sulfide of bismuth is an important ore. Much of the bismuth recovered is used in medicines and cosmetics.

SULFOSALTS

Sulfosalts are a group of about 200 minerals, structurally related to sulfides and with many of the same properties. In these compounds sulfur is combined with a metallic element—commonly silver, copper, lead, or iron—and a semimetal, often antimony or arsenic. Sulfosalts frequently occur in hydrothermal veins, usually in small amounts.

PYRARGYRITE
Trigonal
H 2½ • SG 5.85

Also called ruby silver, this sulfide of silver and antimony is red-black, but thin splinters appear deep ruby-red.

POLYBASITE
Monoclinic
H 2½–3 • SG 6.1

Somewhat uncommon, polybasite is a sulfide of silver, copper, antimony, and arsenic. It can yield worthwhile quantities of silver locally.

BOULANGERITE
Monoclinic
H 2½–3 • SG 6.2

A bluish-gray sulfide of lead and antimony, boulangerite is one of the few sulfide minerals to form fine, hairlike crystals.

STEPHANITE
Orthorhombic
H 2–2½ • SG 6.26

Stephanite is an opaque black sulfide of silver and antimony. It can be an important silver ore—for example, in Nevada.

striated prismatic crystals

bright, metallic lustre

radiating, needlelike crystals

JAMESONITE
Monoclinic • H 2½ • SG 5.63

A sulfide of lead, iron, and antimony, jamesonite's dark gray crystals can be fine and hairlike or larger and prismatic.

PROUSTITE
Trigonal • H 2–2½ • SG 5.55–5.64

This sulfide of silver and arsenic is also called light ruby silver. The transparent crystals are bright scarlet-red.

ZINKENITE
Hexagonal
H 3–3½ • SG 5.25–5.35

Zinkenite is a sulfide of lead and antimony. It occurs as steel-gray hairlike or needlelike crystals.

TETRAHEDRITE
Cubic
H 3–4½ • SG 4.6–5.1

This sulfide of copper, iron, and antimony is named after its tetrahedron-shaped crystals (crystals with four triangular faces).

TENNANTITE
Cubic • H 3–4½ • SG 4.59–4.75

Tennantite, a sulfide of copper, iron, and arsenic, is dark gray or black. It is very similar to tetrahedrite.

ENARGITE
Orthorhombic • H 3 • SG 4.45

A steel-gray colored sulfide of copper and arsenic, enargite has a metallic luster. Crystals are usually small, tabular, or prismatic.

BOURNONITE
Orthorhombic • H 2½–3 • SG 5.83

Black or steel-gray bournonite is a sulfide of lead, copper, and antimony. Crystals are tabular to prismatic.

OXIDES

Oxides are compounds of oxygen and other elements. Many oxides are very hard, some have a high specific gravity, and a number are brightly colored and are sought after as gems. The group includes the chief ores of iron, manganese, aluminum, tin, and chromium. Oxides can occur in hydrothermal veins, igneous and metamorphic rocks, and also, because they can be resistant to weathering and transport, in sands and gravels.

CUPRITE
Cubic • H 3½–4 • SG 6.14
Found in various shades of red, this copper oxide forms near Earth's surface by the oxidation of copper minerals.

PEROVSKITE
Orthorhombic
H 5½ • SG 3.98–4.26
Discovered in Russia in 1839, this dark-colored oxide of calcium and titanium forms in igneous and metamorphic rocks.

ILMENITE
Trigonal
H 5–6 • SG 4.68–4.76
This iron titanium oxide is the principal ore of titanium, a high-strength, low-density metal used in aircraft and rocket construction.

FRANKLINITE
Cubic
H 5½–6 • SG 5.07–5.22
This black or brown zinc iron oxide is found in metamorphosed limestones, notably those at Franklin, New Jersey.

octahedral franklinite crystal

URANINITE
Cubic • H 5–6 • SG 10.63–10.95
This highly radioactive, black- or brown-colored uranium oxide is the main ore of uranium, which is used in nuclear reactors to produce electricity, and in the construction of nuclear weapons.

CASSITERITE
Tetragonal • H 6–7 • SG 6.98–7.01
Almost the sole source of the world's tin, this tin oxide is mainly found in hydrothermal veins and pegmatites and sometimes as small grains among river gravels.

striated crystal face

vitreous luster

SAMARSKITE
Monoclinic
H 5–6 • SG 5–5.69
Minerals known as samarskite—a radioactive oxide of various metals including yttrium, iron, tantalum, and niobium—occur in igneous rocks and alluvial sands.

GAHNITE
Cubic • H 7½–8 • SG 4.62
A rare oxide of aluminum and zinc found mainly in metamorphic rocks, gahnite can form dark green or blue to black crystals.

CORUNDUM
Trigonal
H 9 • SG 3.98–4.1
Corundum is an aluminum oxide, second only in hardness to diamond. Ruby (red) and sapphire (usually blue) varieties are used as gemstones.

CHROMITE
Cubic • H 5½ • SG 4.5–4.8
This iron chromium oxide is the only important source of chromium, an element used in making chrome and stainless steel.

bright, metallic luster

HEMATITE
Trigonal • H 5−6 • SG 5.26
Widespread and abundant, this iron oxide is mined extensively for iron. Different forms are colored black, metallic gray, or brownish red.

HYDROXIDES

Hydroxide minerals are compounds of a metallic element and the hydroxyl radical (OH). They are common minerals and often form through a chemical reaction between an existing oxide and fluids rich in water, seeping through Earth's crust. Hydroxides tend to occur in the altered parts of hydrothermal veins and in metamorphic rocks.

GIBBSITE
Monoclinic
H 2½−3 • SG 2.38−2.42
Gibbsite is one of the essential aluminum hydroxides in the aluminum ore bauxite. It also occurs in hydrothermal veins.

STIBICONITE
Cubic • H 5½−7 • SG 3.5−5.5
An uncommon hydroxide of antimony, stibiconite is white or yellowish brown, and forms by the alteration of other antimony minerals, especially stibnite.

FERGUSONITE
Tetragonal
H 5½−6½ • SG 4.2−5.8
Fergusonite is the group name for an oxide of many metals, including yttrium, lanthanum niobium, and cerium.

 prismatic crystal

PYROLUSITE
Tetragonal
H 2−6½
SG 5.04−5.08
Pyrolusite is a common manganese oxide. It is the primary ore of manganese, an essential element in steel production.

LEPIDOCROCITE
Orthorhombic • H 5
SG 4.05−4.13
This relatively rare iron hydroxide may occur with goethite. It is reddish brown and can form irregular and fibrous shapes.

DIASPORE
Orthorhombic • H 6½−7 • SG 3.2−3.5
Diaspore is an aluminum hydroxide which is found in bauxite. It also occurs in marble and altered igneous rocks.

MASSIVE ROMANÈCHITE

RUTILE
Tetragonal
H 6−6½ • SG 4.23
A source of titanium, this oxide of titanium often forms impressive displays of thin, translucent needles in quartz crystals.

LIMONITE

ROMANÈCHITE
Monoclinic • H 5−6 • SG 6.45
Dark and opaque, this oxide of manganese with barium is usually found in aggregates or in massive forms. Crystals are rare.

BOTRYOIDAL GOETHITE

GOETHITE
Orthorhombic • H 5−5½ • SG 4.27−4.29
Goethite is a common iron hydroxide. A non-crystalline hydrated variety—limonite—gives a yellow-brown color to soils and rocks.

BAUXITE
Amorphous mixture
H 1−3 • SG 2.3−2.7
The primary aluminum ore, bauxite is not a single mineral, but rather an assemblage of aluminum hydroxides and iron oxides.

CHRYSOBERYL
Orthorhombic • H 8½ • SG 3.75
Chrysoberyl is beryllium aluminum oxide. It is a prized gemstone, known for its exceptional hardness and yellow-brown color.

ZINCITE
Hexagonal • H 4 • SG 5.64−5.68
Zincite is a rare oxide of zinc. Its only significant deposit in the US has been exhausted.

BRUCITE
Trigonal • H 2½−3 • SG 2.39
Brucite is a magnesium hydroxide, and its color is white, gray, blue, or green. It occurs in metamorphic rocks.

HALIDES

When metallic elements combine with halogen elements, halides are formed. The halogens are iodine, fluorine, chlorine, and bromine. Halide minerals are commonly very soft and of low specific gravity, often having crystals classified in the cubic system. Many of these minerals, such as halite and sylvite, form in evaporite sequences by the drying out of saline waters. Other halides—for example, fluorite—occur in hydrothermal veins.

YELLOW FLUORITE

PURPLE FLUORITE

cubic crystal

FLUORITE
Cubic • H 4 • SG 3.18
This calcium fluoride often forms transparent to translucent crystals of various colors. Large quantities are used for making hydrofluoric acid.

GREEN FLUORITE

SYLVITE
Cubic • H 2 • SG 1.99
Sylvite is a potassium chloride, a salt similar to halite. It occurs with halite in evaporite deposits. Sylvite is used to make potash fertilizers.

granular carnallite

CARNALLITE
Orthorhombic • H 2½ • SG 1.6
Carnallite is a hydrated chloride of magnesium and potassium, formed by the evaporation of saline water. It is important in fertilizer manufacture.

vitreous luster

transparent cubic crystal

ORANGE HALITE

DIABOLEITE
Tetragonal • H 2½ • SG 5.42
A lead copper chloride hydroxide with a light to dark blue color, diaboleite forms by the alteration of other minerals.

BOLEITE
Cubic • H 3–3½ • SG 5.05
Deep blue boleite is a rare hydroxide of lead, silver, copper, and chlorine. It occurs where lead and copper deposits have been altered.

HALITE
Cubic • H 2 • SG 2.17
Halite (common salt), a sodium chloride, occurs in extensive beds formed by the evaporation of saline water. It can be colored or colorless.

HALITE CRYSTALS

JARLITE
Monoclinic
H 4–4½ • SG 3.78–3.93
Usually white, jarlite is found in igneous rocks. It is a rare sodium strontium magnesium aluminum fluoride hydroxide.

chlorargyrite crust

CARBONATES

Carbonate minerals are compounds of metallic or semimetallic elements combined with the carbonate radical (CO_3). More than 70 carbonate minerals are known, but calcite, dolomite, and siderite account for most of the carbonate material in Earth's crust. Carbonates usually form as "good" crystals with regular shapes and no foreign substances enclosed within them. Many carbonates are pale colored, but some, such as rhodochrosite, smithsonite, and malachite, are brightly colored.

SMITHSONITE
Trigonal
H 4–4½ • SG 4.42–4.44
A zinc carbonate found in the upper oxidized zones of zinc ore deposits, smithsonite is also mined for zinc.

CHLORARGYRITE
Cubic • H 2½ • SG 5.55
This silver chloride is typically scaly, platelike, or, in masses, resembling wax. It occurs where silver deposits are altered.

dark green, tabular crystals of atacamite

NAILHEAD SPAR

CALCITE
Trigonal
H 3 • SG 2.71
Most calcite (a calcium carbonate), one of the most abundant minerals, occurs in the rocks limestone and marble. It can also form outstanding crystals.

DOGTOOTH SPAR

prismatic crystals

limestone groundmass

BARYTOCALCITE
Monoclinic • H 4 • SG 3.66–3.71
This barium calcium carbonate is white to yellowish, and often found in hydrothermal veins within limestone.

curved crystal faces

DOLOMITE
Trigonal
H 3½–4
SG 2.84–2.86
Dolomite is a calcium magnesium carbonate widespread in altered limestones. Dolomite rock, formed exclusively of massive dolomite, is used as a building stone.

ATACAMITE
Orthorhombic • H 3–3½ • SG 3.76
Green atacamite is a copper chloride hydroxide, occurring where copper deposits have been oxidized. It is a minor copper ore.

TRONA
Monoclinic • H 2½ • SG 2.14
A hydrated sodium carbonate, trona is gray, yellowish, or brown. It forms at Earth's surface, especially in saline desert environments.

CALOMEL
Tetragonal
H 1½–2 • SG 7.15
Rarely found, this mercury chloride is a white to gray or brown mineral. Its color deepens when exposed to light.

CRYOLITE
Monoclinic
H 2½ • SG 2.97
A rare aluminum sodium fluoride, cryolite often has an icelike appearance. It is found in granitic pegmatites and granites.

WITHERITE
Orthorhombic
H 3–3½ • SG 4.29
This fairly uncommon carbonate of barium is usually white or gray, and occurs in hydrothermal veins.

MAGNESITE
Trigonal • H 3½–4½
SG 2.98–3.02
A magnesium carbonate, magnesite usually occurs as white to brownish dense masses. It is made into furnace bricks and magnesia cement.

STRONTIANITE
Orthorhombic • H 3½
SG 3.74–3.78
This strontium carbonate is found in hydrothermal veins and limestones. The strontium is used in sugar refining and fireworks.

»

FLOS FERRI
("FLOWERS OF IRON")
ARAGONITE

ARAGONITE
Orthorhombic
H 3½–4 • SG 2.95
Aragonite is a calcium carbonate,
chemically identical to calcite, but
in a different crystal system and
much less common.

BOTRYOIDAL SIDERITE

SIDERITE
Trigonal
H 3½–4½ • SG 3.96
A brown-colored carbonate of iron,
named from the Greek for iron, *sideros*,
siderite occurs in a variety of forms.

RHOMBOHEDRAL
SIDERITE

short,
prismatic
crystals

PHOSGENITE
Tetragonal • H 2–3 • SG 6.12–6.15
This rare lead carbonate chloride is
formed close to the Earth's surface by the
reaction of lead-rich minerals with water.

ARTINITE
Monoclinic • H 2½ • SG 2.01–2.03
A hydrated magnesium carbonate
hydroxide, artinite has a
distinctive habit, with sprays
of white, needle-shaped crystals.
It occurs in serpentine rocks.

HYDROZINCITE
Monoclinic
H 2–2½ • SG 3.5–4
Hydrozincite, a zinc
carbonate hydroxide, is pale
gray, white, pink, or yellowish.
It fluoresces bluish white
under ultraviolet light.

TWINNED
ARAGONITE CRYSTALS

patch of green
malachite
around margins

limonite matrix

AZURITE
Monoclinic • H 3½–4 • SG 3.77
Azurite is a hydrous copper carbonate. Its
rich blue color and frequent association
with green malachite in hydrothermal
veins are distinctive features.

botryoidal
habit

LEADHILLITE
Monoclinic • H 2½–3 • SG 6.55
This lead sulfate carbonate hydroxide
usually occurs as well-formed crystals in
the oxidized zones of lead deposits.

CLUSTER OF
TWINNED
CRYSTALS

CERUSSITE
CRYSTALS

CERUSSITE
Orthorhombic • H 3–3½ • SG 6.53–6.57
A carbonate of lead occurring where
lead-bearing veins have been altered, cerussite
is the most common lead ore after galena.

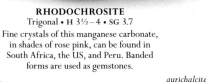

rhombohedral
crystals

RHODOCHROSITE
Trigonal • H 3½–4 • SG 3.7
Fine crystals of this manganese carbonate,
in shades of rose pink, can be found in
South Africa, the US, and Peru. Banded
forms are used as gemstones.

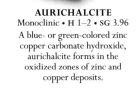

ANKERITE
Trigonal
H 3½–4 • SG 2.97
Ankerite is a carbonate of
calcium with lesser iron,
magnesium, and manganese.
It is sometimes found in
gold-bearing quartz veins.

auric halcite
crystals

characteristic
green coloring

AURICHALCITE
Monoclinic • H 1–2 • SG 3.96
A blue- or green-colored zinc
copper carbonate hydroxide,
aurichalcite forms in the
oxidized zones of zinc and
copper deposits.

MALACHITE ON
CHRYSOCOLLA

associated
azurite

MALACHITE
Monoclinic
H 3½–4 • SG 3.6–4.05
This striking green copper carbonate
often occurs as botryoidal masses.
It is used for ornamentation and
as a source of copper.

BOTRYOIDAL
MALACHITE

BORATES

Borates occur when metallic elements combine
with the borate radical (BO_3). There are more
than 100 borate minerals, the most common
being borax, kernite, ulexite, and colemanite.
Borates tend to be pale colored, relatively soft,
and have low specific gravity. Many occur in
evaporite sequences where saline waters have
dried out and minerals are then precipitated
among layers of sedimentary rocks.

BORACITE
Orthorhombic
H 7–7½ • SG 2.91–3.1
Crystals of this magnesium
borate chloride are pale
green or white, with a
vitreous luster and high
hardness. Boracite occurs
in salt deposits.

translucent,
prismatic
crystals

COLEMANITE
Monoclinic • H 4½ • SG 2.42
This hydrated calcium borate hydroxide forms when
saline water evaporates. It was the main source of
boron until the discovery of kernite.

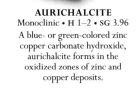

BORAX
Monoclinic • H 2–2½ • SG 1.71
A chalky-white sodium borate
hydrate, borax has many
applications, including in
medicines, laundry detergents,
glasses, and textiles.

KERNITE
Monoclinic • H 2½ • SG 1.91
A colorless or white sodium
borate hydrate, kernite has less
water than borax. The two
minerals often occur together.

ULEXITE
Triclinic • H 2½ • SG 1.95
A hydrated sodium calcium borate
hydroxide, ulexite's white, fibrous
crystals transmit light down their
length. It has uses similar to borax.

HOWLITE
Monoclinic • H 3½–6½ • SG 2.6
Howlite is a calcium borosilicate
hydroxide. It generally forms as
chalky, rounded masses.

NITRATES

Nitrates are a small group of
compounds where metallic
elements combine with the nitrate
radical (NO_2). These minerals
are usually very soft and have
low specific gravity. Many
dissolve readily in water and they
only rarely occur as crystals. They
are generally confined to arid
regions, forming coatings on the
land surface that often cover wide
areas. Commercially, nitrates can
be used as fertilizers or explosives.

NITRATINE
Trigonal • H 1½–2 • SG 2.26
This sodium nitrate typically occurs as crusts
on the ground surface in arid regions, especially
in Chile. It is white, gray, brown, or yellow.

SULFATES

Sulfates are composed of metals combined with the sulfate radical (SO_4). There are about 200 sulfates and many are rare. Some, such as gypsum, form in evaporite deposits where minerals are precipitated from drying saline solutions. Others occur as weathering products, or as primary minerals in hydrothermal veins. Many are economically important—barite is used to lubricate drills on oil rigs.

GYPSUM
Monoclinic • H 2 • SG 2.31–2.32
A widespread mineral, gypsum, a hydrated calcium sulfate, produces plaster of Paris when heated and mixed with water.

SATIN SPAR

RADIATING GYPSUM

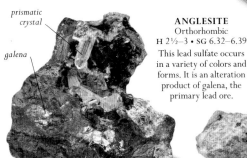

prismatic crystal

galena

ANGLESITE
Orthorhombic
H 2½–3 • SG 6.32–6.39
This lead sulfate occurs in a variety of colors and forms. It is an alteration product of galena, the primary lead ore.

THENARDITE
Orthorhombic
H 2½–3 • SG 2.66
A pale gray or brownish mineral, thenardite is a sodium sulfate. It is found on lava flows and around salt lakes.

crystalline chalcanthite

CHALCANTHITE
Triclinic
H 2½ • SG 2.29
Rich blue or green chalcanthite is a hydrated copper sulfate. It forms through the oxidation of chalcopyrite and other copper sulfates.

radiating, hairlike crystals

rock groundmass

LINARITE
Monoclinic
H 2½ • SG 5.35
Bright blue linarite is a hydrous copper lead sulfate. It occurs in the oxidation zones of copper and lead ores.

mass of needle-shaped brochantite crystals

CYANOTRICHITE
Monoclinic • H 1–3 • SG 2.76
This hydrated copper aluminum sulfate is named after the Greek for "blue" and "hair," referring to its clusters of fine, blue crystals.

GLAUBERITE
Monoclinic
H 2½–3 • SG 2.75–2.85
Glauberite is a sodium and calcium sulfate. Colorless, gray, or yellowish, it forms where saline water evaporates.

ALUNITE
Trigonal
H 3½–4 • SG 2.6–2.9
A hydrous sulfate of potassium and aluminum, alunite may be found at volcanic vents where rocks are altered by sulfur vapors.

CHROMATES

Chromate minerals form when metallic elements combine with the chromate radical (CrO_4). They are rare minerals—crocoite is the only reasonably well-known chromate. Generally brightly colored, they are sought after by mineral collectors. Chromates often form when hydrothermal veins are altered by fluids.

RED CROCOITE

slender, elongate crystals with striations

ORANGE CROCOITE

CROCOITE
Monoclinic • H 2½–3 • SG 5.97–6.02
Orange or red lead chromate forms in the oxidized zones of lead deposits. Good specimens come from Australia.

MELANTERITE
Monoclinic
H 2 • SG 1.89
White, green, or blue
melanterite is hydrated
iron sulfate. It is used for
purifying water supplies,
and as a fertilizer.

JAROSITE
Trigonal
H 2½–3½ • SG 2.9–3.26
Jarosite is a hydrous sulfate of iron and
potassium, occurring as brown coatings
on pyrite and other iron minerals.

prismatic crystals

EPSOMITE
Orthorhombic
H 2–2½ • SG 1.68
This hydrated magnesium
sulfate occurs in arid
regions and limestone cave
walls. It is the source of
the laxative Epsom salts.

CELESTINE
Orthorhombic • H 3–3½ • SG 3.96–3.98
This strontium sulfate is sought after not only
as the main source of strontium, but also for the
beautiful, transparent, pale-colored crystals.

COPIAPITE
Triclinic • H 2½–3 • SG 2.08–2.17
A yellow or green hydrated sulfate of
iron first described from Copiapó,
Chile, copiapite occurs where other
minerals are altered.

*iron oxide
groundmass*

BROCHANTITE
Monoclinic
H 3½–4 • SG 3.97
A copper sulfate
hydroxide, brochantite
forms emerald-green
crystals, crusts, or masses.

ANHYDRITE
Orthorhombic
H 3–3½ • SG 2.98
A form of calcium sulfate,
anhydrite occurs with gypsum but
is less common. It alters to gypsum
in humid conditions.

BARITE
Orthorhombic
H 3 • SG 4.5
The most common
barium mineral,
barium sulfate is
unusually heavy for a
pale-colored mineral.

POLYHALITE
Triclinic • H 2½–3½ • SG 2.78
Polyhalite is a hydrated potassium
calcium magnesium sulfate. Colorless,
white, pink, or red, it is widespread
in many marine salt deposits.

MOLYBDATES

Molybdates form when metals combine with
the molybdate radical (MoO_4). These minerals
are rare and tend to be dense and brightly
colored. Molybdate minerals occur in mineral
veins that have been altered by circulating water.
Wulfenite is the best-known molybdate mineral.
It is prized for its fine crystals and brilliant
orange or yellow colors.

TUNGSTATES

Tungstate minerals are compounds with metallic
elements joined to the tungstate radical (WO_4).
These minerals are rare and usually have a high
specific gravity. Some form as fine crystals.
Tungstates occur in hydrothermal veins and
pegmatites—very coarse-grained granitic
rocks where minerals form from fluids
permeating the rock.

HÜBNERITE
Monoclinic
H 4–4½ • SG 7.12–7.18
This manganese tungstate is a
major source of tungsten,
a metal used in steel alloys,
abrasives, and light bulbs.

quartz
*hübnerite
crystal*

*thin, tabular
wulfenite
crystal*

SCHEELITE
Tetragonal
H 4½–5 • SG 6.1
Mined for tungsten, this
calcium tungstate is found
in hydrothermal veins,
metamorphic and igneous
rocks, and alluvial sands.

*bipyramidal
scheelite crystal*

FERBERITE
Monoclinic
H 4–4½ • SG 7.58
Opaque black
ferberite is found in
hydrothermal veins and
granitic pegmatites. It is
an iron tungstate, mined
as a source of tungsten.

WULFENITE
Tetragonal • H 2½–3 • SG 6.5–7.5
This lead molybdate is found in oxidized
zones of lead and molybdenum deposits.
It is a minor source of molybdenum.

PHOSPHATES

When metals combine with the phosphate radical (PO₄), phosphate minerals are produced. These minerals form a large group of more than 200—however, many are very rare. Some minerals in this group are brightly colored. They usually form by the alteration of sulfide minerals, but they are sometimes primary minerals. A number of phosphates are radioactive.

HYDROXYLHERDERITE
Monoclinic • H 5–5½ • SG 2.95
Hydroxylherderite is a calcium beryllium phosphate. It occurs as pale yellow or greenish crystals, with a glassy luster, in granitic pegmatites.

DUFRÉNITE
Monoclinic
H 3½–4½ • SG 3.10–3.34
This hydrated phosphate of iron and calcium mainly occurs as green to black masses or crusts in altered veins and iron ores.

aggregate of xenotime crystals

XENOTIME-(Y)
Tetragonal
H 4–5 • SG 4.4–5.1
This widely distributed yttrium phosphate is yellow-brown, gray, or greenish, and forms in igneous and metamorphic rocks.

AUTUNITE
Orthorhombic
H 2–2½ • SG 3.05–3.2
Radioactive autunite is a lemon-yellow or pale green hydrated phosphate of calcium and uranium. It occurs where uranium minerals are altered.

tabular torbernite crystal

TURQUOISE
Triclinic • H 5–6 • SG 2.6–2.8
A sought-after gem for thousands of years, this hydrated phosphate of copper and aluminum is found in altered igneous rocks.

PYROMORPHITE
Hexagonal • H 3½–4 • SG 7.04
A lead phosphate chloride, variably colored greenish, orange, yellowish or brownish, pyromorphite forms in the oxidized zones of lead deposits.

prismatic apatite crystal

APATITE
Hexagonal
H 5 • SG 3.16–3.22
Apatite is the group name used for three similar calcium phosphate minerals: fluorapatite, chlorapatite, and hydroxylapatite.

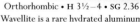

TORBERNITE
Tetragonal • H 2–2½ • SG 3.22
This copper uranium phosphate hydrate is related to autunite and occurs in similar geological environments. It is a radioactive mineral.

WAVELLITE
Orthorhombic • H 3½–4 • SG 2.36
Wavellite is a rare hydrated aluminum phosphate hydroxide. Colorless, gray, or greenish, glassy, needlelike crystals form radiating aggregates in altered rock.

sliced nodule

waxy luster

VARISCITE
Orthorhombic
H 3½–4½ • SG 2.56–2.61
This hydrated aluminum phosphate usually occurs as green microcrystalline masses in nodules, veins, or crusts.

TRIPLITE
Monoclinic • H 5–5½ • SG 3.5–3.9
Triplite is a phosphate of manganese and fluorine, sometimes with iron. It forms in granitic pegmatites.

AMBLYGONITE
Triclinic • H 5½–6 • SG 3.04–3.11
This rare lithium aluminum fluorophosphate mainly forms as masses, but crystals occur in Zimbabwe and Brazil.

VIVIANITE
Monoclinic • H 1½–2 • SG 2.67–2.69
Vivianite is a hydrated iron phosphate. It commonly forms as clusters of dark prismatic crystals in altered iron deposits.

radiating acicular (needle-shaped) crystals

libethenite crystal

LIBETHENITE
Orthorhombic
H 4 • SG 3.97
Libethenite is a light to dark green copper phosphate hydroxide. It forms in the upper oxidized zone of copper deposits.

MONAZITE
Monoclinic • H 5–5½ • SG 5–5.5
Phosphate minerals containing either cerium, lanthanum, or neodymium are all referred to as monazite. They are mined for the various elements.

BRAZILIANITE
Monoclinic • H 5½ • SG 2.98
First discovered in Brazil, this sodium aluminum phosphate hydroxide is yellow or greenish and forms in cavities in granitic pegmatites.

LAZULITE
Monoclinic
H 5½–6 • SG 3.12–3.24
A relatively rare, semiprecious blue gemstone, this magnesium aluminum phosphate hydroxide occurs in metamorphic and igneous rocks.

bipyramidal crystal

VANADINITE
Hexagonal • H 2½–3 • SG 6.88
This relatively rare lead vanadate chloride forms crystals in altered lead deposits. It is an important source of vanadium, used in steel alloys.

VANADATES

Vanadates are formed by the combination of metallic elements and the vanadate radical (VO_4). This group contains many rare minerals, which may be dense and brightly colored. Vanadates often form when hydrothermal veins are altered by permeating fluids. Most vanadates have no commercial value; however, carnotite is an important source of uranium.

powdery crust on sandstone

CARNOTITE
Monoclinic • H 2 • SG 4.75
Generally occurring as powdery yellow crusts in uranium deposits, radioactive carnotite is a hydrated vanadate of potassium and uranium.

TYUYAMUNITE
Orthorhombic
H 1½–2 • SG 3.57–4.35
This rare hydrated vanadate of calcium and uranium looks similar to carnotite, and also occurs in altered uranium deposits.

ARSENATES

Arsenates are mostly rare minerals composed of metallic elements and the arsenate radical (AsO_3 or AsO_4). They generally have a low hardness. Many arsenates are brightly colored—adamite is yellow or green, and clinoclase is green or blue. This group of minerals form in a variety of geological situations, but many arsenates occur in altered metal deposits.

ADAMITE
Orthorhombic
H 3½ • SG 4.32–4.48
This zinc arsenate hydroxide occurs in altered arsenic and zinc deposits, sometimes as exceptional crystals.

ERYTHRITE
Monoclinic • H 1½–2½ • SG 3.06
This hydrated cobalt arsenate forms purple-pink crystals or coatings. Excellent examples occur in Canada and Morocco.

radiating clusters of clinoclase crystals

BAYLDONITE
Monoclinic
H 4½ • SG 5.24–5.65
This hydrous arsenate of copper and lead is usually found as green or yellow crusts in altered hydrothermal veins.

olivenite crystals

OLIVENITE
Monoclinic
H 3 • SG 4.46
Olivenite is a copper arsenate hydroxide. It can be greenish, brownish, yellow, or gray and occurs in altered copper deposits.

quartz

CLINOCLASE
Monoclinic • H 2½–3 • SG 4.38
Clinoclase is a dark blue-green copper arsenate hydroxide that has a variety of forms in altered copper sulfide deposits.

MIMETITE
Hexagonal • H 3½–4 • SG 7.24
Unusual, barrel-shaped crystals characterize this lead arsenate chloride, although other forms also exist. Mimetite occurs in altered lead deposits.

CHALCOPHYLLITE
Trigonal • H 2 • SG 2.67–2.69
Bright blue-green chalcophyllite is a hydrated copper aluminum arsenate sulfate. It forms in oxidized copper deposits.

SILICATES

Silicates are the most common and largest group of minerals. The fundamental building blocks are tetrahedra of silicon and oxygen (SiO_4) together with other elements. They are subdivided into six groups based on the arrangement of the silica tetrahedra. Some form as isolated tetrahedra (nesosilicates), others occur in pairs (sorosilicates), and yet others have a three-dimensional network of tetrahedra (tectosilicates). Some silicates occur as chains of tetrahedra (inosilicates), while others form as sheets (phyllosilicates) or rings (cyclosilicates).

NESOSILICATES

ANDRADITE
Cubic • H 6½–7 • SG 3.8–3.9
Yellowish green, brown, or black, andradite garnet is a calcium iron silicate. Cut gems are excellent at separating white light into colors.

MASSIVE DUMORTIERITE

DUMORTIERITE
Orthorhombic • H 7–8 • SG 3.21–3.41
Dumortierite is a silicate of aluminum, iron, and boron. It usually forms fibrous aggregates of radiating crystals, but it can also be massive.

EUCLASE
Monoclinic
H 7½ • SG 2.99–3.1
Euclase is a beryllium aluminum silicate hydroxide. It may form white, colorless, green, or blue prismatic, striated crystals.

HUMITE
Orthorhombic • H 6 • SG 3.2–3.32
A magnesium iron silicate fluorohydroxide, humite generally occurs as yellow to orange granular masses in metamorphosed limestones and dolomites.

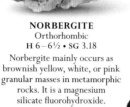

NORBERGITE
Orthorhombic
H 6–6½ • SG 3.18
Norbergite mainly occurs as brownish yellow, white, or pink granular masses in metamorphic rocks. It is a magnesium silicate fluorohydroxide.

KYANITE
Triclinic • H 5½–7 • SG 3.53–3.67
Kyanite is an aluminum silicate. Its bladed crystals in schist and gneiss form at high pressures in the Earth.

DATOLITE
Monoclinic
H 5–5½ • SG 2.96–3
Datolite is a hydrous calcium boron silicate. Not very common, it is mainly found in veins or cavities in igneous rocks.

PYROPE
Cubic • H 7–7½ • SG 3.58
Pyrope garnet is a dark red magnesium aluminum silicate. It forms at high pressures in metamorphic and some igneous rocks.

rhombic crystal faces

ALMANDINE
Cubic • H 7–7½ • SG 4.32
The most common garnet, pinkish red almandine is an iron aluminum silicate. It is widely used as a gem.

green coloring due to vanadium

vitreous luster

red coloring due to iron

GREEN GROSSULAR

GROSSULAR
Cubic • H 6½–7 • SG 3.59
Grossular garnet is a silicate of calcium and aluminum that sometimes forms in marble. It occurs in a wide range of colors.

RED GROSSULAR

OLIVINE
Orthorhombic
H 7 • SG 3.27–4.39

Common in igneous rocks, nesosilicate minerals varying in composition from magnesium silicate to iron silicate are called olivine.

pinkish brown topaz

TOPAZ
Orthorhombic • H 8 • SG 3.4–3.6

Topaz is an aluminum silicate fluoride hydroxide. Crystal size is usually small, but a giant crystal weighing 596 lb (271 kg) was found in Brazil.

translucent

typical green color

wedge-shaped crystal

TWINNED TITANITE CRYSTALS

TITANITE
Monoclinic • H 5–5½ • SG 3.48–3.6

Variable in color, titanite is a calcium titanium silicate. It is excellent at dispersing light—even better than diamond.

CRYSTALS IN MATRIX

CHLORITOID
Monoclinic • H 6½ • SG 3.4–3.8

Widespread in metamorphic and volcanic rocks, chloritoid is a dark green or black hydrous aluminum silicate of iron, magnesium, and manganese.

ANDALUSITE
Orthorhombic
H 6½–7½ • SG 3.13–3.21

Andalusite is an aluminum silicate. It occurs mainly in low-grade metamorphic rocks as coarse prismatic crystals with a square cross-section.

prismatic crystal

ZIRCON
Tetragonal
H 7½ • SG 4.6–4.7

Zircon, a zirconium silicate, is a gem used extensively in jewelry. It is also the main source of the metal zirconium, used in nuclear reactors.

short, prismatic willemite crystal

WILLEMITE
Trigonal
H 5½ • SG 3.89–4.19

Willemite is a zinc silicate. White, green, yellow, or reddish, and usually massive, it occurs in altered zinc deposits and metamorphosed limestone.

long, parallel, fibrous crystals

SILLIMANITE
Orthorhombic • H 6½–7½ • SG 3.23–3.27

Sillimanite is an aluminum silicate with long, slender crystals. Its composition is identical to that of andalusite, but it forms at higher temperatures and pressures.

EPIDOTE
Monoclinic
H 6 • SG 3.38–3.49

Epidote is an abundant mineral. Crystals of this hydrous calcium aluminum iron silicate are prismatic or tabular, green, and striated.

AXINITE
Triclinic
H 6½–7 • SG 3.25–3.28

Axinite is a hydrous calcium iron manganese aluminum boron silicate, with axe-head-shaped crystals.

HEMIMORPHITE
Orthorhombic
H 4½–5 • SG 3.48

This hydrated silicate of zinc occurs in altered zinc deposits. It is very variable in both its color and form.

rounded aggregates

DANBURITE
Orthorhombic
H 7–7½ • SG 2.93–3.02

Danburite is a calcium boron silicate. Its variably colored crystals resemble those of topaz, but it can also be granular.

VESUVIANITE
Tetragonal
H 6½ • SG 3.32–3.43

Vesuvianite, also known as idocrase, is a hydrous calcium sodium magnesium iron aluminum silicate, with fluorine. It occurs in marble and igneous rocks.

CYCLOSILICATES

BENITOITE
Hexagonal
H 6–6½ · SG 3.64–3.65
This usually blue barium titanium silicate occurs in serpentinite and veins in schist. Gem-quality crystals come from California.

six-sided crystal

TOURMALINE
Trigonal
H 7 · SG 2.9–3.1
Tourmaline is the name for a group of 11 hydrous boron silicate minerals with the same crystal structure but varying chemistry.

prismatic crystal

AQUAMARINE   EMERALD

BERYL
Hexagonal
H 7½–8 · SG 2.63–2.92
This beryllium aluminum silicate is both a source of beryllium and a gemstone. Gem varieties include emerald (green) and aquamarine (greenish blue).

MORGANITE
Hexagonal
H 7½–8 · SG 2.63–2.92
Morganite is a pink variety of beryl, colored by additional cesium or manganese. It forms tabular crystals in pegmatites.

SUGILITE
Hexagonal
H 6–6½ · SG 2.74–2.79
This rare silicate of potassium, sodium, iron, and lithium occurs in metamorphic rock.

columnar, six-sided prismatic crystal

HELIODOR
Hexagonal
H 7½–8 · SG 2.63–2.92
Named after the Greek for "sun," heliodor is a yellow variety of beryl. Fine examples come from Russia.

rock groundmass

INOSILICATES

ACTINOLITE
Monoclinic
H 5–6 · SG 3.03–3.24
Actinolite is a more iron-rich, darker colored form of the amphibole tremolite. It is one of the asbestos minerals.

TREMOLITE
Monoclinic · H 5–6 · SG 2.99–3.03
A widespread amphibole, this hydrous silicate of calcium and magnesium forms in metamorphic rocks. It has been used as asbestos.

vitreous luster

PECTOLITE
Triclinic · H 4½–5 · SG 2.48–2.9
This sodium calcium silicate hydroxide forms in cavities within basalt. It is common in Canada, the US, and England.

NEPHRITE
Monoclinic · H 6½ · SG 2.99–3.24
This very tough, cream to dark green form of the amphiboles tremolite and actinolite is commonly known as jade.

AEGIRITE
Monoclinic · H 6 · SG 3.5–3.6
This brown, green, or black pyroxene is a sodium iron silicate. It forms in metamorphic and dark igneous rocks.

HORNBLENDE
Monoclinic · H 5–6 · SG 3–3.4
Common in igneous and metamorphic rocks, this amphibole mineral is a dark, hydrous silicate of calcium, magnesium, iron, and aluminum.

long, prismatic crystal

fibrous mass

WOLLASTONITE
Triclinic · H 4½–5 · SG 2.86–3.09
This calcium silicate, found in marble and other metamorphic rocks, is used in ceramics, paints, and as a form of asbestos.

RHODONITE
Triclinic
H 5½–6½ · SG 3.57–3.76
Rose-red or pink rhodonite is a manganese calcium silicate, occurring as crystals, masses, and grains. It is used in jewelry making.

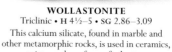

slender prismatic crystal

prismatic diopside crystal

quartz

SPODUMENE
Monoclinic • H 6½–7 • SG 3.1–3.2
This pyroxene mineral is a lithium aluminum silicate. Some enormous crystals have been found; the largest weighed almost 110 tons.

DIOPSIDE
Monoclinic • H 5½–6½ • SG 3.22–3.38
This pyroxene is a generally green-colored calcium magnesium silicate. It occurs in metamorphic and igneous rocks.

richterite crystal

PIGEONITE
Monoclinic
H 6 • SG 3.3–3.46
This brown to purplish black uncommon pyroxene is a magnesium iron calcium silicate. It occurs in igneous rocks and in meteorites.

RICHTERITE
Monoclinic
H 5–6 • SG 3.1
The amphibole richterite is a hydrous silicate of sodium, calcium, and magnesium. It occurs in metamorphosed limestones and igneous rocks.

AUGITE
Monoclinic
H 5½–6 • SG 3.19–3.56
Augite is a common pyroxene. This silicate of calcium, magnesium, and iron occurs in igneous and metamorphic rocks.

ASTROPHYLLITE
Triclinic • H 3 • SG 3.2–3.4
This complex hydrous silicate of potassium, sodium, iron, and titanium, with fluorine, occurs in gneiss and cavities in igneous rocks.

jadeite

JADEITE
Monoclinic
H 6 • SG 3.25–3.35
One of two carvable materials commonly referred to as jade, this pyroxene mineral is a sodium aluminum iron silicate.

long prismatic crystal

rock groundmass

long, striated crystals

RIEBECKITE
Monoclinic • H 5–5½ • SG 3.26–3.44
This amphibole is a hydrous sodium iron silicate, occurring in igneous rocks. The variety crocidolite, or "blue asbestos," occurs in metamorphosed ironstone.

PHYLLOSILICATES

spherical mass of radiating crystals

PREHNITE
Orthorhombic
H 6–6½ • SG 2.8–2.95
Prehnite is a hydrous silicate of calcium and aluminum. It occurs in cavities in basalt.

OKENITE
Triclinic • H 4½–5 • SG 2.28–2.33
This hydrated calcium silicate has fibrous or bladelike crystals, usually white or tinted blue or yellow. It occurs in basalt.

CLINOCHLORE
Monoclinic
H 2–2½ • SG 2.6–3.02
This hydrous silicate of iron, magnesium, and aluminum forms as green tabular crystals. It occurs in a variety of rock types.

tabular crystal

radiating crystal groups

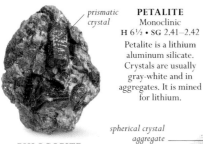

prismatic crystal

PETALITE
Monoclinic
H 6½ • SG 2.41–2.42
Petalite is a lithium aluminum silicate. Crystals are usually gray-white and in aggregates. It is mined for lithium.

MUSCOVITE
Monoclinic
H 2½ • SG 2.77–2.88
Muscovite, or white mica, is a potassium aluminum aluminosilicate hydroxide, with fluorine. It is very common in metamorphic rocks and granite.

PHLOGOPITE
Monoclinic
H 2–3 • SG 2.78–2.85
The colorless, yellow, or brown mica phlogopite is a potassium magnesium aluminosilicate hydroxide.

spherical crystal aggregate

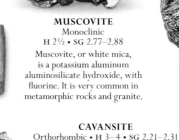

CAVANSITE
Orthorhombic • H 3–4 • SG 2.21–2.31
Cavansite is a hydrated calcium vanadium silicate. It is blue or greenish blue, and occurs in cavities in basalt.

tabular lepidolite crystal

SEPIOLITE
Orthorhombic • H 2 • SG 2–2.2
This pale-colored clay mineral, a hydrous magnesium silicate, usually occurs as earthy masses in altered rocks. It is used for ornamental carving.

typical blue coloring

LEPIDOLITE
Monoclinic • H 2½–3½ • SG 2.80–2.9
Lepidolite is the term used for mica minerals that are potassium lithium aluminum aluminosilicate hydroxides, with fluorine.

TECTOSILICATES

prismatic crystal

SMOKY QUARTZ
Trigonal • H 7 • SG 2.65
Smoky quartz is a brown variety of quartz, a silicon dioxide. It occurs in igneous rocks and hydrothermal veins.

ROSE QUARTZ
Trigonal • H 7 • SG 2.65
Rose quartz is a prized, translucent pink variety of quartz. Crystals are very rare; it usually has a massive habit.

MILKY QUARTZ
Trigonal
H 7 • SG 2.65
This very common, milky white variety of quartz occurs in all types of rocks and in hydrothermal veins.

CITRINE
Trigonal
H 7 • SG 2.65
Citrine is a yellow to brownish variety of quartz. It resembles topaz and is often used as a gemstone.

AMETHYST
Trigonal
H 7 • SG 2.65
Amethyst is a purple variety of quartz, prized since ancient times. It is found in hydrothermal veins and in lava cavities.

slender crystal

ZINNWALDITE
Monoclinic • H 3½–4
SG 2.9–3.1
This brown, gray, or green mica is
potassium lithium iron aluminum
aluminosilicate hydroxide, with
fluorine.

CHRYSOCOLLA
Orthorhombic
H 2½–3½ • SG 1.93–2.4
This blue or greenish blue
hydrated silicate of copper
and aluminum forms in
altered copper deposits.
Crystals do not occur.

VERMICULITE
Monoclinic
H 1½ • SG 2.4–2.7
This green or yellow clay
mineral often occurs where
micas have been altered. It is a
hydrated silicate of magnesium,
iron, and aluminum.

GLAUCONITE
Monoclinic • H 2 • SG 2.4–2.95
A mica, glauconite is a potassium
sodium magnesium aluminum iron
aluminosilicate hydroxide. It occurs
in marine sedimentary rocks.

_tabular biotite
crystal_

CHRYSOTILE
Monoclinic • H 2½ • SG 2.53
Chrysotile is a hydrous silicate of
magnesium, forming fibrous, silky, white
crystals in serpentinite rock. It is the most
abundant of the asbestos minerals.

BIOTITE
Monoclinic • H 2½–3 • SG 3.3
Biotite, or black mica, is a potassium
iron magnesium aluminosilicate
hydroxide, with fluorine. It is abundant
in igneous and metamorphic rocks.

TALC
Triclinic
H 1 • SG 2.58–2.83
The softest mineral, white, gray, or
greenish talc is magnesium silicate
hydroxide. Its many uses include
toiletries, paint, and ceramics.

PYROPHYLLITE
Triclinic
H 1–2 • SG 2.65–2.9
This aluminum silicate hydroxide,
of variable form and color, occurs
in low-grade metamorphic rocks.
It has good insulating properties.

ALLOPHANE
Amorphous
H 3 • SG 2.8
A clay mineral,
this aluminosilicate
hydrate is an alteration
product of feldspars
and other minerals. It
forms crusty masses.

_vitreous
luster_

_prismatic
crystals_

ROCK CRYSTAL
Trigonal
H 7 • SG 2.65
Rock crystal is a transparent,
colorless variety of quartz, which
is widely used as an ornament or
a gemstone.

JASPER
Trigonal • H 7 • SG 2.6
Jasper is a variety of chalcedony, or
microcrystalline quartz, and is used for jewelry. It
is opaque, and usually colored red by impurities.

_white
quartz
vein_

AGATE
Trigonal
H 7 • SG 2.6
Forming in cavities in lavas,
agate is a type of chalcedony.
It is characterized by
concentric color bands
caused by impurities.

≫

» TECTOSILICATES

alternating bands of different colors

CHALCEDONY
Trigonal
H 7 • SG 2.65
Chalcedony is microcrystalline quartz, a silicon dioxide. Pure chalcedony is white. It forms in veins and cavities in many rock types.

ONYX
Trigonal
H 7 • SG 2.7
Onyx is a parallel-banded gemstone variety of chalcedony. It is not particularly common; notable locations are India and South America.

OPAL
Amorphous
H 5½–6½ • SG 1.9–2.3
Opal is a hydrated silicon dioxide, occurring as nodules, encrustations, or masses in most rock types. Impurities impart a variety of colors. Opal is treasured as a gemstone.

CARNELIAN
Trigonal
H 7 • SG 2.7
Carnelian is a variety of chalcedony colored red to orange by iron oxide. The finest quality carnelian comes from India.

BLOODSTONE
Trigonal • H 7 • SG 2.7
Bloodstone is a variety of chalcedony colored dark green by traces of iron silicates. Flecks of red jasper throughout resemble blood.

CHRYSOPRASE
Trigonal • H 7 • SG 2.7
Chrysoprase is a variety of chalcedony containing nickel, which imparts a green color. It is the most valuable of the chalcedony varieties.

precious opal

streaks of yellow opal

yellow cancrinite

prismatic crystal

ironstone matrix

CANCRINITE
Hexagonal
H 5–6 • SG 2.42–2.51
The feldspathoid cancrinite is a variously colored hydrated sodium calcium aluminosilicate carbonate sulfate.

SCAPOLITE
Tetragonal
H 5½–6 • SG 2.5–2.78
The name scapolite encompasses a series of complex sodium calcium silicates, occurring mainly in metamorphic rocks.

MICROCLINE
Triclinic
H 6–6½ • SG 2.54–2.57
A very common alkali feldspar, this potassium aluminosilicate is usually white or pinkish. The green variety is called amazonstone.

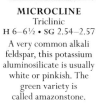

ANORTHITE
Triclinic
H 6–6½ • SG 2.74–2.76
This plagioclase feldspar is a calcium aluminosilicate. It forms pale-colored crystals, grains, or masses.

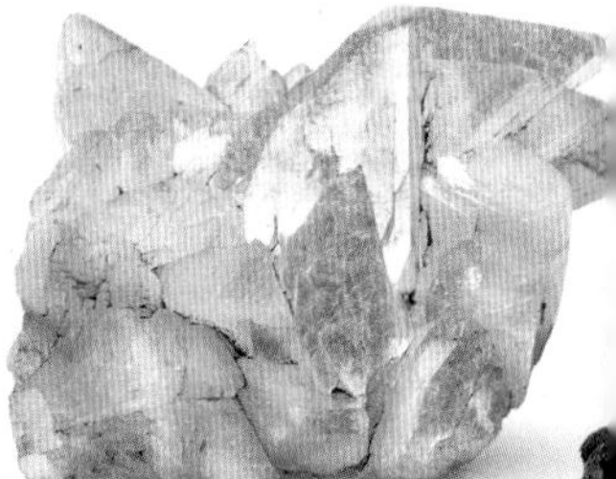

HEULANDITE
Monoclinic
H 3–3½ • SG 2.2
This zeolite is a hydrated sodium calcium aluminosilicate. It is used in petroleum refining as a molecular sieve.

SCOLECITE
Monoclinic
H 5–5½ • SG 2.25–2.29
This zeolite is a hydrated calcium aluminosilicate. Usually colorless or white, it is common in igneous and metamorphic rocks.

long, slender, needlelike crystals

ANDESINE
Triclinic
H 6–6½ • SG 2.66–2.68
A plagioclase feldspar, andesine is a gray or white sodium calcium aluminosilicate. It is widespread in igneous rocks.

NATROLITE
Orthorhombic
H 5–5½ • SG 2.2–2.26
One of the most widespread zeolites, this hydrated sodium aluminosilicate occurs in cavities in basalts, and in hydrothermal veins.

HYALOPHANE
Monoclinic
H 6–6½ • SG 2.81
A relatively rare barium feldspar, this potassium barium aluminosilicate is colorless, white, yellow, or pink.

massive sodalite

SODALITE
Cubic
H 5½–6 • SG 2.27–2.33
The feldspathoid mineral sodalite is a sodium aluminum silicate chloride. Rare crystals have been found in Canada.

STILBITE
Monoclinic
H 3½–4 • SG 2.19
This widespread zeolite is a hydrated sodium calcium potassium aluminosilicate. It forms sheaflike crystals in a variety of rock types.

HARMOTOME
Monoclinic
H 4–5 • SG 2.41–2.47
This widespread zeolite is a pale hydrated barium calcium potassium sodium aluminosilicate. It occurs in hydrothermal veins and volcanic rocks.

ANALCIME
Triclinic
H 5–5½ • SG 2.24–2.29
This pale zeolite, a hydrated sodium aluminosilicate, occurs in igneous, metamorphic, and some sedimentary rocks.

ALBITE
Triclinic
H 6–6½ • SG 2.6–2.65
A plagioclase feldspar, albite is a pale-colored sodium aluminosilicate. It is an abundant mineral.

ANORTHOCLASE
Triclinic
H 6–6½ • SG 2.57–2.6
An alkali feldspar, anorthoclase is a sodium potassium aluminosilicate occurring as prismatic or tabular crystals.

lazurite crystal

calcite

LAZURITE
Cubic
H 5–5½ • SG 2.38–2.45
The intense blue feldspathoid mineral lazurite is a sodium calcium aluminosilicate sulfate. It is the main mineral in lapis lazuli gems.

THOMSONITE
Orthorhombic
H 5–5½ • SG 2.23–2.29
A pale-colored zeolite, thomsonite is a hydrated aluminosilicate of sodium and calcium, widespread in cavities in basalts.

LAUMONTITE
Monoclinic
H 3½–4 • SG 2.23–2.41
A widespread zeolite, laumontite is a hydrated calcium aluminosilicate. It occurs in igneous, metamorphic, and sedimentary rocks.

short, prismatic orthoclase crystal

POLLUCITE
Cubic
H 6½ • SG 2.9
A rare zeolite, this complex hydrated aluminosilicate of cesium and sodium commonly contains other elements, such as calcium. It is a source of cesium.

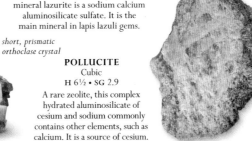

CHABAZITE
Trigonal
H 4 • SG 2.05–2.2
This common zeolite is a hydrated sodium calcium aluminosilicate. Crystals are colorless, white, yellow, or pink.

ORTHOCLASE
Monoclinic • H 6 • SG 2.55–2.63
Orthoclase, an alkali feldspar, is a potassium aluminosilicate. It is a major component of many igneous and metamorphic rocks.

hairlike tufts of mesolite crystals

MESOLITE
Orthorhombic • H 5 • SG 2.26
This white or colorless zeolite occurs in igneous and metamorphic rocks. It is a hydrated sodium calcium aluminosilicate.

HAÜYNE
Cubic • H 5½–6 • SG 2.44–2.5
A feldspathoid mineral, haüyne is a sodium potassium calcium aluminosilicate, with sulfate and chlorine. It occurs mainly in silica-deficient volcanic rocks.

pseudocubic rhombohedral crystal

ROCKS

The rocks that form Earth's crust are composed of different combinations of minerals. Thought of as the epitome of strength and solidity, they are actually in a state of constant change, with rock being destroyed and reformed over immense periods of time. Rocks are classified into three main groups, depending on how they form.

Igneous rocks can be formed by the cooling of magma deep underground, like this granite, or by volcanic eruption.

Sedimentary rocks such as sandstone are formed by the erosion of existing rocks and the deposition of their fragments.

Metamorphic rocks such as schist form when changes in pressure, temperature, or both alter existing rocks.

The world's oldest known rocks, from Canada's Northwest Territories, have existed for about 4 billion years, but most rocks are much younger. The chalk cliffs facing the English Channel date back to the Cretaceous Period, which ended 66 million years ago (see timeline, below), and the European Alps are younger still. Even in the Grand Canyon, the oldest rocks date back 2 billion years—less than half the lifetime of the planet as a whole. The reason for this is that Earth is tectonically active, with new rock being formed by the planet's internal heat. At the same time, existing rocks are broken down, in an endless cycle that started when Earth's crust first solidified.

ROCK GROUPS

Geologists classify rocks into three groups—igneous, sedimentary, and metamorphic—reflecting the different ways in which they are formed. Igneous rocks are created by lava from volcanic eruptions on Earth's surface and from the cooling of molten magma underground. The heat in lava and magma rises from the mantle beneath Earth's crust. The commonest kind, a black volcanic rock called basalt, forms most of the sea floor. Igneous activity can also create plutonic rock, which cools and solidifies beneath the surface, often in huge masses called batholiths. This is how most of the world's granite originates.

Sedimentary rocks are created on Earth's surface. Their key feature is their layers, or strata, built up over long periods of time. Some sedimentary rocks—such as sandstone and shale—form when existing rocks are eroded, releasing particles that are then washed or blown away, producing rock elsewhere. Others, such as rock salt and rock gypsum, are created when salt water evaporates, leaving behind its dissolved minerals, known as evaporites. Sedimentary rock can also have a biological origin: chalk is formed from the microscopic skeletons of marine organisms, and coal is derived from the remains of plants, compressed over millions of years.

Metamorphism takes place deep beneath Earth's surface, when rocks are altered by heat, pressure, or both. For example, marble is created when limestone is heated by lava or by magma. Recrystallization of the limestone may remove any bedding in the original rock, enabling some marbles to be cut without splitting apart, a quality prized in sculpture. Metamorphism can take place over very wide areas. If temperature and pressure are sufficiently high, all rocks can be metamorphosed and ultimately melted, forming magma and completing the rock cycle.

GRAND CANYON, ARIZONA >
This view of the Grand Canyon shows the almost horizontal layering of sedimentary rocks and the effects of river erosion.

GEOLOGICAL TIMELINE

4,600	4,000	2,500	541	485	444	419	359	299	252	201	145	66	56	34	23	5.3	2.58 MYA	11,700 years ago

Hadean	Archean	Proterozoic		Phanerozoic		**Eon**
			Paleozoic	Mesozoic	Cenozoic	**Era**

Cambrian · Ordovician · Silurian · Devonian · Carboniferous · Permian · Triassic · Jurassic · Cretaceous · Paleogene · Neogene · Quaternary **Period**

Paleocene · Eocene · Oligocene · Miocene · Pliocene · Pleistocene · Holocene **Epoch**

Trilobites were a group of marine animals that became extinct at the end of the Paleozoic Era. *Dalmanites caudatus* was a common trilobite during the Silurian Period.

Geologists divide Earth's history into periods of time. The transition from one to the next is defined by a global event verified by fossil evidence. Eons begin with a major change, such as the evolution of multicellular organisms that ushered in the Phanerozoic. Epochs are divided by events such as the last ice age, after which the Holocene begins.

MYA: million years ago

IGNEOUS ROCKS

Rocks that solidify from a molten state are called igneous rocks and are broadly divided into extrusive rocks and intrusive rocks. Extrusive rocks form from lava on Earth's surface, while intrusive rocks form underground, from magma. Lava and magma are rich in silica and metallic elements. As they cool, minerals such as feldspars, micas, amphiboles, and pyroxenes form. Different combinations of these minerals make up many of the igneous rocks.

RHYOLITE
This fine-grained, pale-colored lava contains much quartz, mica, and feldspar. It frequently includes visible phenocrysts (larger crystals).

vesicle (gas bubble cavity)

BANDED RHYOLITE
Similar in composition to granite, rhyolite also has tiny crystals because it forms from rapidly cooled lava. Banding shows the direction of lava flow.

BASALT
A dark, fine-grained volcanic rock, basalt is the most common rock forming the oceanic crust.

VESICULAR BASALT
Dominated by plagioclase feldspars, pyroxenes, and olivine, this lava is dark-colored, with numerous vesicles (former gas cavities).

dark, fine-grained rock

PAHOEHOE
Common in Hawaii, pahoehoe is named after the Hawaiian word *hoe*, meaning "to swirl." This basaltic lava is also known as ropy lava.

AMYGDALOIDAL BASALT
Amygdaloidal means almond-shaped and refers to the gas bubble cavities found in basaltic lavas, which are typically infilled with secondary minerals such as zeolites, carbonates, and agate.

PORPHYRITIC BASALT
This dark-colored rock has large crystals, typically of olivine or plagioclase feldspars, set in a fine-grained matrix.

PÉLÉ'S HAIR
Named after a Hawaiian goddess, and composed of numerous slender, brown, glassy strands, Pélé's hair is formed by lava sprays being blown in the wind.

PUMICE
Formed from frothy lava, pumice contains small crystals of feldspar in a very fine-grained matrix. It is of such low density that it floats on water.

IGNIMBRITE
A very fine-grained, pale-colored volcanic tuff, ignimbrite often shows banding caused by the flow of lava when molten.

LITHIC TUFF
This rock contains fragments of previously formed rock in a very fine-grained matrix. Usually pale, lithic tuff is formed by violent volcanic eruptions.

SPINDLE BOMB
Molten, low-viscosity basaltic lava blasted through the air can assume aerodynamic forms such as this one. These masses cool to form "bombs."

BREADCRUST BOMB
This variety of volcanic bomb is characterized by a cracked crust, due to continued expansion of the interior after the exterior has solidified.

AGGLOMERATE

Composed of relatively large rock fragments set in a finer matrix, agglomerate forms after volcanic explosions.

PITCHSTONE

This dense volcanic rock has variable composition and color, and a pitchlike, resinous luster.

PORPHYRITIC TRACHYTE

This rock has a complex mineralogy of alkali feldspars, quartz, micas, pyroxenes, and hornblende. Its fine-grained matrix contains larger crystals.

PORPHYRITIC ANDESITE

Commonly composed of plagioclase feldspars, pyroxenes, and amphiboles, porphyritic andesite contains larger crystals set in its fine-grained matrix.

DACITE

A pale- to medium-colored rock with very fine grains, dacite is mainly made of plagioclase feldspars and quartz, with pyroxenes, biotite mica, and hornblende.

ANDESITE

Named after the Andes mountains, andesite is common in many of the world's subduction-related volcanic arcs—chains of islands and mountains formed by tectonic plate movement. It is typically about 60 percent silica.

AMYGDALOIDAL ANDESITE

Brown, gray, purple, or red, this is a fine-grained volcanic rock containing infilled gas bubbles called amygdales.

SPILITE

This fine-grained, brownish volcanic rock contains augite and plagioclase feldspars. It forms by the alteration of basalt lava in contact with sea water.

TRACHYTE

Trachytes are a group of fine-grained volcanic rocks containing alkali feldspars and dark-colored mafic minerals such as biotite, hornblende, and pyroxenes.

curved fracture

OBSIDIAN

Typically very dark, obsidian is formed by the rapid cooling of highly viscous rhyolitic lava before individual minerals have had time to form crystals. This gives it a glassy texture. Obsidian has been used as a cutting tool since ancient times.

pale patch of devitrified glass

SNOWFLAKE OBSIDIAN

The "snowflakes" in this glassy, black volcanic rock, with a high silica content, are areas of devitrified (crystalline) glass.

RHOMB PORPHYRY

This rock is characterized by large crystals of feldspars with a rhomb-shaped cross-section set in its dark, fine-grained matrix. Unlike all the extrusive rocks on this spread, rhomb porphyry is an intrusive rock—formed from magma underground—like all the rocks on the following spread.

>>

DIORITE
Composed of plagioclase feldspars, with amphiboles and pyroxenes, diorite contains little or no quartz. It is a coarse-grained plutonic rock.

GRANODIORITE
This is probably the most common intrusive igneous rock in the continental crust. It is over 65 percent plagioclase feldspar.

pale plagioclase feldspar

LARVIKITE
A form of syenite, larvikite is blue-black in color and composed of large amounts of feldspar that can have a shimmering blue schiller effect.

NEPHELINE SYENITE
Coarse-grained, pale-colored, and composed of feldspars, micas, and hornblende, this rock also contains nepheline but usually no quartz.

SYENITE
Gray or pinkish, syenite is a plutonic rock that occurs in large intrusions. Coarse-grained, it contains feldspars, micas, and hornblende, and little or no quartz.

LAMPROPHYRE
Lamprophyre's matrix is medium-grained and studded with distinct crystals of micas and amphiboles. It forms in dykes and sills.

GRANITE
Coarse-grained and of varying color, granite contains more than ten percent quartz. Attractive when polished, it is often used on the facade of buildings.

light plagioclase feldspar

GABBRO
This dark, plutonic rock contains plagioclase feldspars, pyroxenes, and olivine. Coarse-grained, it forms from the slow cooling of basaltic magma at depth.

black tourmaline

OLIVINE GABBRO
Gabbro is a dark, coarse-grained rock with much pyroxene and plagioclase feldspar. Olivine occurs in significant amounts in this variety of gabbro.

LAYERED GABBRO
This type of gabbro shows banding caused by the settling of minerals of different density in the magma.

PORPHYRITIC MICROGRANITE
Larger crystals are set into the matrix of this medium-grained rock, which is composed mainly of quartz, micas, and feldspars.

WHITE MICROGRANITE

DOLERITE
Typically occurring in sills and dykes, dolerite is a dark, medium-grained rock composed of plagioclase feldspars, pyroxenes, and iron oxides.

BOJITE
A dark-colored rock, bojite is a general term for hornblende-rich gabbro. It is coarse-grained and forms from magma.

MICROGRANITE
This medium-grained granite—often with a porphyritic texture—occurs in sills and dykes, which are sheets of intrusive igneous rock.

black biotite mica

gray quartz

crystal of red garnet

GARNET PERIDOTITE
Peridotite has a composition similar to that of Earth's upper mantle. This dense, greenish variety is made of dark minerals including garnets, olivine, and pyroxenes.

DUNITE
Composed almost entirely of olivine, this rock is dark green or brown, with a medium-grained texture. It usually includes a little chromite.

green olivine crystals

PERIDOTITE
A dark, dense rock composed mainly of olivine and pyroxenes, peridotite is coarse-grained and forms slowly at great depth.

KIMBERLITE
Dark-colored and coarse-grained, kimberlite has a very low silica content. Of variable composition, it is the world's main source of diamonds.

pink orthoclase feldspar

ADAMELLITE
This type of granite forms at depth with crystals of quartz, micas, and feldspars. One-third to two-thirds of the total feldspar content is plagioclase.

GRAPHIC GRANITE
This coarse-grained rock contains quartz and feldspars, whose intergrowth results in a graphic texture, vaguely resembling runic writing. It also contains micas.

FELSITE
Fine-grained felsite forms in sheetlike intrusions, sills, and dykes. Pale in color, it is composed mainly of feldspars and quartz.

dark hornblende

PEGMATITE
Pegmatites are very coarse-grained rocks, formed from residual liquid magma after most of a granite intrusion has cooled and crystallized. Some pegmatites are an important source of gemstones.

PORPHYRITIC GRANITE
A pale-colored rock composed of feldspars, quartz, and micas, porphyritic granite has large, well-formed crystals set into the matrix.

HORNBLENDE GRANITE
Granite typically contains quartz, feldspars, and micas. This variety also contains hornblende, a member of the amphibole mineral group.

ANORTHOSITE
Pale-colored, anorthosite is composed mainly of large crystals of plagioclase feldspars. It can also contain olivine and augite.

METAMORPHIC ROCKS

When existing rocks are subjected to heat or pressure or both within the Earth, they are changed into different assemblages of minerals. Contact metamorphism occurs when intense, localized heat radiating from a body of igneous rock causes the surrounding rocks to recrystallize. Regional metamorphism takes place over wide areas, often at considerable depth, as a result of intense heat and pressure. Movements of the Earth's crust can result in rocks being pulverized by dynamic metamorphism.

GRANULITE
Formed at very high temperatures and pressures, granulite is dark, coarse-grained, and rich in pyroxenes, garnets, micas, and feldspars.

SCHIST
Schist is characterized by its parallel planes of similarly oriented minerals within the rock. The corrugated pattern seen on the specimen is called crenulation.

HALLEFLINTA
Originally volcanic tuff, rhyolite, or quartz porphyry, halleflinta is fine-grained, pale-colored, and rich in quartz. It is a type of hornfels.

streaked texture

MYLONITE
When rocks are pulverized deep within a thrust fault zone, the rock dust and fragments produced form a fine-grained rock called mylonite.

SPOTTED SLATE
This is a dark, fine-grained rock characterized by black spots (porphyroblasts) of minerals, such as cordierite and andalusite.

GARNET SCHIST
The red garnets in this variety of schist indicate that it developed at relatively high temperatures and pressures deep within the continental crust.

BIOTITE SCHIST
Formed at relatively high pressures and temperatures, biotite schist contains feldspars, quartz, and much dark-colored biotite mica.

MUSCOVITE SCHIST
A typical schist with pale, glittery muscovite mica, this rock also contains quartz and feldspars.

KYANITE SCHIST
Mainly composed of feldspars, micas, and quartz, this schist also contains crystals of the blue mineral kyanite.

SLATE
Dark and very fine-grained, slate is a compact rock with parallel cleavage planes. It is formed by low-pressure metamorphism.

tubular structure

pink calcite

PHYLLITE
Phyllite is formed at lower temperatures and pressures than schist, but higher than slate. It is fine-grained and splits into slabs with a characteristic sheen.

SKARN
Formed when rocks rich in carbonate minerals are altered by high-temperature contact metamorphism, skarn contains minerals rich in calcium, magnesium, and iron.

FULGURITE
When lightning strikes in deserts or on beaches, sand can melt, forming small, tube-shaped structures, called sand fulgurites.

METAQUARTZITE
With a high percentage of quartz, this rock is harder than most metamorphic rocks. It is formed from sandstone altered at high temperature.

bladed crystal

CORDIERITE HORNFELS
Dark and splintery, cordierite hornfels is produced by the heat of nearby igneous intrusions. It is fine- to medium-grained.

GARNET HORNFELS
Hornfels is a tough, dark, flinty rock formed by the heat from an igneous intrusion. Reddish garnets occur in this particular variety.

GNEISS
A coarse-grained rock formed at very high temperatures and pressures, gneiss is recognized by its alternating dark and light crystalline bands.

AUGEN GNEISS
Gneiss contains quartz, feldspars, and micas, often in parallel bands. Augen gneiss has crystals resembling lens-shaped eyes (*augen* in German).

CHIASTOLITE HORNFELS
Hornfels forms at very high temperatures near magma intrusions. Bladed crystals of pale chiastolite give this particular variety its name.

PYROXENE HORNFELS
Fine- to medium-grained, tough, and flinty, this variety of hornfels contains quartz, micas, and pyroxenes. It forms close to igneous intrusions.

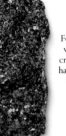

AMPHIBOLITE
Formed by moderate heat and varying pressure deep in the crust, this coarse-grained rock has an abundance of hornblende and plagioclase feldspar, as well as other minerals.

FOLDED GNEISS
Gneiss becomes plastic and folded at great depth. The dark bands are hornblende-rich and the paler ones are of quartz and feldspars.

ECLOGITE
Composed of two main minerals, green omphacite pyroxene and red garnet, eclogite is coarse-grained and forms at very high temperatures and pressures.

MIGMATITE
Formed at the highest temperatures and pressures, coarse-grained migmatite contains bands of dark basaltic minerals and pale granitic minerals. It is often folded.

GRANULAR GNEISS
This gneiss has a granular texture with equal-sized grains. It has dark bands of hornblende and biotite mica, and pale bands of quartz and feldspars.

marble fragment

MARBLE BRECCIA

GRAY MARBLE

MARBLE
Formed by contact or regional metamorphism, marble is rich in calcite, often with colorful veins of other minerals. It is prized as a carving material.

GREEN MARBLE

SERPENTINITE
Commonly banded, speckled, or streaked, serpentinite is a dense, but soft, metamorphic rock derived from peridotite. It is found in convergence zones between tectonic plates.

SEDIMENTARY ROCKS

Sedimentary rocks are generally characterized by stratification, or bedding, and they can contain fossils. These rocks are divided into three groups according to their origin. Clastic rocks are made up of fragments of rocks and minerals eroded from previously formed rocks; organic rocks are derived from the remains of plants and animals; chemical rocks result from the precipitation of chemicals.

MILLET-SEED SANDSTONE

The medium-sized grains of this rock are often coated with reddish iron oxides. The well-rounded, equal-sized quartz grains have been shaped by the wind.

GREENSAND

Colored greenish by the silicate mineral glauconite, greensand is a quartz-rich sandstone formed in the sea.

MICACEOUS SANDSTONE

Rich in quartz, this sandstone also contains glittery flakes of mica. It usually has medium-sized grains.

LIMONITIC SANDSTONE

This rock is colored red-brown or yellowish by the iron oxide mineral limonite, which coats its medium quartz grains.

TRAVERTINE

A pale-colored and often layered rock, travertine is virtually pure calcite. It is formed around hot springs and volcanic vents.

iron oxide gives red color

ROCK SALT

Composed of halite, rock salt is often colored by inclusions of iron oxides and clay minerals. It is soluble in water, with a low hardness and a salty taste.

ROCK GYPSUM

A crystalline rock formed when salt water evaporates, rock gypsum is pale colored, often fibrous, and very soft. It occurs with other evaporite minerals.

SANDSTONE

SANDSTONE

This rock typically occurs as layers of sand grains, held together by various mineral cements that impart different colors. Sandstones are quartz-rich.

BOULDER CLAY

Gray to brownish in color, boulder clay, or till, has a very fine-grained clay matrix with glacially derived angular and rounded rock fragments.

CLAYSTONE

Of varying color, this very fine-grained rock is composed mainly of silicate clay minerals such as kaolinite—mostly derived from the weathering of feldspars.

RED SANDSTONE

quartz grains colored by iron oxides

bands of hematite and chert

BANDED IRONSTONE

This rock, deposited in marine or fresh water, has alternating bands of black hematite and red chert. It is one of the best ores of iron.

OOLITIC IRONSTONE

This rock is composed of small, rounded grains (ooliths) of iron minerals, such as siderite, cemented by other iron minerals, as well as by calcite and quartz.

LOESS

A clay with very fine, dustlike grains lifted by the wind from dry land surfaces, loess is crumbly and lacks obvious layering.

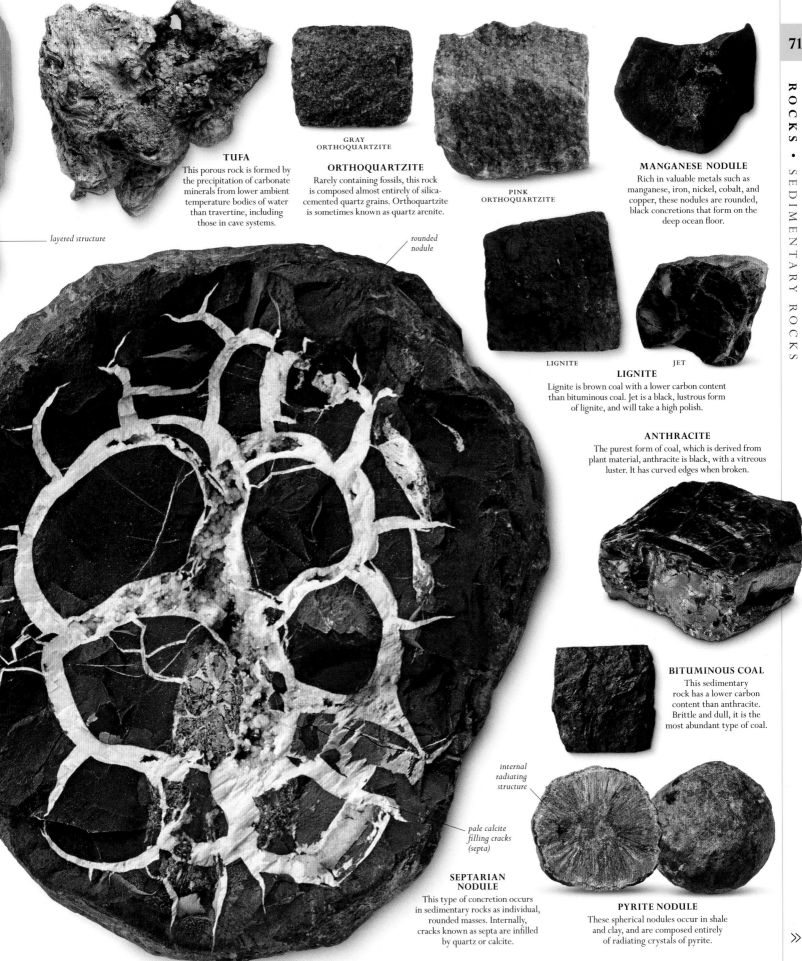

TUFA

This porous rock is formed by the precipitation of carbonate minerals from lower ambient temperature bodies of water than travertine, including those in cave systems.

layered structure

GRAY ORTHOQUARTZITE

ORTHOQUARTZITE

Rarely containing fossils, this rock is composed almost entirely of silica-cemented quartz grains. Orthoquartzite is sometimes known as quartz arenite.

PINK ORTHOQUARTZITE

MANGANESE NODULE

Rich in valuable metals such as manganese, iron, nickel, cobalt, and copper, these nodules are rounded, black concretions that form on the deep ocean floor.

rounded nodule

LIGNITE

JET

LIGNITE

Lignite is brown coal with a lower carbon content than bituminous coal. Jet is a black, lustrous form of lignite, and will take a high polish.

ANTHRACITE

The purest form of coal, which is derived from plant material, anthracite is black, with a vitreous luster. It has curved edges when broken.

BITUMINOUS COAL

This sedimentary rock has a lower carbon content than anthracite. Brittle and dull, it is the most abundant type of coal.

internal radiating structure

pale calcite filling cracks (septa)

SEPTARIAN NODULE

This type of concretion occurs in sedimentary rocks as individual, rounded masses. Internally, cracks known as septa are infilled by quartz or calcite.

PYRITE NODULE

These spherical nodules occur in shale and clay, and are composed entirely of radiating crystals of pyrite.

» SEDIMENTARY ROCKS

fossilized crinoid, or sea lily, stalk

CRINOIDAL LIMESTONE

Crinoids are echinoderms that are attached to the seabed by a flexible stalk. Crinoidal limestone is a mass of broken stalks cemented by hardened lime mud.

FRESHWATER LIMESTONE

This limestone is a pale, calcite-rich rock with some quartz and clay. It contains fossils of freshwater organisms, which indicate the environment of deposition.

pisolith cemented by calcite

NUMMULITIC LIMESTONE

A marine foraminiferan, *Nummulites* is the main fossil in this rock. The cement is calcite, originally lime mud.

CORAL LIMESTONE

This rock is a mass of fossilized corals cemented by fine-grained calcite. It is gray to white or brownish in color.

PISOLITIC LIMESTONE

This rock is composed of pisoliths—pea-sized grains slightly larger than ooliths, often flattened, and loosely cemented by calcite.

OOLITIC LIMESTONE

This limestone is composed of ooliths—small, rounded, concentrically banded grains rolled by seabed currents and cemented by carbonate mud.

BRYOZOAN LIMESTONE

This is a gray or reddish organic limestone, which contains fossils of bryozoans in a matrix of hardened, calcite-rich mud.

LIMESTONE BRECCIA

Often forming below limestone cliffs, this breccia comprises large, angular fragments of limestone and other rocks, cemented by calcite.

FELDSPATHIC GRITSTONE

Coarse-grained and pale- to dark-colored, this gritstone contains much quartz and up to 25 percent feldspar.

QUARTZ GRITSTONE

This gritstone is composed of quartz with some feldspar and mica, all of coarse grain size.

DOLOMITE

This yellowish-brown or gray rock contains a high percentage of dolomite (calcium magnesium carbonate). It is also called dolostone, to distinguish it from this mineral.

GRAYWACKE

This dark rock contains quartz, rock fragments, and feldspars, set in a mass of finer clay and chlorite. It forms in marine basins.

ARKOSE

Variable in color and medium-grained, arkose is a sandstone with a high percentage of feldspar.

FOSSILIFEROUS SHALE

Fine-grained marine sedimentary rocks such as shale often contain large numbers of well-preserved fossils.

brachiopod fossil

SHALE

This fine-grained, layered rock varies in composition, usually containing silt, clay minerals, organic materials, iron oxides, and minute crystals of minerals such as pyrite and gypsum.

POLYGENETIC CONGLOMERATE
A coarse-grained sedimentary rock, polygenetic conglomerate has many different, rounded rock and mineral fragments in a finer grained matrix.

rounded quartz pebble

sandstone matrix

iron oxide gives red color

RED CHALK

QUARTZ CONGLOMERATE
Varying in color, this rock typically has dirty white, pebble-sized, rounded quartz fragments set in a finer, darker matrix.

CHALK
Pure calcite, chalk is very fine-grained and powdery. It is composed of minute fossilized organisms, including coccoliths and radiolarians.

WHITE CHALK

BRECCIA
Containing large, angular fragments of rocks and minerals set in a finer matrix of sand or silt, breccia rarely forms in layers.

SILTSTONE
This dark-colored rock comprises particles, mainly of quartz, which are smaller than fine sand but larger than clay grains. It can also contain organic material and calcite.

CHERT
A very fine-grained form of silica, chert occurs as bands and nodules in rocks such as limestone. It is usually gray in color.

MARL
Intermediate between clay and limestone, marl is a fine-grained, calcite-rich, layered rock. Chlorite and glauconite can give it a green coloring.

MUDSTONE
Composed of clay, with fine-grained quartz and feldspars, mudstone lacks the bedding found in shale.

FLINT
Usually found as nodules in chalk, flint is very hard, black, compact silica. It breaks leaving sharp, curved edges.

FOSSILS

Fossils are evidence of past life that has been buried and preserved in the rocks of the Earth's crust. They give scientists important clues about how life has evolved; they can also be used to date rocks, and create a timeline of the events that have shaped our world.

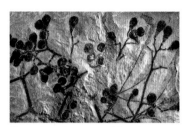

Over time the tissues of this plant have decayed. Only the outline remains, covered by a thin film of carbon.

An insect became stuck in resin oozing from a tree. The resin turned into amber, perfectly preserving the creature.

This fish skeleton has been fossilized in shale. Here all the pores in the original skeleton have been infilled by minerals.

Life has existed on Earth for at least 3.7 billion years. Initially, living things were small and soft-bodied, and left few obvious traces of their existence. But during the past billion years, life gradually changed. Organisms evolved hard body parts, which—given enough time—could fossilize. The significance of this change is hard to overstate. It turned the world's sedimentary rocks into a global data bank, teeming with an incredible array of fossilized species, arranged in the exact order in which they appeared. These fossils show the path evolution has followed. They also highlight mass extinctions—when enormous numbers of species have been wiped out in a relatively short space of time.

DEAD AND BURIED

Fossilization is a lottery, and only a small fraction of living organisms end up being preserved. On land, it is usually triggered by chance events—for example, when animals are overwhelmed by landslides or flash floods, or when they drown in lakes. Marine animals have a much better chance of being fossilized, because sediment routinely accumulates over their dead remains. Fine sediment can preserve soft bodies, but the best fossils are left by animals with hard body parts, such as shells or bones. After burial, the remains are slowly infiltrated by dissolved minerals, which literally turns them to stone. Once formed, many fossils are destroyed by heat, pressure, or geological movements while still deep underground. But if a fossil survives all this, uplift may eventually bring it back to the surface, where erosion can release it from its parent rock (see panel, below). There, it has to be found before it eventually breaks apart.

These body fossils can be breathtaking objects, particularly when they are complete skeletons several yards long. However, they are not the only kind of fossilized remains. Rocks can also yield trace fossils, which are fossilized footprints, burrows, or other signs of animal activity. Trace fossils provide indirect but fascinating evidence of how animals lived: for example, dinosaur footprints can show how fast they moved, how they interacted in herds, and even how they put on weight as they grew.

Much farther back in time, rocks sometimes contain chemical fossils—ancient carbon-based compounds that have been produced by biological processes. Although unspectacular, these chemical smudges are key evidence in the hunt for Earth's earliest living things.

SUDDEN DEATH >
Trilobites from the Late Ordovician Period are fossilized together, suggesting the sudden burial of these animals by sediment.

INDEX FOSSILS

The geological timescale was largely established using fossils. Species which lived across a wide geographical range, but only existed for a short time, are known as index fossils. They can be used to identify particular strata and link them from place to place—the presence of the same index fossil in different places shows that the strata were laid down at the same time. Index fossils therefore help geologists to date rocks and build a relative time sequence. Mesozoic ammonites (a group of extinct marine mollusks) are among the best index fossils—an ammonite zone may be as little as a million years.

HOW FOSSILS FORM

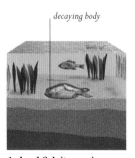

decaying body

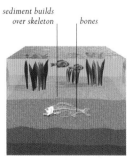

sediment builds over skeleton — *bones*

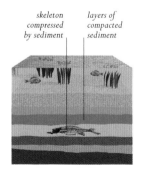

skeleton compressed by sediment — *layers of compacted sediment*

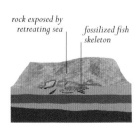

rock exposed by retreating sea — *fossilized fish skeleton*

A dead fish lies on the seabed, where its flesh may rot or be eaten. To be preserved, it must be buried rapidly. As the mud turns into shale, this change will compress and flatten the body.

The fish skeleton has been covered by sediment. To become fossilized, the bones must undergo a chemical change, called permineralization, where the pores in the bone are replaced by minerals.

More sediment has accumulated on the seabed and compressed the lower layers. The sediment is compressed and the fossil may become flattened, distorted, or destroyed.

Millions of years later the seabed sediment, now turned to rock, has been exposed by retreating seas. Weathering has further denuded the rock around the fossilized fish skeleton.

FOSSIL PLANTS

Plants are among the first organisms to appear in the fossil record. Algae are found in rocks from the Precambrian. Vascular plants (plants with tissues for conducting water and nutrients) evolved in the Silurian Period, and by Carboniferous times, Earth was colored green by vast, coal-forming swamp forests. Flowering plants developed later, in the Mesozoic Era.

EARLY LAND PLANT
Cooksonia hemisphaerica
Found in Silurian and Devonian rocks, *Cooksonia* was one of the earliest vascular plants. It had a stiff stem and leafless branches.

CALAMOPHYTON STEMS
Calamophyton primaevum
A primitive, leafless plant, probably related to ferns, *Calamophyton* is found in Devonian and Early Carboniferous rocks.

branch —

CLADOXYLON STEMS
Cladoxylon scoparium
Fossilized in Devonian and Carboniferous rocks, *Cladoxylon* was a low-growing plant, with a tough central stem and leafless, light-absorbing branches.

tough stem —

SEED FERN LEAF
Alethopteris serlii
A seed fern from Carboniferous and Permian strata, *Alethopteris* had compound pinnate fronds, consisting of thick, strongly veined leaflets.

CYCLOPTERIS LEAFLETS
Cyclopteris orbicularis
Oval leaflets from a seed fern called *Neuropteris* are given the scientific name *Cyclopteris*. Its fossils belong to Carboniferous strata.

SEED FERN SEEDS
Trigonocarpus adamsi
The name *Trigonocarpus* is given to fossil seeds from seed ferns found in Carboniferous strata. Each seed has three ribs.

HORSETAIL FOLIAGE
Asterophyllites equisetiformis
Found in Carboniferous and Permian strata, *Asterophyllites* had needle-shaped leaves and a structure similar to that of modern horsetails.

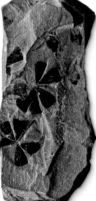

CLIMBING HORSETAIL
Sphenophyllum emarginatum
Found fossilized in rocks of Carboniferous to Permian age, this horsetail had wedge-shaped leaves and long, soft stems adapted for climbing.

SIGILLARIA STEM
Sigillaria aeveolaris
Found in Carboniferous and Permian rocks, *Sigillaria* was a giant relative of clubmosses that grew to over 100 ft (30 m). It had a narrow stem, and its leaves grew in clumps.

vertical rib —

LEPIDODENDRON ROOT
Stigmaria ficoides
Occurring in Carboniferous to Permian rocks, *Stigmaria* is the name for the fossil roots of the club moss relative, *Lepidodendron*.

SALVINIA RHIZOME
Salvinia formosa
A floating water fern from the tropics, *Salvinia* is found fossilized in rocks from Cretaceous to recent strata.

PERMIAN FERN
Oligocarpia gothanii
This ground-hugging fern is found in Carboniferous and Permian strata. It occupied a wetland habitat.

pinnate (featherlike) leaf

CRETACEOUS FERN
Weichselia reticulata
Found fossilized in Cretaceous strata, *Weichselia* was similar to modern bracken and had fronds that were divided twice.

SEED FERN LEAFLETS
Dicrodium sp.
A seed fern from the Triassic Period, *Dicrodium* had pinnate leaves and its fronds were about 3 in (7.5 cm) long.

PALEOZOIC CONIFER
Lebachia piniformis

A cone-bearing plant from Carboniferous and Permian strata, *Lebachia* is an ancestor of modern conifers.

CONIFER SEED CONES
Taxodium dubium

Found in Jurassic strata, *Taxodium* is related to modern cypress trees. It grew in damp habitats and had needlelike leaves.

JURASSIC CONIFER
Araucaria mirabilis

This extinct species of monkey puzzle tree, *A. mirabilis*, bore characteristic female cones with spirally arranged scales attached to a central axis.

SECTION THROUGH CONE

COAST REDWOOD CONE
Sequoia dakotensis

Cones of the giant evergreen tree *Sequoia* have been found in Cretaceous and recent rocks. Some living members of *Sequoia* are more than 2,000 years old.

CRETACEOUS CONIFER
Glyptostrobus sp.

This conifer grew in swamps during the Cretaceous Period and into the Cenozoic Era. *Glyptostrobus* was an important coal-forming tree.

SUBFOSSIL TREE RESIN
Kauri pine amber

Amber is the hardened resin from pine trees, such as Kauri pines. First occurring in the Early Cretaceous, it often contains fossils of insects that perished on the fragrant, sticky resin.

seed

CARBONIFEROUS GYMNOSPERM
Cordaites sp.

An ancestor of the conifers, *Cordaites* grew during the Carboniferous and Permian Periods. It was a tree-sized plant, which reproduced by seed.

GIGANTOPTERID LEAVES
Gigantopteris nicotianaefolia

A flowerless plant from Permian times, this species was so named because its leaves resembled those of tobacco plants.

PERMIAN GINKGO LEAVES
Psygmophyllum multipartitum

Still found in China, ginkgos first appeared in the Permian Period. The fan-shaped leaves can be identified in fossils of *Psygmophyllum*, a precursor of the modern-day ginkgo.

TRIASSIC GINKGO
Baiera munsteriana

Growing up to 6 in (15 cm) in length, the fan-shaped leaves of *Baiera* were split into separate ribs. In living ginkgos, the leaves are almost entire.

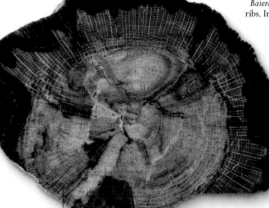

growth rings

STEMLESS PALM FRUIT
Nypa burtinii

Fossils of *Nypa* date from the Eocene Period onwards. This palm holds its woody seeds in a 10 in (25 cm) ball-shaped fruit.

OAK TREE TRUNK
Quercus sp.

The well-known oak genus, *Quercus*, first occurs as fossils in Cretaceous strata. There are over 500 species of oak living today.

MAGNOLIA LEAF
Magnolia longipetiolata

One of the earliest genera of flowering plants, magnolias first appeared in the Cretaceous Period. Early insects fed on their nectar.

central axis

FOSSIL INVERTEBRATES

Invertebrates, animals without a solid internal skeleton, are among the most common fossils found. They first appeared in the Precambrian, but it is only in Early Cambrian times that complex invertebrates such as trilobites become numerous in the fossil record. Fossils of invertebrates such as arthropods, mollusks, brachiopods, echinoderms, and corals are especially common, because they had hard external structures and lived in the sea, where most fossil-bearing rocks form.

ARCHAEOCYATHID
Metaldetes taylori
These reef-building organisms are known only from the Cambrian Period. *Metaldetes* had a cuplike structure, not unlike that of a coral.

corallite

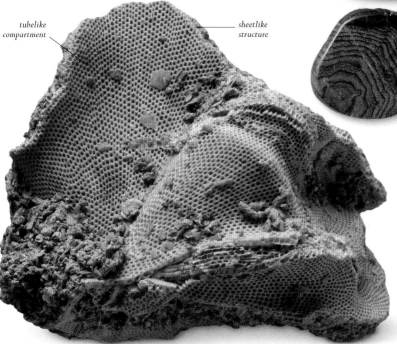

tubelike compartment

sheetlike structure

STROMATOPOROID
Stromatopora concentrica
Found in rocks ranging from the Ordovician to the Permian, often in reef limestone, the fossil sponges in this group were made of porous, calcium-rich tubes.

CALCAREOUS SPONGE
Peronidella pistilliformis
Characterized by needlelike spicules (in the form of calcite) fused together, *Peronidella* is found in Triassic and Cretaceous rocks.

TREPOSTOME BRYOZOAN
Diplotrypa sp.
A bryozoan from Ordovician strata, *Diplotrypa* was a small invertebrate, not unlike coral, that lived in dome-shaped colonies.

CHEILOSTOME BRYOZOAN
Biflustra sp.
Found in Cenozoic rocks, the *Biflustra* genus is extant. These bryozoans have minute compartments, which house zooids—the soft-bodied individuals of a colony.

LACE CORAL
Schizoretepora notopachys
A lace coral, *Schizoretepora* is found in Eocene to Pleistocene strata. It lived on rocky seabeds.

BRANCHING BRYOZOAN
Constellaria sp.
A bryozoan that built branching colonies on the seabed, *Constellaria* occurs in Ordovician strata.

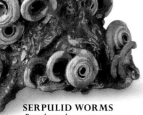

SERPULID WORMS
Rotularia bognoriensis
Found in Jurassic to Eocene rocks, *Rotularia* is a genus of serpulid worm. Like all serpulids, each worm protected its soft body by producing coiled tubes made of calcium carbonate.

SPRIGGINA
Spriggina floundersi
A very early fossil found in Ediacaran rocks, *Spriggina* had a long, wormlike body. Its classification is uncertain.

thecae housed soft-bodied individuals of the colony

single, curved branch

network of branches

"TUNING FORK" GRAPTOLITE
Didymograptus murchisoni
A graptolite (extinct colonial invertebrate) with two stipes (branches), *Didymograptus* is found in Ordovician rocks. It grew from ¾–23½ in (2–60 cm) in length.

BRANCHING GRAPTOLITE
Rhabdinopora socialis
Until recently, this graptolite was called *Dictyonema*. It had numerous thin, radiating stipes, and is from Ordovician strata.

SPIRAL GRAPTOLITE
Monograptus convolutus
A single branch, with thecae (cuplike structures) on one side, characterizes *Monograptus*, which is found in Early Silurian strata. *M. convolutus* had unusual coiling.

TABULATE CORAL
Catenipora sp.
A simple, tabulate coral
with a chainlike structure,
Catenipora lived in warm,
shallow seas during
Ordovician and
Silurian times.

*chainlike
colonial
structure*

SCLERACTINIAN CORAL
Meandrina sp.
Shaped like the human brain, this
colonial coral has ridges and valleys
on its surface. First found in Eocene
rocks, *Meandrina* still survives today.

*thick
corallite wall*

RUGOSE CORAL
Goniophyllum pyramidale
A solitary coral occurring in
Silurian rocks, *Goniophyllum*
has a cone-shaped structure
in which the polyp lived.

**CAMBRIAN
TRILOBITE**
Paradoxides bohemicus
Some *Paradoxides*
trilobites grew to nearly
3¼ ft (1 m) in length.
This species had long
spines on its body and is
from Cambrian strata.

SILURIAN TRILOBITE
Dalmanites caudatus
Common in the Silurian
Period, *Dalmanites* had a
segmented thorax and
a pointed tail spine.

**DEVONIAN
ENROLLED TRILOBITE**
Phacops sp.
Characterized by compound eyes, *Phacops* is
found in Devonian strata. Trilobites could
roll up, like many modern arthropods.

**ORDOVICIAN
TRILOBITE**
Eodalmanitina macrophtalma
An Ordovician trilobite, *Eodalmanitina*
had large, crescent-shaped eyes. Its
thorax was composed of 11 segments.

pincers

**HORSESHOE
CRAB RELATIVE**
Euproops rotundatus
A Carboniferous relative of
horseshoe crabs, *Euproops* had
a crescent-shaped headshield
and long tail spine.

LOBSTER
Eryma leptodactylina
A fossil lobster from rocks of
Jurassic and Cretaceous times,
Eryma was 2¼ in (6 cm) long,
and similar to modern species.

CRAB
Avitelmessus grapsoideus
Covered with many spines, this crab is
from the Cretaceous Period. It grew
to 10 in (25 cm) in width.

COCKROACH RELATIVE
Archimylacris eggintoni
A relative of cockroaches from the
Carboniferous Period, *Archimylacris* had
hind wings with a distinctive vein pattern.

>>

ARTICULATE BRACHIOPOD
Leptaena rhomboidalis

A brachiopod from Ordovician, Silurian, and Devonian strata, *Leptaena* grew to about 2 in (5 cm) in width. Its shell had concentric and radial ribs.

RHYNCHONELLID BRACHIOPOD
Homeorhynchia acuta

Found in Early Jurassic strata, *Homeorhynchia* was a small brachiopod that grew to about ⅜ in (1 cm) in width.

SPIRIFERID BRACHIOPOD
Spiriferina walcotti

A common brachiopod from Triassic and Jurassic strata, *Spiriferina* had a rounded shell up to 1¼ in (3 cm) wide, with clearly visible growth lines.

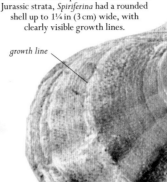

growth line

SWAMP CLAM
Carbonicola pseudorobusta

Occurring in nonmarine rocks of the Carboniferous Period, *Carbonicola* had a tapered shell. Its fossils have been used in the relative dating of these rocks.

ribbed valve

SCALLOP
Pecten maximus

Found in Paleogene to recent rocks, *Pecten* is a genus of bivalve mollusks. *P. maximus* is extant and swims by flapping its valves.

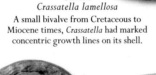

CLAM
Crassatella lamellosa

A small bivalve from Cretaceous to Miocene times, *Crassatella* had marked concentric growth lines on its shell.

MUSSEL RELATIVE
Ambonychia sp.

Found in Ordovician strata, *Ambonychia*, an early bivalve mollusk, grew up to 2½ in (6 cm) wide. Both valves had radial ribs on the surface.

DEVIL'S TOENAIL
Gryphaea arcuata

This fossil oyster occurs in Triassic and Jurassic rocks. *Gryphaea* had one large, hooked valve and a smaller, flat one.

DEVONIAN GASTROPOD
Murchisonia bilineata

A gastropod mollusk from Silurian to Permian strata, *Murchisonia* grew up to 2 in (5 cm) tall, with ridges around its whorls.

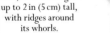

JURASSIC GASTROPOD
Pleurotomaria anglica

Found in Jurassic and Cretaceous rocks, this gastropod mollusk had a broad shell with a combination of radial and spiral patterns.

ROSTROCONCH
Conocardium sp.

Occurring in Devonian and Carboniferous strata, *Conocardium* resembled clams, but its shell had no functional hinge.

NAUTILOID
Vestinautilus cariniferus

This early *Nautilus* relative had very open coiling and little shell ornamentation. *Vestinautilus* is found in Carboniferous rocks.

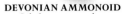

whorl with ridges

shell aperture

CARBONIFEROUS AMMONOID
Goniatites crenistria

An ammonoid mollusk from Devonian and Carboniferous rocks, *Goniatites* had angular sutures where the chamber walls joined the shell.

TRIASSIC AMMONOID
Ceratites nodosus

Found in Triassic strata, *Ceratites* was an ammonoid mollusk. Its heavily ornamented shell had open coiling and strong ribs.

simple rib

DEVONIAN AMMONOID
Soliclymenia paradoxa

An Early Devonian ammonoid, *Soliclymenia* had thin ribs across the shell. Some species had an unusual, triangular shell.

AMMONITE
Mortoniceras rostratum

This ammonite mollusk from the Cretaceous Period grew up to 4 in (10 cm) in diameter, with ribs across the shell.

BELEMNITE
Pachyteuthis abbreviata
A squid relative from the Jurassic, *Pachyteuthis* had a calcite guard. It was about 4 in (10 cm) long.

respiratory structure

CYSTOID
Pseudocrinites bifasciatus
Characterized by rhomboid-shaped respiratory structures, *Pseudocrinites* lived in Silurian and Devonian times. It was attached to the seabed by its stem.

DEVONIAN CRINOID
Cupressocrinites crassus
Up to 1¼ in (3 cm) in diameter, this crinoid from Devonian strata had a tall, five-sided cup at the end of its stem.

flexible arm

stem

JURASSIC CRINOID
Pentacrinites sp.
Named for its five-sided ossicles (stem segments), *Pentacrinites* grew to more than 3¼ ft (1 m) in height. It is often found attached to fossil wood.

densely packed branches

BRITTLESTAR
Lapworthura miltoni
An early fossil brittlestar, *Lapworthura* is from Ordovician and Silurian strata. It grew to 4 in (10 cm) in diameter. This species had five relatively short, thick arms.

tubercle where spine was attached

STARFISH
Tropidaster pectinatus
This extinct starfish from the Early Jurassic was about 1 in (2.5 cm) wide. It had five thick arms.

BLASTOID
Pentremites pyriformis
This echinoderm from the blastoid group lived during Carboniferous times. It had long, armlike structures that were used for feeding.

SEA URCHIN
Hemicidaris intermedia
This common sea urchin from Jurassic strata grew to about 1½ in (4 cm) in diameter. Its many bumps (tubercles) supported stout spines.

HEART URCHIN
Lovenia sp.
A heart-shaped, burrowing sea urchin, *Lovenia* is known from the Palaeocene and still exists today. It is up to 2 in (5 cm) in diameter.

FOSSIL VERTEBRATES

Fossilized vertebrate remains are not as common as those of invertebrates, since many vertebrates lived on land, where fewer fossils form, and they evolved much later than the invertebrates. Fishes were the earliest vertebrates to evolve, some dating back to Cambrian times. Their rapid evolution in the Silurian and Devonian led to amphibians, which first appeared in the Devonian Period. Dinosaurs flourished during the Mesozoic Era, toward the end of which mammals began to diversify.

ZENASPID FISH
Zenaspis sp.
Found in Devonian strata, *Zenaspis* had a massive headshield. Up to 10 in (25 cm) long, its body was covered in bony scales.

finlike structure aided movement

EARLY FISHLIKE VERTEBRATE
Loganellia sp.
A primitive, jawless, flattened "fish," *Loganellia* was covered with toothlike scales. Up to 4¾ in (12 cm) long, it is found in Devonian rocks.

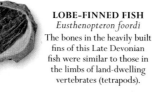

LOBE-FINNED FISH
Eusthenopteron foordi
The bones in the heavily built fins of this Late Devonian fish were similar to those in the limbs of land-dwelling vertebrates (tetrapods).

PSAMMOSTEID FISH
Drepanaspis sp.
A jawless, primitive fish, *Drepanaspis* had a flattened headshield. It is found only in Devonian strata.

PLACODERM
Bothriolepis canadensis
A Devonian placoderm (an extinct group of jawless fishes), *Bothriolepis* had large head- and trunk-shields and spinelike pectoral fins.

SHARK TOOTH
Otodus sokolovi
The serrated edges of the teeth of this Cenozoic shark could easily cut through flesh.

eye socket

fine vertebral column

SHOAL OF DACE
Leuciscus pachecoi
Found in Miocene strata, extinct species of *Leuciscus* or dace resembled modern bony fishes. *L. pachecoi* grew to 2½ in (6 cm) in length.

STINGRAY
Heliobatis radians
Found in Eocene strata, *Heliobatis* was a freshwater stingray that grew to about 12 in (30 cm) in length and had a skeleton of cartilage.

PRIMITIVE FROG
Rana pueyoi
Dating back to Miocene times, *Rana* is a genus of frogs. *R. pueyoi* grew to 6 in (15 cm) in length and shared features such as long hind limbs with modern frogs.

SKULL OF LARGE, PREDATORY BONY FISH
Xiphactinus sp.
A bony fish from Late Cretaceous strata, *Xiphactinus* was a marine predator with a muscular body and large front teeth.

long, sharp tooth

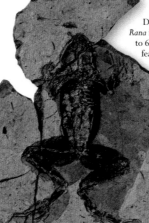

DIPLOCAULID AMPHIBIAN
Diplocaulus magnicornis
A salamanderlike amphibian from the Permian Period, *Diplocaulus* had protrusions on the sides of its skull. It grew up to 3¼ ft (1 m) in length.

DIMETRODON SKULL
Dimetrodon loomisi
Noted for a saillike structure on its back, *Dimetrodon* was an early relative of mammals from the Permian Period. A high skull and short snout translated into a powerful bite.

eye socket

DICYNODONT SKULL
Pelanomodon sp.
This tuskless herbivore was a dicynodont—a member of a group of mammal relatives that lived in the Permian and Triassic Periods.

MARINE TURTLE SKULL
Puppigerus crassicostata
Fossilized marine turtles are found in rocks ranging from the Mesozoic Era to Recent times. *Puppigerus* had a heavy shell and is found in Eocene strata.

PLESIOSAUR FLIPPER
Cryptoclidus eurymerus
Growing up to 26 ft (8 m) in length, *Cryptoclidus* was a long-necked plesiosaur from the Jurassic Period.

GIANT MONITOR LIZARD VERTEBRA
Varanus priscus
A huge monitor lizard, *Varanus priscus* grew to 23 ft (7 m) in length. It is found in rocks of Pleistocene age.

CYNODONT SKULL
Cynognathus crateronotus
A carnivore with a stout skull and large canine teeth, *Cynognathus* belonged to a group of mammal precursors called cynodonts. It is found in Triassic strata.

EARLIEST BIRD
Archaeopteryx lithographica
Archaeopteryx was thought to be the earliest bird. However, recent finds in Jurassic rocks in China are challenging that view.

GIANT GROUNDBIRD SKULL
Phorusrhacos inflatus
A carnivore up to 8¼ ft (2.5 m) tall, *Phorusrhacos* was a flightless bird with a powerful beak. It is found in Miocene rocks.

vertebrae

EARLY HORSE TEETH
Protorohippus sp.
A dog-sized, many-toed ancestor of the modern horse, *Protorohippus* is found in Eocene strata. It had low-crowned molars.

SABRE-TOOTHED CAT SKULL
Smilodon sp.
Large, curved canine teeth were typical of *Smilodon*, which was the size of a tiger and lived in the Pleistocene Period.

EARLY ELEPHANT JAW
Phiomia serridens
Found in Eocene to Oligocene strata, *Phiomia* was 8¼ ft (2.5 m) tall. It had tusks in its upper jaw, and a small trunk.

low, sloping forehead

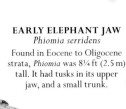

ANTHROPOID SKULL
Proconsul africanus
The first fossil anthropoid (apelike primate) to be found in Africa, *Proconsul* comes from Miocene strata.

NOTOUNGULATE
Toxodon platensis
Growing to about 8¾ ft (2.7 m), *Toxodon* had a sturdy body with a hippopotamuslike head. It lived from Pliocene to Pleistocene times.

≫ FOSSIL VERTEBRATES: DINOSAURS

PLATEOSAURUS SKULL
Plateosaurus sp.

A bulky plant-eater from the Late Triassic, *Plateosaurus* grew to about 26 ft (8 m) in length and had a very small head.

bony spike

DIPLODOCUS TAIL VERTEBRA
Diplodocus longus

Found in Jurassic strata, *Diplodocus* was a giant plant-eater that grew up to 89 ft (27 m) in length. Its tail was long and whiplike.

BRACHIOSAURUS THIGH BONE
Brachiosaurus sp.

A huge, plant-eating dinosaur that grew to 82 ft (25 m) in length, *Brachiosaurus* lived in the Jurassic and Cretaceous Periods.

COELOPHYSIS SKELETON
Coelophysis bauri

Fossils of *Coelophysis* are found in Triassic rocks. Only 10 ft (3 m) long, this carnivore had a birdlike skeleton.

PROCERATOSAURUS PARTIAL SKULL
Proceratosaurus bradleyi

A carnivore from Middle Jurassic strata in Gloucestershire, England, *Proceratosaurus* had a bony crest on its head.

skull

long tail used for balance

MEGALOSAURUS SACRAL VERTEBRAE
Megalosaurus bucklandi

Found in Middle Jurassic strata, *Megalosaurus* was 30 ft (9 m) long, with a large head and strong hind limbs. It was carnivorous.

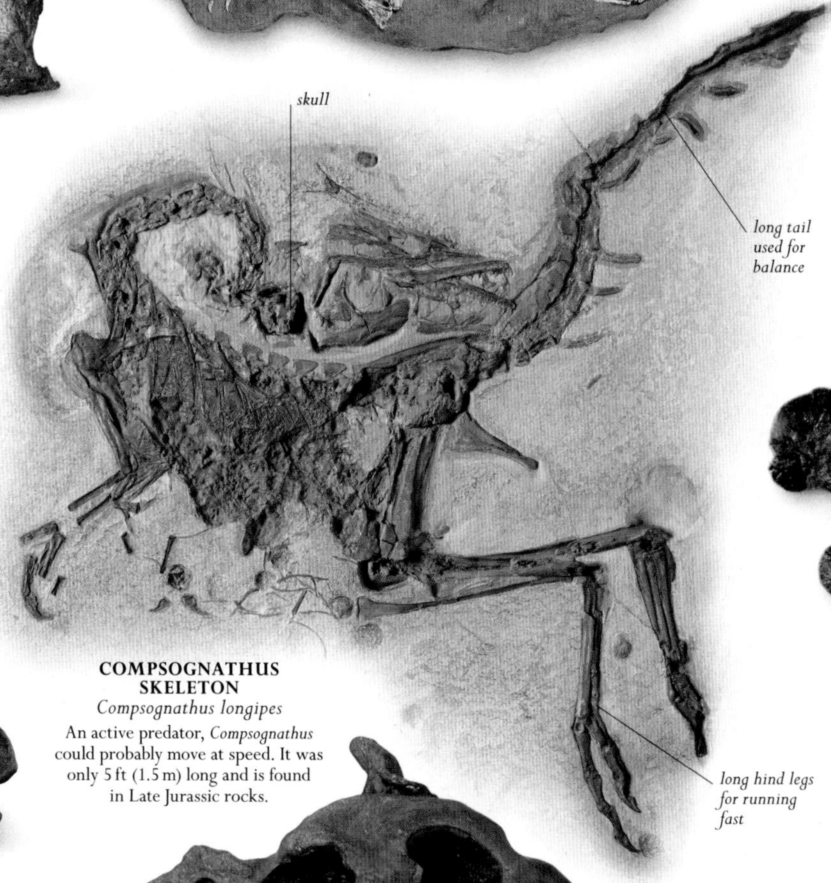

COMPSOGNATHUS SKELETON
Compsognathus longipes

An active predator, *Compsognathus* could probably move at speed. It was only 5 ft (1.5 m) long and is found in Late Jurassic rocks.

long hind legs for running fast

GALLIMIMUS SKULL
Gallimimus bullatus

Growing up to 20 ft (6 m) long, *Gallimimus* had a birdlike, beaked skull, and long neck and legs.

small brain cavity

strong, serrated teeth

ALBERTOSAURUS SKULL
Albertosaurus sp.

A predator and close relative of *Tyrannosaurus rex*, *Albertosaurus* grew to 26 ft (8 m) in length and is found in Late Cretaceous rocks.

DASPLETOSAURUS JAW
Daspletosaurus torosus

This Cretaceous dinosaur had massive hind legs and small arms, and it grew to a length of 30 ft (9 m). It had a powerful jaw with the formidable teeth of a carnivore.

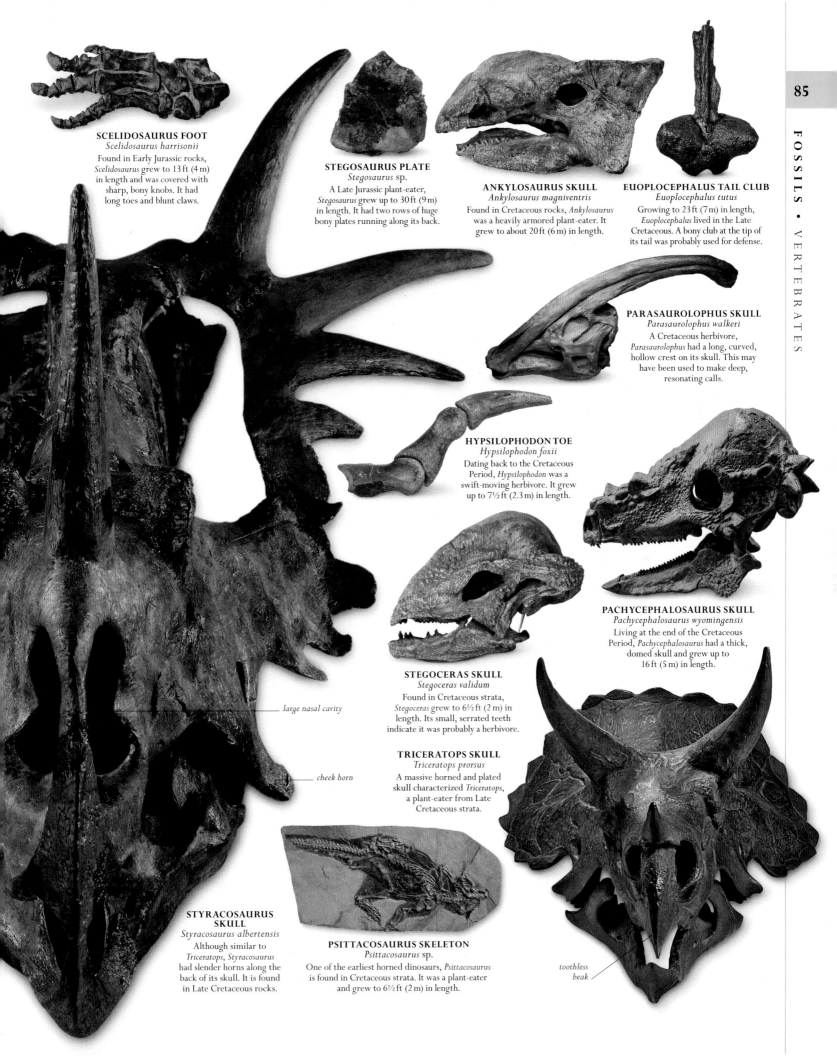

SCELIDOSAURUS FOOT
Scelidosaurus harrisonii
Found in Early Jurassic rocks, *Scelidosaurus* grew to 13 ft (4 m) in length and was covered with sharp, bony knobs. It had long toes and blunt claws.

STEGOSAURUS PLATE
Stegosaurus sp.
A Late Jurassic plant-eater, *Stegosaurus* grew up to 30 ft (9 m) in length. It had two rows of huge bony plates running along its back.

ANKYLOSAURUS SKULL
Ankylosaurus magniventris
Found in Cretaceous rocks, *Ankylosaurus* was a heavily armored plant-eater. It grew to about 20 ft (6 m) in length.

EUOPLOCEPHALUS TAIL CLUB
Euoplocephalus tutus
Growing to 23 ft (7 m) in length, *Euoplocephalus* lived in the Late Cretaceous. A bony club at the tip of its tail was probably used for defense.

PARASAUROLOPHUS SKULL
Parasaurolophus walkeri
A Cretaceous herbivore, *Parasaurolophus* had a long, curved, hollow crest on its skull. This may have been used to make deep, resonating calls.

HYPSILOPHODON TOE
Hypsilophodon foxii
Dating back to the Cretaceous Period, *Hypsilophodon* was a swift-moving herbivore. It grew up to 7½ ft (2.3 m) in length.

PACHYCEPHALOSAURUS SKULL
Pachycephalosaurus wyomingensis
Living at the end of the Cretaceous Period, *Pachycephalosaurus* had a thick, domed skull and grew up to 16 ft (5 m) in length.

STEGOCERAS SKULL
Stegoceras validum
Found in Cretaceous strata, *Stegoceras* grew to 6½ ft (2 m) in length. Its small, serrated teeth indicate it was probably a herbivore.

TRICERATOPS SKULL
Triceratops prorsus
A massive horned and plated skull characterized *Triceratops*, a plant-eater from Late Cretaceous strata.

large nasal cavity

cheek horn

STYRACOSAURUS SKULL
Styracosaurus albertensis
Although similar to *Triceratops*, *Styracosaurus* had slender horns along the back of its skull. It is found in Late Cretaceous rocks.

PSITTACOSAURUS SKELETON
Psittacosaurus sp.
One of the earliest horned dinosaurs, *Psittacosaurus* is found in Cretaceous strata. It was a plant-eater and grew to 6½ ft (2 m) in length.

toothless beak

EUOPLOCEPHALUS

Euoplocephalus belonged to a family of dinosaurs called the ankylosaurids, which were characterized by an armored head and bony plating on the body. This dinosaur grew to about 20 ft (6 m) in length and weighed about two tons. Its tail, body, and neck were covered by plates and bands of tough, leathery skin set with bony studs. Two rows of larger spikes ran along its back. Even its eyes were protected by bony eyelids. Fused bony knobs on the end of the long tail formed a tail club, which the dinosaur could swing at attacking predators. *Euoplocephalus* was a herbivore—its beaked mouth was ideally suited for grazing on vegetation in the thick forests of the Late Cretaceous. It may even have used the blunt hoofs at the end of each toe to dig in the ground for roots and tubers. *Euoplocephalus* was probably a solitary dinosaur, though juveniles may have lived in herds.

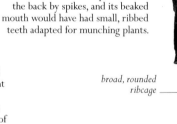

SIZE 20 ft (6 m)
DATE Late Cretaceous
DISTRIBUTION N. America
GROUP Ankylosaurids

ARMORED HEAD >
The head had a massive skull, with protective spikes at the back and a beaked mouth. The name *Euoplocephalus* means "well-armored head."

˅ NECK VERTEBRAE
Although the head was relatively small and the neck short, the neck vertebrae had to be strong to support the head's weighty, studded armor plating.

WALKING TANK >
Euoplocephalus had a wide, low-slung body, supported by short, stout limbs. In cross-section it would have been almost round. Its small head was protected at the back by spikes, and its beaked mouth would have had small, ribbed teeth adapted for munching plants.

short shoulder blade

broad, rounded ribcage

< STUDDED PLATES
One of the most important features of *Euoplocephalus* was its armor plating. This was made of tough plates of skin studded with ridged ovals of bone.

˄ TAIL CLUB
The massive tail club was made of fused bones—two large and several smaller bones. This weapon was probably used in defense.

˄ TAIL VERTEBRAE
About halfway along the tail, the typical tail vertebrae—armed with spikes—gave way to a fused, bony structure. This rigid structure would have supported the club at the end of the tail. The tail would have been quite muscular.

˄ FRONT FOOT
This dinosaur's limbs were short, stocky, and strong. The front feet had short, sturdy toes, which helped to support the considerable body weight.

in life, bony studs were covered with scaly horn

elbow joint

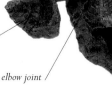

bony studs and
spines along back

head spike

massive leg bones
supported the weight
of the armored body

long, convoluted nasal
passages inside the skull
suggest this dinosaur had
a keen sense of smell

hind feet had three
toes, each tipped
with a blunt hoof

MICROS
LIFE

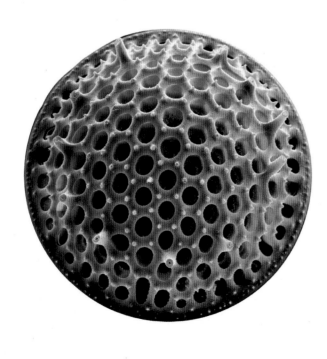

COPIC

Despite their tiny size, microscopic organisms dominate life on Earth. They were the first living things to evolve, and they underpin all the world's ecosystems, capturing and releasing nutrients that other forms of life need to survive. From the simplest bacteria to the most complex protists, they make up a constellation of varied and versatile life-forms that largely goes unseen.

» 90

ARCHAEA AND BACTERIA

Life at its most basic, archaea and bacteria are single-celled organisms without nuclei. Most live as individuals, though some form colonies such as filaments and chains.

» 94

PROTISTS

Among the most numerous living things, protists are also some of the most diverse. Some have no fixed shape, but many have elaborate mineral skeletons or shells. They are typically single-celled.

ARCHAEA AND BACTERIA

An alien visiting Earth might conclude that its true masters were the archaea and the bacteria. They outnumber, and are more diverse than, the more complex life-forms known as the eukaryotes, and thrive in every nook and cranny of the planet.

DOMAIN	ARCHAEA
PHYLA	12
ORDERS	18
FAMILIES	28
SPECIES	Probably millions

DOMAIN	BACTERIA
PHYLA	About 50
ORDERS	180
FAMILIES	430
SPECIES	Probably millions

Hydrothermal vents at the bottom of oceans are home to various thermophile archaea living at high temperatures.

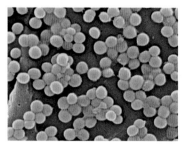

Staphylococcus **bacteria** are common agents of food poisoning in humans, forming microscopic clusters in food.

DEBATE

CAULDRON OF LIFE

Many archaea are adapted for extremes; in hot water their fragile DNA is lagged by protective protein. Similiar protein braces the longer chromosomes of eukaryotes (fungi, plants, and animals). So perhaps what began as insulation in primeval pools evolved into a "scaffold" for the extra DNA needed for more complex life.

The archaea and bacteria are single-celled organisms and were the first forms of life on Earth. They form two of life's three fundamental divisions, along with the eukaryotes. While all living cells possess DNA, eukaryote cells contain it within a membrane-bound nucleus, and most also have energy-generating mitochondria. Archaea and bacteria have neither a nucleus nor mitochondria. They are only distantly related to each other, having evolved from separate, as yet unknown genetic origins. Archaea cells are contained by chemically unique membranes overlain by a tough outer cell wall, and within the cell their DNA is often protein-covered. Bacteria cells have very different physical and chemical makeups, particularly in their cell wall. Their different properties make the typical archaea especially suited to harsh habitats, while bacteria are ubiquitous and thrive in every environment.

SMALLEST LIVING THINGS

All archaea and bacteria are tiny, their size measured in microns or micrometers (µm); 1 µm equals one-thousandth of a millimeter. A human hair is about 80 µm thick; most archaea and bacteria are just 1–10 µm in length. Yet they survive in almost every corner of the biosphere, from the outer atmosphere to deep within Earth's crust, and from ocean depths to inside the human body. Some archaea and bacteria thrive in boiling water or freezing ice, survive radiation, and even live off poisonous gases and corrosive acids. Most get nourishment from dead material, others from infesting living bodies. While some make food in darkness, by using the energy in minerals, others use photosynthesis, using the energy in light to convert carbon dioxide and water into food and oxygen. Although notorious for causing infectious diseases, bacteria are actually essential to human health, with humans depending on their gut bacteria to help break down food and to manufacture essential nutrients. For almost 4 billion years, archaea and bacteria have profoundly influenced Earth's climate, rock formation, and the evolution of other forms of life.

MICROSCOPIC CYANOBACTERIA COLONY >
Although bacteria have single cells, the cells of some, such as cyanobacteria, can join together in spectacular, long filaments.

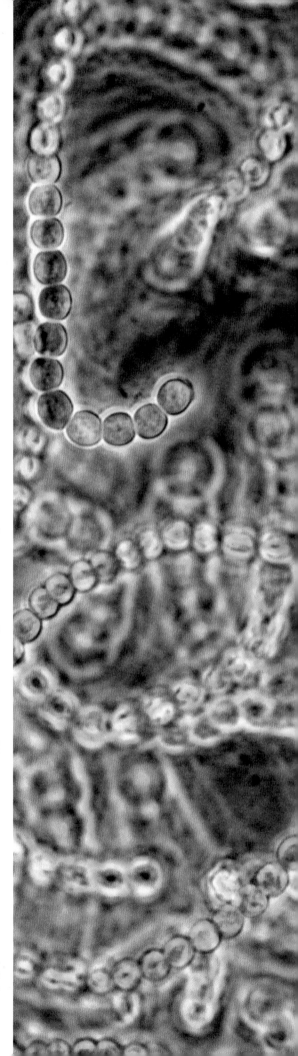

ARCHAEA

protein-making granules

flexible cell wall

1.2 μm

METHANOCOCCOIDES BURTONII

This methane producer was discovered living at the bottom of Ace Lake in Antarctica, where there is no oxygen and the average temperature is 33 °F (0.6 °C).

0.5–15 μm

STAPHYLOTHERMUS MARINUS

Discovered in a hydrothermal vent on the sea floor, this species grows best at 184–198 °F (85–92 °C). It forms grapelike clusters, and can grow comparatively large.

SULFOLOBUS ACIDOCALDARIUS

Like many archaea, this thermophile has a heat-resistant cell wall. It thrives in the high temperatures of hot springs in Yellowstone National Park.

1–5 μm

80 μm

THERMOPROTEUS TENAX

This species forms rod-shaped cells of varying lengths but constant diameter. Its heat-resistant outer envelope is ancient in evolutionary terms.

1 μm

DESULFUROCOCCUS MOBILIS

This anaerobe thrives where oxygen is limited or absent and uses sulfur-containing compounds in place of oxygen. An extreme thermophile, it grows best at 184 °F (85 °C).

0.8–2 μm

8 μm

METHANOSPIRILLUM HUNGATEI

Discovered in human sewage, this archaean generates large amounts of methane during sewage disposal. Each cell is confined within a hollow sheath.

PYROCOCCUS FURIOSUS

Pyrococcus means "fireberry," reflecting this archaean's shape and its tolerance of extreme temperatures: it grows best at 212 °F (100 °C).

BACTERIA

1–4 μm

2–3 μm

1–2 μm

ACETOBACTER ACETI

Used to make vinegar, this species is also a common contaminant wherever alcohol is brewed, causing discoloration and souring of beer in particular.

BACILLUS SUBTILIS

Up to one billion of these bacteria can be found in a single gram of soil. When food is scarce or the environment becomes harsh it can live as an inactive spore.

BACILLUS THURINGIENSIS

This bacterium produces toxic crystals insoluble in the human gut but lethal to insects, and is therefore commercially marketed as an insecticide.

DEINOCOCCUS RADIODURANS

This species is known as the world's toughest bacterium. It was discovered in meat that had been exposed to doses of radiation in an experiment.

1.62 μm

BORDETELLA PERTUSSIS

This species causes whooping cough, or pertussis, a respiratory infection marked by severe, spasmodic coughing episodes, and breakdown of blood cells.

0.25 μm

ESCHERICHIA COLI

An organism much observed and used by scientists, the rod-shaped *E. coli* lives harmlessly in the gut, although virulent strains are a major cause of food poisoning.

1–3 μm

BACTEROIDES FRAGILIS

Forming a large part of the human gut's flora, this bacterium is usually harmless to its host. But it can also cause disease, invading tissues to form pus-filled abscesses.

1.5–4.5 μm

individual rod-shaped E. coli bacterium

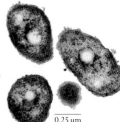

0.6–4 μm

NITROBACTER SP.

This genus of bacteria plays a role in the nitrogen cycle by oxidizing nitrites into nitrates. This helps purify water and is central to fertilizing the oceans and the soil.

4—8 µm

2—3 µm

3—8 µm

cell dividing

NOSTOC SP.
This cyanobacteria forms gelatinous filaments in colonies that are hardy enough to survive from the poles to the tropics.

3—7 µm

1—3 µm

DNA

CLOSTRIDIUM BOTULINUM
This bacterium thrives where oxygen is limited or absent. Living in soil, it produces the neurotoxins that cause botulism, but which also have medical and cosmetic uses.

CLOSTRIDIUM TETANI
A soil organism that can grow in dead tissue in wounds or burns, C. tetani produces tetanospasmin, the neurotoxin that causes tetanus.

SALMONELLA ENTERICA
Belonging to the same family as E. coli, some subspecies of Salmonella bacteria cause gastroenteritis; others cause typhoid fever.

SHIGELLA DYSENTERIAE
This gut bacteria produces shiga toxin that can cause epidemic dysentery. As few as ten bacteria are needed to cause an infection.

STREPTOCOCCUS PNEUMONIAE
This bacterium, present inside humans, can cause pneumonia. In children and the elderly, it is the leading cause of invasive infection that moves into all parts of the body.

0.9 µm

1.5—6 µm

1 µm

pigmented membranes for photosynthesis

LACTOBACILLUS ACIDOPHILUS
Found in the intestine and vagina, this bacterium has nutritional and anti-microbial properties. It is used in probiotic drinks and supplements.

STAPHYLOCOCCUS EPIDERMIDIS
This spherical bacterium can be found on the skin as part of normal flora, but can cause infections in immune-compromised patients.

flagellum helps propel organism

0.4—0.5 µm

1.5—2 µm

1—3 µm

VIBRIO CHOLERAE
Highly mobile curved rods with a single flagellum at one end, this species secretes a potent enterotoxin that causes cholera.

FUSOBACTERIUM NUCLEATUM
This species makes its home in the human mouth, and forms a major component of plaque. It can also cause premature birth.

PSYCHROBACTER URATIVORANS
This species is a psychrophile, or cryophile, which means it thrives at very low temperatures, thanks to natural antifreeze molecules in its cytoplasm within the cell membrane.

numerous rods of bacteria grouped together

NITROSOSPIRA SP.
Filling a vital ecological niche, this nitrifying soil bacteria oxidizes ammonia to form nitrites as part of the nitrogen cycle.

1 µm

cell wall

cell contents, or cytoplasm

1—3 µm

ENTEROCOCCUS FAECALIS
Normally a harmless inhabitant of the human digestive tract and vagina, this species can invade wounds and is resistant to many antibiotics.

0.5—1 µm

YERSINIA PESTIS
This species is the agent that causes bubonic plague. Transmitted to humans via fleas that live on rats, it causes up to 3,000 cases worldwide each year.

PROTISTS

From tiny amoebas to giant kelp, the protists defy simple description, yet this informal grouping of eukaryotes included the first life-forms to evolve that were more complex than the archaea and bacteria. It still produces most of Earth's food and oxygen.

DOMAIN	EUKARYOTA
KINGDOM	PROTISTA
CLADES AND DIVISIONS	More than 9
FAMILIES	About 841
SPECIES	About 73,500

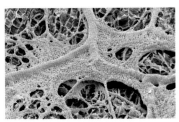

In some slime molds, which are a group of amoebas, thousands of individual nuclei coexist within one giant cell.

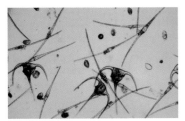

Many single-celled protists have extraordinary shapes, such as these grapplelike dinoflagellates.

Some protists cause devastating disease, such as this *Giardia*, which infects intestines of humans and other animals.

Protists are mainly single-celled creatures that, unlike archaea and bacteria, have cell nuclei. Their basic cell nature separates them from the higher eukaryotes—plants, fungi, and animals—that later emerged from them. The protists include an incredible range of organisms with diverse lifestyles and ecological niches. Most are microscopic, ranging in size from 10 to 100 µm, and some are tiny enough to infest red blood cells. Others, however, are multicellular organisms such as kelp, a seaweed that grows up to tens of yards in length, or the strange, funguslike slime molds, which form creeping puddles of slime that are essentially one giant cell. Typical protists include amoebas, which move around and capture food particles with their pseudopods (cell extensions), or plankton drifting in the sea, such as the beautiful diatoms, which have intricate silicon-based skeletons.

A HIDDEN REALM OF LIFE

Protists are among the most numerous creatures on Earth. Vast numbers of them live in the oceans and rivers, in sediments at the bottom of seas and lakes, and in the soil, while many spend all or part of their life cycles as parasites inside other organisms. They play a vital role in the planet's ecosphere, in particular as primary photosynthesizers, using the energy in light to convert carbon dioxide and water into food, while releasing oxygen into the atmosphere. They also act as predators and recyclers of matter. A few species are well known because they cause major diseases: the parasitic *Plasmodium* causes malaria, one of the biggest killers of humans; another parasite, *Trypanosoma brucei*, causes sleeping sickness. Also familiar are the dinoflagellates, a group of plankton organisms that can cause "red tides"—huge blooms of toxic creatures that kill fishes and poison humans.

The taxonomy of the protists is complex, as they don't form a natural kingdom. However, molecular and genetic analysis has allowed most protists to be placed in a few large clades, groups that each share a common ancestor, such as the Amoebozoa, Rhizaria, and Alveolata. The red algae, green algae, and the stoneworts each have their own division.

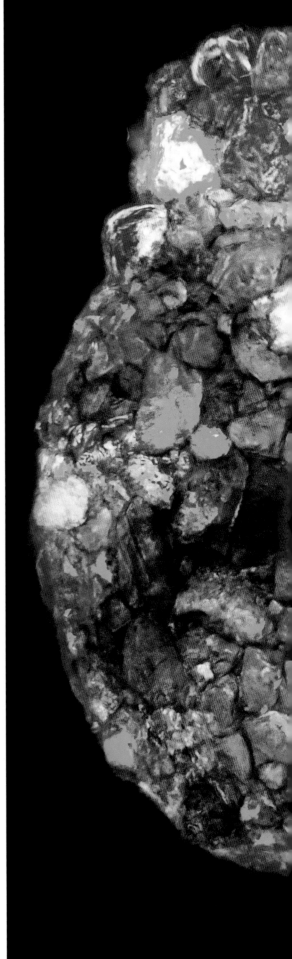

SMALL WONDERS >
This polarized light micrograph reveals an *Arcella* shelled amoeba and a *Micrasterias* desmid in all their beauty.

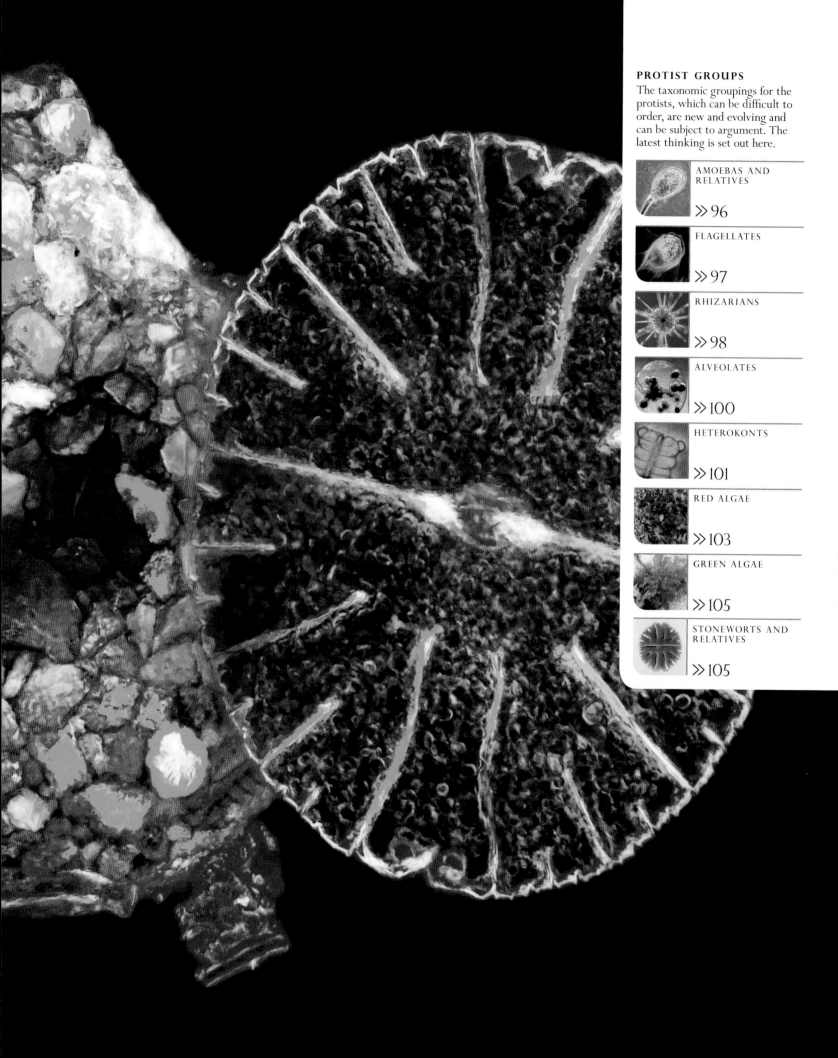

AMOEBAS AND RELATIVES

Two of the protist clades, the Amoebozoa and the Opisthokonta, have evolved different ways of moving around and obtaining food.

Amoebas of the clade Amoebozoa can change shape by oozing projections, called pseudopods, of their single cells. They use these "false feet" to creep forward and hunt smaller organisms, enveloping victims within a sac of fluid and digesting them while still alive. Some amoebas are giant cells, and are just visible to the naked eye. A few are gut parasites, causing amoebic dysentery in humans. One group, the slime molds, have a remarkable strategy for avoiding starvation. When prey runs low their cells are attracted to one another by a chemical distress signal, and aggregate into a tiny creeping slug. This then sprouts upward a number of stalks, which burst to scatter spores. Each spore hatches into a new amoeba ready to hunt on new ground.

ANIMAL-FUNGUS ORIGINS

Most protists of the clade Opisthokonta have developed a single whiplike flagellum for propulsion in open water. Early in the history of life, some of these protists may have given rise to animals, and the single flagellum is still seen in the tail of animal sperm today. Others, called nucleariids, lost their flagellum and actually reverted to the amoebalike state. The nucleariids may be closely related to fungi, as fungi also lack flagella and fertilize one another without the involvement of free-swimming sperm.

DOMAIN	EUKARYOTA
KINGDOM	PROTISTA
CLADES	2
FAMILIES	About 50
SPECIES	About 4,000

DEBATE
LIFE'S FIRST BRANCHES?

One theory of evolution says that the eukaryotes divided into two branches: unikonts (cells with one flagellum), such as opisthokonts, and bikonts (cells with two). Unikonts evolved into animals and fungi, while bikonts gave rise to plants. However, the DNA evidence for the theory is equivocal.

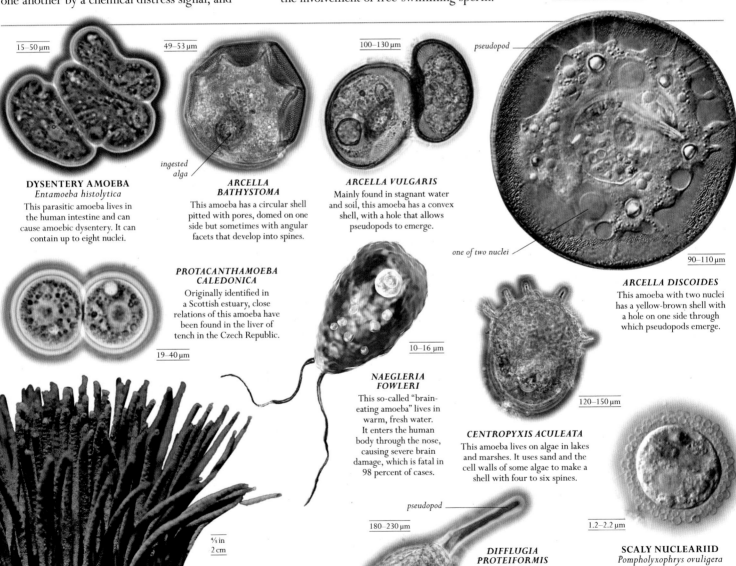

15–50 µm

DYSENTERY AMOEBA
Entamoeba histolytica
This parasitic amoeba lives in the human intestine and can cause amoebic dysentery. It can contain up to eight nuclei.

49–53 µm

ingested alga

ARCELLA BATHYSTOMA
This amoeba has a circular shell pitted with pores, domed on one side but sometimes with angular facets that develop into spines.

100–130 µm

ARCELLA VULGARIS
Mainly found in stagnant water and soil, this amoeba has a convex shell, with a hole that allows pseudopods to emerge.

pseudopod

one of two nuclei

90–110 µm

ARCELLA DISCOIDES
This amoeba with two nuclei has a yellow-brown shell with a hole on one side through which pseudopods emerge.

PROTACANTHAMOEBA CALEDONICA
Originally identified in a Scottish estuary, close relations of this amoeba have been found in the liver of tench in the Czech Republic.

19–40 µm

10–16 µm

NAEGLERIA FOWLERI
This so-called "brain-eating amoeba" lives in warm, fresh water. It enters the human body through the nose, causing severe brain damage, which is fatal in 98 percent of cases.

120–150 µm

CENTROPYXIS ACULEATA
This amoeba lives on algae in lakes and marshes. It uses sand and the cell walls of some algae to make a shell with four to six spines.

1.2–2.2 µm

⅕ in
2 cm

STEMONITIS SP.
This "chocolate tube" or "pipe cleaner" slime mold begins as a mass of cell content with many nuclei from which numerous spore-bearing stalks sprout up.

pseudopod

180–230 µm

DIFFLUGIA PROTEIFORMIS
This pond-ooze amoeba constructs a shell by sticking together minuscule grains of sand and the cell walls of some algae.

SCALY NUCLEARIID
Pompholyxophrys ovuligera
Once assigned to the heliozoans clade, this flagellate of the Opisthokonta is covered in hollow scales or beads.

FLAGELLATES

The whiplike propulsion of single-celled organisms has evolved in several groups of protists, some unrelated. But it predominates in the Excavata.

Flagellates are powerful swimming microbes that move by the whiplike action of one or more external threads called flagella. Many are predators, preying on smaller organisms such as bacteria. However, unlike changeable amoebas, these flagellates have a fixed, rigid shape and direct food into a cellular "mouth" at the base of the flagella. Remarkably, however, some, such as *Euglena*, have a degree of behavioral versatility that provides them with plant- and animal-type nutrition, depending upon circumstance. When exposed to bright light these organisms can photosynthesize; but in darkness their light-absorbing cell organelles, known as chloroplasts, shrivel and they revert to being a predator.

LIVING INSIDE ANIMALS

Many flagellates of this group lack the ordinary cellular features that would enable them to use oxygen in their respiration and survive inside the oxygen-poor environment of animal guts. Many are extraordinarily specialized, subsisting on the partially-digested food inside the abdomens of insects, but causing their host no harm. Others are parasites that cause devastating disease, including in humans. One notorious group, the trypanosomes, are transmitted by biting insects and are responsible for sleeping sickness and leishmaniasis in the tropics.

DOMAIN	EUKARYOTA
KINGDOM	PROTISTA
CLADE	EXCAVATA
FAMILIES	40
SPECIES	About 2,500

DEBATE
EVOLUTION DOWN-SIZING

The flagellates inside termite guts lack mitochondria, the cell organelles most other protists use to respire with oxygen. Some think these flagellates are primitive protists. Others say they are more advanced organisms, and lost their mitochondria by evolving in an oxygen-poor environment.

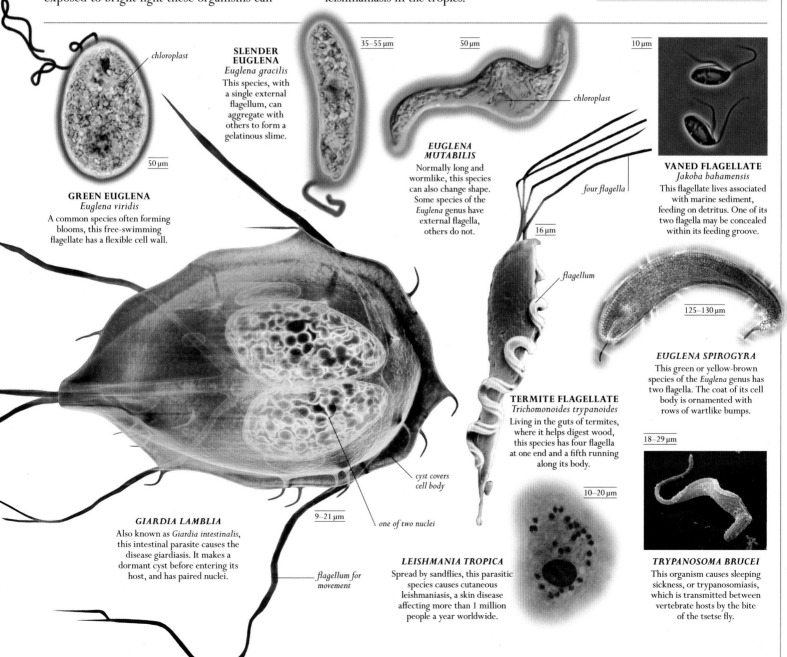

GREEN EUGLENA
Euglena viridis
A common species often forming blooms, this free-swimming flagellate has a flexible cell wall.

chloroplast

50 μm

SLENDER EUGLENA
Euglena gracilis
This species, with a single external flagellum, can aggregate with others to form a gelatinous slime.

35–55 μm

50 μm

EUGLENA MUTABILIS
Normally long and wormlike, this species can also change shape. Some species of the *Euglena* genus have external flagella, others do not.

chloroplast

10 μm

VANED FLAGELLATE
Jakoba bahamensis
This flagellate lives associated with marine sediment, feeding on detritus. One of its two flagella may be concealed within its feeding groove.

four flagella

16 μm

125–130 μm

EUGLENA SPIROGYRA
This green or yellow-brown species of the *Euglena* genus has two flagella. The coat of its cell body is ornamented with rows of wartlike bumps.

flagellum

TERMITE FLAGELLATE
Trichomonoides trypanoides
Living in the guts of termites, where it helps digest wood, this species has four flagella at one end and a fifth running along its body.

18–29 μm

cyst covers cell body

one of two nuclei

9–21 μm

GIARDIA LAMBLIA
Also known as *Giardia intestinalis*, this intestinal parasite causes the disease giardiasis. It makes a dormant cyst before entering its host, and has paired nuclei.

flagellum for movement

10–20 μm

LEISHMANIA TROPICA
Spread by sandflies, this parasitic species causes cutaneous leishmaniasis, a skin disease affecting more than 1 million people a year worldwide.

TRYPANOSOMA BRUCEI
This organism causes sleeping sickness, or trypanosomiasis, which is transmitted between vertebrate hosts by the bite of the tsetse fly.

RHIZARIANS

The clade Rhizaria includes two phyla that are among the most beautiful of all of the miniature protists: the radiolarians and the forams.

The unique, intricately sculptured shells of both phyla make them distinctive members of the microscopic world, and some have left an impressive fossil record. Most radiolarians build a glassy shell with silica, stripping the ocean surface of this otherwise abundant mineral when they bloom in large numbers. Their body radiates long pseudopods that poke through their shell like the rays of the sun, supplemented in some by hardened spines. Radiolarians use their pseudopods for catching food, but some also harbor living algae, gaining nourishment from the sugars their photosynthetic partners make as they drift in sunlit tropical oceans. Their silica dependence binds radiolarians to the sea, but their allies, the cercozoans, have also exploited soil and fresh water. Cercozoans typically retain the long pseudopods, but have shelled, and shell-less forms, and some have flagella, depending on habitat demands.

The forams, or foraminiferans, have flourished in the oceans for hundreds of millions of years and their calcium carbonate shells litter the seabed to form layers of chalky sediment. Foram shells are so distinctive, even when fossilized, that geologists can use them to date deposits and identify hidden reservoirs of oil. When alive, each shell harbors a tiny amoeba that hunts with pseudopods—like the radiolarians—although some are big enough to catch animal larvae.

DOMAIN	EUKARYOTA
KINGDOM	PROTISTA
CLADE	RHIZARIA
FAMILIES	108
SPECIES	About 14,000

CONVERGING GIANTS

The pseudopods of radiolarians and forams enmesh to form nets around their cells. This perhaps supports the idea that both forms descend from a common ancestor. But it is possible that this net evolved independently in the two groups, perhaps a result of both having large single cells.

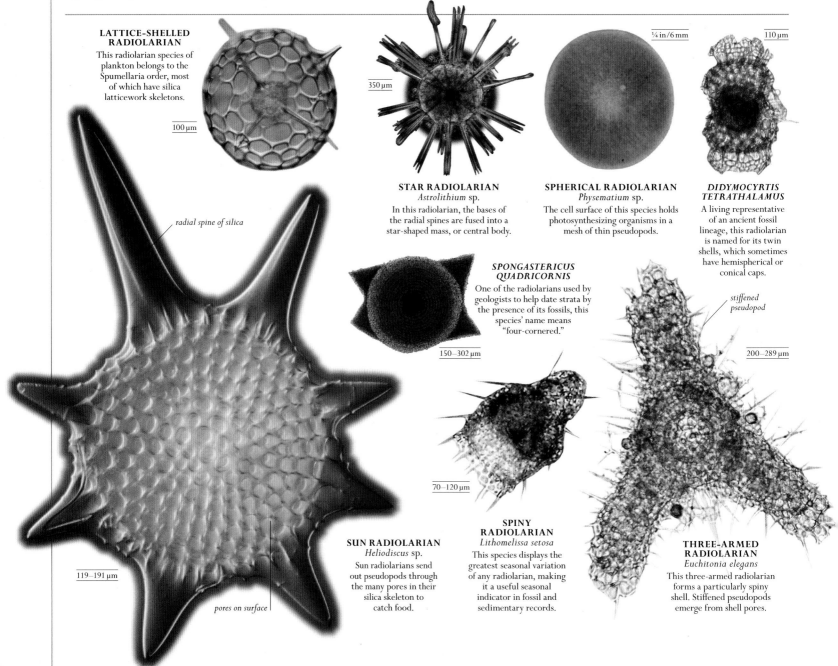

LATTICE-SHELLED RADIOLARIAN
This radiolarian species of plankton belongs to the Spumellaria order, most of which have silica latticework skeletons.

100 μm

radial spine of silica

350 μm

STAR RADIOLARIAN
Astrolithium sp.
In this radiolarian, the bases of the radial spines are fused into a star-shaped mass, or central body.

¼ in/6 mm

SPHERICAL RADIOLARIAN
Physematium sp.
The cell surface of this species holds photosynthesizing organisms in a mesh of thin pseudopods.

110 μm

DIDYMOCYRTIS TETRATHALAMUS
A living representative of an ancient fossil lineage, this radiolarian is named for its twin shells, which sometimes have hemispherical or conical caps.

SPONGASTERICUS QUADRICORNIS
One of the radiolarians used by geologists to help date strata by the presence of its fossils, this species' name means "four-cornered."

150–302 μm

stiffened pseudopod

200–289 μm

119–191 μm

pores on surface

70–120 μm

SUN RADIOLARIAN
Heliodiscus sp.
Sun radiolarians send out pseudopods through the many pores in their silica skeleton to catch food.

SPINY RADIOLARIAN
Lithomelissa setosa
This species displays the greatest seasonal variation of any radiolarian, making it a useful seasonal indicator in fossil and sedimentary records.

THREE-ARMED RADIOLARIAN
Euchitonia elegans
This three-armed radiolarian forms a particularly spiny shell. Stiffened pseudopods emerge from shell pores.

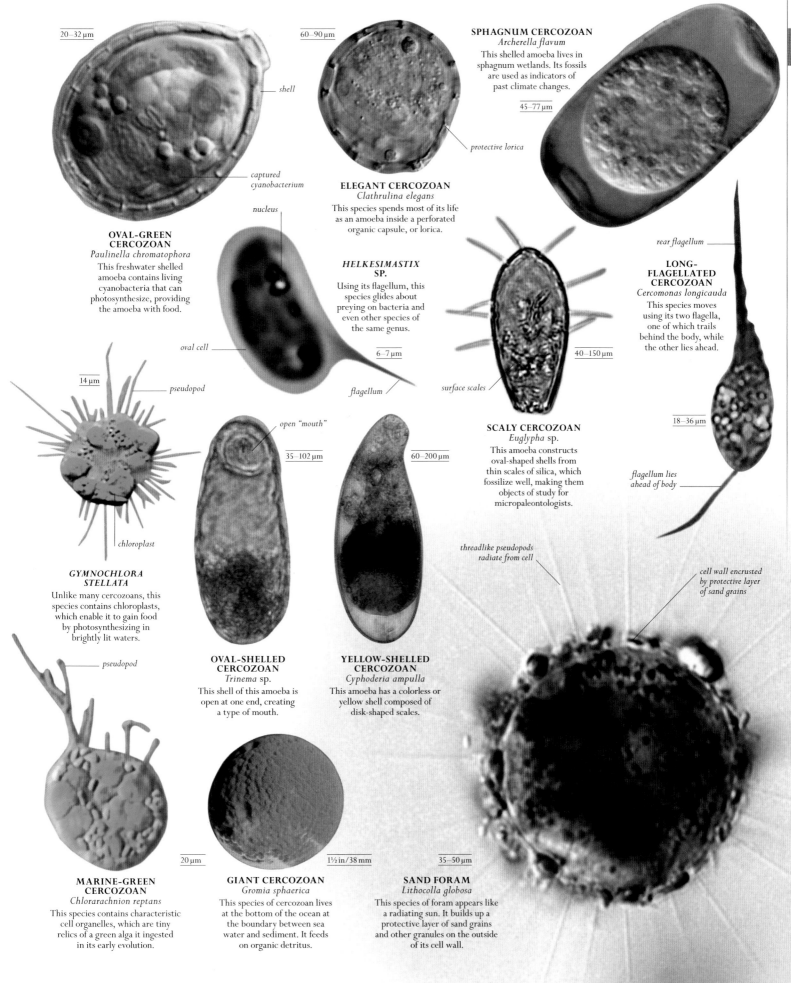

20—32 µm

shell

captured
cyanobacterium

**OVAL-GREEN
CERCOZOAN**
Paulinella chromatophora
This freshwater shelled
amoeba contains living
cyanobacteria that can
photosynthesize, providing
the amoeba with food.

60—90 µm

SPHAGNUM CERCOZOAN
Archerella flavum
This shelled amoeba lives in
sphagnum wetlands. Its fossils
are used as indicators of
past climate changes.

45—77 µm

protective lorica

ELEGANT CERCOZOAN
Clathrulina elegans
This species spends most of its life
as an amoeba inside a perforated
organic capsule, or lorica.

nucleus

oval cell

**HELKESIMASTIX
SP.**
Using its flagellum, this
species glides about
preying on bacteria and
even other species of
the same genus.

6—7 µm

flagellum

rear flagellum

**LONG-
FLAGELLATED
CERCOZOAN**
Cercomonas longicauda
This species moves
using its two flagella,
one of which trails
behind the body, while
the other lies ahead.

18—36 µm

flagellum lies
ahead of body

14 µm

pseudopod

chloroplast

**GYMNOCHLORA
STELLATA**
Unlike many cercozoans, this
species contains chloroplasts,
which enable it to gain food
by photosynthesizing in
brightly lit waters.

open "mouth"

35—102 µm

surface scales

40—150 µm

SCALY CERCOZOAN
Euglypha sp.
This amoeba constructs
oval-shaped shells from
thin scales of silica, which
fossilize well, making them
objects of study for
micropaleontologists.

60—200 µm

threadlike pseudopods
radiate from cell

cell wall encrusted
by protective layer
of sand grains

pseudopod

**OVAL-SHELLED
CERCOZOAN**
Trinema sp.
This shell of this amoeba is
open at one end, creating
a type of mouth.

**YELLOW-SHELLED
CERCOZOAN**
Cyphoderia ampulla
This amoeba has a colorless or
yellow shell composed of
disk-shaped scales.

20 µm

1½ in/38 mm

35—50 µm

**MARINE-GREEN
CERCOZOAN**
Chlorarachnion reptans
This species contains characteristic
cell organelles, which are tiny
relics of a green alga it ingested
in its early evolution.

GIANT CERCOZOAN
Gromia sphaerica
This species of cercozoan lives
at the bottom of the ocean at
the boundary between sea
water and sediment. It feeds
on organic detritus.

SAND FORAM
Lithocolla globosa
This species of foram appears like
a radiating sun. It builds up a
protective layer of sand grains
and other granules on the outside
of its cell wall.

ALVEOLATES

Alveolates are united by possession of the same anatomical quirk—a fringe of tiny sacs around the cell called alveoli, after which the clade is named.

Three groups of superficially different but still single-celled protists are included in the alveolates: the dinoflagellates, the ciliates, and the apicomplexans. Predatory dinoflagellates live in the ocean, most using two whiplike flagella that emerge from grooves in their cell coat at right angles to one another to swim. Some can discharge stinging barbs to immobilize their prey. Others release poisons, and sudden blooms of dinoflagellates are responsible for toxic red tides in parts of the world. A few are bioluminescent, sparkling with light when disturbed at night.

Most ciliates feed on smaller organisms. Their smooth graceful motion is due to the coordinated beating of countless tiny hairs called cilia, which can cover their entire single-celled body. They also use these cilia to waft their food into a furrow that approximates to a mouth. Ciliates are virtually ubiquitous; some even live in the stomachs of mammalian herbivores, where they help in digesting the tough cellulose in plants.

In contrast, all apicomplexans are parasites. They are named for an arrangement of structures, called the apical complex, which helps them penetrate the living cells of animals, from which they derive their nourishment. Infamous among the apicomplexans are the malaria parasites, which penetrate the red blood cells of animals for food, destroying the cells in the process.

DOMAIN	EUKARYOTA
KINGDOM	PROTISTA
CLADE	ALVEOLATA
FAMILIES	222
SPECIES	About 20,000

PROTOZOANS

Early classifications placed food-eating microbes into different classes of the animal phylum Protozoa. Some microbes, such as the ciliates and the dinoflagellates, are now grouped together in modern protist taxonomy, although the exact nature of their relationships is debatable.

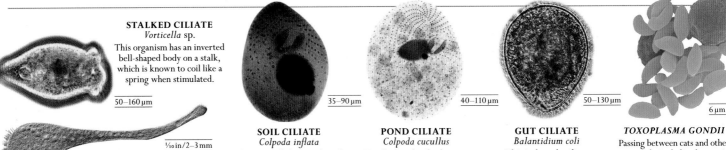

STALKED CILIATE
Vorticella sp.
This organism has an inverted bell-shaped body on a stalk, which is known to coil like a spring when stimulated.

50–160 µm

1/10 in/2–3 mm

TRUMPET CILIATE
Stentor muelleri
This algae-feeding species, with a horn-shaped cell body, is large for a single-celled organism.

SOIL CILIATE
Colpoda inflata
This normally kidney-shaped species plays an important role in soil ecology but is vulnerable to pesticides.

35–90 µm

POND CILIATE
Colpoda cucullus
Usually found in fresh water among decaying plants, this ciliate has food-containing organelles called vacuoles within its cell.

40–110 µm

GUT CILIATE
Balantidium coli
This is the only ciliate known to parasitize humans. Infection can cause intestinal ulcers, or serious intestinal infections.

50–130 µm

TOXOPLASMA GONDII
Passing between cats and other mammals (including humans), this apicomplexan causes the disease toxoplasmosis, which is dangerous to the unborn child in pregnancy.

6 µm

225 µm

THREE-HORNED DINOFLAGELLATE
Tripos muelleri
Formerly known as *Ceratium tripos* and among the most distinctive dinoflagellates, blooms of this plankton can cause dangerous red tides.

CHAIN-FORMING DINOFLAGELLATE
Gymnodinium catenatum
This species of dinoflagellate forms long, swimming chains of up to 32 cells.

38–50 µm

MAHOGANY-TIDE DINOFLAGELLATE
Karlodinium veneficum
This dinoflagellate plankton can undergo a population explosion leading to "mahogany tides" that are lethal to fishes.

19–17 µm

CRYPTOSPORIDIUM PARVUM
This apicomplexan can cause cryptosporidiosis, a diarrhoeal disease transmitted by ingestion of its spores via feces, usually in contaminated water.

4–6 µm

PLASMODIUM FALCIPARUM
The most deadly of the malaria-causing *Plasmodium* species, this apicomplexan parasite is responsible for the deaths of over 1 million people a year worldwide.

9–14 µm

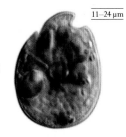

organelle

200–2,000 µm

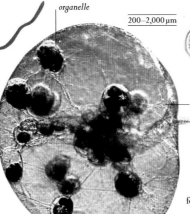

gas bag

AKASHIWO SANGUINEA
A large pentagonal species responsible for several recorded instances of harmful algal blooms, this dinoflagellate can both photosynthesize and prey on other plankton.

40–74 µm

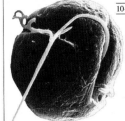

10–100 µm

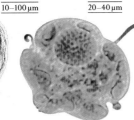

20–40 µm

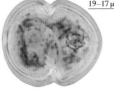

GYMNODINIUM SP.
Producing a neurotoxin, blooms of this dinoflagellate, found in both fresh and salt water, cause red tides and shellfish poisoning.

KARENIA BREVIS
Formerly known as *Gymnodinium brevis*, this dinoflagellate plankton is a major cause of red tides in the Gulf of Mexico.

SEASPARKLE
Noctiluca scintillans
This bioluminescent, plankton-forming species has a gas bag allowing it to float just below the surface.

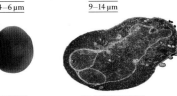

AMPHIDINIUM CARTERAE
This plankton-forming species causes ciguatera, a fish poisoning that can cause human fatalities if contaminated fishes are consumed.

11–24 µm

HETEROKONTS

Heterokonts include some types of algae, which are photosynthesizing protists growing in or near water that do not have true leaves or roots.

Heterokonts are mostly defined as having two different types of flagella on the sperm they use in reproduction. One flagellum is covered in tiny, bristly hairs, or mastigonemes, and the other is smooth and whiplike. Heterokonts include diatoms and brown algae, as well as water molds. Diatoms are single-celled algae with finely sculptured silica walls divided into two parts called valves. They are the dominant eukaryotic phytoplankton—plankton communities of small organisms that drift in open waters and engage in photosynthesis. Here, near the surface,

their pigments absorb energy from sunlight to make food. Diatoms contain green chlorophyll—as do plants—but also a brown pigment called fucoxanthin, which serves to broaden the spectrum of light that can be used, making their photosynthesis significantly more efficient.

Brown algae occupy coastal habitats around the world. They also make use of fucoxanthin, and have evolved into complex multicellular seaweeds that superficially resemble a plant. Rather than true roots and leaves, their body consists of a holdfast for clasping to rock and a creeping thallus that lacks the internal veins of a true leaf. Nevertheless, some brown seaweeds, such as kelps, can grow to extraordinary lengths and make extensive underwater forests on some offshore coastlines.

DOMAIN	EUKARYOTA
KINGDOM	PROTISTA
CLADE	HETEROKONTOPHYTA
FAMILIES	177
SPECIES	About 20,000

DEBATE

FROM ALGA TO MOLD

Water molds grow and feed like mold, but unlike true molds, which are fungi, have plantlike cell walls and heterokont flagella, while some also cause plant diseases. DNA analysis links them with diatoms and brown algae, so perhaps they evolved from algae that abandoned their chloroplasts and turned parasitic.

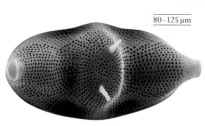

80–125 µm

EPIPHYTIC DIATOM
Biddulphia sp.
The cause of brown film on fish tanks, in the wild this species grows on seaweed and rocks.

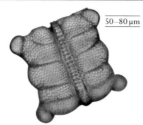

50–80 µm

BARREL DIATOM
Biddulphia pulchella
This image clearly displays the different parts of a diatom: the two valves bound together by a narrow girdle.

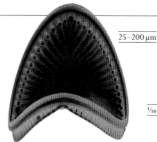

25–200 µm

SADDLE DIATOM
Campylodiscus sp.
This genus of diatom shows a groove between the tubular lips, known as canals, around the edge of its valves.

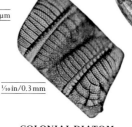

¹⁄₁₀ in/0.3 mm

COLONIAL DIATOM
Isthmia nervosa
This species grows on other algae, specifically seaweeds, forming branching colonies.

60–240 µm

GYROSIGMA **SP.**
The name of this diatom refers to the sigmoidal curve of its cell, meaning that it follows a very slight S-shape.

SLIMY DIATOM
Lyrella lyra
This species secretes a gluey slime from its central groove to help the cells to glide along their host's surface.

125 µm

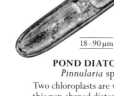

18–90 µm

POND DIATOM
Pinnularia sp.
Two chloroplasts are visible in this pen-shaped diatom, found in ponds and wet ground.

STEPHANODISCUS **SP.**
With its disk shape, areolae (open circles), and a ring of spines, this diatom occurs individually or in chains.

12–20 µm

disk-shaped girdle

spine

areola

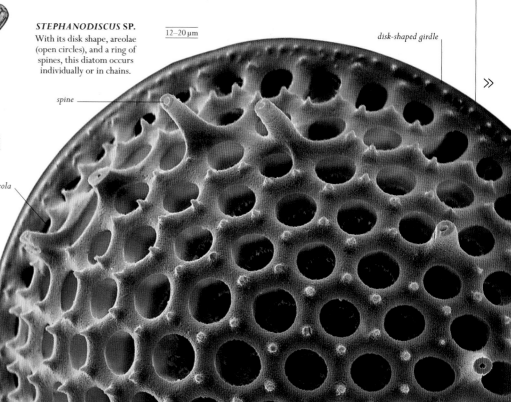

200–1,000 µm

URCHIN HELIOZOAN
Actinosphaerium sp.
This diatom resembles a sea urchin. It moves by shifting cell contents into its thin pseudopods.

10–100 µm

GROOVED DIATOM
Diploneis sp.
The valves of this species of diatom appear as two heavily pronounced lips, known as canals, either side of a slitlike groove known as a raphe.

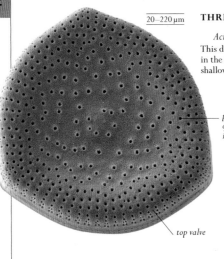

20–220 µm

THREE-CORNERED DIATOM
Actinoptychus sp.
This diatom can be found in the neritic zone on the shallow continental shelf.

pores allow intake of minerals for use in photosynthesis

top valve

¹⁄₃₂ in
0.9 mm

SPIDER'S-WEB DIATOM
Arachnoidiscus sp.
Strong radial ribs and web-patterned valves are features of this disk-shaped diatom. It can grow very large.

protective cell wall

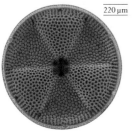

220 µm

20–220 µm

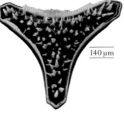

140 µm

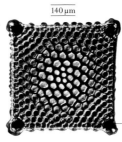

140 µm

TRICERATIUM FAVUS
The heavily silicified cell walls of this species can leave a trail of silica in the water as it moves. This can show how seawater penetrates into freshwater environments.

geometric cell wall

SUN-RAY DIATOM
Actinoptychus heliopelta
A distinctive pattern of alternating raised and depressed sectors on its top and bottom valves characterize this diatom species.

FIVE-ARMED DIATOM
Actinoptychus sp.
A five-arm appearance is produced by the alternating sectors on the valves of this species of the *Actinoptychus* genus.

TRICERATIUM SP.
More than 400 species of this genus of marine diatom are known. Diatoms of this genus are often, but not always, three sided.

fronds are leaflike structures used in photosynthesis

air bladder on frond helps keep seaweed bouyant

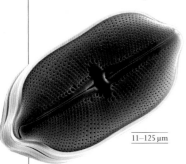

11–125 µm

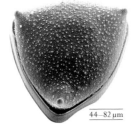

44–82 µm

DAPPLED DIATOM
Stictodiscus sp.
Seen under a scanning electron microscope, this diatom shows pores dappled on its valve surface, above a surrounding girdle.

BOAT-SHAPED DIATOM
Navicula sp.
This species belongs to the largest genus of diatoms, which has thousands of species.

shrublike fronds

6½ ft
2 m

23½ in
60 cm

13 ft
4 m

6½–11 ft
2–3.5 m

3¼–10 ft
1–3 m

12–39 in
30–100 cm

THONGWEED
Himanthalia elongata
During its reproductive phase, this brown seaweed of the northern hemisphere produces long straplike fronds, which are types of divided leaves.

TOOTHED WRACK
Fucus serratus
This robust, shrubby, olive-brown seaweed grows on lower shores all around the North Atlantic.

SUGAR KELP
Saccharina latifolia
Found on rocky shores and at depths of 26–100 ft (8–30 m) in northern seas, this brown seaweed is also often known as sugar wrack.

CUVIE
Laminaria hyperborea
This alga is commercially important in iodine production. It grows at depths of 26–100 ft (8–30 m), mainly in the northern hemisphere.

WIREWEED
Sargassum muticum
Originally from Japan, this invasive brown seaweed has now spread to Europe. It can grow as much as 4 in (10 cm) in a day.

SEA OAK
Halidrys siliquosa
A large brown seaweed, common in rock pools in Europe, this species has a distinctive zigzag stem and air-filled pods known as air bladders.

RED ALGAE

Although some are microscopic organisms, the most familiar red algae are multicellular seaweeds, which are among the largest of the protists. They have evolved into an extraordinary diversity of forms, ranging from rock-hugging encrustations that look more like lichens to bristly or leafy clumps.

The great majority of the 6,500 or so species of red algae are marine, but there are a few freshwater species. Unlike brown and green algae, they do not produce flagellated sperm and rely instead on water currents to transport male sex cells to the female organs. The cells then merge, but subsequent development is highly variable between species.

RED FOR SURVIVAL

Like land plants, most algae use green chlorophyll to trap the energy of sunlight and manufacture food in the process of photosynthesis. However, the cells of red algae also contain a red pigment, called phycoerythrin, which masks the green color of the chlorophyll and gives them their red color—and their name. This pigment allows them to continue photosynthesizing even at depths in the water where only a little blue light can penetrate. This allows red algae to live much deeper in the oceans than brown or green seaweeds, and some have been reported growing as deep as 650 ft (200 m).

Some red algae become calcified, forming encrustations on rocks or hard, upright, antler-like forms. And, in spite of their name, their pigments can make red algae appear olive or gray.

DOMAIN	EUKARYOTA
KINGDOM	PROTISTA
DIVISION	RHODOPHYTA
FAMILIES	92
SPECIES	About 6,500

DEBATE
CLOSE RELATIVES?

Red algae were previously grouped with green algae, but it is now recognized that green algae are two separate and very diverse groups. Some species may be close to red algae, but others are closer to land plants. Red algae branched off the taxonomic tree before the land plants began to evolve.

DULSE
Palmaria palmata
In N. Atlantic coastal communities, this edible seaweed is a traditional source of protein and vitamins.

20 in
50 cm

CORAL WEED
Corallina officinalis
This red seaweed is common in rock pools worldwide, forming branching, featherlike fronds.

²⁄₅–6 in
1–15 cm

AGARDHIELLA SUBULATA
Originally from the western Atlantic, Caribbean, and Gulf of Mexico, this fleshy red alga has invaded some parts of Europe.

16 in
40 cm

SCHMITZIA HISCOCKIANA
Found in areas swept almost bare by the tide, this red seaweed is fleshy and gelatinous, with flattened branches and fingerlike projections.

3¼ in
8 cm

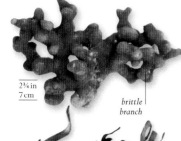

2¾ in
7 cm

brittle branch

MAERL
Phymatolithon calcareum
This species, an encrusting coralline alga of the British Isles, is commercially dredged and ground up as a calcium-rich soil additive.

EYELASH WEED
Calliblepharis ciliata
This northern hemisphere seaweed has flat fronds, fringed with numerous "branchlets."

12 in
30 cm

12 in
30 cm

BLACK CARRAGEEN
Furcellaria lumbricalis
This northern hemisphere seaweed has cylindrical, brown-black fronds that branch into masses of fleshy fingers.

tubular frond

fanlike branch

frond

4–12 in
10–30 cm

GRACILARIA FOLIIFERA
Rising as slender dull purple stems, this red seaweed branches sparsely to give straplike fronds. It grows abundantly in shallow lagoons worldwide.

12 in
30 cm

GRACILARIA BURSA-PASTORIS
A long, thin red seaweed with forked or alternating branches, this species is found from southern England to the Pacific and the Caribbean.

6½ in
17 cm

MASTOCARPUS STELLATUS
This red seaweed has conspicuous papillae, which are reproductive bodies, on the fronds. It is found in the North Atlantic.

8¾ in
22 cm

IRISH MOSS
Chondrus crispus
Also known as carrageen moss, this seaweed of the British Isles is important as a source of carrageen, which is a gelling agent.

»

PTEROCLADIELLA CAPILLACEA
With feathery branches that taper toward the tip, this red seaweed, found in pools worldwide, often develops a Christmas-treelike appearance.

¹⁄₁₀ in
2.5 mm

SPOROLITHON PTYCHOIDES
This red algae forms a crust using calcareous deposits in its cell walls. It is found worldwide in rock pools and on current-swept rocky seabeds.

12 in
30 cm

CERAMIUM VIRGATUM
This small red seaweed grows worldwide on rocks and other seaweeds. From a tiny holdfast it grows into filamentlike fronds with forked tips.

¹⁄₁₀–³⁄₅ in
2–15 mm

GELIDIUM PUSILLUM
Growing worldwide from an extensive creeping base that incorporates shells and small snails, this turf-forming red seaweed has flattened, leaflike fronds.

flat, leaflike structure

6½ in
17 cm

blades on frond

MELANAMANSIA FIMBRIFOLIA
Found in N. America and Australia, this seaweed grows on sediment-covered reefs at depths of up to 180 ft (55 m).

CHONDRIA DASYPHYLLA
This species, found worldwide, has feathery fronds ending in clublike branchlets that sprout spore-bearing branchlets and urn-shaped fruit bodies.

4–8½ in
10–21 cm

8 in
20 cm

BROAD WEED
Lenormandiopsis nozawae
On both sides of this broad-bladed seaweed, found in temperate areas, are clusters of spore-bearing bodies, themselves home to tiny, parasitic algae.

stolon

LEATHER WEED
Ptilophora leliaertii
First described in 2004 after its discovery on a reef off the coast of South Africa, this red seaweed has feathery compound branches.

14 in
35 cm

GELIDIELLA ACEROSA
An important source of agar, used in the food and pharmaceutical industries, this red seaweed, found in India, has a slender, cylindrical shoot, known as a stolon.

3¼ in
8.5 cm

6 in
15 cm

BUSH WEED
Ahnfeltia sp.
A source of the gelatinous substance agar, used in petri dishes for cell culture, this northern-hemisphere seaweed forms dense tufts of fronds.

2¾–8¾ in
7–22 cm

LAURENCIA OBTUSA
This tropical species is a source of halogenated terpenoids— chemical defenses against grazing crabs and urchins—which also make useful antifouling agents.

12 in
30 cm

SEA BEECH
Delesseria sanguinea
Known for its characteristic beechlike leaves, this red seaweed grows in the understory of kelp forests in Europe.

¾–4 in
2–10 cm

POLYSIPHONIA LANOSA
Growing on other seaweeds in tufts like pom-poms in the northern hemisphere, this red alga has branching filaments made up of long, tubular cells.

GREEN ALGAE

Green algae are a loose taxonomic aggregation of very diverse species. Some live in freshwater ponds and streams, some form green mats over damp, shady rocks or tree trunks, and others grow as leafy seaweeds, attached to rocks in shallow sea water.

Many green algae are microscopic and free-floating; others are multicellular with variably complex structures, including the blanketweeds that sometimes clog ponds. Some reproduce asexually by splitting, budding, or producing motile spores, but sexual reproduction is common in larger species. These typically produce sperm with two identical flagella, but also have spore-producing stages in their life cycle. The cells of green algae use the same chlorophyll pigments as land plants for photosynthesis, and they are sometimes included with them in an informal grouping called the Viridiplantae (green plants).

DOMAIN	EUKARYOTA
KINGDOM	PROTISTA
DIVISION	CHLOROPHYTA
FAMILIES	127
SPECIES	4,300

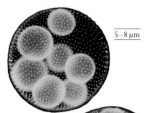
5–8 μm

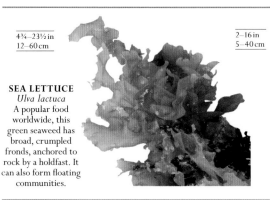
4¾–23½ in
12–60 cm

SEA LETTUCE
Ulva lactuca
A popular food worldwide, this green seaweed has broad, crumpled fronds, anchored to rock by a holdfast. It can also form floating communities.

2–16 in
5–40 cm

GREEN SEA FINGERS
Codium fragile
This tubular green alga occurs worldwide, in seashore rock pools and coastal waters to 6½ ft (2 m) deep.

6½–16 ft
2–5 m

SEA GRAPE
Caulerpa lentillifera
This edible seaweed produces many upright stalks from a long, creeping runner. In the Philippines, the succulent stalks are harvested and eaten raw in salads.

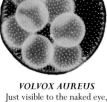

VOLVOX AUREUS
Just visible to the naked eye, spherical colonies of this freshwater alga consist of thousands of microscopic individuals. Threadlike flagellae spin the colony through the water.

STONEWORTS AND RELATIVES

Although related to green algae, the stoneworts and their relatives have more complex structures and more advanced chemistry in their cells, suggesting that they are the ancestors of the true plants.

The Streptophyta (sometimes called Charophyta) includes desmids, which are classed as "conjugating green algae." These typically have a single cell, divided into two symmetrical semicells.

However, the stoneworts and brittleworts are the best known group of streptophytes, and are often regarded as "honorary plants." They grow in the mud of shallow, freshwater or brackish habitats, rooted by cell filaments, called rhizoids, like the roots of simple plants. Their branched form and reproductive structures are also reminiscent of true plants. In fact, some classification systems regard the Streptophyta as an "infra-kingdom" that also includes all of the land plants.

DOMAIN	EUKARYOTA
KINGDOM	PROTISTA
DIVISION	STREPTOPHYTA
FAMILIES	16
SPECIES	2,700

prong

350 μm

egg-bearing capsule

12–23½ in
30–60 cm
long stem is plantlike

NITELLA TRANSLUCENS
This delicate "plant" with translucent green leaves grows in the clear water of ponds, streams, and fens around W. and S. Europe.

4–20 in
10–50 cm
stem made of elongated cells

COMMON STONEWORT
Chara vulgaris
Also known as skunkweed, this streptophyte of the northern hemisphere emits an unpleasant smell.

32–70 μm

PENIUM SP.
This desmid, found in N. America, is symmetrically divided into cylindrical semicells with blunt, oval ends and girdle bands.

CLOSTERIUM SP.
This crescent-shaped desmid, found worldwide, displays semicells, each housing a chloroplast, joined at the isthmus, where the nucleus is situated.

cell divide
semicell
100–460 μm

MICRASTERIAS SP.
This desmid, found in temperate areas, has multiple spined arms, tipped with prongs that allow the semicells to interlock.

PLANTS

By harnessing the energy in sunlight, and using it to grow, green plants play a central role in life on Earth. They generate food for animals and other organisms, and often form their habitats as well. Some plants are small and simple, but they also include giant conifers, and a dazzling wealth of flowering species that have evolved a remarkable array of shapes and strategies for survival.

≫ 111
HORNWORTS • LYCOPHYTES

Hornworts are small plants, related to mosses, which produce spores from an elongated, hornlike structure. Lycophytes were previously thought to belong with the ferns, and include clubmosses.

≫ 112
FERNS AND RELATIVES

Ferns are the largest plants that reproduce with spores instead of seeds. Many are low-growing, but some form trees, which have trunks made of fibrous roots rather than wood.

≫ 116
CYCADS, GINKGOS, AND GNETOPHYTES

These plants do not have flowers, but they do form seeds. Before flowering plants evolved, cycads formed an important part of the world's vegetation.

≫ 118
CONIFERS

Although far fewer in species than flowering plants, conifers dominate the landscape in some parts of the world. All of them are trees or shrubs, and they typically form seeds in woody cones.

≫ 122
FLOWERING PLANTS

This is by far the largest group of living plants, making up the bulk of the world's vegetation. They all grow flowers—often inconspicuously—and spread by forming seeds.

MOSSES

Mosses are nonflowering plants that typically form mats or cushion-shaped clumps. Despite their small size, they are remarkably resilient plants. They are able to grow in a wide range of habitats, from woodland to deserts, and are found on every continent, including Antarctica.

DIVISION	BRYOPHYTA
CLASSES	8
ORDERS	30
FAMILIES	110
SPECIES	About 10,000

In the far north sphagnum mosses form raised bogs—here interspersed with gray patches of reindeer moss lichen.

Mosses have thin leaves that usually spiral around slender, wiry stems, and they reproduce by scattering spores. Like liverworts, they need moist conditions to grow. They can be extremely abundant in habitats that are always damp, and some species—particularly sphagnum mosses—form extensive soggy blankets, dominating the ground in cold parts of the world. Others, however, are able to remain dormant during droughts. They look gray and lifeless, but become green again within minutes if it rains.

Mosses have a two-stage life cycle—a feature shared by all other plants. In mosses, the dominant stage is the gametophyte, which produces male and female cells. After the female cells are fertilized, they produce the spore-forming stages or sporophytes, which remain attached to the parent plant. Most moss sporophytes have a capsule on the end of a long stalk. Once ripe—a process that can take many months—the capsule splits open, releasing up to 50 million spores into the air.

Because moss spores are so small and light, they are carried large distances by the slightest breeze. This makes mosses very good at dispersing and enables them to colonize all kinds of microenvironments, from crevices in the bark of trees to damp walls and roofs.

RELATIVES AND LOOK-ALIKES

The division name Bryophyta is sometimes used to encompass the liverworts and hornworts (see pp.110–111), but these are now regarded as separate, but closely related, taxonomic divisions. Clubmosses are more complex than true mosses, and belong in the division Lycophyta (see p.111). Other moss look-alikes include reindeer moss—actually a form of lichen—and Spanish moss, a flowering plant in the bromeliad family, that grows festooned on trees.

∨ **DEEP COVER**
The cool, moist climate in New Zealand's Fjordland National Park enables a wide variety of mosses and liverworts to thrive.

1¼ in/3 cm

BLACK ROCK MOSS
Andreaea rupestris
F: Andreaeaceae

Widespread on mountains, this dark-colored moss grows on bare rock. Unlike most mosses, its capsules release their spores through four microscopic slits.

1¼ in/3 cm

FIRE MOSS
Ceratodon purpureus
F: Ditrichaceae

Found worldwide, particularly on burned or disturbed ground, this low-growing moss is also common on roofs and walls. It produces a dense carpet of spore capsules in spring.

¾ in/2 cm

COMMON POCKET MOSS
Fissidens taxifolius
F: Fissidentaceae

This widespread moss has short, spreading stems bearing two ranks of pointed leaves. It grows in shaded ground and on rocks.

VELVET FEATHER-MOSS
Brachythecium velutinum
F: Brachytheciaceae

With its much-branched stems, this abundant moss forms creeping mats on dead wood and in poorly drained grass. It has a worldwide range.

spore capsule with a conical lid

4 in 10 cm

6 in/15 cm

WHITE FORK MOSS
Leucobryum glaucum
F: Dicranaceae

Growing in large, neatly rounded cushions, this woodland moss has a characteristic gray-green hue, turning almost white in dry weather.

2 in/5 cm

MOUNTAIN FORK MOSS
Dicranum montanum
F: Dicranaceae

A moss that forms compact feathery mounds, this lowland species has narrow leaves that curl up when dry, sometimes breaking off to form new plants.

2 in/5 cm

COMMON TAMARISK MOSS
Thuidium tamariscinum
F: Thuidiaceae

Its finely divided "fronds" make this woodland moss resemble a miniature fern. It grows on rotting wood and rocks in Europe and N. Asia.

1½ in 4 cm

CYPRESS-LEAVED PLAIT MOSS
Hypnum cupressiforme
F: Hypnaceae

This highly variable mat-forming moss has crowded, overlapping leaves. Found worldwide, it is common on rocks and walls, and beneath trees.

4 in 10 cm

OSTRICH-PLUME FEATHER MOSS
Ptilium crista-castrensis
F: Hypnaceae

Found mainly in northern forests, this moss has feather-shaped, symmetrically branched stems, often forming extensive patches below spruces and pines.

32 in 80 cm

GREATER WATER-MOSS
Fontinalis antipyretica
F: Fontinalaceae

A submerged freshwater species, greater water-moss trails from rocks in slow-flowing rivers and streams. It has three rows of keeled, dark green leaves.

1¼ in/3 cm

PURPLE FORK MOSS
Grimmia pulvinata
F: Grimmiaceae

Growing on rocks, roofs, and walls, this widespread moss has leaves that end in long, silvery hairs. Its spore capsules grow on curved stalks.

SWAN'S NECK THYME MOSS
Mnium hornum
F: Mniaceae

Brilliant green in spring, this is a common woodland moss in Europe and N. America. Each spore capsule has a curved stalk resembling a swan's neck.

spore capsule is horizontal when ripe

⅜ in/1 cm

CAPE THREAD-MOSS
Orthodontium lineare
F: Bryaceae

This moss from the southern hemisphere was introduced to Europe in the early 20th century and has become an invasive species.

10 in 25 cm

BLUNT-LEAVED BOG MOSS
Sphagnum palustre
F: Sphagnaceae

Like its relatives, this peat-forming moss grows on wet ground, and holds large amounts of water. Each stem ends in a flat-topped rosette of small branches.

16 in 40 cm

COMMON HAIR-CAP MOSS
Polytrichum commune
F: Polytrichaceae

This tall, tussock-forming moss is common in wet moorland throughout the northern hemisphere. Its stems are stiff and unbranched, with narrow, pointed leaves.

½ in 1.5 cm

BONFIRE MOSS
Funaria hygrometrica
F: Funariaceae

One of the world's most widespread mosses, this mat-forming species is particularly common on disturbed ground and sites of recent fires. Its ripe spore capsules have orange stalks.

LIVERWORTS

Found mainly in damp, shaded habitats, liverworts are thought to be the simplest of all the existing groups of land plants. They come in two distinct forms: some are flat and ribbonlike; others are more like mosses, with slender stems flanked by tiny leaves.

DIVISION	MARCHANTIOPHYTA
CLASSES	3
ORDERS	16
FAMILIES	88
SPECIES	About 7,500

DEBATE

PLANTS APART

Liverworts have unique features that distinguish them from mosses and hornworts, although they were once all grouped together as bryophytes. They are the only land plants without stomata, or breathing pores, in their sporophyte (spore-producing) stage. Also, their hairlike "roots" are single-celled. They are therefore classed in their own division, Marchantiophyta.

Ribbonlike liverworts look unlike any other plants. Instead of stems and leaves, they have a flat body, or thallus, that repeatedly forks as it grows. Many species have a glistening upper surface with deeply divided lobes. Medieval herbalists noted this liverlike shape, giving the plants their common name. Leafy liverworts look quite different, with trailing or spreading stems. They usually have two ranks of leaves, with a third row of smaller leaves on their underside. Some kinds form loose mats in damp grass, and others grow on rocks. They are much more numerous than ribbonlike liverworts, especially in the tropics, where many species live epiphytically on rainforest trees.

SPREADING THEIR KIND

Unlike most flowering plants, liverworts have no preset limits to their growth. They can form patches up to several yards across, which gradually fragment as they grow, forming multiple plants. In addition, some species have cell clusters, called gemmae, in depressions on their upper surface. Raindrops scatter these gemmae, which grow into new plants.

Liverworts also reproduce by scattering spores, microscopic cells that form new plants. Spore formation requires damp conditions, because a liverwort's female cells must first be fertilized by swimming male sperm. Sperm cells often reach egg cells with the help of raindrops, which splash them from one plant to another. Many liverworts produce their sperm or egg cells in structures that look like tiny parasols. After fertilization, the developed spores are scattered into the air.

female reproductive structure with star-shaped rays

GREATER WHIPWORT
Bazzania trilobata
F: Lepidoziaceae
A native of damp woodland, this mound-forming liverwort has main leaves that overlap and curve downward, giving its stems a caterpillarlike shape.

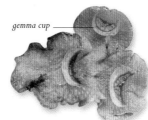
gemma cup

CRESCENT-CUP LIVERWORT
Lunularia cruciata
F: Lunulariaceae
Common in gardens and greenhouses, this ribbon-shaped liverwort is bright green, with distinctive reproductive structures, gemma cups, that look like tiny fingernails.

two ranks of main leaves

OVERLEAF PELLIA
Pellia epiphylla
F: Pelliaceae
Growing on wet peat and rocks, this ribbon-shaped liverwort often forms a tufted mat. It produces black spore capsules on slender white stalks.

GREATER FEATHERWORT
Plagiochila asplenioides
F: Plagiochilaceae
A delicate liverwort with translucent leaves, this species forms spreading, mosslike tufts. It grows on shaded rocks and soil, particularly on chalk and limestone.

EVEN SCALEWORT
Radula complanata
F: Radulaceae
Ranging in color from light green to brown, this scaly-leaved liverwort creeps over flat, unshaded surfaces as varied as tree trunks and coastal rocks.

MALE REPRODUCTIVE STRUCTURES

glossy thallus

COMMON LIVERWORT
Marchantia polymorpha
F: Marchantiaceae
In spring and summer this ribbon-shaped liverwort produces conspicuous spore-forming structures that look like tiny parasols. It is widespread in damp habitats and gardens.

HORNWORTS

Hornworts are the third division in the informal grouping called bryophytes. They are small, easily overlooked, leafy plants found in damp places worldwide.

Like mosses and liverworts, hornworts have two generations in their life cycle: a vegetative gametophyte and a spore-bearing sporophyte. The gametophytes form flattened, leafy lobes, rather like some liverworts, up to 2 in (5 cm) in diameter. In their sporophytes, the spores are released from a characteristic, elongated hornlike structure, which gives them their name. Some species grow as tiny weeds on the soil of gardens or cultivated fields. Other, rather larger, tropical and subtropical species grow mainly on the bark of trees. These hornworts do not produce flowers, unlike the small order of flowering water plants that, confusingly, are also called hornworts (see p.124).

DIVISION	ANTHOCEROTOPHYTA
CLASSES	2
ORDERS	5
FAMILIES	5
SPECIES	About 220

hornlike sporophyte

leafy gametophyte

DENDROCEROS
Dendroceros sp.
F: Dendrocerotaceae
The ribbonlike lobes of dendroceros gametophytes spread across damp rocks or tree bark in tropical and subtropical forests. The hornlike sporophyte grows from this, up to 2 in (5 cm) tall.

FIELD HORNWORT
Anthoceros agrestis
F: Anthocerotaceae
Damp stubble fields, trampled ground, and the sides of ditches are home to this inconspicuous hornwort with distinctive, frilly leaf lobes. It is found across temperate Europe and N. America.

CAROLINA HORNWORT
Phaeoceros carolinianus
F: Notothyladaceae
Growing in similar places to field hornwort, this species has separate male and female gametophytes. Raindrops carry sex cells from leaf pits on males to fertilize females, allowing sporophytes to develop.

LYCOPHYTES

Although their taxonomy is still under review, lycophytes include clubmosses and quillworts, plants which are closely related to, but distinct from, ferns.

Lycophytes are the first plants in this book that have a vascular system, a network of conducting tissues that transport water, minerals, and nutrients around their roots, stems, and leaves. This allows lycophytes to grow taller and survive in drier habitats than bryophytes. Fossil lycophytes date back 425 million years. They were dominant plants during the Carboniferous era. Some then were tree-sized, unlike their smaller modern relatives. Lycophytes spread by spores, produced in sporangia (spore cases) at the base of simple leaves.

DIVISION	LYCOPHYTA
CLASSES	1
ORDERS	3
FAMILIES	3
SPECIES	About 170

STAGHORN CLUBMOSS
Lycopodiella cernua
F: Lycopodiaceae
Lycopodiella species are generally more delicate than other clubmosses. Looking like a miniature pine tree, staghorn clubmoss grows in bogs throughout the tropics.

IBERIAN QUILLWORT
Isoetes longissimi
F: Isoetaceae
Quillworts grow in slow-moving streams and clear ponds. They have densely tufted, hollow leaves, with spore cases hidden in their swollen bases.

INTERRUPTED CLUBMOSS
Lycopodium annotinum
F: Lycopodiaceae
This clubmoss of northern moors and mountains produces spores from conelike clusters at the tip of its upright stems.

HICKEY'S TREE-CLUBMOSS
Lycopodium hickeyi
F: Lycopodiaceae
Native to hardwood forests and scrub in eastern N. America, this species spreads by underground stems which produce treelike, fertile fronds.

FERNS AND RELATIVES

Most ferns can be recognized by their graceful fronds, which unroll as they grow. Together with horsetails and whisk ferns, they make up a diverse and ancient group of nonflowering plants that reproduce by means of spores. Ferns grow in a wide variety of habitats, although the majority thrive where there is moisture and shade.

DIVISION	POLYPODIOPHYTA
CLASSES	1
ORDERS	11
FAMILIES	40
SPECIES	About 12,000

DEBATE

LIVING FOSSILS

Horsetails have been around for over 300 million years. One of the largest extinct genera, called *Calamites*, grew up to 65 ft (20 m) tall, and had massive ridged stems over 23½ in (60 cm) across. The 20 or so species of horsetails that survive today are much smaller plants, growing in water, woodland, and disturbed ground. They look very different from ferns, producing their spores in conelike structures at the tips of some of their stems, but most botanists now agree that they should be classified as an order within the ferns.

Some ferns are small and easily overlooked, but the group also includes large, treelike species that can be over 50 ft (15 m) high. Many are compact plants, with a single cluster of fronds, but others have creeping stems that sprout fronds as they spread. Bracken fern—one of the most widespread creeping species—is particularly vigorous. Over many years, it can produce clumps, or clones, over ½ mile (800 m) across. Ferns also include species that live in fresh water and a wide variety of epiphytic species, which live on other plants.

Fern fronds are often finely divided, and usually develop from tightly coiled buds, known as crosiers or fiddleheads. Spores are formed on the underside of the fronds, in structures that look like raised dots or lines. In many fern species, all the fronds produce spores, but some have two kinds of fronds: fertile fronds form spores, while sterile fronds are primarily adapted to collecting sunlight so the fern can grow. When a fern spore germinates, it develops into an intermediate stage known as a gametophyte.

Tiny, flat, and paper-thin, this eventually produces a new sporophyte—the spore-forming plant we recognize as a fern.

FERN RELATIVES

Fern taxonomy is in a state of flux, but all the scientific evidence suggests that ferns are more closely related to seed plants (gymnosperms and angiosperms) than to the lycophytes with which they were once grouped.

The fern division also includes horsetails, which are upright plants with hollow, cylindrical stems, surrounded by whorls of slender branches. Silica granules give them a rough texture, and they were once used for scouring pots and pans. Adder's tongues and whisk ferns are distinctive groups, with twig-shaped stems or a single, divided frond, that are also classed within the ferns.

∨ **CHARACTERISTIC COIL**
The delicate tips of young fern fronds are protected inside tight coils. These unwind as they stretch upward toward the light.

HORSETAILS

32 in/80 cm

COMMON HORSETAIL
Equisetum arvense
F: Equisetaceae

Widespread in the northern hemisphere, this horsetail can be a troublesome weed. Its black underground rhizomes sprout hollow shoots, bearing symmetrical rings of bright green branches.

TRUE FERNS

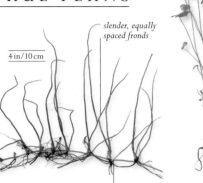

slender, equally spaced fronds

4 in/10 cm

PILLWORT
Pilularia globulifera
F: Marsileaceae

creeping stem

This marshland fern from W. Europe forms clumps resembling tufts of grass. Its spores develop at ground level, in green capsules shaped like pills.

WATER CLOVER
Marsilea quadrifolia
F: Marsileaceae

4 in/15 cm

With its four-lobed fronds, this clump-forming aquatic fern looks deceptively like a flowering plant. It is widely distributed across the northern hemisphere.

5 ft
1.5 m

ROYAL FERN
Osmunda regalis
F: Osmundaceae

Often cultivated, this stately fern from the northern hemisphere has a rosette of spreading fronds with several narrower, spore-bearing fronds in the center.

¾ in
2 cm

FLOATING FERN
Salvinia natans
F: Salviniaceae

Often forming a dense carpet, this water fern has small, oval leaves covered with water-repellent hairs. It is common throughout the tropics.

½ in
1.5 cm

MOSQUITO FERN
Azolla filiculoides
F: Salviniaceae

A floating plant with mat-forming fronds, mosquito fern spreads rapidly across lakes and ponds. It is widespread in warm parts of the world.

WHISK FERNS, MOONWORTS, AND ADDER'S TONGUES

23½ in
60 cm

WHISK FERN
Psilotum nudum
F: Psilotaceae

A primitive relative of true ferns, this largely tropical plant has leafless, brushlike stems, bearing berrylike capsules that produce its spores.

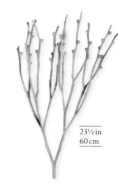

12 in
30 cm

ADDER'S TONGUE FERN
Ophioglossum vulgatum
F: Ophioglossaceae

This unusual fern has a single, oval frond, clasping a slender, spore-producing stalk. It grows in grassy places throughout the northern hemisphere.

COMMON MOONWORT
Botrychium lunaria
F: Ophioglossaceae

Found worldwide in temperate regions, moonwort has a single vegetative frond and a fertile frond topped by a cluster of spore capsules.

8 in
20 cm

lobed vegetative frond

3¼ ft
1 m

UMBRELLA FERN
Sticherus cunninghamii
F: Gleicheniaceae

This distinctive fern from New Zealand has a pencil-thin stem topped by a horizontal crown of narrow, radiating fronds. It spreads by creeping rhizomes.

33 ft
10 m

SILVER TREE FERN
Alsophila dealbata
F: Cyatheaceae

This tree fern gets its name from its fronds, which have a silvery underside. Native to New Zealand, it grows in open forest and scrub.

20 ft
6 m

fronds are evergreen in mild climates

TASMANIAN TREE FERN
Dicksonia antarctica
F: Dicksoniaceae

This stout-stemmed tree fern is widespread in Tasmania and S.E. Australia. It needs shady conditions, and usually grows among trees.

59 ft
18 m

BLACK TREE FERN
Sphaeropteris medullaris
F: Cyatheaceae

From New Zealand, this tall and slender tree fern has a sooty-black trunk and dark leaf stalks, contrasting with its bright green fronds.

26 ft
8 m

SOFT TREE FERN
Alsophila smithii
F: Cyatheaceae

Native to New Zealand and sub-Antarctic islands, this is the most southerly tree fern. It often has a collar of dead fronds hanging beneath its crown.

»

long terminal leaflet

BLACK MAIDENHAIR FERN
Adiantum capillus-veneris
F: Pteridaceae

Found in crevices on limestone, this widespread fern has bright green, translucent leaflets that contrast with its slender black stems.

12 in
30 cm

SILVER BRAKE
Pteris argyraea
F: Pteridaceae

This shade-loving fern is named for its distinctive fronds, which have a silvery white central stripe. Originally from S.E. Asia, it is widely grown as a houseplant.

3¼ ft
1 m

purple stem and veins

23½ in
60 cm

PAINTED BRAKE
Pteris tricolor
F: Pteridaceae

From Malaysia, painted brake has an unusual range of colors: its metallic-looking, copper-colored young fronds turn green as they mature.

CLIFF BRAKE
Pellaea viridis
F: Pteridaceae

Native to South Africa, this drought-resistant fern has shining green fronds with wiry, black stems. It grows in open woodland on mountains.

20 in
50 cm

SILVER LIP FERN
Cheilanthes argentea
F: Pteridaceae

From E. Asia, this small, evergreen fern has wedge-shaped fronds with black veins and distinctive silvery markings on their underside.

10 in
25 cm

LADDER BRAKE
Pteris vittata
F: Pteridaceae

Widespread in warm regions, this fern has upright or arching fronds with narrow, linear leaflets. It usually grows on limestone and in alkaline soils.

3¼ ft
1 m

16 in
40 cm

NORTHERN BEECH FERN
Phegopteris connectilis
F: Thelypteridaceae

Found as far north as Greenland, this compact fern grows in a variety of habitats from woodland to rocky tundra.

30 in
75 cm

MOUNTAIN FERN
Oreopteris limbosperma
F: Thelypteridaceae

Found in damp habitats on acid soil, this clump-forming European fern gives off a characteristic lemony smell if its fronds are bruised.

6½ ft
2 m

BRACKEN FERN
Pteridium aquilinum
F: Dennstaedtiaceae

Found on every continent except Antarctica, bracken spreads vigorously by growing underground rhizomes. It often dies back in winter, sprouting new fronds in spring.

purplish-black stalk

20 in
50 cm

SQUIRREL'S FOOT FERN
Davallia trichomanoides
F: Davalliaceae

This epiphytic fern from Malaysia creeps over trees and other plants. Its rhizomes have furlike scales and tips that resemble a squirrel's foot.

6 ft
1.8 m

EUROPEAN CHAIN FERN
Woodwardia radicans
F: Blechnaceae

Native to S.W. Europe and Algeria, this is a luxuriant fern of damp, shaded habitats. Its arching fronds sometimes produce bulbils (small bulbs) at their tips.

30 in
75 cm

DEER FERN
Blechnum spicant
F: Blechnaceae

This evergreen fern has spreading sterile fronds, and upright fertile fronds with leaflets like narrow ribs. It is native to the temperate northern hemisphere.

SOFT SHIELD FERN
Polystichum setiferum
F: Dryopteridaceae
Native to damp European woodlands, soft shield fern has feather-shaped fronds, with leaflets turned at an angle, instead of lying flat.

4 ft
1.2 m

16 in
40 cm

OAK FERN
Gymnocarpium dryopteris
F: Cystopteridaceae
Delicate bright green fronds, sprouting separately, distinguish this northern fern. It grows in shaded screes as well as in woodland.

MALE FERN
Dryopteris filix-mas
F: Dryopteridaceae
One of the most common woodland ferns in Europe, this species has a crown of concentric fronds shaped like the tail of a shuttlecock.

spore capsules on underside of fronds

4 ft
1.2 m

BRITTLE BLADDER-FERN
Cystopteris fragilis
F: Cystopteridaceae
Named for its swollen spore capsules borne on the underside of its fronds, this fern grows in temperate regions worldwide.

23½ in
60 cm

16 in
40 cm

23½ in
60 cm

toothed, pale green leaflets

winged midrib

fronds form cone-shaped crown

5 ft
1.5 m

SENSITIVE FERN
Onoclea sensibilis
F: Onocleaceae
The fronds of this wetland fern from N. America and E. Asia seem very sensitive to cold, dying soon after the first frosts.

HART'S TONGUE FERN
Asplenium scolopendrium
F: Aspleniaceae
With its glossy, strap-shaped fronds, this fern is often cultivated as an ornamental plant. It grows wild in Europe, W. Asia, and N. America.

6 in
15 cm

WALL RUE
Asplenium ruta-muraria
F: Aspleniaceae
Widespread across the northern hemisphere, this small tuft-forming fern grows on limestone rocks and on walls containing lime-rich mortar.

OSTRICH FERN
Matteuccia struthiopteris
F: Onocleaceae
Native to the northern hemisphere, this tall waterside fern has a symmetrical crown of sterile fronds in summer, followed by brown fertile fronds in the winter.

12 in
30 cm

8 in
20 cm

MAIDENHAIR SPLEENWORT
Asplenium trichomanes
F: Aspleniaceae
Occurring from the tropics to subarctic regions, this small clump-forming fern grows in rocky places. Its fronds have paired elliptical leaflets.

COMMON STAGHORN FERN
Platycerium bifurcatum
F: Polypodiaceae
Found in Indonesia and Australasia, this impressive epiphyte grows on tree trunks. It has kidney-shaped sterile fronds and spreading fertile fronds shaped like antlers.

35 in
90 cm

COMMON POLYPODY
Polypodium vulgare
F: Polypodiaceae
Instead of forming clumps, this fern sprouts separate fronds along its creeping rhizome. It is common on rocks, trees, and leaf litter in northern temperate regions.

CYCADS, GINKGOS, AND GNETOPHYTES

Found mainly in warm parts of the world, the cycads, ginkgos, and gnetophytes are three ancient divisions of nonflowering plants with strikingly varied shapes. They range from climbers and low-growing shrubs to plants that are easily mistaken for palms.

DIVISION	CYCADOPHYTA
CLASS	1
ORDER	1
FAMILIES	2
SPECIES	330

DIVISION	GINKGOPHYTA
CLASS	1
ORDER	1
FAMILY	1
SPECIES	1

DIVISION	GNETOPHYTA
CLASS	1
ORDERS	1
FAMILIES	3
SPECIES	70

∨ BUILT TO LAST
Like palms, cycads usually have a single crown of compound leaves, surrounding a central growing point, or apical meristem. Their tough leaves are able to withstand strong sunshine and drying winds.

These three groups are traditionally classed, along with conifers, as gymnosperms—a word that means "naked seed." Unlike flowering plants, gymnosperms form their seeds on exposed surfaces, which are typically specialized scales. Flowering plants produce their seeds in closed chambers, or ovaries.

Scientists have not yet resolved the exact relationship between cycads, ginkgos, and gnetophytes. Nor can they say where they stand in relation to conifers and flowering plants. At the cellular level, some features of gnetophytes suggest they are more closely related to conifers than to cycads and ginkgos, but they also have features in common with flowering plants.

Apart from the nature of their seeds, the three groups have little in common, and hardly ever grow in the same place. Cycads grow mainly in the tropics and subtropics, while the single surviving species of ginkgo comes from China. Gnetophytes are more diverse, including trees and climbers from the tropics, densely branched shrubs from dry habitats, and the bizarre *Welwitschia*, which grows only in the Namib Desert of southwest Africa.

VARYING FORTUNES

Cycads have a long history, dating back nearly 300 million years. Once an important part of the world's vegetation, they have gradually lost ground in the face of competition from flowering plants. Today, nearly a quarter of cycad species are endangered, under threat from illegal plant collection and habitat change.

Gnetophytes face fewer problems, and the ginkgo has seen a dramatic improvement in its status. For centuries it was preserved by Buddhist monks, who grew it in temple gardens long after it had vanished in the wild. Introduced to Europe in the 1700s, it proved easy to grow and highly tolerant of polluted air. It is now planted worldwide in parks and city streets.

CYCADS

stiff, upswept leaves

10 ft/3 m

trunk branches when mature

JAPANESE SAGO PALM
Cycas revoluta
F: Cycadaceae

Unrelated to true sago, this palmlike cycad from S. Japan has a thick stem and glossy leaves. It is widely cultivated as an ornamental plant.

spiny leaves with recurved tips

EASTERN CAPE BLUE CYCAD
Encephalartos horridus
F: Zamiaceae

Unlike most cycads, this low-growing semidesert species from South Africa has gray-blue leaves, composed of stiff leaflets armed with sharp spines.

4½ ft/1.4 m

20 ft/6 m

PRICKLY CYCAD
Encephalartos altensteinii
F: Zamiaceae

Found near South Africa's eastern coast, this tall, subtropical cycad is named for the toothed edges of its leaves. Its yellow cones produce bright red seeds.

6 ft
1.8 m

4 ft
1.2 m

23 ft
7 m

10 ft/3 m

MEXICAN FERN PALM
Dioon edule
F: Zamiaceae

A cycad rather than a true palm, this slow-growing plant from E. Mexico has oval seed cones up to 12 in (30 cm) long.

MEXICAN HORNCONE
Ceratozamia mexicana
F: Zamiaceae

6½ ft/2 m

This robust cycad from E. Mexico has a large crown of spreading leaves and gray-green cones with distinctive horns on their scales.

COONTIE
Zamia pumila
F: Zamiaceae

Once used as a source of starch, this dwarf cycad from the Caribbean has a short, half-buried stem and upright reddish-brown cones.

MACROZAMIA MOOREI
F: Zamiaceae

One of Australia's tallest cycads, this palmlike plant has giant seed cones up to 36 in (90 cm) long. It grows in dry woodland.

BURRAWONG
Macrozamia communis
F: Zamiaceae

Native to Australia's southeast coast, this cycad has large cones with red, fleshy seeds. It often grows in dense stands.

GNETOPHYTES

LONGLEAF JOINT-FIR
Ephedra trifurca
F: Ephedraceae

Also known as Mormon tea, this mound-forming plant has stems ringed by scalelike leaves. It grows in deserts in Mexico and the southern US.

6½ ft/2 m

MELINJO
50 ft/15 m
Gnetum gnemon
F: Gnetaceae

A nonflowering tree from S.E. Asia and the Pacific, melinjo has evergreen leaves and nutlike seeds. Both are used in cooking.

GINKGO

MAIDENHAIR TREE
Ginkgo biloba
F: Ginkgoaceae

100 ft/30 m

This distinctive tree is easily recognized by its fan-shaped leaves, which turn bright yellow in the fall. Originally restricted to S. China, it is now cultivated worldwide.

SEEDS WITH FLESHY COAT

EDIBLE KERNELS

shredded leaf

18 in/45 cm

CHINESE EPHEDRA
Ephedra sinica
F: Ephedraceae

This straggling plant from E. Asia contains powerful alkaloids and has a long history of use as a medicinal herb.

male cones attached to central stem

WELWITSCHIA
Welwitschia mirabilis
F: Welwitschiaceae

Endemic to the Namib Desert, this extremely long-lived plant has a single pair of strap-shaped leaves. Over the centuries they crack and split, creating a tangled mound of foliage.

3¼ ft/1 m

CONIFERS

Conifers evolved more than 300 million years ago, long before the world's first broad-leaved trees. With their tough waxy leaves, conifers thrive in harsh climates. They are less diverse than broad-leaved trees, but they dominate forests on cold mountains, and in the far north.

DIVISION	PINOPHYTA
CLASS	1
ORDER	2
FAMILIES	6
SPECIES	About 600

Between areas of open tundra, conifers grow far to the north of the Arctic Circle, forming the largest forest on Earth.

Although relatively few in species, conifers include the world's tallest trees, as well as the heaviest, the longest-lived, and some of the most widespread. Conifers are traditionally placed in a group known as gymnosperms, along with cycads, ginkgos, and gnetophytes. Unlike broad-leaved trees, they do not have flowers and produce pollen and seeds in cones.

LEAVES AND CONES

Most conifers are evergreen, with highly resinous leaves that are good at withstanding cold winds and strong sunshine. Pines have slender needles, arranged singly or in bundles, while other conifers often have linear leaves or flat scales. A few conifers—including larches and some redwoods—are deciduous, with soft leaves that are shed each year.

Conifers have two types of cones, which normally grow on the same tree. Male cones produce pollen. Small, soft, and often very numerous, they typically appear in spring, and wither soon after shedding their pollen into the air. The larger female cones contain one or more seeds. They can take several years to develop, and usually become hard and woody when mature.

The cones of some species, including firs and cedars, release their seeds by slowly disintegrating on the tree. Pine cones remain intact, and often remain on the tree long after they are ripe. Most of them shed their seeds in dry weather, when their scales open wide, but some hold onto their seeds until they are scorched. This is a specialized adaptation that allows pines to recolonize areas that have been burned by forest fires.

A few conifers, the yews, junipers, and podocarps, have small, berrylike cones with fleshy scales that do not fully enclose the seed. These are scattered by birds as they feed.

∨ **CONIFERS IN CONTROL**
Conifers often form extensive single-species stands. These pines are growing in California's Yosemite National Park in California.

115 ft / 35 m

FOUR-SIDED RIGID LEAVES

CONE

COLORADO BLUE SPRUCE
Picea pungens
F: Pinaceae
Bright blue-gray leaves, with prickly tips, make this a popular ornamental tree. Native to western N. America, it typically grows on mountains.

260 ft / 80 m

GRAND FIR
Abies grandis
F: Pinaceae
The giant fir is a tall, fast-growing tree that is native to western N. America. Its leaves have a citruslike scent when crushed.

245 ft / 75 m

CALIFORNIA RED FIR
Abies magnifica
F: Pinaceae
This drought-resistant fir is found on dry mountain slopes in California. It has up-curved leaves and upright cones up to 8 in (20 cm) long.

200 ft / 60 m

EUROPEAN SILVER FIR
Abies alba
F: Pinaceae
Named after the silver bands on the underside of its leaves, this fir has upright resinous cones. They disintegrate in order to scatter their seeds.

200 ft / 60 m

CAUCASIAN FIR
Abies nordmanniana
F: Pinaceae
Originally from mountains in the Black Sea region, this is a popular Christmas tree in Europe because it retains its needles indoors.

SITKA SPRUCE
Picea sitchensis
F: Pinaceae
Thriving in cold, wet conditions, the sitka spruce is often grown as a forestry tree. It comes from the coastal belt of western N. America.

lower side of leaf has two blue-white bands

245 ft / 75 m

cylindrical cone with toothed scales

165 ft / 50 m

NORWAY SPRUCE
Picea abies
F: Pinaceae
An important lumber tree, this fast-growing conifer has spiky leaves and cylindrical cones. Its natural range is N. and C. Europe.

115 ft / 35 m

SCOTCH PINE
Pinus sylvestris
F: Pinaceae
Ranging from the British Isles to China, the Scots pine is the world's most widespread conifer after common juniper. Its upper branches have beautiful orange-red bark.

LONG PAIRED NEEDLES

65 ft / 20 m

CONE WITH EDIBLE SEEDS

50 ft / 15 m

SINGLE-LEAF PINYON
Pinus monophylla
F: Pinaceae
Unique among pines, this species has leaves that grow singly, rather than in pairs or clusters. It comes from rocky slopes in Mexico and the American Southwest.

100 ft / 30 m

LODGEPOLE PINE
Pinus contorta
F: Pinaceae
Growing in coastal dunes and bogs, this N. American pine has paired leaves and prickly cones. The cones release their seeds after being scorched by fire.

STONE PINE
Pinus pinea
F: Pinaceae
Prized for its edible seeds (pine nuts), this Mediterranean pine has large, oval cones, and an elegant umbrella-shaped outline when mature.

150 ft / 45 m

male pollen-producing cones

stiff, sharp needles

65 ft / 20 m

AROLLA PINE
Pinus cembra
F: Pinaceae
Found on European mountains, this slow-growing tree produces small cones that fall intact. Birds such as nutcrackers feed on the cones and disperse the seeds.

115 ft / 35 m

MARITIME PINE
Pinus pinaster
F: Pinaceae
Native to the W. Mediterranean, the maritime pine grows rapidly on poor sandy soils. It has glossy brown cones that are up to 8 in (20 cm) long.

130 ft / 40 m

AUSTRIAN PINE
Pinus nigra
F: Pinaceae
Tall, rangy, and open-crowned, this pine has long leaves arranged in pairs. Despite its name, it is widespread throughout Europe, typically growing on limestone.

82 ft / 25 m

CHINESE PINE
Pinus tabuliformis
F: Pinaceae
With age, this oriental pine develops a distinctive flat, spreading crown. It grows on mountains, and produces small, egg-shaped cones.

MONTEREY PINE
Pinus radiata
F: Pinaceae
Originally restricted to a small area of California, this fast-growing pine is now widely planted for its lumber, particularly in the southern hemisphere.

»

male cone withers after shedding pollen

immature cone

200 ft/60 m

DOUGLAS FIR
Pseudotsuga menziesii
F: Pinaceae
One of the world's tallest conifers, the Douglas fir grows in western N. America. Its cones have projecting bracts formed from modified scales.

165 ft/50 m

mature cone with three-pronged bracts

130 ft 40 m

200 ft 60 m

WESTERN HEMLOCK
Tsuga heterophylla
F: Pinaceae
From western N. America, this is the largest hemlock. It thrives in cool, damp conditions, and can live to be over 1,000 years old.

GOLDEN LARCH
Pseudolarix amabilis
F: Pinaceae
Native to E. China, the golden larch turns brilliant yellow in the fall before losing its leaves. Its cones break up as they scatter their seeds.

DEODAR
Cedrus deodara
F: Pinaceae
Native to the W. Himalayas, this fast-growing cedar has branches that droop at their tips. Its cones are purple-brown when ripe.

130 ft/40 m

130 ft 40 m

ATLAS CEDAR
Cedrus atlantica
F: Pinaceae
This N. African cedar has short needles and upright barrel-shaped cones. When ripe, the cones slowly break up to release their seeds.

CEDAR OF LEBANON
Cedrus libani
F: Pinaceae
A majestic conifer with spreading branches, the cedar of Lebanon is now rare in the wild, but is widely planted as an ornamental tree.

98 ft/30 m

JAPANESE LARCH
Larix kaempferi
F: Pinaceae
A deciduous tree—like all larches—this has soft leaves and persistent woody cones with downcurved scales. It grows on mountains in N. Japan.

165 ft/50 m

MONKEY PUZZLE
Araucaria araucana
F: Araucariaceae
Originally from the mountains of Chile, this primeval-looking tree has spirally arranged, sharply pointed leaves. Mature trees develop an umbrellalike crown.

82 ft/25 m

FRAGILE RIPE CONE

JAPANESE UMBRELLA PINE
Sciadopitys verticillata
F: Sciadopityaceae
The only member of its family, this ancient Japanese species is distinguished from other conifers by having 10–13 needles clustered in whorls.

WESTERN RED-CEDAR
Thuja plicata
F: Cupressaceae

165 ft/50 m

A large tree with flat sprays of small, scalelike leaves, this species grows in northwest N. America. Its wood is highly rot-resistant, even when dead.

LAWSON CYPRESS
Chamaecyparis lawsoniana
F: Cupressaceae

230 ft/70 m

Like other cypresses, this has small cones and sprays of tiny, scalelike leaves. Originally from western N. America, it exists in many cultivated forms.

JAPANESE CEDAR
Cryptomeria japonica
F: Cupressaceae

100 ft
30 m

A cypress rather than a true cedar, this tree has slender leaves and small, rounded cones. It grows on mountains in China and Japan.

WESTERN JUNIPER
Juniperus occidentalis
F: Cupressaceae

This long-lived tree grows on rocky mountain slopes in W. US. Like other junipers, it produces seeds inside berrylike cones.

65 ft
20 m

82 ft/25 m

CHINESE JUNIPER
Juniperus chinensis
F: Cupressaceae

Widespread in temperate E. Asia, this shrub or small tree has prickly leaves when young, and scalelike leaves when mature.

linear leaves in opposite ranks

377 ft
115 m

COAST REDWOOD
Sequoia sempervirens
F: Cupressaceae

Native to the coasts of N. California and Oregon, this is the world's tallest tree. Mature trees have a soaring trunk and relatively sparse branches, and can be as much as 2,000 years old.

312 ft/95 m

GIANT SEQUOIA
Sequoiadendron giganteum
F: Cupressaceae

This Californian redwood is the world's most massive tree. The largest living specimen is estimated to weigh 2,200 tons, with fireproof bark more than 2 ft (60 cm) thick.

RIPENING
CONE

DAWN REDWOOD
Metasequoia glyptostroboides
F: Cupressaceae

Native to C. China, this deciduous redwood is extremely rare in the wild. Until the 1940s it was thought to be extinct, being known only from fossils.

130 ft/40 m

82 ft/25 m

MONTEREY CYPRESS
Cupressus macrocarpa
F: Cupressaceae

Although widely cultivated, the wild Monterey cypress is restricted to a small region of California's coast. Mature trees typically have an irregular spreading shape.

aril turns red when ripe

66 ft/20 m

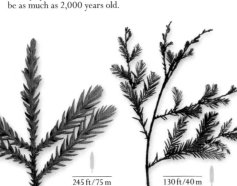

245 ft/75 m

TAIWANIA
Taiwania cryptomerioides
F: Cupressaceae

One of Asia's largest conifers, this tropical species has a trunk up to 10 ft (3 m) thick. It has spiny-tipped leaves, and small, rounded cones.

SWAMP CYPRESS
Taxodium distichum
F: Cupressaceae

130 ft/40 m

Also known as the bald cypress, this deciduous conifer grows in swamps in S.E. US. Its trunk often has a buttressed base.

CALIFORNIA NUTMEG
Torreya californica
F: Taxaceae

98 ft/30 m

This rare conifer is restricted to canyons and mountains in California. It is unrelated to the true nutmeg, although it has nutlike seeds.

CHINESE PLUM-YEW
Cephalotaxus fortunei
F: Taxaceae

66 ft/20 m

This small, densely branched conifer has fleshy cones, which turn purple-brown when ripe. It grows in mountain forests in C. and E. China.

EUROPEAN YEW
Taxus baccata
F: Taxaceae

66 ft/20 m

This long-lived tree produces seeds in fleshy arils—modified cone scales. Found wild in Europe and S.W. Asia, it is commonly planted.

FLOWERING PLANTS

With more than 300,000 species, flowering plants—or angiosperms—are by far the largest group of plants, as well as the most diverse. They play a vital role in most land-based ecosystems, producing food and shelter for animals and many other living things.

DIVISION	ANGIOSPERMAE
CLADES	10
ORDERS	64
FAMILIES	416
SPECIES	About 304,000

Wind-pollinated flowers such as hazel catkins shed clouds of pollen grains into the air. They are rarely brightly colored.

Animal-pollinated flowers are usually conspicuous, and have sticky pollen grains. Hummingbirds transfer pollen as they feed.

Fleshy fruits evolved to attract animals. Wild cucumbers are eaten by antelopes, which scatter the seeds in their droppings.

Dry fruits often burst open when their seeds are ripe. This willowherb is scattering its fluffy seeds into the wind.

The first flowering plants evolved 140 million years ago, which makes them relative latecomers to life on Earth. Since then, they have become the dominant forms of plant life. The smallest flowering plants are not much bigger than a pinhead, but this group also includes all broadleaved trees, and a variety of other species, from cacti and grasses to orchids and palms.

Flowering plants share several key features that help to make them successful. Foremost among these are flowers—actually a collection of highly modified leaves. In most flowers, the outermost layers are the sepals, then the petals. These surround the male stamens, which make pollen, and the female carpels, which collect pollen arriving from similar flowers, so that their female ovules can develop into seeds. Some flowers spread their pollen through the air, but many more use animals to transfer pollen for them. In these showy species, flowers lure animals to sip the sugary nectar in the innermost part of the flower, brushing past the pollen on the way. These plants mostly attract insects, but birds and bats are also important pollinators.

STRATEGIES FOR SPREADING

Flowering plants are not unique in making seeds, but they are the only plants that produce them in fruits. A fruit develops from a flower's ovaries—the closed chambers that house the seeds, close to the flower's central stem. Fruits have a double function: to protect the seeds and to help them spread. Fleshy fruits do this by attracting animals, which feed and scatter the seeds in their excrement. Dry fruits work in a variety of ways. Some burst open when their seeds are ripe. Others have hooks that latch onto clothing, skin, or fur. Still more drift in water or through the air. Many flowering plants spread by the way they grow, often producing new plants from creeping stems. This can create vast, interconnected clones. The largest examples, formed by North American aspen trees, stretch over 100 acres (40 ha), and may be 10,000 years old.

ATTRACTING ATTENTION >
Hellebores use colorful sepals, rather than petals, to attract pollinating insects—or multiple sepals, in this garden variety.

FLOWERING PLANTS GROUPS

DNA sequencing is revolutionizing our understanding of relationships between flowering plants. The groupings below represent the current thinking (see p.150 for subdivisions within the Eudicots).

BASAL ANGIOSPERMS

≫124

MAGNOLIIDS

≫128

MONOCOTS

≫130

EUDICOTS

≫150

BASAL ANGIOSPERMS

Of the more than 60 orders of flowering plants, or angiosperms, three, known as basal angiosperms, evolved early on and still exist today.

Modern DNA sequencing is a powerful way of identifying groups of species that are closely related to one another. However, the interrelationship between these groups is sometimes less clear, with different DNA markers suggesting different affinities. In flowering plants, there seems to be a pattern of some flourishing groups that have diversified hugely, leaving behind a few, remnant groups that have survived for millions of years but diversified little.

"Basal angiosperm" is the term used for three of these remnant orders. They are primitive species, apparently with little in common, growing in widely scattered parts of the world. They include trees, shrubs, and climbers, as well as water plants. The Amborellales order survives today as a single species of shrub on one South Pacific island. The Nymphaeales are aquatic plants with primitive but often showy flowers. They include more than 70 species of water lilies, found all over the globe. The Austrobaileyales order contains nearly 100 species of woody plants, mainly in the tropics.

Two other orders feature on this page as "orphans," whose place in the phylogeny (evolutionary history) of angiosperm families is unclear. The most recent review suggests that the Chloranthales might be related to magnoliids, and the Ceratophyllales to eudicots, but for now that remains uncertain and unproven.

DIVISION	ANGIOSPERMAE
GROUPS	BASAL ANGIOSPERMS
ORDERS	3
FAMILIES	7
SPECIES	190

MYSTERIOUS ORIGINS

Angiosperms may have evolved from seed ferns, a group that died out over 50 million years ago, or more probably from gnetophytes (see p.117). DNA and fossil evidence suggest that the Amborellales were the first angiosperms to appear, about 140 million years ago. The Nymphaeales and Austrobaileyales were the next two groups to evolve.

CHLORANTHALES

The Chloranthaceae, the only family in the order Chloranthales, contains four genera and some 70 living species, but its fossil record dates back over 100 million years. Its members are mainly tropical shrubs and trees with small, stalkless flowers.

SARCANDRA GLABRA
F: Chloranthaceae

With a range of medicinal uses, this evergreen shrub inhabits damp ground, especially wooded banks of streams, in S.E. Asia, China, and Japan.

5 ft
1.5 m

berry clusters in winter

CERATOPHYLLALES

A single family, the Ceratophyllaceae, makes up this order. Called hornworts, these plants are completely unrelated to the nonflowering hornworts (see p.111). The four species are free-floating, rootless aquatics with whorls of finely-divided leaves, tiny male and female flowers, and spiny fruits.

3¼ ft
1 m

RIGID HORNWORT
Ceratophyllum demersum
F: Ceratophyllaceae

This is a submerged species with no roots, inhabiting ponds and ditches in non-arctic Europe. It has tiny flowers and whorls of leaves.

6½ ft
2 m

AMBORELLALES

This order of primitive evergreen shrub contains a single family, which has a single genus with only one species, *Amborella trichopoda*. It bears small flowers, males and females being produced on separate plants, and red berries, each containing a single seed.

AMBORELLA TRICHOPODA
F: Amborellaceae

Found only on the island of New Caledonia in the South Pacific, this sprawling, white-flowered shrub is quite common in mountain forests.

AUSTROBAILEYALES

The order Austrobaileyales is made up of only three families. Members of this order are trees, shrubs, and climbers. The flowers of most species are single and have many petals. Perhaps the best known species is the spice, star anise.

many-petaled single flower

50 ft/15 m

STEM, LEAVES, AND FLOWERS

59 ft/18 m

FRUIT

STAR ANISE
Illicium verum
F: Schisandraceae

Widely used for flavoring, star anise fruits are woody and star-shaped. This woodland species is native to China and Vietnam.

AUSTROBAILEYA SCANDENS
F: Austrobaileyaceae

The flowers of this rare primitive climber, found only in rainforests in Queensland, Australia, smell of rotting fish to attract pollinating flies.

NYMPHAEALES

This primitive order encompasses three families of aquatic plants with floating, submerged, or, more rarely, emergent leaves. The family Nymphaeaceae includes water lilies, which are grown in ornamental ponds throughout the world for their showy flowers.

star-shaped semidouble flower

10 ft/3 m

FLOWER

RIMMED LEAVES

AMAZON WATER LILY
Victoria amazonica
F: Nymphaeaceae
Native to deep Amazonian backwaters, this water lily has enormous round leaves with upright rims. The flowers open at night.

6½ ft/2 m

CAROLINA FANWORT
Cabomba caroliniana
F: Cabombaceae
From still, freshwater habitats in US and S.E. South America, this species has submerged and floating leaves. It can be invasive.

NYMPHAEA 'SUNRISE'
F: Nymphaeaceae
With large, showy, fragrant flowers and mid-green floating foliage, this hybrid, thought to be of American origin, is one of the largest flowering water lilies.

20 in
50 cm

WHITE WATER LILY
Nymphaea alba
F: Nymphaeaceae
This European species has fragrant flowers and its fruits ripen under water, releasing floating seeds. It inhabits lakes, ponds, and slow-flowing rivers.

5 ft/1.5 m

starlike, pure white flower

PRICKLY FOXNUT
Euryale ferox
F: Nymphaeaceae
This inhabitant of deep, slow-flowing water and still water in parts of Asia has prickly leaves, flowers, and many-seeded berries.

5 ft/1.5 m

sharp prickles protect stems, flowers, and floating leaves from animals

young stems, fruits, and seeds are all edible

bright purple flower

grass-green to olive-green shoots

∨ **FLOATING FLOWERS**
Water lily flowers are hermaphrodite—they have both male and female organs. Pollen from the male parts can fertilize ovules in the female reproductive parts, so they can "self-pollinate." However, female parts mature slightly before the flower starts releasing pollen. This increases the chances of flowers being fertilized by pollen carried by insects.

stamen

petal

∧ **FLOWER**
The flowers have numerous white petals and bright yellow stamens, and can grow to 8 in (20 cm) in diameter.

sepal

seeds

< **OVARY SECTION**
The female reproductive organs (consisting of ovules, ovary, and stigma) are fused together. Cavities contain ovules, which turn into seeds when fertilized.

∨ **LEAVES**
The water lily has large leaves, up to 14 in (35 cm) in diameter. Unlike most leaves, their stomata (breathing pores) are on the upper surface, which has a water-repellent coating.

furled young leaf

aerenchyma

∧ **STEM SECTION**
Stems have longitudinal air spaces (aerenchyma), as well as structural tissue, creating buoyancy and allowing oxygen to circulate.

WHITE WATER LILY
Nymphaea alba

One of about 75 wild species of water lily, this handsome plant lives in still or slow-flowing water, casting a deep shade with its rounded, glossy leaves. It grows in water up to 5 ft (1.5 m) deep, and flowers in mid- to late summer, producing a succession of pure white blooms. Each flower lasts for three to four days, opening in the morning and closing by late afternoon. They attract pollinating beetles, which often spend the night inside the flowers, before being released at dawn. White water lilies are useful to water animals: pond snails glue their eggs to the underside of the leaves, and fishes hide beneath them, avoiding predatory birds. After the flowers are pollinated, they produce buoyant seeds. These float for several weeks, before sinking into the mud.

SIZE Leaf diameter 4–14 in (10–35 cm)
HABITAT Ponds, streams, lakes, backwaters
DISTRIBUTION Europe
LEAF TYPE Simple, orbicular with a basal notch

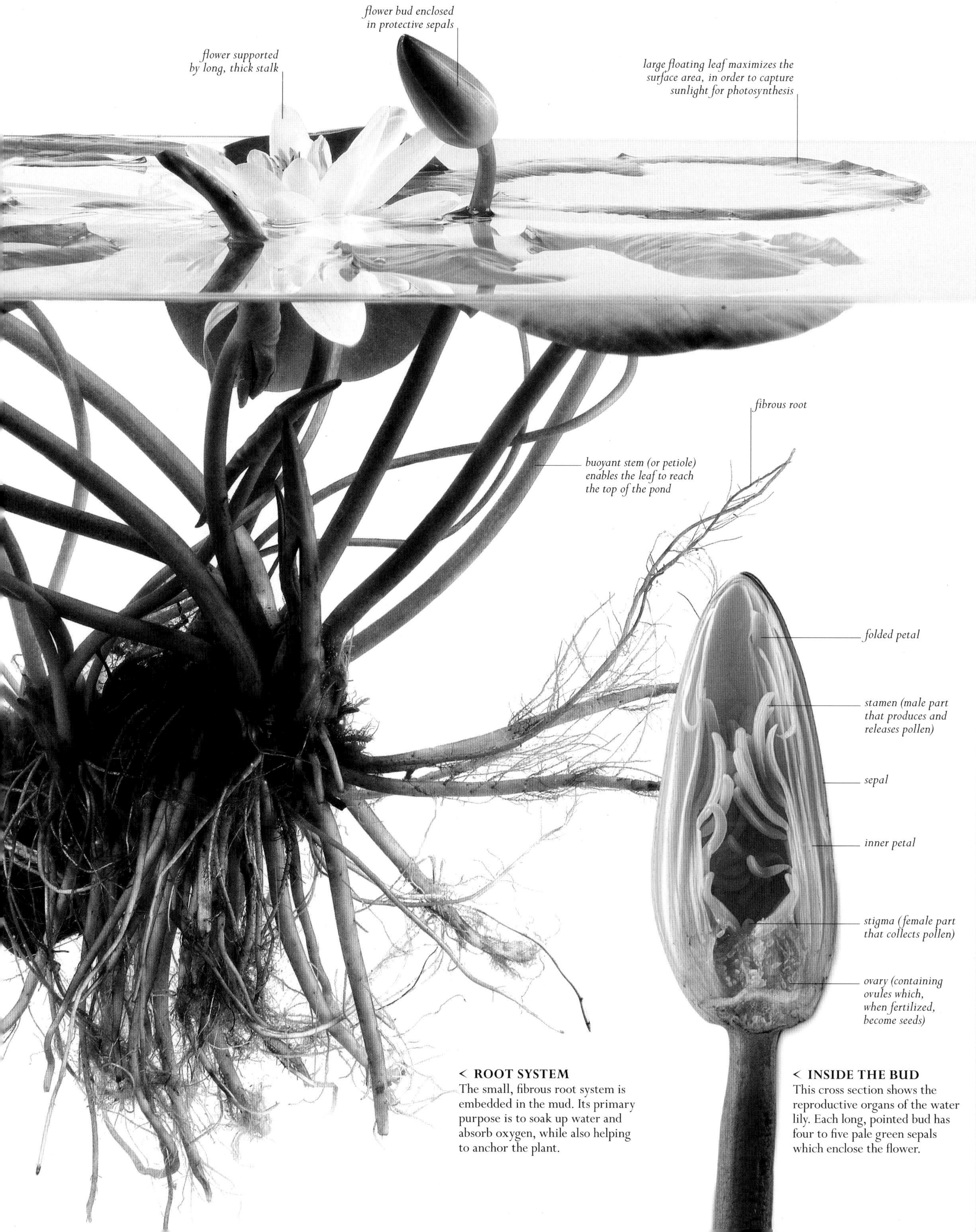

flower supported
by long, thick stalk

flower bud enclosed
in protective sepals

large floating leaf maximizes the
surface area, in order to capture
sunlight for photosynthesis

fibrous root

buoyant stem (or petiole)
enables the leaf to reach
the top of the pond

folded petal

stamen (male part
that produces and
releases pollen)

sepal

inner petal

stigma (female part
that collects pollen)

ovary (containing
ovules which,
when fertilized,
become seeds)

< ROOT SYSTEM

The small, fibrous root system is
embedded in the mud. Its primary
purpose is to soak up water and
absorb oxygen, while also helping
to anchor the plant.

< INSIDE THE BUD

This cross section shows the
reproductive organs of the water
lily. Each long, pointed bud has
four to five pale green sepals
which enclose the flower.

MAGNOLIIDS

The magnoliids are a large group of primitive flowering plants that in botanical terms are between basal angiosperms and monocots.

Found in tropical and temperate regions, the magnoliids form a major plant group that emerged early in the history of flowering plants. They take their name from the magnolia family, which is one of the largest in the group. Plants in this family nearly all have woody stems, and some kinds grow into large trees. However, the group also includes herbaceous or nonwoody species, both as free-standing plants and as climbers.

Some herbaceous magnoliids have highly specialized flowers. In birthworts and pipevines, for example, the flower is shaped like a flaring tube, lined with backward-pointing hairs. Flowers like these act as temporary traps for pollinating flies, which are attracted by a powerful scent. But these are an exception. Most magnoliid flowers are structurally simple, with numerous spirally arranged parts, attached separately to a central stem. Instead of sepals and petals (see p.122), they have a single layer of flaps that are similar in color, size, and shape, known as tepals. The fossil record shows that similar flowers existed more than 100 million years ago.

COMMON CHARACTERISTICS

Magnoliids are grouped together mainly by genetic evidence, but they also share features that can be seen with a microscope, or with the naked eye. At a microscopic level, their pollen grains have a single pore. This small but significant feature links them with monocots, but distinguishes them from eudicots—by far the largest group of flowering plants—which have three-pored pollen grains.

Most magnoliids have leaves with smooth margins, and a network of branching veins. Their fruits may be soft and fleshy, or hard and conelike, and can contain one or several seeds. Species with fleshy fruits use animals to disperse their seeds: many of them are swallowed whole and then scattered by birds. In prehistoric times, wild avocados may have been dispersed by giant ground sloths. Now that ground sloths are extinct, avocados depend on human cultivation to disperse their seeds and ensure survival.

DIVISION	ANGIOSPERMAE
CLADE	MAGNOLIIDAE
ORDERS	4
FAMILIES	18
SPECIES	10,000

DEBATE

PIONEERING PLANTS

Scientists have differing opinions about the growth habit, or shape and lifestyles, of the first flowering plants. According to the "woody hypothesis," these plants were trees or shrubs, like many magnoliids are today. However, a rival view—known as the "paleoherb hypothesis"— suggests that the first angiosperms were herbaceous, or nonwoody, plants with relatively fast life cycles. This meant that they were good at colonizing disturbed ground such as riverbanks. Currently, neither view has proved decisive, although molecular analysis does favor the idea that they were woody plants.

CANELLALES

There are two families in the order Canellales: the Canellaceae and the Winteraceae. These are aromatic trees and shrubs with leathery entire leaves. The flowers in most species have both male and female parts and the fruit is a berry. The leaves and bark of some species can be used medicinally. The Winteraceae is a primitive family— its members have woody stems that contain no water-conducting vessels.

WINTER'S BARK
Drimys winteri
F: Winteraceae

36 ft
11 m

Native to the coastal rainforest in Chile and Argentina, this tree has aromatic bark and leaves and fragrant flowers.

PIPERALES

The order Piperales includes three families of herbaceous plants, trees, and shrubs, widely distributed in tropical regions. The stems have scattered bundles of vascular tissue, a characteristic of monocots. Members of the family Piperaceae have tiny flowers which lack petals and are clustered in spikes. Many species are aromatic.

3¼ ft
1 m

4 in
10 cm

BIRTHWORT
Aristolochia clematitis
F: Aristolochiaceae

A foul-smelling and poisonous perennial, birthwort was grown for medicinal use. It is native to Europe, growing in damp places.

ASARABACCA
Asarum europaeum
F: Aristolochiaceae

This creeping species grows in European woodlands. Its glossy evergreen leaves hide insignificant flowers.

fruits, or peppercorns, used dried

13 ft
4 m

BLACK PEPPER
Piper nigrum
F: Piperaceae

An evergreen climber from shady habitats in S. India and Sri Lanka, this species is widely cultivated for its aromatic fruits.

MAGNOLIALES

Consisting almost exclusively of trees and shrubs, the Magnoliales are a primitive order, widely distributed in the fossil record. Although very variable, most have simple alternately arranged leaves and flowers with both male and female parts. There are six families in the order, of which the Magnoliaceae are the best known, being widely grown in gardens for their spectacular flowers.

tepals protect stamen

100 ft
30 m

100 ft
30 m

TULIP TREE
Liriodendron tulipifera
F: Magnoliaceae

A native of eastern N. America, the tulip tree grows in woodland. The leaves turn yellow in the fall before dropping.

CAMPBELL'S MAGNOLIA
Magnolia campbellii
F: Magnoliaceae

Several of the 210 or so species of Magnolia are grown as garden ornamentals. This species is native to mountain forests in China, India, and Nepal.

LAURALES

The order Laurales consists of seven families of trees, shrubs, and woody climbing plants. A few genera grow in temperate parts of the world, but most are found in tropical and subtropical regions. Their classification is based on genetic analysis rather than morphological characteristics. Many plants are aromatic and are used in perfumery, cookery, and medicine. Others provide lumber or are grown ornamentally.

BAY LAUREL
Laurus nobilis
F: Lauraceae
Widely cultivated for its aromatic leaves used in flavoring, this species grows in woods, scrub, and rocky habitats around the Mediterranean.

unripe fruit

50 ft
15 m

SASSAFRAS
Sassafras albidum
F: Lauraceae
This suckering deciduous tree of woodland in eastern N. America has aromatic leaves which are brightly colored in the fall.

80 ft
25 m

FRUITS

CINNAMON
Cinnamomum verum
F: Lauraceae
The spice cinnamon comes from the aromatic bark of this species, found in lowland forests in Sri Lanka.

CINNAMON STICK

60 ft
18 m

LEAVES

CAROLINA ALLSPICE
Calycanthus floridus
F: Calycanthaceae
Native to woods and streamsides in the S.E. US, Carolina allspice has aromatic leaves and bark and large, fragrant flowers.

CALIFORNIA LAUREL
Umbellularia californica
F: Lauraceae
Sometimes called "headache tree," as scent from the crushed leaves can cause headaches, this species from the W. US flowers in winter.

60 ft
18 m

LEAF CLUSTER

EDIBLE FRUIT

8 ft/2.5 m

AVOCADO TREE
Persea americana
F: Lauraceae
Probably originating in S. Mexico, in well-drained parts of rainforests, the avocado tree is now widely cultivated for its edible, pear-shaped fruits.

60 ft
18 m

white to dark pink flowers open early in spring before the leaves

SCENTED FLOWERS

EVERGREEN LEAVES

65 ft
20 m

YLANG-YLANG
Cananga odorata
F: Annonaceae
Oil from the fragrant flowers of ylang-ylang are used in perfumery. This evergreen tree comes from parts of Asia and Australia.

SWEETSOP
Annona squamosa
F: Annonaceae
Believed to have Caribbean origins, sweetsop, or sugar or custard apple, is widely cultivated. The fruits have edible flesh, which look and taste like custard.

26 ft
8 m

GLOSSY LEAVES

NUTMEG
Myristica fragrans
F: Myristicaceae
The spices nutmeg and mace are both obtained from the seeds of this evergreen tree, a native of the Molucca—or Spice—Islands of Indonesia.

mace

SEED, OR NUTMEG, WITH MACE COVERING

60 ft
18 m

26 ft
8 m

LEAVES

EDIBLE FRUITS

PAWPAW
Asimina triloba
F: Annonaceae
This deciduous tree is native to damp woods of eastern N. America. Its singly borne flowers produce edible fruits.

MONOCOTS

Defined by their unique internal anatomy, monocots include grasses and palms, plus lilies, orchids, and many other ornamental plants.

Early in the evolution of flowering plants, their family tree developed two major branches. A smaller but still substantial branch became the monocots, while the larger branch became the eudicots. Monocots get their name from their seeds, which have a single prepacked seed-leaf, or cotyledon. Apart from this, there is no foolproof way of recognizing a monocot, although there are strong clues. Most have long, narrow leaves with parallel veins, and the flower parts (petals, stamens, and so on) are arranged in threes or multiples of three. The pollen grains have a single opening, unlike those of eudicots which have three. In many monocot flowers, such as tulips, the sepals and petals are almost identical and are known as tepals, but many magnoliids share this feature.

Below ground, monocots usually have a cluster of fibrous roots, instead of a main taproot with smaller roots leading off it. Another key characteristic—revealed under a microscope—involves the structure of their stems. Monocots have scattered bundles of vascular tissue—the specialized cells that carry water and sap—unlike eudicots, which have theirs in concentric rings. This makes monocot stems more flexible than those of a typical eudicot, but it also makes it harder for them to evolve into trees. Treelike monocots, which principally include palms, have a very different way of growing to typical broadleaved trees and conifers. Monocot trunks grow taller, but not thicker, and are usually topped by a single rosette of leaves.

SURVIVAL STRATEGIES

Smaller monocots vary greatly, from climbers to aquatic plants that survive harsh periods as underground storage organs, such as bulbs or tubers. Grasses thrive on the grazing effects of animals, and are the only plant family that forms an entire habitat: grassland. In the tropics, many monocots are epiphytic, growing high up in trees. This group includes bromeliads, and many orchids—by far the largest monocot family, with more than 25,000 species worldwide.

DIVISION	ANGIOSPERMAE
CLADE	MONOCOTS
ORDERS	11
FAMILIES	77
SPECIES	60,000

DID MONOCOTS HAVE AQUATIC ORIGINS?

Living monocots include many freshwater plants, and also some of the few flowering plants that live in the sea. According to a longstanding theory, monocots may have originally evolved in fresh water, before diversifying and taking up life on land. This would account for the long, thin leaves found in many species, and also the internal structure of their stems (see left). Having spread to land, some terrestrial monocots then came full circle and evolved into aquatic forms—duckweeds, for example, originated in this way.

ACORALES

Only one genus and as few as two species make up the order Acorales. Called sweet flags, these waterside and wetland plants have a fleshy flower spike of small flowers and were once classified as aroids (see right). Botanists now believe they represent the earliest branch of the monocots family tree, and may hold clues as to what the first monocots looked like.

3¼ ft
1 m

SWEET FLAG
Acorus calamus
F: Acoraceae
Once cut to strew on floors because of its fresh citrus odor, this waterside plant grows across the northern hemisphere.

ALISMATALES

This order includes many common aquatic plants as well as the mainly land-based aroid family (Araceae). Aroids, which can be dramatic in appearance, have a distinctive reproductive anatomy consisting of a fleshy spike of tiny flowers called a spadix and a leaflike surround called a spathe. The other families in the order include many freshwater species, as well as several families of seagrasses.

WATER-PLANTAIN
Alisma plantago-aquatica
F: Alismataceae
Found in waterside habitats, this plant is common across Europe, Asia, and N. Africa. It has white, pinkish, or pale purple flowers that last only a day.

OVAL FLOATING LEAF

3¼ ft
1 m

BRANCHED FLOWER CLUSTER

WATER HAWTHORN
Aponogeton distachyos
F: Aponogetonaceae
Native to South Africa, this plant has become widely naturalized elsewhere. Its vanilla-scented flowers open just above the water surface.

3¼ ft
1 m

elliptical to oval leaves

male flower

FLOWER STEM

ARROW-SHAPED LEAF

whorl of three female flowers

ARROWHEAD
Sagittaria sagittifolia
F: Alismataceae
This European wetland plant grows arrowhead-shaped leaves above water, ribbonlike ones below water, and sometimes floating leaves too.

3¼ ft
1 m

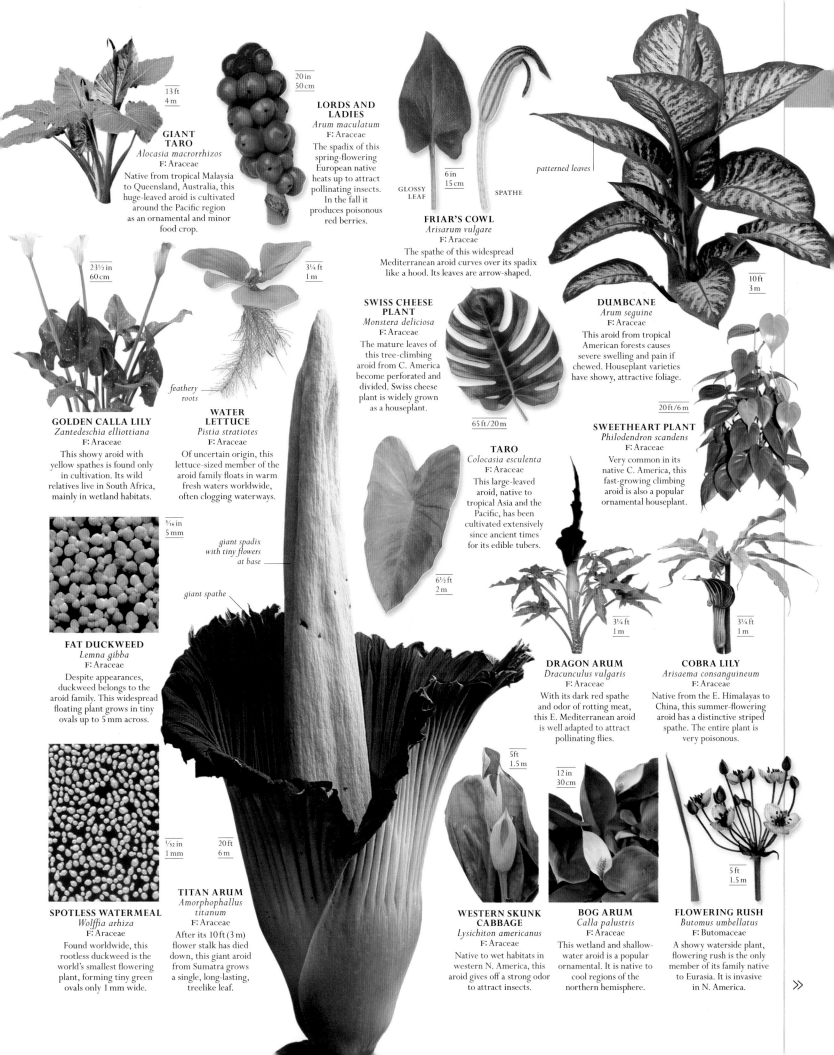

GIANT TARO
Alocasia macrorrhizos
F: Araceae
Native from tropical Malaysia to Queensland, Australia, this huge-leaved aroid is cultivated around the Pacific region as an ornamental and minor food crop.

13 ft
4 m

20 in
50 cm

LORDS AND LADIES
Arum maculatum
F: Araceae
The spadix of this spring-flowering European native heats up to attract pollinating insects. In the fall it produces poisonous red berries.

6 in
15 cm

GLOSSY LEAF   SPATHE

FRIAR'S COWL
Arisarum vulgare
F: Araceae
The spathe of this widespread Mediterranean aroid curves over its spadix like a hood. Its leaves are arrow-shaped.

patterned leaves

10 ft
3 m

DUMBCANE
Arum seguine
F: Araceae
This aroid from tropical American forests causes severe swelling and pain if chewed. Houseplant varieties have showy, attractive foliage.

23½ in
60 cm

3¼ ft
1 m

feathery roots

SWISS CHEESE PLANT
Monstera deliciosa
F: Araceae
The mature leaves of this tree-climbing aroid from C. America become perforated and divided. Swiss cheese plant is widely grown as a houseplant.

65 ft / 20 m

20 ft / 6 m

GOLDEN CALLA LILY
Zantedeschia elliottiana
F: Araceae
This showy aroid with yellow spathes is found only in cultivation. Its wild relatives live in South Africa, mainly in wetland habitats.

WATER LETTUCE
Pistia stratiotes
F: Araceae
Of uncertain origin, this lettuce-sized member of the aroid family floats in warm fresh waters worldwide, often clogging waterways.

TARO
Colocasia esculenta
F: Araceae
This large-leaved aroid, native to tropical Asia and the Pacific, has been cultivated extensively since ancient times for its edible tubers.

SWEETHEART PLANT
Philodendron scandens
F: Araceae
Very common in its native C. America, this fast-growing climbing aroid is also a popular ornamental houseplant.

³⁄₁₆ in
5 mm

giant spadix with tiny flowers at base

6½ ft
2 m

FAT DUCKWEED
Lemna gibba
F: Araceae
Despite appearances, duckweed belongs to the aroid family. This widespread floating plant grows in tiny ovals up to 5 mm across.

giant spathe

3¼ ft
1 m

3¼ ft
1 m

DRAGON ARUM
Dracunculus vulgaris
F: Araceae
With its dark red spathe and odor of rotting meat, this E. Mediterranean aroid is well adapted to attract pollinating flies.

COBRA LILY
Arisaema consanguineum
F: Araceae
Native from the E. Himalayas to China, this summer-flowering aroid has a distinctive striped spathe. The entire plant is very poisonous.

¹⁄₃₂ in
1 mm

20 ft
6 m

5 ft
1.5 m

12 in
30 cm

5 ft
1.5 m

SPOTLESS WATERMEAL
Wolffia arhiza
F: Araceae
Found worldwide, this rootless duckweed is the world's smallest flowering plant, forming tiny green ovals only 1 mm wide.

TITAN ARUM
Amorphophallus titanum
F: Araceae
After its 10 ft (3 m) flower stalk has died down, this giant aroid from Sumatra grows a single, long-lasting, treelike leaf.

WESTERN SKUNK CABBAGE
Lysichiton americanus
F: Araceae
Native to wet habitats in western N. America, this aroid gives off a strong odor to attract insects.

BOG ARUM
Calla palustris
F: Araceae
This wetland and shallow-water aroid is a popular ornamental. It is native to cool regions of the northern hemisphere.

FLOWERING RUSH
Butomus umbellatus
F: Butomaceae
A showy waterside plant, flowering rush is the only member of its family native to Eurasia. It is invasive in N. America.

»

» ALISMATALES

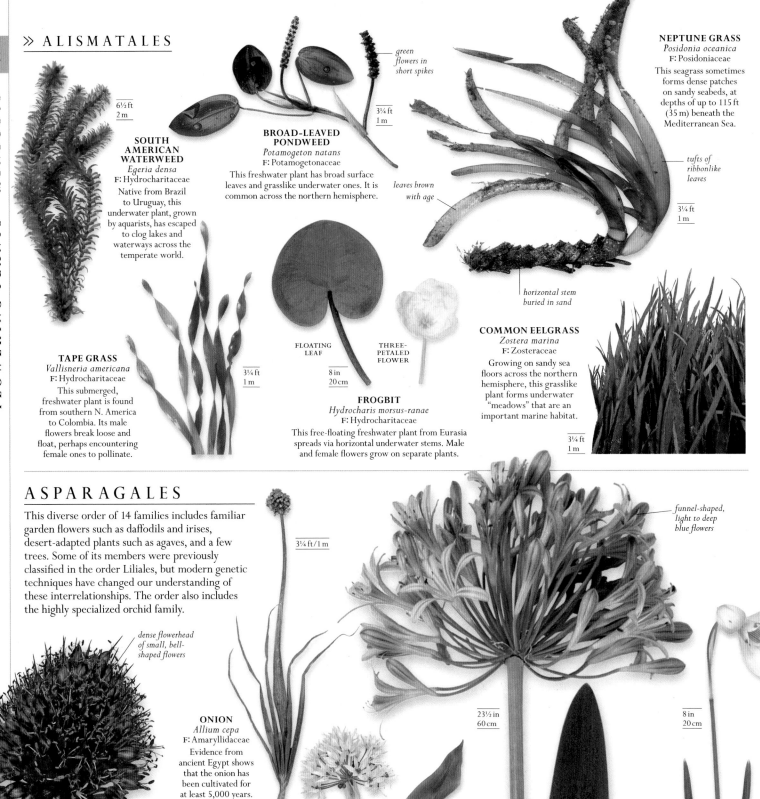

SOUTH AMERICAN WATERWEED
Egeria densa
F: Hydrocharitaceae
Native from Brazil to Uruguay, this underwater plant, grown by aquarists, has escaped to clog lakes and waterways across the temperate world.

6½ ft
2 m

green flowers in short spikes

BROAD-LEAVED PONDWEED
Potamogeton natans
F: Potamogetonaceae
This freshwater plant has broad surface leaves and grasslike underwater ones. It is common across the northern hemisphere.

3¼ ft
1 m

NEPTUNE GRASS
Posidonia oceanica
F: Posidoniaceae
This seagrass sometimes forms dense patches on sandy seabeds, at depths of up to 115 ft (35 m) beneath the Mediterranean Sea.

tufts of ribbonlike leaves

leaves brown with age

3¼ ft
1 m

horizontal stem buried in sand

TAPE GRASS
Vallisneria americana
F: Hydrocharitaceae
This submerged, freshwater plant is found from southern N. America to Colombia. Its male flowers break loose and float, perhaps encountering female ones to pollinate.

3¼ ft
1 m

FLOATING LEAF

THREE-PETALED FLOWER

8 in
20 cm

FROGBIT
Hydrocharis morsus-ranae
F: Hydrocharitaceae
This free-floating freshwater plant from Eurasia spreads via horizontal underwater stems. Male and female flowers grow on separate plants.

COMMON EELGRASS
Zostera marina
F: Zosteraceae
Growing on sandy sea floors across the northern hemisphere, this grasslike plant forms underwater "meadows" that are an important marine habitat.

3¼ ft
1 m

ASPARAGALES

This diverse order of 14 families includes familiar garden flowers such as daffodils and irises, desert-adapted plants such as agaves, and a few trees. Some of its members were previously classified in the order Liliales, but modern genetic techniques have changed our understanding of these interrelationships. The order also includes the highly specialized orchid family.

3¼ ft / 1 m

funnel-shaped, light to deep blue flowers

dense flowerhead of small, bell-shaped flowers

ONION
Allium cepa
F: Amaryllidaceae
Evidence from ancient Egypt shows that the onion has been cultivated for at least 5,000 years.

edible bulb

20 in
45 cm

23½ in
60 cm

8 in
20 cm

32 in
80 cm

ROUND-HEADED LEEK
Allium sphaerocephalon
F: Amaryllidaceae
This European relative of the onion favors limestone soils. Gardeners grow it for its colorful, tightly packed flowerheads.

RAMSONS
Allium ursinum
F: Amaryllidaceae
Related to and smelling of garlic, this plant often carpets the floors of European woodlands in spring with its broad, green leaves.

AFRICAN LILY
Agapanthus africanus
F: Amaryllidaceae
Native to South Africa, this plant has a tall, flowering stem and narrow, arching leaves. It can survive fires in its native habitat, regrowing from a fleshy underground stem.

SNOWDROP
Galanthus nivalis
F: Amaryllidaceae
The early spring flowers of this shade-tolerant European native have three white sepals that are much longer than the three inner petals.

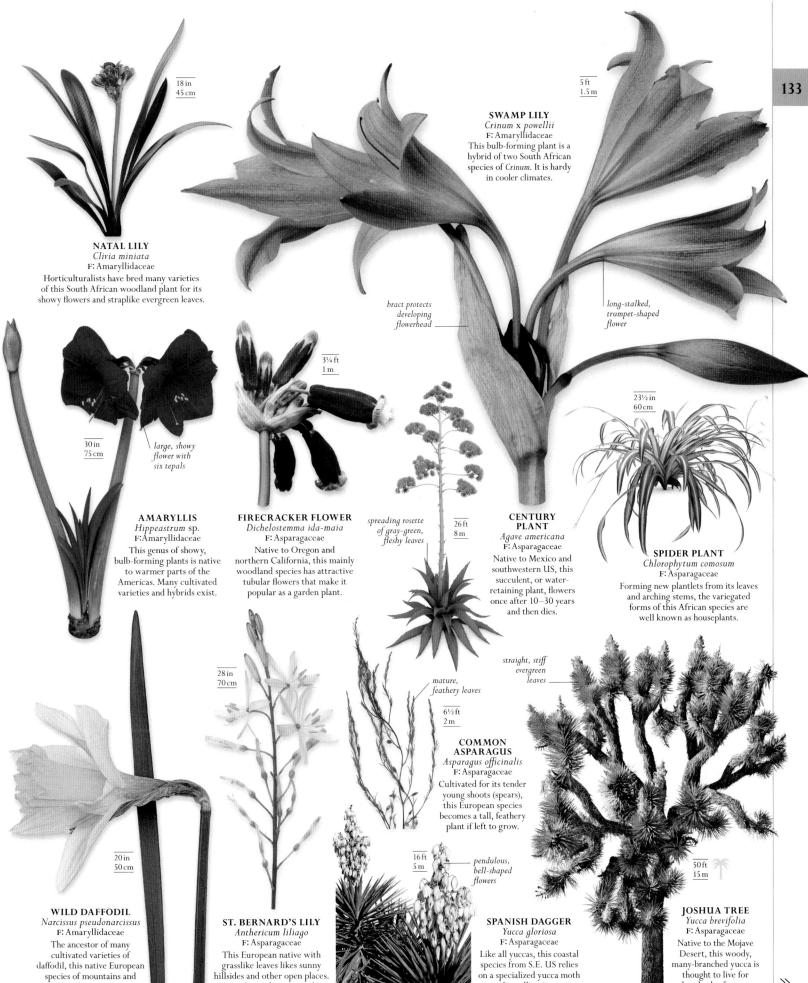

18 in
45 cm

NATAL LILY
Clivia miniata
F: Amaryllidaceae
Horticulturalists have bred many varieties
of this South African woodland plant for its
showy flowers and straplike evergreen leaves.

5 ft
1.5 m

SWAMP LILY
Crinum x *powellii*
F: Amaryllidaceae
This bulb-forming plant is a
hybrid of two South African
species of *Crinum*. It is hardy
in cooler climates.

bract protects
developing
flowerhead

long-stalked,
trumpet-shaped
flower

30 in
75 cm

large, showy
flower with
six tepals

AMARYLLIS
Hippeastrum sp.
F:Amaryllidaceae
This genus of showy,
bulb-forming plants is native
to warmer parts of the
Americas. Many cultivated
varieties and hybrids exist.

3¼ ft
1 m

FIRECRACKER FLOWER
Dichelostemma ida-maia
F: Asparagaceae
Native to Oregon and
northern California, this mainly
woodland species has attractive
tubular flowers that make it
popular as a garden plant.

spreading rosette
of gray-green,
fleshy leaves

26 ft
8 m

CENTURY
PLANT
Agave americana
F: Asparagaceae
Native to Mexico and
southwestern US, this
succulent, or water-
retaining plant, flowers
once after 10–30 years
and then dies.

23½ in
60 cm

SPIDER PLANT
Chlorophytum comosum
F: Asparagaceae
Forming new plantlets from its leaves
and arching stems, the variegated
forms of this African species are
well known as houseplants.

straight, stiff
evergreen
leaves

mature,
feathery leaves

6½ ft
2 m

COMMON
ASPARAGUS
Asparagus officinalis
F: Asparagaceae
Cultivated for its tender
young shoots (spears),
this European species
becomes a tall, feathery
plant if left to grow.

28 in
70 cm

20 in
50 cm

WILD DAFFODIL
Narcissus pseudonarcissus
F: Amaryllidaceae
The ancestor of many
cultivated varieties of
daffodil, this native European
species of mountains and
woods is becoming rarer.

ST. BERNARD'S LILY
Anthericum liliago
F: Asparagaceae
This European native with
grasslike leaves likes sunny
hillsides and other open places.
It prefers rich, moist soils.

16 ft
5 m

pendulous,
bell-shaped
flowers

SPANISH DAGGER
Yucca gloriosa
F: Asparagaceae
Like all yuccas, this coastal
species from S.E. US relies
on a specialized yucca moth
for pollination.

50 ft
15 m

JOSHUA TREE
Yucca brevifolia
F: Asparagaceae
Native to the Mojave
Desert, this woody,
many-branched yucca is
thought to live for
hundreds of years.

FLOWERING PLANTS · MONOCOTS

flower stem holds
up to 20 flowers

10 in
25 cm

**LILY-OF-THE-
VALLEY**
Convallaria majalis
F: Asparagaceae
A sweetly scented plant
from temperate Eurasia,
lily-of-the-valley has
poisonous red berries.

23½ in
60 cm

CAST-IRON PLANT
Aspidistra elatior
F: Asparagaceae
Native to Japan, this woodland plant is now
a popular houseplant. Its small, purple
flowers bloom close to the soil.

28 in
70 cm

SOLOMON'S-SEAL
Polygonatum multiflorum
F: Asparagaceae
A mainly woodland plant, this
Eurasian species has leafy, arched
stems and tube-shaped,
scentless flowers.

3¼ ft
1 m

BUTCHER'S BROOM
Ruscus aculeatus
F: Asparagaceae
Flowers and fruits grow
straight out of the "leaves"
(actually flattened stems)
of this shrubby
European plant.

5 ft
1.5 m

SEA SQUILL
Drimia maritima
F: Asparagaceae
Native to Mediterranean
coasts, this large-
bulbed plant grows
a tall spike of white flowers
in late summer after its
leaves have withered.

bell-shaped flower

12 in
30 cm

STAR-OF-BETHLEHEM
Ornithogalum angustifolium
F: Asparagaceae
This widespread European plant
closes its flowers in dull weather.
Its narrow, grooved leaves have
a central white stripe.

35 in
90 cm

**MOTHER-IN-
LAW'S TONGUE**
Dracaena trifasciata
F: Asparagaceae
This tropical
W. African species
with stiff patterned
leaves is a popular
houseplant. It is also
grown for its fibers.

20 in
50 cm

PORTUGUESE SQUILL
Scilla peruviana
F: Asparagaceae
From S.W. Europe, this showy,
bulb-forming perennial has long,
broad leaves growing from
the plant's base.

12 in
30 cm

APHYLLANTHES
Aphyllanthes monspeliensis
F: Asparagaceae
When not flowering, this
Mediterranean species looks
like a clump of rushes, with
many slender stems that are
almost leafless.

18 in
45 cm

spike of up to 50
fragrant flowers

12 in
30 cm

HYACINTH
Hyacinthus orientalis
"Blue jacket"
F: Asparagaceae
This is one of the many
fragrant, colorful varieties
derived from a S.W. Asian
species over many centuries.

CAPE COWSLIP
Lachenalia aloides
F: Asparagaceae
This bulb-forming species,
native to the southwestern
tip of Africa, has only one
or two strap-shaped leaves.

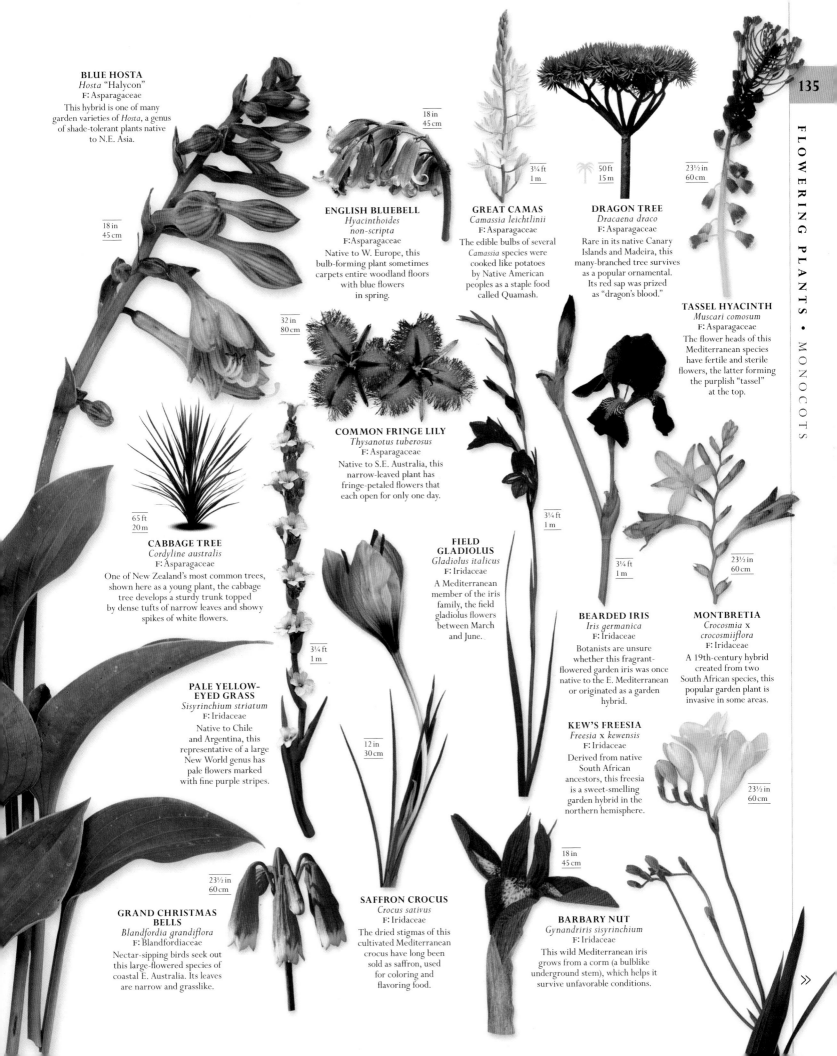

BLUE HOSTA
Hosta "Halycon"
F: Asparagaceae
This hybrid is one of many garden varieties of *Hosta*, a genus of shade-tolerant plants native to N.E. Asia.

18 in
45 cm

ENGLISH BLUEBELL
Hyacinthoides non-scripta
F: Asparagaceae
Native to W. Europe, this bulb-forming plant sometimes carpets entire woodland floors with blue flowers in spring.

18 in
45 cm

GREAT CAMAS
Camassia leichtlinii
F: Asparagaceae
The edible bulbs of several *Camassia* species were cooked like potatoes by Native American peoples as a staple food called Quamash.

3¼ ft
1 m

DRAGON TREE
Dracaena draco
F: Asparagaceae
Rare in its native Canary Islands and Madeira, this many-branched tree survives as a popular ornamental. Its red sap was prized as "dragon's blood."

50 ft
15 m

23½ in
60 cm

TASSEL HYACINTH
Muscari comosum
F: Asparagaceae
The flower heads of this Mediterranean species have fertile and sterile flowers, the latter forming the purplish "tassel" at the top.

32 in
80 cm

COMMON FRINGE LILY
Thysanotus tuberosus
F: Asparagaceae
Native to S.E. Australia, this narrow-leaved plant has fringe-petaled flowers that each open for only one day.

65 ft
20 m

CABBAGE TREE
Cordyline australis
F: Asparagaceae
One of New Zealand's most common trees, shown here as a young plant, the cabbage tree develops a sturdy trunk topped by dense tufts of narrow leaves and showy spikes of white flowers.

3¼ ft
1 m

FIELD GLADIOLUS
Gladiolus italicus
F: Iridaceae
A Mediterranean member of the iris family, the field gladiolus flowers between March and June.

3¼ ft
1 m

BEARDED IRIS
Iris germanica
F: Iridaceae
Botanists are unsure whether this fragrant-flowered garden iris was once native to the E. Mediterranean or originated as a garden hybrid.

3¼ ft
1 m

MONTBRETIA
Crocosmia x *crocosmiiflora*
F: Iridaceae
A 19th-century hybrid created from two South African species, this popular garden plant is invasive in some areas.

23½ in
60 cm

PALE YELLOW-EYED GRASS
Sisyrinchium striatum
F: Iridaceae
Native to Chile and Argentina, this representative of a large New World genus has pale flowers marked with fine purple stripes.

3¼ ft
1 m

KEW'S FREESIA
Freesia x *kewensis*
F: Iridaceae
Derived from native South African ancestors, this freesia is a sweet-smelling garden hybrid in the northern hemisphere.

23½ in
60 cm

12 in
30 cm

GRAND CHRISTMAS BELLS
Blandfordia grandiflora
F: Blandfordiaceae
Nectar-sipping birds seek out this large-flowered species of coastal E. Australia. Its leaves are narrow and grasslike.

23½ in
60 cm

SAFFRON CROCUS
Crocus sativus
F: Iridaceae
The dried stigmas of this cultivated Mediterranean crocus have long been sold as saffron, used for coloring and flavoring food.

18 in
45 cm

BARBARY NUT
Gynandriris sisyrinchium
F: Iridaceae
This wild Mediterranean iris grows from a corm (a bulblike underground stem), which helps it survive unfavorable conditions.

» ASPARAGALES

FAIRY SLIPPER ORCHID
Calypso bulbosa
F: Orchidaceae
Widespread across cooler parts of the northern hemisphere, this fragrant, single-flowered orchid prefers damp woodland and marshland.

8 in
20 cm

ONCIDIUM SP.
F: Orchidaceae
Native to the tropical Americas, this genus of broad-lipped orchids includes about 330 tree-living species of varying sizes.

3¼ ft
1 m

LESSER BUTTERFLY ORCHID
Platanthera bifolia
F: Orchidaceae
Found in various habitats across temperate Eurasia, this sweetly scented, pale-flowered orchid is visited and pollinated by night-flying moths.

12 in
30 cm

TRACY'S CYMBIDIUM
Cymbidium tracyanum
F: Orchidaceae
This epiphytic orchid from Burma, Thailand, and S.W. China produces its strongly fragrant flowers in the fall.

3¼ ft
1 m

PHRAGMIPEDIUM X SEDENII
F: Orchidaceae
This fragrant, ground-living orchid is a hybrid created from species of the genus *Phragmipedium*, slipper orchids native to tropical regions of the Americas.

23½ in
60 cm

HYACINTH ORCHID
Dipodium squamatum
F: Orchidaceae
A leafless, ground-living Australian orchid, this species depends on associating with underground fungi to survive. It lives in woodland habitats.

32 in
80 cm

PYRAMIDAL ORCHID
Anacamptis pyramidalis
F: Orchidaceae
This chalk-loving orchid from temperate Eurasia sticks its pollen sacs onto visiting butterflies and moths, as first described by Charles Darwin.

23½ in
60 cm

PINK LADY'S SLIPPER
Cypripedium acaule
F: Orchidaceae
Widespread in eastern N. America, this two-leaved slipper orchid favors acid soil as in pine forests.

14 in
35 cm

HAIRY SLIPPER ORCHID
Paphiopedilum villosum
F: Orchidaceae
Native to S. China and parts of S.E. Asia, this slipper orchid has been used to create many ornamental hybrids.

12 in
30 cm

flowers have three outer sepals and three inner petals

12 in
30 cm

ORANGE GUARIANTHE
Guarianthe aurantiaca
F: Orchidaceae
This tree-living species is native to tropical C. America. Many ornamental varieties and hybrids have been bred by horticulturalists.

WINDOWSILL ORCHID
Pleione formosana
F: Orchidaceae
Native to parts of China, this small ground-living orchid dies back during the winter months.

6 in
15 cm

central section has distinctive frilly edge

WAGENER'S MASDEVALLIA
Masdevallia wageneriana
F: Orchidaceae
The sepals of this small epiphytic orchid from the mountains of N. Venezuela have narrow "tails" (spurs), a feature typical of the genus.

¾ in
2 cm

RED HELLEBORINE
Cephalanthera rubra
F: Orchidaceae
This ground-living species with rose-pink or purplish flowers grows in open woodlands from Europe eastward to W. Asia.

23½ in
60 cm

VIOLET BIRD'S-NEST ORCHID
Limodorum abortivum
F: Orchidaceae
This S. European orchid lacks green leaves. For food, it depends entirely on *Russula* fungi living in its roots.

32 in / 80 cm

DENDROBIUM SP.
F: Orchidaceae
Including more than 1,000 species of diverse shapes, colors, and sizes, this genus of epiphytic orchids is found from S.E. Asia to New Zealand.

6½ ft / 2 m

LIPPEROSE MOTH ORCHID
Phalaenopsis 'Lipperose'
F: Orchidaceae
This cultivated hybrid is one of many developed using the genus *Phalaenopsis* from S.E. Asia with its broad, flattened flowers.

3¼ ft / 1 m

12 in / 30 cm

TONGUE ORCHID
Serapias lingua
F: Orchidaceae
The lip of this unusual Mediterranean orchid's flower hangs down like a tongue, acting as a landing platform for visiting insects.

35 in / 90 cm

LIZARD ORCHID
Himantoglossum hircinum
F: Orchidaceae
The long lips of this orchid from southern Europe have a fanciful resemblance to tiny lizards, hence its common name.

three sepals and two petals on each flower

23½ in / 60 cm

MILITARY ORCHID
Orchis militaris
F: Orchidaceae
This chalk-loving Eurasian orchid may have acquired its name because the individual flowers resemble helmeted human figures.

23½ in / 60 cm

COMMON DONKEY ORCHID
Diuris corymbosa
F: Orchidaceae
Native to S.E. and S.W. Australia, this orchid gets its common name from the shape of its side petals, which resemble donkeys' ears.

drooping sepals

12 in / 30 cm

50 ft / 15 m

SEED POD

LEATHERY EVERGREEN LEAVES

ROTHCHILD'S VANDA
Vanda 'Rothschildiana'
F: Orchidaceae
This cultivated hybrid was created by crossing species of the genus *Vanda*, which are tree-living orchids native to tropical Asia.

FLAT-LEAVED VANILLA ORCHID
Vanilla planifolia
F: Orchidaceae
Native to Mexico and C. America, this climbing orchid is the source of the flavoring vanilla.

BEARDED GREENHOOD ORCHID
Pterostylis sp.
F: Orchidaceae
Scent from these ground-living Australian orchids attracts small male flies. A feathery trap then spurs the fly to collect the pollen sacs.

lanceolate leaves on each stalk

AUTUMN LADY'S-TRESSES
Spiranthes spiralis
F: Orchidaceae
This small, mainly grassland orchid from temperate Eurasia is notable for its flowers being arranged spirally up the flower stalk.

8 in / 20 cm

flower resembles a fat bumblebee

sepal and petals form protective hood over inner flower

flower attacts pollinating insects

flower trigger snaps shut to trap insect

3¼ ft / 1 m

16 in / 40 cm

20 in / 50 cm

DARK-RED HELLEBORINE
Epipactis atrorubens
F: Orchidaceae
The long roots of this fragrant Eurasian orchid help it grow successfully even within the cracks of limestone cliffs.

23½ in / 60 cm

NUN'S ORCHID
Phaius tankervilleae
F: Orchidaceae
This fragrant ground-living orchid is widely cultivated. Its native range is tropical and subtropical S.E. Asia and the South Pacific.

lower sepals provide landing stage for insects

BEE ORCHID
Ophrys apifera
F: Orchidaceae
Although in a genus whose flowers attract pollinating male insects by mimicking females, this Eurasian orchid in practice often pollinates itself.

∨ RAMBLING STEMS
The many-bulbed orchid grows all over its chosen support structure, whether it be a rock or another plant, and forms a dense mat of stolons—the horizontal stems that reach out from the base of the plant.

broad, whitish petal forms a lip, which acts as a flag to attract pollinating insects

outer three sepals enclose the flower bud

MANY-BULBED ORCHID
Dinema polybulbon

This diminutive orchid is the only species in the genus *Dinema*. Rambling and prolific, it is an epiphytic or lithophytic plant, meaning that it grows on trees or rocks, using them for support. It gains its nutrients from the air, animal detritus, or other plant remains, and absorbs water from rain or mist. Moisture collects on the orchid's leaves, which are thick and waxy, and have reduced stomata—all features which help to reduce water loss. The leaves funnel water into upright, swollen outgrowths from the stems, called pseudobulbs. These act as water-storage organs to keep the orchid hydrated during tropical dry seasons. The creeping stems, with the pseudobulbs, in turn connect to aerial and (sometimes) underground roots. These absorb dissolved nutrients through a spongy, outer layer of cells called velamen. In winter, each pseudobulb produces a single flower. Plants can survive temperatures down to 45°F (7°C), and are a favorite of orchid collectors.

SIZE Height 3 in (7.5 cm)
HABITAT Humid, mixed forests
DISTRIBUTION Mexico, C. America, Jamaica, Cuba
LEAF TYPE Parallel veins

flattened leaf tip has a notch

PSEUDOBULB >
These small, oval, bulblike outgrowths from the creeping stems store water for dry periods. Each produces one to three glossy leaves.

∧ **FLOWER**
The pseudobulb produces a small, sweet-smelling flower with narrow, yellow-brown sepals, purple petals, and a spreading, white lip.

< **POLLEN SACS (POLLINIA)**
Orchid pollen grains are contained in two sacs. A sticky mass glues the sacs onto pollinating insects, which carry the pollen to another plant.

< **AERIAL ROOTS**
The orchid's aerial roots do not have to connect to the soil. The older parts of the root have a protective layer of dead cells that act like blotting paper to absorb and retain water.

» ASPARAGALES

16 ft
5 m

tall spike of tiny, white or cream flowers

GRASS TREE
Xanthorrhoea australis
F: Asphodelaceae

Perhaps living for up to 450 years, this bushfire-resistant Australian plant develops a sturdy trunk and a spherical crown of narrow leaves.

COMMON ASPHODEL
Asphodelus aestivus
F: Asphodelaceae

This common narrow-leaved Mediterranean plant produces many pinkish white flowers on a tall stem.

buds at top of spike open last

18 in
45 cm

16 ft
5 m

NEW ZEALAND FLAX
Phormium tenax
F: Asphodelaceae

This common New Zealand species grows large leaves at ground level and a tall flower stalk bearing tubular reddish flowers.

sap from leaves has healing properties

3¼ ft
1 m

MEDICINAL ALOE
Aloe vera
F: Asphodelaceae

This spiny succulent adapted to dry areas has been cultivated since ancient times for its reputed medicinal properties.

orange-red flower spike fades to yellow near base

5 ft
1.5 m

TORCH LILY
Kniphofia uvaria
F: Asphodelaceae

Native to the S.W. tip of Africa, this multiflowered lily with long, narrow leaves is now a popular ornamental specimen.

5 ft
1.5 m

ORANGE DAY LILY
Hemerocallis fulva
F: Asphodelaceae

Each flower of this ornamental species lasts just one day. Native to Asia, it can be an invasive weed in N. America.

LILIALES

Many plants once considered part of the lily family, including onions and hyacinths, have been moved into the order Asparagales (see pp.132–39). The ten families in the Liliales include the true lilies and tulips, the sarsaparilla and colchicum families, and the colorful alstroemeria family from the New World. Beware that several lily species are highly toxic to cats.

18 in
45 cm

WILD TULIP
Tulipa sylvestris
F: Liliaceae

Still widespread as a native plant in southern Europe, this yellow-flowered relative of the cultivated garden tulip grows wild in meadows, open woodland, and rocky places.

16 in
40 cm

HERB PARIS
Paris quadrifolia
F: Liliaceae

Found in ancient woodland in temperate Eurasia, this four-leaved plant has a single green flower that develops into a black inedible fruit.

20 in
50 cm

SNAKESHEAD FRITILLARY
Fritillaria meleagris
F: Liliaceae

This bulb-forming European species has distinctive checkered flowers. It grows wild in damp meadows but is more common in cultivation.

DIOSCOREALES

This order contains just three families. It is dominated by the yams, a family of mainly tropical climbing plants. Several species of yam have been cultivated since ancient times for their edible tubers. The small bog asphodel family is mainly north temperate.

BOG ASPHODEL
Narthecium ossifragum
F: Nartheciaceae

This European plant lives in nutrient-poor upland habitats. After fertilization, its seed capsules develop a fiery orange color.

YAM
Dioscorea sp
F: Dioscoreaceae

Widely cultivated in the tropics, yams grow large, starch-rich tubers. Plants have stems that trail or climb and heart-shaped leaves.

edible tuber

16 ft
5 m

13 ft/4 m

BLACK BRYONY
Tamus communis
F: Dioscoreaceae

This poisonous European climbing plant of the yam family is named for its black underground tuber. Its globular fruits turn bright red when ripe.

unripe fruits are green

heart-shaped leaf

MADONNA LILY
Lilium candidum
F: Liliaceae
Often encountered in Christian art as a symbol of purity, this widely cultivated lily is native to the E. Mediterranean.

6½ ft
2 m

female stigma receives the pollen

male stamen releases the pollen

leaves arranged spirally up stem

GAGEA RETICULATA
F: Liliaceae
This small, bulb-forming plant grows in open habitats across much of temperate Asia, as well as S.E. Europe and N. Africa.

6 in
15 cm

TURK'S CAP LILY
Lilium martagon
F: Liliaceae
This widespread Eurasian lily has the reflexed (backward-bending) tepals that form the flower shape characteristic of the species.

6½ ft
2 m

GIANT HIMALAYAN LILY
Cardiocrinum giganteum
F: Liliaceae
Found from the Himalayas to China, this giant lily grows for several seasons before flowering, after which the main plant dies.

10 ft
3 m

FLAME LILY
Gloriosa superba
F: Colchicaceae
This striking plant, found from S. Africa to S.E. Asia, is a climber, clambering upward with the aid of tendrils.

6½ ft/2 m

FLOWERHEADS

6 in
15 cm

MEADOW SAFFRON
Colchicum autumnale
F: Colchicaceae
The crocuslike flowers of this European native appear in the fall, months before its leaves. Despite being poisonous, it is cultivated widely.

TYPICAL LEAF

flower with six tepals (lobes)

leaf twists from stem so underside faces upwards

up to 40 tubular flowers

TRAILING LILY
Bomarea multiflora
F: Alstroemeriaceae
This many-stemmed climbing plant, with clusters of orange-red flowers, is native to S. America.

13 ft
4 m

PERUVIAN LILY
Alstroemeria sp.
F: Alstroemeriaceae
This lilylike S. American genus includes many popular ornamentals. The leaves twist during growth to become anatomically upside down.

4 ft
1.2 m

≫ LILIALES

CHILEAN BELLFLOWER
Lapageria rosea
F: Philesiaceae
Cultivated for its spectacular bell-shaped flowers, this twining climber is native to humid woodlands of Chile.

30 ft
10 m

5 ft
1.5 m

BERRIES

50 ft
15 m

FLOWER CLUSTER

SMILAX
Smilax aspera
F: Smilacaceae
Native to the Mediterranean and S.W. Asia, this vigorous climber bears male and female flowers on separate plants.

FALSE HELLEBORINE
Veratrum sp.
F: Melanthiaceae
Native to the northern hemisphere, this genus of poisonous plants produces branched flower heads of greenish white flowers.

PANDANALES

Found mainly in the tropics, the five families in this order include over 1,300 species of trees, shrubs, climbers, and smaller plants. Many look superficially like palms, except that they have simpler, strap-shaped leaves. About half of them belong to the single genus *Pandanus* (screw pines).

MULTI-SEEDED FRUIT

THATCH SCREW PINE
Pandanus tectorius
F: Pandanaceae
No relation of real pines, this tropical coastal tree has long been a vital source of materials for Pacific Island cultures.

MATURE TREE

60 ft
18 m

ARECALES

Since 2016, a family of 16 endemic Australian trees has been added to the order Arecales, which is dominated by more than 2,000 species in the palm family. Typically growing from one central bud, palms vary from towering trees to slender climbing rattans. Their huge leaves are either feather-shaped or fanlike. Often associated with desert islands, most species actually live in tropical rainforests.

100 ft
30 m

80 ft
25 m

pinnate leaf

COCONUT PALM
Cocos nucifera
F: Arecaceae
Now widely cultivated, this species probably originated in the W. Pacific, colonizing new islands via its floating fruits.

MATURE TREE

SINGLE-SEEDED FRUIT

MATURE TREE

RIPE FRUIT

BETEL NUT PALM
Areca catechu
F: Arecaceae
Cultivated for its seeds, which release a psychoactive substance when chewed, this palm originates from S.E. Asia.

65 ft
20 m

SUGAR PALM
Arenga pinnata
F: Arecaceae
This native of India and S.E. Asia grows showy yellow flower heads. It yields many products, including sugar from its sap and fibers.

DATE PALM
Phoenix dactylifera
F: Arecaceae
Seen here as a young tree, this cultivated palm from the Middle East grows as separate male and female trees.

100 ft
30 m

DESERT FAN PALM
Washingtonia filifera
F: Arecaceae
Dead leaves hanging below this palm's crown provide shelter for birds and insects in its native deserts of S.W. US.

60 ft
18 m

PETTICOAT PALM
Copernicia macroglossa
F: Arecaceae
Native to Cuba, this relatively small palm is named for the skirt of dead leaves that it retains below its crown.

23 ft
7 m

PINNATE LEAF

65 ft
20 m

RIPE FRUIT

AFRICAN OIL PALM
Elaeis guineensis
F: Arecaceae

A plant of humid tropical lowlands, this palm is extensively cultivated beyond its native Africa for the oils contained in its fruits.

20 ft
6 m

BOTTLE PALM
Hyophorbe lagenicaulis
F: Arecaceae

This swollen-based palm comes from tiny Round Island near Mauritius. It is widely grown as an ornamental.

50 ft
15 m

SAGO PALM
Metroxylon sagu
F: Arecaceae

Shown here as a young plant, this swamp-living palm probably originated in New Guinea but is now grown throughout S.E. Asia.

PALMATE LEAF

65 ft
20 m

CHUSAN PALM
Trachycarpus fortunei
F: Arecaceae

From C. China, this cold-tolerant species has separate male and female plants. The females produce round, blue-black fruits.

100 ft
30 m

SEED

COCO DE MER
Lodoicea maldivica
F: Arecaceae

The seeds of this palm from the Seychelles are the biggest in the plant kingdom. They take six years to mature.

ROYAL PALM
Roystonea regia
F: Arecaceae

Often planted in stately avenues in the tropics, this handsome, smooth-trunked species is native from C. America to Florida and Cuba.

80 ft
25 m

30 m
10 m

RAFFIA PALM
Raphia farinifera
F: Arecaceae

At 65 ft (20 m) long, the leaves of this African palm are the largest of any tree.

65 ft
20 m

PALMYRA PALM
Borassus flabellifer
F: Arecaceae

This tall-trunked species from S. Asia favors drier habitats. It is cultivated for its fruits and sugar-yielding sap.

fan-shaped leaf up to 3¼ ft (1 m) across

BRAZILIAN WAX PALM
Copernicia prunifera
F: Arecaceae

A waxy coating on the leaves of this palm from N.E. Brazil helps it resist drought. The harvested wax is used in polish and soap.

50 ft
15 m

10 ft
3 m

EUROPEAN FAN PALM
Chamaerops humilis
F: Arecaceae

Often trunkless in the wild, cultivated specimens grow several short trunks from a single base. It originates from Mediterranean countries.

80 ft
25 m

CHILEAN WINE PALM
Jubaea chilensis
F: Arecaceae

Also known as the coquito palm, this massive-trunked, cold-tolerant palm is native only to C. Chile, where it is now a protected species.

COMMELINALES

This order contains five families, two of which are minor groups of about five species each, plus the three larger families illustrated here. Most are low-growing plants, typically from warmer regions. Many have attractive, three-petaled blue flowers (reduced to two petals in some species), making them popular as ornamentals.

23½ in
60 cm

BLUE SPIDERWORT
Commelina coelestis
F: Commelinaceae
Sometimes planted as ground cover, this sprawling plant is native to Mexico and C. America.

QUEEN'S SPIDERWORT
Dichorisandra reginae
F: Commelinaceae
A popular garden flower grown in warmer regions, this tropical woodland species is from Peru. It has blue flowers with white centers.

12 in
30 cm

TAHITIAN BRIDAL VEIL
Callisia procumbens
F: Commelinaceae
Native to C. and S. America, this species has weak stems and three-petaled white flowers.

6 in
15 cm

4 in
10 cm

TURTLE VINE
Callisia repens
F: Commelinaceae
This succulent plant from forest edges in the American tropics spreads by rooting from creeping stems.

2¼ ft
70 cm

trailing stem

SILVER INCH PLANT
Tradescantia zebrina
F: Commelinaceae
With striped, succulent foliage, this ground-living species from the tropical Americas is also a popular houseplant.

POALES

The 14 families in the Poales include wind-pollinated families that dominate certain ecosystems: grasses in prairies and savannas, sedges and rushes in bogs and heathlands, especially in the northern hemisphere, and restios in heathlands and swamps in the southern hemisphere. Bromeliads mainly live as epiphytes growing on tropical trees.

SLENDER CLUB-RUSH
Isolepis cernua
F: Cyperaceae
With narrow, green stems and silvery flower heads, this widespread temperate sedge is also dubbed "fiber-optics grass."

12 in
30 cm

SCARLET STAR
Guzmania lingulata
F: Bromeliaceae
This tree-living bromeliad has a wide native range from C. America to Brazil. It is a popular ornamental species.

18 in
45 cm

PINK QUILL
Wallisia cyanea
F: Bromeliaceae
Native to Ecuador and Peru, pink quill grows on rainforest trees at up to 2,800 ft (850 m) above sea level.

20 in
50 cm

arching stalk of flower cluster

5 ft
1.5 m

long, narrow leaf

PINEAPPLE
Ananas comosus
F: Bromeliaceae
Christopher Columbus introduced Europeans to this cultivated bromeliad with its large, seedless fruits. Its wild origins are uncertain.

12 in
30 cm

12 in
30 cm

red bracts surround flowers

orange-red bracts protect tiny, white flowers

BIRD'S NEST BROMELIAD
Nidularium innocentii
F: Bromeliaceae
The small, white flowers of this Brazilian bromeliad nestle within the chamber formed by the surrounding colored bracts.

spiny-leaved rosette

vaselike rosette of leaves collects rainwater

14 in
35 cm

10 in
25 cm

DYER'S TILLANDSIA
Racinaea dyeriana
F: Bromeliaceae
This epiphytic bromeliad is endangered in the wild due to destruction of its native mangrove forests in Ecuador.

LORENTZ'S BROMELIAD
Deuterocohnia lorentziana
F: Bromeliaceae
Native to the Andes of Argentina and Bolivia, this is a ground-dwelling species.

BLUSHING BROMELIAD
Neoregelia carolinae
F: Bromeliaceae
At flowering time, the central leaves of this Brazilian bromeliad turn crimson, and it produces blue or violet flowers.

yellow blotch on
uppermost petal
of each flower

3¼ ft / 1 m

10 ft
3 m

18 in
45 cm

TALL KANGAROO PAW
Anigozanthos flavidus
F: Haemodoraceae
From sandy areas of S.W. Australia,
this species gets its common name
from the appearance of its
woolly-haired flower buds.

flowers in compact
conical cluster

spike of male flowers

swollen leaf stalk

COMMON WATER
HYACINTH
Pontederia crassipes
F: Pontederiaceae
Native to Amazonia, this
floating plant is a major
tropical pest, but can be
used to absorb pollution.

PICKEREL WEED
Pontederia cordata
F: Pontederiaceae
Fast-growing and with
striking blue flower heads,
this water-edge species is
common from eastern N.
America south to Argentina.

PENDULOUS SEDGE
Carex pendula
F: Cyperaceae
Like other true sedges (genus *Carex*), this
European species has triangular stems and
separate male and female flower clusters.

4½ ft
1.4 m

5 ft
1.5 m

16 ft / 5 m

spike of female
flowers

crimson
central leaves

giant spike of
more than
3,000 flowers

CHINESE WATER
CHESTNUT
Eleocharis dulcis
F: Cyperaceae
Native to Asia, this
wetland sedge has
clumps of tubular
stems. It is cultivated
for its edible
underwater tubers.

PAPYRUS SEDGE
Cyperus papyrus
F: Cyperaceae
This tall African perrenial is a
wetland plant. The Ancient
Egyptians wrote on a paperlike
material called papyrus made
from its leaves.

evergreen
variegated foliage

33 ft
10 m

3¼ ft
1 m

spine-edged,
striped leaf

yellow bracts
around small,
white flowers

stiff, spiny
leaves

ZEBRA PLANT
Aechmea chantinii
F: Bromeliaceae
This large, strap-leaved
epiphytic bromeliad comes
from the rainforests of
S. America. It is pollinated
by hummingbirds.

HAHN'S CATOPSIS
Catopsis paniculata
F: Bromeliaceae
This epiphytic
bromeliad is native to
the cloud forests of
S. Mexico and
C. America.

QUEEN OF THE ANDES
Puya raimondii
F: Bromeliaceae
Native to the C. Andes, the
world's largest bromeliad
produces a single colossal
flower spike after many years'
growth and then dies.

gray-green
leaves with
waxy texture

20 in
50 cm

rosette of
strap-shaped leaves

» POALES

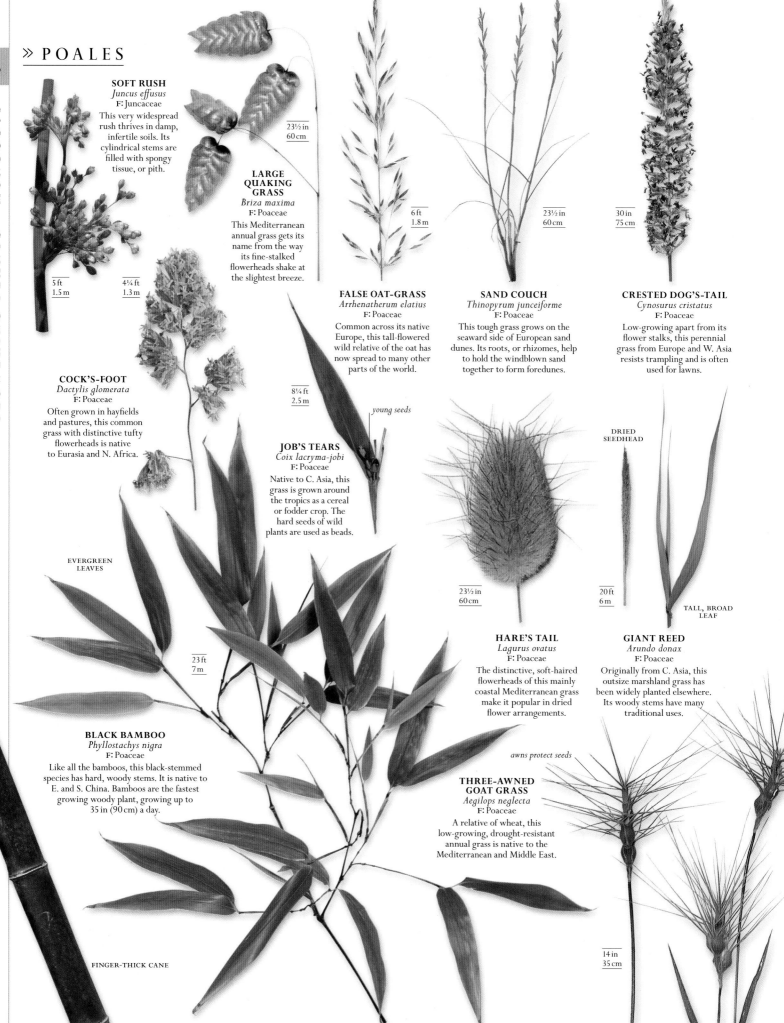

SOFT RUSH
Juncus effusus
F: Juncaceae
This very widespread rush thrives in damp, infertile soils. Its cylindrical stems are filled with spongy tissue, or pith.

5 ft
1.5 m

4¼ ft
1.3 m

LARGE QUAKING GRASS
Briza maxima
F: Poaceae
This Mediterranean annual grass gets its name from the way its fine-stalked flowerheads shake at the slightest breeze.

23½ in
60 cm

COCK'S-FOOT
Dactylis glomerata
F: Poaceae
Often grown in hayfields and pastures, this common grass with distinctive tufty flowerheads is native to Eurasia and N. Africa.

EVERGREEN LEAVES

FALSE OAT-GRASS
Arrhenatherum elatius
F: Poaceae
Common across its native Europe, this tall-flowered wild relative of the oat has now spread to many other parts of the world.

6 ft
1.8 m

8¼ ft
2.5 m

JOB'S TEARS
Coix lacryma-jobi
F: Poaceae
Native to C. Asia, this grass is grown around the tropics as a cereal or fodder crop. The hard seeds of wild plants are used as beads.

young seeds

SAND COUCH
Thinopyrum junceiforme
F: Poaceae
This tough grass grows on the seaward side of European sand dunes. Its roots, or rhizomes, help to hold the windblown sand together to form foredunes.

23½ in
60 cm

CRESTED DOG'S-TAIL
Cynosurus cristatus
F: Poaceae
Low-growing apart from its flower stalks, this perennial grass from Europe and W. Asia resists trampling and is often used for lawns.

30 in
75 cm

DRIED SEEDHEAD

23½ in
60 cm

HARE'S TAIL
Lagurus ovatus
F: Poaceae
The distinctive, soft-haired flowerheads of this mainly coastal Mediterranean grass make it popular in dried flower arrangements.

GIANT REED
Arundo donax
F: Poaceae
Originally from C. Asia, this outsize marshland grass has been widely planted elsewhere. Its woody stems have many traditional uses.

20 ft
6 m

TALL, BROAD LEAF

23 ft
7 m

BLACK BAMBOO
Phyllostachys nigra
F: Poaceae
Like all the bamboos, this black-stemmed species has hard, woody stems. It is native to E. and S. China. Bamboos are the fastest growing woody plant, growing up to 35 in (90 cm) a day.

awns protect seeds

THREE-AWNED GOAT GRASS
Aegilops neglecta
F: Poaceae
A relative of wheat, this low-growing, drought-resistant annual grass is native to the Mediterranean and Middle East.

14 in
35 cm

FINGER-THICK CANE

COMMON REED
Phragmites australis
F: Poaceae

Widespread in both temperate and tropical regions, this shallow-water grass can colonize large areas via its creeping horizontal stems.

20 ft
6 m

SWEET VERNAL GRASS
Anthoxanthum odoratum
F: Poaceae

Native to Eurasia, this early-flowering grass contains a chemical called coumarin that makes it smell pleasantly of new-mown hay.

3¼ ft
1 m

MARRAM GRASS
Calamagrostis arenaria
F: Poaceae

This tough grass from Europe flourishes on sand dunes, which its long underground stems and roots help to stabilize.

4 ft
1.2 m

OAT
Avena sativa
F: Poaceae

This cultivated grass, grown widely as a cereal crop for livestock and human consumption, thrives in cool, wet climates.

6 ft
1.8 m

awn

BARLEY
Hordeum vulgare
F: Poaceae

Originating in the ancient Near East, this cultivated cereal grass is notable for the long, hairlike bristles (awns) on its flowerheads.

32 in
80 cm

BREAD WHEAT
Triticum aestivum
F: Poaceae

The world's top-tonnage cereal, this species originated in the ancient Near East. It is a hybrid between the wild and earlier cultivated varieties.

3¼ ft
1 m

CREEPING BENT-GRASS
Agrostis stolonifera
F: Poaceae

This common perennial grass with feathery flowerheads is found worldwide. It spreads using creeping horizontal stems, or stolons.

16 in
40 cm

CITRONELLA
Cymbopogon nardus
F: Poaceae

A type of lemongrass, citronella originates from tropical Asia. It yields an oil used in perfumes and for repelling insects.

10 ft
3 m

PERENNIAL RYE-GRASS
Lolium perenne
F: Poaceae

Often used in pastures, lawns, and sports fields, this common Eurasian grass has now spread around the world.

single stem

RICE
Oryza sativa
F: Poaceae

Native to E. Asia, this major grain crop of warmer regions is usually cultivated in shallow water or in flood-prone areas.

6 ft
1.8 m

CORN
Zea mays
F: Poaceae

This major food crop was first cultivated in ancient Mexico. It has male or female flowerheads – edible corn is the female flower.

10 ft
3 m

SUGARCANE
Saccharum officinarum
F: Poaceae

Possibly related to corn, this tropical cultivated grass may have originated in New Guinea. Sugar is extracted from its thick stems.

20 ft
6 m

VETIVER
Chrysopogon zizanioides
F: Poaceae

Native to India, this tropical grass is widely cultivated for its fragrant oil and to bind soils to prevent erosion.

10 ft
3 m

35 in
90 cm

32 in
80 cm

YORKSHIRE FOG
Holcus lanatus
F: Poaceae

A common European grass found in damp meadows, Yorkshire fog has leaves with soft, velvety hairs.

dense white flowerheads

PAMPAS GRASS
Cortaderia selloana
F: Poaceae

This tall grass from southern S. America is a popular ornamental species. It has become invasive in some areas.

10 ft
3 m

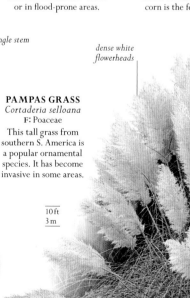

FLOWERING PLANTS · MONOCOTS

» POALES

GIANT FEATHER GRASS
Stipa gigantea
F: Poaceae
Gardeners cultivate this tall grass for its showy flowerheads, which persist until winter. It is native to Spain, Portugal, and Morocco.

8 ft
2.5 m

female flower

male flowerhead

5 ft
1.5 m

BRANCHED BUR-REED
Sparganium erectum
F: Typhaceae
Separate ball-shaped clusters of male and female flowers develop on the same flower stalk in this widespread wetland species of the northern hemisphere.

10 ft
3 m

CATTAIL
Typha latifolia
F: Typhaceae
This widespread wetland plant has a distinctive, cigar-shaped female flowerhead. The male flowers grow above it in a tuft.

YELLOW-EYED GRASS
Xyris sp.
F: Xyridaceae
This genus of grasslike plants, widespread in warmer regions of the world, bears small, yellow flowers on slender stems.

12 in
30 cm

ZINGIBERALES

Many species in this mainly tropical order of eight families grow giant leaves at the end of stalks. Although there are no true woody trees, some species such as the banana plant grow very large. Many Zingiberales have showy flowers and foliage and have become ornamentals. The ginger family, Zingiberaceae, the largest in the order, includes several other important spice plants beside ginger itself.

16 in
40 cm

NEVER-NEVER PLANT
Ctenanthe amabilis
F: Marantaceae
Native to forest floors in the Brazilian tropics, this warmth-loving plant requires high humidity if grown in cultivation.

20 in
50 cm

ETERNAL FLAME
Goeppertia crocata
F: Marantaceae
Related to *Ctenanthe* and *Maranta*, this Brazilian species has more striking flowers than its relatives, although living in similar forest habitats.

6½ ft
2 m

INDIAN SHOT
Canna indica
F: Cannaceae
This tropical American plant has unusual flowers in which some of the petals are fused with pollen-producing stamens. There are many cultivated varieties.

23½ in
60 cm

SPIRAL FLAG
Chamaecostus cuspidatus
F: Costaceae
This orange-flowered plant with dark green, glossy leaves is native to tropical E. Brazil. It is also grown as an ornamental.

12 in
30 cm

SILVER-VEINED PRAYER PLANT
Maranta leuconeura
F: Marantaceae
This Brazilian forest plant folds its leaves together at night to conserve moisture. Cultivated varieties have strikingly patterned foliage.

40 ft / 12 m

ENSET
Ensete ventricosum
F: Musaceae
This African-relative of the banana has long been cultivated in Ethiopia for its nutritious rootstock and stem, not its (inedible) fruits.

30 ft
9 m

BANANA
Musa acuminata
F: Musaceae
The seedless hybrid of Asian wild ancestors, cultivated bananas grow sterile male flowers at the end of their fruiting branches.

5 ft / 1.5 m

each yellow bract protects four or five tiny flowers

CHINESE YELLOW BANANA
Musella lasiocarpa
F: Musaceae
Possibly extinct in the wild, this mountain species from China produces a yellow flower spike that can last for months.

VIOLET-STRIPED WHITE PETAL

LANCEOLATE LEAF

18 ft / 5.5 m

SEED POD

CARDAMOM
Elettaria cardamomum
F: Zingiberaceae
This widely cultivated
tropical plant is native
to forests in S. India and
Sri Lanka. Its dried,
unripe fruits provide
the spice cardamom.

WHITE FLOWER

5 ft / 1.5 m

LESSER GALANGAL
Alpinia officinarum
F: Zingiberaceae
Native to E. Asia, this
relative of ginger
develops a swollen
underground stem and is
similarly used as a spice.

RHIZOME AND ROOTS

AROMATIC GINGER
Kaempferia galanga
F: Zingiberaceae
Native to tropical Asia,
this short-stemmed
species develops small
flowers and is grown as
an ornamental.

12 in / 30 cm

*paddle-shaped
leaf, green at tip,
yellow at leaf stem*

3¼ ft / 1 m

FLESHY RHIZOME

TURMERIC
Curcuma longa
F: Zingiberaceae
Native to S.E. Asia, this large-leaved
plant develops a swollen underground
stem from which the yellow spice
turmeric is obtained.

STEM WITH FLOWERHEAD

*symmetrical
fan-shaped tree*

EDIBLE RHIZOME

3¼ ft / 1 m

GINGER
Zingiber officinale
F: Zingiberaceae
The spice ginger comes from
the underground stems of this
cultivated S.E. Asian plant,
which no longer grows wild.

50 ft / 15 m

TRAVELER'S TREE
Ravenala madagascariensis
F: Strelitziaceae
Native to open forests in
Madagascar, this relative
of *Strelitzia* is pollinated
by lemurs in its
natural habitat.

6½ ft / 2 m

BIRD-OF-PARADISE
Strelitzia reginae
F: Strelitziaceae
Orange-and-blue bird-pollinated
flowers open one at a time from a
beaklike sheath in this South
African plant.

*purple flowers
develop in summer*

10 in / 25 cm

HUME'S ROSCOEA
Roscoea humeana
F: Zingiberaceae
This relative of the
ginger plant grows
unusual orchidlike
flowers. Its native
habitat is the mountains
of S.W. China.

12 ft / 4 m

HELICONIA STRICTA
F: Heliconiaceae
In its natural habitat, this
large-leaved tropical plant
from northern areas of
S. America is pollinated
by nectar-sipping
hummingbirds.

EUDICOTS

Almost three-quarters of the world's known flowering plants are classified in the eudicot grouping, which evolved more than 125 million years ago.

Eudicots get their name from the cotyledons, or seed-leaves, in the seed before it germinates. Unlike monocots, which have a single seed-leaf, eudicots have two. They include an enormous variety of plants, from agricultural weeds to towering rainforest trees, and they are of huge economic importance, as well as being prized as garden flowers. Many eudicots are annuals, with short lifespans measured in months or even weeks. Others are biennials and perennials, which live from two years to a century or more.

Despite their great diversity, eudicots have many physical features in common. Their leaves often have a netlike tracery of veins, unlike the parallel veins of monocots, and their stems have a well-developed vascular system, arranged in rings, for transporting water and sap. As well as getting taller, woody species thicken and strengthen their stems as they grow. This "secondary growth" is absent in most monocots, which is why eudicots make up most of the world's flowering shrubs and trees. Below ground level, most eudicots have a taproot, with smaller roots branching off it.

Eudicot flowers usually have parts in fours or fives, rather than the threes found in monocots, with sepals and petals that look different, both in color and in shape. Each species has its own distinctive kind of pollen, but the grains always have three pores; monocot pollen grains only have a single opening.

A VITAL ROLE

The fossil pollen record suggests that the eudicots diverged from other flowering plants some 125 million years ago. Since then, they have colonized every habitat on land, although they are less common as waterplants. Their importance to animals is impossible to overestimate. Countless species—with the notable exception of grazing animals that feed on grasses (monocots)—depend on them for food and shelter, and in return, many act as pollinators, or help to spread their seeds.

DIVISION	ANGIOSPERMAE
CLADE	EUDICOTS
ORDERS	44
FAMILIES	312
SPECIES	Over 210,000

FOUR-WAY SPLIT

For many years, all flowering plants were classified in two groups—the monocotyledons (monocots) and the dicotyledons (dicots)—based on the number of their seed-leaves. However, DNA analysis, combined with palynology—the study of pollen—has shown this classification does not fully reflect plant evolution. As a result, flowering plants are now split into four overall groups: basal angiosperms, magnoliids, monocots, and the eudicots, or true dicots.

NEW PERSPECTIVES

Computer analysis of DNA data has suggested new ways of subdividing the eudicots. These new hierarchical groupings, known as clades (see pp.30–31), give insights into the relationships between species and how they evolved, but they are difficult to interpret in terms of physical characteristics of the resulting groups and do not fit neatly into the traditional taxonomic hierarchy. This book is organized around orders and families, which still provide the clearest way to understand the diversity of flowering plants. However, the table below shows how they "nest" together in the proposed new system of clades.

EUDICOTS (pp.150–209)
(BUXALES—DIPSACALES)

CORE EUDICOTS (pp.155–209)
(GUNNERALES—DIPSACALES)

SUPERROSIDS (pp.156–181)
(SAXIFRAGALES—SAPINDALES)

ROSIDS (pp.158–181)
(VITALES—SAPINDALES)

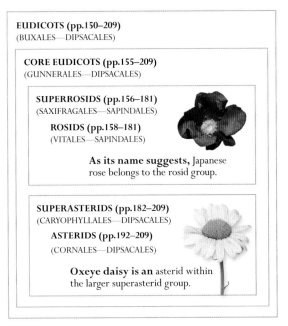

As its name suggests, Japanese rose belongs to the rosid group.

SUPERASTERIDS (pp.182–209)
(CARYOPHYLLALES—DIPSACALES)

ASTERIDS (pp.192–209)
(CORNALES—DIPSACALES)

Oxeye daisy is an asterid within the larger superasterid group.

BUXALES

Containing just a single family of about 120 species, the order Buxales is found in temperate, subtropical, and tropical regions. Most of the members are trees or shrubs, with simple, evergreen leaves and separate male and female flowers on the same plant. Many species are grown ornamentally and box wood is used for carving.

33 ft
10 m

5 ft
1.5 m

SWEET BOX
Sarcococca hookeriana
F: Buxaceae
This species bears clusters of small, fragrant flowers in winter, and is found in shady habitats in W. China.

BOXWOOD
Buxus sempervirens
F: Buxaceae
Native to rocky woods and scrub from Europe to N. Africa, this evergreen shrub or small tree is often grown in gardens for topiary.

PROTEALES

Of the four Proteales families, the largest is the Proteaceae, evergreen trees and shrubs from the southern hemisphere. The Platanaceae are northern hemisphere deciduous trees, and the Nelumbonaceae are aquatic plants from Asia, Australia, and North America.

outer bract

SEED HEADS

LOTUS LEAF

3 ft
1 m

SACRED LOTUS
Nelumbo nucifera
F: Nelumbonaceae
Growing in shallow freshwater habitats in parts of Asia and Australia, sacred lotus carries large, fragrant flowers on long stalks above the water.

SEED POD SEEN FROM ABOVE

collection of flowers at center of flowerhead

6½ ft
2 m

KING PROTEA
Protea cynaroides
F: Proteaceae
This species is native to hillsides and scrub in South Africa. Clusters of small flowers are enclosed by petallike bracts.

23 ft/7 m

RED SILKY OAK
Grevillea banksii
F: Proteaceae
Cultivated as an ornamental species for its striking "bottle-brush" flowers, this native of N.E. Australia grows in woods and open habitats.

styles

20 ft
6 m

HONEYSUCKLE GREVILLEA
Grevillea juncifolia
F: Proteaceae
Found in drier inland regions across Australia, this erect shrub has narrow, gray-green leaves and tight heads of orange flowers with long, yellow styles.

10 ft
3 m

WARATAH
Telopea speciosissima
F: Proteaceae
From dry woodland in New South Wales, Australia, common waratah is also known as Sydney waratah.

6½ ft / 2 m

MOUNTAIN DEVIL
Lambertia formosa
F: Proteaceae
From coastal heathland to mountain forests in New South Wales, Australia, this species has red-tinged bracts surrounding the flower clusters.

5 ft
1.5 m

PINCUSHION PROTEA
Leucospermum cordifolium
F: Proteaceae
This shrub of sandstone soils in South Africa's Cape Province is noted for its brightly colored, spherical flowerheads.

simple, evergreen leaf

50 ft
15 m

SAW BANKSIA
Banksia serrata
F: Proteaceae
Saw Banksia grows in woodland and scrub in E. Australia. Its bark is fire-resistant, enabling it to survive bushfires.

NARROW-LEAF DRUMSTICKS
Isopogon anemonifolius
F: Proteaceae
Spherical, yellow flowerheads above feathery leaves characterize this shrub, native to dry woodland and heathland in New South Wales, Australia.

6½ ft / 2 m

fruit takes about six months to ripen

thick, stiff-textured, broad leaf, similar to maple

157 ft/48 m

LONDON PLANETREE
Platanus x hispanica
F: Platanaceae
Planted in London since the 17th century, this deciduous hybrid arose from the cross-fertilization of two species of *Platanus* in Spain. Tolerant of pollution, it is widely planted in urban parks and streets.

leafy spike of red, or occasionally yellow or white flowers

30 ft
10 m

CHILEAN FIRE BUSH
Embothrium coccineum
F: Proteaceae
From forest and open habitats in S. Chile and Argentina, this species is cultivated for its flame-colored, 30-ft flowers, and thrives in sheltered gardens.

UNRIPE NUTS

MACADAMIA NUT
Macadamia integrifolia
F: Proteaceae
Native to coastal rainforest in E. Australia, macadamias are cultivated for their edible nuts.

FLOWER SPIKE

50 ft
15 m

RANUNCULALES

The order Ranunculales, consists of seven families. Four are relatively obscure, with few species, but the buttercup family, Ranunculaceae—after which the order is named—contains about 3,200 species. Together with members of the poppy and barberry families, these include some of the most common agricultural weeds worldwide and many familiar garden plants.

oval, toothed leaf

10 ft
3 m

— *oblong berry*

OREGON GRAPE
Berberis aquifolium
F: Berberidaceae

Flowering in springtime, Oregon grape is a suckering evergreen shrub from N.W. US. It grows in shady habitats.

SACRED BAMBOO
Nandina domestica
F: Berberidaceae

Also called heavenly bamboo, this shrub from mountain valleys in China and Japan is unrelated to true bamboos.

BRIGHT RED BERRIES

BARBERRY
Berberis vulgaris
F: Berberidaceae

A European species of hedges and scrub, barberry has distinctive triple spines, hanging flower clusters, and red berries.

5 ft
1.5 m

BISHOP'S HAT
Epimedium davidii
F: Berberidaceae

This evergreen species from W. China grows in woods and scrub. The young leaves are coppery, later turning green.

EVERGREEN LEAVES

6½ ft
2 m

12 in
30 cm

33 ft
10 m

20 ft
6 m

clustered flowers

20 in
50 cm

twining stem

MOONSEED
Menispermum canadense
F: Menispermaceae

Although the fruits of this climber resemble black grapes, they are extremely poisonous. The species inhabits woodland and banks of streams in Canada and the US.

CAROLINA SNAILSEED
Cocculus carolinus
F: Menispermaceae

This climbing woodland species from S.E. US has tiny flowers; the males and females are borne on separate plants.

13 ft
4 m

fragrant flower

CHOCOLATE VINE
Akebia quinata
F: Lardizabalaceae

Bearing sprays of scented flowers in spring, chocolate vine inhabits forest edges in China, Korea, and Japan.

30 in
75 cm

CREAMY-COLORED FLOWER

PINNATE LEAF

16 in/40 cm

33 ft/10 m

LEONTICE
Leontice leontopetalum
F: Berberidaceae

Native to cultivated ground and dry hillsides in N. Africa and E. Mediterranean countries, leontice grows from a tuber.

MAY APPLE
Podophyllum peltatum
F: Berberidaceae

A native of N. America, May apple—also known as American mandrake—grows in open woodland.

JAPANESE STAUNTON VINE
Stauntonia hexaphylla
F: Lardizabalaceae

Native to woodland in Japan and South Korea, this vigorous evergreen climber has woody stems and fragrant flowers.

CLIMBING CORYDALIS
Ceratocapnos claviculata
F: Papaveraceae

This scrambling annual species from W. Europe grows in woods and shady habitats on acid soils, supporting itself by leaf tendrils.

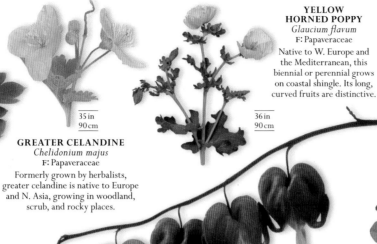

YELLOW HORNED POPPY
Glaucium flavum
F: Papaveraceae
Native to W. Europe and the Mediterranean, this biennial or perennial grows on coastal shingle. Its long, curved fruits are distinctive.

35 in
90 cm

36 in
90 cm

GREATER CELANDINE
Chelidonium majus
F: Papaveraceae
Formerly grown by herbalists, greater celandine is native to Europe and N. Asia, growing in woodland, scrub, and rocky places.

12 in
30 cm

4 ft
1.2 m

BLEEDING HEART
Lamprocapnos spectabilis
F: Papaveraceae
Named for its heart-shaped flowers, this species grows in damp woodland edges in Siberia, N. China, and Korea.

YELLOW CORYDALIS
Pseudofumaria lutea
F: Papaveraceae
This European species grows on walls and in stony and rocky places. It spreads vigorously by seed.

spurred petal

CALIFORNIA POPPY
Eschscholzia californica
F: Papaveraceae
Native to open habitats in western US and Mexico, this poppy is grown in gardens for its brightly colored flowers.

MATILIJA POPPY
Romneya coulteri
F: Papaveraceae
A species of scrub and grassland in California and Mexico, this poppy has fragrant flowers and is often grown in gardens.

12 in
30 cm

6½ ft
2 m

20 in
50 cm

COMMON FUMITORY
Fumaria officinalis
F: Papaveraceae
Found in Europe and N. Africa on cultivated and waste ground, this species usually grows on light soils.

pollen produced by a ring of dark cental anthers

YELLOW OR ORANGE FLOWER

16 in
40 cm

23½ in
60 cm

23½ in
60 cm

PINNATE LEAF

cut stem exudes poisonous latex

SICKLEFRUIT HYPECOUM
Hypecoum imberbe
F: Papaveraceae
This Mediterranean native grows in cultivated and waste ground and on walls.

COMMON POPPY
Papaver rhoeas
F: Papaveraceae
Arable and waste ground are home to this poppy, native to Europe, N. Africa, and parts of Asia. It is used as a symbol to commemorate World War I.

OPIUM POPPY
Papaver somniferum
F: Papaveraceae
Grown as the source of opium, heroin, and poppy seed, this inhabits cultivated and disturbed ground in Eurasia.

5 ft
1.5 m

WELSH POPPY
Papaver cambrica
F: Papaveraceae
Inhabiting shady, rocky places in hilly areas and often grown in gardens, Welsh poppy is native to W. Europe.

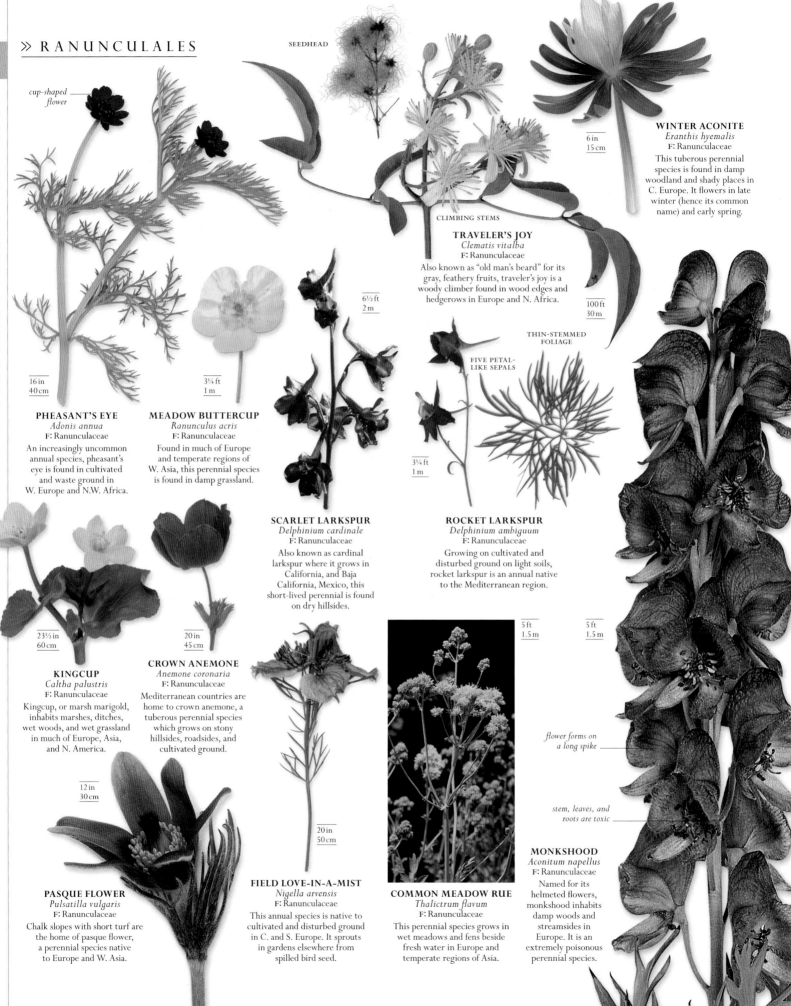

» RANUNCULALES

cup-shaped flower

SEEDHEAD

CLIMBING STEMS

WINTER ACONITE
Eranthis hyemalis
F: Ranunculaceae

This tuberous perennial species is found in damp woodland and shady places in C. Europe. It flowers in late winter (hence its common name) and early spring.

6 in
15 cm

TRAVELER'S JOY
Clematis vitalba
F: Ranunculaceae

Also known as "old man's beard" for its gray, feathery fruits, traveler's joy is a woody climber found in wood edges and hedgerows in Europe and N. Africa.

6½ ft
2 m

100 ft
30 m

THIN-STEMMED FOLIAGE

FIVE PETAL-LIKE SEPALS

3¼ ft
1 m

16 in
40 cm

PHEASANT'S EYE
Adonis annua
F: Ranunculaceae

An increasingly uncommon annual species, pheasant's eye is found in cultivated and waste ground in W. Europe and N.W. Africa.

3¼ ft
1 m

MEADOW BUTTERCUP
Ranunculus acris
F: Ranunculaceae

Found in much of Europe and temperate regions of W. Asia, this perennial species is found in damp grassland.

SCARLET LARKSPUR
Delphinium cardinale
F: Ranunculaceae

Also known as cardinal larkspur where it grows in California, and Baja California, Mexico, this short-lived perennial is found on dry hillsides.

ROCKET LARKSPUR
Delphinium ambiguum
F: Ranunculaceae

Growing on cultivated and disturbed ground on light soils, rocket larkspur is an annual native to the Mediterranean region.

23½ in
60 cm

20 in
45 cm

KINGCUP
Caltha palustris
F: Ranunculaceae

Kingcup, or marsh marigold, inhabits marshes, ditches, wet woods, and wet grassland in much of Europe, Asia, and N. America.

CROWN ANEMONE
Anemone coronaria
F: Ranunculaceae

Mediterranean countries are home to crown anemone, a tuberous perennial species which grows on stony hillsides, roadsides, and cultivated ground.

5 ft
1.5 m

5 ft
1.5 m

12 in
30 cm

flower forms on a long spike

stem, leaves, and roots are toxic

20 in
50 cm

PASQUE FLOWER
Pulsatilla vulgaris
F: Ranunculaceae

Chalk slopes with short turf are the home of pasque flower, a perennial species native to Europe and W. Asia.

FIELD LOVE-IN-A-MIST
Nigella arvensis
F: Ranunculaceae

This annual species is native to cultivated and disturbed ground in C. and S. Europe. It sprouts in gardens elsewhere from spilled bird seed.

COMMON MEADOW RUE
Thalictrum flavum
F: Ranunculaceae

This perennial species grows in wet meadows and fens beside fresh water in Europe and temperate regions of Asia.

MONKSHOOD
Aconitum napellus
F: Ranunculaceae

Named for its helmeted flowers, monkshood inhabits damp woods and streamsides in Europe. It is an extremely poisonous perennial species.

MOUSETAIL
Myosurus minimus
F: Ranunculaceae
The "tail" of mousetail is the elongated fruit that develops from its flower. This species grows in bare, damp ground across Europe, Asia, and N. America.

4 in
10 cm

leafless stem topped by a single flower

fruit developing from a flower

narrow, threadlike leaf

28 in
70 cm

GLOBEFLOWER
Trollius europaeus
F: Ranunculaceae
Native to Europe and W. Asia, globeflower is a perennial species found in damp mountain pasture.

6 in
15 cm

HEPATICA
Hepatica nobilis
F: Ranunculaceae
Native to much of Europe, this perennial woodland species has distinctive semievergreen, three-lobed leaves.

green flower

23½ in
60 cm

toothed leaf

LONG, DISTINCTIVE SPUR

3¼ ft
1 m

LEAVES

HELLEBORE
Helleborus lividus
F: Ranunculaceae
The Balearic Islands, particularly Mallorca, are home to this perennial species. Hellebore inhabits woods and rocky hillsides.

COLUMBINE
Aquilegia vulgaris
F: Ranunculaceae
A perennial species native to much of Europe, N. Africa, and temperate regions of Asia, columbine is found in shady and damp chalky habitats.

GUNNERALES

The two families which make up the Gunnerales order were previously classified in different orders, as they are visually very different. However, genetic analysis has recently shown the two families to be closely related. The family Gunneraceae consists of a single genus of large herbaceous plants growing in damp habitats, whereas Myrothamnaceae species are inhabitants of African deserts. *Gunnera* species are often grown in gardens as ornamental plants.

8 ft
2.5 m

GUNNERA
Gunnera manicata
F: Gunneraceae
Massive leaves and tall flower spikes characterize this perennial species. It is native beside fresh water in S. Brazil.

DILLENIALES

The order Dilleniales contains only the Dilleniaceae, a mainly tropical family of around 330 species of trees, shrubs, and climbers. Most species have alternately arranged leaves and bisexual flowers (with male and female parts), with five sepals, five petals, and numerous stamens. Some produce dry fruits that open to shed their seeds, others have berries. A few are grown as ornamentals or for lumber used in construction and boat building.

10 ft
3 m

GOLDEN GUINEA VINE
Hibbertia scandens
F: Dilleniaceae
A vigorous climber or scrambler, this evergreen shrub grows near the E. coast of Australia and in New Guinea.

23 ft
7 m

SIMPOH AIR
Dillenia suffruticosa
F: Dilleniaceae
A large and vigorous evergreen shrub endemic to Malaysia, Sumatra, and Borneo, simpoh air is found on swampy ground and forest edges.

SAXIFRAGALES

Of the 15 diverse families in the Saxifragales, five have just two members each and only three have more than 500 species. Best known is the saxifrage family, after which the order is named. The Latin *saxifraga* literally means "rock breaker," as these plants often grow in cracks in rocks and walls. The largest family is the stonecrops, Crassulaceae, which includes many succulent or water-retaining plants, adapted to dry conditions.

MEXICAN FIRECRACKER
Echeveria setosa
F: Crassulaceae
Named for its flame-colored flowers, this Mexican species has rosettes of succulent, blue-green, variably downy leaves.

4 in
10 cm

COMMON HOUSELEEK
Sempervivum tectorum
F: Crassulaceae
Originating from mountains in C. Europe, houseleek is often planted on roofs and walls. This plant forms dense mats of succulent rosettes.

20 in
50 cm

SAUCER PLANT
Aeonium tabuliforme
F: Crassulaceae
This native of cliffs on the N. coast of Tenerife, Canary Islands, forms a flat, succulent leaf rosette.

23½ in
60 cm

ORPINE
Hylotelephium telephium
F: Crassulaceae
Rocky terrain, woods, and hedgebanks are the habitats of orpine, a species native to Eurasia and introduced to N. America.

23½ in
60 cm

NAVELWORT
Umbilicus rupestris
F: Crassulaceae
This species has round, fleshy leaves, each with a central dimple. Navelwort inhabits rocks and walls in S. Europe and N. Africa.

20 in
50 cm

16 in
40 cm

FLAMING KATY
Kalanchoe blossfeldiana
F: Crassulaceae
Arid areas of Madagascar are home to this bushy species with glossy, succulent leaves and brightly colored flowers.

16 in
40 cm

ROSEROOT
Rhodiola rosea
F: Crassulaceae
Inhabiting mountain rocks and coastal cliffs, roseroot is a fleshy species found in the Arctic and alpine regions in Europe, N. America, and Asia.

BLACKCURRANT
Ribes nigrum
F: Grossulariaceae
This species grows in damp woods in much of Europe and C. Asia. Blackcurrant is cultivated for its flavorful berries.

6½ ft
2 m

BUFFALO CURRANT
Ribes aureum
F: Grossulariaceae
Scented flowers and thornless stems characterize buffalo currant, an inhabitant of rocky and sandy areas in C. US.

6½ ft / 2 m

WESTERN WATER MILFOIL
Myriophyllum hippuroides
F: Haloragaceae
This aquatic species has finely divided leaves. It is native to freshwater habitats in western N. America.

3¼ ft / 1 m

CHINESE SWEETGUM
Liquidambar formosana
F: Altingiaceae
A deciduous tree of damp woodland in
S.W. Asia, this species is noted for its
brightly colored fall foliage.

40 ft
12 m

PARROTIOPSIS
*Parrotiopsis
jacquemontiana*
F: Hamamelidaceae
Instead of petals, the
flowers of this forest
species from the
W. Himalayas have
white bracts, or
modified leaves.

20 ft/6 m

LEAVES CHANGING
COLOR
BUDS AND
FLOWERS

33 ft/10 m

VIRGINIAN WITCH HAZEL
Hamamelis virginiana
F: Hamamelidaceae
The flowers of this witch hazel are fragrant
and the leaves turn yellow in the fall. It
inhabits woodland in eastern N. America.

CHINESE ASTILBE
Astilbe rubra
F: Saxifragaceae
With distinctive plumelike clusters of flowers, this
moisture-loving plant grows in damp woodland and
by streams from Myanmar to Siberia.

3¼ ft/1 m

8 in
20 cm

50 ft/15 m

PERSIAN IRONWOOD
Parrotia persica
F: Hamamelidaceae
Native to forests in the Caucasus region and northern
Iran, Persian ironwood is a winter-flowering deciduous
tree with brilliantly colored fall leaves.

**MOTHER
OF THOUSANDS**
Saxifraga stolonifera
F: Saxifragaceae
Native to shady habitats
in China and Japan, this
perennial spreads by
producing rooting
plantlets at the end
of threadlike
runners.

**YELLOW
SAXIFRAGE**
Saxifraga aizoides
F: Saxifragaceae
Streamsides and wet
stony ground on
mountains are home to
this cushion-forming
species from Europe,
N. America, and
W. Asia.

12 in
30 cm

23½ in/60 cm

ALUMROOT
Heuchera americana
F: Saxifragaceae
Inhabiting rocky woodland in N. America,
alumroot has glossy leaves which are
mottled when young.

*flowers form
in a flattish-
topped cluster*

ELEPHANT'S EARS
Bergenia stracheyi
F: Saxifragaceae
An inhabitant of damp woodland and
meadows in the W. Himalayas and
Afghanistan, elephant's ears has fragrant
flowers and large, glossy leaves.

6 in/15 cm

28 in
70 cm

COMMON PEONY
Paeonia officinalis
F: Paeoniaceae
A herbaceous species
found in woodland,
meadows, and scrub
in parts of S. Europe,
common peony is
known for its
showy flowers.

*large,
glossy leaf*

12 in
30 cm

**OPPOSITE-LEAVED
GOLDEN-SAXIFRAGE**
Chrysosplenium oppositifolium
F: Saxifragaceae
The procumbent stems of this
species form extensive patches
in damp, shady habitats in
W. and C. Europe.

**PIGGYBACK
PLANT**
Tolmiea menziesii
F: Saxifragaceae
Young plantlets growing on
leaf bases give piggyback
plant its name. It is a hairy
perennial species found in
damp, shady places
in N. America.

28 in
70 cm

VITALES

This order consists of a single family, the Vitaceae or grape family. It contains 14 genera and 850 species, including the important grapevine and the ornamental Virginia creeper. Members of the Vitaceae are mainly native to the tropics or warm temperate regions. Mostly vines or lianas, they usually have swollen nodes (where leaves fork from the stem) and tendrils for climbing. Their flowers are usually held in flat-topped clusters.

foliage turns red in the fall

GRAPEVINE
Vitis vinifera
F: Vitaceae

115 ft
35 m

Humans have used grapes to make wine, food, and medicine since neolithic times. Very widespread, the grapevine is native to the Mediterranean, Europe, and Asia.

fruits smaller than domesticated varieties

CRIMSON GLORY VINE
Vitis coignetiae
F: Vitaceae

Cultivated for its giant, dimpled leaves (12 in/30 cm across) and fall color, this deciduous climber is native to temperate Asia.

VIRGINIA CREEPER
Parthenocissus quinquefolia
F: Vitaceae

This prolific climber from N. and C. America clings onto smooth surfaces using the adhesive pads on its small, forked tendrils.

100 ft
30 m

50 ft
15 m

GERANIALES

Two families comprise the Geraniales order. The Geraniaceae or geranium family contains about 800 species, including 400 cranesbills in the genus *Geranium*. Many of the 200 species in the African genus *Pelargonium* are of horticultural importance, including the garden plants called geraniums. The new family Francoaceae encompasses four previously recognized families, mainly of trees and shrubs from Africa.

8 ft
2.5 m

GIANT HONEYBUSH
Melianthus major
F: Francoaceae

Nectar drips from the bronze flower spikes of this South African native. Touching the leaves causes them to give off a strong odor.

12 in
30 cm

20 in
50 cm

32 in
80 cm

8 in
20 cm

APPLE PELARGONIUM
Pelargonium odoratissimum
F: Geraniaceae

This perennial originates from South Africa and has spreading flower stalks. It is cultivated for geranium oil, which has a strong scent of apple and rose.

HERB ROBERT
Geranium robertianum
F: Geraniaceae

Widespread in the northern hemisphere, this sprawling species has red stems and long flower stalks. It has a strong, mousey smell.

MEADOW CRANESBILL
Geranium pratense
F: Geraniaceae

Preferring grasslands on chalky soil, this perennial is native to Europe and Asia. Its flowers are pollinated by honey bees and other wild bee species.

ROCK STORKSBILL
Erodium foetidum
F: Geraniaceae

Storksbills take their name from the shape of their seeds. This French native is cultivated in rock gardens.

MYRTALES

The nine families in the Myrtales are most common in warmer regions. The 5,800 species in the Myrtaceae or myrtle family provide essential oils, spices, and fruits such as guava. They include more than 700 species of eucalypts from Australia and New Guinea. The Lythraceae or loosestrife family consists mainly of tropical trees and shrubs, producing fruits such as pomegranate and various dyes, and the Melastomataceae is a pan-tropical family of about 4,500 trees, shrubs, and climbers.

SWAMP LOOSESTRIFE
Decodon verticillatus
F: Lythraceae
Native to N.E. America, this shrub grows in swamps. It has arching stems with leaves in whorls of three, and red or purple flowers up to 1 in (2.5 cm) across.

8 ft
2.5 m

PURPLE LOOSESTRIFE
Lythrum salicaria
F: Lythraceae
Numerous purple-red, square stems grow from the woody, creeping rootstock of this perennial. Native to Europe, Asia, S.E. Australia, and N.W. Africa, it can be invasive.

23 ft
7 m

5 ft
1.5 m

triangular leaf

multi-seeded fruit develops from flower

20 ft
6 m

CRAPE MYRTLE
Lagerstroemia indica
F: Lythraceae
Blooming for up to 120 days, this tree is native to China, Korea, and Japan. Its bark is smooth, mottled, pinkish-gray, and is shed each year.

20 ft
6 m

HENNA
Lawsonia inermis
F: Lythraceae
Native to N. Africa and the Middle East, henna's leaves make a reddish-brown dye and its fragrant flowers yield essential oils.

30 in
75 cm

WATER CHESTNUT
Trapa natans
F: Lythraceae
This floating plant is from Europe and Asia. It has an edible, starchy seed in a nut with four hornlike barbed spines.

POMEGRANATE
Punica granatum
F: Lythraceae
This spiny, shrubby tree, from S.W. Asia, is widely cultivated in the Mediterranean for its pulpy fruits, containing numerous seeds.

35 in
90 cm

CIGAR FLOWER
Cuphea ignea
F: Lythraceae
A densely branched, perennial shrub, cigar flower is a garden and house ornamental, native to Mexico. Its fruits are paperlike capsules.

59 ft
18 m

RANGOON CREEPER
Combretum indicum
F: Combretaceae
This climber from tropical Asia has clusters of tubular red flowers. Its ellipsoidal fruits, each with five wings, taste of almonds.

100 ft
30 m

INDIAN ALMOND
Terminalia catappa
F: Combretaceae
Found on coasts of the Indo-Pacific oceans, this tree has horizontal branches. Its corky fruit, dispersed by water, contains an edible, almond-tasting nut.

flower has five or six petals

ROSE GRAPE
Medinilla magnifica
F: Melastomataceae
This ornamental from the Philippines is an epiphyte, growing on trees. Its leaves have the longitudinal veins characteristic of the melastomes.

10 ft
3 m

16 ft
5 m

BRAZILIAN SPIDER FLOWER
Tibouchina urvilleana
F: Melastomataceae
In warm regions this Brazilian ornamental blooms for most of the year. Its velvety leaves have red edges and three to five obvious longitudinal veins.

23½ in
60 cm

VIRGINIA MEADOW BEAUTY
Rhexia virginica
F: Melastomataceae
This hairy, perennial herb from E. US and Canada grows on wetlands. It has square stems and stalkless leaves with toothed margins.

»

FLOWERING PLANTS · EUDICOTS

stamens form bottlebrush-shaped flowers

woody seed capsule

TASMANIAN SNOW GUM
Eucalyptus coccifera
F: Myrtaceae
With gray-white bark, peeling in long strips to reveal a creamy-white trunk, this species of Tasmanian uplands has unstalked, opposite leaves.

120 ft
36 m

CIDER GUM
Eucalyptus gunnii
F: Myrtaceae
This hardy Tasmanian tree forms juvenile leaves that are round and silvery, maturing to scythe-shaped and blue-gray. It is used in paper production.

82 ft
25 m

TONGHI BOTTLEBRUSH
Melaleuca subulata
F: Myrtaceae
A spreading shrub with small, woody fruits containing hundreds of seeds. It is found mainly in New South Wales and Victoria, Australia.

10 ft
3 m

RED GUM
Eucalyptus camaldulensis
F: Myrtaceae
Widespread in Australia, with smooth, pale bark and blue-green leaves, this riverside tree has durable lumber and its nectar makes good honey.

130 ft
40 m

GREEN BOTTLEBRUSH
Melaleuca virens
F: Myrtaceae
Resisting snow, frost, and drought, this sprawling subalpine shrub from Tasmania has leaves with sharp, pointed ends. It attracts birds and butterflies.

10 ft
3 m

elliptical leaf smells of peppermint

82 ft
25 m

16 ft
5 m

40 ft
12 m

65 ft
20 m

CAJUPUT
Melaleuca cajuputi
F: Myrtaceae
Native from S.E. Asia to N. Australia, this tree has aromatic leaves that contain a pale yellow medicinal oil.

COMMON MYRTLE
Myrtus communis
F: Myrtaceae
Native from the Mediterranean to Pakistan, myrtle produces fragrant flowers and blue-black berries. Its aromatic leaves yield essential oils.

ALLSPICE
Pimenta dioica
F: Myrtaceae
From the Caribbean, S. Mexico, and C. America, this unisexual tree has small, white flowers that form brown, berrylike stoned fruits. Unripe fruits are dried and ground to make "allspice."

URN-FRUITED GUM
Eucalyptus urnigera
F: Myrtaceae
This tree from S.E. Tasmania has urn-shaped fruits, blue-gray juvenile leaves, and white flowers in clusters of three with numerous stamens.

50 ft
15 m

10 ft
3 m

EVERGREEN LEAVES

SCARLET KUNZEA
Kunzea baxteri
F: Myrtaceae
Kunzea species are unusual in having colorful stamens much longer than their petals, which adds to the showy appearance that attracts pollinating birds.

40 ft
12 m

65 ft
20 m

CHILEAN MYRTLE
Luma apiculata
F: Myrtaceae
This slow-growing tree has a contorted trunk and smooth gray-orange peeling bark. Its fruit is a black berry.

EDIBLE FRUIT

POHUTAKAWA
Metrosideros excelsa
F: Myrtaceae
Native to North Island, New Zealand, this tree produces red flowers in December. Old trees have beardlike masses of hanging aerial roots.

65 ft
20 m

CLOVE
Syzygium aromaticum
F: Myrtaceae
The dried flower buds of this native of Indonesian islands are used as a spice. Its cream-colored flowers have red stamens and ripen to purple, single-seeded berries.

GUAVA
Psidium guajava
F: Myrtaceae
This tree from tropical and subtropical America has flaky, coppery bark. The fruits have a sweet, musky smell when ripe, and contain numerous hard, yellow seeds.

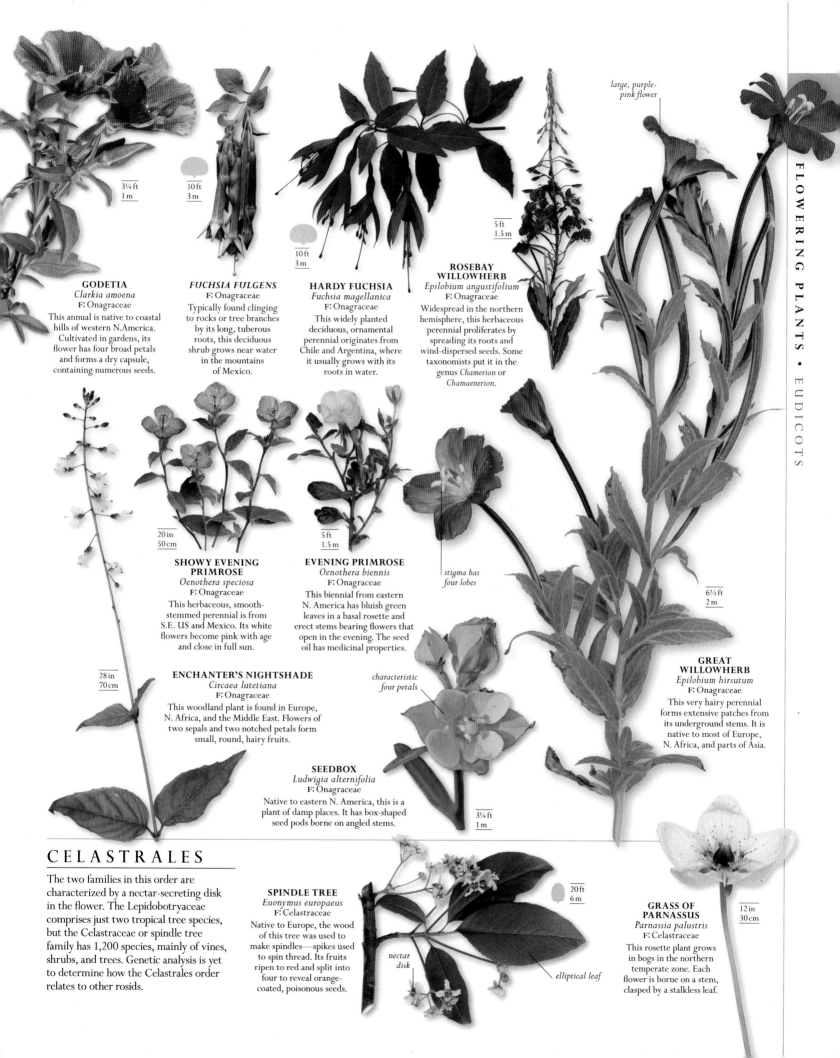

GODETIA
Clarkia amoena
F: Onagraceae
This annual is native to coastal hills of western N.America. Cultivated in gardens, its flower has four broad petals and forms a dry capsule, containing numerous seeds.

3¼ ft
1 m

FUCHSIA FULGENS
F: Onagraceae
Typically found clinging to rocks or tree branches by its long, tuberous roots, this deciduous shrub grows near water in the mountains of Mexico.

10 ft
3 m

HARDY FUCHSIA
Fuchsia magellanica
F: Onagraceae
This widely planted deciduous, ornamental perennial originates from Chile and Argentina, where it usually grows with its roots in water.

10 ft
3 m

ROSEBAY WILLOWHERB
Epilobium angustifolium
F: Onagraceae
Widespread in the northern hemisphere, this herbaceous perennial proliferates by spreading its roots and wind-dispersed seeds. Some taxonomists put it in the genus *Chamerion* or *Chamaenerion*.

5 ft
1.5 m

large, purple-pink flower

SHOWY EVENING PRIMROSE
Oenothera speciosa
F: Onagraceae
This herbaceous, smooth-stemmed perennial is from S.E. US and Mexico. Its white flowers become pink with age and close in full sun.

20 in
50 cm

EVENING PRIMROSE
Oenothera biennis
F: Onagraceae
This biennial from eastern N. America has bluish green leaves in a basal rosette and erect stems bearing flowers that open in the evening. The seed oil has medicinal properties.

5 ft
1.5 m

stigma has four lobes

ENCHANTER'S NIGHTSHADE
Circaea lutetiana
F: Onagraceae
This woodland plant is found in Europe, N. Africa, and the Middle East. Flowers of two sepals and two notched petals form small, round, hairy fruits.

28 in
70 cm

characteristic four petals

SEEDBOX
Ludwigia alternifolia
F: Onagraceae
Native to eastern N. America, this is a plant of damp places. It has box-shaped seed pods borne on angled stems.

3¼ ft
1 m

GREAT WILLOWHERB
Epilobium hirsutum
F: Onagraceae
This very hairy perennial forms extensive patches from its underground stems. It is native to most of Europe, N. Africa, and parts of Asia.

6½ ft
2 m

CELASTRALES

The two families in this order are characterized by a nectar-secreting disk in the flower. The Lepidobotryaceae comprises just two tropical tree species, but the Celastraceae or spindle tree family has 1,200 species, mainly of vines, shrubs, and trees. Genetic analysis is yet to determine how the Celastrales order relates to other rosids.

SPINDLE TREE
Euonymus europaeus
F: Celastraceae
Native to Europe, the wood of this tree was used to make spindles—spikes used to spin thread. Its fruits ripen to red and split into four to reveal orange-coated, poisonous seeds.

nectar disk

20 ft
6 m

elliptical leaf

GRASS OF PARNASSUS
Parnassia palustris
F: Celastraceae
This rosette plant grows in bogs in the northern temperate zone. Each flower is borne on a stem, clasped by a stalkless leaf.

12 in
30 cm

CUCURBITALES

Eight mainly tropical families of trees, shrubs, herbs, and climbers make up this order. Six have few members, but the Begoniaceae or begonia family has 1,400 species, 130 of which are horticultural plants. The 850 species in the Cucurbitaceae or gourd family include notable food plants such as squash and pumpkins. Both of these families have male and female flowers on the same plant (monoecious).

LARGE LEAF

EDIBLE FLOWER

13 ft
4 m

NUTRITIOUS FRUIT

3¼ ft
1 m

SQUASH
Cucurbita pepo
F: Cucurbitaceae
With five-sided prickly stems and large, yellow-orange flowers, pumpkins, marrows, and zucchini are all varieties of *Cucurbita pepo* from C. America.

WHITE BRYONY
Bryonia cretica ssp. *dioica*
F: Cucurbitaceae
Common in hedgerows and scrub on calcium-rich soils in S. Europe and N. Africa, this poisonous climber has a swollen, tuberous root.

6½ ft
2 m

HAIRY LEAVES

CUCUMBER
Cucumis sativus
F: Cucurbitaceae
Originally from tropical Asia, cucumber has been cultivated for 3,000 years and there are many varieties, including the gherkin. The plant has tubular yellow flowers.

IMMATURE FRUIT

15 ft/4.5 m

FRUIT   LEAF SPRIG

SMOOTH LOOFAH
Luffa cylindrica
F: Cucurbitaceae
This climbing, annual vine, originally from Asia, produces cylindrical fruits. Their fibrous interior can be eaten young, or dried and used as a sponge.

large, spherical fruit narrows above middle

fruit stem

CALABASH
Lagenaria siceraria
F: Cucurbitaceae

16 ft
5 m

This vine produces whitish, crinkly flowers. Its edible, hard-shelled fruit floats in the sea for months and is used as a container.

3¼ ft
1 m

SQUIRTING CUCUMBER
Ecballium elaterium
F: Cucurbitaceae
The ripe fruit of this Mediterranean plant becomes so swollen with juices that it explodes when touched, squirting its seeds up to 20 ft (6 m).

3¼ ft/1 m

WATERMELON
Citrullus lanatus
F: Cucurbitaceae
With yellow-green flowers, this annual, prostrate vine originated in southern Africa and its fruits are cultivated throughout warmer climates.

16 in/40 cm

BEGONIA LISTADA
F: Begoniaceae
Native to Paraguay, this sprawling species has thick, velvety leaves with a red underside and white-pinkish flowers. It was first described in 1981.

poisonous, exploding fruit

leaf is up to 4 in (10 cm) long

BALSAM PEAR
Momordica charantia
F: Cucurbitaceae
Also called bitter gourd, this tropical vine produces bitter-tasting fruits. Their tips split into three to reveal yellow pulp and seeds encased in scarlet coats (arils).

10 ft
3 m

150 ft
45 m

TETRAMELES NUDIFLORA
F: Datiscaceae
Props at the trunk base support this tall tree from Asia and N. Australia. Male flowers grow on different trees than the female flowers, which form round seed capsules.

FABALES

Found all over the world, except for Antarctica, the Fabales have compound leaves with tiny outgrowths (stipules) at the base and seed pods that open when mature. Called legumes, these plants have nodules—swellings that contain bacteria—on their roots that help them take nitrogen from the air into soil. The Fabaceae, the largest of the four families, includes plants like the pea with flowers that have a large upper petal with adjacent smaller petals.

SENSITIVE PLANT
Mimosa pudica
F: Fabaceae
This prickly S. American plant's leaves close when touched. Its flower heads (³⁄₈ in/1 cm) usually have pink flowers and long, pink-purple stamens.

20 in/50 cm

PEANUT
Arachis hypogaea
F: Fabaceae
Native to Brazil, this legume has yellow, reddish-veined, pealike flowers. Its seed pod matures in the ground, producing peanuts.

20 in / 50 cm

pinnate leaf

SEED POD   HERBACEOUS PLANT

GOLDEN SHOWER TREE
Cassia fistula
F: Fabaceae
This slender tree from S.E. Asia has pendulous, pealike flowers and pinnate leaves with three to eight pairs of leaflets. Its seeds are poisonous.

FLOWER RACEME

65 ft / 20 m

DECIDUOUS LEAFLET

GLOSSY LEAVES

POD HAS EDIBLE PULP

130 ft / 40 m

WEST INDIAN LOCUST TREE
Hymenaea courbaril
F: Fabaceae
This hardwood tree has a straight, thick trunk and stems of white-purplish flowers with large petals and long stamens. Orange gum oozes from its trunk, which is used in incense and perfumes.

flowers are clusters of stamens with reduced petals

SILVER WATTLE
Acacia dealbata
F: Fabaceae
This tree has fragrant flowers and silvery bark, aging to black. Native to S.E. Australia, silver wattle is often called mimosa elsewhere.

65 ft / 20 m

SAINFOIN
Onobrychis viciifolia
F: Fabaceae
This fodder crop from S. Europe has green, oval, pinnate leaves with 6–14 pairs of leaflets. Densely packed pink flowers appear on a main stem.

32 in / 80 cm

4 ft / 1.2 m

23½ in / 60 cm

LEAF SPRIG

KIDNEY VETCH
Anthyllis vulneraria
F: Fabaceae
A plant of dry, grassy places throughout Europe, this perennial has silky-haired stems and clusters of yellow to red flowers, with a ruff of downy bracts beneath.

TAMARIND
Tamarindus indica
F: Fabaceae
This evergreen tree with drooping branches from East Africa and Asia has yellow-orange flowers on stems and edible pulp inside the seed pod.

65 ft / 20 m

SEED PODS

FLAT WATTLE
Acacia glaucoptera
F: Fabaceae
The globular flowers of this spreading shrub from S.W. Australia grow from a modified stem looking like a twisted leaf.

40 ft / 12 m

leaf with 6–12 pairs of pinnate leaflets

SILK TREE
Albizia julibrissin
F: Fabaceae
This deciduous tree is native to S.W. Asia. It has dark green bark, striped vertically with age, and is fast-growing but short lived.

WILD LICORICE
Astragalus glycyphyllos
F: Fabaceae

With similar leaves to true licorice, this European grassland herb has curved pods. It can be used as a tea.

35 in / 90 cm

BLUE FALSE INDIGO
Baptisia australis
F: Fabaceae

The sap of this herbaceous perennial, native to woods and streamsides in E. US, turns purple when exposed to air, and is used as an indigo dye substitute.

5 ft / 1.5 m

GLOSSY PINNATE LEAF

BLACK BEAN TREE
Castanospermum australe
F: Fabaceae

The orange-and-red flowers of this Australian lumber tree develop into woody pods containing three to five beanlike seeds.

130 ft / 40 m

HARD SEED

EVERGREEN LEAVES

CAROB TREE
Ceratonia siliqua
F: Fabaceae

This Mediterranean tree has a thick trunk, dense foliage, and small, dull-green flowers. Pulp from the seed pods is a chocolate substitute.

50 ft / 15 m

SEEDS

PULPY PODS

JUDAS TREE
Cercis siliquastrum
F: Fabaceae

This deciduous, ornamental and lumber tree from the E. Mediterranean flowers profusely in the spring, then later develops flat 4 in (10 cm) seed pods.

33 ft / 10 m

young seed pod

CHICK PEA
Cicer arietinum
F: Fabaceae

One of the earliest cultivated vegetables from the Middle East, its seed pod contains one to three seeds, which can be cooked to make hummus.

20 in / 50 cm

GOAT'S-RUE
Galega officinalis
F: Fabaceae

Naturalized in temperate areas, this perennial, with long, cylindrical, red-brown seed pods, is believed by some to improve lactation and reduce fevers and diabetes.

5 ft / 1.5 m

MOUNT ETNA BROOM
Genista aetnensis
F: Fabaceae

Native to Sicily, around the slopes of Mount Etna, and Sardinia, this small tree has few leaves; flattened green shoots assist photosynthesis.

10 ft / 3 m

DORYCNIUM
Lotus hirsutus
F: Fabaceae

This Mediterranean perennial forms gray-green shrubby mounds. Its red-brown cylindrical pods can be mistaken for small berries.

20 in / 50 cm

5 ft / 1.5 m

PURPLISH-PINK FLOWER

10 ft / 3 m

SMALL-FLOWERED GORSE
Ulex parviflorus
F: Fabaceae

This dense, spiny, perennial shrub from the W. Mediterranean has yellow leaves and short, brown-black pods. Fire and brush clearance aid seed germination.

BROAD-LEAVED EVERLASTING PEA
Lathyrus latifolius
F: Fabaceae

Distributed through S. Europe and N. Africa, this vigorous, climbing perennial has winged stems and clusters of 5–15 pinkish flowers. Its seeds form in long pods.

POD CONTAINS 10–15 SEEDS

COMMON BIRD'S-FOOT TREFOIL
Lotus corniculatus
F: Fabaceae

Found in grassland in Europe, Asia, and Africa, it has yellow flowers and five lobed leaves, three of which sit above the others. Its seed pods resemble a bird's foot.

12 in / 30 cm

COMMON VETCH
Vicia sativa
F: Fabaceae

This scrambling, widespread annual is native to Europe and the Mediterranean. An animal-feed crop, its flowers are normally paired and its tendrils can be branched.

5 ft / 1.5 m

CROWN VETCH
Securigera varia
F: Fabaceae
This rapidly spreading
perennial of S. Europe and
W. Asia has thick leaves and
deep roots, good for binding
soil and controlling erosion.

3¼ ft
1 m

BROOM
Cytisus scoparius
F: Fabaceae
Found in heaths across Europe,
this shrub has strongly scented
flowers and many slender, ridged,
and angled branches traditionally
used for brooms.

6½ ft
2 m

PEALIKE
FLOWERS

12 in
30 cm

TWISTED SEED
PODS

HORSESHOE VETCH
Hippocrepis comosa
F: Fabaceae
This prostrate perennial, from S. Europe
and the Mediterranean, is a major food
plant for blue butterfly caterpillars.
Its seed pods are twisted into
horseshoelike segments.

3¼ ft
1 m

COMPOUND
LEAVES

ROOT

BLUE-VIOLET
FLOWERS

3¼ ft
1 m

RED CLOVER
Trifolium pratense
F: Fabaceae
A food crop for grazing
animals, this perennial
has long-stalked basal
leaves, and an oblong
fruit pod hidden within
its flowerhead.

LICORICE
Glycyrrhiza glabra
F: Fabaceae
This feathery perennial has smooth, small,
oblong pods and deep roots. Extract
from the roots is 50 times sweeter than
sugar and is medicinal.

ALFALFA
Medicago sativa
F: Fabaceae
This deeply rooting perennial
grows on chalk grassland
throughout S.W. Asia, Europe, and
the US. It is a useful animal-feed
crop and has medicinal properties.

32 in
80 cm

5 ft
1.5 m

*densely packed
flowers*

strong stem

narrow leaflet

**SCARLET
RUNNER BEAN**
Phaseolus coccineus
F: Fabaceae
Native to C. American
mountains, the shoot of this
perennial twists clockwise
around its support. Its seeds
(beans) are variously colored.

12 ft
3.7 m

**COMMON
LABURNUM**
Laburnum anagyroides
F: Fabaceae
This deciduous tree is
native to C. and
S. Europe. Clusters
of brown hairy pods
(3 in/7.5 cm) contain
toxic black seeds.

40 ft
12 m

82 ft/25 m

BLACK LOCUST
Robinia pseudoacacia
F: Fabaceae
This vigorous deciduous tree,
originally from S.E. USA, spreads
by suckers. It has poisonous bark
and leaves, and flat brown seed pods.

16 in
40 cm

NICEAN MILKWORT
Polygala nicaeensis
F: Polygalaceae
This perennial from France
and Italy has flowers with two
sepals and three joined petals,
one of which is fringed, and a
small capsule as a fruit.

GARDEN LUPIN
Lupinus polyphyllus
F: Fabaceae
Originating from
western N. America
and naturalized
throughout Europe,
this popular
ornamental has hairy,
black pods and
undersides of leaves. Its
flowers are fragrant.

FAGALES

The seven families in this order include some of the world's best-known trees, which dominate the woodlands in which they grow: beech trees in the Fagaceae, birches in the Betulaceae, southern beeches in the Nothofagaceae, walnut trees in the Juglandaceae, and Australian she-oaks in the Casuarinaceae. Most have simple leaves and small, wind-pollinated flowers.

65 ft
20 m

FOREST OAK
Allocasuarina torulosa
F: Casuarinaceae
The lumber of this W. Australian tree is prized by wood-turners. Its pendulous branches carry needlelike, green twigs with tiny leaves and warty cones.

fruit and seed develop from catkins

80 ft
25 m

JAPANESE HOP HORNBEAM
Ostrya japonica
F: Betulaceae
This species from the Far East has gray-brown scaly bark. Its nutlet seeds are enclosed in husks, like hops, and hang in long clusters.

80 ft
25 m

EUROPEAN HORNBEAM
Carpinus betulus
F: Betulaceae
Frequenting hedgerows, this European native has small, green, male and female catkin flowers that develop into nuts surrounded by three-lobed bracts.

bract

male catkin

leaf bud

female flower

43 ft
13 m

EUROPEAN HAZELNUT
Corylus avellana
F: Betulaceae
This European shrubby tree is harvested for its edible nuts. Its male catkins and female flowers appear in early spring.

100 ft
30 m

RED ALDER
Alnus rubra
F: Betulaceae
A native of western N. America, this tree prefers damp slopes and margins of watercourses. It has pale gray bark that becomes red when bruised or scraped.

catkin is a cylindrical cluster of male flowers

80 ft
25 m

RAULI
Nothofagus alpina
F: Nothofagaceae
A lumber tree, rauli is native to Argentina and Chile. The young leaves are bronze. Greenish female flowers grow in clusters and form bristly husks enclosing small nuts.

PINNATE LEAVES

RIPE PECAN

130 ft
40 m

PECAN
Carya illinoinensis
F: Juglandaceae
The pecan is native to N. America and cultivated for its edible nuts. The nuts are enclosed in a husk that splits four ways when ripe.

PLATYCARYA STROBILACEA
F: Juglandaceae
Native to E. Asia, this tree has clusters of erect male catkins and a fruit, like a conifer cone, with winged seeds.

50 ft
15 m

BUTTERNUT
Juglans cinerea
F: Juglandaceae
This deciduous tree, native to N. America, has yellow-green catkins. Small bunches of egg-shaped nuts contain sweet seeds.

100 ft
30 m

ERMAN'S BIRCH
Betula ermanii
F: Betulaceae
Native to mountain areas from Siberia to Japan, this tree is widely planted in parks and gardens. It has peeling, whitish bark.

nut clusters form in leaf axil

6½ ft
2 m

SWEET GALE
Myrica gale
F: Myricaceae
This sweet-smelling shrub, traditionally used as an insect repellent, grows in peat bogs in northern temperate zones. It has either male or female reddish catkins.

100 ft
30 m

ENGLISH WALNUT
Juglans regia
F: Juglandaceae
A species from the mountains of S.W. Asia, the walnut tree is valued for its fruits and lumber. It has smooth gray bark and male catkins up to 6 in (15 cm) long.

100 ft
30 m

RIPE WALNUT

LEAF SPRIG WITH UNRIPE FRUITS

deciduous leaf

SCALY ACORN CUP

130 ft
40 m

ENGLISH OAK
Quercus robur
F: Fagaceae
Long-lived, with valuable lumber, this tree is common in W. Europe, especially Britain. It has male flowers in long, hanging catkins, and long-stalked acorns.

SCARLET OAK
Quercus coccinea
F: Fagaceae
The leaves of this N. American tree turn a characteristic deep red in the fall. This oak bears long catkins of yellowish-green male flowers and its acorns have a glossy cup.

80 ft
25 m

FALL LEAF

SPRING COLOR

CATKIN SPRIG

toothed leaf

100 ft
30 m

spiny burr encases nuts

30 ft
10 m

KERMES OAK
Quercus coccifera
F: Fagaceae
An evergreen, shrubby tree from the Mediterranean area, this species has hollylike leaves that are bronze when young, and yellow-brown male catkins.

100 ft
30 m

50 ft
15 m

JAPANESE STONE OAK
Lithocarpus edulis
F: Fagaceae
This Japanese evergreen tree has upright creamy-white catkins, which have female flowers at the base and males above. The edible acorns ripen over two years.

AMERICAN BEECH
Fagus grandifolia
F: Fagaceae
This broadly spreading tree from eastern N. America has gray-brown bark. It has glossy deciduous leaves and two fruits (beechnuts) in each husk.

EUROPEAN CHESTNUT
Castanea sativa
F: Fagaceae
Cultivated for 3,000 years for its nuts, this tree is from S.E. Europe and W. Asia. It bears catkins with male flowers in the upper part and female flowers beneath.

MALPIGHIALES

One of the largest and most diverse orders, Malpighiales comprises 36 mainly tropical families with more than 16,000 species, grouped together by their DNA but widely different in form. Half the families, including one only recognized in 2016, have fewer than 100 members and many are unfamiliar outside their native regions, but they also include the huge spurge family (Euphorbiaceae), with 6,300 species.

32 in
80 cm

POISONOUS FRUITS

TUTSAN
Hypericum androsaemum
F: Hypericaceae
This small shrub from W. Europe has reddish, two-ridged stems. Its aromatic leaves have medicinal uses, but its berries are poisonous.

coarsely toothed, ovate leaf

PAIRED LEAVES

100 ft
30 m

32 in
80 cm

CEYLON IRONWOOD
Mesua ferrea
F: Calophyllaceae
This Asian tree yields heavy lumber, hence its common name. It has large flowers with four white petals and bright red young leaves.

PERFORATE ST JOHN'S-WORT
Hypericum perforatum
F: Hypericaceae
This Eurasian perennial is a common plant of banks, fields, and roadsides. It has round stems with ridges on either side, leaves with translucent dots, and many-seeded capsules.

» MALPIGHIALES

TRILOBED
LEAF

50 ft
15 m

16 ft
5 m

NUT CONTAINING
HARD SEED

CANDLE-NUT TREE
Aleurites moluccanus
F: Euphorbiaceae
Oil from the nuts of this
tropical tree was burned in
lamps. It has stems of small,
creamy flowers and its variable
leaves are gray-green
when young.

PALMATE
LEAF

FLOWER HEAD

CASTOR OIL PLANT
Ricinus communis
F: Euphorbiaceae
Castor oil is extracted from the
poisionous seeds of this species
from tropical Africa. With
upright flower heads, the upper
female flowers have red stigmas
and lower males yellow anthers.

13 ft
4 m

CASSAVA
Manihot esculenta
F: Euphorbiaceae
The tuberous roots of this
S. American plant are boiled and
ground as food. It has small
flower clusters that grow on
secondary branches.

13 ft
4 m

STICKY
LEAVES

OIL-YIELDING
SEEDS

JATROPHA GOSSYPIIFOLIA
F: Euphorbiaceae
This invasive, toxic plant,
banned in parts of Australia, is
native to tropical America. It
has sticky leaves, watery sap,
and flowers with purple bracts.

4 ft
1.2 m

LARGE MEDITERRANEAN SPURGE
Euphorbia characias
F: Euphorbiaceae
This ornamental, Mediterranean
perennial has upright, purplish,
woolly stems, bare and smooth at the
base. The fruit, or seed capsule, is
hairy and berrylike.

CROWN OF THORNS
Euphorbia milii
F: Euphorbiaceae
This semi-succulent, climbing
shrub from Madagascar has
very spiny stems. Its leaves are
mainly on the new shoots.

6 ft
1.8 m

8 in
20 cm

BASEBALL PLANT
Euphorbia obesa
F: Euphorbiaceae
Rare in the wild, this ball-shaped
succulent originates from the Great
Karoo, South Africa. Its tiny flowers
grow from "eyes" at its top.

20 ft
6 m

CROTON
Croton tiglium
F: Euphorbiaceae
Used in Chinese medicine,
this tree from S.E. Asia has
malodorous leaves and produces
fruit capsules containing three
poisonous seeds.

12 in
30 cm

130 ft
40 m

RUBBER TREE
Hevea brasiliensis
F: Euphorbiaceae
A native of Brazil, this tree is famous
for the milky sap (latex) under its
bark used for rubber. It has leaves
with three leaflets and pungent
yellow flowers.

16 in
40 cm

DOG'S MERCURY
Mercurialis perennis
F: Euphorbiaceae
This downy perennial has a
single upright stem. Plants are
either male or female and
produce tiny green flowers
on long, thin stalks.

16 in
40 cm

PETTY SPURGE
Euphorbia peplus
F: Euphorbiaceae
This poisonous, weedy annual,
native to Europe, N. Africa, and
W. Asia, has dividing stems and
flower stalks with three branches.

SAUSAGE SPURGE
Euphorbia guentheri
F: Euphorbiaceae
This evergreen succulent from
Kenya has white flower bracts
with purple markings. It has
fleshy, sickle-shaped leaves
mainly on young growth.

23½ in
60 cm

PERENNIAL FLAX
Linum perenne
F: Linaceae
This slender, often
spreading perennial is
native from C. Europe to
China. Its flowers form
buds at intervals at the
top of the stem.

10 ft
3 m

COCA
Erythroxylum coca
F: Erythroxylaceae
The leaves of this evergreen shrub
from northwestern S. America yield
cocaine. It has oval leaves and clusters
of small, yellowish white flowers.

FIVE-LOBED
LEAF AND
FLOWER

65 ft
20 m

BERRY WITH
NUMEROUS SEEDS

*leaves have twining
tendril at base*

BLUE PASSION FLOWER
Passiflora caerulea
F: Passifloraceae
This ornamental S. American vine has
scented flowers associated with
Christian symbolism. Flowers have ten
very similar sepals and petals, five
stamens, and three purple stigmas.

AYAHUASCA
Banisteriopsis caapi
F: Malpighiaceae

This woody vine, native to the Amazon, is traditionally used to make a sacred, medicinal drink. It produces stalks of pink flowers and winged seed pods.

33 ft
10 m

RED MANGROVE
Rhizophora mangle
F: Rhizophoraceae

Found throughout the tropics, especially in swampy salt marshes, the red mangrove has prop roots and its seeds germinate before leaving their parent tree.

80 ft
25 m

RAFFLESIA
Rafflesia arnoldii
F: Rafflesiaceae

This leafless parasite on S.E Asian rainforest vines has the world's largest flowers, up to 3¼ ft (1 m) in diameter. Their foul smell attracts flies.

23½ in
60 cm

PUSSY WILLOW
Salix caprea
F: Salicaceae

This shrubby tree, native to Europe and Asia, has downy, oval, toothed, alternate leaves. Female catkins form capsules, releasing cottony seeds.

80 ft
25 m

40 ft
12 m

IDESIA
Idesia polycarpa
F: Salicaceae

A mountain tree from E. Asia, this has smooth, gray bark. Small yellow-green, fragrant flowers produce deep red berries.

70 ft
21 m

WHITE WILLOW
Salix alba
F: Salicaceae

This waterside tree is native to Europe and Asia. Trees have either male or female catkins, never both. The bark is a source of salicin, the active ingredient of aspirin.

80 ft
25 m

GIANT GRANADILLA
Passiflora quadrangularis
F: Passifloraceae

flower is up to 5 in (12 cm) across

This passion flower is a perennial native of S. America. It bears oblong fruits on four-sided stems.

fragrant flower, may be white, red, or purple

ASPEN
Populus tremula
F: Salicaceae

Native to Europe and Asia, the gray bark of young trees has diamond-shaped scars. Flattened stalks cause the leaves of aspen to quiver.

WHITE POPLAR
Populus alba
F: Salicaceae

This deciduous tree, native to C. Europe and C. Asia, tolerates salt water. It is unisexual, and most trees are female. Catkins form capsules that release fluffy seeds.

100 ft
30 m

50 ft
15 m

AZARA MICROPHYLLA
F: Salicaceae

This evergreen tree from Argentina and Chile has small, vanilla-scented flowers with yellow stamens. Each leaf has a circular growth (stipule) at the base.

33 ft
10 m

WILD PANSY
Viola tricolor
F: Violaceae

This sprawling, short-lived perennial is native to Europe and W. Asia, growing in grassy habitats with neutral to acid soils. Used in herbal remedies, it is also known as heart's-ease.

12 in/30 cm

SHRUB VIOLET
Pigea floribundus
F: Violaceae

This Australian woody perennial accumulates nickel. Shrub violet has bluish petals with a yellow patch and small, dark green, lance-shaped leaves.

4 ft/1.2 m

OXALIDALES

This order contains about 2,000 species, grouped within seven families. Of these, the Cephalotaceae contains only one species—the carnivorous Albany pitcher plant. The Cunoniaceae are woody plants that produce woody fruit capsules containing small seeds. The wood sorrel family (Oxalidaceae) is the largest, with 800 species within five genera. They are characterized by divided leaves that exhibit or open out by day and then close up at night.

ALBANY PITCHER PLANT
Cephalotus follicularis
F: Cephalotaceae

This carnivorous plant is native to the coast of S.W. Australia. It has basal, oval leaves, and a liquid-filled pitcher that traps prey but it is not a member of the pitcher family (see p.194).

8 in
20 cm

ROSALES

This order of plants contains nine families, among them the Rosaceae (rose family), Cannabaceae (hemp family), Moraceae (mulberry family), Rhamnaceae (buckthorn family), Ulmaceae (elm family), and Urticaceae (nettle family). Members of the Rosales are often grown for their fruits or other products. Plants in this order tend to have five sepals and numerous stamens. Mostly insect-pollinated, they are often thorny or hairy.

flower has five petals

13 ft
4 m

CHRISTMAS BUSH
Ceratopetalum gummiferum
F: Cunoniaceae

From coastal E. Australia, this shrub produces insignificant white flowers in spring. The pink and red sepals enlarge in winter, enclosing the fruit.

14 in
35 cm

PINK SORREL
Oxalis articulata
F: Oxalidaceae

Growing from a swollen rhizome, or root stalk, this plant from S. America forms leafy mounds topped with clusters of flowers. These produce seed capsules that explode.

FLOWER CLUSTER

40 ft
12 m

BLACK WATTLE
Callicoma serratifolia
F: Cunoniaceae

The wood of this shrubby tree from the east coast of Australia was used by early settlers to make wattle-and-daub shelters. Its young foliage is bronze.

oil-rich berry

LEAFLET CLUSTER

RIPE FRUIT

FLOWER SPRIG

STAR FRUIT
Averrhoa carambola
F: Oxalidaceae

This bushy tree, native to S.E. Asia, is widely cultivated for its edible star-shaped fruits. It can flower four times a year.

50 ft/15 m

SEA BUCKTHORN
Hippophae rhamnoides
F: Elaeagnaceae

Widespread through Asia and Europe, this shrub bears tiny flowers before the leaves. The small, bright-orange berries are rich in vitamin C.

33 ft
10 m

20 ft / 6 m

OLEASTER
Elaeagnus angustifolia
F: Elaeagnaceae

Native to Eurasia, this spreading, deciduous tree has spiny shoots, covered in silver scales. Its egg-shaped, yellow-red fruits are edible.

6½ ft / 2 m

HEMP
Cannabis sativa
F: Cannabaceae

Originating from C. and W. Asia, this annual is a source of cannabis from its leaves, fiber for making rope, and oil is extracted from its seeds.

23 ft / 7 m

FEMALE PLANT AND FLOWERS

MALE PLANT AND FLOWERS

HOP
Humulus lupulus
F: Cannabaceae

This climbing perennial from N. Europe and N. Asia is widely naturalized elsewhere. The conelike female flower or hop is used to flavor and preserve beer.

65 ft / 20 m

young fruit

JACKFRUIT
Artocarpus heterophyllus
F: Moraceae

From lowland S.E. Asia, this tree yields the largest tree-borne fruit, weighing up to 110 lb (50 kg) and 36 in (90 cm) long when ripe.

43 ft / 13 m

BLACK MULBERRY
Morus nigra
F: Moraceae

This deciduous, spreading tree, widely cultivated for its rich-flavored fruits, is native to the Middle East. It has bumpy, fissured orange bark.

33 ft / 10 m

GLOSSY LEAF

FLESHY FRUIT

COMMON FIG
Ficus carica
F: Moraceae

As with most figs, flowers of this Mediterranean and Asian species develop inside a flask-shaped receptacle and are pollinated by a single species of wasp.

100 ft / 30 m

SACRED FIG
Ficus religiosa
F: Moraceae

Under this sacred tree, Buddha is said to have gained enlightenment. Native to S.E. Asia, its flowers, then its fruits, are enclosed within purple, flecked figs.

50 ft / 15 m

CONICAL FRUITS

SPRIG OF FEMALE FLOWERS

PAPER MULBERRY
Broussonetia papyrifera
F: Moraceae

Fine paper is made from the inner bark of this tree from Japan and S.E. China. Male catkins produce large quantities of wind-dispersed pollen.

65 ft / 20 m

LEAF SPRIG

INEDIBLE FRUIT

OSAGE ORANGE
Maclura pomifera
F: Moraceae

Native to the S.E. US, this tree is used for hedging and its roots and wood were valued by native Americans.

33 ft / 10 m

JUJUBE
Ziziphus jujuba
F: Rhamnaceae

This thorny, shrubby tree is widely cultivated for its fruits in China and Korea. The immature, smooth, oval, green stoned fruit tastes like apple.

COMMON BUCKTHORN
Rhamnus cathartica
F: Rhamnaceae

A shrubby, invasive tree that has spine-tipped shoots, tiny yellow-green flowers, and black berries. It is native to Europe, Asia, and N.W. Africa.

26 ft / 8 m

30 in / 75 cm

NEW JERSEY TEA
Ceanothus americanus
F: Rhamnaceae

This bush from eastern N. America has purple, three-lobed capsules, containing seeds. Its red roots and hairy leaves have been used to make tea.

≫

FLOWERING PLANTS • EUDICOTS

CHINESE PHOTINIA
Photinia serratifolia
F: Rosaceae
A plant of Chinese forests, this tree is commonly cultivated for ornamental purposes. Its dense wood is used in furniture making.

26 ft/8 m

SILVERWEED
Potentilla anserina
F: Rosaceae
This silky-haired, creeping perennial grows in wasteland, pastures, and dunes in Europe, Asia, and N. America, and has been widely introduced elsewhere.

32 in
80 cm

SNOWY MESPILUS
Amelanchier lamarckii
F: Rosaceae
Frothy clusters of star-shaped flowers in spring are followed by dark red berries in summer in this showy tree from eastern N. America.

40 ft/12 m

LEAVES AND
EDIBLE FRUITS

80 ft
25 m

SPRING
FLOWERS

WILD CHERRY
Prunus avium
F: Rosaceae
The wild ancestor of orchard cherries grows in woods and hedges in Europe, Asia, and N. Africa and is naturalized in N. America.

35 ft/11 m

PRAIRIE CRAB APPLE
Malus ioensis
F: Rosaceae
One of several crab apples native to N. America, this species is cultivated for its rather tart fruits.

SUMMER
FLOWER

EDIBLE
FRUITS

STRAWBERRY
Fragaria vesca
F: Rosaceae
The wild strawberry is a perennial in European and N. American woodlands. Tiny edible fruits form from the swollen flower receptacle.

12 in
30 cm

THREE-LOBED
LEAF

40 ft
12 m

ORCHARD PLUM
Prunus domestica
F: Rosaceae
These trees originated as a cross between the cherry plum of China and the European blackthorn, but without the sharp thorns of the latter.

double flower garden variety

6½ ft/2 m

JAPANESE ROSE
Rosa rugosa
F: Rosaceae
Tolerant of salt spray, this rose from E. Asia is often planted as a hedge near the sea. It has prickly stems and pink flowers with wrinkled petals.

WHITE OR
PINK FLOWERS

32 in
80 cm

10 ft
3 m

DOG ROSE
Rosa canina
F: Rosaceae
The thorny, arching branches of this rose are familiar in hedgerows in Europe, Asia, and N. Africa. This plant is also naturalized in N. America.

6½ ft
2 m

SWEET BRIAR
Rosa rubiginosa
F: Rosaceae
One of the most deeply colored wild roses, sweet briar enlivens hedgerows and scrub around Europe, Asia, and Africa. Glands on its leaves produce an apple scent when crushed.

fragrant, deep pink flower

APOTHECARY'S ROSE
Rosa gallica var. *officinalis*
F: Rosaceae
A popular garden plant, this European parent of many floribunda and hybrid tea roses has a long pedigree. "Attar of roses" is a fragrant oil distilled from its petals.

20 in / 50 cm

OAKLIKE LEAF

EIGHT-PETALED FLOWER

MOUNTAIN AVENS
Dryas octopetala
F: Rosaceae

The flowers of this Arctic and mountain undershrub turn to follow the sun to warm their centers and attract pollinating insects.

COTONEASTER
Cotoneaster horizontalis
F: Rosaceae

This creeping Chinese shrub is commonly grown in gardens for its flowers and fruits, and occasionally escapes. The semievergreen leaves form flattened clumps.

3¼ ft / 1 m

10 ft / 3 m

FIRETHORN
Pyracantha rogersiana
F: Rosaceae

This thorny evergreen shrub from E. China belongs to a genus often planted for their attractive but inedible, orange, berrylike fruits.

EDIBLE FRUITS

WHITE OR PINK FLOWERS

8 ft / 2.5 m

BLACKBERRY
Rubus fruticosus
F: Rosaceae

Familiar in Europe as scrambling hedgerow bushes bearing edible fruits in the fall, blackberries form a group of very closely related "microspecies."

80 ft / 25 m

WILD SERVICE TREE
Sorbus torminalis
F: Rosaceae

This rare tree grows in ancient woods in Europe, Asia Minor, and N. Africa. Its name is derived from the brew, *cerevisia*, made from its fruits.

many small flowers

AMERICAN MOUNTAIN ASH
Sorbus americana
F: Rosaceae

This deciduous woodland tree is native to eastern N. America. Its orange berries last into winter, providing food for thrushes and jays.

40 ft / 12 m

FIVE-PETALED WHITE FLOWER

20 ft / 6 m

MEDLAR
Mespilus germanica
F: Rosaceae

Originally from C. and S. Europe, medlar produces a hard, yellowish-brown fruit. This turns soft and edible in the fall, then decays.

FRUITS WITH REMAINS OF SEPALS

five-petalled white flower

SALAD BURNET
Sanguisorba minor
F: Rosaceae

Native from Europe to Iran, and introduced in N. America, this perennial of lime-rich grassland has edible leaves, hence its name.

TOOTHED LEAFLETS

FLOWER HEADS

23½ in / 60 cm

lobed leaf

WILLOW-LEAVED PEAR
Pyrus salicifolia
F: Rosaceae

Cultivated for its pendulous, silvery foliage, not its inedible fruits, this Middle Eastern tree is endangered in the wild in Turkey.

40 ft / 12 m

red fruits with single stem

showy five-petaled flower

23½ in / 60 cm

NODDING SUMMER FLOWERS

BURRLIKE FRUIT

WATER AVENS
Geum rivale
F: Rosaceae

This downy perennial grows in damp places in Europe, Asia Minor, and N. America. Hooked hairs on the fruits latch onto animals for dispersal.

23½ in / 60 cm

LADY'S MANTLE
Alchemilla vulgaris
F: Rosaceae

This name encompasses several closely related grassland species from Europe, Asia, and eastern N. America. Their leaves are said to resemble a trailing gown.

RED FRUITS, OR HAWS

COMMON HAWTHORN
Crataegus monogyna
F: Rosaceae

Occurring in woods and hedges in the wild from Europe and N. Africa to Afghanistan, this densely branched small tree produces massed white blossom in spring, followed by deep red fruits. It is also cultivated in gardens.

FRAGRANT BLOSSOM

52 ft / 16 m

>>

SWEETBRIAR
Rosa rubiginosa

Sweetbriar, or eglantine rose, is one of the most attractive of a range of wild roses that brighten hedgerows and scrub across Europe into Asia and Africa. They have five-petaled flowers that are usually white or pale pink. To nonbotanists, they seem rather similar, and are often called "dog roses." This derogatory name is said to be because they are less showy and strongly scented than the garden roses bred by horticulturalists over centuries, although some would argue they have a more subtle beauty and delicate scent than their fancy relatives. Many are prickly shrubs with scrambling, arched stems, for which "briar" is a descriptive term. Sweetbriar has deep pink flowers and leaves that emit the scent of apples when crushed. It has been introduced to the Americas and Australia, where it has become invasive.

SIZE Erect stems to 6½ ft (2 m) tall
HABITAT Open scrub, hedgerows, parks, and roadsides
DISTRIBUTION Europe to Asia, introduced elsewhere

prickly, spreading stem

⌄ DEVELOPING HIPS
The flower base swells to form a succulent hip, technically a "false fruit" because it is not formed from the ovary. This contains numerous, small, one-seeded "pips" which are the true fruits.

sepals remain as a leafy tuft

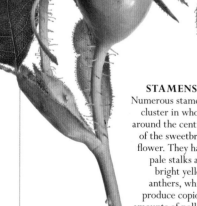

⌄ ROSEBUD
Sepals around the flower bud protect the developing petals and fertile parts. The stubble of glandular hairs may insulate the bud and repel pests.

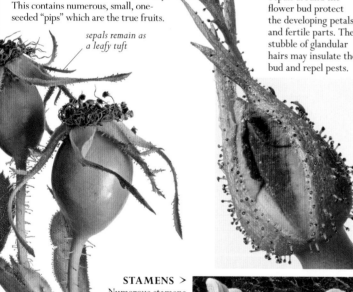

leaflet teeth fringed with stalked glands

⌄ UNDERSIDE OF LEAF
The underside of the leaf is densely hairy, mixed with many brownish, stalked glands. These release sweet-smelling terpene chemicals that may deter grazers.

STAMENS ＞
Numerous stamens cluster in whorls around the centrer of the sweetbriar flower. They have pale stalks and bright yellow anthers, which produce copious amounts of pollen.

THORNS ＞
Hooked thorns, interspersed with strong bristles, discourage large animals from grazing the leaves.

leaflet with saw-toothed edge

＜ RIPE HIPS
Hips shed their sepal tuft before ripening. Ripe, red hips are rich in vitamin C and are eaten by thrushes and pigeons, which spread the seeds in their droppings.

showy flower up to 1½ in (4 cm) in diameter

five overlapping petals, slightly notched at tip

5- or 7-lobed leaf, paler beneath

flower base

∧ SIDE VIEW
Five sepals spread beneath the petals. After pollination by bees or butterflies, the flower base swells to form a hip.

pollen-producing stamen

∧ THE SWEET SMELL OF SUCCESS
Scent plays different roles in plants, to attract pollinators or repel pests. We may enjoy the smell of sweetbriar leaves, but old herbals say leaf extracts were "harsh on the throat" and would "purge the chest," so the apple-scented terpenes in the leaves may well be distasteful to grazing animals.

cluster of sticky, pollen-receiving stigmas

≫ ROSALES

HACKBERRY
Celtis occidentalis
F: Cannabaceae
Native to N. America, hackberry
has bright green, elmlike leaves,
and red fruits eaten by many
birds and mammals.

130 ft
40 m

red berry

JAPANESE ZELKOVA
Zelkova serrata
F: Ulmaceae
In the US elms killed by Dutch
elm disease are sometimes
replaced by this valuable Asian
lumber tree. The Japanese grow
it in bonsai form.

100 ft
30 m

toothed leaf

EUROPEAN FIELD ELM
Ulmus minor
F: Ulmaceae
This tree used to be a major
feature in European
landscapes, but was decimated
by Dutch elm disease.

120 ft
36 m

FRIENDSHIP PLANT
Pilea involucrata
F: Urticaceae
Several *Pilea* species make valuable
houseplants, including this species
from C. and S. America with
strongly veined leaves.

12 in
30 cm

STINGING NETTLE
Urtica dioica
F: Urticaceae
Stinging hairs discourage animals
from eating nettle leaves. This species
grows in disturbed ground in Europe,
Asia, N. Africa, and N. America.

6½ ft/2 m

BRASSICALES

The order Brassicales has 17 families. Many members have bitter or
fragrant oils in their leaves, stems, or swollen roots. Although these oils
evolved to deter grazers, they make many species pleasantly edible for
humans, and have culinary, perfumery, or herbal uses. The cabbage
family (Brassicaceae), is the largest group, with 3,300 species.

WILD CABBAGE
Brassica oleracea
F: Brassicaceae
Humans have grown this
W. European plant for
millennia. Cauliflower,
broccoli, and brussels
sprouts are cultivated forms
of the same species.

3¼ ft
1 m

FLOWER HEAD

STALKLESS BLUE-
GREEN LEAF

male
flower

30 ft
10 m

PAPAYA
Carica papaya
F: Caricaceae
This C. and S. American
treelike plant produces yellow
flowers. The female flowers
develop into large orange-
fleshed fruits
of the same name.

HORSERADISH
TREE LEAF

FLOWER
STEM

30 ft/10 m

HORSERADISH TREE
Moringa oleifera
F: Moringaceae
This tropical Asian tree has corky gray
bark and fernlike leaves. A condiment
can be made from its crushed roots.

NASTURTIUM
Tropaeolum majus
F: Tropaeolaceae
A colorful annual from
C. and S. America, this is a
popular garden plant. The
flowers and leaves can be
eaten in salads.

10 ft/3 m

5 ft
1.5 m

CAPER
Capparis spinosa
F: Capparaceae
A perennial spiny bush, caper is
native to the Mediterranean. Its
flower buds are used salted and
pickled for culinary use.

20 in/50 cm

**COMMON
MIGNONETTE**
Reseda odorata
F: Resedaceae
Originally from N. Africa, common
mignonette is grown in gardens
across S. Europe. An oil from its
fragrant flowers is used in perfumery.

23½ in
60 cm

WALLFLOWER
Erysimum x *cheiri*
F: Brassicaceae
Probably first developed
as a hybrid in Greece, this
species has been cultivated
throughout Europe since
medieval times.

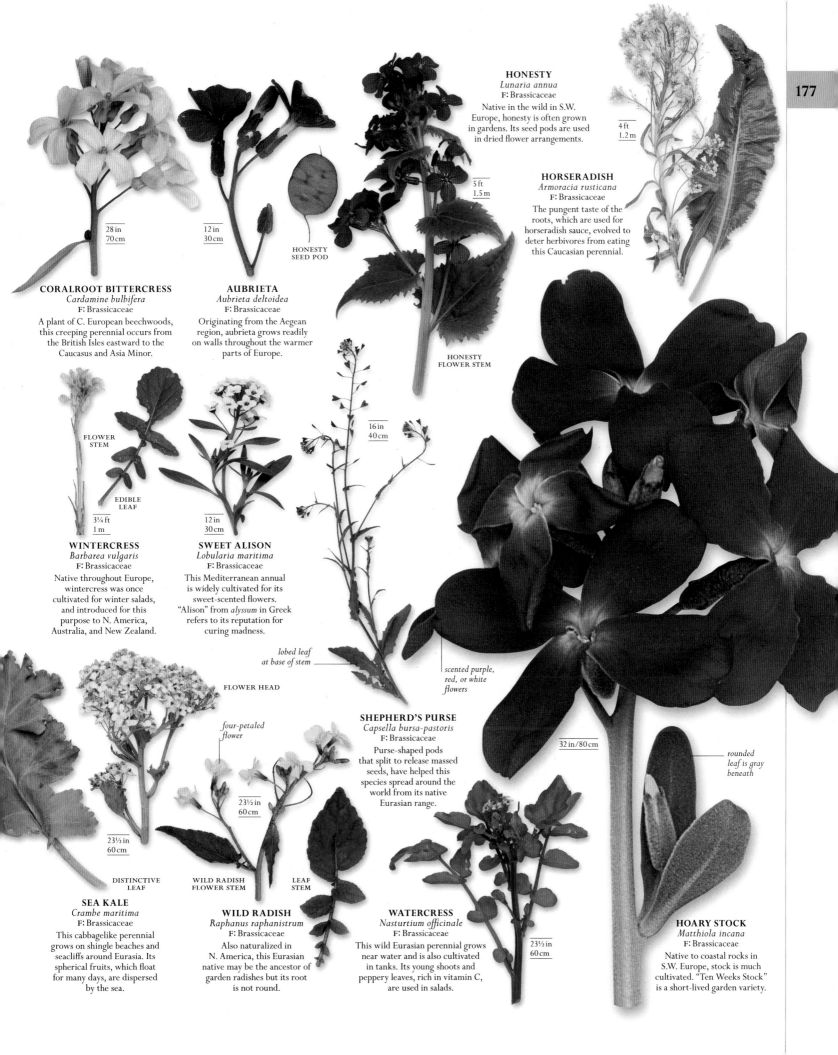

CORALROOT BITTERCRESS
Cardamine bulbifera
F: Brassicaceae
A plant of C. European beechwoods, this creeping perennial occurs from the British Isles eastward to the Caucasus and Asia Minor.

28 in
70 cm

AUBRIETA
Aubrieta deltoidea
F: Brassicaceae
Originating from the Aegean region, aubrieta grows readily on walls throughout the warmer parts of Europe.

12 in
30 cm

HONESTY
SEED POD

HONESTY
Lunaria annua
F: Brassicaceae
Native in the wild in S.W. Europe, honesty is often grown in gardens. Its seed pods are used in dried flower arrangements.

4 ft
1.2 m

HORSERADISH
Armoracia rusticana
F: Brassicaceae
The pungent taste of the roots, which are used for horseradish sauce, evolved to deter herbivores from eating this Caucasian perennial.

5 ft
1.5 m

HONESTY
FLOWER STEM

FLOWER
STEM

EDIBLE
LEAF

3¼ ft
1 m

WINTERCRESS
Barbarea vulgaris
F: Brassicaceae
Native throughout Europe, wintercress was once cultivated for winter salads, and introduced for this purpose to N. America, Australia, and New Zealand.

12 in
30 cm

SWEET ALISON
Lobularia maritima
F: Brassicaceae
This Mediterranean annual is widely cultivated for its sweet-scented flowers. "Alison" from *alyssum* in Greek refers to its reputation for curing madness.

16 in
40 cm

lobed leaf at base of stem

scented purple, red, or white flowers

FLOWER HEAD

four-petaled flower

SHEPHERD'S PURSE
Capsella bursa-pastoris
F: Brassicaceae
Purse-shaped pods that split to release massed seeds, have helped this species spread around the world from its native Eurasian range.

32 in/80 cm

rounded leaf is gray beneath

23½ in
60 cm

DISTINCTIVE
LEAF

WILD RADISH
FLOWER STEM

LEAF
STEM

23½ in
60 cm

SEA KALE
Crambe maritima
F: Brassicaceae
This cabbagelike perennial grows on shingle beaches and seacliffs around Eurasia. Its spherical fruits, which float for many days, are dispersed by the sea.

WILD RADISH
Raphanus raphanistrum
F: Brassicaceae
Also naturalized in N. America, this Eurasian native may be the ancestor of garden radishes but its root is not round.

WATERCRESS
Nasturtium officinale
F: Brassicaceae
This wild Eurasian perennial grows near water and is also cultivated in tanks. Its young shoots and peppery leaves, rich in vitamin C, are used in salads.

23½ in
60 cm

HOARY STOCK
Matthiola incana
F: Brassicaceae
Native to coastal rocks in S.W. Europe, stock is much cultivated. "Ten Weeks Stock" is a short-lived garden variety.

MALVALES

The ten families in the Malvales include many shrubs and trees, found mainly in tropical and warm temperate regions, but extending into cooler parts. The main members are the rockrose family (Cistaceae), mostly shrubs from the northern hemisphere; the more widespread mallow family (Malvaceae) of herbs, shrubs, and massive trees; and the pantropical Dipterocarpaceae, which includes some of the most important tropical lumber trees.

COMMON ROCKROSE
Helianthemum nummularium
F: Cistaceae
Found on sunny banks throughout most of Europe, this low shrub prefers lime-rich soils. Different subspecies grow in Europe's mountains.

20 in
50 cm

FLOWERING SPRIG

3¼ ft/1 m

HAIRY ROCKROSE
Cistus incanus
F: Cistaceae
Several scientific names have been given to this widespread Mediterranean shrub, because its leaves vary greatly in hairiness and size.

spiny fruit

33 ft/10 m

ANNATTO
Bixa orellana
F: Bixaceae
The food dye annatto comes from the spiny fruits of this pink-flowered shrub or small tree, native to tropical America.

13 ft/4 m

CALIFORNIAN FLANNELBUSH
Fremontodendron californicum
F: Malvaceae
The flowers of this spreading shrub produce showy masses in early summer. It lives high up in the granite mountains of California.

five-petaled rose-red flower

CHINESE HIBISCUS
Hibiscus rosa-sinensis
F: Malvaceae
Possibly native to India, despite its name, this tropical shrub is one of several *Hibiscus* species cultivated for their showy blossoms.

OBLONG LEATHERY LEAVES

RIBBED POD

40 ft
12 m

COCOA
Theobroma cacao
F: Malvaceae
Originally from Brazilian rainforests, this tree is cultivated around the tropics. Cocoa comes from the seeds, known as beans, inside its fruit pods.

15 ft/4.5 m

toothed leaf

6½ ft/2 m

strongly scented flowers

FRUITING STEM

32 in
80 cm

CAFFEINE-CONTAINING SEED

82 ft/25 m

NUTRITIOUS BAOBAB FRUIT

MATURE TREE

GLOSSY OVAL LEAVES

82 ft
25 m

MEZEREON
Daphne mezereum
F: Thymelaeaceae
Damp woods and shady gorges are the typical habitat of this deciduous shrub, which is found across most of Europe.

MUSK MALLOW
Malva moschata
F: Malvaceae
Native to N. Africa and S. Europe, and a garden plant farther north, this tall perennial grows in grassy and bushy places.

COLA NUT TREE
Cola nitida
F: Malvaceae
The caffeine-rich seeds (called cola nuts) from the pods of this W. African tree are chewed for their stimulant effects.

BAOBAB TREE
Adansonia digitata
F: Malvaceae
This massive African tree is called an upside-down tree because, when leafless, its branches look like roots. It can live for 1,500 years.

pollen-collecting stigma

pollen is attached to filaments on central stamen tube

TRAILING ABUTILON
Callianthe megapotamicum
F: Malvaceae
Native to dry mountain valleys of Brazil, this spreading shrub is also popular as a colorful ornamental in warm, sunny gardens.

6 ft
1.8 m

GLOSSY LEAVES

SPINY FRUIT

DURIAN
Durio zibethinus
F: Malvaceae
The spiny fruits of this Asian rainforest tree smell of sweaty socks. This attracts animals to eat the fruit then disperse the seeds in their droppings.

130 ft
40 m

120 ft
36 m

6 ft
1.8 m

AMERICAN LIME
Tilia americana
F: Malvaceae
This medium to large deciduous tree adds to fall colors in woodlands across eastern N. America. It often grows alongside sugar maple.

WILD HOLLYHOCK
Alcea pallida
F: Malvaceae
This tall evergreen perennial, a close relative of garden hollyhocks, comes from the E. Mediterranean. It grows in rocky places and scrubland.

FLOWER SPRIG

PALMATE LEAVES

230 ft
70 m

DOWN-FILLED POD

KAPOK
Ceiba pentandra
F: Malvaceae
Massive kapok trees grow wild in W. Africa, and C. and S. America. Fiber from their fruits, or pods, is used as a stuffing for toys.

OPEN BOLL

5 ft
1.5 m

UPLAND COTTON
Gossypium hirsutum
F: Malvaceae
This C. American shrub is the most common cultivated cotton species. The cotton fibers protect seeds inside the fruit, or boll.

SAPINDALES

The Sapindales is an important order of nine families, mostly of trees, shrubs, and woody vines, often with divided leaves. It includes many dominant woodland species and commercially important species such as citrus fruits. More than half its members belong to two families: the maple family (Sapindaceae), with about 1,900 species; and the rue family (Rutaceae), with 1,700 species, which mostly originate in Australia and South Africa.

EVERGREEN LEAVES

130 ft/40 m

RIPE FRUIT

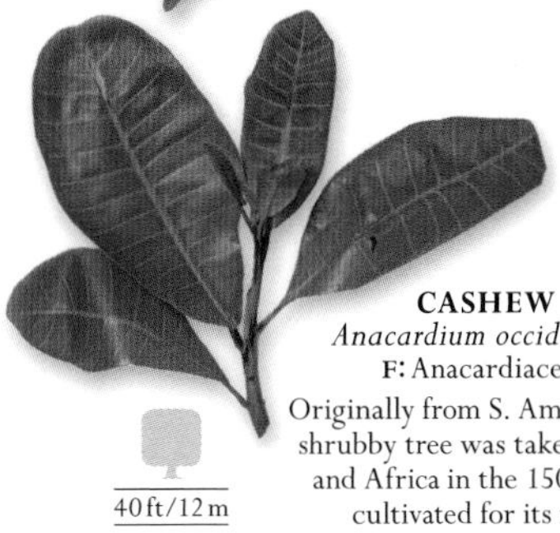

16 ft/5 m

CASHEW
Anacardium occidentale
F: Anacardiaceae
Originally from S. America, this shrubby tree was taken to Asia and Africa in the 1500s, and cultivated for its nuts.

40 ft/12 m

33 ft/10 m

STAG'S HORN SUMACH
Rhus typhina
F: Anacardiaceae
This deciduous shrub or small tree grows on forest edges and wasteland in eastern N. America. It has spiky clusters of red berries.

SMOKE BUSH
Cotinus coggygria
F: Anacardiaceae
Finely branched clusters of pale green flowers give this bush a smoky appearance. It is found in S. Europe and Asia.

MANGO
Mangifera indica
F: Anacardiaceae
A native of Asia, mango is one of the most widely cultivated fruits in the tropical world, and is rich in vitamin A.

>>

» SAPINDALES

CRABWOOD
Carapa guianensis
F: Meliaceae

180 ft
55 m

The dark lumber of this tropical S. American tree is sometimes sold as Brazilian mahogany. Soap is made from its seeds.

130 ft
40 m

RIPE FRUITS

PINNATE LEAVES

NEEM TREE
Azadirachta indica
F: Meliaceae

Valued for its lumber, medicinal oils, and edible shoots, this tree is grown throughout the Old World tropics. Indian farmers produce insecticides from the oil and leaves.

80 ft
25 m

CHINESE MAHOGANY
Toona sinensis
F: Meliaceae

Chinese people eat the leaves of this E. Asian tree as a vegetable. Furniture is made from its hard, reddish timber.

6½ ft
2 m

SCENTED BORONIA
Boronia megastigma
F: Rutaceae

This erect shrub of wet sandy sites in W. Australia has bell-like flowers, brownish on the outside and golden-green inside.

23½ in
60 cm

FRINGED RUE
Ruta chalepensis
F: Rutaceae

A native of rocky habitats in S. Europe and S.W. Asia, this is thought to be the rue mentioned in the Bible.

23½ in
60 cm

PEPPER AND SALT
Philoteca spicata
F: Rutaceae

Widespread in sandy and gravelly places in S.W. Australia, this low shrub has narrow leaves and pink, white, or bluish flowers.

5 ft
1.5 m

SYDNEY ROCKROSE
Boronia serrulata
F: Rutaceae

This small shrub, in a genus confined to Australia, grows in coastal heaths near Sydney. It has bright pink, cup-shaped flowers.

prominent midrib

6½ ft
2 m

MEXICAN ORANGE
Choisya ternata
F: Rutaceae

Originating from Mexico but common in gardens elsewhere, this irregular, bushy, evergreen shrub has branched clusters of sweet-scented, white flowers.

20 ft
6 m

HOP TREE
Ptelea trifoliata
F: Rutaceae

This small tree, native to eastern N. America, is grown for ornament. Its fruits were once used as a substitute for hops in brewing.

3¼ ft / 1 m

SALMON CORREA
Correa pulchella
F: Rutaceae

A native of South Australia, this small shrub is grown in gardens for its delicate, pendulous, tubular flowers.

33 ft
10 m

COMMON PRICKLY ASH
Zanthoxylum americanum
F: Rutaceae

This spiny N. American tree grows as far north as Quebec, Canada. Native Americans chewed its bark to help soothe toothaches.

6½ ft / 2 m

SKIMMIA
Skimmia japonica
F: Rutaceae

Much planted in gardens, parks, and amenity areas, this evergreen, aromatic shrub from E. Asia produces red berries in late summer.

GOLF BALL-SIZED FRUIT

26 ft / 8 m

JAPANESE BITTER ORANGE
Citrus trifoliata
F: Rutaceae

The small, inedible yellow fruits of this spiny shrub resemble oranges with a downy skin. They have several medicinal uses.

STEM

20 ft
6 m

EVERGREEN LEAVES

RIPENING FRUIT

LEMON
Citrus x *limon*
F: Rutaceae

Thought to have originated by hybridization in Assam or China, this evergreen tree is now widely cultivated for its fruits.

unripe fruit

30 ft
9 m

SEVILLE ORANGE
Citrus x *aurantium*
F: Rutaceae

Unlike the sweet orange we eat raw (*C.* x *sinensis*), this species' bitter fruits are only good for cooking. Both originated as Asian hybrids.

52 ft / 16 m

JAPANESE MAPLE
Acer palmatum
F: Sapindaceae
Centuries of breeding have produced many cultivated varieties of this native Japanese tree, with varying leaf shapes and spectacular fall colors.

SYCAMORE
Acer pseudoplatanus
F: Sapindaceae
Native to mountain woods in Europe and Asia, sycamore is widely planted elsewhere. Its winged seeds spread in the wind.

FLOWERING STEM

WINGED SEEDS

SUGAR MAPLE
Acer saccharum
F: Sapindaceae
This tree is native to N.E. US and S.E. Canada. Maple syrup is made by boiling the sap collected from the tree in spring.

115 ft / 35 m

GOLDEN-RAIN TREE
Koelreuteria paniculata
F: Sapindaceae
This showy tree from E. Asia is much planted in temperate regions for its cascading yellow blossoms and distinctive bladderlike seed pods.

40 ft / 12 m

100 ft / 30 m

CONKER IN SPIKY SHELL

young flowers have yellow "eye"

100 ft / 30 m

LYCHEE
Litchi chinensis
F: Sapindaceae
Probably originating in S. China, this tree is cultivated for its fruits. Their sweet flesh is encased within a tough shell.

130 ft / 40 m

WHITE SPRING FLOWERS

HORSE CHESTNUT
Aesculus hippocastanum
F: Sapindaceae
Native to S.E. Europe, this tree is often planted for shade along city streets. Its fruits resemble chestnuts but are hard and inedible.

130 ft / 40 m

BRIGHT GREEN LEAVES

EDIBLE NUTS

JAVA ALMOND
Canarium indicum
F: Burseraceae
Native to rainforests in the Pacific islands, this is one of the region's most useful trees, providing lumber, oil, and edible nuts.

toothed leaflet

26 ft / 8 m

65 ft / 20 m

redder eye of older flower

PINNATE LEAF

FLOWER SPIKE

26 ft / 8 m

26 ft / 8 m

YELLOWHORN
Xanthoceras sorbifolium
F: Sapindaceae
This small tree grows wild in China. Its scientific name records its leaves, which resemble those of rowan (*Sorbus*).

TREE OF HEAVEN
Ailanthus altissima
F: Simaroubaceae
Hailing from China, this slightly rank-smelling tree is planted along city streets because it endures pollution and most soil types.

BITTERWOOD
Quassia amara
F: Simaroubaceae
Boiled extracts from the bark and leaves of this tree are used to make tonics against malaria in tropical America.

FRANKINCENSE
Boswellia sacra
F: Burseraceae
Frankincense, a gum resin used in incense and to fix perfumes, oozes as milky juice from cuts in trunks of these Arabian trees.

CARYOPHYLLALES

The Caryophyllales is a diverse order of 38 families of trees, shrubs, climbers, succulents, and herbaceous plants, ranging from carnations to cacti. Many grow in difficult environments, but have evolved special adaptations for survival. Some have fleshy leaves to store water in drought conditions. Another extreme example is the ability of carnivorous species to trap and digest insects for extra nutrients.

2 in
5 cm

CONOPHYTUM MINUTUM
F: Aizoaceae

A minute, clump-forming species, with fleshy pebblelike leaves, this succulent perennial inhabits semidesert areas in South Africa.

1 in
2.5 cm

WARTY TIGER JAWS
Faucaria tuberculosa
F: Aizoaceae

Native to semidesert areas of South Africa, this species has warty leaves that resemble an open mouth, hence the name.

12 in
30 cm

PURPLE DEWPLANT
Disphyma crassifolium
F: Aizoaceae

This procumbent species—with stems that trail along ground—has fleshy leaves and daisylike flowers. Native to South Africa, Australia, and New Zealand, it lives on saline soils.

many-petaled yellow or light pink flower

fleshy leaf

4 in
10 cm

16 in
40 cm

SCHWANTESIA RUEDEBUSCHII
F: Aizoaceae

A clump-forming species with unequal pairs of keeled, fleshy leaves, this succulent grows on hillsides in Namibia and South Africa.

LAMPRANTHUS AURANTIACUS
F: Aizoaceae

This succulent plant grows in sandflats along the west coast of South Africa, where it is known locally as "vygie." Its daisylike flowers are about 2 in (5 cm) across.

1¼ in
3 cm

TITANOPSIS CALCAREA
F: Aizoaceae

The encrusted leaves of this South African succulent disguise it among desert limestone rocks. It flowers in late summer and fall.

12 in
30 cm

HOTTENTOT FIG
Carpobrotus edulis
F: Aizoaceae

This sprawling, fleshy South African species with showy flowers and edible figlike fruits grows in open dry habitats, where it can be invasive.

LIVING STONES
Lithops aucampiae
F: Aizoaceae

Growing among pebbles in semidesert regions of South Africa, this dwarf clump-forming species has bulbous leaves.

1¼ in
3 cm

paired leaves

carmine-red flower

12 in
30 cm

ICE PLANT
Mesembryanthemum crystallinum
F: Aizoaceae

Popularly named for the tiny glistening swellings covering the entire plant, this species inhabits saline areas in parts of Africa, Europe, and W. Asia.

4 in
10 cm

bulbous leaf

FAIRY ELEPHANT'S FEET
Frithia pulchra
F: Aizoaceae

This small succulent grows in montane grassland in South Africa. In drought its leaves shrink, pulling the plant into the soil to protect it from desiccation.

3¼ in
8 cm

GIBBAEUM VELUTINUM
F: Aizoaceae

Pairs of unequal-sized fleshy leaves, joined near their bases, characterize this mat-forming species. It is found in semidesert areas in South Africa.

LOVE-LIES-BLEEDING
Amaranthus caudatus
F:Amaranthaceae
Believed to originate in
S. America, this annual
plant has long been
cultivated for its edible
leaves and seeds.

8 ft/2.5 m

6½ ft
2 m

POLPALA
Aerva lanata
F:Amaranthaceae
With catkinlike flower clusters, this
perennial from tropical regions of
Asia and Africa grows on open,
disturbed ground.

6 ft
1.8 m

*panicle of
tiny flowers*

**ACHYRANTHES
BIDENTATA**
F:Amaranthaceae
Forest edges, streamsides, and
moist shady places in China,
Japan, India, and Nepal are
home to this species.

ANNUAL SEABLITE
Suaeda maritima
F:Amaranthaceae
Mainly coastal, inhabiting salt
marshes in Europe, but growing
inland in parts of Asia and
N. America, annual seablite has
green leaves which turn red.

12 in
30 cm

5 ft
1.5 m

HEART-
SHAPED
LEAF

TALL,
LEAFY
STEM

SEA BEET
Beta vulgaris
F:Amaranthaceae
The wild ancestor of the
beet, sea beet is a fleshy
inhabitant of bare
ground by the sea in
parts of Europe,
N. Africa, and Asia.

keel, or ridge

3¼ ft
1 m

SEA PURSLANE
Atriplex portulacoides
F:Amaranthaceae
Salt marshes, especially edges
of tidal channels and pools, are
home to sea purslane, a sprawling,
silvery species from S. Europe
and the Mediterranean.

12 in
30 cm

COMMON GLASSWORT
Salicornia europaea
F:Amaranthaceae
Common glasswort is a bushy
species of muddy salt marshes
in W. Europe. Its succulent
stems are sometimes cooked
as a vegetable.

*white, pink, or
lilac flower*

SPIKE OF REDDISH
FLOWERS

EDIBLE
SALAD
LEAF

MEXICAN TEA
Dysphania ambrosioides
F:Amaranthaceae
A short-lived inhabitant of
cultivated and waste ground
in tropical America, this
aromatic species is used as
a seasoning and in tea.

3¼ ft
1 m

GARDEN ORACHE
Atriplex hortensis
F:Amaranthaceae
Probably originally from
seashores in S.W. Asia, this
spinachlike species has
long been cultivated for its
edible leaves.

4 ft
1.2 m

6½ ft
2 m

paired fleshy leaf

PLUMED COCKSCOMB
Celosia argentea
F:Amaranthaceae
This showy plant is native to dry
slopes and stony ground in tropical
Africa, but is widely planted
elsewhere for its colorful foliage.

»

≫ CARYOPHYLLALES

PARODIA HASELBERGII
F: Cactaceae

This cactus has a spherical stem and funnel-shaped flowers. It inhabits mountainous parts of Brazil.

6 in
15 cm

spines protect slow-growing plant

ERIOSYCE SUBGIBBOSA
F: Cactaceae

This spherical species grows in dry, stony places, often on the coast in its native Chile.

35 in
90 cm

funnel-shaped flower

23 ft
7 m

PERUVIAN OLD MAN CACTUS
Espostoa lanata
F: Cactaceae

Identified by long, white hairs on the columnar stem, this is a slow-growing species from the hilly areas of Peru and S. Ecuador.

16 in
40 cm

BARREL CACTUS
Echinocactus sp.
F: Cactaceae

The barrel-shaped members of this genus are confined to the deserts of S.W. US and Mexico.

flesh in leaf stores water

4 in/10 cm

23½ in
60 cm

CLEISTOCACTUS BROOKEAE
F: Cactaceae

With semi-erect or spreading fleshy single stems, this species inhabits mountainous areas of Bolivia.

23½ in/60 cm

LEUCHTENBERGIA PRINCIPIS
F: Cactaceae

The stem of this species is either spherical or short and cylindrical, and bears fragrant flowers. It inhabits semidesert regions of N. Mexico.

OLD MAN CACTUS
Cephalocereus senilis
F: Cactaceae

Long white hairs on the stem give this cactus its common name. It is native to rocky areas in Mexico.

40 ft
12 m

REBUTIA HELIOSA
F: Cactaceae

A clump-forming species with brightly colored flowers, this native of Bolivia grows in partially shaded, mountainous habitats.

GOLDEN COLUMN
Weberbauerocereus johnsonii
F: Cactaceae

This is a tall species from Peru, where it inhabits sandy soils.

20 ft
6 m

52 ft
16 m

arm, or branch, supports extra flowers

SAGUARO CACTUS
Carnegiea gigantea
F: Cactaceae

Living for up to 150 years in desert areas of Mexico, Arizona, and California, saguaro cactus is an extremely tall species.

13 ft/4 m

MISTLETOE CACTUS
Rhipsalis baccifera
F: Cactaceae

The only cactus found naturally outside the Americas, this epiphyte was perhaps carried to tropical Africa, Madagascar, and Sri Lanka by migrating birds.

7¾ ft
2.4 m

62 ft
19 m

GIANT CARDÓN
Pachycereus pringlei
F: Cactaceae

The tallest living cactus species, giant cardón grows in the desert of Baja California, Mexico. Its flowers open at night.

QUEEN OF THE NIGHT
Harrisia jusbertii
F: Cactaceae

Possibly of natural hybrid origin in the hills of Argentina and Paraguay, this columnar cactus opens its flowers at night, hence its name.

LOPHOCEREUS SCHOTTII
F: Cactaceae

A tall, slow-growing species from Mexico and S. Arizona, this species has unpleasant-smelling flowers that open at night.

16 ft
5 m

12 in
30 cm

MATUCANA INTERTEXTA
F: Cactaceae

A mountain valley region of Peru is the only home of this clump-forming species with a spherical or short cylindrical stem.

MONK'S HOOD CACTUS
Astrophytum ornatum
F: Cactaceae

A spherical or columnar stem with long, brownish yellow protective spines characterizes this species. It grows in arid areas of Mexico.

28 in/70 cm

CLARET CUP HEDGEHOG
Echinocereus triglochidiatus
F: Cactaceae

Growing in deserts, scrub, and on rocky slopes in S. US and N. Mexico, this variable cactus is pollinated by hummingbirds.

flower bud

RAT'S-TAIL CACTUS
Aporocactus flagelliformis
F: Cactaceae

Living on trees or rocks in wooded areas of Mexico, rat's-tail cactus has trailing fleshy stems and colorful flowers.

5 ft
1.5 m

long, thin, trailing stem

8 in
20 cm

TURK'S CAP CACTUS
Melocactus salvadorensis
F: Cactaceae

With a characteristic flower-bearing structure on top of a spherical stem when mature, turk's cap cactus inhabits open, rocky ground in N.E. Brazil.

8 in
20 cm

OLD LADY CACTUS
Mammillaria hahniana
F: Cactaceae

Growing in semidesert regions, old lady cactus is a Mexican species with grayish hairs on the spherical stem.

BRAIN CACTUS
Stenocactus multicostatus
F: Cactaceae

This spherical species inhabits dry grasslands in N.E. Mexico. Funnel-shaped buds open into pink striped flowers.

4 in
10 cm

GYMNOCALYCIUM HORSTII
F: Cactaceae

This spherical, clump-forming cactus probably grows wild only in rocky grassland near Rio Grande do Sul in Brazil.

15 cm
6 in

12 in
30 cm

CRAB CACTUS
Schlumbergera truncata
F: Cactaceae

An epiphytic winter-flowering species of tropical rainforest in S.E. Brazil, crab cactus is often grown as a houseplant.

newly formed prickly pear fruit

16 ft/5 m

BRIGHT YELLOW FLOWER

PADDLELIKE GREEN STEM

PRICKLY PEAR
Opuntia ficus-indica
F: Cactaceae

With flattened jointed stems and edible egg-shaped fruits, prickly pear is a native of rocky hillsides and dry ground in Mexico, and is widely naturalized elsewhere.

cluster of spines

8 in
20 cm

GLORY OF TEXAS
Thelocactus bicolor
F: Cactaceae

Native to arid regions of Texas and N.E. Mexico, this is a spherical species.

»

∨ **SPIKY HAIRY STAR**
Viewed from above, this cactus
appears star shaped, with five to ten
(usually eight) outstretched ribs, and
crossbands of woolly, white scales
across the surface between them.

*clusters of 5—11
spines grow from
a small hollow, or
areole, surrounded
by white hair*

MONK'S HOOD CACTUS
Astrophytum ornatum

The genus name for this cactus, *Astrophytum*, means "star plant" in ancient Greek. It was first collected in 1827 by Thomas Coulter, an Irish doctor and botanist, who sent it to Professor de Candolle at the Botanic Garden of Geneva. When de Candolle unpacked the cactus he thought it was covered in fungus, but discovered that the white spots were actually tufts of hair or trichomes. These woolly scales may help the cactus collect water and protect it against the sun, but they also provide camouflage. This species is the most densely covered of all the *Astrophytum* cacti, and also the spiniest. It is now rare in the wild.

SIZE 5 ft (1.5 m)
HABITAT Semidesert scrub
DISTRIBUTION N.E. Mexico
LEAF TYPE Spines

spine protects the slow-growing plant from being eaten

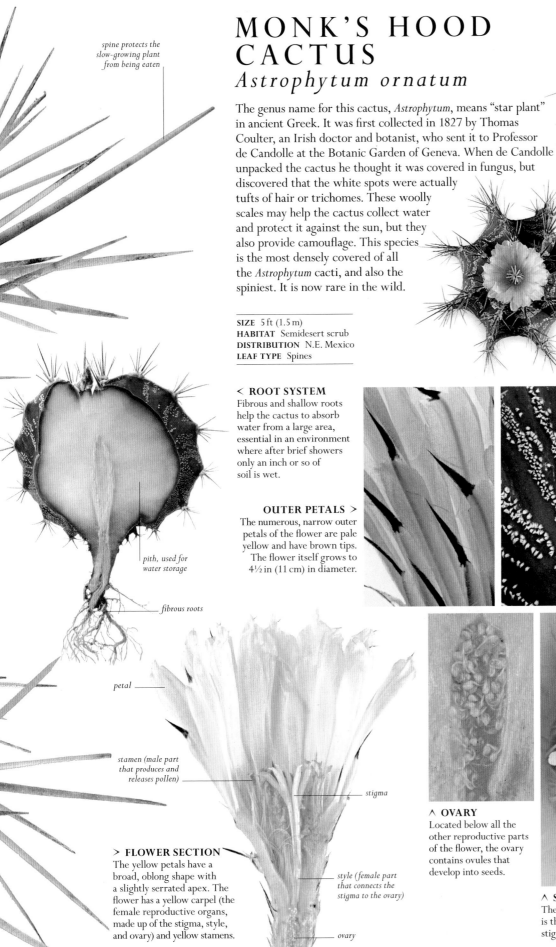

pith, used for water storage

fibrous roots

petal

stamen (male part that produces and releases pollen)

< ROOT SYSTEM
Fibrous and shallow roots help the cactus to absorb water from a large area, essential in an environment where after brief showers only an inch or so of soil is wet.

OUTER PETALS >
The numerous, narrow outer petals of the flower are pale yellow and have brown tips. The flower itself grows to 4½ in (11 cm) in diameter.

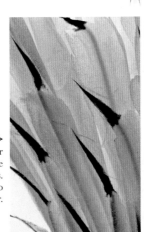

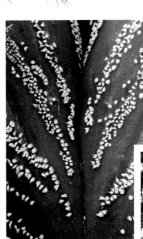

< ∨ WOOLLY SCALES
This cactus has white, scalelike tufts of trichomes arranged in bands between its ribs. They are densely woolly in young plants but become sparse with age.

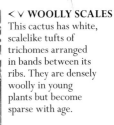

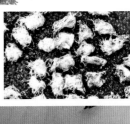

> FLOWER SECTION
The yellow petals have a broad, oblong shape with a slightly serrated apex. The flower has a yellow carpel (the female reproductive organs, made up of the stigma, style, and ovary) and yellow stamens.

stigma

style (female part that connects the stigma to the ovary)

ovary

∧ OVARY
Located below all the other reproductive parts of the flower, the ovary contains ovules that develop into seeds.

∧ STIGMA
The flower has a single stigma with 7–12 lobes. This is the part of the flower used to catch pollen. The stigma is about ½ in (1.5 cm) long.

23½ in / 60 cm

DEPTFORD PINK
Dianthus armeria
F: Caryophyllaceae
Inhabiting dry grassland, especially on light, sandy soils in much of Europe, Deptford pink has starry flowers with toothed petals.

3¼ ft / 1 m

CORNCOCKLE
Agrostemma githago
F: Caryophyllaceae
A native of E. Mediterranean countries, corncockle was formerly a widespread cornfield weed, but is now much rarer.

23½ in / 60 cm

COW BASIL
Vaccaria hispanica
F: Caryophyllaceae
An increasingly rare weed of grain fields in Europe and Asia, this annual herb has pink flowers and bluish green leaves.

10 in / 25 cm

SEA SANDWORT
Honckenya peploides
F: Caryophyllaceae
A fleshy, prostrate species, sea sandwort is found on coastal sand and shingle in Europe, Asia, and N. America.

12 in / 30 cm

THYME-LEAVED SANDWORT
Arenaria serpyllifolia
F: Caryophyllaceae
Found in bare or disturbed ground in Europe, temperate Asia, and N. America, this species gets its common name from its tiny leaves.

ALPINE CAMPION
Viscaria alpina
F: Caryophyllaceae
This native of the European Alps, Pyrenees, and subarctic areas of Europe, W. Asia, and N. America grows on mineral-rich rocks.

32 in / 80 cm

12 in / 30 cm

pink, or occasionally white, flower

flower stalk lengthens as flower matures

8 in / 20 cm

four or five small green leaves per shoot

4 in / 10 cm

FIVE-PETALED FLOWER

LANCEOLATE LEAVES

RAGGED ROBIN
Silene flos-cuculi
F: Caryophyllaceae
Divided petals give ragged robin its name. It is a European species found in marshy fields and damp ground.

FIELD MOUSE-EAR
Cerastium arvense
F: Caryophyllaceae
This is a native of dry grassland in Europe, N. Africa, N. America, and W. temperate Asia.

inflated calyx

32 in / 80 cm

central cushion has long tap root

MOSS CAMPION
Silene acaulis
F: Caryophyllaceae
Cushion-forming and mosslike, moss campion grows on mountains in W., C., and N. Europe, as well as Asia, and N. America.

BLADDER CAMPION
Silene vulgaris
F: Caryophyllaceae
Named for its inflated calyx, bladder campion inhabits open grassy places in Europe, N. Africa, and temperate regions of Asia.

sheath around flower

20 in
50 cm

CHILDING PINK
Petrorhagia nanteuilii
F: Caryophyllaceae
The flowers of childing pink open one at
a time. This species inhabits dry, grassy
places on sandy soils in W. Europe.

3¼ ft
1 m

SOAPWORT
Saponaria officinalis
F: Caryophyllaceae
Once used to make soap, this species
is found along streams and damp
ground in S. Europe and W. Asia.

20 in
50 cm

COMMON CHICKWEED
Stellaria media
F: Caryophyllaceae
This sprawling species is found on cultivated
and open ground worldwide and is
sometimes used as a salad vegetable.

**ROUND-LEAVED
SUNDEW**
Drosera rotundifolia
F: Droseraceae
Inhabiting open, boggy
moorland and heathland
in Europe, N. Asia, and
N. America, this
insectivorous species has
leaves with numerous
sticky hairs covered
in insect-dissolving
enzymes.

*leaf hairs
trap insects*

4 in
10 cm

*white flowers
surrounded by bracts*

*brightly colored
bract*

*trap triggered by hairs
on leaf surface*

VENUS FLY-TRAP
Dionaea muscipula
F: Droseraceae
An insectivorous perennial
species from coastal bogs in
N. and S. Carolina, Venus
fly-trap has hinged,
two-lobed leaves.

26 ft/8 m

*when triggered,
leaf snaps shut,
trapping insect*

6 in
15 cm

*"teeth" prevent
insect from escaping*

BOUGAINVILLEA
Bougainvillea glabra
F: Nyctaginaceae
What seem to be petals on
bougainvillea are actually
bracts. This slow-growing
evergreen climber from Brazil
is widely cultivated.

*nectar glands on
lid attract insects*

*insects slither
off slippery rim
into pitcher*

40 ft
12 m

12 in/30 cm

SEA HEATH
Frankenia laevis
F: Frankeniaceae
A mat-forming species, sea heath
grows on bare sandy ground in
the drier parts of salt marshes in
W. Europe and N.W. Africa.

*fluid in base of pitcher
digests insects*

FOUR O'CLOCK FLOWER
Mirabilis jalapa
F: Nyctaginaceae
From dry, open habitats in
tropical C. and S. America, this
species has fragrant flowers which
open during the late afternoon.

3¼ ft/1 m

PITCHER PLANT
Nepenthes stenophylla
F: Nepenthaceae
In the forests of Borneo,
this scrambling vine gets
extra nutrients by trapping
insects in its pitchers.

*pitcher
grows from
tendril
at end
of leaf*

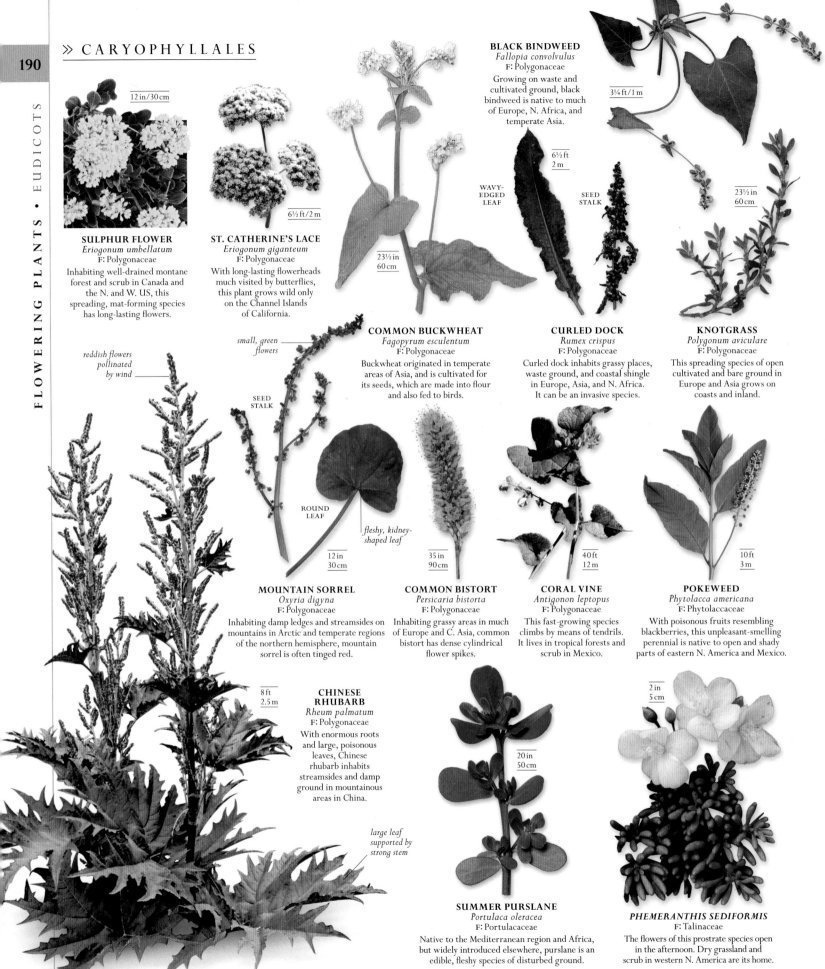

BLACK BINDWEED
Fallopia convolvulus
F: Polygonaceae
Growing on waste and cultivated ground, black bindweed is native to much of Europe, N. Africa, and temperate Asia.

3¼ ft/1 m

6½ ft
2 m

WAVY-EDGED LEAF

SEED STALK

23½ in
60 cm

12 in/30 cm

SULPHUR FLOWER
Eriogonum umbellatum
F: Polygonaceae
Inhabiting well-drained montane forest and scrub in Canada and the N. and W. US, this spreading, mat-forming species has long-lasting flowers.

6½ ft/2 m

ST. CATHERINE'S LACE
Eriogonum giganteum
F: Polygonaceae
With long-lasting flowerheads much visited by butterflies, this plant grows wild only on the Channel Islands of California.

23½ in
60 cm

small, green flowers

COMMON BUCKWHEAT
Fagopyrum esculentum
F: Polygonaceae
Buckwheat originated in temperate areas of Asia, and is cultivated for its seeds, which are made into flour and also fed to birds.

CURLED DOCK
Rumex crispus
F: Polygonaceae
Curled dock inhabits grassy places, waste ground, and coastal shingle in Europe, Asia, and N. Africa. It can be an invasive species.

KNOTGRASS
Polygonum aviculare
F: Polygonaceae
This spreading species of open cultivated and bare ground in Europe and Asia grows on coasts and inland.

reddish flowers pollinated by wind

SEED STALK

ROUND LEAF

fleshy, kidney-shaped leaf

12 in
30 cm

35 in
90 cm

40 ft
12 m

10 ft
3 m

MOUNTAIN SORREL
Oxyria digyna
F: Polygonaceae
Inhabiting damp ledges and streamsides on mountains in Arctic and temperate regions of the northern hemisphere, mountain sorrel is often tinged red.

COMMON BISTORT
Persicaria bistorta
F: Polygonaceae
Inhabiting grassy areas in much of Europe and C. Asia, common bistort has dense cylindrical flower spikes.

CORAL VINE
Antigonon leptopus
F: Polygonaceae
This fast-growing species climbs by means of tendrils. It lives in tropical forests and scrub in Mexico.

POKEWEED
Phytolacca americana
F: Phytolaccaceae
With poisonous fruits resembling blackberries, this unpleasant-smelling perennial is native to open and shady parts of eastern N. America and Mexico.

8 ft
2.5 m

CHINESE RHUBARB
Rheum palmatum
F: Polygonaceae
With enormous roots and large, poisonous leaves, Chinese rhubarb inhabits streamsides and damp ground in mountainous areas in China.

large leaf supported by strong stem

2 in
5 cm

20 in
50 cm

SUMMER PURSLANE
Portulaca oleracea
F: Portulacaceae
Native to the Mediterranean region and Africa, but widely introduced elsewhere, purslane is an edible, fleshy species of disturbed ground.

PHEMERANTHIS SEDIFORMIS
F: Talinaceae
The flowers of this prostrate species open in the afternoon. Dry grassland and scrub in western N. America are its home.

LEWISIA BRACHYCALYX
F: Montiaceae

Native to damp, stony meadows in mountainous areas of S.W. US, this species has a basal rosette of fleshy leaves.

1½ in
4 cm

12 in
30 cm

SPRING BEAUTY
Claytonia perfoliata
F: Montiaceae

Also known as miner's lettuce, this native of cultivated and bare ground in western N. America, Mexico, and Guatemala has two fused leaves beneath the flowers.

FRINGED REDMAIDS
Calandrinia ciliata
F: Montiaceae

Native to grasslands from S. Canada to Argentina, this species has escaped to the Falkland Islands and is also grown in gardens.

12 in
30 cm

CEYLON LEADWORT
Plumbago zeylanica
F: Plumbaginaceae

Growing mainly in open, human-created habitats, this scrambling shrub is found throughout tropical and subtropical regions.

5 ft
1.5 m

STATICE
Limonium sinuatum
F: Plumbaginaceae

Statice has winged, branching stems and is found in rocky and sandy coastal habitats and in saline areas inland in Mediterranean countries.

16 in
40 cm

SEA PINK
Armeria maritima
F: Plumbaginaceae

In much of W. Europe, thrift, or sea pink, inhabits coastal rocks and cliffs, salt marshes, and mountains. It is a cushion-forming species.

10 in
25 cm

TAMARISK
Tamarix gallica
F: Tamaricaceae

Usually coastal, tamarisk also grows in saline areas inland. It originates from S. Europe, N. Africa, and the Canary Islands, and is planted elsewhere.

10 ft
3 m

JOJOBA
Simmondsia chinensis
F: Simmondsiaceae

Deserts in Arizona, California, and Mexico are home to jojoba, a species often cultivated for its oil.

6½ ft
2 m

SANTALALES

Found mainly in the tropics and subtropics, the seven families in the order Santalales include several important lumber trees. They also feature many parasitic and semiparasitic plants, notably the 900 species in the showy mistletoe family (Loranthaceae) of the southern hemisphere. These live attached to other plants, from which they obtain all or most of the water and nutrients they need for growth.

FIRE TREE
Nuytsia floribunda
F: Loranthaceae

This species is semiparasitic, deriving moisture and nutrients from the roots of surrounding plants in woodland in S.W. Australia.

33 ft/10 m

berry can be poisonous

3¼ ft
1 m

MISTLETOE
Viscum album
F: Santalaceae

Mistletoe forms spherical masses on tree branches. A semiparasitic species with white berries, it inhabits much of Europe, N. Africa, and Asia.

30 ft
9 m

SANDALWOOD
Santalum album
F: Santalaceae

Cultivated for its wood and fragrant oil, this semiparasitic tree grows in dry, rocky areas in Asia and Australia.

OSYRIS
Osyris alba
F: Santalaceae

This broomlike, semiparasitic species with fragrant flowers is native to S. Europe, N. Africa, and S.W. Asia, inhabiting dry, rocky places.

4 ft
1.2 m

CORNALES

Seven families constitute the revised order Cornales. Five of these are small and relatively insignificant, one with just a single species of evergreen tree from Africa. The main family is the dogwood family (Cornaceae), a loose taxonomic grouping of shrubs and small trees found in temperate zones and on tropical mountains. The hydrangea family (Hydrangeaceae) includes several popular garden plants.

HYDRANGEA
Hydrangea macrophylla
F: Hydrangeaceae
Native to Japan, this showy shrub is cultivated for its rounded clusters of rose, lavender, blue, and even white flowers.

5 ft / 1.5 m

FLOWERING DOGWOOD
Cornus florida
F: Cornaceae
The showy white "petals" of this N. American dogwood tree are bracts (specialized leaves) around clusters of tiny flowers.

40 ft
12 m

DOVE TREE
Davidia involucrata
F: Nyssaceae
This small, flowering tree from China is impressive in bloom, with flowerheads ¾ in (2 cm) across, surrounded by creamy bracts.

80 ft
25 m

MOCK ORANGE
Philadelphus sp.
F: Hydrangeaceae
About 45 species of this shrubby genus, named for their orange-blossom fragrance, are native to Asia, western N. America, and Mexico.

15 ft
4.5 m

ERICALES

Comprising 22 families, the Ericales is a major order, both economically and as one of the world's main herbaceous groups. It is named after its largest member, the heather family (Ericaceae), with more than 4,000 species of flowering shrubs mainly in acid soils. It also includes 900 species in the primrose family (Primulaceae), largely confined to north temperate mountain regions; herbs and shrubs in the Jacob's ladder family (Polemoniaceae); and carnivorous plants in the pitcher plant family (Sarraceniaceae).

3¼ ft
1 m

JACOB'S LADDER
Polemonium caeruleum
F: Polemoniaceae
This tall perennial has cup-shaped lavender or white flowers. It grows in rocky and grassy places in Europe and N. Asia.

3¼ ft / 1 m

PERENNIAL PHLOX
Phlox paniculata
F: Polemoniaceae
A pyramidal cluster of trumpet-shaped, pink or lavender flowers top the stem of this perennial, native to open woods in the S.E. US.

HEATHER
Calluna vulgaris
F: Ericaceae
This evergreen shrub, with spikes of pale purple flowers, dominates vast areas of moorland in N. Europe, eastward into Asia.

3¼ ft
1 m

DORSET HEATH
Erica ciliaris
F: Ericaceae
One of several similar spiky ericas, this species is known from S. England (including Dorset), as well as W. Ireland, France, Iberia, and Morocco.

23½ in
60 cm

STRAWBERRY TREE
Arbutus unedo
F: Ericaceae
Although related to heathers, the warty, red fruits of this evergreen Mediterranean tree look remarkably like strawberries but are tasteless.

40 ft
12 m

scarlet, bell-shaped flower

100 ft / 30 m

TREE RHODODENDRON
Rhododendron arboreum
F: Ericaceae
There are about 1,000 species of *Rhododendron*, most with leathery, evergreen leaves and showy flowers. Many, like this one, come from the Himalayas.

WHITE, BELL-SHAPED FLOWERS

16 ft
5 m

12 in
30 cm

ACIDIC BERRIES

COWBERRY
Vaccinium vitis-idaea
F: Ericaceae
This neat heather, with leathery leaves and glossy red berries in the fall, is found across N. Europe, Asia, and N. America.

HIMALAYAN PIERIS
Pieris formosa
F: Ericaceae
Formosa means "beautiful," a reference to the hanging clusters of white, urn-shaped flowers on this Asian bush or small tree.

thick trunk covered in small leaves

6 in
15 cm

CHECKERBERRY
Gaultheria procumbens
F: Ericaceae
This creeping, aromatic shrub forms patches beneath oaks and conifers in eastern N. America. Its red fruits last through the winter.

10 ft
3 m

SWEET PEPPER BUSH
Clethra alnifolia
F: Clethraceae
This deciduous shrub grows in wet forests and bogs in eastern N. America. Its leaves turn yellow or orange in the fall.

DECIDUOUS LEAVES

ORANGE FRUITS

65 ft
20 m

PERSIMMON
Diospyros virginiana
F: Ebenaceae
This native N. America tree has yellowish white, bell-shaped flowers and produces globular orange fruits, about 1½ in (4 cm) across.

6½ ft / 2 m

INDIAN BALSAM
Impatiens glandulifera
F: Balsaminaceae
Exploding fruit pods, which shoot their seeds outward, have helped this Himalayan species become established by rivers in most of Europe.

65 ft
20 m

ROUND, WOODY FRUIT

33 ft
10 m

FRUITS

black seeds

green flesh

OPEN FRUIT

KIWI
Actinidia chinensis
F: Actinidiaceae
Also known as Chinese gooseberry, this woody climber was introduced from China to New Zealand, where the berries were sold as "kiwi fruits."

165 ft
50 m

BRAZIL NUT
Bertholletia excelsa
F: Lecythidaceae
Brazil "nuts" are actually seeds inside this S. American tree's hard, cannonball-like fruits. In the wild, agoutis (large rodents) open the fallen fruits.

MULTIPLE "NUTS" INSIDE

BOOJUM TREE
Fouquieria columnaris
F: Fouquieriaceae
This strange tree is mainly confined to Baja, California. Its stem and upright, spiny branches are green and can photosynthesize.

≫

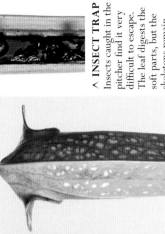

∧ **INSECT TRAP**
Insects caught in the pitcher find it very difficult to escape. The leaf digests the soft parts, but the skeletons remain.

FLOWER ∨
The flowers have five sepals, five yellow petals (which redden as they mature), and a whitish, umbrella-like style. They hang upside down.

∧ **"WINDOWS"**
Insects fly toward the bright translucent "windows" or areoles on the upper part of the pitcher, but then slide down into the tube.

SEED CAPSULE ∨
The broad, rough capsules crack open and scatter many tiny knobbly seeds, which are about ⅛in (3 mm) long.

RHIZOME ∨
The plant's domed pitchers rise up from a horizontally branching, underground stem known as a rhizome.

ROOTS >
The fibrous roots are 8–12 in (20–30 cm) long and grow along the length of the rhizome.

HOODED PITCHER PLANT
Sarracenia minor

This insectivorous plant is highly adapted to living in nutrient-poor, acidic swamps and bogs, digesting insects to obtain phosphorus and nitrogen from their bodies. Insects are attracted by nectar into the pitcher, the long, thin, tubular, modified leaf from which the plant gets its name. The interior of the pitcher is covered in a slippery, waxy coating and downward-pointing hairs that become coarser nearer the bottom of the tube. The insects fall in and, unable to get out, eventually die from exhaustion. Digestive glands in the tube walls exude enzymes which break down the insects, resulting in a liquid slurry which is then absorbed into the leaf, providing the pitcher plant with valuable nutrients.

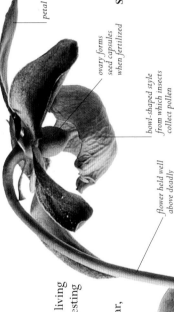

petal

ovary forms seed capsules when fertilized

bowl-shaped style from which insects collect pollen

flower held well above deadly pitchers

SIZE Up to 12 in (30 cm) tall
HABITAT Swampy areas in pinewoods and savannas
DISTRIBUTION S.E. US
LEAF TYPE Hooded, narrow, funnel-shaped "pitcher"

< IN BLOOM
The yellow, odorless flowers are produced from late March to mid-May. Bees are the main pollinator, and will favor this species over other nearby plants.

< HOODS
The concave hoods prevent rainwater from filling up the pitcher and also shade the opening, making it harder for trapped insects to find the exit.

nectar glands at the back of the hood

reddish purple colors of the hood attract insects to the pitcher

< CARNIVORE AT RISK
The hooded pitcher plant is now a rare and threatened species due to habitat loss. It is found in North Carolina, South Carolina, Georgia, and Florida. There are two varieties: *S. minor* grows to 12 in (30 cm), while *S. minor* var. *okefenokeensis*, with pitchers up to 4 ft (1.2 m) long, is found only at Okefenokee Swamp in Georgia.

GRANTHAM'S CAMELLIA
Camellia granthamiana
F: Theaceae
One of well over 100 *Camellia* species,
all from Asia, this endangered tree
from China was discovered in 1955.

30 ft/10 m

EPAULETTE TREE
Pterostyrax hispidus
F: Styracaceae
Native to E. Asia, this
large-leaved, deciduous tree is
sometimes cultivated for its
hanging clusters of fragrant,
creamy-white flowers.

40 ft/12 m

15 ft
4.5 m

SILKY CAMELLIA
Stewartia malacodendron
F: Theaceae
Named after its silkily
haired young stems,
this deciduous shrub or
tree grows in woodland
in S.E. US.

TEA
Camellia sinensis
F: Theaceae
The dried, fermented leaves and buds of
this evergreen Asian shrub are used to
make tea. Their astringent tannins
deter grazing animals.

56 ft/17 m

*paired oval leaf
with rounded tips*

23½ in/60 cm

SPANISH CHERRY
Mimusops elengi
F: Sapotaceae
This evergreen Indian tree is
planted in tropical countries
for its fragrant flowers. Its
durable wood is used in
shipbuilding and construction.

100 ft/30 m

**HOODED
PITCHER PLANT**
Sarracenia minor
F: Sarraceniaceae
Pitcher plants trap and
digest insects to supplement
their nutrient-poor soils.
This species is native
to the S.E. US.

12 in
30 cm

**PARROT
PITCHER PLANT**
Sarracenia psittacina
F: Sarraceniaceae
Sarracenia species grow wild only
in N.E. America. The pitchers
of this species, found from
Florida to Louisiana, form
horizontal traps.

14 in
35 cm

CORALBERRY
Ardisia crenata
F: Primulaceae
A greenhouse favorite
for its long-lasting, bright
berries, this Asian shrub
has become an invasive
pest in Hawaii, Florida,
and Texas.

6 ft
1.8 m

*horizontal,
creeping stem*

3¼ ft/1 m

*hood resembles
cobra ready to strike*

COBRA PLANT
Darlingtonia californica
F: Sarraceniaceae
The only species of
Darlingtonia, this carnivorous
plant grows in coastal
bogs and beside mountain
streams in W. US.

SCARLET PIMPERNEL
Anagallis arvensis
F: Primulaceae
Just as likely to have deep blue as
scarlet flowers, this sprawling
plant now grows in disturbed
ground virtually worldwide,
except the tropics.

12 in
30 cm

HAIRY ANDROSACE
Androsace villosa
F: Primulaceae
In a genus of arctic-alpines, this
perennial of mountains from Europe
to the Himalayas is known for its dense
white flowerheads and silky leaves.

6 in
15 cm

CREEPING JENNY
Lysimachia nummularia
F: Primulaceae
From Sweden to the Caucasus,
the spreading stems of this
perennial creep among
vegetation in damp grassy
places, shady hedgerows,
and streamsides.

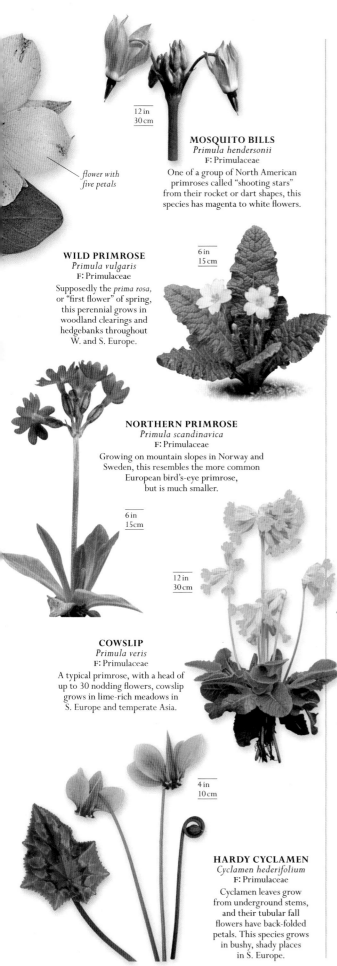

MOSQUITO BILLS
Primula hendersonii
F: Primulaceae
One of a group of North American primroses called "shooting stars" from their rocket or dart shapes, this species has magenta to white flowers.

flower with five petals

12 in
30 cm

WILD PRIMROSE
Primula vulgaris
F: Primulaceae
Supposedly the *prima rosa,* or "first flower" of spring, this perennial grows in woodland clearings and hedgebanks throughout W. and S. Europe.

6 in
15 cm

NORTHERN PRIMROSE
Primula scandinavica
F: Primulaceae
Growing on mountain slopes in Norway and Sweden, this resembles the more common European bird's-eye primrose, but is much smaller.

6 in
15 cm

COWSLIP
Primula veris
F: Primulaceae
A typical primrose, with a head of up to 30 nodding flowers, cowslip grows in lime-rich meadows in S. Europe and temperate Asia.

12 in
30 cm

HARDY CYCLAMEN
Cyclamen hederifolium
F: Primulaceae
Cyclamen leaves grow from underground stems, and their tubular fall flowers have back-folded petals. This species grows in bushy, shady places in S. Europe.

4 in
10 cm

BORAGINALES

Because of uncertainties over how the forget-me-not family (Boraginaceae) relates to other families, a new order, Boraginales, was proposed in 2016, just for this one family of about 2,700 species. They range from small annual herbs to large trees, often with conspicuous hairs and swollen bases on their stems or leaves. Some species are edible, and dyes are produced from others.

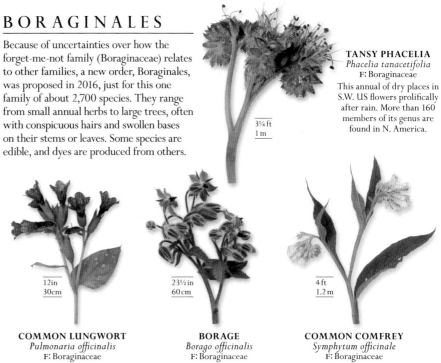

3¼ ft
1 m

TANSY PHACELIA
Phacelia tanacetifolia
F: Boraginaceae
This annual of dry places in S.W. US flowers prolifically after rain. More than 160 members of its genus are found in N. America.

12 in
30 cm

COMMON LUNGWORT
Pulmonaria officinalis
F: Boraginaceae
With blotched leaves supposedly resembling unhealthy lungs, and once thought to be efficacious against tuberculosis, this perennial grows in shady places throughout C. Europe.

23½ in
60 cm

BORAGE
Borago officinalis
F: Boraginaceae
Stiff hairs cover the stem and leaves of this S. European wayside plant, also cultivated for ornament and its oil-rich seeds.

4 ft
1.2 m

COMMON COMFREY
Symphytum officinale
F: Boraginaceae
Symphytum means "growing together," named because this perennial of damp ground across Eurasia was traditionally used in the treatment of injuries.

28 in
70 cm

WATER FORGET-ME-NOT
Myosotis scorpioides
F: Boraginaceae
This surface-breaking plant of streams and ponds is widely distributed in Eurasia, and introduced to the Americas and New Zealand.

23½ in
60 cm

VIPER'S-BUGLOSS
Echium vulgare
F: Boraginaceae
Widespread in grassy places in Europe and temperate Asia, this rough-haired biennial has pinkish flower buds but deep blue flowers.

rough, hairy, lanceolate leaves

PURPLE GROMWELL
Lithospermum purpureocaeruleum
F: Boraginaceae
Upright flower stems grow from creeping ground stems in this woodland perennial found on chalky soils in S. Europe and S.W. Asia.

3¼ ft
1 m

GARRYALES

Any classification system raises anomalies, and the order Garryales is one of these. Once placed in the Cornales, modern genetic investigation separated this order of two families and about 20 species. The silky tassel family (Garryaceae) includes two genera: *Garrya* from N. America and *Aucuba* from E. Asia. The Eucommiaceae covers just one species: *Eucommia ulmoides,* a tree of Chinese mountain forests.

16 ft/5 m

SPOTTED LAUREL
Aucuba japonica
F: Garryaceae
Some cultivars of this ornamental Japanese shrub grow yellow-blotched leaves. Plants are either male, with upright flower spikes, or female with small, clustered flowerheads.

SILK-TASSEL BUSH
Garrya elliptica
F: Garryaceae
Growing in coastal scrub in California and Oregon, this small tree has showy, gray-green male catkins and shorter, silvery female catkins.

16 ft
5 m

FLOWERING PLANTS • EUDICOTS

GENTIANALES

The order name celebrates the gentians of mountains and gardens, but the Gentianaceae family has only about 1,600 members. The madder family (Rubiaceae) is the largest in the order, with more than 13,000 species, including tropical shrubs such as coffee. The oleander family (Apocyanaceae), strychnine family (Loganiaceae), and jessamine family (Gelsemiaceae) complete the order.

MADAGASCAR PERIWINKLE
Catharanthus roseus
F: Apocynaceae
The leaves of this garden ornamental, endangered in its native Madagascar, contain tiny amounts of alkaloids used to treat childhood leukaemias. This use has saved it from extinction.

3¼ ft
1 m

40 ft
12 m

FIVE-PETALED FLOWER

OVAL LEAF

COMMON FRANGIPANI
Plumeria rubra
F: Apocynaceae
Native from Mexico to Venezuela, the flowers of this ornamental tree are fragrant at night and attract sphinx moths for pollination.

6 in/15 cm

BEAD PLANT
Nertera granadensis
F: Rubiaceae
Named for its small, beadlike fruits, this perennial with tiny, green flowers is native to Australia, New Zealand, Pacific islands, and S. America.

GREATER PERIWINKLE
Vinca major
F: Apocynaceae
Long, rooting stems of this evergreen perennial creep across woodland floors in S. and C. Europe and N. Africa.

5 ft
1.5 m

13 ft/4 m

OLEANDER
Nerium oleander
F: Apocynaceae
All parts of this evergreen shrub are poisonous. It grows by streams from the Mediterranean to China, producing clusters of fragrant pink flowers.

30 in
75 cm

BUTTERFLY WEED
Asclepias tuberosa
F: Apocynaceae
Native Americans chewed the roots of this showy perennial as a cure for pleurisy. It grows in fields and on roadsides in N. America.

incurved, rudimentary leaf

succulent stem

CLEAVERS
Galium aparine
F: Rubiaceae
Prickles on the stem and leaf margins attach to animal fur, dispersing parts of this agricultural weed of Europe and Asia.

10 ft
3 m

glossy, evergreen leaf

CROSSWORT
Cruciata laevipes
F: Rubiaceae
With stem leaves in crosslike whorls and clustered, honey-scented flowers, this perennial grows in grassy places across Europe and Asia.

23½ in
60 cm

50 ft
15 m

QUININE TREE
Cinchona calisaya
F: Rubiaceae
The anti-malaria drug quinine comes from the bark of this S. American tree. Seeds were smuggled to Asia in the mid-19th century for cultivation there.

33 ft/10 m

6½ ft
2 m

COFFEE SHRUB
Coffea arabica
F: Rubiaceae
Native to Ethiopia but transported elsewhere for cultivation, this evergreen shrub produces crimson drupes (stone fruits) within which are two seeds (coffee beans).

fruits red when ripe

CAPE JASMINE
Gardenia jasminoides
F: Rubiaceae
Originating in Asia, this evergreen shrub has fragrant, waxy, tubular white flowers that turn yellow with age. These ripen to berrylike fruits.

OVAL
LEAVES

82 ft
25 m

STRYCHNINE
SEEDS

STRYCHNINE
Strychnos nux-vomica
F: Loganiaceae
The poison strychnine comes
from seeds of this evergreen
tree from S.E. Asia.

12 in
30 cm

PERSIAN VIOLET
Exacum affine
F: Gentianaceae
Grown as an indoor ornamental but native to
Yemen and nearby Socotra Island, this evergreen
biennial has purple flowers with yellow centers.

SPRING GENTIAN
Gentiana verna
F: Gentianaceae
With ground-level leaf
rosettes and deep blue, tubular
flowers, this perennial grows
in Arctic and mountain
regions of Europe and W. Asia.

5 in
12 cm

20 in
50 cm

COMMON
CENTAURY
Centaurium erythraea
F: Gentianaceae
Grassy places and
dunes from Europe to
S.W. Asia are home
to this annual with
funnel-shaped,
five-lobed pink flowers.

12 in
30 cm

*flower's center emits
stench of rotting meat*

*basal rosette
of leaves*

CARRION FLOWER
Stapelia gettlifei
F: Apocynaceae
Carrion flower is a succulent,
spiny plant from southern Africa.
The stench of its barred or
blotched flowers attracts carrion
flies as pollinators.

*hairy edge
to petals*

20 ft
6 m

MADAGASCAR
JASMINE
Stephanotis floribunda
F: Apocynaceae
Originally from Madagascar,
this woody, twining vine is
a popular greenhouse plant
with leathery leaves and
clusters of fragrant, waxy,
white flowers.

LAMIALES

Modern taxonomy has expanded the
scope of this order to include 25 families,
typically with tubular flowers and
unequal petal lobes. The largest are
the mint family (Lamiaceae, often
known by its older name Labiatae) and
the figwort family (Scrophulariaceae),
each with 5,000–6,000 species. Others
include the olive family (Oleaceae)
and plantain family (Plantaginaceae).

5 ft
1.5 m

STOUT
FLOWER
SPIKE

CRINKLY
LOBED
LEAF

3¼ ft
1 m

BEAR'S
BREECHES
Acanthus mollis
F: Acanthaceae
This robust perennial of rocky places in
the W. Mediterranean has spikes of
purple-veined white flowers with a
three-lobed lower lip.

3¼ ft
1 m

*dark green leaf
with creamy veins*

SHRIMP BUSH
Justicia brandeegeana
F: Acanthaceae
This popular ornamental
from Mexico has spikes of
white flowers surrounded
by overlapping red-
bronze bracts (specialized
leaves), looking like a
large shrimp.

ZEBRA PLANT
Aphelandra squarrosa
F: Acanthaceae
Originating from coastal Brazilian
forests, this popular houseplant
has pale-veined leaves and spikes
of yellow flowers surrounded
by yellow bracts.

50 ft
15 m

BLACK
MANGROVE
Avicennia germinans
F: Acanthaceae
This tree forms thickets beside tidal
estuaries along tropical Atlantic
coasts. Pointed fruits, falling into
the mud, sprout new plants.

»

BUGLE
Ajuga reptans
F: Lamiaceae

Growing around Europe, N. Africa, and S.W. Asia, this woodland and meadow perennial produces its flower stems from far-creeping, rooting runners.

12 in
30 cm

tubular flower with large, three-lobed lip

oval leaf, often bronzy underneath

SAGE
Salvia officinalis
F: Lamiaceae

This grayish shrub from S.W. Europe and the Balkans is widely cultivated for its pungent leaves, used to flavor foods.

23½ in/60 cm

ROSEMARY
Salvia rosmarinus
F: Lamiaceae

From dry Mediterranean habitats, the oils in the leaves of this shrub reduce potential water loss from transpiration. Rosemary is a well-known culinary herb.

6½ ft/2 m

LATE SUMMER FLOWERS

BASIL
Ocimum basilicum
F: Lamiaceae

Also known as sweet basil, this is an annual from India and Iran. In cultivation, its pungent leaves are used as a culinary herb.

32 in
80 cm

EDIBLE LEAVES

COMMON LAVENDER
Lavandula angustifolia
F: Lamiaceae

This is another evergreen shrub of dry Mediterranean scrub habitats. The water-saving oils in its leaves are valued in perfumery.

32 in
80 cm

DENSELY PACKED FLOWERS

BETONY
Betonica officinalis
F: Lamiaceae

A common hedgerow and grassland herb in most of Europe, the Caucasus, and N. Africa, betony has reddish purple or white flowers.

32 in
80 cm

CHASTE TREE
Vitex agnus-castus
F: Lamiaceae

Chaste tree grows in damp places from S. Europe to Pakistan. It was once thought to preserve chastity, and is used to regulate hormonal functions in herbal medicine.

20 ft/6 m

JERUSALEM SAGE
Phlomis fruticosa
F: Lamiaceae

Native to dry, rocky habitats in the E. Mediterranean, this evergreen shrub with gray-felted leaves is a widely grown garden ornamental.

5 ft
1.5 m

WILD BERGAMOT
Monarda fistulosa
F: Lamiaceae

A minty tea is made from the leaves of this showy American perennial of dry fields and thickets from New England to Texas.

LANCEOLATE LEAF

CLUSTER OF FLOWERS

4 ft
1.2 m

WILD MARJORAM
Origanum vulgare
F: Lamiaceae

A perennial aromatic herb of grassy and stony habitats in Europe and W. Asia, marjoram is cultivated for many culinary uses.

branched cluster of pinkish-purple flowers

3¼ ft/1 m

stalked, oval leaf

ROUND-LEAVED MINT BUSH
Prostanthera rotundifolia
F: Lamiaceae

This aromatic Australian shrub with circular leaves and pink to purple spring flowers grows in open forests from New South Wales to Tasmania.

10 ft/3 m

COMMON THYME
Thymus vulgaris
F: Lamiaceae

Native to dry, rocky places in the W. Mediterranean, this densely branched shrub is used as a culinary herb, and an ingredient in perfumes and soaps.

16 in/40 cm

WATER MINT
Mentha aquatica
F: Lamiaceae

Growing part-submerged in ponds and ditches in Europe, Africa, and S.W. Asia, this herb is one parent of the hybrid peppermint.

3¼ ft
1 m

FLOWER
SPRIG

OLIVE
Olea europaea
F: Oleaceae
The fleshy fruits of this evergreen
Mediterranean tree contain 40 percent
unsaturated oils. Before eating, the
olives must be pickled in brine.

EGG-SHAPED
FRUIT

50 ft / 15 m

*heart-
shaped leaf*

GOLDEN BELL
Forsythia suspensa
F: Oleaceae
Also called weeping forsythia,
this deciduous shrub has
hollow, pendulous stems and
yellow flowers. It is native to
China and perhaps Japan.

10 ft / 3 m

MANNA ASH
Fraxinus ornus
F: Oleaceae
Unusually for an ash, this
S. European tree has clusters
of showy, white flowers.
Manna, a gum from its bark,
is used medicinally.

65 ft / 20 m

*leaf up
to 5 in
(12 cm) long*

23 ft / 7 m

*dense, pyramid-
shaped cluster of
scented flowers*

12 in
30 cm

LILAC
Syringa vulgaris
F: Oleaceae
Gardeners have developed
many hybrids and varieties
of this showy, deciduous
tree, found wild on scrubby
hill slopes in S.E. Europe.

TRUMPET-
SHAPED
FLOWER

ROSETTE
OF STICKY
LEAVES

**FALSE AFRICAN
VIOLET**
Streptocarpus saxorum
F: Gesneriaceae
Originating from Kenya and
Tanzania, this evergreen
perennial has whorls of
small, hairy, almost succulent
leaves and five-lobed,
trumpet-shaped flowers.

40 ft
12 m

COMMON JASMINE
Jasminum officinale
F: Oleaceae
Widely cultivated for its
showy, scented flowers,
this scrambling, deciduous
shrub, found from the
Caucasus to China,
twists counterclockwise
as it climbs.

**COMMON
BUTTERWORT**
Pinguicula vulgaris
F: Lentibulariaceae
This bog plant from
N. Europe, Asia,
and N. America
gains nutrients by
trapping and digesting
insects on its
sticky leaves.

7 in
18 cm

40 ft
12 m

6 in
15 cm

AFRICAN VIOLET
Streptocarpus ionanthus
F: Gesneriaceae
Endangered in the rainforests of
tropical E. Africa but popular as
houseplants, African violets were once
grouped in the genus *Saintpaulia*.

60 ft / 18 m

BEANLIKE
PODS

SPRIG WITH
FLOWER
CLUSTER

INDIAN BEAN TREE
Catalpa bignonioides
F: Bignoniaceae
Despite its common name,
this woodland tree grows in
the S. US. It is planted farther
north and in Europe for
its showy flowers.

23½ in
60 cm

HARDY GLOXINIA
Incarvillea delavayi
F: Bignoniaceae
A hardy garden perennial
with rosy-purple trumpet
flowers, this herbaceous
plant grows wild in
mountain grasslands
of China.

50 ft
15 m

JACARANDA
Jacaranda mimosifolia
F: Bignoniaceae
Originally from Argentina
to Bolivia, this tropical tree
with hanging clusters of
lavender-colored flowers is
widely planted for shade
and ornament.

TRUMPET VINE
Campsis x tagliabuana
F: Bignoniaceae
A garden hybrid between
a N. American and Asian
species, this vigorous
climbing shrub has clusters
of orange-red, trumpet-
shaped flowers.

FLOWERING PLANTS · EUDICOTS

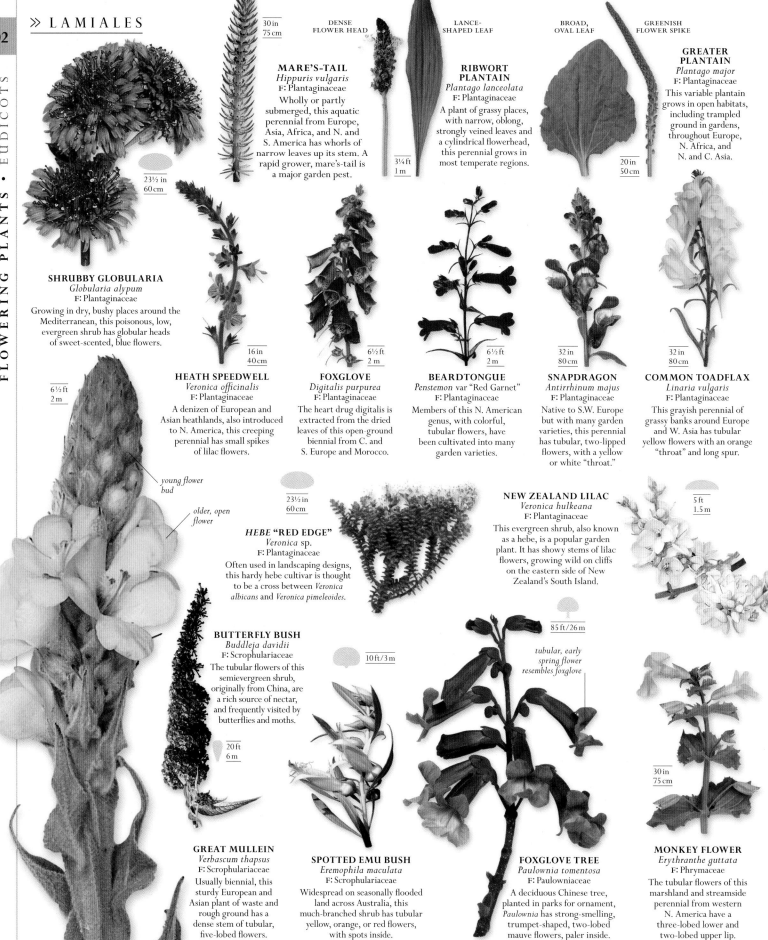

SHRUBBY GLOBULARIA
Globularia alypum
F: Plantaginaceae
Growing in dry, bushy places around the
Mediterranean, this poisonous, low,
evergreen shrub has globular heads
of sweet-scented, blue flowers.

30 in
75 cm

DENSE
FLOWER HEAD

MARE'S-TAIL
Hippuris vulgaris
F: Plantaginaceae
Wholly or partly
submerged, this aquatic
perennial from Europe,
Asia, Africa, and N. and
S. America has whorls of
narrow leaves up its stem. A
rapid grower, mare's-tail is
a major garden pest.

23½ in
60 cm

LANCE-
SHAPED LEAF

**RIBWORT
PLANTAIN**
Plantago lanceolata
F: Plantaginaceae
A plant of grassy places,
with narrow, oblong,
strongly veined leaves and
a cylindrical flowerhead,
this perennial grows in
most temperate regions.

3¼ ft
1 m

BROAD,
OVAL LEAF

GREENISH
FLOWER SPIKE

**GREATER
PLANTAIN**
Plantago major
F: Plantaginaceae
This variable plantain
grows in open habitats,
including trampled
ground in gardens,
throughout Europe,
N. Africa, and
N. and C. Asia.

20 in
50 cm

16 in
40 cm

HEATH SPEEDWELL
Veronica officinalis
F: Plantaginaceae
A denizen of European and
Asian heathlands, also introduced
to N. America, this creeping
perennial has small spikes
of lilac flowers.

6½ ft
2 m

FOXGLOVE
Digitalis purpurea
F: Plantaginaceae
The heart drug digitalis is
extracted from the dried
leaves of this open-ground
biennial from C. and
S. Europe and Morocco.

6½ ft
2 m

BEARDTONGUE
Penstemon var "Red Garnet"
F: Plantaginaceae
Members of this N. American
genus, with colorful,
tubular flowers, have
been cultivated into many
garden varieties.

32 in
80 cm

SNAPDRAGON
Antirrhinum majus
F: Plantaginaceae
Native to S.W. Europe
but with many garden
varieties, this perennial
has tubular, two-lipped
flowers, with a yellow
or white "throat."

32 in
80 cm

COMMON TOADFLAX
Linaria vulgaris
F: Plantaginaceae
This grayish perennial of
grassy banks around Europe
and W. Asia has tubular
yellow flowers with an orange
"throat" and long spur.

6½ ft
2 m

young flower
bud

older, open
flower

23½ in
60 cm

HEBE "RED EDGE"
Veronica sp.
F: Plantaginaceae
Often used in landscaping designs,
this hardy hebe cultivar is thought
to be a cross between *Veronica
albicans* and *Veronica pimeleoides*.

NEW ZEALAND LILAC
Veronica hulkeana
F: Plantaginaceae
This evergreen shrub, also known
as a hebe, is a popular garden
plant. It has showy stems of lilac
flowers, growing wild on cliffs
on the eastern side of New
Zealand's South Island.

5 ft
1.5 m

BUTTERFLY BUSH
Buddleja davidii
F: Scrophulariaceae
The tubular flowers of this
semievergreen shrub,
originally from China, are
a rich source of nectar,
and frequently visited by
butterflies and moths.

20 ft
6 m

10 ft / 3 m

85 ft / 26 m

tubular, early
spring flower
resembles foxglove

GREAT MULLEIN
Verbascum thapsus
F: Scrophulariaceae
Usually biennial, this
sturdy European and
Asian plant of waste and
rough ground has a
dense stem of tubular,
five-lobed flowers.

SPOTTED EMU BUSH
Eremophila maculata
F: Scrophulariaceae
Widespread on seasonally flooded
land across Australia, this
much-branched shrub has tubular
yellow, orange, or red flowers,
with spots inside.

FOXGLOVE TREE
Paulownia tomentosa
F: Paulowniaceae
A deciduous Chinese tree,
planted in parks for ornament,
Paulownia has strong-smelling,
trumpet-shaped, two-lobed
mauve flowers, paler inside.

30 in
75 cm

MONKEY FLOWER
Erythranthe guttata
F: Phrymaceae
The tubular flowers of this
marshland and streamside
perennial from western
N. America have a
three-lobed lower and
two-lobed upper lip.

YELLOW RATTLE
Rhinanthus minor
F: Orobanchaceae

A partial parasite whose roots tap grasses for nutrition, this variable, yellow-flowered annual grows in grasslands in northern temperate regions.

20 in
50 cm

3 in
8 cm

PURPLE TOOTHWORT
Lathraea clandestina
F: Orobanchaceae

Lacking green leaves, this parasitic W. European perennial taps the roots of willows and poplars for food. Its flowerheads emerge directly from underground shoots.

flowers attract butterflies for pollination

opposite, oval shiny leaf

LANTANA
Lantana camara
F: Verbenaceae

The tubular flowers of this prickly shrub open yellow or orange, then turn red. Native to C. and S. America, it is one of the most invasive weeds in warm regions.

5 ft
1.5 m

VERVAIN
Verbena officinalis
F: Verbenaceae

Widely distributed in rough grassland in temperate and tropical regions, this stiffly hairy perennial has slender spikes of two-lipped, lilac flowers.

3¼ ft
1 m

BODINIER'S BEAUTYBERRY
Callicarpa bodinieri
F: Lamiaceae

Related to American beautyberry, this Chinese ornamental shrub is grown in gardens for its decorative purple berries, which are bitter but not poisonous.

10 ft
3 m

FLAMING GLORYBOWER
Clerodendrum splendens
F: Lamiaceae

This African vine, with clusters of tubular, red flowers, twines around trees in its native forest habitat, or trellises in gardens.

12 ft
3.7 m

SOLANALES

The potato family (Solanaceae), an economically important family of up to 4,000 species, dominates the order Solanales. Many species contain poisonous alkaloids. The bindweed family (Convolvulaceae) includes tropical climbers and low-growing herbs. The three other families in this order are the Hydrolaceae, with one genus found only in the Americas; five African trees in the Montiniaceae; and a pan-tropical herb in the Sphenocleaceae.

GREAT BINDWEED
Calystegia sylvaticus
F: Convolvulaceae

A plant of hedgerows and waste ground, even in cities, around the temperate northern hemisphere, this large-flowered perennial spreads by far-creeping rhizomes (underground plant stems).

10 ft
3 m

trumpet-shaped flower with white or yellow center

three-lobed leaf

13 ft
4 m

MORNING GLORY
Ipomoea tricolor
F: Convolvulaceae

This twining herbaceous climber from Central America has lobed leaves and funnel-shaped flowers that open only in the morning.

COMMON DODDER
Cuscuta epithymum
F: Convolvulaceae

23½ in
60 cm

A Eurasian annual forming a dense network of threadlike, twining stems, parasitic dodder taps into gorse, heather, and other species for food.

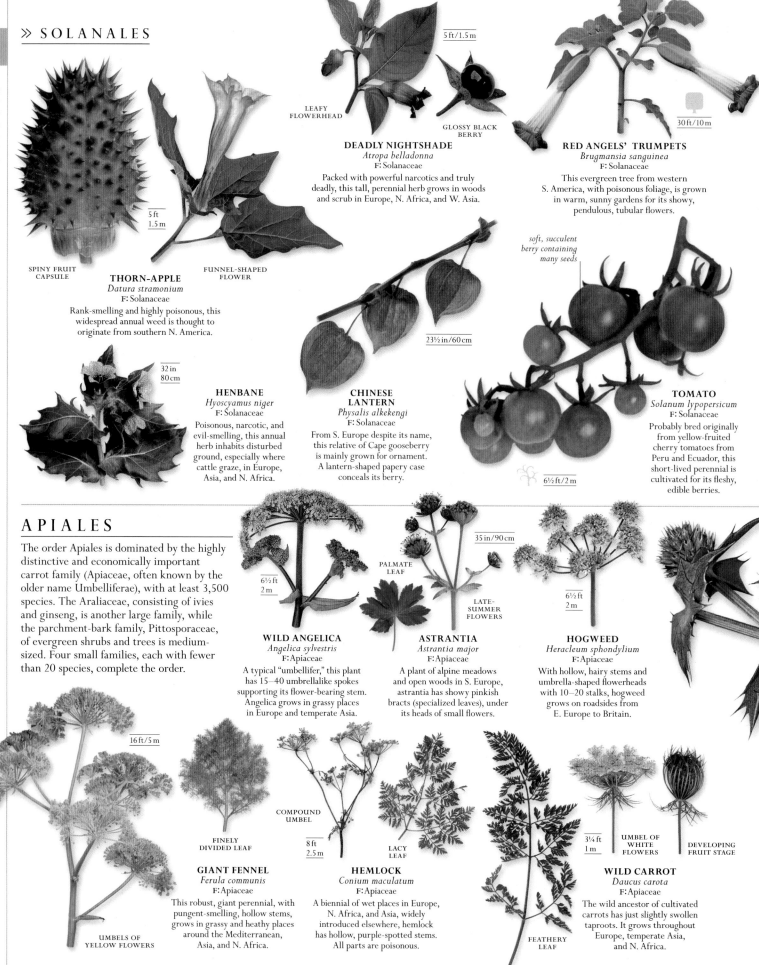

DEADLY NIGHTSHADE
Atropa belladonna
F: Solanaceae

Packed with powerful narcotics and truly deadly, this tall, perennial herb grows in woods and scrub in Europe, N. Africa, and W. Asia.

LEAFY FLOWERHEAD

GLOSSY BLACK BERRY

5 ft/1.5 m

RED ANGELS' TRUMPETS
Brugmansia sanguinea
F: Solanaceae

This evergreen tree from western S. America, with poisonous foliage, is grown in warm, sunny gardens for its showy, pendulous, tubular flowers.

30 ft/10 m

SPINY FRUIT CAPSULE

5 ft 1.5 m

FUNNEL-SHAPED FLOWER

THORN-APPLE
Datura stramonium
F: Solanaceae

Rank-smelling and highly poisonous, this widespread annual weed is thought to originate from southern N. America.

32 in 80 cm

HENBANE
Hyoscyamus niger
F: Solanaceae

Poisonous, narcotic, and evil-smelling, this annual herb inhabits disturbed ground, especially where cattle graze, in Europe, Asia, and N. Africa.

23½ in/60 cm

CHINESE LANTERN
Physalis alkekengi
F: Solanaceae

From S. Europe despite its name, this relative of Cape gooseberry is mainly grown for ornament. A lantern-shaped papery case conceals its berry.

soft, succulent berry containing many seeds

6½ ft/2 m

TOMATO
Solanum lypopersicum
F: Solanaceae

Probably bred originally from yellow-fruited cherry tomatoes from Peru and Ecuador, this short-lived perennial is cultivated for its fleshy, edible berries.

APIALES

The order Apiales is dominated by the highly distinctive and economically important carrot family (Apiaceae, often known by the older name Umbelliferae), with at least 3,500 species. The Araliaceae, consisting of ivies and ginseng, is another large family, while the parchment-bark family, Pittosporaceae, of evergreen shrubs and trees is medium-sized. Four small families, each with fewer than 20 species, complete the order.

6½ ft 2 m

WILD ANGELICA
Angelica sylvestris
F: Apiaceae

A typical "umbellifer," this plant has 15–40 umbrellalike spokes supporting its flower-bearing stem. Angelica grows in grassy places in Europe and temperate Asia.

PALMATE LEAF

LATE-SUMMER FLOWERS

35 in/90 cm

ASTRANTIA
Astrantia major
F: Apiaceae

A plant of alpine meadows and open woods in S. Europe, astrantia has showy pinkish bracts (specialized leaves), under its heads of small flowers.

6½ ft 2 m

HOGWEED
Heracleum sphondylium
F: Apiaceae

With hollow, hairy stems and umbrella-shaped flowerheads with 10–20 stalks, hogweed grows on roadsides from E. Europe to Britain.

16 ft/5 m

COMPOUND UMBEL

FINELY DIVIDED LEAF

GIANT FENNEL
Ferula communis
F: Apiaceae

This robust, giant perennial, with pungent-smelling, hollow stems, grows in grassy and heathy places around the Mediterranean, Asia, and N. Africa.

8 ft 2.5 m

LACY LEAF

HEMLOCK
Conium maculatum
F: Apiaceae

A biennial of wet places in Europe, N. Africa, and Asia, widely introduced elsewhere, hemlock has hollow, purple-spotted stems. All parts are poisonous.

FEATHERY LEAF

3¼ ft 1 m

UMBEL OF WHITE FLOWERS

DEVELOPING FRUIT STAGE

WILD CARROT
Daucus carota
F: Apiaceae

The wild ancestor of cultivated carrots has just slightly swollen taproots. It grows throughout Europe, temperate Asia, and N. Africa.

UMBELS OF YELLOW FLOWERS

CHILI PEPPER
Capsicum frutescens
F: Solanaceae
Native to Brazil and Bolivia, this small shrub is much cultivated in tropical Asia and equatorial America. Its fruits are chilies.

4 ft
1.2 m

TOBACCO
Nicotiana tabacum
F: Solanaceae
Originally S. American, this herb is the most common of two species grown for smoking tobacco. The leaves contain nicotine.

10 ft
3 m

POTATO
Solanum tuberosum
F: Solanaceae
Potatoes are a major food crop. The swollen tubers turn green (and poisonous) in sunlight. The plant was bred from ancestors in the S. American Andes.

3¼ ft
1 m

FLOWERS AND PINNATE LEAVES

EDIBLE TUBERS

mostly spineless upper leaf

berrylike red fruit

COMMON HOLLY
Ilex aquifolium
F: Aquifoliaceae
An evergreen, woodland understory shrub or small tree from Europe, N. Africa, and N.W. Asia, holly develops spiny leaves to discourage grazers.

80 ft / 24 m

egg-shaped flowerhead

33 ft
10 m

KOHUHU
Pittosporum tenuifolium
F: Pittosporaceae
Growing wild in forests and scrub on both islands of New Zealand, this evergreen tree has almost black branches and honey-scented spring flowers.

23½ in / 60 cm

leathery, spiny leaf

SEA HOLLY
Eryngium maritimum
F: Apiaceae
Leathery leaves help this sturdy perennial reduce water loss and repel salt spray in sand dunes around Europe, N. Africa, and S.W. Asia.

32 in / 80 cm

ORIENTAL GINSENG
Panax ginseng
F: Araliaceae
The root of this Asian herb is used as a traditional remedy. Its species name, *Panax*, comes from the Greek word *panacea*, meaning a cure-all for diseases.

IVY
Hedera helix
F: Araliaceae
An evergreen shrub from Europe and S.W. Asia, ivy climbs up other plants or forms creeping carpets in woods.

100 ft
30 m

AQUIFOLIALES

The mainly tropical holly family (Aquifoliaceae) includes trees and shrubs with characteristically toothed leaves, and is the main representative of the Aquifoliales order. Other members of this small order are an odd family of twining herbs, the Cardiopteridaceae; three Asian shrubs in the Helwingiaceae; four South American shrubs and trees in the Phyllonomaceae; and a family of tropical trees, the Stemonuraceae.

ASTERALES

Eleven families make up the Asterales. The largest is the daisy family (Asteraceae) with some 25,000 species. Typically, each flowerlike head (capitula) has many individual flowers, called florets, surrounded by showy rays. Some of the bellflower family (Campanulaceae) show similar characteristics. This order also includes the bogbean (Menyanthaceae) and fan-flower (Goodeniaceae) families, and seven smaller families.

ray florets

disk florets

bract

5 ft
1.5 m

MICHAELMAS DAISY
Symphotrichum novi-belgii
F: Asteraceae
Previously classified as a species of *Aster*, this showy, variable garden plant is a native of eastern N. America.

SALSIFY
Tragopogon porrifolius
F: Asteraceae
A biennial herb of grassy places around the Mediterranean, salsify has lilac or reddish-purple flower heads, surrounded by longer, pointed bracts.

4 ft
1.25 m

COMMON RAGWORT
Jacobaea vulgaris
F: Asteraceae
Toxic to farm animals and avoided by rabbits, this native perennial from Europe and W. Asia has invaded grasslands almost worldwide.

5 ft
1.5 m

capitula of disk florets which mature into seeds

lobed ray florets, usually yellow or orange

oval, toothed leaf

11 ft
3.5 m

3¼ ft
1 m

SERRATED LEAF

SUNFLOWER
Helianthus annuus
F: Asteraceae
Probably Mexican in origin, this tall annual is grown as an ornamental as well as commercially for its seeds that contain 27–40 percent polyunsaturated oil and 13–20 percent protein.

CORNFLOWER
Centaurea cyanus
F: Asteraceae
Probably native to S. Europe and W. Asia, cornflower seeds were inseparable from corn seeds, making it a pestilential agricultural weed before herbicides.

SMALL, WHITE FLOWER HEAD

OXEYE DAISY
Leucanthemum vulgare
F: Asteraceae
One of the most common white-flowered daisies of Europe and W. Asia, this variable patch-forming perennial quickly seeds into disturbed ground.

30 in
75 cm

UPRIGHT OR HORIZONTAL LEAF

5 in
12 cm

20 in
50 cm

YELLOW FLOWER HEAD

PINK OR PURPLE RAY FLORETS

4 ft
1.2 m

tall, robust stem

DAISY
Bellis perennis
F: Asteraceae
Originally from Europe and W. Asia, this short perennial of grazed grasslands and lawns has been transported almost worldwide. It is commonly considered a weed.

COMMON DANDELION
Taraxacum officinale
F: Asteraceae
The name for this familiar weed covers an amazing diversity of minutely differing forms, with perhaps 1,000 recognized "microspecies" in Europe and Asia.

PURPLE CONEFLOWER
Echinacea purpurea
F: Asteraceae
Named after the cone shape of its central disk florets, this ornamental perennial from eastern N. America is cultivated for herbal remedies against colds and flu.

ROUGH, TOOTHED LEAF

MILK THISTLE
Silybum marianum
F:Asteraceae

This weakly spiny biennial, with white-blotched leaves, grows in waste and cultivated ground around S. Europe, N. Africa, and W. Asia.

8 ft
2.5 m

CANADIAN GOLDENROD
Solidago canadensis
F:Asteraceae

Native to N. America but grown in many gardens, this downy perennial has yellow flower heads borne in curved, spreading flower spikes.

8 ft
2.5 m

GLOBE THISTLE
Echinops bannaticus
F:Asteraceae

Native to grassy places in S.E. Europe, this hairy-stemmed perennial has a spherical head of bluish, tubular "disk" florets.

4 ft
1.2 m

SPINY HEAD

6½ ft
2 m

CARDOON
Cynara cardunculus
F:Asteraceae

The fleshy flower base, eaten as globe artichoke, comes from the *scolymus* variety of cardoon, a wasteground perennial from the E. Mediterranean.

DEEPLY CUT LEAF

6½ ft
2 m

EARLY SPRING FLOWER

BUTTERBUR
Petasites hybridus
F:Asteraceae

Plants of butterbur have stems with either male or female flowers. It grows in wet meadows and beside streams throughout Europe to Iran.

20 in
50 cm

COTTONWEED
Otanthus maritimus
F:Asteraceae

This shrubby perennial grows in coastal habitats in S. Europe, N. Africa, and S.W. Asia. The common name derives from its woolly stems and leaves.

6 ft
1.8 m

DENSE BLAZING STAR
Liatris spicata
F:Asteraceae

A plant of damp ground in eastern N. America, this plant is named after its dense, elongated, spike of rose-purple flower heads.

5 ft
1.5 m

SPEAR THISTLE
Cirsium vulgare
F:Asteraceae

Common in disturbed ground in Europe and W. Asia, this robust perennial with spine-tipped leaves has been carried worldwide by agriculture.

20 in
50 cm

outer ray florets

GARDEN MARIGOLD
Calendula officinalis
F:Asteraceae

This species has been cultivated so long that its origins are unknown. Marigold extract from the flowers is used to treat skin problems.

leaf is hairy on both sides

23½ in
60 cm

YARROW
Achillea millefolium
F:Asteraceae

Found in grassy places throughout Europe, W. Asia, and N. America, this feathery-leaved perennial has been introduced to Australia and New Zealand.

BLUE OR LAVENDER FLOWERS

4 ft
1.2 m

OBLONG ROSETTE LEAF

CHICORY
Cichorium intybus
F:Asteraceae

Introduced almost worldwide from Europe, W. Asia and N. Africa, chicory is cultivated for salad leaves and for a coffee substitute from its roots.

6½ ft/2 m

FRENCH TARRAGON
Artemisia dracunculus
F:Asteraceae

Leaves of this bushy, aromatic herb from S.E. Europe, Asia, and N. America are used to flavor fish and other foods.

COTTON LAVENDER
Santolina chamaecyparissus
F:Asteraceae

Widely grown in gardens, this strongly aromatic, dwarf, evergreen shrub with gray, hairy leaves is native to rocky places around the W. Mediterranean.

23½ in
60 cm

20 in
50 cm

GARDEN GAZANIA
Gazania sp.
F:Asteraceae

Seventeen species of *Gazania* grow in southern Africa, most in dry habitats. Cultivars like these can survive in gardens with little watering.

8 in/20 cm

STRAND GAZANIA
Gazania rigens
F:Asteraceae

This sprawling, mat-forming perennial often flowers prolifically on dunes and rocky outcrops along the southern Cape coast of South Africa.

≫

≫ ASTERALES

4 ft
1.2 m

PERENNIAL
LEAF

FLOWER
SPIKE

GREAT BLUE LOBELIA
Lobelia siphilitica
F: Campanulaceae
A showy N. American perennial of
woods and meadows from Manitoba
to Alabama, this plant was once
supposed to cure syphilis.

tubular flower

16 in
40 cm

ROYAL BLUEBELL
Wahlenbergia gloriosa
F: Campanulaceae
The state plant of the Australian
Capital Territory, this erect
perennial with deep blue flowers
grows in mountain grasslands
in southern Australia.

28 in/70 cm

BELL-SHAPED
FLOWER

OVAL
LEAF

BALLOON FLOWER
Platycodon grandiflorus
F: Campanulaceae
The only member of its genus,
this Asian perennial with
blue or white, bell-shaped
flowers also comes in many
garden varieties.

5 ft
1.5 m

BUCKBEAN
Menyanthes trifoliata
F: Menyanthaceae
This far-creeping perennial
produces its beanlike fruits in
bogs and shallow water in
N. America, Greenland,
N. Europe, and Asia.

20 in
50 cm

CLUSTERED
FLOWER HEAD

3¼ ft
1 m

ROUND-HEADED
RAMPION
Phyteuma orbiculare
F: Campanulaceae
With dark blue flowers in rounded
heads, this unbranched perennial
grows in chalk grassland from
S. England to Greece.

LANCEOLATE
LEAF

NETTLE-LEAVED
BELLFLOWER
Campanula trachelium
F: Campanulaceae
Found in hedgebanks
around Europe, Iran, and
N. Africa, this roughly
hairy perennial has
nettlelike leaves and
blue, bell-shaped flowers.

6 in
15 cm

ONE-SIDED
FLOWER

SWAMPWEED
Selliera radicans
F: Goodeniaceae
This creeping herb is found on
coastal sandflats around Chile,
Australia, and New Zealand—
where it also grows alongside
mountain streams.

FLESHY
LEAF

5 ft
1.5 m

FRINGED WATER LILY
Nymphoides peltata
F: Menyanthaceae
Unlike true water lilies, this
aquatic perennial has smaller
yellow flowers with five fringed
petals. It is found from
Europe to Japan.

DIPSACALES

The order Dipsacales is found worldwide, but
especially in the northern hemisphere. Its members
generally have small flowers, often in compact
heads, and include many popular ornamental plants.
It consists of two families. The moschatel and
viburnum family (Adoxaceae) contains about 200
species, and the honeysuckle family (Caprifoliaceae)
has some 860 species, including valerians and teasels
that were previously placed in their own families.

6 in
15 cm

MUSKROOT
Adoxa moschatellina
F: Adoxaceae
Inflorescences of this
delicate woodland
perennial from Europe,
Asia, and N. America
have five flowers at
right angles (with one
facing upward).

ELDER FLOWER HEAD

32 in
80 cm

RED VALERIAN
Centranthus ruber
F: Caprifoliaceae
Butterflies appreciate
the flowers of this
grayish perennial. It is
found on coastal rocks
and old walls around
the Mediterranean,
and is naturalized
elsewhere.

lobed leaf

ELDER
BERRIES

ELDERBERRY
Sambucus nigra
F: Adoxaceae
A hedgerow shrub or small tree with
lobed leaves and flat-topped flower heads,
elder grows throughout Europe,
W. Asia, and N. Africa.

40 ft/12 m

*shiny, red
berry*

6½ ft
2 m

EUROPEAN
CRANBERRYBUSH
Viburnum opulus
F: Adoxaceae
Native across Europe to Asia, this
deciduous hedgerow shrub has
flattened flower heads with large,
sterile outer flowers, around
smaller, fertile ones.

GARDEN VALERIAN
Valeriana officinalis
F: Caprifoliaceae
This perennial is found in grassy places
from N. Europe into Asia. It has flattened
heads of pale pink, five-petaled flowers,
which are darker when in bud.

SCENTED
FLOWERS

PINNATE
LEAF

13 ft
4 m

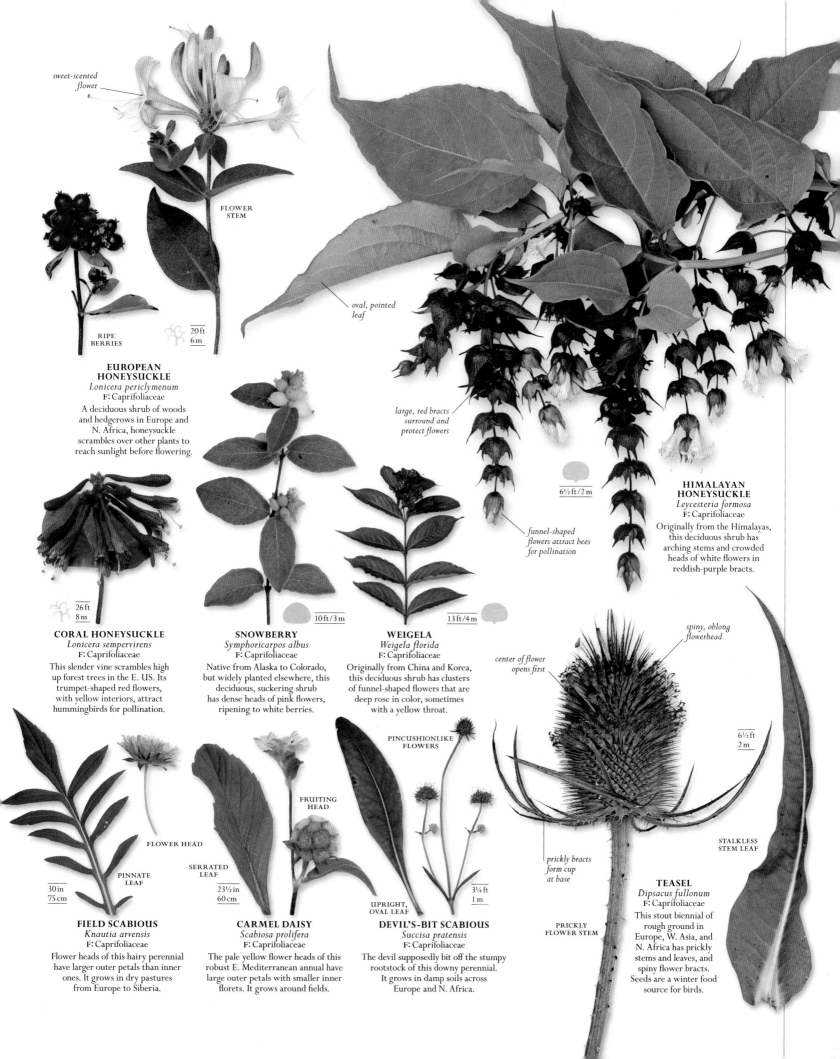

sweet-scented flower

FLOWER STEM

RIPE BERRIES

20 ft
6 m

EUROPEAN HONEYSUCKLE
Lonicera periclymenum
F: Caprifoliaceae

A deciduous shrub of woods and hedgerows in Europe and N. Africa, honeysuckle scrambles over other plants to reach sunlight before flowering.

oval, pointed leaf

large, red bracts surround and protect flowers

6½ ft / 2 m

funnel-shaped flowers attract bees for pollination

HIMALAYAN HONEYSUCKLE
Leycesteria formosa
F: Caprifoliaceae

Originally from the Himalayas, this deciduous shrub has arching stems and crowded heads of white flowers in reddish-purple bracts.

26 ft
8 m

CORAL HONEYSUCKLE
Lonicera sempervirens
F: Caprifoliaceae

This slender vine scrambles high up forest trees in the E. US. Its trumpet-shaped red flowers, with yellow interiors, attract hummingbirds for pollination.

10 ft / 3 m

SNOWBERRY
Symphoricarpos albus
F: Caprifoliaceae

Native from Alaska to Colorado, but widely planted elsewhere, this deciduous, suckering shrub has dense heads of pink flowers, ripening to white berries.

13 ft / 4 m

WEIGELA
Weigela florida
F: Caprifoliaceae

Originally from China and Korea, this deciduous shrub has clusters of funnel-shaped flowers that are deep rose in color, sometimes with a yellow throat.

spiny, oblong flowerhead

center of flower opens first

6½ ft
2 m

FLOWER HEAD

PINNATE LEAF

30 in
75 cm

FIELD SCABIOUS
Knautia arvensis
F: Caprifoliaceae

Flower heads of this hairy perennial have larger outer petals than inner ones. It grows in dry pastures from Europe to Siberia.

SERRATED LEAF

23½ in
60 cm

CARMEL DAISY
Scabiosa prolifera
F: Caprifoliaceae

The pale yellow flower heads of this robust E. Mediterranean annual have large outer petals with smaller inner florets. It grows around fields.

FRUITING HEAD

PINCUSHIONLIKE FLOWERS

UPRIGHT, OVAL LEAF

3¼ ft
1 m

DEVIL'S-BIT SCABIOUS
Succisa pratensis
F: Caprifoliaceae

The devil supposedly bit off the stumpy rootstock of this downy perennial. It grows in damp soils across Europe and N. Africa.

prickly bracts form cup at base

PRICKLY FLOWER STEM

STALKLESS STEM LEAF

TEASEL
Dipsacus fullonum
F: Caprifoliaceae

This stout biennial of rough ground in Europe, W. Asia, and N. Africa has prickly stems and leaves, and spiny flower bracts. Seeds are a winter food source for birds.

FUNGI

Ranging from mushrooms to microscopic molds, fungi were once classified as plants. Today, they are recognized as a distinct kingdom of living things. Growing in or through their food, they digest organic matter, and often become visible only when they reproduce. Fungi are both allies and enemies to other forms of life: vital recyclers and mutually benefiting partners, they also include parasites and pathogens.

MUSHROOMS

As well as mushrooms, this large group includes bracket fungi, puffballs, and many other species. Their fruit bodies vary in shape, but all make their spores on microscopic cells called basidia.

SAC FUNGI

These fungi produce their spores in microscopic sacs, which often form a felty layer on their fruit bodies. Many are cup-shaped, but they also include truffles, morels, and single-celled yeasts.

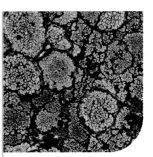

LICHENS

Living partnerships between fungi and algae, lichens colonize all kinds of bare surfaces. Some form flat crusts, while others are like tiny plants. Most grow slowly, and are exceptionally long-lived.

MUSHROOMS

The phylum Basidiomycota includes the majority of what are commonly called mushrooms and toadstools. Found in almost every major habitat type, nearly all share the ability to form sexual spores externally on special cells called basidia.

PHYLUM	BASIDIOMYCOTA
CLASSES	16
ORDERS	52
FAMILIES	177
SPECIES	About 32,000

These typical mushrooms show the cap with radiating gills on the underside, supported by the central stem.

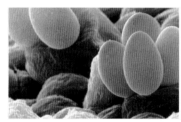

The microscopic spores of the field mushroom are shown here still attached to the basidia, or reproductive cells.

This stinkhorn is covered with a dark spore mass, which rapidly liquifies and emits a foul odor, usually of rotting flesh.

Few fungi are as diverse in form as the Basidiomycota. Their fruit bodies (the structures in which spores are produced) have evolved to provide varied and efficient mechanisms for spore dispersal. These fruit bodies often take the form of a stem, cap, and gills, but many are simple, crustlike sheets, and others are more complex, bracketlike forms. More exotic still are the totally enclosed, spherical types such as the puffballs and the earthstars. Then there are the beautiful, corallike structures of the club and coral fungi, and the often weird, animallike shapes of the stinkhorns. Each structure supports and produces special spore-producing cells called basidia, and—depending on the fungus concerned—uses a variety of mechanisms to disperse these spores. For example, gilled fungi forcibly eject their spores, to be carried off by air currents. The puffballs, however, depend on a combination of wind and raindrops to puff out their spores. Stinkhorns, on the other hand, use foul odors and bright colors to attract animals, such as insects and other invertebrates, to eat their spore mass and to pass the spores unharmed through their digestive tracts. In this way the spores are scattered as far as the animals travel before excreting them.

FUNGAL FEEDING

The main body of the fungus is usually underground, formed of fine fungal threads called hyphae, which make up a mycelium (fungal body). These threads permeate the substance on which the fungus grows, be it soil, old leaf litter, fallen trees, living plant tissues, or even dead and decaying animals. The mycelium may spread over a large area, and in many fungi it also forms a mutually beneficial relationship with the roots of plants. In this relationship—called a mycorrhiza—the mycelium wraps around and penetrates the plant's roots, providing the fungus with access to carbohydrates from the plant. In return, the plant benefits from the mycelium's greater ability to absorb water and mineral nutrients. Other mushrooms break down dead organic matter, or devour living organisms for food—both processes help to renew the soil and to improve the growing conditions for other living things.

MUSHROOMS AMONG FALLEN LEAVES >
Many fungi feed on dead and decaying plant tissues and may spread in huge numbers where nutrients and moisture occur.

AGARICALES

Many of the most familiar mushrooms and toadstools belong to this order.
They include fungi with fleshy, not woody, fruitbodies (the body supporting
the spore-forming cells). Many have caps and stems with gills; some also
have pores. Other forms include bird's-nest, brackets, crusts, trufflelike,
and puffballs. Most live on leaf litter, soil, or wood; others are parasitic or
live in association with plant roots (mycorrhizal).

1½–4 in
4–10 cm

FIELD MUSHROOM
Agaricus campestris
F: Agaricaceae
Common in meadows of
N. America and Eurasia, this
fungus has a rounded cap;
bright pink gills turning
brown with maturity; and a
short stem with a small ring.

2–4 in
5–10 cm

CULTIVATED MUSHROOM
Agaricus bisporus
F: Agaricaceae
Selling in the millions, this familiar
fungus is common worldwide.
Its cap varies from white to dark
brown with scales.

THE PRINCE
Agaricus augustus
F: Agaricaceae
A large, orange-brown,
scaly cap and a white,
woolly stem with a floppy
ring characterize this
common species
from Eurasia and
N. America.

3¼–6 in
8–15 cm

2–6 in
5–15 cm

FRECKLED
DAPPERLING
Echinoderma asperum
F: Agaricaceae
From woods and
gardens in Eurasia and
N. America, this uncommon
fungus's brown cap has
pyramidal scales that rub off.
The brown stem has a ring.

*slightly
fibrous
surface*

2–4¾ in
5–12 cm

YELLOW STAINER
Agaricus xanthodermus
F: Agaricaceae
This species has a flattened
cap center, chrome-yellow
stains on the stem base,
and an unpleasant odor. It
is common in Eurasia and
western N. America.

2¾–6 in
7–15 cm

HORSE MUSHROOM
Agaricus arvensis
F: Agaricaceae
Common in parks in
N. America and Eurasia,
this species stains dull
brassy yellow. The pendent
ring on its stem has a
cogwheellike pattern
on its underside.

⅜–1½ in
1–4 cm

STINKING
DAPPERLING
Lepiota cristata
F: Agaricaceae
A rubberlike smell is a
characteristic feature
of this species of *Lepiota*
(most common in woods
and meadows) from
N. America and Eurasia.

yellow stains

1¼–3¼ in
3–8 cm

MATURE
CAP

¾–2¼ in
2–6 cm

YOUNG
CAP

BLUSHING
DAPPERLING
Leucoagaricus badhamii
F: Agaricaceae
Found in Eurasia, this
rare fungus grows in woods
and gardens with rich soils.
White when young, it
bruises blood-red when
handled, and then blackens.

LAWYER'S WIG
Coprinus comatus
F: Agaricaceae
This fungus is common on
disturbed soils and pathsides in
N. America and Eurasia. Its tall,
shaggy cap is unmistakable, and
dissolves into inky fluid.

1½–4½ in
4–11 cm

ring

ORANGE-GIRDLED
DAPPERLING
Lepiota ignivolvata
F: Agaricaceae
Found in Eurasia, this
uncommon species has
an orange-margined ring
set low down on the
club-shaped, white stem.

2–3¼ in
5–8 cm

WHITE DAPPERLING
Leucoagaricus leucothites
F: Agaricaceae
Common in meadows and grassy
roadsides in Eurasia and
N. America, this fungus's ivory
fruitbody turns grey-brown
with age. Its gills are white
to pale pink.

2–6 in
5–15 cm

SHAGGY PARASOL
Chlorophyllum rhacodes
F: Agaricaceae
Common in Eurasia and
N. America, this fungus
has a shaggy, brown cap;
thick, double-edged ring;
reddening flesh; and a
swollen stem base.

⅜–2 in
1–5 cm

PLANTPOT DAPPERLING
Leucocoprinus birnbaumii
F: Agaricaceae
Common worldwide, this
species grows on soils of potted
plants. Its features include a
delicate golden-yellow cap and
a slender stem with a ring.

⅜–1¼ in
1–3 cm

INNER
LAYER

GRAY PUFFBALL
Bovista plumbea
F: Agaricaceae
This ball-shaped fungus is
smooth when young. As it
matures, its skin flakes to reveal a
papery inner layer. It is common
in N. America and Eurasia.

OUTER SKIN

8–20 in
20–50 cm

GIANT PUFFBALL
Calvatia gigantea
F: Agaricaceae
Common in hedgerows, fields, and
gardens in N. America and Eurasia,
this mushroom's large, smooth,
white fruit body with a white to
yellow interior is unmistakable.

*bulbous
stem base*

MEADOW PUFFBALL
Lycoperdon pratense
F: Agaricaceae
This short-stemmed species has an internal membrane separating the spore-mass from the stem. It is common in meadows throughout Eurasia.

¾–1¾ in
2–4.5 cm

¾–1½ in
2–4 cm

COMMON PUFFBALL
Lycoperdon perlatum
F: Agaricaceae
The most common white puffball, this species is found in N. America and Eurasia. Its granular spines rub off to leave circular scars in a regular pattern.

2–4 in
5–10 cm

⅜–1¼ in
1–3 cm

contrasting brown scales

4–12 in
10–30 cm

PARASOL
Macrolepiota procera
F: Agaricaceae
Common in meadows in N. America and Eurasia, this tall species has a scaly cap and snakeskin-patterned stem with a thick ring.

snakeskin pattern on the stem

SPINY PUFFBALL
Lycoperdon echinatum
F: Agaricaceae
This species is found in beech woods in Eurasia and N. America. It has groups of long spines that meet at their tips. Its spores are purplish brown.

PESTLE PUFFBALL
Lycoperdon excipuliforme
F: Agaricaceae
One of the tallest puffballs, this fungus has a buff-white stem that is left behind once the spores are released. It is found in Eurasia.

⅜–1 in
1–2.5 cm

STUMP PUFFBALL
Apioperdon pyriforme
F: Agaricaceae
Common in N. America and Eurasia, this pear-shaped species grows on wood. It has prominent cords at its base, and is firm when young.

¾–2 in
2–5 cm

STINKING PUFFBALL
Lycoperdon foetidum
F: Agaricaceae
Common in acidic heaths and woods in Eurasia, this yellowish brown puffball has dark brown spines. Its flesh has an unpleasant smell.

2¼–6 in
6–15 cm

⅜–1½ in
1–3.5 cm

1¼–2¾ in
3–7 cm

POPLAR FIELDCAP
Cyclocybe cylindracea
F: Bolbitiaceae
Found in Eurasia, this rare species grows on poplar wood. The cap cracks when dry and the stem has a ring.

COMMON FIELDCAP
Agrocybe pediades
F: Bolbitiaceae
Usually found in lawns in Eurasia, this fungus's smooth, yellowish cap and slender stem have a mealy smell and no veil.

SPRING FIELDCAP
Agrocybe praecox
F: Bolbitiaceae
Common in Eurasia and N. America during spring, this fungus's cap may have marginal veil fragments. The stem may have a fragile ring.

bulbous base

2¼–4 in
6–10 cm

1¼–2 in
3–5 cm

⅜–½ in
1–1.5 cm

⅜–1¼ in
1–3 cm

WINTER STALKBALL
Tulostoma brumale
F: Agaricaceae
Typically found in sandy soils in dune slacks in Eurasia and N. America, this fungus has tiny, whitish yellow, round heads on slender, pale brown stalks.

SANDY STILTBALL
Battarreoides digueti
F: Tulostomataceae
Found in very dry, sandy soils in N. America, this tall, stalked species emerges from a leathery "egg." Its brown cap contains the spores.

MILKY CONECAP
Conocybe apala
F: Bolbitiaceae
This species is frequently found in lawns in Eurasia and N. America. Its conical, ivory cap with orange-brown gills sits on a tall stem.

YELLOW FIELDCAP
Bolbitius titubans
F: Bolbitiaceae
Growing in meadows, this fungus is common in N. America and Eurasia. It has a delicate, sticky cap that lasts only a day or so.

»

3¼–8 in
8–20 cm

volva

CAESAR'S MUSHROOM
Amanita caesarea
F: Amanitaceae
Found under oaks, mainly
from the Mediterranean to
C. Eurasia, this mushroom
has an orange cap and stem,
and a white volva.

3¼–6 in
8–15 cm

DEATH CAP
Amanita phalloides
F: Amanitaceae
Common in Eurasia and parts
of N. America, this species has
a fragile ring, large volva, and
sickly smell. Its cap may be
greenish, yellowish, or white.

2–4 in
5–10 cm

FALSE DEATHCAP
Amanita citrina
F: Amanitaceae
Found in Eurasia and
eastern N. America, this
species has a stem with a
distinctly rimmed,
rounded bulb, and flesh
with a potato odor. Its cap
is white or pale yellow.

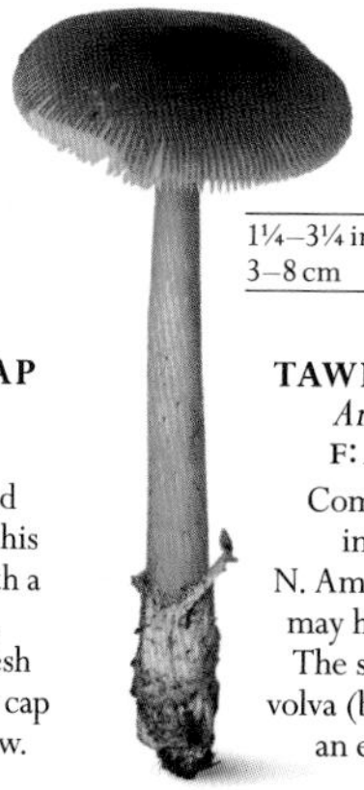

1¼–3¼ in
3–8 cm

TAWNY GRISETTE
Amanita fulva
F: Amanitaceae
Common in woods
in Eurasia and
N. America, this fungus
may have veil patches.
The stem has a white
volva (baglike remains of
an enclosing veil).

buff warts

2¼–7 in
6–18 cm

BLUSHER
Amanita rubescens
F: Amanitaceae
This species is
common in mixed
woods in Eurasia and
N. America. Its cap
varies from cream
to brown, and its flesh
bruises pinkish red.

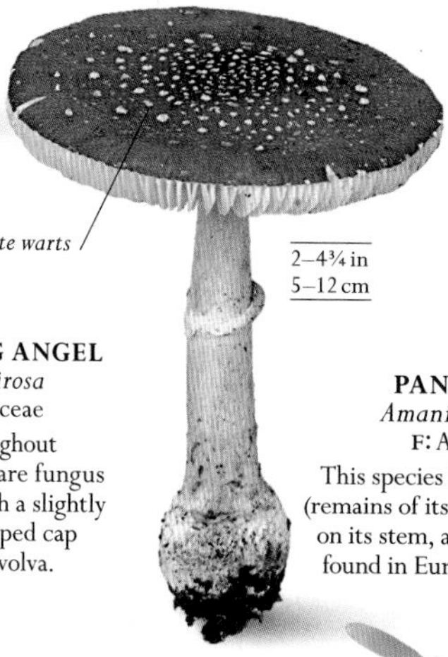

rounded bulb

2¼–6 in
6–15 cm

FLY AGARIC
Amanita muscaria
F: Amanitaceae
Common in Eurasia and
N. America, this distinctive
mushroom is especially
common under birches.
Its white warts may
wash off after rain.

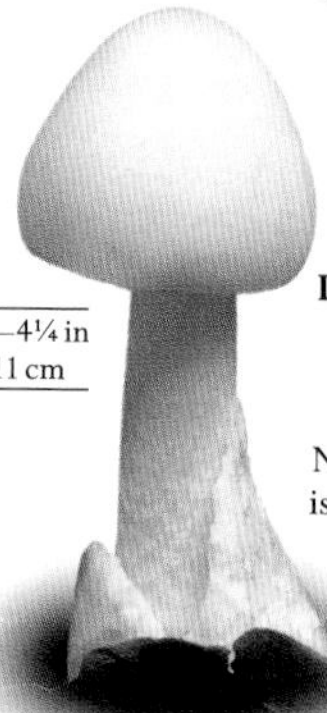

2¼–4¼ in
6–11 cm

DESTROYING ANGEL
Amanita virosa
F: Amanitaceae
Found throughout
N. Eurasia, this rare fungus
is pure white with a slightly
sticky, bell-shaped cap
and a white volva.

pure white warts

2–4¾ in
5–12 cm

PANTHER CAP
Amanita pantherina
F: Amanitaceae
This species has pure white
warts (remains of its protective
veil), a ring on its stem, and a
ridged bulb. It is found in
Eurasia and N. America.

2–6 in
5–15 cm

GOLDEN SPINDLES
Clavulinopsis corniculata
F: Clavariaceae
This golden-yellow club
with antlerlike branches is
relatively common in acidic,
unimproved meadows
and grassy clearings in
woods in Eurasia.

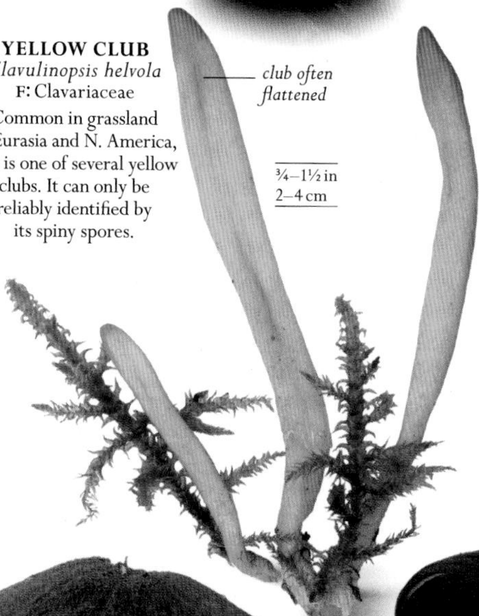

YELLOW CLUB
Clavulinopsis helvola
F: Clavariaceae
Common in grassland
in Eurasia and N. America,
this is one of several yellow
clubs. It can only be
reliably identified by
its spiny spores.

club often
flattened

¾–1½ in
2–4 cm

1½–3¼ in
4–8 cm

VIOLET CORAL
Clavaria zollingeri
F: Clavariaceae
A rare species found in mossy
meadows and woodlands in Eurasia
and N. America, this species has
violet, corallike branches.

1¼–4 in
3–10 cm

WHITE SPINDLES
Clavaria fragilis
F: Clavariaceae
This fungus forms dense
clusters of pure white,
simple clubs in woods
and meadows in Eurasia.

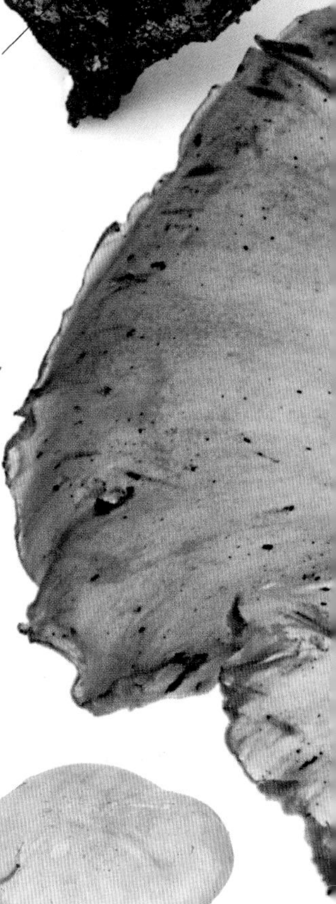

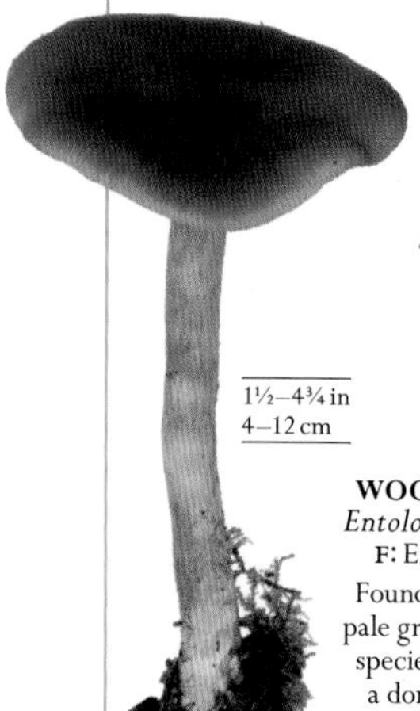

1½–4¾ in
4–12 cm

WOOD PINKGILL
Entoloma rhodopolium
F: Entolomataceae
Found in Eurasia, this
pale gray to gray-brown
species has a cap with
a domed center and
slender stem. It may
smell of bleach.

⅜–1¼ in
1–3 cm

MOUSEPEE PINKGILL
Entoloma incanum
F: Entolomataceae
Smelling of mice, this pinkgill
has a hollow stem and a bright
grass-green cap that fades
to brown with age. It grows
in unimproved meadows
in Eurasia.

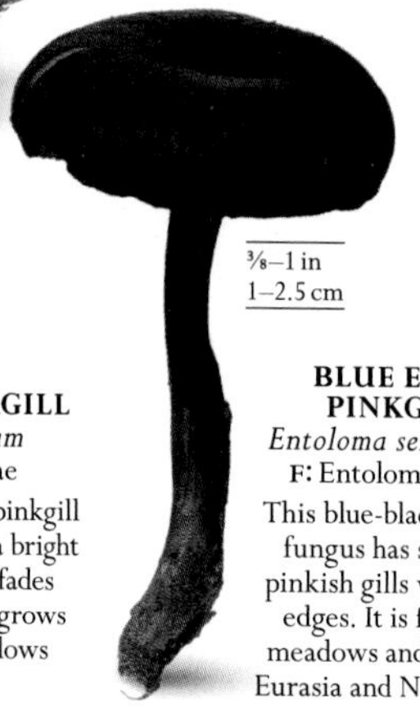

⅜–1 in
1–2.5 cm

BLUE EDGE PINKGILL
Entoloma serrulatum
F: Entolomataceae
This blue-black capped
fungus has serrated
pinkish gills with dark
edges. It is found in
meadows and parks in
Eurasia and N. America.

1½–3¼ in
4–8 cm

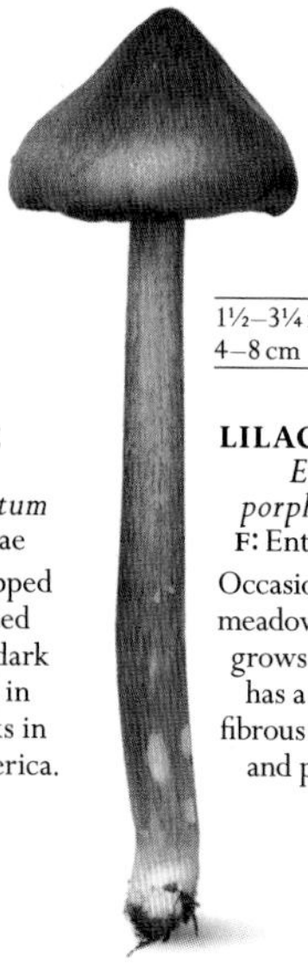

LILAC PINKGILL
*Entoloma
porphyrophaeum*
F: Entolomataceae
Occasionally found in
meadows, this fungus
grows in Eurasia. It
has a gray-purple
fibrous cap and stem,
and pinkish gills.

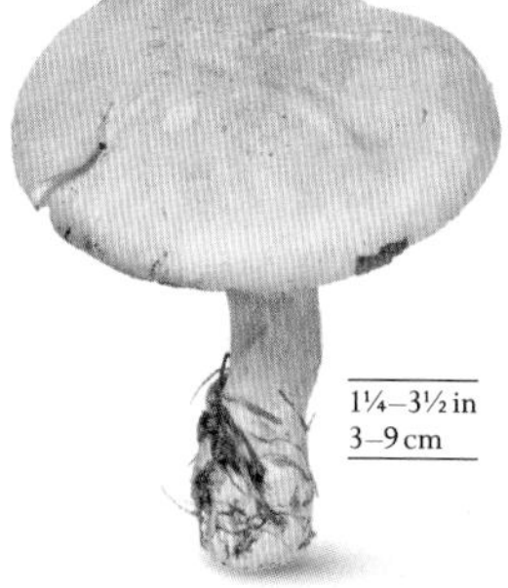

1¼–3½ in
3–9 cm

THE MILLER
Clitopilus prunulus
F: Entolomataceae
Common in mixed woods in
Eurasia and N. America, this
fungus has a cap that changes
shape with age, starting convex
then becoming depressed.

WOOD BLEWIT
Lepista nuda
F: Tricholomataceae

2–8 in
5–20 cm

contrasting cream gills

Common in mixed woods in Eurasia and N. America, this species has a violet cap that fades to brown, but the stem and gills remain violet.

BLUE LEGS
Lepista personata
F: Tricholomataceae

2–8 in
5–20 cm

The blue legs has a buff cap and a bright blue-lilac stem with perfumed flesh. It is common in open meadows in Eurasia.

CLOUDED FUNNEL
Clitocybe nebularis
F: Tricholomataceae

3¼–8 in
8–20 cm

This fleshy mushroom has crowded gills that often run down the stem. It frequently grows in large circles in Eurasia and N. America.

ANISEED FUNNEL
Clitocybe odora
F: Tricholomataceae

1¼–2¼ in
3–6 cm

Common in Eurasia and N. America, this fungus is unmistakable due to its strong aniseed odor. Its sea-green color fades as it matures to grayish green.

IVORY FUNNEL
Clitocybe dealbata
F: Tricholomataceae

¾–2¼ in
2–6 cm

This species grows in rings in lawns and meadows in Eurasia and N. America. It has a frosted appearance and slightly decurrent gills.

SULPHUR KNIGHT
Tricholoma sulphureum
F: Tricholomataceae

¾–3¼ in
2–8 cm

Characterized by its bright sulphur-yellow color and repulsive smell of coal gas, this fungus is common in mixed woods in Eurasia and parts of N. America.

decurrent gills (running down stem)

GIANT FUNNEL
Leucopaxillus giganteus
F: Tricholomataceae

inrolled cap (curls inward at margins)

4¾–16 in
12–40 cm

short stem

Often found in huge fairy rings in Eurasia and N. America, this giant fungus is white, with decurrent gills, an inrolled cap margin, and a short stem.

DINGY AGARIC
Tricholoma portentosum
F: Tricholomataceae

2–4¾ in
5–12 cm

Usually found with conifers in northern N. America and N. Eurasia, this species has a smooth, bell-shaped cap and a stout, whitish stem that flushes yellowish.

SOAPY KNIGHT
Tricholoma saponaceum
F: Tricholomataceae

1½–4 in
4–10 cm

Variable in color—gray-brown, greenish or pinkish gray to mottled—with a soapy smell, this species is common in mixed woods in Eurasia and N. America.

PIPE CLUB
Typhula fistulosa
F: Typhulaceae

2–10 in / 5–25 cm

Often found in large groups in Eurasia and parts of N. America, this species has tall, slender clubs attached to fallen branches and hardwood litter.

»

FLY AGARIC
Amanita muscaria

The fly agaric is arguably the most famous of all fungi, and it appears in illustrations in children's books all over the world. The bright scarlet cap, usually scattered with white spots of veil tissue, make it one of the easiest fungi to identify. Found originally throughout most of Europe, northern Asia, and North America, it now grows wherever humans have planted the host trees—usually birch—with which the fungus forms a mutually beneficial relationship. It is now commonly found in parts of Africa, India, and Australasia. All parts of the fly agaric are poisonous, although rarely fatal.

large floppy ring is easily torn

SIZE Cap diameter 2¼–6 in (6–15 cm)
HABITAT Birch and pine woods
DISTRIBUTION Almost worldwide
SPORE COLOR White

gills are pure white

yellow flesh under outer skin

white warts loosely attached to cap

∨ SLICE THROUGH CAP
In cross section, the yellowish orange flesh immediately beneath the red skin of the cap is revealed. The gills, or lamellae, are bright white, in contrast.

stem flesh

∨ SCALES
The white or pale yellow warty scales are all that remain of an enveloping veil of tissue that once enclosed the young cap.

crowded gills

red skin peels off easily

< FAMILIAR FUNGUS
The fly agaric changes shape dramatically as it grows and expands, but the color and key characters usually remain constant. The cap may fade to yellow if there has been persistent rain. Its common name refers to an old folk remedy—the red skin of the cap was used in a saucer of milk to poison annoying houseflies.

STEM BASE >
The base is swollen and clublike, with warty ridges encircling the upper part of the bulb. Threads of mycelium (fungal tissue) connect to fine tree roots below the soil.

∧ GILLS
The fly agaric's radiating gills are of different lengths, some stopping well short of the stem. This allows them to fill the available space, ensuring that spore production is maximized.

GROWTH STAGES OF A FLY AGARIC

veil warts

warts dispersed

partial veil beneath cap

veil starting to tear

ring now formed

fully exposed gills

upturned cap

∧ BABY BUTTON
The young button is completely enclosed in a warty universal veil.

∧ BROKEN VEIL
The stem has started to grow, and the universal veil on the cap has split apart.

∧ GROWING CAP
The cap expands, with the partial veil still attached underneath.

∧ GILLS APPEAR
The partial veil starts to tear, forming the ring and exposing the gills above.

∧ SPORE DISPERSAL
Spores grow on the exposed gills and are then released into the air.

∧ OLD AGE
At the end of its life the cap coloration fades and the cap may turn upward.

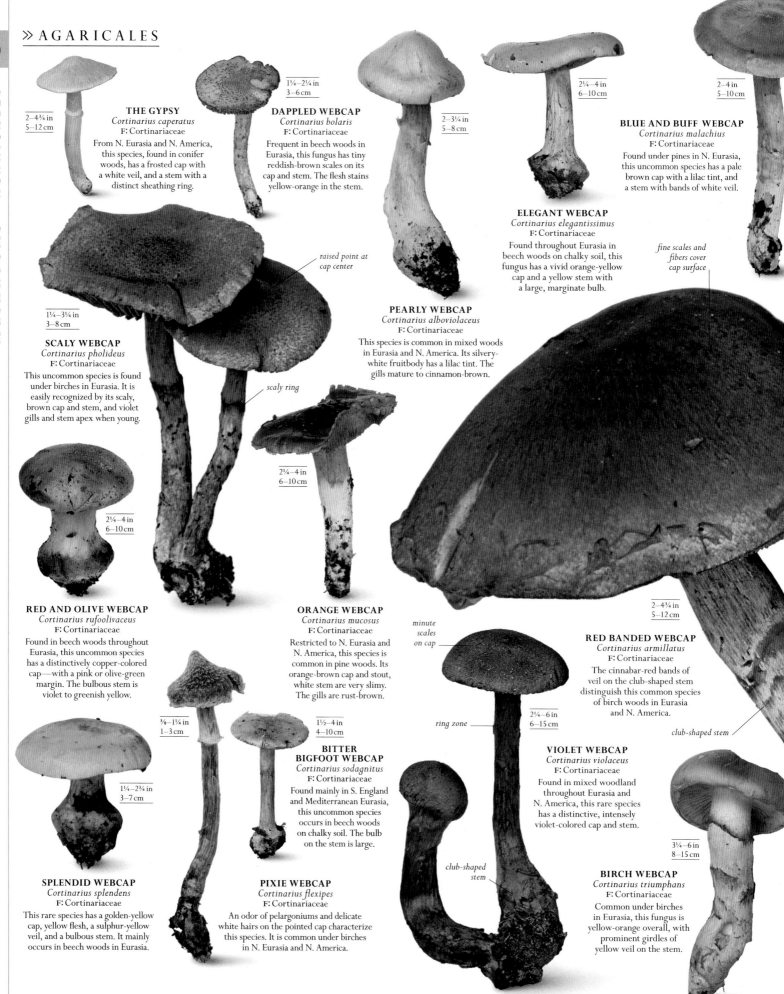

2–4¾ in
5–12 cm

THE GYPSY
Cortinarius caperatus
F: Cortinariaceae
From N. Eurasia and N. America, this species, found in conifer woods, has a frosted cap with a white veil, and a stem with a distinct sheathing ring.

1¼–2¼ in
3–6 cm

DAPPLED WEBCAP
Cortinarius bolaris
F: Cortinariaceae
Frequent in beech woods in Eurasia, this fungus has tiny reddish-brown scales on its cap and stem. The flesh stains yellow-orange in the stem.

2–3¼ in
5–8 cm

2¼–4 in
6–10 cm

BLUE AND BUFF WEBCAP
Cortinarius malachius
F: Cortinariaceae
Found under pines in N. Eurasia, this uncommon species has a pale brown cap with a lilac tint, and a stem with bands of white veil.

2–4 in
5–10 cm

ELEGANT WEBCAP
Cortinarius elegantissimus
F: Cortinariaceae
Found throughout Eurasia in beech woods on chalky soil, this fungus has a vivid orange-yellow cap and a yellow stem with a large, marginate bulb.

fine scales and fibers cover cap surface

raised point at cap center

1¼–3¼ in
3–8 cm

SCALY WEBCAP
Cortinarius pholideus
F: Cortinariaceae
This uncommon species is found under birches in Eurasia. It is easily recognized by its scaly, brown cap and stem, and violet gills and stem apex when young.

PEARLY WEBCAP
Cortinarius alboviolaceus
F: Cortinariaceae
This species is common in mixed woods in Eurasia and N. America. Its silvery-white fruitbody has a lilac tint. The gills mature to cinnamon-brown.

scaly ring

2¼–4 in
6–10 cm

2¼–4 in
6–10 cm

RED AND OLIVE WEBCAP
Cortinarius rufoolivaceus
F: Cortinariaceae
Found in beech woods throughout Eurasia, this uncommon species has a distinctively copper-colored cap—with a pink or olive-green margin. The bulbous stem is violet to greenish yellow.

ORANGE WEBCAP
Cortinarius mucosus
F: Cortinariaceae
Restricted to N. Eurasia and N. America, this species is common in pine woods. Its orange-brown cap and stout, white stem are very slimy. The gills are rust-brown.

minute scales on cap

2–4¾ in
5–12 cm

RED BANDED WEBCAP
Cortinarius armillatus
F: Cortinariaceae
The cinnabar-red bands of veil on the club-shaped stem distinguish this common species of birch woods in Eurasia and N. America.

club-shaped stem

⅜–1¼ in
1–3 cm

1½–4 in
4–10 cm

BITTER BIGFOOT WEBCAP
Cortinarius sodagnitus
F: Cortinariaceae
Found mainly in S. England and Mediterranean Eurasia, this uncommon species occurs in beech woods on chalky soil. The bulb on the stem is large.

ring zone

2¼–6 in
6–15 cm

VIOLET WEBCAP
Cortinarius violaceus
F: Cortinariaceae
Found in mixed woodland throughout Eurasia and N. America, this rare species has a distinctive, intensely violet-colored cap and stem.

1¼–2¾ in
3–7 cm

SPLENDID WEBCAP
Cortinarius splendens
F: Cortinariaceae
This rare species has a golden-yellow cap, yellow flesh, a sulphur-yellow veil, and a bulbous stem. It mainly occurs in beech woods in Eurasia.

PIXIE WEBCAP
Cortinarius flexipes
F: Cortinariaceae
An odor of pelargoniums and delicate white hairs on the pointed cap characterize this species. It is common under birches in N. Eurasia and N. America.

club-shaped stem

3¼–6 in
8–15 cm

BIRCH WEBCAP
Cortinarius triumphans
F: Cortinariaceae
Common under birches in Eurasia, this fungus is yellow-orange overall, with prominent girdles of yellow veil on the stem.

CAP (TOPSIDE)   **CAP (UNDERSIDE)**

¾–2¾ in
2–7 cm

PEELING OYSTERLING
Crepidotus mollis
F: Crepidotaceae

This fungus occurs in Eurasia and N. America. Its small, fan-shaped, pale cap has a gelatinous skin, which peels easily. The stem is absent or very short.

³⁄₁₆–1¼ in
0.5–3 cm

VARIABLE OYSTERLING
Crepidotus variabilis
F: Crepidotaceae

One of several similar species in Eurasia and N. America, its cap is dry with fine fibers. It usually lacks a stem.

⅛–⁵⁄₁₆ in
0.3–0.8 cm

TINY PIXIE CAP
Galerina calyptrata
F: Hymenogastraceae

Found in Eurasia, this fungus is one of many species only identifiable under a microscope. Its rounded cap is orange-brown and striated.

CAP NOT FULLY EXPANDED

2–4¾ in
5–12 cm

ROOTING POISONPIE
Hebeloma radicosum
F: Hymenogastraceae

The deeply rooting stem with a large ring, buff cap with flat scales, and strong smell of marzipan characterize this Eurasian species.

attached to wood

⅜–2 in
1–5 cm

DEADLY GALERINA
Galerina marginata
F: Hymenogastraceae

This species has an orange-brown cap and a small ring on the stem. It grows on fallen trees in Eurasia and N. America.

1½–3½ in
4–9 cm

POISONPIE
Hebeloma crustuliniforme
F: Hymenogastraceae

From Eurasia and N. America, this species strongly smells of radish. Its cap is ivory-white to buff, and slimy when wet. In damp weather, the gills exude droplets.

1¼–2¾ in
3–7 cm

SPLIT FIBERCAP
Inocybe rimosa
F: Inocybaceae

Common in mixed woodland in Eurasia and N. America, its pointed, fibrous cap is straw-yellow and the stem is tall and slender.

red band

1¼–2¾ in
3–7 cm

STAR FIBERCAP
Inocybe asterospora
F: Inocybaceae

A flattened bulb at the stem's base and star-shaped spores distinguish this small, brown fibrous-capped species from Eurasia.

radial fibers

1¼–3½ in
3–9 cm

stout stem

⅜–1½ in
1–4 cm

WHITE FORM

WHITE FIBERCAP
Inocybe geophylla
F: Inocybaceae

This is one of the most common *Inocybe* species in woods in Eurasia and N. America. It has a conical, silky cap and a slender stem.

VIOLET FORM

⅜–1¾ in
1–4.5 cm

TORN FIBERCAP
Inocybe lacera
F: Inocybaceae

Found in Eurasia and N. America, this fungus is distinguished by its cylindrical spores. It has a scaly, fibrous cap and a slender brown stem.

DEADLY FIBERCAP
Inocybe erubescens
F: Inocybaceae

Found in mixed woodland on chalk in Eurasia, this uncommon species has a fibrous cap that discolors with age, and stout stem that bruises red.

⁵⁄₁₆–1½ in
0.8–4 cm

LILAC LEG FIBERCAP
Inocybe griseolilacina
F: Inocybaceae

A common species of beech woods in Eurasia, it has a scaly cap and a pale violet, slender stem.

gills are broadly attached

3/8–2¼ in
2.5–6 cm

½–2 in
1.5–5 cm

3/8–1½ in
1–4 cm

green stem apex

MEADOW WAXCAP
Cuphophyllus pratensis
F: Hygrophoraceae
One of the larger meadow waxcaps, it is found in Eurasia and N. America. It has a stout stem and a fleshy cap with gills running down the stem.

SNOWY WAXCAP
Cuphophyllus virgineus
F: Hygrophoraceae
The most common waxcap in all types of grassland in Eurasia, its waxy cap and slender stem are translucent. It has decurrent gills.

PARROT WAXCAP
Gliophorus psittacinus
F: Hygrophoraceae
This fungus occurs in Eurasia and N. America. Its slimy cap is vivid green to orange and the upper part of the stem is bright green.

fleshy, waxy, pink gills

1¼–2¾ in
3–7 cm

PINK WAXCAP
Porpolomopsis calyptriformis
F: Hygrophoraceae
This highly recognizable but rare fungus has a pointed pink cap, fragile pale stem, and waxy gills. Found in Eurasia, it grows in unimproved meadows.

3/8–2 in
1–5 cm

½–2¼ in
1.5–6 cm

1½–4¾ in
4–12 cm

pale yellow margin

BLACKENING WAXCAP
Hygrocybe conica
F: Hygrophoraceae
Quite common in grasslands and woodlands in Eurasia and N. America, this species has a conical reddish orange cap and fibrous stem that blacken with age or bruising.

SCARLET WAXCAP
Hygrocybe coccinea
F: Hygrophoraceae
The scarlet, waxy cap, gills, and stem make this a very striking fungus of unimproved grasslands in Eurasia and N. America.

1½–3¼ in
4–8 cm

CLUB FOOT
Ampulloclitocybe clavipes
F: Hygrophoraceae
Common in mixed woods in late fall, this species from Eurasia and N. America has decurrent gills and a swollen, spongy base.

½–2¾ in
1.5–7 cm

CRIMSON WAXCAP
Hygrocybe punicea
F: Hygrophoraceae
Found in unimproved grassland in Eurasia and N. America, this uncommon species is the largest blood-red waxcap. Its stem is dry and fibrous with a white base.

GOLDEN WAXCAP
Hygrocybe chlorophana
F: Hygrophoraceae
Found in Eurasia, this is the commonest waxcap found in meadows. It has a bright yellow-orange, slightly sticky cap.

1¼–3¼ in
3–8 cm

IVORY WOODWAX
Hygrophorus eburneus
F: Hygrophoraceae
This species is found in beech woods in Eurasia and N. America. It has a sticky cap and stem, thick gills, and a flowery smell.

¾–2 in
2–5 cm

3/8–2 in
1–5 cm

3/16–¾ in
0.5–2 cm

AMETHYST DECEIVER
Laccaria amethystina
F: Hydnangiaceae
This highly recognizable, slender fungus from Eurasia and N. America has an intense violet coloration when fresh, and powdery spores dusting the gills.

CONIFERCONE CAP
Baeospora myosura
F: Cyphellaceae
Found in Eurasia and N. America, this is one of a few fungi growing mostly on pine cones. It has crowded, narrow gills and a velvety stem.

pine cone

1¼–2 in
3–5 cm

DECEIVER
Laccaria laccata
F: Hydnangiaceae
Variable in color—from brick-red to flesh-pink—with a dry cap and thick gills, this species is abundant in temperate woods in Eurasia and N. America.

HERALD OF WINTER
Hygrophorus hypothejus
F: Hygrophoraceae
Appearing in pine woods after frost in Eurasia and N. America, this sticky fungus has an olive-brown cap and yellow stem.

cap frequently splits as it ages

⅜–1½ in
1–4 cm

³⁄₁₆–½ in
0.5–1.5 cm

1¼–4¾ in
3–12 cm

ST. GEORGE'S MUSHROOM
Calocybe gambosa
F: Lyophyllaceae
This ivory-buff species from Eurasia often grows in fairy rings at woodland margins in late spring. The mushroom smells strongly of newly ground meal.

PINK DOMECAP
Calocybe carnea
F: Lyophyllaceae
This species grows in short grass in Eurasia and N. America. Its smooth cap and fibrous stem are rose-pink and the gills are white.

2–4 in
5–10 cm

SILKY PIGGYBACK
Asterophora parasitica
F: Lyophyllaceae
One of two well-known parasitic species to live on rotten fruitbodies of brittle-caps, this fungus from Eurasia and N. America has a rounded, silky cap.

CLUSTERED DOMECAP
Lyophyllum decastes
F: Lyophyllaceae
Common on roadsides, pathsides, and disturbed soils in Eurasia and N. America, this fungus forms clumps of tough caps with stout stems.

HORSEHAIR PARACHUTE
Gymnopus androsaceus
F: Marasmiaceae
Found in Eurasia, this fungus has a distinctive black hairlike stem, while its radially grooved cap is pale pinkish brown.

⅛–⅜ in
0.3–1 cm

HAIRY PARACHUTE
Crinipellis scabella
F: Marasmiaceae
Found on dead stems of grasses, this recognizable fungus is from Eurasia. It has a small cap and a long, slender stem covered with dense, bristling brown hair.

³⁄₁₆–½ in
0.5–1.5 cm

cap depressed at center

radial groove

⅜–2 in
1–5 cm

FAIRY RING CHAMPIGNON
Marasmius oreades
F: Marasmiaceae
This classic fairy-ring mushroom is common in open grass in Eurasia and N. America. It has fleshy, beige caps with thick gills.

³⁄₁₆–⅜ in
0.5–1 cm

COMMON BIRD'S NEST
Crucibulum laeve
F: Nidulariaceae
Forming miniature nests with spore-filled, egglike cases, this common species may be hard to spot on woody debris. It occurs in Eurasia and N. America.

³⁄₁₆–¾ in
0.5–2 cm

½–1½ in
1.5–4 cm

GARLIC PARACHUTE
Mycetinis alliaceus
F: Marasmiaceae
This species grows in beech woods in Eurasia. It has a tall, slender, blackish stem and a strong odor of rancid garlic.

wiry stem

COLLARED PARACHUTE
Marasmius rotula
F: Marasmiaceae
Like little parachutes, this mushroom's rounded, depressed caps are radially grooved. Its tough stems are attached to woody debris. It occurs in Eurasia and N. America.

³⁄₁₆–⅜ in
0.5–1 cm

FLUTED BIRD'S NEST
Cyathus striatus
F: Nidulariaceae
This fungus is recognized by its hairy, brown, and tall fluted nests containing 10–15 egglike cases. It occurs on woody debris across Eurasia and N. America, but is not common.

>>

¾–2 in
2–5 cm
wavy margin
purple underside

SILVERLEAF FUNGUS
Chondrostereum purpureum
F: Cyphellaceae
Common on cherry and plum trees, this fungus is found in Eurasia and N. America. Violet below when young, it darkens to purple-brown with age.

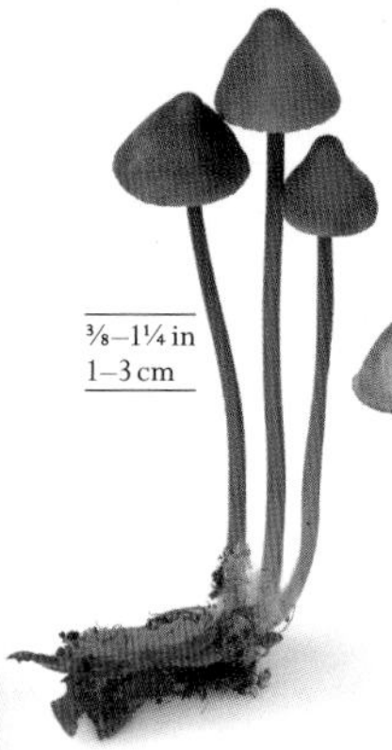
⅜–1¼ in
1–3 cm

SAFFRONDROP BONNET
Mycena crocata
F: Mycenaceae
Common in woods with chalky soils in Eurasia, this species exudes a bright saffron-orange juice when it is broken.

⅜–2¼ in
1–6 cm

COMMON BONNET
Mycena galericulata
F: Mycenaceae
Abundant in temperate woods in Eurasia and N. America, this fungus is variable in color. Its pinkish-gray gills have linking veins and cross-ridges.

⅛–⅜ in
0.3–1 cm

ORANGE BONNET
Mycena acicula
F: Mycenaceae
Common in leaf litter and debris in broadleaf woods in Eurasia and N. America, this fungus has a tiny, translucent cap that is striped almost to the center.

³⁄₁₆–1 in
0.5–2.5 cm

YELLOWLEG BONNET
Mycena epipterygia
F: Mycenaceae
Occurring in Eurasia and N. America, this fungus grows on acidic soils in woods and heaths. Its cap and stem have a sticky, peelable layer.

⅜–1½ in
1–4 cm

CLUSTERED BONNET
Mycena inclinata
F: Mycenaceae
Found in dense clusters on wood in Eurasia and N. America, this fungus has a cap with toothed margins and a strong, soapy smell.

1¼–2¼ in
3–6 cm

BLACKEDGE BONNET
Mycena pelianthina
F: Mycenaceae
Found in Eurasia, this fungus is characterized by a strong smell of radish. It has purplish black-edged gills and a pale lilac to gray-brown cap.

domed
cap center
¾–2¼ in
2–6 cm
club-shaped
stem

ROSY BONNET
Mycena rosea
F: Mycenaceae
Found in beech woods in Eurasia, this common species has a robust pink cap and stem, and smells strongly of radish.

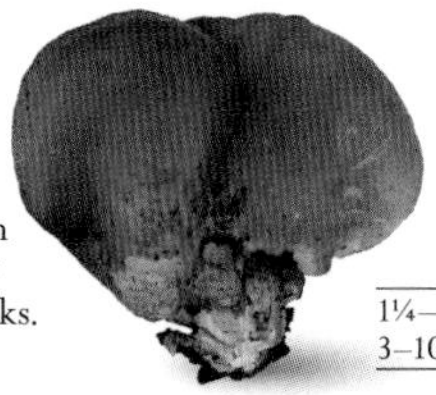
1¼–4 in
3–10 cm

OLIVE OYSTERLING
Sarcomyxa serotina
F: Mycenaceae
From Eurasia and N. America, this mushroom fruits in winter, often near water on hardwood tree trunks. Its cap is slimy when wet.

1½–3¼ in
4–8 cm

SPINDLE TOUGHSHANK
Gymnopus fusipes
F: Omphalotaceae
Abundant in Eurasian oak woods from early summer onward, this fungus forms large clumps of tough fruit bodies on tree roots.

1–2¼ in
2.5–6 cm

WOOD WOOLLYFOOT
Gymnopus peronatus
F: Omphalotaceae
The common name of this species from Eurasia refers to the stiff, fuzzy hairs at the base of the stem.

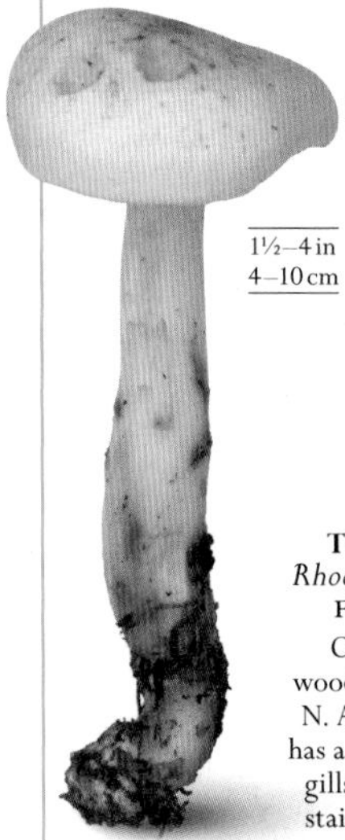
1½–4 in
4–10 cm

SPOTTED TOUGHSHANK
Rhodocollybia maculata
F: Omphalotaceae
Common in mixed woods across Eurasia and N. America, this fungus has a white cap, stem, and gills. The crowded gills stain rust-red with age.

1¼–2¼ in
3–6 cm

BUTTER CAP
Rhodocollybia butyracea
F: Omphalotaceae
This fungus is abundant in woods in Eurasia and N. America. Varying from blackish or reddish brown to dark ocher, its cap is greasy to the touch.

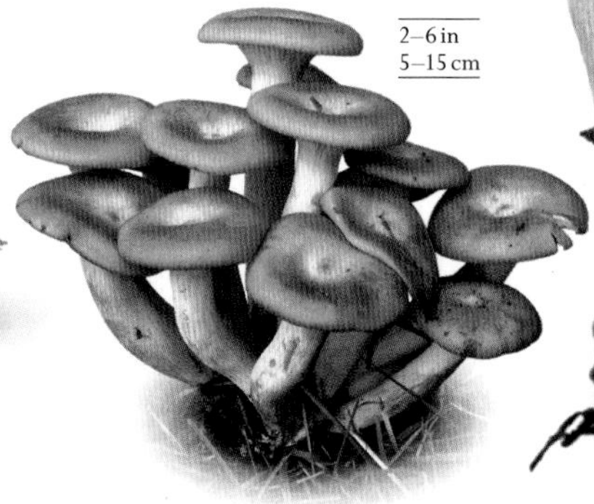
2–6 in
5–15 cm

JACK O'LANTERN
Omphalotus illudens
F: Omphalotaceae
Found across Eurasia and N. America, this bright orange, poisonous fungus is famous for gills that glow in the dark with an eerie, greenish light.

1¼–4 in
3–10 cm

HONEY FUNGUS
Armillaria lutea
F: Physalacriaceae
Found in woods across
Eurasia, this species is
usually found on the
ground. It is a weak
parasite on the trees
around which it grows.

1–4 in
2.5–10 cm

WRINKLED PEACH
Rhodotus palmatus
F: Physalacriaceae
Found on fallen logs of mainly
elm, this uncommon species of
Eurasia and N. America has an
unusual, peach-pink, wrinkled
cap and a fruity odor.

⅜–2¼ in
1–6 cm

VELVET SHANK
Flammulina velutipes
F: Physalacriaceae
Growing throughout winter
in Eurasia and N. America,
this fungus is characterized
by its often sticky cap and
velvety stem.

¾–6 in
2–15 cm

**PORCELAIN
FUNGUS**
Mucidula mucida
F: Physalacriaceae
Usually found on beech logs
in Eurasia, this fungus has
gray-white caps that are
slimy when wet. The tough
stem has a thin ring.

1–4 in
2.5–10 cm

ROOTING SHANK
Hymenopellis radicata
F: Physalacriaceae
Found across Eurasia
and N. America, this
deep-rooted fungus has a
stiff, tall stem. The cap is
slimy when wet, and the
gills are widely spaced.

BRANCHING OYSTER
Pleurotus cornucopiae
F: Pleurotaceae
Growing in clusters on fallen logs, usually
of elm, this fungus has trumpet-shaped
caps with gills running down the short,
frequently branching stems. It occurs
in Eurasia and N. America.

1½–4¾ in
4–12 cm

*caps often
overlap*

2¼–8 in
6–20 cm

tiny glistening granules

¾–1¼ in
2–3 cm

GLISTENING INKCAP
Coprinellus micaceus
F: Psathyrellaceae
Usually found in clusters, this
mushroom's rounded, grooved caps
are dusted with micalike flakes of
veil. It is common on wood in
Eurasia and N. America.

OYSTER MUSHROOM
Pleurotus ostreatus
F: Pleurotaceae
From Eurasia and N. America,
this species is found on trees
and logs. Its shelflike caps vary
from blue-green to pale buff.
The stem is almost absent.

*white veil
breaks into
patches*

1–3 in
2.5–7.5 cm

³⁄₁₆–⅜ in
0.5–1 cm

FAIRIES' BONNETS
Coprinellus disseminatus
F: Psathyrellaceae
Often found in clusters on rotted
stumps in Eurasia and N. America,
it has tiny, umbrellalike caps with
deeply fluted grooves. The gills
are blackish when mature.

WEEPING WIDOW
*Lacrymaria
lacrymabunda*
F: Psathyrellaceae
Common on pathsides
and disturbed soils in
Eurasia and N. America,
it is named after its black
gills that weep droplets
from their edges.

½–2¾ in
1.5–7 cm

PALE BRITTLESTEM
Psathyrella candolleana
F: Psathyrellaceae
Usually found in clusters on
woody debris in early summer,
this fungus has fragile, slender
stems and pale or buff caps. It
occurs in Eurasia and N. America.

⁵⁄₁₆–1½ in
0.8–4 cm

**CLUSTERED
BRITTLESTEM**
Psathyrella multipedata
F: Psathyrellaceae
The densely clustered stems
of this species are joined at
the base. It is usually found
in open grass across Eurasia.

1–3¼ in
2.5–8 cm

COMMON INKCAP
Coprinopsis atramentaria
F: Psathyrellaceae
Occurring in Eurasia and N. America, this
species has egg-shaped caps that dissolve to
a blackish fluid as the spores are released.

2–3¼ in
5–8 cm

MAGPIE INKCAP
Coprinopsis picacea
F: Psathyrellaceae
Found on chalky soils in
Eurasian woodlands, this
uncommon species has white,
fluffy scales that contrast with
the dark gray-brown cap.

»

1½–4 in
4–10 cm

DEER SHIELD
Pluteus cervinus
F: Pluteaceae

Found in Eurasia and
N. America, this species is variable in
color, with a radially fibrous cap, often
dome-centered; pink gills not attached
to the stem; and a fibrous stem.

⅜–2¼ in
1–6 cm

YELLOW SHIELD
Pluteus chrysophlebius
F: Pluteaceae

A golden to greenish-yellow
cap, yellow gills turning
pink, and a whitish stem
characterize this species.
Found in Eurasia, it grows
on decayed wood.

1–3¼ in
2.5–8 cm

WILLOW SHIELD
Pluteus salicinus
F: Pluteaceae

Common in broad-leaved
woodlands in Eurasia, this
species can be identified by
the bluish-gray staining at the
base of its slender stem.

4–10 in
10–25 cm

BEEFSTEAK FUNGUS
Fistulina hepatica
F: Schizophyllaceae

Resembling a fleshy steak,
with bloodlike red juice,
this species is particularly
common in warmer parts
of Eurasia and N. America.

CAP (UNDERSIDE)

*fine, silky-hairy
surface*

4–10 in
10–25 cm

SILKY ROSEGILL
Volvariella bombycina
F: Pluteaceae

Growing on deciduous trees in Eurasia and
N. America, this rare species has a white to pale
lemon cap and a whitish stem with a baglike,
thin veil enclosing the base of the stem.

⅜–2 in
1–5 cm

COMMON PORECRUST
Schizophyllum commune
F: Schizophyllaceae

Found in Eurasia and
N. America, this fan-shaped
fungus is identified by the
gill-like spore surface on
the underside of the cap.

CAP
(UPPER SIDE)

*greenish
yellow gills*

*orange cap
center*

2¼–5½ in
6–14 cm

STUBBLE ROSEGILL
Volvopluteus gloiocephalus
F: Pluteaceae

Quite common in fields and old
stubble, and on woodchip mulch,
in Eurasia and N. America, this
species has a sticky, gray cap.

1¼–2¾ in
3–7 cm

CONIFER TUFT
Hypholoma capnoides
F: Strophariaceae

Found in Eurasia
and N. America, this
uncommon species
occurs on conifer wood.
Its whitish gills mature
to grayish lilac.

1¼–2¾ in
3–7 cm

SULPHUR TUFT
Hypholoma fasciculare
F: Strophariaceae

Abundant in temperate woods
in Eurasia and N. America, this
species has greenish yellow gills
that mature to dark purple, and
are easy to recognize in the field.

2–4 in
5–10 cm

BRICK TUFT
Hypholoma lateritium
F: Strophariaceae

This species is characterized by a
fleshy cap with veil fragments and
pale yellow gills that mature to
lavender. It grows on hardwoods
in Eurasia and N. America.

SHEATHED WOODTUFT
Kuehneromyces mutabilis
F: Strophariaceae

¾–2¾ in
2–7 cm

ring on stem

Often confused with the deadly *Galerina marginata*, this species can be identified by its sticky cap, scaly stem, and brown gills. It occurs in Eurasia and N. America.

ALDER SCALYCAP
Pholiota alnicola
F: Strophariaceae

Found in Eurasia, this species is characterized by a sticky cap and tufted habit. Despite its common name, it is usually seen at the base of birches.

½–2¼ in
1.5–6 cm

1¼–2¾ in
3–7 cm

REDLEAD ROUNDHEAD
Leratiomyces ceres
F: Strophariaceae

Previously known as *Stropharia aurantiaca*, this species has a bright red cap and reddish, flushed stem. It grows among woodchips in Eurasia and N. America.

1¼–4¾ in
3–12 cm

GOLDEN SCALYCAP
Pholiota aurivella
F: Strophariaceae

This rather common species is found on beech logs or trunks in Eurasia and N. America. Its sticky golden caps have dark orange-brown scales.

DUNG ROUNDHEAD
Protostropharia semiglobata
F: Strophariaceae

³⁄₁₆–1½ in
0.5–4 cm

Occurring on animal dung or in grass where animals graze, this species is found in Eurasia and N. America. It has a sticky cap and a slender stem with a ring.

2–6 in
5–15 cm

SHAGGY SCALYCAP
Pholiota squarrosa
F: Strophariaceae

Growing in Eurasia and N. America, this fungus has a distinctive dry, sharply scaly cap and stem; pale yellow gills; and a smell of corn or radish.

scaly, furry surface

LIBERTY CAP
Psilocybe semilanceata
F: Strophariaceae

³⁄₁₆–¾ in
0.5–2 cm

This well-known agaric has a conical, dull yellow cap with a distinct point. Common in meadows in late fall, it occurs in Eurasia and N.America.

1¼–2¾ in
3–7 cm

BLUE-GREEN SLIMEHEAD
Stropharia cyanea
F: Strophariaceae

This fungus has a blue-green cap that fades to yellow. It is found in Eurasia and N. America.

2¼–6 in
6–15 cm

1½–2¾ in
4–7 cm

2–4 in
5–10 cm

2–6 in
5–15 cm

cap margin with veil remnants

WHITELACED SHANK
Megacollybia platyphylla
F: Incertae sedis

Found in Eurasia and N. America, this species has a radially fibrous, pale gray-brown cap; widely spaced, deep gills; and a stem base with rootlike strands.

COMMON CAVALIER
Melanoleuca polioleuca
F: Incertae sedis

Common in grass in Eurasia, this species has a gray-brown cap with white gills. Its stem has blackish flesh at the base.

PLUMS AND CUSTARD
Tricholomopsis rutilans
F: Incertae sedis

Common on pine stumps, this species has a reddish purple cap and stem, and golden-yellow gills. It grows in Eurasia and N. America.

SPECTACULAR RUSTGILL
Gymnopilus junonius
F: Incertae sedis

Occurring in Eurasia, this fungus is found in clumps, usually at the base of a tree. It has a dry cap with crowded, shallow yellowish gills.

⅜–1½ in
1–4 cm

2–8 in
5–20 cm

root attaches to conifer wood

PETTICOAT MOTTLEGILL
Panaeolus papilionaceus
F: Incertae sedis

The minute toothed remains of the veil at the cap margin and mottled black gills characterize this species. It is found in Eurasia and N. America.

⅜–2¼ in
1–6 cm

¾–3¼ in
2–8 cm

EGGHEAD MOTTLEGILL
Panaeolus semiovatus
F: Incertae sedis

Found on animal dung, this species occurs in Eurasia and N. America. It is characterized by a sticky, gray cap and a ring around its tall stem.

GOBLET
Pseudoclitocybe cyathiformis
F: Incertae sedis

1¼–2¾ in
3–7 cm

This distinctive species has a very dark coloration and a tall, fibrous stem. Found in Eurasia, it is common in late fall and winter.

CINNABAR POWDERCAP
Cystodermella cinnabarina
F: Incertae sedis

Found in Eurasia and N. America, this species has a brick-red cap and pale cream gills. The cap and stem have a granular surface.

TROOPING FUNNEL
Infundibulicybe geotropa
F: Incertae sedis

This fungus has a funnel-shaped, fleshy cap; a tall stem; and pale leather-brown coloring. It occurs in Eurasia.

BOLETALES

This order contains fleshy fungi and includes those with both pored and gilled fruit bodies, most with cap and stem, but some are crusts, puffballs, or trufflelike. The majority live in association with trees (mycorrhizal), but some feed on dead wood and cause a brown rot, while others are parasitic. The spore-producing layer—the hymenium—is easily loosened from the flesh.

wrinkled cap surface

orange-brown cap

1½–6 in
4–15 cm

cylindrical stem

4–10 in
10–25 cm

2¾–6 in
7–15 cm

BAY BOLETE
Imleria badia
F: Boletaceae

A common species found among conifers or beech trees in Eurasia and N. America, its color varies from orange-brown to bay.

PENNY BUN
Boletus edulis
F: Boletaceae

Found worldwide, this species has fine white markings on its stem, unchanging cream flesh, and white pores that mature to yellow.

SUMMER BOLETE
Boletus reticulatus
F: Boletaceae

A cracking, matte, brown cap and a stem with a fine white net extending to its base distinguish this species from Eurasia and eastern N. America.

PARASITIC BOLETE
Pseudoboletus parasiticus
F: Boletaceae

This small mushroom from Eurasia and N. America grows only on the common earth-ball, causing its host to become hollow.

2¼–5½ in
6–14 cm

BITTER BEECH BOLETE
Caloboletus calopus
F: Boletaceae

From Eurasia and western N. America, its cap varies from white to buff. Its yellow pores and cream flesh bruise blue.

sticky cap

black scales on face

1¼–2 in
3–5 cm

2¼–6 in
6–15 cm

BITTER BOLETE
Tylopilus felleus
F: Boletaceae

This species, found in Eurasia and N. America, is characterized by pores that turn pink with age and a strongly netted stem.

¾–2¾ in
2–7 cm

PEPPERY BOLETE
Chalciporus piperatus
F: Boletaceae

Common under conifers and also found with fly agaric under birches in Eurasia and N. America, it has cinnamon pores and yellow flesh.

common earthball

2–4 in
5–10 cm

OLD MAN OF THE WOODS
Strobilomyces strobilaceus
F: Boletaceae

Found in Eurasia and N. America, this rare species is unmistakable with black woolly scales on its cap and stem, and white tubes.

3¼–6 in
8–15 cm

ORANGE BIRCH BOLETE
Leccinum versipelle
F: Boletaceae
Found in Eurasia, this species has a yellow-orange fleshy cap. The stem has black, woolly flecks and the flesh stains lilac-black.

2¼–6 in
6–15 cm

BROWN BIRCH BOLETE
Leccinum scabrum
F: Boletaceae
One of many similar species, this fungus is from Eurasia and N. America. Its flesh may flush pink when cut, and the cap is sticky when wet.

⅜–¾ in
1–2 cm

STALKED PUFFBALL-IN-ASPIC
Calostoma cinnabarinum
F: Calostomataceae
From N. America, this species' bright cinnabar-red ball on a stem emerges from a gelatinous layer.

2–39 in
5–100 cm

WET ROT
Coniophora puteana
F: Coniophoraceae
Found worldwide, this fungus forms a brown sheet of tissue, often warty or wrinkled, on wet lumber. It causes serious damage to buildings.

2–3½ in
5–9 cm

BAROMETER EARTHSTAR
Astraeus hygrometricus
F: Diplocystidiaceae
Frequent in Eurasia and N. America, this species' starlike arms peel back to reveal the inner spore-filled ball, but close again in dry weather.

2–3¼ in
5–8 cm

CORNFLOWER BOLETE
Gyroporus cyanescens
F: Gyroporaceae
Growing on acidic soils in Eurasia and eastern N. America, this uncommon species is distinguished by its brittle, hollow stem.

½–2 in
1.5–5 cm

ROSY SPIKE
Gomphidius roseus
F: Gomphidiaceae
Growing in association with the bolete *Suillus bovinus* under pines in Eurasia, this fungus has a slimy, rose-pink cap with grayish gills.

COPPER SPIKE
Chroogomphus rutilus
F: Gomphidiaceae
Common under pines in Eurasia and western N. America, this species has a pointed, coppery brown cap.

1½–3¼ in
4–8 cm

BROWN ROLLRIM
Paxillus involutus
F: Paxillaceae
Common in mixed woods in Eurasia and N. America, the downy, inrolled cap margin and soft yellow gills staining brown characterize this species.

2¼–6 in
6–15 cm

1½–4 in
4–10 cm

COMMON EARTH-BALL
Scleroderma citrinum
F: Sclerodermataceae
This potatolike fungus is common in damp woodland in Eurasia and N. America. Its thick skin has dark scales and a spore-filled black interior.

¾–2 in
2–5 cm

POTATO EARTHBALL
Scleroderma bovista
F: Sclerodermataceae
Common in woods in Eurasia and N. America, its smooth outer skin cracks into a fine mosaic. The inner purple-black spore-mass dries brownish.

cap skin peels off

¾–3¼ in
2–8 cm

FALSE CHANTERELLE
Hygrophoropsis aurantiaca
F: Hygrophoropsidaceae
Sometimes mistaken for a chanterelle, this species from Eurasia and N. America is distinguished by its crowded, multi-forked, soft gills.

4–12 in
10–30 cm

VELVET ROLLRIM
Tapinella atrotomentosa
F: Tapinellaceae
Common on pine stumps in Eurasia and N. America, it has a cap with an inrolled margin; soft, thick gills; and a velvety stem.

2–4 in
5–10 cm

DYEBALL
Pisolithus arhizus
F: Sclerodermataceae
This species is found worldwide on poor, sandy soils associated with pines. The egglike spore-sacs in its interior are embedded in blackish jelly.

1½–4 in
4–10 cm

dots can weep milky latex

WEEPING BOLETE
Suillus granulatus
F: Suillaceae
Common under pines in Eurasia and N. America, its ringless stem is covered with glandular dots.

2–4 in
5–10 cm

SLIPPERY JACK
Suillus luteus
F: Suillaceae
Found on pines in Eurasia and N. America, this fungus has a slimy cap and yellow pores. The large ring on its stem has a lavender underside.

1¼–2¾ in
3–7 cm

BOVINE BOLETE
Suillus bovinus
F: Suillaceae
Common under pines in Eurasia and N. America, this species has a sticky cap and angular, irregular pores.

yellow-brown cap surface is dry and rough

2–4 in
5–10 cm

LARCH BOLETE
Suillus grevillei
F: Suillaceae
Confined to larch woods in Eurasia and N. America, this species is yellow-orange to brick-red in color. The stem has an ascending ring.

VELVET BOLETE
Suillus variegatus
F: Suillaceae
Frequent under pines in Eurasia, this species has a felty-scaly cap, with dark cinnamon-brown pores, and a ringless stem.

2¾–5 in
7–13 cm

CANTHARELLALES

The species of the Cantharellales order may look like agarics (members of the Agaricales order), but differ in several important respects. They may have fleshy fruitbodies with a cap and stem, but they lack true gills, having instead a smooth, wrinkled, or folded gill-like spore-producing surface on the underside. The spores are smooth and usually white to cream.

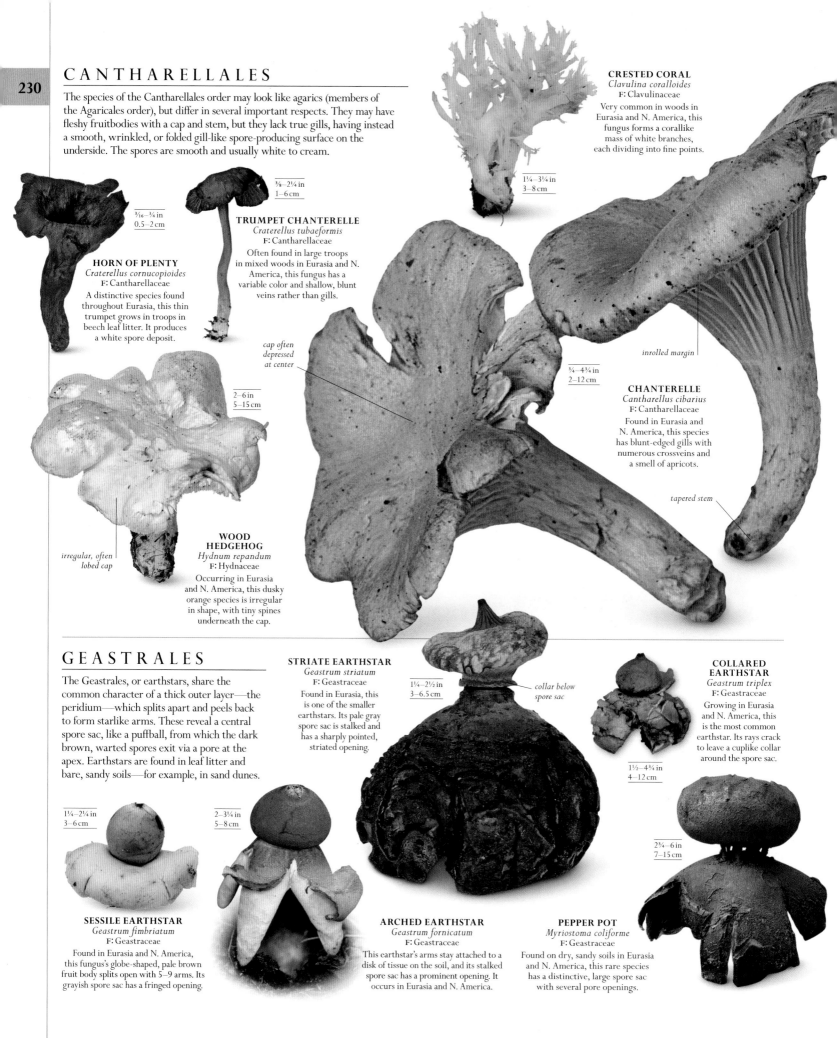

CRESTED CORAL
Clavulina coralloides
F: Clavulinaceae

Very common in woods in Eurasia and N. America, this fungus forms a corallike mass of white branches, each dividing into fine points.

1¼–3¼ in
3–8 cm

³⁄₁₆–¾ in
0.5–2 cm

HORN OF PLENTY
Craterellus cornucopioides
F: Cantharellaceae

A distinctive species found throughout Eurasia, this thin trumpet grows in troops in beech leaf litter. It produces a white spore deposit.

⅜–2¼ in
1–6 cm

TRUMPET CHANTERELLE
Craterellus tubaeformis
F: Cantharellaceae

Often found in large troops in mixed woods in Eurasia and N. America, this fungus has a variable color and shallow, blunt veins rather than gills.

cap often depressed at center

2–6 in
5–15 cm

¾–4¾ in
2–12 cm

inrolled margin

CHANTERELLE
Cantharellus cibarius
F: Cantharellaceae

Found in Eurasia and N. America, this species has blunt-edged gills with numerous crossveins and a smell of apricots.

tapered stem

irregular, often lobed cap

WOOD HEDGEHOG
Hydnum repandum
F: Hydnaceae

Occurring in Eurasia and N. America, this dusky orange species is irregular in shape, with tiny spines underneath the cap.

GEASTRALES

The Geastrales, or earthstars, share the common character of a thick outer layer—the peridium—which splits apart and peels back to form starlike arms. These reveal a central spore sac, like a puffball, from which the dark brown, warted spores exit via a pore at the apex. Earthstars are found in leaf litter and bare, sandy soils—for example, in sand dunes.

STRIATE EARTHSTAR
Geastrum striatum
F: Geastraceae

Found in Eurasia, this is one of the smaller earthstars. Its pale gray spore sac is stalked and has a sharply pointed, striated opening.

1¼–2½ in
3–6.5 cm

collar below spore sac

COLLARED EARTHSTAR
Geastrum triplex
F: Geastraceae

Growing in Eurasia and N. America, this is the most common earthstar. Its rays crack to leave a cuplike collar around the spore sac.

1½–4¾ in
4–12 cm

1¼–2¼ in
3–6 cm

2–3¼ in
5–8 cm

2¾–6 in
7–15 cm

SESSILE EARTHSTAR
Geastrum fimbriatum
F: Geastraceae

Found in Eurasia and N. America, this fungus's globe-shaped, pale brown fruit body splits open with 5–9 arms. Its grayish spore sac has a fringed opening.

ARCHED EARTHSTAR
Geastrum fornicatum
F: Geastraceae

This earthstar's arms stay attached to a disk of tissue on the soil, and its stalked spore sac has a prominent opening. It occurs in Eurasia and N. America.

PEPPER POT
Myriostoma coliforme
F: Geastraceae

Found on dry, sandy soils in Eurasia and N. America, this rare species has a distinctive, large spore sac with several pore openings.

GOMPHALES

Although some species were included with the chanterelles in Cantharellales, DNA analysis of the Gomphales fungi suggests that they are more closely related to the stinkhorn fungi in the Phallales. They often form large fruit bodies, varying in shape from simple clubs (in genus *Clavariadelphus*) to trumpet- or chanterellelike structures with a complex spore-producing surface (in genus *Gomphus*).

¾–2¼ in
2–6 cm

GIANT CLUB
Clavariadelphus pistillaris
F: Clavariadelphaceae
A rare species of Eurasia and N. America, this fungus forms a large, swollen club, with a smooth to slightly wrinkled surface, which bruises purple-brown.

1¼–3¼ in
3–8 cm

UPRIGHT CORAL
Ramaria stricta
F: Gomphaceae
A fairly common species in Eurasia and N. America, it is always attached to decaying wood or woodchip mulch. The branches are pale brown, bruising reddish.

2–4 in
5–10 cm

SCALY VASE CHANTERELLE
Gomphus floccosus
F: Gomphaceae
Common in N. America, this fungus resembles a fleshy, trumpet-shaped vase with scales at the top and wrinkled "gills" on its underside.

2¾–6 in
7–15 cm

ROSSO CORAL
Ramaria botrytis
F: Gomphaceae
Found in beech woods in Eurasia and N. America, this uncommon coral fungus has pinkish-white branches with deep red tips.

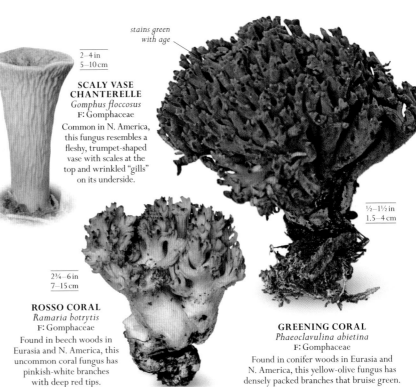

stains green with age

½–1½ in
1.5–4 cm

GREENING CORAL
Phaeoclavulina abietina
F: Gomphaceae
Found in conifer woods in Eurasia and N. America, this yellow-olive fungus has densely packed branches that bruise green.

GLOEOPHYLLALES

This is an order of wood-decay fungi that is characterized by the ability to produce a brown rot of wood. The order Gloeophyllales has a single family, the Gloeophyllaceae, which includes the genus *Gloeophyllum*. Some well-known bracket fungi on conifer wood are members of this genus.

ANISE MAZEGILL
Gloeophyllum odoratum
F: Gloeophyllaceae
This fungus, which grows on decayed conifer wood, is found in Eurasia and N. America. It has irregular brackets with yellow pores and an aniseed odor.

2–8 in
5–20 cm

DACRYMYCETALES

This order forms very simple, rounded or branched gelatinous fruit bodies, usually with bright orange coloration. Smooth or wrinkled, the Dacrymycetales have unusual spore-producing cells (basidia), which typically have two stout stalks (sterigmata), each bearing a spore. They mainly feed on dead wood.

YELLOW STAGSHORN
Calocera viscosa
F: Dacrymycetaceae
Attached to conifer wood, this fungus is found in Eurasia and N. America. Its clubs usually divide into branches with a gelatinous, rubbery texture.

³⁄₁₆–1½ in
0.5–4 cm

HYMENOCHAETALES

This group contains a number of diverse types of fungi including some crust fungi, polypores, such as *Mensularia* and *Phellinus*, as well as several agaric species, such as *Rickenella*. The Hymenochaetales are defined through molecular studies and have few uniting physical characteristics. Many feed on wood and may cause a white rot of wood.

4–16 in
10–40 cm

thick, pale margin

TIGER'S EYE
Coltricia perennis
F: Hymenochaetaceae
Frequently found in acidic heathlands of Eurasia and N. America, this fungus's goblet-shaped, thin fruitbodies are concentrically zoned.

¾–4 in
2–10 cm

³⁄₈–2¼ in
1–6 cm

OAK CURTAIN CRUST
Hymenochaete rubiginosa
F: Hymenochaetaceae
Found across Eurasia, this fungus forms overlapping brackets and crusts mainly on fallen oaks. Its tough fruit bodies have concentric markings.

WILLOW BRACKET
Phellinus igniarius
F: Hymenochaetaceae
This gray to almost black perennial bracket from Eurasia and N. America grows over many years. It is hoof-shaped and extremely woody.

1¼–3¼ in
3–8 cm

ALDER BRACKET
Mensularia radiata
F: Hymenochaetaceae
This deep reddish-brown fungus with a paler margin often forms vertical chains of brackets on alder and other trees in Eurasia and N. America.

⅛–³⁄₈ in
0.3–1 cm

ORANGE MOSSCAP
Rickenella fibula
F: Incertae sedis in Rickenella clade
This tiny species is common in mossy grasslands of Eurasia and N. America. Its bright orange cap is radially marked with a darker center.

POLYPORALES

The Polyporales form a large group of diverse fungi. Most of these are polypores—wood decomposers whose spores are formed in tubes (rather like the tubes of the boletes) or sometimes on spines. Most lack fully developed stems and grow shelf-, bracket-, or crustlike fruit bodies on wood, but some have more or less central stems and grow at the base of trees. A few Polyporales appear to grow from the soil.

UPPER SURFACE

4—12 in
10—30 cm

UNDERSIDE

OAK MAZEGILL
Daedalea quercina
F: Fomitopsidaceae

Found on fallen oaks in Eurasia and N. America, this perennial species has tough brackets with long, mazelike pores on the underside.

RAZORSTROP FUNGUS
Fomitopsis betulinas
F: Fomitopsidaceae

This large, kidney-shaped bracket is pale brown to white. A damaging parasite of birches, it is found in Eurasia and N. America.

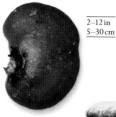

2—12 in
5—30 cm

4—23½ in
10—60 cm

ARTIST'S BRACKET
Ganoderma applanatum
F: Ganodermataceae

Found in Eurasia and N. America, this woody perennial bracket grows for many years to reach huge sizes. Its fallen spores are a rich cinnamon-brown.

bracket grows
in layers

6—12 in
15—30 cm

DYER'S MAZEGILL
Phaeolus schweinitzii
F: Fomitopsidaceae

Usually found at the base of conifers in Eurasia and N. America, this large, furry, cushionlike bracket is used for the manufacture of dye.

 hairy surface

CHICKEN OF THE WOODS
Laetiporus sulphureus
F: Fomitopsidaceae

These large brackets, found in Eurasia and N. America, usually grow on oaks and sometimes on other trees.

LACQUERED BRACKET
Ganoderma lucidum
F: Ganodermataceae

This deep reddish to purple-brown bracket with a shiny surface may have a long, lateral stem. It is found in Eurasia and N. America.

4—12 in
10—30 cm

GIANT POLYPORE
Meripilus giganteus
F: Meripilaceae

One of the largest polypores, its overlapping brackets are thick and fleshy. It grows around beech and other trees in Eurasia and N. America.

4—10 in
10—25 cm

RED BELTED POLYPORE
Fomitopsis pinicola
F: Fomitopsidaceae

Occurring in Eurasia and N. America, this hoof-shaped, woody bracket is usually found on pines, and sometimes on birches.

4—20 in
10—50 cm

margin is wavy
and lobed

4—20 in
10—50 cm

surface
bruises black

4—8 in
10—20 cm

ZONED ROSETTE
Podoscypha multizonata
F: Meruliaceae

Growing on soil from buried oak roots, this rare species occurs in Eurasia. It has densely packed lobes that form a circular mass.

JELLY ROT
Phlebia tremellosa
F: Meruliaceae

Found on logs in Eurasia, this species has a pale, velvety upper side and a yellow-to-orange underside with dense ridges.

1½—6 in
4—15 cm

1¼—2¾ in
3—7 cm

SMOKY BRACKET
Bjerkandera adusta
F: Meruliaceae

This common bracket, found in Eurasia and N. America, can be identified by the ash-gray pore surface on its underside.

TINDER BRACKET
Fomes fomentarius
F: Polyporaceae

2–12 in
5–30 cm

This gray-brown, hoof-shaped, perennial bracket grows on birches and other deciduous trees. It is found in Eurasia and N. America.

BLUSHING BRACKET
Daedaleopsis confragosa
F: Polyporaceae

3¼–6 in
8–15 cm

One of the most common brackets, especially on willows, in Eurasia and N. America, this semicircular species has cream pores that bruise pinkish red.

BIRCH MAZEGILL
Lenzites betulina
F: Polyporaceae

Usually found on birches in Eurasia and N. America, this tough, leathery species has pores that can be extremely elongated in gill-like ridges.

UPPER SURFACE

1¼–4 in
3–10 cm

UNDERSIDE

DRYAD'S SADDLE
Cerioporus squamosus
F: Polyporaceae

4–23½ in
10–60 cm

Found in early summer in Eurasia and N. America, these circular or fan-shaped brackets have concentric scales and pores on the underside.

funnel-shaped cap

BLACK FOOTED POLYPORE
Polyporus durus
F: Polyporaceae

2–8 in
5–20 cm

velvety stem base

This fungus has a funnel-shaped, leathery cap and its stem is black at the base. It grows on fallen beech logs in Eurasia and N. America.

TUBEROUS POLYPORE
Polyporus tuberaster
F: Polyporaceae

2–8 in
5–20 cm

Growing on fallen branches in Eurasia and N. America, this species may root into the ground and form a large, tuberous mass.

WINTER POLYPORE
Lentinus brumalis
F: Polyporaceae

1¼–3¼ in
3–8 cm

This small species grows on fallen branches in Eurasia and N. America. It has rather large, decurrent (running down the stem) pores. The stem is central or off-center.

CINNABAR BRACKET
Pycnoporus cinnabarinus
F: Polyporaceae

1¼–4 in
3–10 cm

Growing on dead deciduous timber in Eurasia and N. America, this rare species has a bright reddish-orange, leathery, annual bracket.

PURPLEPORE BRACKET
Trichaptum abietinum
F: Polyporaceae

¾–1½ in
2–4 cm

Found on fallen conifers in Eurasia and N. America, this fan-shaped bracket has concentric zones of pale gray, often tinged green by algae, with a purple margin.

HAIRY BRACKET
Trametes hirsuta
F: Polyporaceae

2–4¾ in
5–12 cm

This semicircular bracket is covered in minute hairs. It grows on dead deciduous wood or sometimes on old gorse stems in Eurasia and N. America.

TURKEYTAIL
Trametes versicolor
F: Polyporaceae

¾–2¾ in
2–7 cm

Found in Eurasia and N. America in a variety of colors, this species has brackets with concentric zones of different colors and a white pore surface.

LUMPY BRACKET
Trametes gibbosa
F: Polyporaceae

This cream-colored bracket is often stained green from algae. Its pores may be long and mazelike. It grows on fallen deciduous logs in Eurasia and N. America.

fleshy, bright yellow-orange bracket

small-lobed bracket

WOOD CAULIFLOWER
Sparassis crispa
F: Sparassidaceae

4–16 in
10–40 cm

4–12 in
10–30 cm

Growing at the base of conifers in Eurasia and N. America, this species has cream lobes that are flattened and fleshy like a cauliflower.

HEN OF THE WOODS
Grifola frondosa
F: Sparassidaceae

¾–2¼ in
2–6 cm

Found in Eurasia and N. America, this species grows at the base of oak, often where lightning has struck. It has a dense cluster of small brackets.

basal stem

RUSSULALES

The best-known genuses within this order are *Russula* and *Lactarius*, which, although resembling typical mushrooms, are not related at all to the true Agaricales. Apart from the cap-and-stem shapes, Russulales produce fruit bodies in a wide range of forms. Most Russulales have spores with warts that stain blue-black in iodine solutions. When cut, the *Lactarius* species produce a latex, which may be white to colored.

BEECH MILKCAP
Lactarius blennius
F: Russulaceae

This is a common species growing with beech in Eurasia. Slimy when wet, the gray-green cap has spots around the margin.

1½–3½ in
4–9 cm

1½–3¼ in
4–8 cm

OAKBUG MILKCAP
Lactarius quietus
F: Russulaceae
Common under oaks in Eurasia, its reddish brown cap has darker zones, and the flesh has a sweet, oily smell.

1¼–2¼ in
3–6 cm

LIVER MILKCAP
Lactarius hepaticus
F: Russulaceae
Associated with pines in Eurasia, this mushroom has a smooth cap with a grooved margin. The gills when cut bleed a white latex that stains yellow.

UGLY MILKCAP
Lactarius turpis
F: Russulaceae
A common species of birch woods in N. America and Eurasia, it is olive-green to almost black in color and has a slimy cap.

2–6 in
5–15 cm

depressed
cap center

2–6 in
5–15 cm

SAFFRON MILKCAP
Lactarius deliciosus
F: Russulaceae
Found with pines in N. America and Eurasia, this fungus has an orange-zoned and spotted cap that stains green with age. The flesh bleeds orange-red latex.

grooved
cap margin

2–6 in
5–15 cm

1¼–2¼ in
3–6 cm

hairy
and rolled-in
margin

WOOLLY MILKCAP
Lactarius torminosus
F: Russulaceae
Common under birches in N. America and Eurasia, this mushroom has a dark pink, hairy cap that is strikingly zoned.

CURRY MILKCAP
Lactarius camphoratus
F: Russulaceae
As the fruit body of this fungus dries, it gives off a smell of curry, which persists for many weeks. It is found in Eurasia and N. America.

2¼–4 in
6–10 cm

SOOTY MILKCAP
Lactarius fuliginosus
F: Russulaceae
Found in deciduous woods in Eurasia, this uncommon species has a dark brown cap and stem that bleed a white latex that rapidly turns pink.

3¼–8 in
8–20 cm

PEPPERY MILKCAP
Lactifluus piperatus
F: Russulaceae
This uncommon species from mixed woods in N. America and Eurasia has a funnel-shaped cap with extraordinarily crowded, narrow gills. It exudes white latex when cut.

dry, smooth cap surface

2–6 in
5–15 cm

CHARCOAL BURNER
Russula cyanoxantha
F: Russulaceae
Growing in mixed woods in N. America and Eurasia, this mushroom's cap varies from purple-lilac to uniformly green. The gills are forked, flexible, and greasy.

2–4 in
5–10 cm

BLOODY BRITTLEGILL
Russula sanguinaria
F: Russulaceae
Associated with pines in N. America and Eurasia, this species has a scarlet cap and red-streaked stem. Its spores are pale ocher.

1¼–2¾ in
3–7 cm

BEECHWOOD SICKENER
Russula mairei
F: Russulaceae
Exclusively found with beech in Eurasia, this species has a scarlet cap and bluish white gills.

GREEN BRITTLEGILL
Russula aeruginea
F: Russulaceae
Common under birches in N. America and Eurasia, this mushroom's pale olive to grass-green cap has tiny rust-colored spots. Its spores are pale cream.

1½–3½ in
4–9 cm

2–4 in
5–10 cm

YELLOW SWAMP BRITTLEGILL
Russula claroflava
F: Russulaceae
A common species on moss in boggy birch woods in N. America and Eurasia, its cap, yellowish cream gills, and white stem all bruise gray-black.

2–4¾ in
5–12 cm

OCHER BRITTLEGILL
Russula ochroleuca
F: Russulaceae
One of the commonest species found in Eurasia, it is identifiable by its matte yellow-ocher or greenish yellow cap and white gills.

1½–4 in
4–10 cm

PRIMROSE BRITTLEGILL
Russula sardonia
F: Russulaceae
Found in Eurasia, this species grows in association with pine. Variable in color, from purple to green or yellow, it has a fruity odor.

SICKENER
Russula emetica
F: Russulaceae
Growing in wet pine woods in N. America and Eurasia, the sickener has a bright scarlet cap that contrasts with its pure white gills and stem.

dry and matte cap surface

1¼–3¼ in
3–8 cm

1½–4¾ in
4–12 cm

gills may have red edges

2¼–6 in
6–15 cm

CRAB BRITTLEGILL
Russula xerampelina
F: Russulaceae
Found in N. America and Eurasia, this fungus is one of many related species that can only be identified under a microscope or by habitat.

3¼–6 in
8–15 cm

STINKING BRITTLEGILL
Russula foetens
F: Russulaceae
This large, orange-brown fungus with a grooved, lumpy cap margin grows in N. America and Eurasia. It has a sour, rancid odor.

stem often flushes red

ROSY BRITTLEGILL
Russula lepida
F: Russulaceae
Found in Eurasia, this fungus has a carmine-red, dry, hard cap that fades rapidly. Its stem may also be red and the flesh has a smell similar to cedarwood.

¾–2¼ in
2–6 cm

EARPICK FUNGUS
Auriscalpium vulgare
F: Auriscalpiaceae
Growing on pine cones in N. America and Eurasia, this unique mushroom looks like a bent-over spoon with minute spines hanging down from the small, furry cap.

3/16–¾ in
0.5–2 cm

2–10 in
5–25 cm

4–16 in
10–40 cm

HAIRY CURTAIN CRUST
Stereum hirsutum
F: Stereaceae
Found in N. America and Eurasia, this mushroom varies in shape from fully crustlike to small overlapping brackets. It is hairy above and smooth below.

4–20 in
10–50 cm

BLEEDING BROADLEAF CRUST
Stereum rugosum
F: Stereaceae
Growing in N. America and Eurasia, this species forms small crusts, and occasionally small brackets on wood. Its upper surface bleeds red where cut.

ROOT ROT
Heterobasidion annosum
F: Bondarzewiaceae
Usually parasitic on conifers in N. America and Eurasia, this species has a pale brown crust that darkens with age.

CORAL TOOTH
Hericium coralloides
F: Hericiaceae
Usually found on beech trees in N. America and Eurasia, this endangered species has white fruit-body branches with pendent spines on the lower surfaces.

AURICULARIALES

Although often grouped with other jelly fungi, the Auriculariales are separated by their unusual basidia (spore-producing cells). These vary in shape but all are partitioned by dividing membranes into four divisions, with each division producing a spore.

WITCHES' BUTTER
Exidia nigricans
F: Auriculariaceae

Found in temperate Eurasia and N. America, frequently on hardwood trees, it resembles a wrinkled mass of gelatinous tar. It shrivels when dry to a hard, black mass.

¾–4 in
2–10 cm

³⁄₈–3¼ in
1–8 cm

broad, thick, jellylike bracket

TRIPE FUNGUS
Auricularia mesenterica
F: Auriculariaceae

1½–6 in
4–15 cm

Found in Eurasia, this species is common on dead wood, especially elm. It resembles a small bracket fungus from the top, and has a wrinkled, rubbery, grayish purple underside.

1½–4¾ in
4–12 cm

JELLY EAR
Auricularia auricula-judae
F: Auriculariaceae

Common on dead wood of deciduous trees in Eurasia and N. America, this fungus has thin, elastic "ears" that are velvety outside and wrinkled inside.

JELLY TOOTH
Pseudohydnum gelatinosum
F: Incertae sedis

This species from Eurasia and N. America has a pale translucent gray to pale brown coloration. It is occasionally found on conifer stumps.

soft, peglike spines on underside, on which spores are formed

THELEPHORALES

This diverse group includes bracket fungi, crust fungi, earthfans, and toothed fungi. Many of these have tough, leathery flesh, and commonly feature knobbed or spiny spores. The group was only identified as a result of molecular studies as the Thelephorales have few physical features in common.

DRAB TOOTH
Bankera fuligineoalba
F: Bankeraceae

2–4 in
5–10 cm

An uncommon species of conifer woods in Eurasia, this fungus has a short, stout stem. Its cap has an underside covered with tiny, grayish-white spines.

thick, fleshy scales

1½–5½ in
4–14 cm

BITTER TOOTH
Hydnellum scabosum
F: Bankeraceae

Found in mixed woods in Eurasia and N. America, this rare species has a centrally depressed cap with irregular scales. The spines beneath are pale buff.

1¼–4 in
3–10 cm

stout, velvety stem

BLACK TOOTH
Phellodon niger
F: Bankeraceae

Growing in mixed woods in Eurasia and N. America, this uncommon species smells of fenugreek when dry. The irregularly shaped cap is gray to purplish black with gray spines underneath.

blue-green stem base

1¼–6 in
3–15 cm

DEVIL'S TOOTH
Hydnellum peckii
F: Bankeraceae

This locally common species from conifer woods in Eurasia and N. America has a flattened, knobbly, woolly cap that often exudes blood-red droplets. The underside has pale brown spines.

ragged margin

1½–4 in
4–10 cm

EARTHFAN
Thelephora terrestris
F: Thelephoraceae

Quite common in woods or on heaths in Eurasia and N. America, this fungus grows on soil or woody debris. Its fan-shaped fruit bodies overlap, forming clumps with paler, fringed margins.

PHALLALES

Named for the phallic shape of many of the species in this group, such as the stinkhorns, this order also contains some false truffles. The stinkhorns "hatch" from an egglike structure, often in just a few hours.

spores on inside of cage

4 in
10 cm

cage bursts from "egg"

RED CAGE FUNGUS
Clathrus ruber
F: Clathraceae

1–5½ in
2.5–14 cm

Found in parks and gardens in Eurasia, this rare species has a red cage with black, foul-smelling spores. The cage hatches from a small, pale "egg."

DEVIL'S FINGERS
Clathrus archeri
F: Clathraceae

Introduced from Australasia, this species is found mainly in S. Eurasia, where it remains rare. Its red arms emerge from a white "egg" and have blackish spores that smell fetid.

EXOBASIDIALES

This small group consists mainly of gall-forming plant parasites whose spore-producing cells form a layer on the leaf surface. Some cause disease in cultivated plants of the *Vaccinium* genus, which includes the common blueberry.

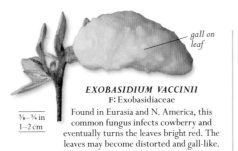

gall on leaf

EXOBASIDIUM VACCINII
F: Exobasidiaceae

Found in Eurasia and N. America, this common fungus infects cowberry and eventually turns the leaves bright red. The leaves may become distorted and gall-like.

⅜–¾ in
1–2 cm

UROCYSTIDIALES

This order contains some well-known smut fungi, in particular species of the genus *Urocystis*. These are parasites of flowering plants such as anemone, onions, wheat, and rye, and often cause serious injury to the host plant.

black, powdery spores on leaf surface

1/16–5/32 in
2–4 mm

ANEMONE SMUT
Urocystis anemones
F: Urocystidaceae

This smut fungus from Eurasia and N. America forms dark brown, powdery, raised pustules on the leaves of anemones and some other plants.

foul-smelling spore mass on cap

hollow, spongy stem

DOG STINKHORN
Mutinus caninus
F: Phallaceae

This common stinkhorn is found in mixed woods in Eurasia and N. America. Its tip, covered in greenish black spores, is joined to the spongy stem that emerges from a white "egg."

⅜–4¾ in
1–12 cm

white skirt drops down from cap

2–8 in
5–20 cm

PHALLUS MERULINUS
F: Phallaceae

This tropical species is mainly found in Australasia. It hatches from a white "egg." There are many similar species, some with brightly colored "skirts."

large, white "egg"

2–8 in
5–20 cm

STINKHORN
Phallus impudicus
F: Phallaceae

Common in mixed woods, this fungus is found in Eurasia. Its spore-covered, honeycomblike cap hatches in a few hours from an "egg." Its foul odor is often evident many yards away.

PUCCINIALES

One of the largest orders of fungi with over 7,000 species, the rust fungi include numerous serious parasites of crop plants. They can have very complex life cycles with multiple hosts and produce different types of spores at different stages in their life.

black, powdery spores in yellow spots on lower leaf surface

RASPBERRY YELLOW RUST
Phragmidium rubi-idaei
F: Phragmidiaceae

This rust from Eurasia and N. America causes pustules to form on the upper surface of leaves. It survives winter with the help of black spores on the underside of leaves.

orange rust damaging rose stem

ROSE RUST
Phragmidium tuberculatum
F: Phragmidiaceae

This common rust is found in N. America and Eurasia. It causes orange pustules on the undersides of leaves and on distorted stems. The pustules turn black in late summer.

ALEXANDERS RUST
Puccinia smyrnii
F: Pucciniaceae

Found across Eurasia, this common rust fungus forms raised plaques or warts on the leaves of Alexanders (*Smyrnium olusatrum*).

yellow warts of rust fungus

leaf surface spotted with rust fungus

raised blister of rust fungus

HOLLYHOCK RUST
Puccinia malvacearum
F: Pucciniaceae

This serious pest of hollyhocks is found in Eurasia and N. America. It covers the leaves with small pustules. Older leaves die and fall off.

PUCCINIA ALLII
F: Pucciniaceae

Common on onions, garlic, and leeks in Eurasia and N. America, this species forms pustules on infected leaves. These break open to release dustlike airborne spores.

rounded bright orange-yellow pustules erupt through leaf surface

powdery pustules on leaf

HYPERICUM RUST
Melampsora hypericorum
F: Melampsoraceae

This common species from Eurasia is visible as scattered, raised pustules on the undersides of *Hypericum* leaves.

FUCHSIA RUST
Pucciniastrum epilobii
F: Pucciniastraceae

A parasite of fuchsias and fireweed in Eurasia, this species infects the leaves and forms pustules on the undersides of the leaves.

SAC FUNGI

The Ascomycota, or sac fungi, are fungi that produce their spores in rounded to elongate sacs called asci located on the fertile surface of the fruit body. They are the largest group of fungi, and include many cup- or saucer-shaped species.

PHYLUM	ASCOMYCOTA
CLASSES	7
ORDERS	56
FAMILIES	226
SPECIES	About 33,000

Many sac fungi display vivid colors, although it is uncertain what biological function such bright colors may have.

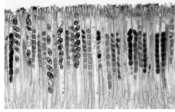

Spore-producing asci are shown here under a microscope. Arranged in dense layers, the asci contain eight spores each.

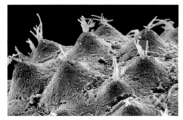

Many species form their asci in special protective chambers called perithecia; from here they discharge their spores.

HEROES OR VILLAINS?

Sac fungi form beneficial associations with plants, algae, and even arthropods such as beetles. But equally they include some of the world's most damaging pathogens: *Cryphonectria*, for example, is responsible for the recent death of millions of chestnut trees. Perhaps no other group of fungus has such contrasting effects on our world.

Ranging in size from the microscopic to about 8 in (20 cm) in height, the sac fungi occupy a wide variety of habitats, growing on dead, dying, and living tissue, and floating in fresh and saltwater environments. Many species are parasitic, and include some of the most serious crop pests. Others form mutually beneficial relationships, called mycorrhizae, with plants. Sac fungi include some of the most important fungi in the history of medicine, such as the source of penicillin, while others are serious pathogens—*Pneumocystis jirovecii*, for instance, can cause lung infections in people with a weak immune system. This phylum also contains the yeasts, which, being vital components in the production of alcohol and bread, have played a pivotal role in human history.

The Ascomycota display a wide range of fruit-body shapes, including cup-shaped, club-shaped, potato-like, simple crusts or sheets, pimplelike, corallike, shield-shaped, or stalked with a spongelike cap. Depending on the type of fruit body, the spore-producing asci may grow externally on a special fertile layer or be contained internally. Not all species have a sexual stage; in fact, many use asexual methods of reproduction. Most yeasts grow and rapidly colonize new areas as a result of asexual division, or budding, where a small bud forms on the outside of a yeast cell, then separates, becoming a new cell.

THE CUP FUNGUS

The common name of cup fungus refers to one of the most conspicuous shapes of fruitbody associated with the Ascomycota. Its open top (which is sometimes disk- or saucer-shaped) allows wind and rain to scatter the spores that line its inner surface. In some varieties, water is absorbed by the asci, which contain the spores, so that pressure builds up and forcibly ejects the spores up to 12 in (30 cm) from the fruit body. Close examination of surfaces such as rotting logs, fallen branches, or leaves may reveal a fascinating world of almost microscopic cups. In the larger cup species, disturbance of the cup can produce such a vigorous ejection of spores that their release is not only visible as a faint cloud of spores, but also audible.

ORANGE PEEL CUP >
The orange peel fungus is a good example of the simple cup-shaped fruit body adopted by many fungi species.

HYPOCREALES

The fungi in this order are usually distinguished by their brightly colored, spore-producing structures. These are usually yellow, orange, or red. The Hypocreales are often parasitic on other fungi and also on insects. The best known among them is the genus *Cordyceps* with club- or branchlike fruit bodies (bodies that support the spore-forming cells). Some species have medicinal uses.

1¼–2¼ in
3–6 cm

2–5 in
5–13 cm

false truffle parasitized by truffleclub

SCARLET CATERPILLAR CLUB
Cordyceps militaris
F: Cordycipitaceae
Found in Eurasia and N. America, this fungus is parasitic on moth pupae. The head of its club bears tiny spore-producing structures.

SNAKETONGUE TRUFFLECLUB
Tolypocladium ophioglossoides
F: Ophiocordycipitaceae
Parasitic on buried false truffles, this species is found in Eurasia and N. America. It forms yellow clubs with an elongated, greenish black heads.

rounded, cushionlike fruitbodies

purple ergot on flower heads

infected bolete fruit body

½ in
1.5 cm

8–12 in
20–30 cm

CORAL SPOT
Nectria cinnabarina
F: Nectriaceae
Abundant on damp wood in Eurasia and N. America, this fungus forms pink "pimples" when sexually immature and red-brown clusters when mature.

ERGOT
Claviceps purpurea
F: Clavicipitaceae
This species has caused outbreaks of mass poisoning. Found in Eurasia and N. America, it is parasitic on grass and cereal crops.

BOLETE EATER
Hypomyces chrysospermus
F: Hypocreaceae
This common mold, found in N. America and Eurasia on bolete, turns bright golden yellow, with a fluffy texture.

XYLARIALES

Members of this order often have their spore-producing cells in chambers, which are embedded in a woody growth called stroma. Although many species live on wood, some also occur on animal dung, fruit, leaves, and soil, or are associated with insects. The order includes many economically important plant parasites.

tips covered in powdery spores

⅜–½ in
1–1.5 cm

⅜–1½ in
1–4 cm

¾–3¼ in
2–8 cm

CANDLESNUFF FUNGUS
Xylaria hypoxylon
F: Xylariaceae
Common on dead wood in Eurasia and N. America, this species resembles a snuffed-out candle with a velvety black stem.

DEAD MAN'S FINGERS
Xylaria polymorpha
F: Xylariaceae
Growing on dead wood in Eurasia and N. America, this fungus forms brittle black clubs with a rough surface, tiny pores, and thick, white flesh.

NAIL FUNGUS
Poronia punctata
F: Xylariaceae
Found on horse dung in Eurasia and N. America, this species is in decline. Its flattened disk has numerous tiny holes from which spores are released.

stemless fruit body has hard, brittle surface

dead trunk of ash tree

¾–4 in
2–10 cm

¼–⅜ in
0.5–1 cm

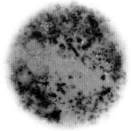

BEECH WOODWART
Hypoxylon fragiforme
F: Xylariaceae
Occurring in clusters on beech logs, this European and N. American species forms hard, rounded fruit bodies with tiny spore-releasing chambers.

CRAMP BALLS
Daldinia concentrica
F: Xylariaceae
Found in Eurasia and N. America, this fungus has rounded fruit bodies that reveal concentric, whitish zones when cut in half. They eject black spores from their outer layer.

ERYSIPHALES

Parasitic on leaves and fruit of flowering plants, the Erysiphales are powdery mildews. The hyphae (filaments) of their mycelium (vegetative part of the fungus) penetrate the cells of the host plant and take up nutrients.

mildew patches

affected apple leaf

feltlike white mycelium covers leaf surface

OAK POWDERY MILDEW
Erysiphe alphitoides
F: Erysiphaceae
Growing on oaks in Eurasia and N. America, this fungus covers young leaves, causing them to shrivel and blacken.

APPLE POWDERY MILDEW
Podosphaera leucotricha
F: Erysiphaceae
Common on apple leaves in Eurasia and N. America, this mildew appears first on the underside of leaves as whitish patches, before spreading rapidly.

POWDERY MILDEW
Golovinomyces cichoracearum
F: Erysiphaceae
Found in Eurasia and N. America, this species appears on members of Asteraceae (the sunflower family), causing patches on leaves, which eventually die.

CAPNODIALES

Commonly called sooty molds, these sac fungi are frequently found on leaves. They feed on the honeydew excreted by insects or on liquid exuded from the leaves. Some cause skin problems in humans.

CLADOSPORIUM CLADOSPORIOIDES
F: Davidiellaceae
This mold, common in Eurasia and N. America, grows on damp bathroom walls. It may cause an allergic reaction in some people.

3/16–1/2 in
0.5–1.5 cm

HELOTIALES

The fungi of this order are distinguished by their disk- or cup-shaped fruit bodies, unlike some similar cup fungi. Their saclike spore-producing cells, or asci, do not have an apical lid (a flap through which they open). Most members live on humus-rich soil, dead logs, and other organic matter. This order also includes some of the most damaging plant parasites.

BROWN CUP
Rutstroemia firma
F: Rutstroemiaceae
Consisting of a light brown cup and a narrow stem, this fungus grows on fallen branches, especially of oaks, in Europe. It turns the host wood black.

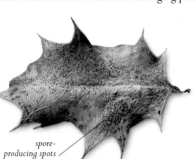
spore-producing spots

TROCHILA ILICINA
F: Dermateaceae
Abundant on fallen holly leaves, this fungus is found in Eurasia and N. America. It forms spore-producing spots on the leaf surface.

merged black spots

ROSE BLACK SPOT
Diplocarpon rosae
F: Dermateaceae
Commonly found on rose leaves in Eurasia and N. America, this fungus causes black spots, which merge together to cause large black patches.

1¼–2¾ in
3–7 cm

spore-producing inner surface

smooth surface of cup

SCALY EARTHTONGUE
Geoglossum fallax
F: Geoglossaceae
Growing in meadows in Eurasia and N. America, this uncommon fungus is one of several species with blackish, flattened clubs, which can only be separated using a microscope.

3/16–1½ in
0.5–4 cm

BLACK BULGAR
Bulgaria inquinans
F: Bulgariaceae
Found in Eurasia and N. America, this species has a brown outer surface. Its spore-producing inner surface is smooth, black, and rubbery.

ANEMONE CUP
Dumontinia tuberosa
F: Sclerotiniaceae
Parasitic on the tubers of anemones, this common Eurasian species has a long, black stem and a simple brown cup.

BEECH JELLYDISK
Neobulgaria pura
F: Helotiaceae
Frequently found on fallen beech logs in Eurasia, this translucent jellydisk varies from pale pink to pale lilac. The disks are often distorted from crowding together.

3/16–1¼ in
0.5–3 cm

3/16–1¼ in
0.5–3 cm

1/16–3/8 in
0.2–1 cm

buff-colored pustules on infected fruit

APPLE BROWN ROT
Monilia fructigena
F: Sclerotiniaceae
Very common in Eurasia, this fungus occurs primarily on apples and pears, although it is also found on *Prunus* sp., causing brown fruit rot.

1/32–1/8 in
1–3 mm

LEMON DISCO
Bisporella citrina
F: Helotiaceae
Very common on dead hardwood in Eurasia, this species occurs in clusters. Its golden-yellow disks sometimes cover entire branches.

⅛–3/8 in
0.3–1 cm

1/16–3/8 in
0.2–1 cm

3/16–¾ in
0.5–2 cm

JELLY BABIES
Leotia lubrica
F: Leotiaceae
Frequently found in mixed woodland in Eurasia and N. America, this species has a lobed head with a margin that rolls back on itself.

GREEN ELFCUP
Chlorociboria aeruginascens
F: Helotiaceae
Found in Eurasia and N. America, the blue-green mature cups are rather uncommon, but the green stains they cause on fallen oak logs are easily visible.

LARGE PURPLE DROP
Ascocoryne cylichnium
F: Helotiaceae
Quite common on fallen beech logs in Eurasia, this species attaches itself centrally to the wood and forms gelatinous, irregular disks when sexually mature.

BOG BEACON
Mitrula paludosa
F: Helotiaceae
This species, found in Eurasia and N. America, grows on plant remains in shallow water in spring and early summer. It has a rounded to tongue-shaped head.

1/16–3/8 in
0.2–1 cm

PEZIZALES

Members of this order produce spores inside saclike structures, or asci, which typically open by rupturing to form an operculum (terminal lid) and eject the spores. The order includes a number of species of economic importance, such as morels, truffles, and desert truffles.

**3/16–3/4 in
0.5–2 cm**

COMMON EARTHCUP
Geopora arenicola
F: Pyronemataceae

Common in Eurasia, this fungus is difficult to spot as it lies buried in sandy soils. It has a smooth, spore-producing inner surface.

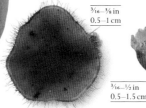

tall orange cups

**2–4 in
5–10 cm**

HARE'S EAR
Otidea onotica
F: Pyronemataceae

This common species is often found in clusters in broadleaved forests in Eurasia and N. America. Its tall cups are split down on one side.

**3/16–3/8 in
0.5–1 cm**

COMMON EYELASH
Scutellinia scutellata
F: Pyronemataceae

This is one of many similar species; its fruit body is a scarlet cup with a fringe of black hairs. This fungus is common on wet, rotten wood in Eurasia and N. America.

**3/16–1/2 in
0.5–1.5 cm**

TOOTHED CUP
Tarzetta cupularis
F: Pyronemataceae

A common species in alkaline soils in woods of Eurasia and N. America, its gobletlike cup has a short stem.

dark brown, wrinkled cap

irregular ridges and pits

ORANGE PEEL FUNGUS
Aleuria aurantia
F: Pyronemataceae

Often found along gravelly dirt tracks in Eurasia and N. America, this is an unmistakable species, with thin, orange caps.

**3/4–4 in
2–10 cm**

**1 1/2–4 in
4–10 cm**

BLEACH CUP
Disciotis venosa
F: Morchellaceae

Growing in spring in damp woodlands in Eurasia and N. America, this short-stemmed species has a chlorinelike smell. Its inner surface is brown and wrinkled and the outer surface is pale.

**2–6 in
5–15 cm**

FALSE MOREL
Gyromitra esculenta
F: Discinaceae

This poisonous species is found throughout N. America and Eurasia, and usually grows under conifers in spring. The shiny brown cap looks like a wrinkled brain.

**2–8 in
5–20 cm**

hollow stem

HALF-FREE MOREL
Morchella semilibera
F: Morchellaceae

This hollow morel looks like a dark, ridged thimble on a scurfy, pale stem. It is common in spring in mixed woods in Eurasia and N. America.

smooth cap surface

**2–4 in
5–10 cm**

**2–6 in
5–15 cm**

THIMBLE MOREL
Verpa conica
F: Morchellaceae

An uncommon species, this fungus grows in woods and hedges on chalky soils in Eurasia and N. America. Its smooth, thimble-like cap sits on a hollow stem.

BLACK MOREL
Morchella elata
F: Morchellaceae

Common in woods during spring in Eurasia and N. America, this species has a pinkish buff to black cap, with cross-connected black ridges, and a hollow stem.

**2–6 in
5–15 cm**

MOREL
Morchella esculenta
F: Morchellaceae

This prized species grows in spring in calcareous woodland in N. America and Eurasia. It has a spongelike, hollow cap and a hollow stem.

**1 1/4–4 in
3–10 cm**

BLISTERED CUP
Peziza vesiculosa
F: Pezizaceae

Common in Eurasia and N. America, this fungus typically occurs in clusters on compost, straw, or manure. Its brittle cups can have ragged edges.

**1–3 in
2.5–7.5 cm**

CELLAR CUP
Peziza cerea
F: Pezizaceae

Found across Eurasia and N. America, this fungus often occurs on damp brickwork. It is dark ocher on the inside and pale outside.

**1/2–2 3/4 in
1.5–7 cm**

BAY CUP
Peziza badia
F: Pezizaceae

One of many similar species, this common fungus of Eurasian and N. American woods has a spore-producing inner surface. Its brown cup develops an olive tint with age.

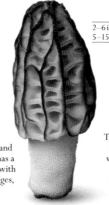
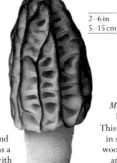

EUROTIALES

Better known for the blue and green molds, this order includes the fungi *Penicillium* (famous for producing penicillin, the first discovered antibiotic) and *Aspergillus* (a significant cause of disease in humans).

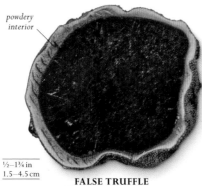

powdery interior

½–1¾ in
1.5–4.5 cm

FALSE TRUFFLE
Elaphomyces granulatus
F: Elaphomycetaceae

Common in sandy soils below conifers in Eurasia and N. America, this reddish-brown truffle has a roughened surface and purple-black inner spore-mass.

RHYTISMATALES

Commonly called tar spot fungi, the species in this order infect plant matter, such as leaves, twigs, bark, female conifer cones, and occasionally berries. Many species attack the needles of conifers, causing needle drop. The tar spot of maple leaves is perhaps the most frequently seen.

brown spots on upper leaf surface

⅜–¾ in
1–2 cm

TAR SPOT
Rhytisma acerinum
F: Rhytismataceae

Abundant on acer or maple trees in N. America and Eurasia, this fungus causes irregular spots, with paler yellow margins, which disfigure the leaves.

1–2 in
2.5–5 cm

YELLOW FAN
Spathularia flavida
F: Cudoniaceae

This common species grows in wet, mossy conifer woods of Eurasia and N. America. It has flattened, pale to darker yellow, rubbery heads.

¾–2¾ in
2–7 cm

PERIGORD TRUFFLE
Tuber melanosporum
F: Tuberaceae

This highly prized truffle grows underground around oaks in Eurasia and the Mediterranean region. The truffles are found using dogs or pigs.

¾–3¼ in
2–8 cm

WHITE TRUFFLE
Tuber magnatum
F: Tuberaceae

Prized in Italy and France, this expensive truffle of alkaline soils in S. Europe can be cultivated on suitable host trees such as oak and poplar, by inoculation.

¾–2 in/2–5 cm

SUMMER TRUFFLE
Tuber aestivum
F: Tuberaceae

Found across S. and C. Europe, this highly valued truffle grows underground, near a variety of broadleaved trees.

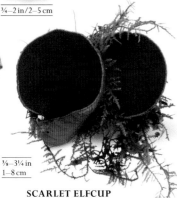

⅜–3¼ in
1–8 cm

SCARLET ELFCUP
Sarcoscypha austriaca
F: Sarcoscyphaceae

Found from winter to early spring on fallen branches in Eurasia and N. America, this species has a scarlet cup contrasting with its pale outer surface.

TAPHRINALES

This order contains many plant parasites, with most species in the genus *Taphrina*. All species have two growth states: in the saprophytic state, they are yeastlike and propagate by budding; but in the parasitic state, they emerge through plant tissues, causing distorted leaves and galls.

BIRCH BESOM
Taphrina betulina
F: Taphrinaceae

Found across Eurasia, this common species causes witches' broom, a disease of birch trees that causes clumps of narrow twigs to grow at branch ends.

8–37 in
20–95 cm

PEACH LEAF CURL
Taphrina deformans
F: Taphrinaceae

This fungus infects most varieties of peach and nectarine trees in Eurasia and N. America. The affected leaves curl and crinkle, and frequently turn reddish purple.

5½–16 in
14–40 cm

red patches caused by infection

PLEOSPORALES

Typical members of this order develop their asci within a flasklike fruiting body. The asci have two wall layers: at maturity, the inner wall protrudes beyond the outer wall, ejecting the spores. Many species grow on plants; some form lichens.

dark cones

LEPTOSPHAERIA ACUTA
F: Leptosphaeriaceae

Found in Eurasia and N. America, this common species infects the dead stems of the stinging nettle, *Urtica dioica*. It forms tiny cones that push through the stem surface of the host in order to eject its spores.

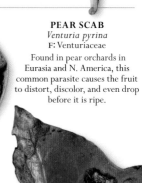

PEAR SCAB
Venturia pyrina
F: Venturiaceae

Found in pear orchards in Eurasia and N. America, this common parasite causes the fruit to distort, discolor, and even drop before it is ripe.

dark, sunken spots

BOEREMIA HEDERICOLA
F: Didymellaceae

Found in Eurasia and N. America, this species causes circular white lesions on the leaves of the ivy plant, which turn brown and die.

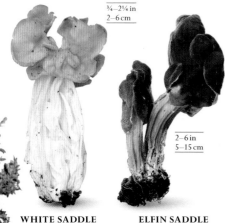

¾–2¼ in
2–6 cm

2–6 in
5–15 cm

WHITE SADDLE
Helvella crispa
F: Helvellaceae

Possibly poisonous, this species is common in mixed woods in Eurasia and N. America. Its thin, saddlelike cap sits on a fragile, ribbed stem.

ELFIN SADDLE
Helvella lacunosa
F: Helvellaceae

Found in mixed woods in Eurasia and N. America, this common species has a lobed, dark cap on a gray, fluted, and columned stem.

LICHENS

From outcrops of rock exposed to the sea to desert areas where the only place they can grow is inside the rock itself, lichens survive in the harshest parts of the world. They are some of nature's pioneers, creating foundations on which others can live.

PHYLA	ASCOMYCOTA BASIDIOMYCOTA
CLASSES	10
ORDERS	15
FAMILIES	40
SPECIES	About 18,000

Asexual soredia are bundles of fungal hyphae (filaments) and algal cells. Here they are budding off and awaiting dispersal.

Asexual isidia are tiny peglike cells on the edge of the lichen, which break off to form new colonies of lichen.

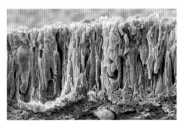

Lichen cells are shown here in microscopic cross-section. Spore-producing asci rise from algal cells.

A lichen is not a single organism, but a composite consisting of a green alga or cyanobacterium and a fungus, living together in a mutually beneficial association. The algal partner provides nutrients through photosynthesis, while the fungal partner aids the alga by retaining water and capturing mineral nutrients. The fungus is normally a member of the Ascomycota (the sac fungi), or, more rarely, of the Basidiomycota (mushrooms)—lichen classification reflects the type of fungus involved. Typically, the fungus surrounds the photosynthetic cells of the alga, enclosing them within special fungus tissues unique to lichens. It appears that although neither partner is capable of surviving on its own, together they can endure the most extreme conditions. Lichens have even been found some 250 miles (400 km) from the South Pole, but they also grow in familiar places, such as dry-stone walls, rocks, and bark.

Lichens broadly fall into three types, based on shape: foliose lichens, which have leaves; crustose lichens, which form a crust; and fruticose lichens, which have branches. However, there are some that defy this categorization, such as filamentous (hairlike) and gelatinous (water-absorbent) lichens.

REPRODUCTION

Many lichens reproduce sexually by means of spores. These are produced by the fungal half of the partnership and are usually formed in special cuplike or disklike structures called apothecia. These spores, once ejected, must land next to a suitable algal partner if they are to form another lichen and survive. Other lichens produce spores inside special chambers called perithecia, which are like microscopic volcanoes, releasing their spores through a hole at the top. Alternatively, lichens may reproduce asexually by budding or breaking off specialized parts of their body. These soredia or isidia contain a mix of fungal and algal cells, which, if they land in a suitable habitat, will go on to form new lichen colonies. Rocky shorelines in North America are home to vast colonies of lichens several miles in length. These will have taken hundreds or even thousands of years to spread to such a huge size.

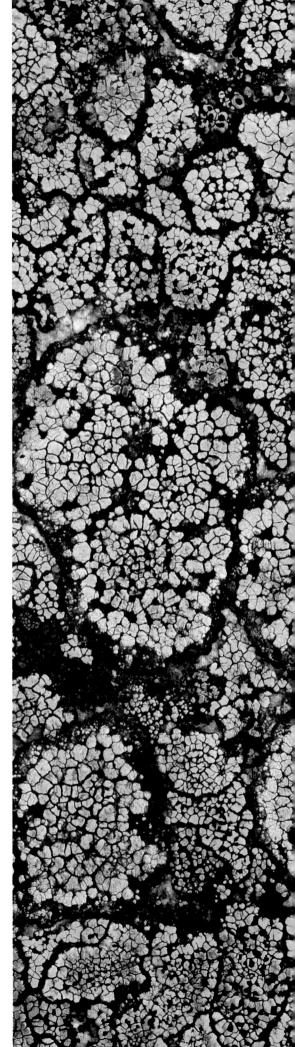

SPREADING MAP LICHEN >
Typical of lichens, *Rhizocarpon* is able to colonize harsh environments such as dry, exposed rock surfaces.

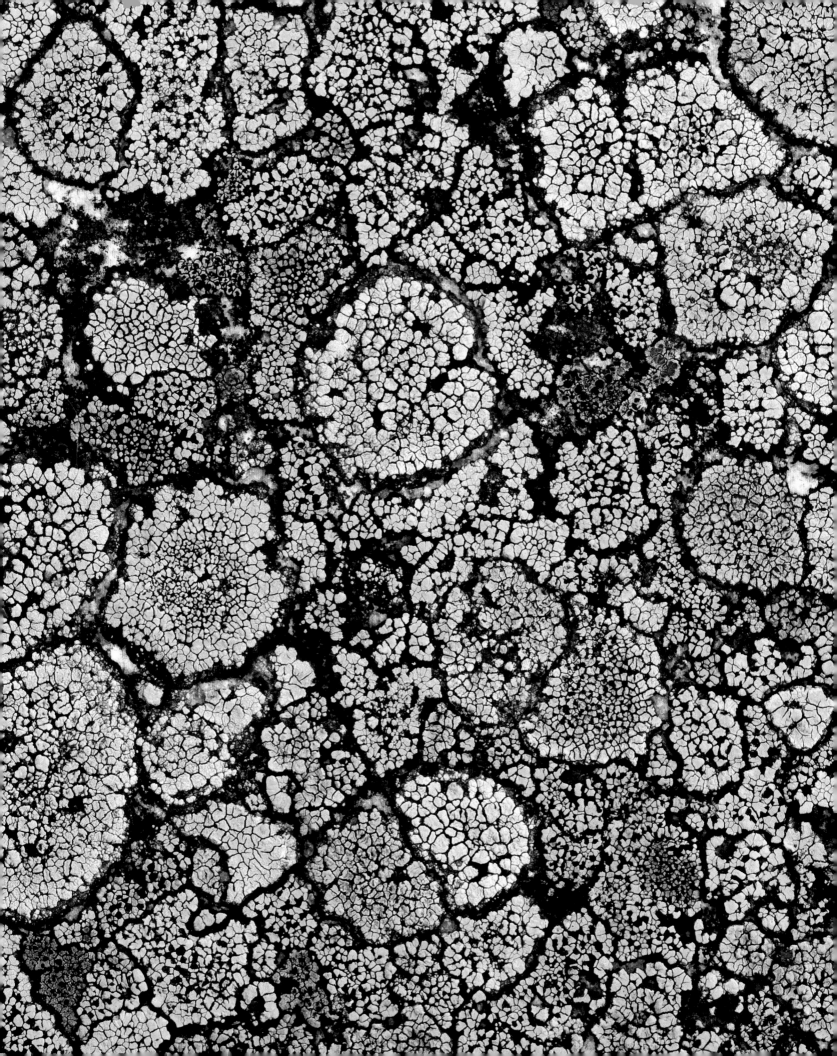

ORANGE LICHEN
Caloplaca verruculifera
F: Teloschistaceae

This lichen has radiating lobes with apothecia (spore-producing disks) at the center. It is found on rocks in coastal areas in N. America and Eurasia, often near bird perches.

2–4 in
5–10 cm

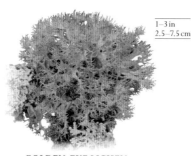

1–3 in
2.5–7.5 cm

GOLDEN-EYE LICHEN
Teloschistes chrysophthalmus
F: Teloschistaceae

A critically endangered species in America, Eurasia, and the tropics, this lichen grows on shrubs and small trees in old orchards and hedgerows. Its branched lobes produce large, orange disks.

1–3 in
2.5–7.5 cm

rounded lobes that lift at edges

COMMON WALL LICHEN
Xanthoria parientina
F: Teloschistaceae

Growing on trees, walls, and roofs in N. America, Eurasia, Africa, and Australia, this lichen has rounded, yellow-orange lobes.

2–6 in/5–15 cm

1–3 in
2.5–7.5 cm

REINDEER MOSS
Flavocetraria nivalis
F: Parmeliaceae

Found on mountain heaths and upland moors in N. America and Eurasia, this species has brownish, flattened leafy fronds with spiny margins.

HAMMERED SHIELD LICHEN
Parmelia sulcata
F: Parmeliaceae

This lichen's flattened lobes are gray-green, with rounded tips and powdery, reproductive structures on the surface. It is commonly found on trees in N. America and Eurasia.

1–3 in
2.5–7.5 cm

1–3 in
2.5–7.5 cm

HOARY ROSETTE LICHEN
Physcia aipolia
F: Physciaceae

Growing on tree bark in the Americas and Eurasia, this gray to brownish-gray species forms rough patches with a lobed margin. It has black apothecia.

3/8–2 in
1–5 cm

BEARD LICHEN
Usnea filipendula
F: Parmeliaceae

Found mainly in northern regions, this species forms green-gray, pendent clumps on trees. Spiny apothecia form at the tips.

1–3 in
2.5–7.5 cm

1–3 in
2.5–7.5 cm

POWDER-HEADED TUBE-LICHEN
Hypogymnia tubulosa
F: Parmeliaceae

Common on twigs and tree trunks, this lichen has lobes that are gray-green above and dark underneath. It is found in N. America and Eurasia.

HOODED TUBE-LICHEN
Hypogymnia physodes
F: Parmeliaceae

Found worldwide on trees, rocks, and walls, this lichen has pale gray-green lobes with wavy margins. The rare apothecia are red-brown with gray margins.

DEVIL'S MATCHSTICK
Cladonia floerkeana
F: Cladoniaceae

This species is common on peaty soil in N. America and Eurasia. It forms a crust of greenish gray scales from which stalks emerge, tipped with scarlet apothecia.

1–5 in
2.5–12.5 cm

CARTILAGE LICHEN
Ramalina fraxinea
F: Ramalinaceae

Growing on trees in N. America and Eurasia, this species forms flattened, gray-green branches dotted with apothecia.

1–4 in
2.5–10 cm

CORAL LICHEN
Sphaerophorus globosus
F: Sphaerophoraceae

Found on rocks in mountainous areas of N. Eurasia and N. America, this fungus forms dense cushions of pinkish brown branches with globular apothecia.

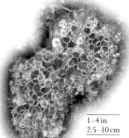

1–4 in
2.5–10 cm

BLACK SHIELDS
Tephromela atra
F: Mycoblastaceae

This lichen has pale gray crustose lobes that look like dried porridge. Its apothecia are black. The lichen is found on exposed rocks in N. America and Eurasia.

1–4 in/2.5–10 cm

STONEWALL RIM-LICHEN
Lecanora muralis
F: Lecanoraceae

This lichen often grows on concrete and on rocks. Its gray-green lobes radiate outward. It is found in N. America and Eurasia.

1–4 in
2.5–10 cm

REINDEER LICHEN
Cladonia portentosa
F: Cladoniaceae

One of several reindeer lichens, this species is common on heath and moorland in N. America and Eurasia. Its thin, hollow branches repeatedly divide.

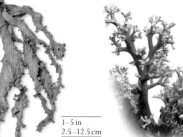

BLACK TAR LICHEN
Verrucaria maura
F: Verrucariaceae

2–20 in
5–50 cm

Occurring on rocks along seashores
in N. America and Eurasia, this
species has a dark gray, cracked
crust that contains apothecia.

DOG LICHEN
Peltigera praetextata
F: Peltigeraceae

8–12 in
20–30 cm

Found on rocks in Eurasia and
N. America, this lichen has large
gray-black lobes, with paler margins
and reddish brown apothecia.

BLISTERED JELLY LICHEN
Collema furfuraceum
F: Collemataceae

1–2 in
2.5–5 cm

This lichen has flat, gelatinous,
wrinkled lobes. It is found on rocks
and trees in areas with high rainfall
in N. America and Eurasia.

*green lobes spread
out from center*

TREE LUNGWORT
Lobaria pulmonaria
F: Lobariaceae

2–6 in
5–15 cm

Mainly found on tree bark in coastal areas in
N. America, Eurasia, and Africa, this species
is declining due to habitat loss. The branching
lobes are pale orange underneath.

ROCK TRIPE
Lasallia pustulata
F: Umbilicariaceae

2–8 in
5–20 cm

This N. American and Eurasian
lichen forms colonies on nutrient-
rich rocks in coastal or upland areas.
It has a gray-brown upper surface
with many oval pustules.

PETALED ROCK-TRIPE
Umbilicaria polyphylla
F: Umbilicariaceae

1–3 in
2.5–7.5 cm

A common species on rocks
in mountains in N. America and
Eurasia, it has smooth, broad
lobes that are dark brown above
and black beneath.

COMMON
SCRIPT LICHEN
Graphis scripta
F: Graphidaceae

2–4 in
5–10 cm

Often seen on tree bark in
N. America and Eurasia, this
lichen forms a thin, gray-green
crust with slitlike openings.

*black, slitlike
openings
release spores*

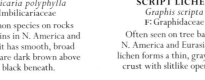

2–8 in
5–20 cm

1–3 in
2.5–7.5 cm

1–4 in
2.5–10 cm

PERTUSARIA PERTUSA
F: Pertusariaceae

Common on tree bark
in N. America and Eurasia,
this lichen forms gray crusts
with a pale margin. The crusts
are covered with groups of
warts with tiny openings.

LECIDEA LICHEN
Lecidea fuscoatra
F: Lecideaceae

This species is common on
siliceous rocks and old brick
walls in N. America and Eurasia.
It forms a gray, cracked crust,
with sunken, black apothecia.

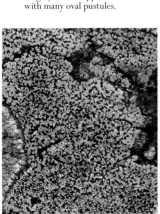

MAP LICHEN
Rhizocarpon geographicum
F: Rhizocarpaceae

³⁄₁₆–2¹⁄₂ in
5–65 mm

Common on rocks in
mountains in northern regions
and Antarctica, this lichen
forms a flat patch bordered
by a black line of spores.
Groups of these lichens have
a patchwork appearance.

CRAB-EYE LICHEN
Ochrolechia parella
F: Ochrolechiaceae

This lichen forms patches on walls and
rocks in N. America and Eurasia. Its surface
usually has many pink-brown apothecia.

*dome-shaped
apothecia*

1–5 in
2.5–12.5 cm

BAEOMYCES
RUFUS
F: Baeomycetaceae

Forming gray-green crusts
on sandy soil and rocks,
this lichen has brown ball-like
apothecia on stalks a few
millimeters high. It is found
in N. America and Eurasia.

ANIMA

LS

Animals make up the largest kingdom of living things. Driven by the need to eat food, and to escape being eaten, they are uniquely responsive to the world around them. The vast majority are invertebrates, but mammals and other chordates often outclass them in size, strength, and speed.

» 250

INVERTEBRATES

Hugely varied in shape, invertebrates also have a wide range of lifestyles. Insects make up the largest group, but others include jellyfish, worms, and animals protected by a hard shell.

» 322

CHORDATES

Most of the world's large animals are chordates. Externally, they can be covered in fur, feathers, or overlapping scales, but internally nearly all have a backbone, which makes up part of a bony skeleton.

INVERTEBRATES

With about 1.4 million species identified, animals make up the largest kingdom of living things on Earth. The vast majority of them are invertebrates—animals without a backbone. Invertebrates are extraordinarily varied; many are microscopic, but the largest are over 33 ft (10 m) long.

Invertebrates were the first animals to evolve. Initially, they were small, soft-bodied, and aquatic—features that many living invertebrates still share. During the Cambrian Period, which ended about 485 million years ago, invertebrates underwent a spectacular burst of evolution, developing a huge range of body forms and some very different ways of life. This evolutionary explosion produced almost all the major phyla, or groups, of invertebrates that exist today.

GREAT DIVERSITY

There is no such thing as a typical invertebrate, and many of the phyla have very little in common. The simplest kinds do not have a head or a brain, and usually rely on internal fluid pressure to keep their shape. At the other extreme, arthropods have a well-developed nervous system and elaborate sense organs, including complex eyes. Crucially, they also have an external body case, or exoskeleton, with legs that bend at flexible joints. This particular body plan has proved

to be remarkably successful, allowing arthropods to invade every natural habitat in water, on land, and in the air. Invertebrates also include animals that have a shell, and ones that are reinforced by mineral crystals or hard plates. Only vertebrates, however, have an internal skeleton made of bone.

SPLIT LIVES

Most invertebrates begin life as an egg. Some look like miniature versions of their parents when they hatch, but many start life with a very different body form. These larvae change shape, food sources, and feeding methods as they grow up. For example, sea urchins have drifting larvae, which filter food from the sea, while the adults scrape algae from rocks. The change in shape—metamorphosis—can be gradual, or it can take place abruptly, with the young animal's body being broken down, and an adult body being assembled in its place. Metamorphosis enables invertebrates to exploit more than one food source, and it also helps them to spread, often over great distances.

SPONGES
Among the simplest animals, sponges have a sievelike body, with an internal skeleton of mineral crystals. Classified as the phylum Porifera, there are more than 9,000 species.

ARTHROPODS
The phylum Arthropoda is the biggest in the animal kingdom, with more than 1.2 million species identified. It includes insects, crustaceans, arachnids, centipedes, and millipedes.

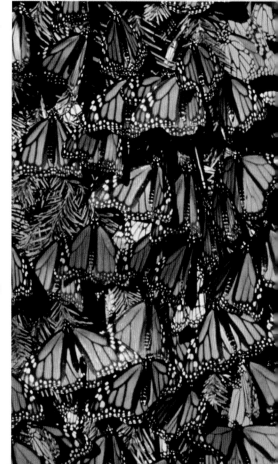

INVERTEBRATE TREE

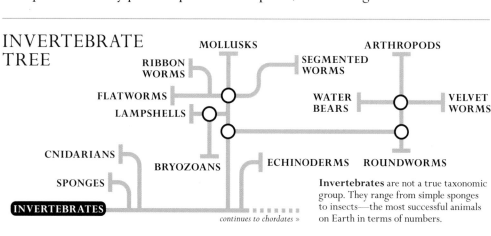

MOLLUSKS — ARTHROPODS
RIBBON WORMS — SEGMENTED WORMS
FLATWORMS — WATER BEARS — VELVET WORMS
LAMPSHELLS
CNIDARIANS
BRYOZOANS — ECHINODERMS — ROUNDWORMS
SPONGES
INVERTEBRATES

continues to chordates »

Invertebrates are not a true taxonomic group. They range from simple sponges to insects—the most successful animals on Earth in terms of numbers.

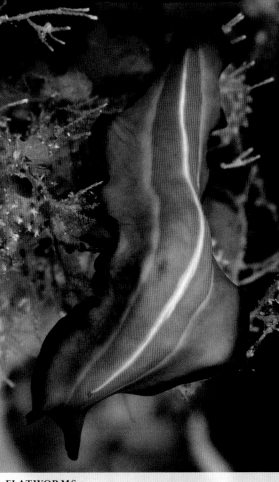

CNIDARIANS
Members of the phylum Cnidaria are soft-bodied invertebrates that kill their prey with stinging cells. Of the 11,947 known species, the vast majority are marine.

FLATWORMS
Numbering about 30,000 species, the phylum Platyhelminthes contains animals with a flat, thin body, and a distinguishable head and tail.

SEGMENTED WORMS
With about 18,000 species, the phylum Annelida contains worms that have a sinuous body, which is divided into ringlike segments. It includes earthworms and leeches.

CRUSTACEANS
Mainly aquatic, crustaceans are arthropods that breathe with gills. Classified as the subphylum Crustacea, there are about 70,000 species, including crabs and lobsters.

MOLLUSKS
One of the most diverse invertebrate groups, the phylum Mollusca contains almost 72,000 species. It includes gastropods, bivalves, and cephalopods.

ECHINODERMS
Recognizable by their five-fold symmetry, members of the phylum Echinodermata have a skeleton of small, chalky plates set in their skin. There are about 7,450 species.

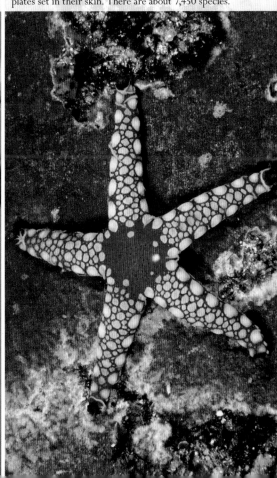

SPONGES

Simple, mostly marine animals, adult sponges live permanently attached to rocks, corals, and shipwrecks. A few species live in fresh water.

The Porifera vary in size and shape, from thin sheets to huge barrels, but all have the same basic structure: different types of specialist cells, but no organs. A system of water canals branches through the sponge, and water is drawn into them through tiny pores in the sponge's surface. Special collar cells lining the water canals trap and engulf bacteria and other small plankton for food, and waste water exits through large openings called oscula.

Many sponges are classically "spongy," others are rock hard, soft, or even slimy depending on the supporting skeleton, which is made up of tiny spicules of silica or calcium carbonate. Spicules vary in shape and number between species and are an important feature in identification.

PHYLUM	PORIFERA
CLASSES	4
ORDERS	32
FAMILIES	144
SPECIES	More than 9,000

3¼ ft
1 m

BLUE SPONGE
Haliclona sp.
F: Chalinidae
One of very few blue-colored sponges, this species commonly grows on the tops of corals and rocks off N. Borneo.

CALCAREOUS SPONGES

The skeleton of Calcarea is composed of densely packed, mostly starlike, calcium carbonate spicules, each with three or four pointed rays. Variable in shape and crunchy to the touch, most species are small and lobe- or tube-shaped.

3¼ in
8 cm

LEMON SPONGE
Leucetta chagosensis
F: Leucettidae
Growing on steep coral reefs in the W. Pacific, this saclike sponge provides a splash of bright color.

⅜–1½ in
1–4 cm

CLATHRINA CLATHRUS
F: Clathrinidae
Made up of many tubes, each only a few millimeters wide, this Mediterranean sponge is a distinctive yellow color.

spicules surrounding osculum

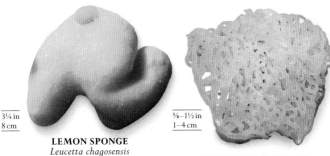

1¾–2 in
2–5 cm

PURSE SPONGE
Sycon ciliatum
F: Sycettidae
This simple, hollow sponge of N.E. Atlantic coasts has spiky, calcareous spicules surrounding its osculum.

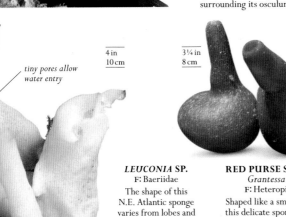

tiny pores allow water entry

4 in
10 cm

LEUCONIA SP.
F: Baeriidae
The shape of this N.E. Atlantic sponge varies from lobes and cushions to crusts. It grows in areas of high water movement.

3¼ in
8 cm

RED PURSE SPONGE
Grantessa sp.
F: Heteropiidae
Shaped like a small gourd, this delicate sponge grows between coral heads on shallow reefs in Malaysia and Indonesia.

DEMOSPONGES

More than 85 percent of all sponges belong to the Demospongiae. Although they are varied in appearance, most have a skeleton made up of scattered silicon dioxide spicules and a flexible organic collagen called spongin. A few encrusting species have no skeleton and some have only spongin.

3¼ ft
1 m

BROWN TUBE SPONGE
Agelas tubulata
F: Agelasidae
This sponge consists of uneven brown tubes arranged in clusters. It is common on deep reefs in the Caribbean and Bahamas.

2–4 in
5–10 cm

ELEPHANT HIDE SPONGE
Pachymatisma johnstonia
F: Geodiidae
Mounds of this tough sponge can cover large areas of rocks and wrecks in clear coastal waters of the N.E. Atlantic.

12–16 in
30–40 cm

RED TREE SPONGE
Negombata magnifica
F: Podospongiidae
Attempts to farm this beautiful Red Sea sponge are meeting with some success. It contains chemicals of potential medical importance.

14 in
35 cm

MEDITERRANAEAN BATH SPONGE
Spongia (Spongia) officinalis
F: Spongiidae
This sponge has an elastic skeleton and no harsh scpicules, so after it has been cleaned and dried, it is ideal to use for washing.

GOLF BALL SPONGE
Paratetilla bacca
F: Tetillidae
One of many tropical ball-shaped sponges, this species grows on sheltered coral reefs in the W. Pacific.

4¾ in
12 cm

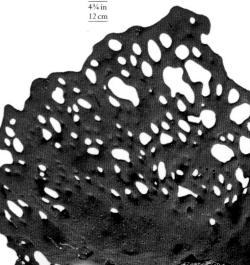

BORING SPONGE
Cliona celata
F: Clionaidae
Although it appears as yellow lumps, much of this European sponge remains hidden as strands growing through shells and calcareous rocks.

20 in
50 cm

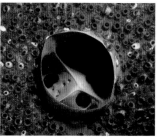

BORING SPONGE
Cliona delitrix
F: Clionaidae
Water leaves this Caribbean sponge via a number of large oscula (one is shown here). This sponge bores through corals by secreting acid.

BREADCRUMB SPONGE
Halichondria panicea
F: Halichondriidae
This N.E. Atlantic sponge grows as crusts on rocky shores and in shallow water. Its color comes from symbiotic algae.

⅜ in/1 cm

SPIRASTRELLA CUNCTATRIX
F: Spirastrellidae
One of many colorful encrusting sponges found in the ocean, this species covers inshore rocks in the Mediterranean and North Atlantic.

⅜–¾ in
1–2 cm

6–12 in
15–30 cm

YELLOW FINGER SPONGE
Callyspongia (Callyspongia) nuda
F: Callyspongiidae
The bright color of this tropical Pacific sponge is due to the chemicals it contains. Sponge extracts are used by the pharmaceutical industry.

12–16 in
30–40 cm

BARREL SPONGE
Xestospongia testudinaria
F: Petrosiidae
Small fishes and many invertebrates live on and in this huge Indo-Pacific sponge, which grows big enough to fit a person inside.

6½ ft/2 m

APLYSINA ARCHERI
F: Aplysinidae
The graceful long tubes of this sponge sway gently with the current on the Caribbean reefs where it grows.

2½–6½ ft
0.8–2 m

1½ ft
45 cm

AZURE VASE SPONGE
Callyspongia plicifera
F: Callyspongiidae
A common Caribbean species, this sponge colors reefs with its pale blue to purple vases. Its surface is sculpted with ridges and valleys.

HOMOSCLEROMORPHA SPONGES

This small class of just under 130 sponges is found in both temperate and tropical waters. They are mostly soft, as they have tiny silica spicules or none at all. Their planktonic larvae are unique in form.

CHICKEN LIVER SPONGE
Plakortis lita
F: Plakinidae
The common name of this sponge reflects its look and feel. Dark brown inside and out and smooth to the touch, it is found on coral reefs in the W. Pacific.

4–6 in
10–15 cm

soft coral polyp

GLASS SPONGES

Hexactinellida are a small group of sponges that live mostly in the deep ocean, where some form reeflike mounds up to 65 ft (20 m) high. The six rayed, starlike, silica spicules of the skeleton are usually fused together as a rigid lattice. After death, the skeleton remains as a ghostly outline.

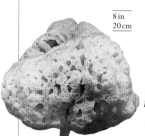

8 in
20 cm

BOLOSOMA GLASS SPONGE
Bolosoma sp.
F: Euplectellidae
Looking like fat fungi, *Bolosoma* sponges live in the deep sea and are held up into the water currents by a long, thin stalk.

14 in
35 cm

VENUS'S FLOWER BASKET
Euplectella aspergillum
F: Euplectellidae
Growing in tropical oceans below about 500 ft (150 m), this sponge was collected and displayed by Victorians for its delicate silica skeleton.

rigid lattice of silica spicules

CNIDARIANS

This phylum includes jellyfish, corals, and anemones. They have stinging tentacles to catch live prey, which they digest in a simple saclike gut.

All cnidarians are aquatic and most are marine. They have two body plans: a free-swimming, bell-like shape called a medusa—as seen in jellyfish—and a static polyp, typical of anemones. Neither medusae nor polyps have a head or front end. Their tentacles encircle a single gut opening, used to take in food and to eliminate waste.

A cnidarian's nervous system consists of a simple network of fibers; there is no brain. This means that the animal's behavior is usually simple. Although they are carnivorous, cnidarians cannot actively pursue their prey, with the possible exception of the box jellyfish. Instead, most species wait for swimming animals to blunder into the reach of their tentacles.

STINGS

Cnidarians' outer skins—and in some species the inner skin as well—are peppered with tiny stinging capsules of a kind that is unique to animals of this phylum. These stinging organs are called cnidocysts (or cnidocytes), and it is from these that cnidarians get their name. Cnidocysts are concentrated on the tentacles, and are triggered by touch or chemical signals, both when the animal comes into contact with potential prey and when it is under attack. Each cnidocyst contains a microscopic sac of venom and discharges a tiny coiled "harpoon" to inject the venom into flesh. Some stings can penetrate human skin and inflict severe pain, but the vast majority of cnidarians are harmless to humans.

ALTERNATING LIFE CYCLE

Many cnidarians have a life cycle that alternates between medusa and polyp—usually with one of these forms dominating, though in some groups one or other form is missing. The free-swimming medusa is normally the sexual stage. In most species fertilization is external: sperm and eggs are released in open water, forming planktonic planula larvae that resemble tiny flatworms. These settle in order to grow into polyps. Specialized polyps then produce new medusae to complete the cycle.

PHYLUM	CNIDARIA
CLASSES	6
ORDERS	22
FAMILIES	278
SPECIES	11,947

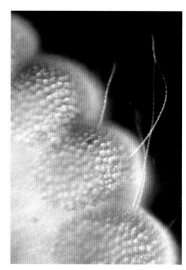

This microscopic view shows cnidocysts (stinging cells) that have been triggered to discharge their venom-laden harpoons.

BOX JELLYFISH

Members of the class Cubozoa are found in tropical and subtropical waters. They differ from true jellyfish in having a greater ability to control the direction and speed of their movement, rather than drifting with the currents. A flapping skirt at the bottom of the bell powers them at considerable speed. Box jellyfish have eyes, set in clusters on the sides of their transparent bell. They can see enough to steer past obstacles and to spot prey.

bell

1–10 ft
0.3–3 m

SEA WASP
Chironex fleckeri
F: Chirodropidae
This Indo-Pacific species— the largest box jellyfish— delivers a particularly painful sting and has caused human fatalities.

STALKED JELLYFISH

The adult is the only easily seen phase in the life cycle of members of the Class Staurozoa. Unlike other jellyfish it does not swim, but is attached to the substrate by a stalk. Eight arms radiate from the mouth, with clumps of tentacles at their tips. The other life stages are the extremely small planula larvae and polyps (called stauropolyps).

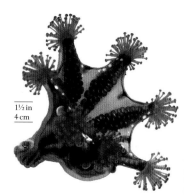

1½ in
4 cm

STALKED JELLYFISH
Haliclystus auricula
F: Haliclystidae
This species is found attached to algae and seagrasses in shallow cooler waters of the Atlantic and Pacific, where it catches food using stinging cells on the tentacles.

TRUE JELLYFISH

The familiar jellyfish bell is the medusa stage of the Scyphozoa life cycle. The polyp is reduced, or in some deep-sea forms, is missing. Jellyfish polyps undergo strobilation, a process that produces new tiny medusae by budding. Rhizostome jellyfish lack a fringe of tentacles around the bell.

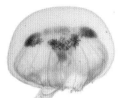

8–16 in
20–40 cm

MOON JELLYFISH
Aurelia aurita
F: Ulmaridae
This globally widespread genus has four long "arms" as well as small marginal tentacles. It swarms to breed near coasts and polyps settle in estuaries.

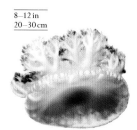

8–12 in
20–30 cm

5½–6½ in
14–16 cm

SPOTTED LAGOON JELLYFISH
Mastigias papua
F: Mastigiidae
Like other rhizostome jellyfish, this algae-containing species traps planktonic food with mucus. It enters South Pacific lagoons, including those of oceanic islands.

UPSIDE-DOWN JELLYFISH
Cassiopea andromeda
F: Cassiopeidae
Superficially resembling an anemone, this Indo-Pacific rhizostome jellyfish lives at the bottom of lagoons, mouth upward, with its bell pulsating to circulate water.

HYDROZOANS

Most animals of the class Hydrozoa live as branching sea-dwelling colonies forming miniature polyp-bearing forests. Some anchor themselves to surfaces by creeping horizontal stems, called stolons. Colonies are supported by a horny translucent sheath and some of the polyps produce sexual medusae. Freshwater hydra lack medusae: instead, sex organs develop directly on solitary polyps.

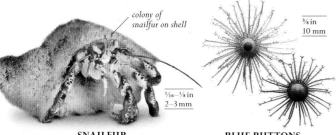

colony of snailfur on shell

1/16–1/8 in
2–3 mm

SNAILFUR
Hydractinia echinata
F: Hydractiniidae
One of a family of colonial spiny hydrozoans, this N.E. Atlantic species grows over the surface of snail shells containing hermit crabs.

3/8 in
10 mm

BLUE BUTTONS
Porpita porpita
F: Porpitidae
This colonial, jellyfishlike hydrozoan of tropical oceans is sometimes regarded as a highly modified individual polyp.

PURPLE LACE "CORAL" HYDROID
Distichopora violacea
F: Stylasteridae
Like some other hydrozoans, colonies of this Indo-Pacific lace coral have polyps specialized for different functions, such as feeding and stinging.

2–4 in
5–10 cm

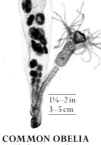

1¼–2 in
3–5 cm

COMMON OBELIA
Obelia geniculata
F: Campanulariidae
This is a worldwide hydrozoan, abundant on intertidal seaweeds. It grows in zig-zag colonies of cup-encased polyps that sprout from creeping horizontal stems.

PORTUGUESE MAN OF WAR
Physalia physalis
F: Physaliidae
Resembling a jellyfish, this is really an ocean-going colony of hydrozoans. Stinging tentacles and specialized polyps are suspended from a gas-filled float.

gas-filled float

33–165 ft
10–50 m

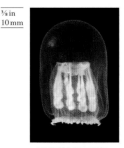

3/8 in
10 mm

EIGHT-RIBBED HYDROMEDUSA
Melicertum octocostatum
F: Melicertidae
This hydrozoan from the North Atlantic and North Pacific oceans belongs to a family related to those with cup-encased polyps. It is known mainly from medusae.

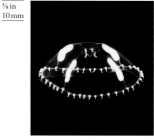

3/8 in
10 mm

CUP HYDROMEDUSA
Phialella quadrata
F: Phialellidae
Like related *Obelia*, branching colonies of this globally widespread hydrozoan release free-living medusae, which reproduce sexually.

16–23½ in
4–6 cm

FEATHER HYDROID
Aglaophenia cupressina
F: Aglaopheniidae
Feather hydroids and the closely related sea ferns belong to a group of colonial hydrozoans with cup-encased polyps. This is an Indo-Pacific species.

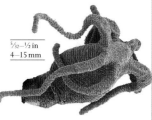

5/32–½ in
4–15 mm

COMMON HYDRA
Hydra vulgaris
F: Hydridae
Lacking the medusa stage, hydras have small freshwater polyps that can reproduce asexually by budding. This is a globally widespread cold-water species, shown here in false color.

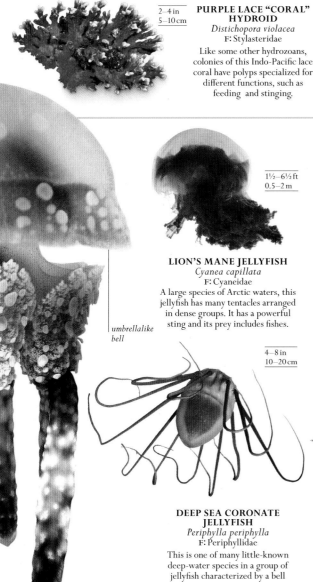

umbrellalike bell

oral arm

1½–6½ ft
0.5–2 m

LION'S MANE JELLYFISH
Cyanea capillata
F: Cyaneidae
A large species of Arctic waters, this jellyfish has many tentacles arranged in dense groups. It has a powerful sting and its prey includes fishes.

4–8 in
10–20 cm

DEEP SEA CORONATE JELLYFISH
Periphylla periphylla
F: Periphyllidae
This is one of many little-known deep-water species in a group of jellyfish characterized by a bell with an encircling groove.

16 in
40 cm

HULA SKIRT SIPHONPHORE
Physophora hydrostatica
F: Physophoridae
This widespread free-floating hydrozoan colony has a smaller float than its relative, the Portuguese man of war, and carries prominent swimming bells.

4–8 in
10–20 cm

PINK-HEARTED HYDROID
Tubularia sp.
F: Tubulariidae
Tubularia hydrozoans have polyps on long stems with two whorls of tentacles, one around the polyp base and one around its mouth.

16–20 in
40–50 cm

FIRE "CORAL" HYDROID
Millepora sp.
F: Milleporidae
Like others of its family, this fiercely stinging colonial hydrozoan has a calcified skeleton and is a reef-builder. It is distantly related to true stony corals.

¾–1 in
2–2.5 cm

FRESHWATER JELLYFISH
Craspedacusta sowerbii
F: Olindiidae
Sporadic in freshwater ponds, lakes, and streams worldwide, minute polyps of this animal develop into dominant medusae.

ANEMONES AND CORALS

Unlike other cnidarians, the Anthozoa lack the free-swimming medusa stage altogether and most remain static throughout their life cycle. The polyps, many of which have a flowerlike appearance, produce sperm and eggs. Members of this class include solitary anemones, colonial sea pens, soft corals, and tropical reef-forming stony corals.

8–12 in
20–30 cm

slender red or
yellow lobe

3¼–5 ft
1–1.5 m

polyp cluster

**COMMON
SEA FAN**
Gorgonia ventalina
F: Gorgoniidae
Sea fans, such as this Caribbean species, have upright fanlike colonies supported by a central axis, which is stiffened with a horny substance called gorgonin.

8–12 in
20–30 cm

3½–20 ft
1–6 m

white
polyps

**TOADSTOOL
LEATHER CORAL**
Sarcophyton trocheliophorum
F: Alcyoniidae
This soft coral can form enormous leathery colonies on tropical reefs of the Indo-Pacific. It is fast-growing, nourished by photosynthetic algae.

4–6 in
10–15 cm

**SEA
STRAWBERRY**
Gersemia rubiformis.
F: Nephtheidae
This stalked soft coral forms bright pink lumpy colonies. It occurs in northern parts of the Pacific and Atlantic Oceans.

**CARNATION
CORAL**
Dendronephthya sp.
F: Nephtheidae
Typical of a family of stalked soft corals with polyps arranged in clusters, this colorful species occurs on Indo-Pacific tropical reefs.

20–40 in
50–100 cm

flexible
body

16–20 in
40–50 cm

4–8 in
10–20 cm

**RED DEAD-MAN'S
FINGERS**
Alcyonium glomeratum
F: Alcyoniidae
This is a more slender, upright ally of dead-man's fingers, found on sheltered rocky European coastlines. It varies from yellow to red.

**COMMON DEAD-MAN'S
FINGERS**
Alcyonium digitatum
F: Alcyoniidae
A typical soft coral, this thickly lobed European species has colonies of polyps, which are carried on a fleshy mass unsupported by a rigid skeleton.

RED CORAL
Corallium rubrum
F: Coralliidae
Colonies of this Mediterranean sea fan (not a true coral) are supported by a skeleton of tiny, enmeshed, calcified needles. It is valued for jewelry.

14–16 in
35–40 cm

KIDNEY SEA PEN
Sarcoptilus grandis
F: Pennatulidae
Widespread in temperate-waters, this sea pen has kidney-shaped branches that occur in rows along each side of its body.

ORANGE SEA PEN
Ptilosarcus gurneyi
F: Pennatulidae
One of many brightly colored sea pens, this species from the Pacific coast of N. America retracts into its burrow when disturbed by predators.

1½–6½ ft
0.5–2 m

calcareous
tubes

6–12 in
15–30 cm

20–40 in
50–100 cm

20–40 in
50–100 cm

tentacle

WHITE SEA WHIP
Junceella fragilis
F: Ellisellidae
Sea whips are threadlike relatives of sea fans, with their supporting horny axis reinforced with calcium. This is an Indonesian reef species.

RED SEA WHIP
Ellisella sp.
F: Ellisellidae
Ellisella sea whips form two-prong branching colonies, some of which may form dense underwater bushes. They occur in tropical and temperate waters.

ORGAN PIPE CORAL
Tubipora musica
F: Tubiporidae
This Indo-Pacific soft coral has polyps in upright, calcified tubes, which are connected to the colony by a rootlike network at the base.

BLUE CORAL
Heliopora coerulea
F: Helioporidae
Despite its hard, calcified skeleton, this species is more closely related to soft coral than stony coral. It is the only member of its order.

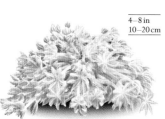

FLOWERPOT CORAL
Goniopora columna
F: Poritidae

This species has daisylike polyps that are greatly lengthened when fully extended. It is an Indo-Pacific relative of the lobe coral.

4–8 in
10–20 cm

LOBE CORAL
Porites lobata
F: Poritidae

One of the most common reef-builders of the Indo-Pacific, this coral forms large encrusting colonies in places buffeted by strong wave action.

13–16 ft
4–5 m

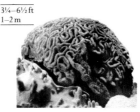

LOBED BRAIN CORAL
Lobophyllia sp.
F: Lobophylliidae

Massive colonies of this brain coral—either flattened or dome-shaped—occur on reefs in tropical waters of the Indo-Pacific.

3¼–6½ ft
1–2 m

DAHLIA ANEMONE
Urticina felina
F: Actiniidae

Sticky swellings on this anemone may carry so much debris that, with tentacles retracted, it resembles a small pile of gravel. It occurs around the North Pole.

4–4¾ in
10–12 cm

STAGHORN CORAL
Acropora sp.
F: Acroporidae

The branching staghorn corals are among the largest tropical reef-builders. Nourished by photosynthetic algae, these species are fast-growing.

3¼–9¾ ft
1–3 m

DAISY CORAL
Goniopora sp.
F: Poritidae

Long-reaching polyps of *Goniopora* stony corals on tropical reefs typically have 24 tentacles. They are among the most flowerlike of corals.

3¼ ft
1 m

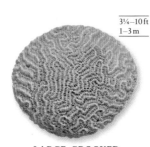

LARGE-GROOVED BRAIN CORAL
Colpophyllia sp.
F: Faviidae

The hemispherical brainlike structure of this coral is typical of its family. It is a tropical reef-builder that contains photosynthetic algae.

3¼–10 ft
1–3 m

MUSHROOM CORAL
Fungia fungites
F: Fungiidae

This is one of many tropical mushroom corals that are not reef-building, but live among other species as solitary polyps crawling over the ocean floor.

4–8 in
10–20 cm

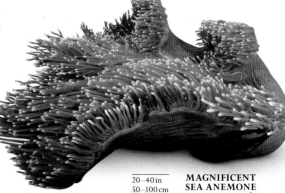

MAGNIFICENT SEA ANEMONE
Heteractis magnifica
F: Stichodactylida

This giant anemone is from Indo-Pacific reefs, where it lives in close partnership with a variety of fish species, including clown fishes.

20–40 in
50–100 cm

ATLANTIC COLD WATER CORAL
Desmophyllum pertusum
F: Caryophylliidae

Unlike many deep sea corals lacking nutrient-supplying algae, this North Atlantic species forms extensive reefs, although only very slowly.

300–600 ft / 100–200 m

⅜–½ in
10–15 mm

PLUMOSE ANEMONE
Metridium senile
F: Metridiidae

This is a globally widespread member of a family of anemones that is characterized by a fuzzy mass of tentacles. It can split to form genetically identical populations.

1–6 in
2.5–15 cm

SNAKELOCKS ANEMONE
Anemonia viridis
F: Actiniidae

The conspicuously long tentacles of this common European intertidal anemone rarely retract, even when exposed at low tide.

2–2¾ in
5–7 cm

DEVONSHIRE CUP CORAL
Caryophyllia (Caryophyllia) smithii
F: Caryophylliidae

This N. E. Atlantic species belongs to a family containing cold water corals, some with large, anemonelike polyps. Barnacles frequently attach to it.

COMMON TUBE ANEMONE
Cerianthus membranaceus
F: Cerianthidae

Tube anemones burrow through sediment, using unique non-stinging cnidocysts to build feltlike tubes from mucus. This species occurs on European offshore mud.

4–6 in
10–15 cm

ATLANTIC BLACK CORAL
Antipathes sp.
F: Antipathidae

Mostly found in deep waters, black corals consist of spiny polyp colonies encased in slender, horny exoskeletons.

20–40 in
50–100 cm

FLATWORMS

Structurally among the simplest of animals, flatworms live wherever a moist habitat can provide their thin, flat body with oxygen and food.

Platyhelminthes, commonly known as flatworms, are a major phylum of animals found in various forms in the ocean, in freshwater ponds, or even within the bodies of other animals. Superficially resembling leeches, flatworms are, in fact, far simpler animals. With no blood system and no organs for breathing, they use their entire body surface to absorb oxygen and release carbon dioxide. Flatworm species that have no gut also absorb their food in this way. In other species the gut has one opening but branches frequently, so that digested food can still reach all tissues,

even without a blood stream to circulate the nutrients. Some free-living flatworms are detritus-eaters, gliding along on a bed of microscopic, beating hairs. Others prey on other invertebrates.

INTERNAL PARASITES

Tapeworms and flukes are parasites. Inside the host animal, their flattened body is perfectly suited for absorbing nutrients. Many have complex ways of passing from host to host and may infect more than one type of animal. They enter the body of the host in infected food or sometimes by penetrating the skin. Once inside, a parasite may go deeper into the body—burrowing through the gut wall to lodge in vital organs.

PHYLUM	PLATYHELMINTHES
CLASSES	6
ORDERS	41
FAMILIES	About 420
SPECIES	About 30,000

A NEW PHYLUM?

Traditionally, the tiny, gutless, and brainless marine animals called acoels are viewed as flatworms. Controversial studies suggest that they belong in a new phylum of their own, and that they are the most primitive living animals to show bilateral symmetry—in contrast to radial (round) cnidarians.

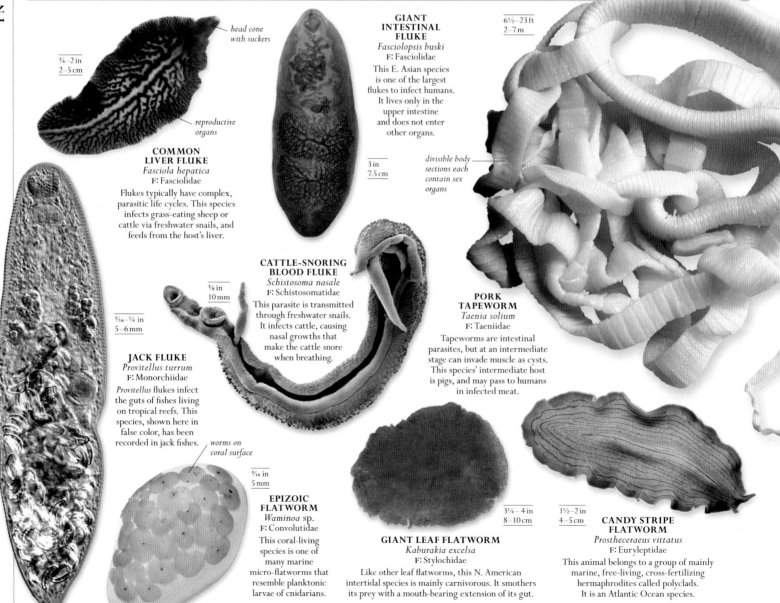

head cone with suckers

³/₄–2 in
2–5 cm

reproductive organs

COMMON LIVER FLUKE
Fasciola hepatica
F: Fasciolidae
Flukes typically have complex, parasitic life cycles. This species infects grass-eating sheep or cattle via freshwater snails, and feeds from the host's liver.

GIANT INTESTINAL FLUKE
Fasciolopsis buski
F: Fasciolidae
This E. Asian species is one of the largest flukes to infect humans. It lives only in the upper intestine and does not enter other organs.

3 in
7.5 cm

6½–23 ft
2–7 m

divisible body sections each contain sex organs

¹/₁₆–¹/₄ in
5–6 mm

JACK FLUKE
Provitellus turrum
F: Monorchiidae
Provitellus flukes infect the guts of fishes living on tropical reefs. This species, shown here in false color, has been recorded in jack fishes.

CATTLE-SNORING BLOOD FLUKE
Schistosoma nasale
F: Schistosomatidae
This parasite is transmitted through freshwater snails. It infects cattle, causing nasal growths that make the cattle snore when breathing.

³/₈ in
10 mm

PORK TAPEWORM
Taenia solium
F: Taeniidae
Tapeworms are intestinal parasites, but at an intermediate stage can invade muscle as cysts. This species' intermediate host is pigs, and may pass to humans in infected meat.

worms on coral surface

³/₁₆ in
5 mm

EPIZOIC FLATWORM
Waminoa sp.
F: Convolutidae
This coral-living species is one of many marine micro-flatworms that resemble planktonic larvae of cnidarians.

GIANT LEAF FLATWORM
Kaburakia excelsa
F: Stylochidae
Like other leaf flatworms, this N. American intertidal species is mainly carnivorous. It smothers its prey with a mouth-bearing extension of its gut.

3¼–4 in
8–10 cm

1½–2 in
4–5 cm

CANDY STRIPE FLATWORM
Prostheceraeus vittatus
F: Euryleptidae
This animal belongs to a group of mainly marine, free-living, cross-fertilizing hermaphrodites called polyclads. It is an Atlantic Ocean species.

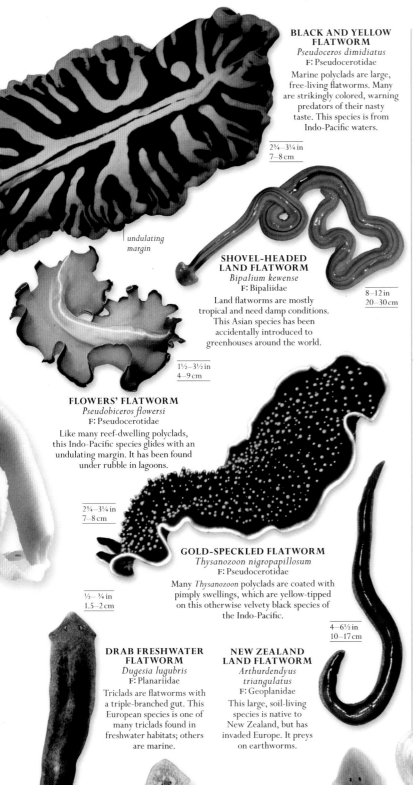

BLACK AND YELLOW FLATWORM
Pseudoceros dimidiatus
F: Pseudocerotidae

Marine polyclads are large, free-living flatworms. Many are strikingly colored, warning predators of their nasty taste. This species is from Indo-Pacific waters.

2¾–3¼ in
7–8 cm

undulating margin

SHOVEL-HEADED LAND FLATWORM
Bipalium kewense
F: Bipaliidae

Land flatworms are mostly tropical and need damp conditions. This Asian species has been accidentally introduced to greenhouses around the world.

8–12 in
20–30 cm

1½–3½ in
4–9 cm

FLOWERS' FLATWORM
Pseudobiceros flowersi
F: Pseudocerotidae

Like many reef-dwelling polyclads, this Indo-Pacific species glides with an undulating margin. It has been found under rubble in lagoons.

2¾–3¼ in
7–8 cm

GOLD-SPECKLED FLATWORM
Thysanozoon nigropapillosum
F: Pseudocerotidae

Many *Thysanozoon* polyclads are coated with pimply swellings, which are yellow-tipped on this otherwise velvety black species of the Indo-Pacific.

½– ¾ in
1.5–2 cm

DRAB FRESHWATER FLATWORM
Dugesia lugubris
F: Planariidae

Triclads are flatworms with a triple-branched gut. This European species is one of many triclads found in freshwater habitats; others are marine.

NEW ZEALAND LAND FLATWORM
Arthurdendyus triangulatus
F: Geoplanidae

This large, soil-living species is native to New Zealand, but has invaded Europe. It preys on earthworms.

4–6½ in
10–17 cm

BROWN FRESHWATER FLATWORM
Dugesia tigrina
F: Planariidae

This species is native to freshwater habitats of N. America, but has been introduced to Europe.

⅜–½ in
1–1.5 cm

STREAM FLATWORM
Dugesia gonocephala
F: Planariidae

Many of the freshwater *Dugesia* flatworms, such as this European species of running water, have earlike flaps for detecting water currents.

¾–1¼ in
2–3 cm

ROUNDWORMS

The simple, cylindrical roundworms are remarkably successful—surviving almost everywhere, resisting drought, and reproducing rapidly.

The phylum Nematoda is ubiquitous. There can be millions of roundworms in a square yard of soil, and they also live in freshwater or marine habitats. Many are parasites. These worms, or nematodes, can be extraordinarily prolific, producing hundreds of thousands of eggs per day. When environmental conditions deteriorate they can survive heat, frost, or drought by producing resistant stages and becoming dormant.

Roundworms have a muscle-lined body cavity and two openings to their gut—the mouth and anus. Their cylindrical body is coated in a tough layer, called a cuticle—functionally similar to that of arthropods—which they molt periodically as they grow.

PHYLUM	NEMATODA
CLASSES	2
ORDERS	17
FAMILIES	About 160
SPECIES	About 26,000

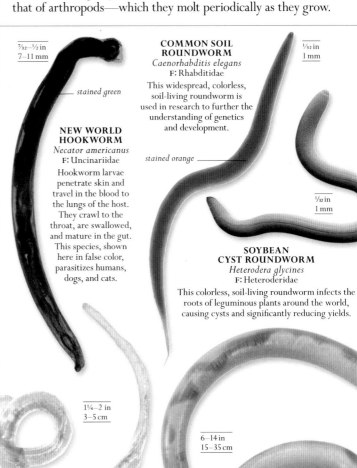

7/32–½ in
7–11 mm

stained green

NEW WORLD HOOKWORM
Necator americanus
F: Uncinariidae

Hookworm larvae penetrate skin and travel in the blood to the lungs of the host. They crawl to the throat, are swallowed, and mature in the gut. This species, shown here in false color, parasitizes humans, dogs, and cats.

COMMON SOIL ROUNDWORM
Caenorhabditis elegans
F: Rhabditidae

This widespread, colorless, soil-living roundworm is used in research to further the understanding of genetics and development.

1/32 in
1 mm

stained orange

1/32 in
1 mm

SOYBEAN CYST ROUNDWORM
Heterodera glycines
F: Heteroderidae

This colorless, soil-living roundworm infects the roots of leguminous plants around the world, causing cysts and significantly reducing yields.

1¼–2 in
3–5 cm

COMMON WHIPWORM
Trichuris trichiura
F: Trichuridae

Like many other gut parasites, the mainly tropical whipworm infects humans who eat food contaminated with feces. It completes its life cycle in the intestines.

6–14 in
15–35 cm

LARGE INTESTINAL ROUNDWORM
Ascaris lumbricoides
F: Acarididae

A common parasite of humans in regions of poor sanitation where people get infected by eating food contaminated with feces, with adult worms then living in the host intestine.

SEGMENTED WORMS

With more complex muscle and organ systems than flatworms, many species of this phylum, called Annelida, are accomplished swimmers or burrowers.

Segmented worms include earthworms, ragworms, and leeches. For most, their blood circulates in vessels and they have a firm sac of fluid (the coelom) running the length of their body, which keeps motion of the gut separate from motion of the body wall. The coelum is split into sections, with each one corresponding to a body segment. Each body segment contains a set of muscles, and coordination of these sets can send a wave of contraction down the length of the body, or make it flex back and forth. This makes many annelids highly mobile both on land and in the water.

Marine segmented worms—the predatory ragworms and their filter-feeding relatives—carry bundles of bristles along their body, typically borne on little paddles that help with swimming, burrowing, or even walking. This group of bristled, segmented worms are called polychaetes or bristle worms.

Land-living earthworms are more sparsely bristled detritus-feeders, and are important recyclers of dead vegetation and aerators of soil. Many leeches are specialized further, and carry suckers to extract blood from a host. Their saliva contains chemicals that stop blood clotting. Other leeches are predatory. Both earthworms and leeches have a saddle-shaped, glandular structure around their body called a clitellum, which they use to make their egg cocoons.

PHYLUM	ANNELIDA
CLASSES	4
ORDERS	17
FAMILIES	About 130
SPECIES	About 18,000

SLUDGE WORM
Tubifex sp.
F: Naididae
This widely distributed worm can be seen in mud polluted by sewage, the front end buried, the rear end wiggling to extract oxygen.

¾–2¾ in
2–7 cm

20 in
50 cm

MEGADRILE EARTHWORM
Glossoscolex sp.
F: Glossoscolecidae
Glossoscolex are large earthworms from tropical C. and S. America. Many are found in rainforest habitats.

4–6 in
10–15 cm

TIGER WORM
Eisenia foetida
F: Lumbricidae
This European inhabitant of rotting vegetation secretes defensive pungent fluid and, like other earthworms, has a "saddle" for producing cocoons of eggs.

6–10 in
15–25 cm

clitellum or "saddle"

COMMON EARTHWORM
Lumbricus terrestris
F: Lumbricidae
This earthworm—native to Europe but introduced elsewhere—drags leaves into its burrow at night as a source of food.

1½–2¾ in
4–7 cm

CHRISTMAS TREE TUBE WORM
Spirobranchus giganteus
F: Serpulidae
Characterized by its spiral whorls of tentacles, used for filter feeding and extracting oxygen, this species is widespread on tropical reefs.

VELVET WORMS

These soft-bodied worms, cousins of arthropods, lumber slowly along dark forest floors like giant caterpillars, but they are extraordinary hunters.

These animals have the body of an earthworm and the multiple limbs of a millipede, but belong in a phylum of their own: Onychophora. Velvet worms are rarely seen in their warm native

rainforests of tropical America, Africa, and Australasia; they shun the open, preferring to hide in crevices and leaf litter. They come out at night or after rainfall to hunt other invertebrates. Velvet worms capture their prey in a unique way: they immobilize their victims by spraying them with sticky slime which is produced by glands that open through pores straddling the mouth.

PHYLUM	ONYCHOPHORA
CLASS	1
ORDER	1
FAMILIES	2
SPECIES	About 200

skin covered in fine hairs

4 in
10 cm

SOUTHERN AFRICAN VELVET WORM
Peripatopsis moseleyi
F: Peripatopsidae
This species belongs to a family with a globally southern distribution.

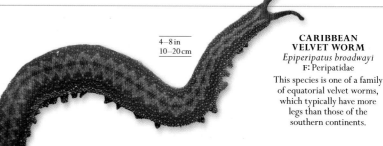

4–8 in
10–20 cm

CARIBBEAN VELVET WORM
Epiperipatus broadwayi
F: Peripatidae
This species is one of a family of equatorial velvet worms, which typically have more legs than those of the southern continents.

FIRE WORM
Hermodice carunculata
F: Amphinomidae

A tropical Atlantic offshore species, this polychaete preys on corals, sucking out the soft flesh from the hard skeleton. Its body extensions have painfully irritating bristles.

2¼–12 in
6–30 cm

SEA MOUSE
Aphrodita aculeata
F: Aphroditidae

This mud-burrowing scale worm is found in the shallow waters of N. Europe. Its scales are covered with a hairlike coat.

4–8 in
10–20 cm

1–1¼ in
2.5–3 cm

SEA CUCUMBER SCALE WORM
Gastrolepidia clavigera
F: Polynoidea

This Indo-Pacific polychaete has flattened back scales. It is parasitic on sea cucumbers.

4¾–10 in
12–25 cm

LUGWORM
Arenicola marina
F: Arenicolidae

An earthwormlike polychaete, the lugworm lives in burrows on beaches and mudflats, where it ingests sediment and feeds on detritus.

3¼–13 ft
1–4 m

HONEYCOMB WORM
Sabellaria alveolata
F: Sabellaridae

This tubeworm builds its tubes from sand and shell fragments. Dense populations form honeycomblike reefs in the Atlantic and Mediterranean.

2–6 in
5–15 cm

GREEN PADDLE WORM
Eulalia viridis
F: Phyllodocidae

Paddleworms are active carnivores with leaf-shaped flaps on their body extensions. This European species lives among intertidal rocks and kelp.

PACIFIC FEATHERDUSTER WORM
Sabellastarte sanctijosephi
F: Sabellidae

A tropical Indo-Pacific tubeworm, this species is common along coastlines, including coral reefs and tidal pools.

3¼–4 in
8–10 cm

10–16 in
25–40 cm

KING RAGWORM
Alitta virens
F: Nereididae

Ragworms, close relatives of paddleworms, have two-pronged body extensions. This Atlantic mud-burrowing species can deliver a painful bite.

2–2¾ in
5–7 cm

RED TUBE WORM
Serpula vermicularis
F: Serpulidae

Serpulid tubeworms build hard, chalky tubes. Like most serpulids, this widespread species has tentacles modified to plug the tube after retraction.

NORTHERN SPIRAL TUBE WORM
Spirorbis borealis
F: Serpulidae

The small coiled tube of this tubeworm is cemented to brown *Fucus* seaweeds and kelps along North Atlantic coastlines.

¹⁄₁₆–⅛ in
2–3 mm

GIANT TUBE WORM
Riftia pachyptila
F: Siboglinidae

This tubeworm lives in the hot, sulfur-rich darkness of ocean-floor volcanic vents in the Pacific. The red plume harbors bacteria that make its food using chemicals from these vents.

body used
as anchor

6½–7¾ ft
2–2.4 m

WATER BEARS

Visible only through a microscope, stubby-legged, agile water bears share their aquatic community with microbes and much simpler invertebrates.

Tiny water bears, or tardigrades (meaning "slow-stepped") clamber through miniature forests of water weed, four pairs of short legs hanging on with clawed feet. Most are less than a millimeter long. They abound among moss or algae, where many use their needlelike jaws to pierce the cells of this vegetation and suck its sap. Many water bear species are known only as females, which reproduce asexually by spawning offspring from unfertilized eggs. If its habitat dries up, a water bear is able to survive by going into a kind of suspended animation called cryptobiosis, shriveling into a husk, sometimes for years at a time, until rainfall revives it.

PHYLUM	TARDIGRADA
CLASSES	3
ORDERS	5
FAMILIES	20
SPECIES	About 1,000

MOSS WATER BEAR
Echiniscus sp.
F: Echiniscidae

Many water bears live in moss, but their ability to survive in a dried state has helped a number of species to disperse worldwide.

0.25 mm

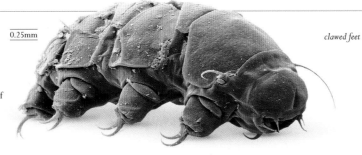

clawed feet

SEAWEED WATER BEAR
Echiniscoides sigismundi
F: Echiniscoididae

One of many little-known marine water bears, this species has been recorded among seaweed in coastal locations around the world.

0.25 mm

ARTHROPODS

Jointed legs and flexible armor have helped a phylum that includes winged insects and underwater crustaceans to evolve unrivaled diversity.

There are more species of arthropods known to scientists than all the other animal phyla put together—and doubtless many more await discovery. They have an extraordinary range of lifestyles including grazing, predation, filtering particles of food in water, and drinking fluids such as nectar or blood.

Arthropods are coated by an exoskeleton made of a tough material called chitin. It is sufficiently flexible around the joints and between segments to allow for mobility, but resists stretching. As arthropods grow, they must periodically molt their exoskeleton, which is replaced by a slightly larger one. The exoskeleton serves as protective armor but also helps reduce water loss in very dry habitats.

BODY PARTS
Arthropods evolved from a segmented ancestor, perhaps a wormlike creature. Body segmentation persists in all arthropods, but is especially obvious in millipedes and centipedes. In other groups, various segments have fused together into discrete body sections. Insects are divided into a sensory head, a muscular thorax with legs and wings, and an abdomen containing most of the internal organs. In arachnids and some crustaceans, the head and thorax are fused into one single section.

GETTING OXYGEN
Aquatic arthropods, such as crustaceans, breathe with gills. The bodies of most land arthropods—insects and myriapods—are permeated by a network of microscopic, air-filled tubes called tracheae. These open out through pores known as spiracles on the sides of the body, usually one pair per body section. Small muscles inside the spiracles regulate the flow of air by means of valves. In this way, oxygen can reach all cells of the body and does not need to be transported in blood. Some arachnids breathe with tracheae; others with book lungs in their abdomen that evolved from the gills of their aquatic ancestors. Most use a combination of the two.

PHYLUM	ARTHROPODA
CLASSES	19
ORDERS	123
FAMILIES	About 2,300
SPECIES	About 1.2 million

DEBATE

CAMOUFLAGE AND MIMICRY

Many arthropods have adapted to their environment by blending imperceptibly into it. Stick insects, for example, look so like twigs that they are very difficult for predators to spot. In contrast, arthropods, such as wasps, have evolved bright colors that warn others they are distasteful or dangerous. The hornet moth is completely harmless but avoids predation because it looks and sounds exactly like a hornet. Despite their similar appearance, other anatomical features separate them into different orders. The hornet moth belongs to the order Lepidoptera and the hornet to the order Hymenoptera.

MILLIPEDES AND CENTIPEDES

Millipedes and centipedes form a group of multi-segmented arthropods called myriapods. Millipedes typically have two legs per body segment; centipedes have only one per segment. Whereas millipedes are vegetarian, centipedes are predatory carnivores.

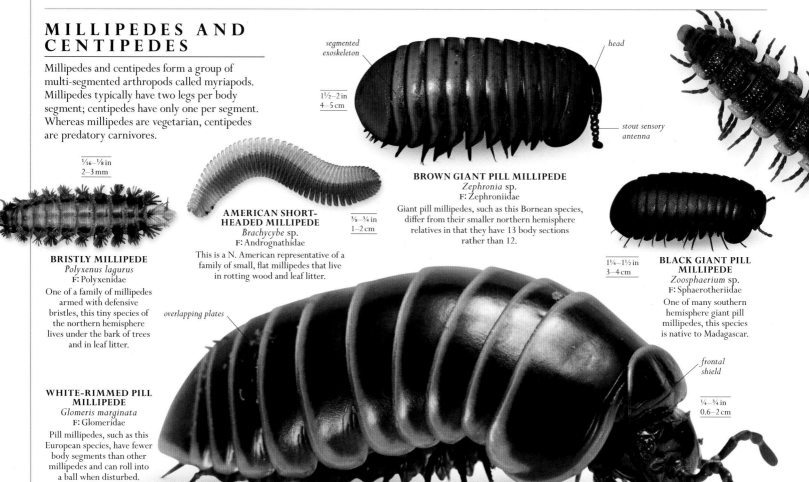

segmented exoskeleton

head

1½–2 in
4–5 cm

stout sensory antenna

BROWN GIANT PILL MILLIPEDE
Zephronia sp.
F: Zephroniidae
Giant pill millipedes, such as this Bornean species, differ from their smaller northern hemisphere relatives in that they have 13 body sections rather than 12.

1/16–1/8 in
2–3 mm

3/8–3/4 in
1–2 cm

AMERICAN SHORT-HEADED MILLIPEDE
Brachycybe sp.
F: Andrognathidae
This is a N. American representative of a family of small, flat millipedes that live in rotting wood and leaf litter.

BRISTLY MILLIPEDE
Polyxenus lagurus
F: Polyxenidae
One of a family of millipedes armed with defensive bristles, this tiny species of the northern hemisphere lives under the bark of trees and in leaf litter.

overlapping plates

1¼–1½ in
3–4 cm

BLACK GIANT PILL MILLIPEDE
Zoosphaerium sp.
F: Sphaerotheriidae
One of many southern hemisphere giant pill millipedes, this species is native to Madagascar.

WHITE-RIMMED PILL MILLIPEDE
Glomeris marginata
F: Glomeridae
Pill millipedes, such as this European species, have fewer body segments than other millipedes and can roll into a ball when disturbed.

frontal shield

¼–¾ in
0.6–2 cm

BLACK SNAKE MILLIPEDE
Tachypodoiulus niger
F: Julidae

¾–2¼ in
2–6 cm

Unlike its close relatives, this distinctive, white-legged millipede from W. Europe spends most of its time above ground—even climbing trees and walls.

3–5 in
7.5–13 cm

AMERICAN GIANT MILLIPEDE
Narceus americanus
F: Spirobolidae

This large, Atlantic coast millipede belongs to a family of mainly American, cylindrical millipedes. Like its relatives, it releases noxious chemicals in defense.

½–1¼ in
1.5–3 cm

BROWN SNAKE MILLIPEDE
Julus scandinavius
F: Julidae

A member of a large family of cylindrical millipedes with ring-encased segments, this species occurs in European deciduous woodland, especially on acid soils.

8–15 in
20–38 cm

AFRICAN GIANT MILLIPEDE
Archispirostreptus gigas
F: Spirostreptidae

One of the largest of all millipedes, this species occurs mainly in tropical E. Africa and is one of many species that release irritating chemicals to defend themselves.

BORING MILLIPEDE
Polyzonium germanicum
F: Polyzoniidae

This somewhat primitive millipede has a scattered distribution in Europe. It lives in woodland and resembles beech bud scales when coiled.

³⁄₁₆–¾ in
0.5–1.8 cm

EASTERN FLAT-BACKED MILLIPEDE
Polydesmus complanatus
F: Polydesmidae

Polydesmids have projections on their exoskeleton that make them look flat-backed. This E. European species is a fast runner.

½–2¼ in
1.5–6 cm

YELLOW EARTH CENTIPEDE
Geophilus flavus
F: Geophilidae

Eyeless geophilids have more segments (and therefore legs) than other centipedes. This European soil-dweller has been introduced to the Americas and Australia.

¾–1¾ in
2–4.5 cm

1½–2¼ in
4–6 cm

TANZANIAN FLAT-BACKED MILLIPEDE
Coromus diaphorus
F: Oxydesmidae

The shiny, pitted surface typical of many of the eyeless flat-backed millipedes is especially prominent in this species from tropical Africa.

¾–1¼ in
2–3 cm

BANDED STONE CENTIPEDE
Lithobius variegatus
F: Lithobiidae

Once thought to be confined to Britain, this centipede is now known in continental Europe, where native populations have unbanded legs.

¾–1¼ in
2–3 cm

BROWN STONE CENTIPEDE
Lithobius forficatus
F: Lithobiidae

Found sheltering beneath bark and rock, stone centipedes characteristically have 15 body segments. This globally widespread species frequents forests, gardens, and seashores.

1–2 in
2.5–5 cm

HOUSE CENTIPEDE
Scutigera coleoptrata
F: Scutigeridae

An unmistakable centipede, this long-legged animal with compound eyes is one of the fastest invertebrate runners. Native to the Mediterranean region, it has been introduced elsewhere.

8–10 in
20–25 cm

TIGER GIANT CENTIPEDE
Scolopendra hardwickei
F: Scolopendridae

Many of the giant *Scolopendra* centipedes have vibrant warning colors. This Indian species is one of several with a "tiger" pattern.

one pair of legs per body segment

"fangs" inject venom

4–6 in
10–15 cm

BLUE-LEGGED CENTIPEDE
Ethmostigmus trigonopodus
F: Scolopendridae

This close relative of the giant *Scolopendra* centipedes is widespread throughout Africa and is one of several species with blue-tinted legs.

ARACHNIDS

A class within the phylum Arthropoda, arachnids include predatory spiders and scorpions, as well as mites and blood-sucking ticks.

Arachnids and related horseshoe crabs are chelicerates—arthropods named for their clawlike mouthparts (chelicerae). The head and thorax of chelicerates are fused into one body section, which carries the sensory organs, brain, and four pairs of walking legs. Unlike other arthropods, chelicerates lack antennae.

EFFECTIVE PREDATORS
Scorpions, spiders, and their relatives are land-living predators that have evolved quick ways of immobilizing and killing prey. Spiders use their

chelicerae as fangs to inject venom and many first ensnare prey by spinning webs. Scorpions inject venom with the stinger in their tail. Between their legs and chelicerae, arachnids also have a pair of limblike pedipalps. These are modified into grasping pincers in scorpions or sperm-transferring clubs in male spiders.

MICROSCOPIC DIVERSITY
Many mites are too small to see with the naked eye. They abound in almost every habitat, where they scavenge on detritus, prey on other tiny invertebrates, or live as parasites. Some lead innocuous lives in skin follicles, feathers, or fur; others cause disease or allergy. One group— the ticks—are blood-suckers and can spread disease-causing microbes.

PHYLUM	ARTHROPODA
CLASS	ARACHNIDA
ORDERS	12
FAMILIES	661
SPECIES	About 103,000

This female wasp spider waits at the center of her dewy orb web to ensnare any insect prey that might fly in.

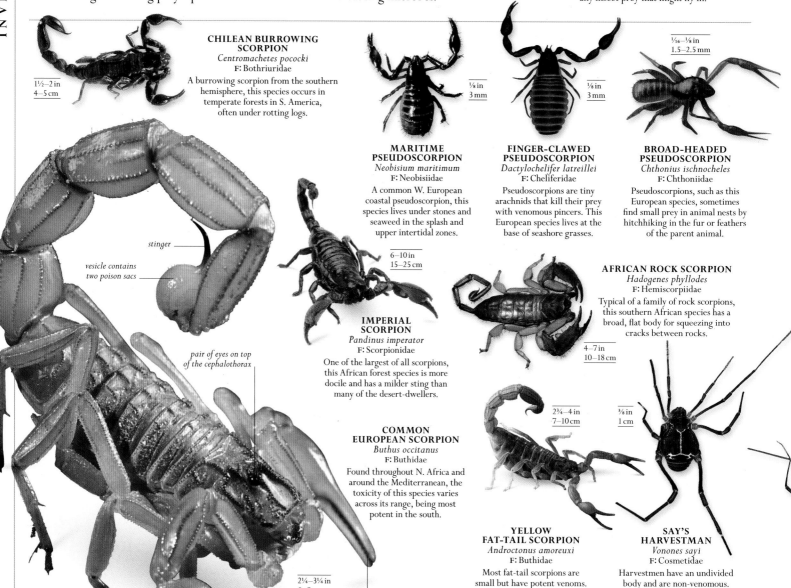

CHILEAN BURROWING SCORPION
Centromachetes pococki
F: Bothriuridae
A burrowing scorpion from the southern hemisphere, this species occurs in temperate forests in S. America, often under rotting logs.

1½–2 in
4–5 cm

MARITIME PSEUDOSCORPION
Neobisium maritimum
F: Neobisiidae
A common W. European coastal pseudoscorpion, this species lives under stones and seaweed in the splash and upper intertidal zones.

⅛ in
3 mm

FINGER-CLAWED PSEUDOSCORPION
Dactylochelifer latreillei
F: Cheliferidae
Pseudoscorpions are tiny arachnids that kill their prey with venomous pincers. This European species lives at the base of seashore grasses.

⅛ in
3 mm

BROAD-HEADED PSEUDOSCORPION
Chthonius ischnocheles
F: Chthoniidae
Pseudoscorpions, such as this European species, sometimes find small prey in animal nests by hitchhiking in the fur or feathers of the parent animal.

1/16–⅛ in
1.5–2.5 mm

stinger

vesicle contains two poison sacs

pair of eyes on top of the cephalothorax

pedipalps modified as pincers

IMPERIAL SCORPION
Pandinus imperator
F: Scorpionidae
One of the largest of all scorpions, this African forest species is more docile and has a milder sting than many of the desert-dwellers.

6–10 in
15–25 cm

AFRICAN ROCK SCORPION
Hadogenes phyllodes
F: Hemiscorpiidae
Typical of a family of rock scorpions, this southern African species has a broad, flat body for squeezing into cracks between rocks.

4–7 in
10–18 cm

COMMON EUROPEAN SCORPION
Buthus occitanus
F: Buthidae
Found throughout N. Africa and around the Mediterranean, the toxicity of this species varies across its range, being most potent in the south.

2¼–3¼ in
6–8 cm

YELLOW FAT-TAIL SCORPION
Androctonus amoreuxi
F: Buthidae
Most fat-tail scorpions are small but have potent venoms. This large species from the Sahara and Middle East has caused human fatalities.

2¾–4 in
7–10 cm

SAY'S HARVESTMAN
Vonones sayi
F: Cosmetidae
Harvestmen have an undivided body and are non-venomous. Many, including this American species, produce distasteful chemicals to deter predators.

⅜ in
1 cm

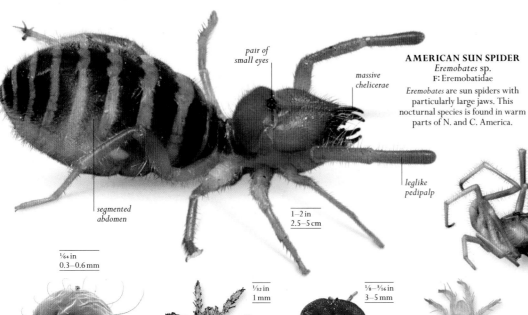

pair of
small eyes

massive
chelicerae

leglike
pedipalp

segmented
abdomen

¹⁄₆₄ in
0.3–0.6 mm

1–2 in
2.5–5 cm

AMERICAN SUN SPIDER
Eremobates sp.
F: Eremobatidae

Eremobates are sun spiders with
particularly large jaws. This
nocturnal species is found in warm
parts of N. and C. America.

3¼–4 in
8–10 cm

PAINTED SUN SPIDER
Metasolpuga picta
F: Solpugidae

Sun spiders, also called camel
spiders, are fast-running, desert-living
relatives of true spiders. This is a
diurnal species from Namibia.

3¼–6 in
8–15 cm

WIND SPIDER
Galeodes arabs
F: Galeodidae

A common species in the largest
genus of sun spiders, this Middle
Eastern species is named for its
ability to tolerate sand storms.

¹⁄₃₂ in
1 mm

⅛–³⁄₁₆ in
3–5 mm

¹⁄₆₄ in
0.4 mm

⁵⁄₁₆–⅜ in
8–10 mm

characteristic white
spot on body

FLOUR MITE
Acarus siro
F: Acaridae

This is an important pest
that feeds on stored cereal
products. Like many
other species, its presence
can cause allergic
reactions in humans.

AUTUMN CHIGGER
Neotrombicula autumnalis
F: Trombiculidae

Adults of chigger mites
are vegetarian, but their
larvae feed on the skin of
other animals, including
humans. Their chewing
causes intense irritation.

COMMON VELVET MITE
Trombidium holosericeum
F: Trombidiidae

This is a widespread Eurasian
species of velvet mite. When
young, it lives parasitically off
other arthropods, but becomes
predatory when mature.

RED SPIDER MITE
Tetranychus urticae
F: Tetranychidae

Spider mites are a
family of mites that suck
the sap of plants. They
weaken the plant and can
transmit viral diseases.

♀

¹⁄₃₂–¹⁄₁₆ in
1–2 mm

¹⁄₁₀–¹⁄₆₄ in
0.3–0.5 mm

¹⁄₆₄–¹⁄₃₂ in
0.5–1 mm

¹⁄₆₄ in
0.5 mm

LONE STAR TICK
Amblyomma americanum
F: Ixodidae

Like other blood-sucking
ticks, this species—common
in woodlands in the US—
can transmit a number of
disease-causing microbes.

VARROA MITE
Varroa sp.
F: Varroidae

Parasitic on honeybees, young
of this mite feed on bee larvae.
When mature, they attach to
and feed on adult bees and
spread to other hives.

MANGE MITE
Sarcoptes scabiei
F: Sarcoptidae

This tiny mite burrows into the
skin of various mammal species,
completing its life cycle there.
It causes scabies in humans and
mange in carnivores.

CHICKEN MITE
Dermanyssus gallinae
F: Dermanyssidae

A blood-sucking parasite of
poultry, this mite lives—and
completes its life cycle—in
crevices away from its host,
but emerges at night to feed.

PERSIAN FOWL TICK
Argas persicus
F: Argasidae

A blood-sucking parasite of fowl,
including domestic chickens, this
oval-shaped, soft-bodied tick can
spread disease between birds and
can cause paralysis.

¾–1½ in
2–4 cm

pincerlike
palps

very long
front legs

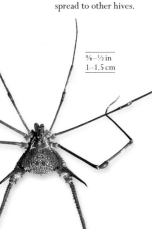

⅜–½ in
1–1.5 cm

SPINY
HARVESTMAN
Discocyrtus sp.
F: Gonyleptidae

This S. American harvestman
has spiny hind legs as a possible
defense against predators.
It lives under stones and
logs in forests.

⁵⁄₃₂–1¹⁄₃₂ in
4–9 mm

HORNED HARVESTMAN
Phalangium opilio
F: Phalangiidae

A common harvestman of Eurasia
and N. America, the male of this
species has a jaw ornamented
with projecting horns.

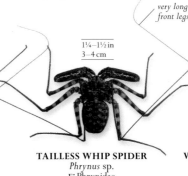

1¼–1½ in
3–4 cm

TAILLESS WHIP SPIDER
Phrynus sp.
F: Phrynidae

Whip spiders are not true
spiders. They have long,
whiplike front legs and pincers
for grabbing prey, but lack
venom. All are tropical.

WHIP SCORPION
Thelyphonus sp.
F: Thelyphonidae

Tropical whip scorpions such
as this species have a whiplike
tail, but no stinger or venom.
However, they can spray acetic
acid from their abdomen.

≫

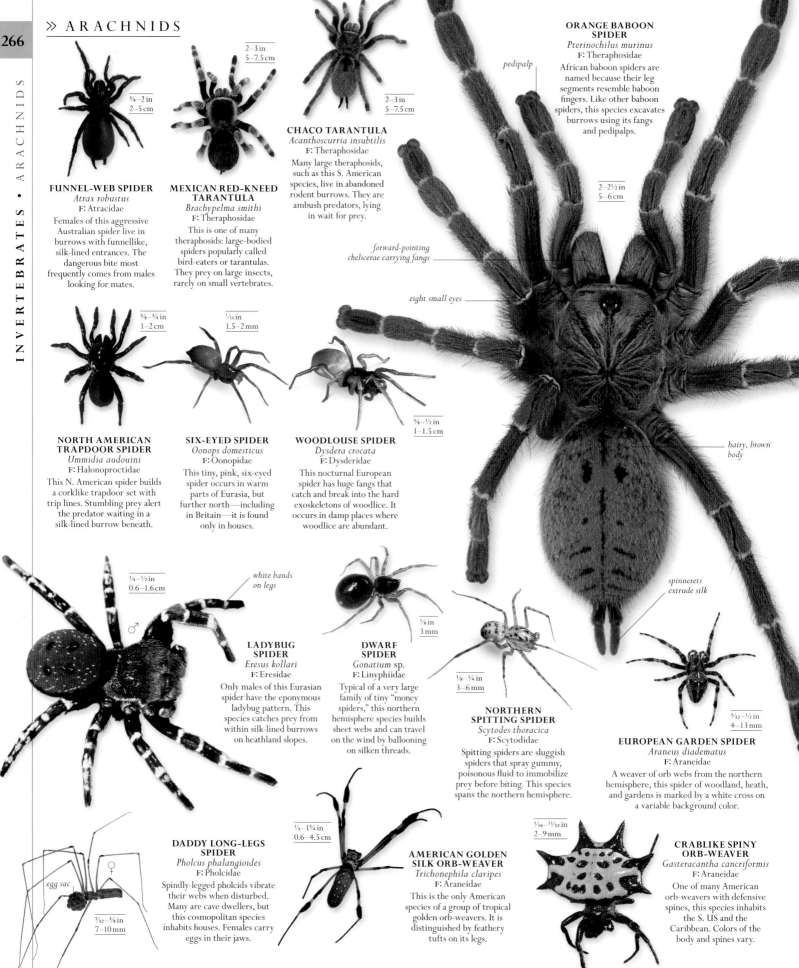

FUNNEL-WEB SPIDER
Atrax robustus
F: Atracidae
Females of this aggressive
Australian spider live in
burrows with funnellike,
silk-lined entrances. The
dangerous bite most
frequently comes from males
looking for mates.

¾–2 in
2–5 cm

**MEXICAN RED-KNEED
TARANTULA**
Brachypelma smithi
F: Theraphosidae
This is one of many
theraphosids: large-bodied
spiders popularly called
bird-eaters or tarantulas.
They prey on large insects,
rarely on small vertebrates.

2–3 in
5–7.5 cm

CHACO TARANTULA
Acanthoscurria insubtilis
F: Theraphosidae
Many large theraphosids,
such as this S. American
species, live in abandoned
rodent burrows. They are
ambush predators, lying
in wait for prey.

2–3 in
5–7.5 cm

**ORANGE BABOON
SPIDER**
Pterinochilus murinus
F: Theraphosidae
African baboon spiders are
named because their leg
segments resemble baboon
fingers. Like other baboon
spiders, this species
excavates
burrows using its fangs
and pedipalps.

pedipalp

2–2½ in
5–6 cm

*forward-pointing
chelicerae carrying fangs*

eight small eyes

*hairy, brown
body*

*spinnerets
extrude silk*

**NORTH AMERICAN
TRAPDOOR SPIDER**
Ummidia audouini
F: Halonoproctidae
This N. American spider builds
a corklike trapdoor set with
trip lines. Stumbling prey alert
the predator waiting in a
silk-lined burrow beneath.

⅜–¾ in
1–2 cm

SIX-EYED SPIDER
Oonops domesticus
F: Oonopidae
This tiny, pink, six-eyed
spider occurs in warm
parts of Eurasia, but
further north—including
in Britain—it is found
only in houses.

1/16 in
1.5–2 mm

WOODLOUSE SPIDER
Dysdera crocata
F: Dysderidae
This nocturnal European
spider has huge fangs that
catch and break into the hard
exoskeletons of woodlice. It
occurs in damp places where
woodlice are abundant.

⅜–½ in
1–1.5 cm

*white bands
on legs*

¼–½ in
0.6–1.6 cm

♂

**LADYBUG
SPIDER**
Eresus kollari
F: Eresidae
Only males of this Eurasian
spider have the eponymous
ladybug pattern. This
species catches prey from
within silk-lined burrows
on heathland slopes.

**DWARF
SPIDER**
Gonatium sp.
F: Linyphiidae
Typical of a very large
family of tiny "money
spiders," this northern
hemisphere species builds
sheet webs and can travel
on the wind by ballooning
on silken threads.

⅛ in
3 mm

⅛–¼ in
3–6 mm

**NORTHERN
SPITTING SPIDER**
Scytodes thoracica
F: Scytodidae
Spitting spiders are sluggish
spiders that spray gummy,
poisonous fluid to immobilize
prey before biting. This species
spans the northern hemisphere.

5/32–½ in
4–13 mm

EUROPEAN GARDEN SPIDER
Araneus diadematus
F: Araneidae
A weaver of orb webs from the northern
hemisphere, this spider of woodland, heath,
and gardens is marked by a white cross on
a variable background color.

egg sac

7/32–⅜ in
7–10 mm

**DADDY LONG-LEGS
SPIDER**
Pholcus phalangioides
F: Pholcidae
Spindly-legged pholcids vibrate
their webs when disturbed.
Many are cave dwellers, but
this cosmopolitan species
inhabits houses. Females carry
eggs in their jaws.

♀

¼–1¾ in
0.6–4.5 cm

**AMERICAN GOLDEN
SILK ORB-WEAVER**
Trichonephila clavipes
F: Araneidae
This is the only American
species of a group of tropical
golden orb-weavers. It is
distinguished by feathery
tufts on its legs.

1/16–11/32 in
2–9 mm

**CRABLIKE SPINY
ORB-WEAVER**
Gasteracantha cancriformis
F: Araneidae
One of many American
orb-weavers with defensive
spines, this species inhabits
the S. US and the
Caribbean. Colors of the
body and spines vary.

⁵⁄₃₂–½ in
4–13 mm

**BLACK
WIDOW SPIDER**
Latrodectus mactans
F: Theridiidae
A small spider with venom
unusually dangerous to humans,
females of this N. American
species frequently eat the smaller
male after mating.

¾–1¼ in
2–3 cm

TARANTULA WOLF SPIDER
Lycosa tarantula
F: Lycosidae
This species is the only true tarantula:
a large member of the wolf spider
family from Mediterranean Europe,
not to be confused with hairy-
bodied "bird-eaters."

³⁄₁₆–⁵⁄₁₆ in
5–8 mm

EUROPEAN WOLF SPIDER
Pardosa amentata
F: Lycosidae
A typical wolf spider, this furry
brown spider hunts on the ground
without a web. Females carry an
egg sac and young ride on
the mother's back.

**NURSERY WEB
SPIDER**
Pisaura mirabilis
F: Pisauridae
Females of this Eurasian
spider carry their egg sac
beneath the body by their
jaws, before weaving a
nursery "tent" of silk to
protect hatching young.

*egg sac
carried by jaws*

⅜–½ in
1–1.5 cm

⅜ in
1.2 cm

CAVE SPIDER
Meta menardi
F: Tetragnathidae
Many spiders of the genus *Meta*, such as
this European species, inhabit caves, where
they suspend their drop-shaped egg sacs.

egg sac

⁷⁄₃₂–⁷⁄₁₀ in
7–18 mm

*silvery abdomen
is due to
trapped air*

WATER SPIDER
Argyroneta aquatica
F: Dictynidae
The only routinely aquatic spider,
this Eurasian pond species builds
a submerged, air-filled chamber,
inside which it eats its prey,
including small fishes.

⅜–¾ in
1–1.8 cm

GIANT HOUSE SPIDER
Eratigena duellica
F: Agelenidae
A member of a group of
spiders that build sheet webs
with tubular retreats, this
species frequents buildings
across the northern hemisphere.

*distinctive
pale stripes
on body*

⅜–⅞ in
1–2.2 cm

*dark,
velvety body*

RAFT SPIDER
Dolomedes fimbriatus
F: Pisauridae
A European swamp spider,
this species preys on
small fishes, which it can
attract by vibrating the water
surface with its legs.

1¼–2 in
3–5 cm

**BRAZILIAN
WANDERING SPIDER**
Phoneutria nigriventer
F: Ctenidae
Wandering spiders get their
name from their nocturnal
roaming habits. This species
from the Brazilian forests is
believed to be the world's
most venomous spider.

³⁄₁₆–⅜ in
5–10 mm

**GOLDENROD
CRAB SPIDER**
Misumena vatia
F: Thomisidae
Females of this northern
hemisphere spider can slowly
change their color from white
to yellow for camouflage in
flowers before ambushing
nectar-feeding insects.

¾–1¼ in
2–3 cm

**BROWN HUNTSMAN
SPIDER**
Heteropoda venatoria
F: Sparassidae
Widespread throughout tropical and
subtropical regions, this large but
harmless spider is sometimes welcomed
in homes for preying on cockroaches.

thickened front legs

**ELEGANT JUMPING
SPIDER**
Chrysilla lauta
F: Salticidae
Jumping spiders are
especially diverse in the
tropics. Many, such as this
E. Asian species, are
strikingly colored,
sometimes to aid
communication.

⅛–⅜ in
3–9 mm

*large forward-facing
eyes for judging distance*

³⁄₁₆–⁷⁄₃₂ in
5–7 mm

**BROWN JUMPING
SPIDER**
Evarcha arcuata
F: Salticidae
This Eurasian grassland species
is one of thousands of species
of jumping spiders with acute
binocular vision and complex
courtship behavior.

MEXICAN RED-KNEED TARANTULA
Brachypelma smithi

With its stout, furry body, the Mexican red-kneed tarantula might seem more mammallike than spiderlike. The female can live up to 30 years, which is unusual for an invertebrate, but the life span of the male is only up to six years. *Brachypelma smithi* is one of a group of spiders that gets the nickname of bird-eater because of their size, and although most of their diet consists of fellow arthropods, they can and do take down small mammals and reptiles when the opportunity arises. In its native Mexico this species lives in burrows in earth banks, where it is safe to molt and lay eggs, and from which it ambushes its prey. Habitat destruction is now a threat, and, as the species makes a popular pet, it is routinely bred in captivity.

SIZE Body length 2–3 in (5–7.5 cm)
HABITAT Tropical deciduous forests
DISTRIBUTION Mexico
DIET Mostly insects

legs are covered in specialized hairs sensitive to air movements and touch

padded foot

< EYES
Like most spiders, the tarantula has eight simple eyes arranged at the front of the head. Even so, it has poor vision, and relies more on touch than sight to sense its surroundings and the presence of prey.

< JOINTS
Like all arthropods, tarantulas have jointed legs. Each leg consists of seven tubular sections of exoskeleton that connect via flexible joints. Muscles run through these joints to move the sections.

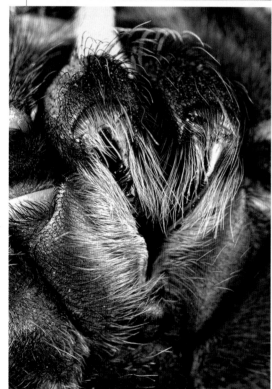

< FOOT
At the tip of each foot are two claws used for grip as the tarantula moves. As in other hunting spiders, there are also pads of tiny hairs for extra gripping on smooth surfaces.

^ VENOMOUS FANGS
Tarantula fangs hinge forward when attacking prey—unlike most other spiders, whose fangs are angled toward each other. Both fangs inject venom from muscular sacs within the head to paralyze the victim.

< SPINNERETS
Glands in the abdomen produce silk in liquid form. Using its hind legs, the tarantula pulls the silk from tubes called spinnerets. The silk solidifies into threads used to make egg sacs and line the spider's burrow.

dark abdomen contains most of the animal's vital internal organs

orange-red patella, or knee

∧ HAIRY WEAPONS
Like many tropical American tarantulas, this spider defends itself by rubbing its hind legs against its abdomen to release the barbed, stinging hairs that grow there. These urticating hairs are tiny and light, and designed to float in a cloud around a predator's face, getting in its eyes, nose, and mouth. They cause intense irritation when lodged in skin.

palps are used for feeling, groping prey, and transferring sperm to the female

UNDERSIDE >
The head and thorax are fused into one body part, which carries the legs and mouthparts. The abdomen has openings for breathing and reproduction, and two pairs of spinnerets at the rear.

SEA SPIDERS

These fragile-looking marine animals live among seaweeds in shallow seas and coral reefs in tropical waters. The largest species live in the deep ocean.

Sea spiders of the class Pycnogonida are not true spiders, and are so remarkably different from other arthropods that some scientists think they belong to an ancient lineage that is not closely related to any groups alive today. Others think they are distant cousins of the arachnids. Most sea spiders are small—less than ⅜ in (1 cm) in body length. They have three or four pairs of legs and the head and thorax are fused together. Instead of clawlike mouthparts they have a stabbing proboscis, which they use like a hypodermic needle to suck fluids from their invertebrate prey. Their spindly body shape means they do not need gills; they rely instead on oxygen seeping directly into all their cells through their body surface.

PHYLUM	ARTHROPODA
CLASS	PYCNOGONIDA
ORDER	1
FAMILIES	13
SPECIES	1,348

¾ in
2 cm

GIANT SEA SPIDER
Colossendeis megalonyx
F: Colossendeidae
One of the largest of all sea spiders, this is a deep-sea animal of sub-Antarctic waters, with a leg span of 28 in (70 cm).

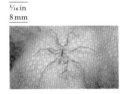

⁵⁄₁₆ in
8 mm

SPINY SEA SPIDER
Endeis spinosa
F: Endeidae
This sea spider—found around European coasts but perhaps ranging elsewhere—has an especially spindly body, with a long, cylindrical proboscis.

³⁄₁₆ in
5 mm

FAT SEA SPIDER
Pycnogonum litorale
F: Pycnogonidae
Unlike most other sea spiders, this European species has a thick body and comparatively short, curved, clawed legs. It feeds on anemones.

YELLOW-KNEED SEA SPIDER
Unknown sp.
F: Callipallenidae
Some species of sea spiders—such as this species of Australian reefs—are strikingly colored, often as camouflage against their colorful surroundings.

⁵⁄₁₆ in
8 mm

GRACEFUL SEA SPIDER
Nymphon brevirostre
F: Nymphonidae
One of the most common sea spiders of the N.E. Atlantic, this species occurs in the intertidal zone and in offshore shallows.

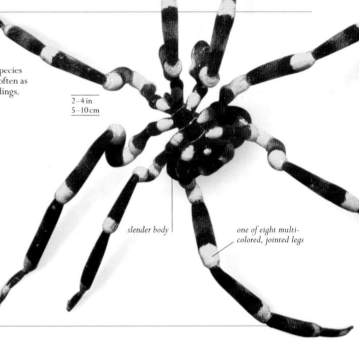

2–4 in
5–10 cm

slender body

one of eight multi-colored, jointed legs

HORSESHOE CRABS

The Merostomata are a small class of sea-dwelling relatives of spiders and scorpions. Of prehistoric origin, they are widely regarded as "living fossils."

Horseshoe crabs were far more diverse in prehistory, when animals like them may well have been the first chelicerates on Earth. Their clawlike mouthparts and lack of antennae indicate that, despite their hard carapace, horseshoe crabs are most closely related to arachnids. The leaflike gills on the underside of their abdomens are the forerunners of similar internal structures—the so-called book lungs—that arachnids use to breathe on land. The pedipalps are used as a fifth pair of legs, one pair more than in spiders. Horseshoe crabs grub around in muddy ocean waters to catch prey. They come ashore in large numbers to breed, laying their spawn in sand.

PHYLUM	ARTHROPODA
CLASS	MEROSTOMATA
ORDER	1
FAMILY	1
SPECIES	4

spiny abdomen

16–23½ in
40–60 cm

JAPANESE HORSESHOE CRAB
Tachypleus tridentatus
F: Limulidae
This horseshoe crab spawns in sandy shores of E. Asia. Habitat destruction and pollution have depleted its numbers in some parts of its range.

fused head and thorax covered by carapace

long, spine-like tail

MANGROVE HORSESHOE CRAB
Carcinoscorpius rotundicauda
F: Limulidae
Associated with muddy-sand habitats in S.E. Asia, this species feeds on insect larvae and other invertebrates, as well as small fishes.

16–23½ in
40–60 cm

CRUSTACEANS

Most crustaceans are aquatic, breathe with gills, and have limbs for crawling or swimming. A few are sedentary as adults, parasitic, or land living.

The basic body plan of a crustacean consists of a head, thorax, and abdomen, though in many groups the head and thorax are fused together. In crabs, lobsters, and prawns a frontal flap grows backward to cover the entire head and thorax with a shell-like carapace. Crustaceans are the only arthropods that have two pairs of antennae and primitively two-pronged limbs. Limbs on the thorax are usually used for locomotion, but in some species they are modified as pincers for feeding or defense. Many crustaceans also have well-developed abdominal limbs, which are

often used for brooding young. Gills tie most crustaceans to water, although woodlice and some crabs have modified breathing organs that enable them to live in damp places on land.

Being permanently under water, where their weight is supported, means that crustaceans can develop a thick, heavy exoskeleton. Many species have an exoskeleton hardened with minerals, which are reabsorbed between molts when the outer coating of the body is shed. With their body buoyed up by water, many crustaceans have become bigger than their land-based relatives. The world's largest arthropod is the deep-sea Japanese spider crab (leg span 13 ft/4 m). At the other extreme, crustaceans make up the bulk of animal life in plankton, either as tiny larvae or as adult shrimps and krill.

PHYLUM	ARTHROPODA
SUBPHYLUM	CRUSTACEA
ORDERS	56
FAMILIES	About 1,000
SPECIES	About 70,000

WATER FLEAS AND RELATIVES

The class Branchiopoda are primarily freshwater crustaceans, which make up part of the plankton that flourish in short-lived pools, surviving long periods of drought as eggs. Their thorax carries leafy limbs for breathing and filter feeding. Water fleas are typically encased in a transparent carapace.

BRINE SHRIMP
Artemia salina
F: Artemiidae

³⁄₈–¹⁄₂ in
1–1.5 cm

This soft-bodied, stalk-eyed animal is found worldwide in salt pools and swims upside down. Its hard-shelled eggs can resist years of drought.

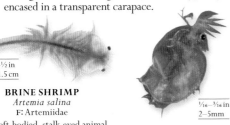

LARGE WATER FLEA
Daphnia magna
F: Daphniidae

¹⁄₁₆–³⁄₁₆ in
2–5 mm

This N. American water flea, like its relatives, can brood eggs inside its carapace. These hatch without fertilization, rapidly populating ponds as a result.

¹⁄₁₆ in
1.5 mm

MARINE WATER FLEA
Evadne nordmanni
F: Podonidae

Most water fleas—named for their jerky swimming—occur in stagnant freshwater pools. This, however, is a saltwater species of oceanic plankton.

VERNAL POOL TADPOLE SHRIMP
Lepidurus packardi
F: Triopsidae

2 in
5 cm

Tadpole shrimp are ancient, bottom-living crustaceans of temporary freshwater pools. Relatives of this Californian species have changed little in 220 million years of evolution.

— two tails

BARNACLES AND COPEPODS

Like other marine crustaceans, the class Maxillopoda start life as tiny planktonic larvae. Barnacle larvae cement themselves, head down, on rocks. Most copepods stay free-swimming, although some become parasites.

³⁄₁₆–¹⁄₂ in
0.5–1.5 cm

COMMON ACORN BARNACLE
Semibalanus balanoides
F: Archaeobalanidae

Sensitive to drying out, this intertidal species is most abundant at the base of the barnacle zone on exposed North Atlantic rocky coastlines.

³⁄₄ in
1.8 cm

SEA LOUSE
Caligus sp.
F: Caligidae

A representative of a group of copepods that are parasitic on marine fishes, this crustacean attacks salmon and related species.

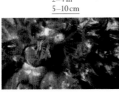

2–4 in
5–10 cm

GIANT ACORN BARNACLE
Balanus nubilus
F: Balanidae

The world's largest barnacle, this species occurs attached to rocks below the intertidal level along the Pacific N. American coast.

¹⁄₄–1¹⁄₄ in
2–3 cm

ASIAN ACORN BARNACLE
Tetraclita squamosa
F: Tetraclitidae

This acorn barnacle lives in the intertidal zone on Indo-Pacific shores. Recent research has led to five subspecies being recognized.

¹⁄₈–³⁄₁₆ in
3–5 mm

GLACIAL COPEPOD
Calanus glacialis
F: Calanidae

Within Arctic Ocean plankton, this copepod is an important part of the food chain.

GIANT COPEPOD
Macrocyclops albidus
F: Cyclopidae

Copepods are tiny predators of plankton; this widespread species even preys on mosquito larvae, offering a potential means of controlling these insects.

¹⁄₃₂–¹⁄₈ in
1–2.5 mm

DEEP SEA GOOSE BARNACLE
Neolepas sp.
F: Eolepadidae

This goose barnacle ally lives around volcanic vents on the ocean floor, where it filter-feeds on other organisms including bacteria.

2–4 in/5–10 cm

COMMON GOOSE BARNACLE
Lepas anatifera
F: Lepadidae

Goose barnacles are attached—mostly to oceanic flotsam—by a flexible stalk. This species occurs in temperate waters of the N.E. Atlantic.

3¹⁄₄–36 in
8–90 cm

COMMON FISH LOUSE
Argulus sp.
F: Argulidae

This is a flattened, fast-swimming crustacean with an oval carapace. It uses suckers to attach to fishes and suck their blood.

³⁄₁₆–³⁄₈ in
0.5–1 cm

SEED SHRIMP

Seed shrimp belong to the class Ostracoda. The entire body of a seed shrimp is enclosed within a two-valve hinged carapace, with just the limbs poking through. If threatened, the animal can shut itself inside. These small crustaceans crawl through vegetation in marine and freshwater habitats; some use their antennae to swim.

¾–1¼ in
2–3 cm

GIANT SWIMMING SEED SHRIMP
Gigantocypris sp.
F: Cypridinidae
Most seed shrimp are tiny crustaceans with two-valved carapaces; this is a large, deep-sea species with big eyes for hunting bioluminescent prey.

¹⁄₆₄–¹⁄₁₆ in
0.5–2 mm

CRAWLING SEED SHRIMP
Cypris sp.
F: Cyprididae
This widespread freshwater crustacean belongs to a group of small, hard-shelled seed shrimp that crawl through detritus.

CRABS AND RELATIVES

The Malacostraca class is the most diverse crustacean group. The basic features are head, thorax, and an abdomen with multiple limbs. Two large orders within this class include the decapods, which have a carapace that curves around the fused head-thorax to contain a gill cavity, and the isopods (woodlice and relatives), which lack a carapace and are the largest group of land-living crustaceans.

ANTARCTIC KRILL
Euphausia superba
F: Euphausiidae
Swarms of these plankton-feeding crustaceans are critical components of the Southern Ocean food chains that support whales, seals, and seabirds.

1½–2¼ in
4–6 cm

RELICT OPOSSUM SHRIMP
Mysis relicta
F: Mysidae
Translucent, feathery-legged opossum shrimp carry their larvae in a brood pouch. Most live in coastal waters, but this species is found in fresh waters in the northern hemisphere.

⅜–¾ in
1–1.8 cm

½–⅞ in
1.5–2.2 cm

⅜–¾ in
1–2 cm

abdomen

COMMON FRESHWATER SHRIMP
Gammarus pulex
F: Gammaridae
This N. European amphipod, abundant in freshwater streams, belongs to a family of detritus-eating freshwater shrimp. Related species inhabit brackish waters.

COMMON SANDHOPPER
Orchestia gammarellus
F: Talitridae
One of many amphipods (sideways-flattened crustaceans), this is a European intertidal sandhopper—so called because it can jump by flipping its abdomen.

⅜–½ in
1–1.5 cm

SPINY SKELETON SHRIMP
Caprella acanthifera
F: Caprellidae
Like other skeleton shrimp, this is a slender, slow-moving, predatory amphipod with few legs. It clings to seaweeds in European rockpools.

½ in
13 mm

COMMON WATER SLATER
Asellus aquaticus
F: Asellidae
A common European member of a family of freshwater woodlice relatives, this species crawls among detritus in stagnant water.

tail fin used for swimming

segmented exoskeleton

GIANT DEEPSEA ISOPOD
Bathynomus giganteus
F: Cirolanidae
A giant marine relative of woodlice, this species crawls along ocean floors, scavenging on dead animals and occasionally taking live prey.

7½–14 in
19–36 cm

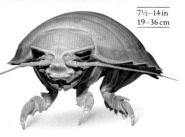

¾–1¼ in
2–3 cm

4–4¾ in
10–12 cm

COMMON SEA SLATER
Ligia oceanica
F: Ligiidae
A large coastal woodlouse, this European species lives in rock crevices above the intertidal zone. It feeds on detritus.

BLACK-HEADED WOODLOUSE
Porcellio spinicornis
F: Porcellionidae
Often living alongside humans, especially in lime-rich habitats, this distinctly marked woodlouse is native to Europe but has invaded N. America.

⅜–¾ in
1–1.8 cm

COMMON PILL WOODLOUSE
Armadillidium vulgare
F: Armadillidiidae
Pill woodlice are characterized by their ability to roll into a ball when disturbed. This species is widespread across Eurasia and introduced elsewhere.

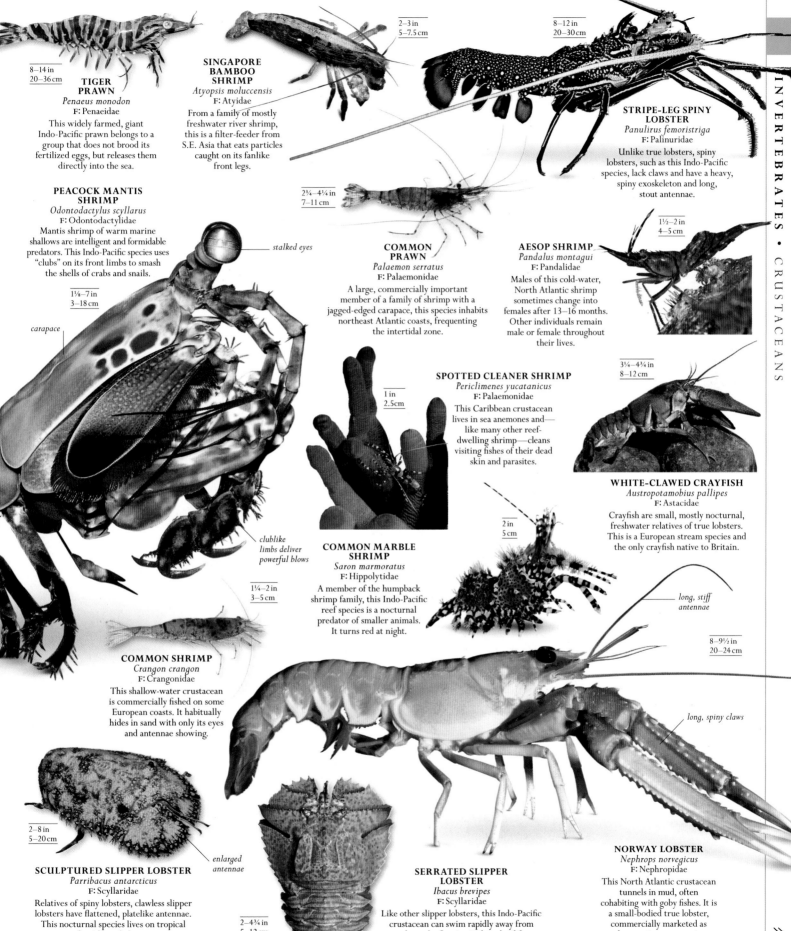

8–14 in
20–36 cm

TIGER PRAWN
Penaeus monodon
F: Penaeidae
This widely farmed, giant
Indo-Pacific prawn belongs to a
group that does not brood its
fertilized eggs, but releases them
directly into the sea.

PEACOCK MANTIS SHRIMP
Odontodactylus scyllarus
F: Odontodactylidae
Mantis shrimp of warm marine
shallows are intelligent and formidable
predators. This Indo-Pacific species uses
"clubs" on its front limbs to smash
the shells of crabs and snails.

1⅛–7 in
3–18 cm

carapace

stalked eyes

clublike
limbs deliver
powerful blows

1¼–2 in
3–5 cm

COMMON SHRIMP
Crangon crangon
F: Crangonidae
This shallow-water crustacean
is commercially fished on some
European coasts. It habitually
hides in sand with only its eyes
and antennae showing.

2–8 in
5–20 cm

enlarged
antennae

SCULPTURED SLIPPER LOBSTER
Parribacus antarcticus
F: Scyllaridae
Relatives of spiny lobsters, clawless slipper
lobsters have flattened, platelike antennae.
This nocturnal species lives on tropical
reefs with sandy bottoms.

SINGAPORE BAMBOO SHRIMP
Atyopsis moluccensis
F: Atyidae
From a family of mostly
freshwater river shrimp,
this is a filter-feeder from
S.E. Asia that eats particles
caught on its fanlike
front legs.

2–3 in
5–7.5 cm

2¾–4¼ in
7–11 cm

COMMON PRAWN
Palaemon serratus
F: Palaemonidae
A large, commercially important
member of a family of shrimp with a
jagged-edged carapace, this species inhabits
northeast Atlantic coasts, frequenting
the intertidal zone.

SPOTTED CLEANER SHRIMP
Periclimenes yucatanicus
F: Palaemonidae
This Caribbean crustacean
lives in sea anemones and—
like many other reef-
dwelling shrimp—cleans
visiting fishes of their dead
skin and parasites.

1 in
2.5 cm

COMMON MARBLE SHRIMP
Saron marmoratus
F: Hippolytidae
A member of the humpback
shrimp family, this Indo-Pacific
reef species is a nocturnal
predator of smaller animals.
It turns red at night.

2 in
5 cm

SERRATED SLIPPER LOBSTER
Ibacus brevipes
F: Scyllaridae
Like other slipper lobsters, this Indo-Pacific
crustacean can swim rapidly away from
predators by flapping its broad tail fan.

2–4¾ in
5–12 cm

8–12 in
20–30 cm

STRIPE-LEG SPINY LOBSTER
Panulirus femoristriga
F: Palinuridae
Unlike true lobsters, spiny
lobsters, such as this Indo-Pacific
species, lack claws and have a heavy,
spiny exoskeleton and long,
stout antennae.

1½–2 in
4–5 cm

AESOP SHRIMP
Pandalus montagui
F: Pandalidae
Males of this cold-water,
North Atlantic shrimp
sometimes change into
females after 13–16 months.
Other individuals remain
male or female throughout
their lives.

3¼–4¾ in
8–12 cm

WHITE-CLAWED CRAYFISH
Austropotamobius pallipes
F: Astacidae
Crayfish are small, mostly nocturnal,
freshwater relatives of true lobsters.
This is a European stream species and
the only crayfish native to Britain.

long, stiff
antennae

8–9½ in
20–24 cm

long, spiny claws

NORWAY LOBSTER
Nephrops norvegicus
F: Nephropidae
This North Atlantic crustacean
tunnels in mud, often
cohabiting with goby fishes. It is
a small-bodied true lobster,
commercially marketed as
langoustine or scampi.

» CRABS AND RELATIVES

12–16 in
30–40 cm

ROBBER CRAB
Birgus latro
F: Coenobitidae
The largest land-living arthropod, this hermit crab is a relative of squat lobsters and inhabits Indo-Pacific island forests, eating coconuts using its massive pincers.

⅜ in
1 cm

PINK SQUAT LOBSTER
Lauriea siagiani
F: Galatheidae
Many tropical squat lobsters are associated with specific reef organisms. This tiny, hairy Indonesian species lives on *Xestospongia* vase sponges.

BLUE-STRIPED SQUAT LOBSTER
Galathea strigosa
F: Galatheidae
Slender-clawed squat lobsters such as this European species are decapods (ten-legged), but the last leg pair is reduced, making them appear eight-legged.

2¾–3½ in
7–9 cm

1¾ in
2 cm

anemone attached to shell

unequal-sized claws used for signaling

ANEMONE PORCELAIN CRAB
Neopetrolisthes maculatus
F: Porcellanidae
Porcelain crabs are tiny, eight-legged decapods more closely related to squat lobsters than true crabs. This Indo-Pacific species lives in giant *Stichodactyla* anemones.

5/16–½ in
8–12 mm

antenna

INDO-PACIFIC PEA CRAB
Pinnotheres sp.
F: Pinnotheridae
Tiny pea crabs complete their life cycle on or in the body of other marine invertebrates. This Philippine species lives on cup corals.

1½ in
4 cm

2¼–4 in
6–10 cm

RED REEF HERMIT CRAB
Paguristes cadenati
F: Diogenidae
Hermit crabs inhabit discarded snail shells that accommodate their soft coiled abdomen. This species lives on Indo-Pacific and E. Atlantic reefs.

5–8 in
13–20 cm

WHITE-SPOTTED HERMIT CRAB
Dardanus megistos
F: Diogenidae
This crustacean of the E. Atlantic and Indo-Pacific coasts is a "left-handed" hermit crab, with an enlarged left claw.

ANEMONE HERMIT CRAB
Dardanus pedunculatus
F: Diogenidae
This hermit crab of Indo-Pacific reefs always carries a *Calliactis* anemone on its shell, which shares its food and provides camouflaged protection.

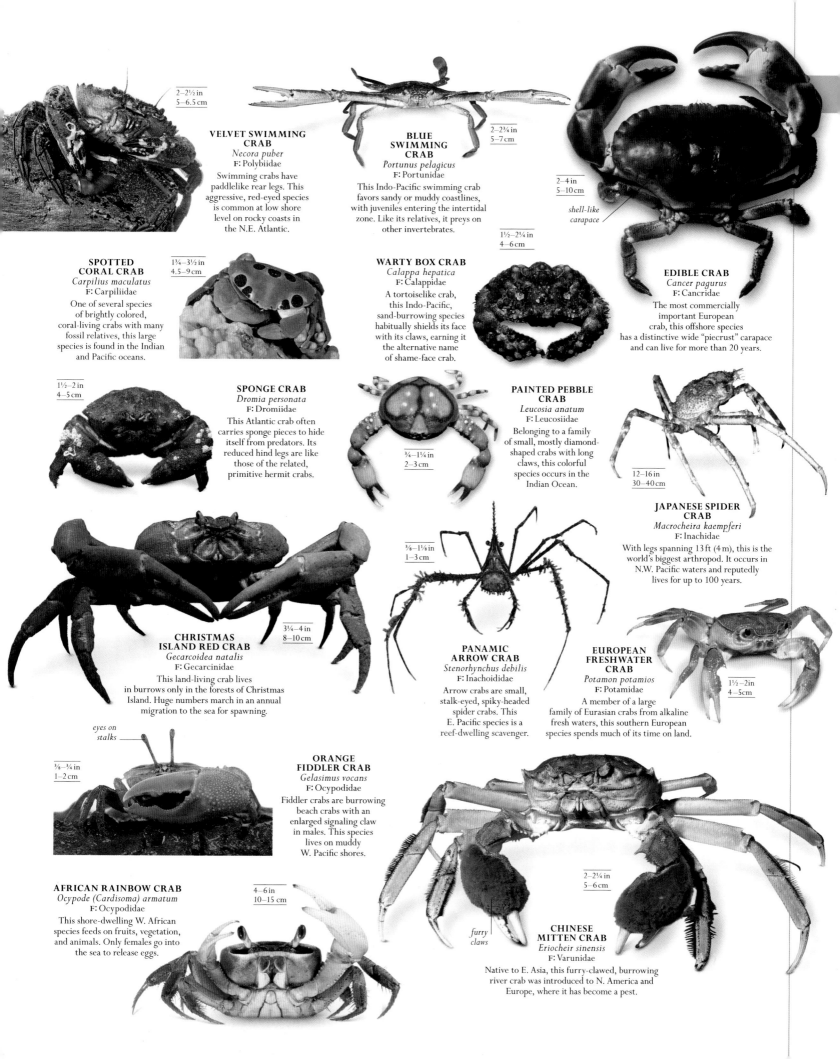

VELVET SWIMMING CRAB
Necora puber
F: Polybiidae
Swimming crabs have paddlelike rear legs. This aggressive, red-eyed species is common at low shore level on rocky coasts in the N.E. Atlantic.

2–2½ in
5–6.5 cm

BLUE SWIMMING CRAB
Portunus pelagicus
F: Portunidae
This Indo-Pacific swimming crab favors sandy or muddy coastlines, with juveniles entering the intertidal zone. Like its relatives, it preys on other invertebrates.

2–2¾ in
5–7 cm

2–4 in
5–10 cm

shell-like carapace

EDIBLE CRAB
Cancer pagurus
F: Cancridae
The most commercially important European crab, this offshore species has a distinctive wide "piecrust" carapace and can live for more than 20 years.

SPOTTED CORAL CRAB
Carpilius maculatus
F: Carpiliidae
One of several species of brightly colored, coral-living crabs with many fossil relatives, this large species is found in the Indian and Pacific oceans.

1¾–3½ in
4.5–9 cm

WARTY BOX CRAB
Calappa hepatica
F: Calappidae
A tortoiselike crab, this Indo-Pacific, sand-burrowing species habitually shields its face with its claws, earning it the alternative name of shame-face crab.

1½–2¼ in
4–6 cm

SPONGE CRAB
Dromia personata
F: Dromiidae
This Atlantic crab often carries sponge pieces to hide itself from predators. Its reduced hind legs are like those of the related, primitive hermit crabs.

1½–2 in
4–5 cm

PAINTED PEBBLE CRAB
Leucosia anatum
F: Leucosiidae
Belonging to a family of small, mostly diamond-shaped crabs with long claws, this colorful species occurs in the Indian Ocean.

¾–1¼ in
2–3 cm

JAPANESE SPIDER CRAB
Macrocheira kaempferi
F: Inachidae
With legs spanning 13 ft (4 m), this is the world's biggest arthropod. It occurs in N.W. Pacific waters and reputedly lives for up to 100 years.

12–16 in
30–40 cm

CHRISTMAS ISLAND RED CRAB
Gecarcoidea natalis
F: Gecarcinidae
This land-living crab lives in burrows only in the forests of Christmas Island. Huge numbers march in an annual migration to the sea for spawning.

3¼–4 in
8–10 cm

⅜–1⅛ in
1–3 cm

PANAMIC ARROW CRAB
Stenorhynchus debilis
F: Inachoididae
Arrow crabs are small, stalk-eyed, spiky-headed spider crabs. This E. Pacific species is a reef-dwelling scavenger.

EUROPEAN FRESHWATER CRAB
Potamon potamios
F: Potamidae
A member of a large family of Eurasian crabs from alkaline fresh waters, this southern European species spends much of its time on land.

1½–2 in
4–5 cm

eyes on stalks

⅜–¾ in
1–2 cm

ORANGE FIDDLER CRAB
Gelasimus vocans
F: Ocypodidae
Fiddler crabs are burrowing beach crabs with an enlarged signaling claw in males. This species lives on muddy W. Pacific shores.

AFRICAN RAINBOW CRAB
Ocypode (Cardisoma) armatum
F: Ocypodidae
This shore-dwelling W. African species feeds on fruits, vegetation, and animals. Only females go into the sea to release eggs.

4–6 in
10–15 cm

2–2¼ in
5–6 cm

furry claws

CHINESE MITTEN CRAB
Eriocheir sinensis
F: Varunidae
Native to E. Asia, this furry-clawed, burrowing river crab was introduced to N. America and Europe, where it has become a pest.

AFRICAN RAINBOW CRAB
Ocypode (Cardisoma) armatum

African rainbow crabs dig deep individual burrows above the shoreline on sandy beaches, dunes, mangrove swamps, and estuaries. They mate in or near the burrows and females then carry the fertilized eggs under their abdomen for 2–3 weeks before releasing them into shallow water. Millions of floating larvae may be produced by each female per spawn, but only a few survive to make their way back to continue their life on land.

∨ BUCCAL CAVITY
Setae on the outer mouthparts scrub food from particles of sand and pass it inward to the mandibles

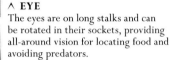

∧ EYE
The eyes are on long stalks and can be rotated in their sockets, providing all-around vision for locating food and avoiding predators.

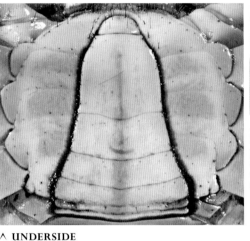

∧ UNDERSIDE
The abdomen in males, as shown here, is narrow. In females it is broad and wide for carrying the egg mass.

∧ LEG JOINT
Leg joints move in one plane, like a human knee, but each one moves in a different plane, making the crab very agile.

∧ SHELL
To increase in size the crab must shed its hard shell, a process called molting. Water is absorbed into the body until the shell opens up and can be discarded, revealing a new, larger shell beneath.

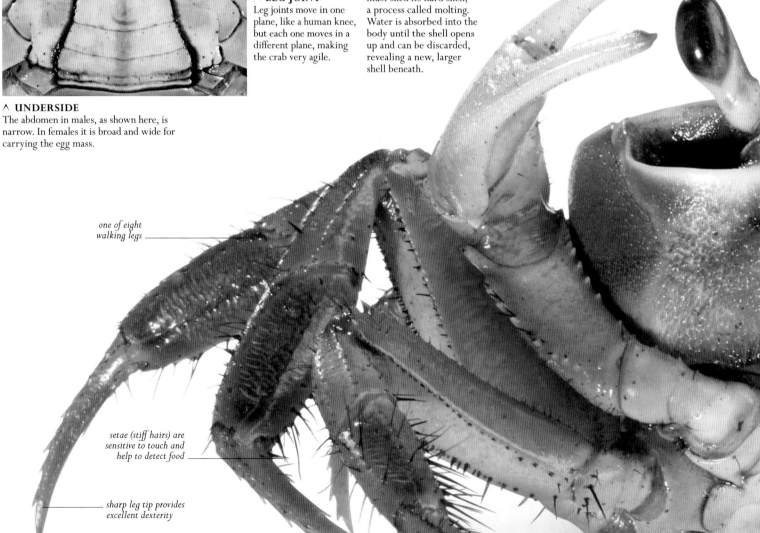

one of eight walking legs

setae (stiff hairs) are sensitive to touch and help to detect food

sharp leg tip provides excellent dexterity

SIZE 4–6 in (10–15 cm)
HABITAT Sandy and muddy shorelines
DISTRIBUTION Atlantic coast of W. Africa
DIET Opportunistic feeders, mainly on decaying plants and animals

∨ BEACHCOMBER

African rainbow crabs are most active at night when they emerge to forage, racing back to their burrows if disturbed. They eat plant and animal detritus, dead fishes, and insects, leaving numerous little balls of sand which they have searched through for food.

moveable dactylus allows the claw to operate as a pincer

static claw against which the dactylus moves

unequal-sized claw-bearing leg used for feeding and social interactions

buccal cavity contains six pairs of specialized mouthparts

eye on thick, elongated stalk

six-jointed walking leg

INSECTS

Insects first appeared on land more than 400 million years ago and today they account for more species than any other class on the planet.

Insects have evolved diverse lifestyles and although most are terrestrial, there are also numerous freshwater species but almost none are marine. They have a number of key features, such as a small size, an efficient nervous system, high reproductive rates, and—in many cases—the power of flight, which have led to their success.

Insects include (among others) beetles, flies, butterflies, moths, ants, bees, and true bugs. Yet for all their diversity, insects are remarkably similar. Evolution has modified the basic insect anatomy many times over to produce a multitude

of variants based on three major body regions—the head, thorax, and abdomen. The head, made up of six fused segments, houses the brain and carries the major sensory organs: compound eyes, secondary light-receptive organs called ocelli, and the antennae. The mouthparts are modified according to diet, allowing the sucking of liquids or the chewing of solid foods.

The thorax is made up of three segments, each bearing a pair of legs. The posterior two thoracic segments usually each bear a pair of wings. The legs, which are each made up of a number of segments, can be greatly modified to serve a variety of functions, from walking and running to jumping, digging, or swimming. The abdomen, which is usually made up of 11 segments, contains the digestive and reproductive organs.

PHYLUM	ARTHROPODA
CLASS	INSECTA
ORDERS	29
FAMILIES	About 1,000
SPECIES	About 1.1 million

DEBATE

HOW MANY SPECIES?

The actual number of insect species likely to exist far exceeds the 1.1 million so far described, and many more are discovered each year. However, research based on sampling in species-rich rainforests suggests that there could be as many as 10–12 million species of insects.

SILVERFISH

These primitive, wingless insects in the order Zygentoma have a lengthy body that is covered with scales. The head has a pair of long antennae and small eyes. The abdominal segments have small appendages (styles).

⅜–½ in
1–1.5 cm

½ in
1.2 cm

SILVERFISH
Lepisma saccharina
F: Lepismatidae
This common domestic species can sometimes be a nuisance in kitchens, where it feeds on tiny scraps of dropped food.

FIREBRAT
Thermobia domestica
F: Lepismatidae
This insect is found worldwide and lives under stones and in leaf litter. Indoors, it favors warm conditions and can be a pest in bakeries.

MAYFLIES

Ephemeroptera are soft-bodied insects with slender legs and two pairs of wings. The head has a pair of short antennae and large, compound eyes. The end of the abdomen has two or three long tail filaments. The life cycle is dominated by the aquatic nymphal stages—the non-feeding adults live for only a few hours or days.

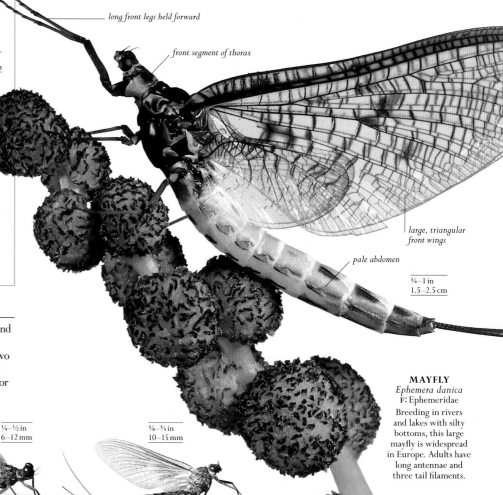

long front legs held forward

front segment of thorax

large, triangular
front wings

pale abdomen

¾–1 in
1.5–2.5 cm

MAYFLY
Ephemera danica
F: Ephemeridae
Breeding in rivers and lakes with silty bottoms, this large mayfly is widespread in Europe. Adults have long antennae and three tail filaments.

³⁄₁₆–⁷⁄₃₂ in
5–7 mm

¼–½ in
6–12 mm

⅜–⅗ in
10–15 mm

POND OLIVE
Cloeon dipterum
F: Baetidae
A widespread European species, this mayfly breeds in a range of habitats from ponds and ditches to troughs and water butts.

BLUE-WINGED OLIVE
Ephemerella ignita
F: Ephemerellidae
Adults of this N. European species have three tail filaments. The males' rounded eyes have two parts, the larger upper portion for spotting females.

SIPHLONURUS LACUSTRIS
F: Siphlonuridae
Very common in upland lakes in N. Europe, this summer mayfly has two long tails and greenish-gray wings. The hindwings are small.

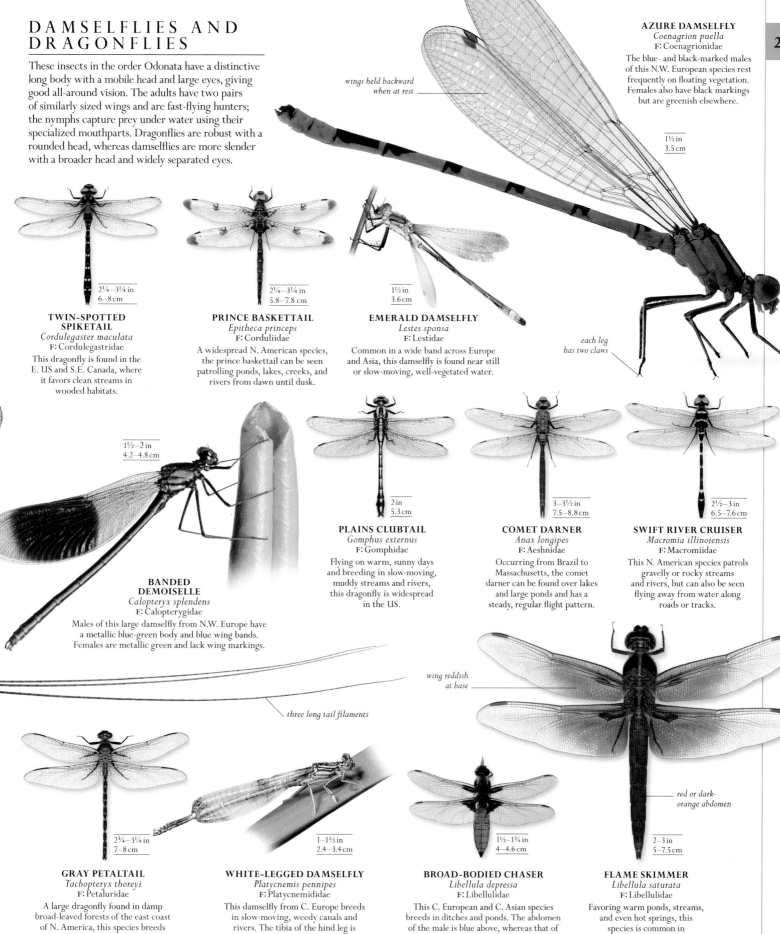

DAMSELFLIES AND DRAGONFLIES

These insects in the order Odonata have a distinctive long body with a mobile head and large eyes, giving good all-around vision. The adults have two pairs of similarly sized wings and are fast-flying hunters; the nymphs capture prey under water using their specialized mouthparts. Dragonflies are robust with a rounded head, whereas damselflies are more slender with a broader head and widely separated eyes.

wings held backward when at rest

AZURE DAMSELFLY
Coenagrion puella
F: Coenagrionidae
The blue- and black-marked males of this N.W. European species rest frequently on floating vegetation. Females also have black markings but are greenish elsewhere.

1½ in
3.5 cm

each leg has two claws

2¼–3¼ in
6–8 cm

TWIN-SPOTTED SPIKETAIL
Cordulegaster maculata
F: Cordulegastridae
This dragonfly is found in the E. US and S.E. Canada, where it favors clean streams in wooded habitats.

2¼–3¼ in
5.8–7.8 cm

PRINCE BASKETTAIL
Epitheca princeps
F: Corduliidae
A widespread N. American species, the prince baskettail can be seen patrolling ponds, lakes, creeks, and rivers from dawn until dusk.

1½ in
3.6 cm

EMERALD DAMSELFLY
Lestes sponsa
F: Lestidae
Common in a wide band across Europe and Asia, this damselfly is found near still or slow-moving, well-vegetated water.

1½–2 in
4.2–4.8 cm

BANDED DEMOISELLE
Calopteryx splendens
F: Calopterygidae
Males of this large damselfly from N.W. Europe have a metallic blue-green body and blue wing bands. Females are metallic green and lack wing markings.

2 in
5.3 cm

PLAINS CLUBTAIL
Gomphus externus
F: Gomphidae
Flying on warm, sunny days and breeding in slow-moving, muddy streams and rivers, this dragonfly is widespread in the US.

3–3½ in
7.5–8.8 cm

COMET DARNER
Anax longipes
F: Aeshnidae
Occurring from Brazil to Massachusetts, the comet darner can be found over lakes and large ponds and has a steady, regular flight pattern.

2½–3 in
6.5–7.6 cm

SWIFT RIVER CRUISER
Macromia illinoiensis
F: Macromiidae
This N. American species patrols gravelly or rocky streams and rivers, but can be seen flying away from water along roads or tracks.

wing reddish at base

three long tail filaments

red or dark-orange abdomen

2¾–3¼ in
7–8 cm

GRAY PETALTAIL
Tachopteryx thoreyi
F: Petaluridae
A large dragonfly found in damp broad-leaved forests of the east coast of N. America, this species breeds in boggy areas and seeps.

1–1⅖ in
2.4–3.4 cm

WHITE-LEGGED DAMSELFLY
Platycnemis pennipes
F: Platycnemididae
This damselfly from C. Europe breeds in slow-moving, weedy canals and rivers. The tibia of the hind leg is expanded and appears slightly feathery.

1½–1¾ in
4–4.6 cm

BROAD-BODIED CHASER
Libellula depressa
F: Libellulidae
This C. European and C. Asian species breeds in ditches and ponds. The abdomen of the male is blue above, whereas that of the female is yellowish brown.

2–3 in
5–7.5 cm

FLAME SKIMMER
Libellula saturata
F: Libellulidae
Favoring warm ponds, streams, and even hot springs, this species is common in the S.W. US.

STONEFLIES

Members of the order Plecoptera are soft, slender-bodied insects with a pair of thin tail filaments and two pairs of wings. The nymphal stages are aquatic.

¾–1 in
2–2.8 cm

PERLA BIPUNCTATA
F: Perlidae

The males of this species, which favors stony streams in upland regions, have much shorter wings and can be half the size of the females.

¹¹⁄₃₂–½ in
0.9–1.3 cm

YELLOW SALLY
Isoperla grammatica
F: Perlodidae

Particularly common in limestone areas, this species prefers clean, gravel-bottomed streams and stony lakes. Males are shorter than females.

STICK AND LEAF INSECTS

These slow-moving, herbivorous insects of the order Pseudophasmatidae have a stick- or leaflike body that may be smooth or spiny. Many are well camouflaged to avoid predators.

TWO-STRIPED STICK INSECT
Anisomorpha buprestoides
F: Pseudophasmatidae

Found in the S. US, this species can eject an acidic, defensive liquid from glands in the thorax.

1½–2½ in
4.2–6.8 cm

JUNGLE NYMPH STICK INSECT
Heteropteryx dilatata
F: Heteropterygidae

This impressive species is found in Malaysia. Females are flightless and green; males are smaller, winged, and brownish.

4–6 in
10–15.5 cm

JAVANESE LEAF INSECT
Phyllium bioculatum
F: Phylliidae

Females of this S.E. Asian insect are large, winged, and very leaflike. The males are smaller, more slender, and brown.

2¾–3¾ in
7–9.4 cm

— large, fan-shaped hindwing

— flattened, leaflike abdomen

EARWIGS

These slender, slightly flattened, scavenging insects belong to the order Dermaptera. They have short front wings with large, fan-shaped hindwings folded beneath. The flexible abdomen ends in a pair of multipurpose forceps.

½–⅗ in
1.2–1.5 cm

COMMON EARWIG
Forficula auricularia
F: Forficulidae

This species can be found under bark and in leaf litter. The female cares for her eggs and feeds the young nymphs.

TAWNY EARWIG
Labidura riparia
F: Labiduridae

The tawny earwig, the largest European species, is especially common along sandy river banks and in coastal areas.

½–1¼ in
1.6–3 cm

PRAYING MANTIDS

These predatory insects of the order Mantodea have a triangular, highly mobile head with large eyes. The enlarged, spined front legs are specially modified for snatching prey. A pair of toughened front wings protects the larger membranous hindwings folded beneath.

large, compound eyes on a triangular head —

extended prothorax —

leaflike front wing —

spiny, enlarged front femur —

2¼–3½ in
6–9 cm

head crest —

2–2¾ in
5–7 cm

COMMON PRAYING MANTIS
Mantis religiosa
F: Mantidae

The strike of a praying mantis lasts a fraction of a second, and impales the prey on the sharp spines of the front legs.

CONEHEAD MANTID
Empusa pennata
F: Empusidae

A slender mantid with a distinctive head crest, this S. European species eats small flies and can be green or brown.

ORCHID MANTIS
Hymenopus coronatus
F: Hymenopodidae

With its flower-mimicking color and legs resembling petals, this S.E. Asian mantis is able to hide among foliage and ambush small insects.

1¼–2¾ in
3–7 cm

CRICKETS AND GRASSHOPPERS

Orthoptera are mainly herbivorous. They have two pairs of wings, although some can be short-winged or wingless. The hind legs are often large and used for jumping. They sing by rubbing their front wings together or their hind legs on a wing edge.

½–1 in
1.4–2.4 cm

2–2¾ in
5–7 cm

½–1 in
1.5–2.6 cm

COMMON FIELD GRASSHOPPER
Chorthippus brunneus
F: Acrididae

This grasshopper is typically found on short, dry, grazed grassland in Europe, N. Africa, and temperate Asia, where it is most active on sunny days.

WELLINGTON TREE WETA
Hemideina crassidens
F: Anostostomatidae

Native to New Zealand, this nocturnal insect lives in rotten wood and tree stumps. It eats plant material as well as small insects.

AFRICAN CAVE CRICKET
Phaeophilacris bredoides
F: Rhaphidophoridae

This C. African species is an omnivorous scavenger. It has very long antennae, an adaptation to life in dark microhabitats.

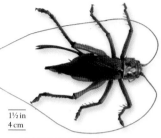
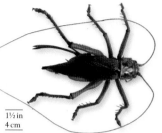

1½ in
4 cm

DESERT LOCUST
Schistocerca gregaria
F: Acrididae

Crowding of nymphs after rain stimulates this African insect to transform from a solitary into a gregarious form. It swarms in the billions, devastating crops.

2–3¼ in
5–8 cm

OAK BUSH CRICKET
Meconema thalassinum
F: Tettigoniidae

Found on a range of broadleaved trees, this European cricket feeds on small insects after dark. The female has a long, curved ovipositor.

½–¾ in
1.3–1.8 cm

ovipositor for laying eggs

LEAF-ROLLING CRICKET
Hyalogryllacris subdebilis
F: Gryllacrididae

Found in Australia, this species has relatively long wings and very long antennae, which may be up to three times the length of the body.

MOLE CRICKET
Gryllotalpa gryllotalpa
F: Gryllotalpidae

This species from Europe uses its strong front legs to burrow, just like a miniature mole. It is found in meadows and banks near rivers, where the soil is damp and sandy.

1½–1¾ in
3.5–4.6 cm

warty surface

HOUSE CRICKET
Acheta domesticus
F: Gryllidae

A nocturnal species, the house cricket makes an attractive chirping song. Originally from S.W. Asia and N. Africa, it has spread into Europe.

½–¾ in
1.4–2 cm

⅝–⅞ in
1.7–2.3 cm

SOUTHERN FIELD CRICKET
Gryllus bimaculatus
F: Gryllidae

This cricket is widespread in S. Europe, parts of Africa, and Asia and lives on the ground under wood or debris.

bright red markings

FOAMING GRASSHOPPER
Dictyophorus spumans
F: Pyrgomorphidae

Bright colors advertise that this South African grasshopper is toxic to predators. It can also produce a noxious foam from its thoracic glands.

2¼–3¼ in
6–8 cm

COCKROACHES

Members of the order Blattodea are scavenging insects, with a flattened oval body. The downward-pointing head is often largely concealed by a shieldlike pronotum and there are usually two pairs of wings. The tip of the abdomen has a pair of sensory organs, or cerci.

2–3¼ in
5–8 cm

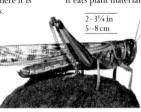

MADAGASCAN HISSING COCKROACH
Gromphadorhina portentosa
F: Blaberidae

This large, wingless cockroach is raised worldwide as a pet. The male has prominent bumps on the thorax for male-to-male combat.

short terminal cerci

11⁄32–7⁄10 in
0.8–1.3 cm

light brown pronotum

DUSKY COCKROACH
Ectobius lapponicus
F: Ectobiidae

Small and fast-running, this European species can be found among leaf litter and occasionally in foliage. It has been introduced into the US.

NYMPH

1⅛–1¾ in
2.7–4.4 cm

AMERICAN COCKROACH
Periplaneta americana
F: Blattidae

This species, which was originally from Africa, is now found worldwide. It lives on ships and in food warehouses.

TERMITES

These nest-building social insects of the order Isoptera live in colonies with different castes: reproductives (the kings and queens), workers, and soldiers. Workers are generally pale and wingless; reproductives have wings that are shed after a nuptial fight; and soldiers have a large head and jaws.

SUBTERRANEAN TERMITE
Coptotermes formosanus
F: Rhinotermitidae

Native to S. China, Taiwan, and Japan, this invasive species has now spread to other parts of the world, where it is a serious pest.

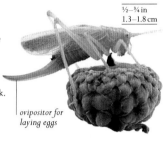

¼–7⁄32 in
6–7 mm

PACIFIC DAMPWOOD TERMITE
Zootermopsis angusticollis
F: Archotermopsidae

This termite, found along the Pacific coastal states of N. America, nests in and feeds on decayed, fungus-infected wood.

¼–½ in
0.8–1.5 cm

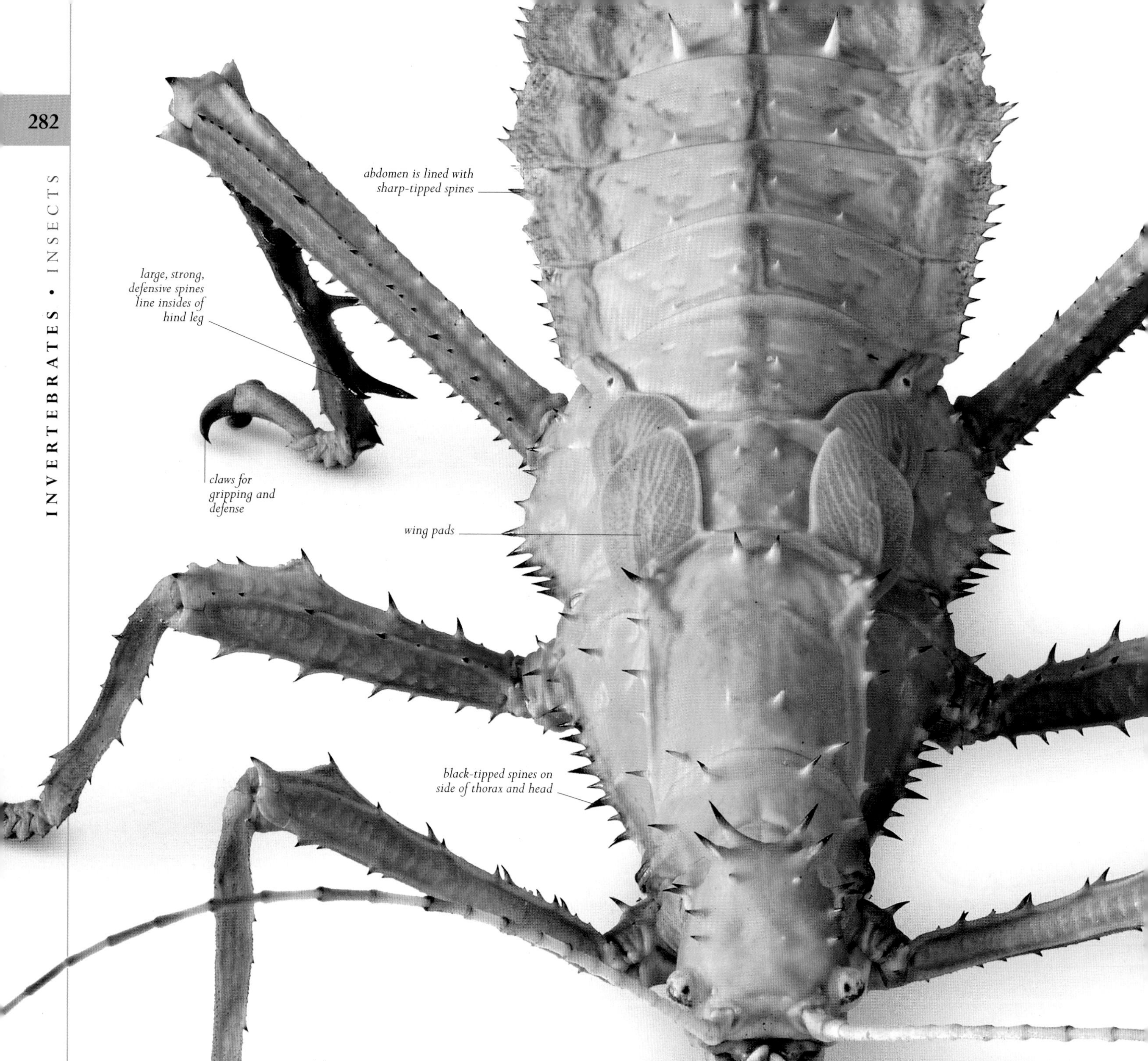

abdomen is lined with
sharp-tipped spines

large, strong,
defensive spines
line insides of
hind leg

claws for
gripping and
defense

wing pads

black-tipped spines on
side of thorax and head

∧ **ADOLESCENT**
The small, nonoverlapping
wing pads show that this
female is still a nymph—
not yet sexually mature.
At the next molt, when it
sheds its skin, the nymph will
become an adult, acquiring
short, stubby wings and a
functional ovipositor. After
mating, the abdomen will
swell as the eggs develop.

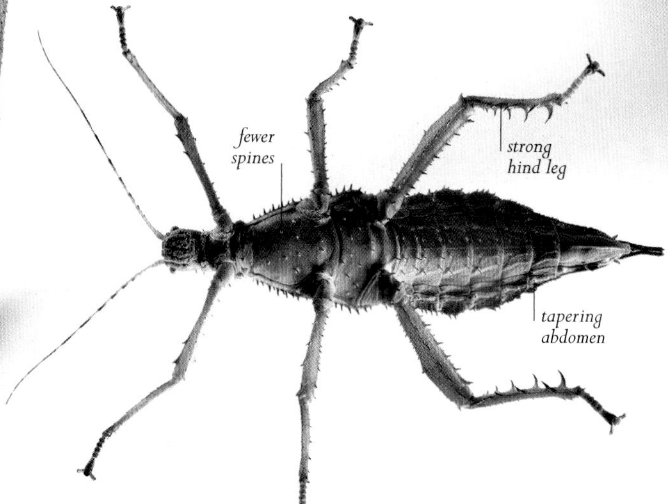

fewer
spines

strong
hind leg

tapering
abdomen

< **UNDERSIDE**
The female's underside is
dark green and, although
it has fewer spines than
the top surface, it is well
protected by the spiny legs.

JUNGLE NYMPH STICK INSECT
Heteropteryx dilatata

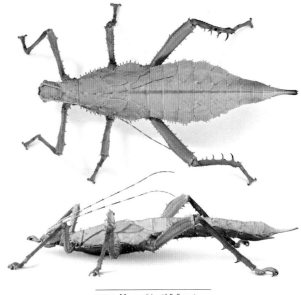

This species is also known as the Malayan jungle nymph. Females are large, with bright green coloration on their upper surface and darker green below. The adult male is much smaller, more slender, and darker in color. Both have wings, but the female is flightless. Nymphs and adults feed on the foliage of a range of different plants, including durian, guava, and mango. Mature females with eggs can be very aggressive. If disturbed, they make a loud hissing sound using their short wings, and splay their strong, spiny hind legs in a defensive posture. If attacked, they will kick out. This lively, nocturnal species has become a popular pet worldwide.

SIZE Up to 6 in (15.5 cm)
HABITAT Tropical forest
DISTRIBUTION Malaysia
DIET Foliage of various plants

MOUTHPARTS >
At rest the jaws are concealed behind two pairs of food-handling appendages called palps. These are covered with sensory organs, which enable the insect to taste the surface of the leaves on which it feeds.

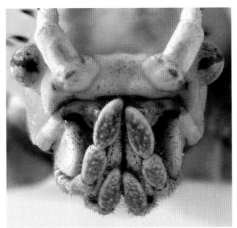

< COMPOUND EYE
Compound eyes—typical of insects—do not need to be as acute as those of many predatory arthropods, but they do need to be able to detect movement and potential enemies.

SEGMENTED BODY >
The tough, spiny plates that make up the body segments are joined by soft membranes that allow flexibility.

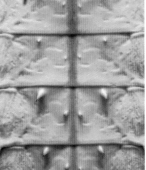

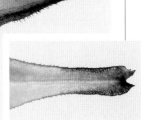

∧ OVIPOSITOR
The female—which can lay up to 150 eggs in her lifetime—uses her ovipositor to lay her large eggs one at a time, concealed in leaf litter or moist soil.

long, segmented antennae sense the immediate environment, including air movements

∧> FOOT
The foot is made up of a number of short tarsal segments and a long, spiny terminal segment, which bears a pair of sharp, curved claws.

TRUE BUGS

Abundant and widespread in terrestrial and aquatic habitats, members of the order Hemiptera range from minute, wingless insects to giant water bugs capable of catching fishes and frogs. The mouthparts are used for piercing and sucking up liquids such as plant sap, dissolved prey tissues, or blood. Many species are plant pests and some transmit disease.

1/32–1/16 in
1–2 mm

GLASSHOUSE WHITEFLY
Trialeurodes vaporariorum
F: Aleyrodidae
This small, mothlike bug is found in temperate regions worldwide and can be a serious pest of greenhouse crops.

1/8–3/16 in
3–5 mm

AMERICAN LUPIN APHID
Macrosiphum albifrons
F: Aphididae
Aphids such as this US species can quickly infest plants because females can produce many offspring without fertilization.

distinct, dark marginal area

5/16–3/8 in
8–10 mm

11/32–3/8 in
9–11 mm

INDIAN CICADA
Angamiana aetherea
F: Cicadidae
This cicada is found in India. As with all cicadas, males produce loud songs both for courtship and to signal aggression.

1 1/4–1 1/2 in
3.5–4 cm

pale area at base of hindwing

SPITTLE BUG
Aphrophora alni
F: Aphrophoridae
Commonly occurring on a wide range of trees and shrubs across Europe, this froghopper can vary in appearance from pale to dark brown.

FROGHOPPER
Cecopis vulnerata
F: Cercopidae
The nymphs of this conspicuously colored European species live communally underground in a protective, frothy mass and feed on plant root sap.

1 1/2 in
35–40 mm

ASSAM CICADA
Platypleura assamensis
F: Cicadidae
Known from N. India, this species is also found in Bhutan and parts of China where it prefers temperate, deciduous forests.

1/4–5/16 in
6–8 mm

HORSE CHESTNUT SCALE
Pulvinaria regalis
F: Coccidae
Although commonly found on the bark of horse chestnut trees, this European scale insect will also attack a range of other deciduous species.

1/2–3/4 in
1.3–1.8 cm

3 1/4 in
8 cm

large eyespot

LEAFHOPPER
Ledra aurita
F: Cicadellidae
Mottled coloration helps this flat-bodied leafhopper from N. Europe to blend with the lichen-covered bark of oak trees in its habitat.

1/4–5/16 in
6–8 mm

CICADELLA VIRIDIS
F: Cicadellidae
Found feeding on grasses and sedges in wet, boggy, or marshy areas in Europe and Asia, this leafhopper can also be found near garden ponds.

PEANUT-HEADED BUG
Fulgora laternaria
F: Fulgoridae
This bug is found in C. and S. America and the West Indies. The bulbous head was once thought to glow.

3/8–1/2 in
1–1.2 cm

THORN BUG
Umbonia crassicornis
F: Membracidae
Found in C. and S. America, almost the whole body of this bug is concealed under the enlarged, thornlike pronotum.

distinctive marking across forewings

elongated head

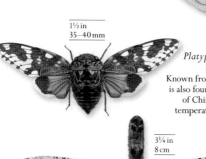

1 1/4 in
3.2 cm

WART-HEADED BUG
Phrictus quinquepartitus
F: Fulgoridae
Also known as the dragon-headed bug, this species can be found in Costa Rica, Panama, Colombia, and parts of Brazil.

1/16–1/8 in
2–3 mm

ASH PLANT LOUSE
Psyllopsis fraxini
F: Psyllidae
Found commonly on ash trees, the nymphs of this species cause the edges of leaves to form red, swollen galls when they feed.

1/8–5/32 in
3–4 mm

brightly colored hindwings

THISTLE LACE BUG
Tingis cardui
F: Tingidae
This bug occurs in most of Europe and feeds on spear, musk, and marsh thistles. The body is covered in powdery wax.

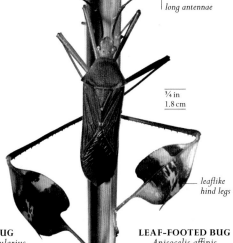

$^3/_8$–$^1/_2$ in
1–1.2 cm

$^3/_8$–$^1/_2$ in
1–1.6 cm

$^5/_{32}$ in
4 mm

$^3/_{16}$ in
5 mm

long antennae

$^3/_4$ in
1.8 cm

EUROPEAN TORTOISE BUG
Eurygaster maura
F: Scutelleridae

This bug feeds on a wide range of grasses but can sometimes be a minor pest of cereal crops.

HAWTHORN SHIELD BUG
Acanthosoma haemorrhoidale
F: Acanthosomatidae

This attractive European species feeds on the buds and berries of hawthorn and occasionally other deciduous trees such as oak.

COMMON FLOWER BUG
Anthocoris nemorum
F: Anthocoridae

This predatory bug can be found on a wide range of plants. Despite its small size it can pierce human skin.

BIRCH BARK BUG
Aradus betulae
F: Aradidae

The flattened body of this European bug allows it to live under the bark of birch trees, where it feeds on fungi.

leaflike hind legs

LEAF-FOOTED BUG
Anisocelis affinis
F: Coreidae

Found in Mexico and C. America, the leg expansions of this herbivorous bug may provide camouflage and protect it from enemies.

strong front leg

single sharp claw

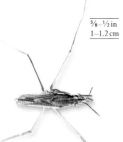

$^5/_{32}$–$^3/_{16}$ in
4–5mm

BED BUG
Cimex lectularius
F: Cimicidae

This widespread species feeds on the blood of humans and other warm-blooded mammals. These wingless, flattened bugs are active after dark.

$^3/_8$–$^1/_2$ in
1–1.2 cm

$^5/_{16}$–$^3/_5$ in
8–15 mm

$^5/_{16}$–$^1/_2$ in
8–11 mm

COMMON POND SKATER
Gerris lacustris
F: Gerridae

Widespread and immediately recognizable as it darts around on the surface of water, the pond skater locates prey by the ripples it sends out.

WATER BOATMAN
Corixa punctata
F: Corixidae

Using its powerful, oarlike rear legs for swimming, this common European species feeds on algae and detritus in ponds.

TOAD BUG
Nerthra grandicollis
F: Gelastocoridae

The warty surface and drab coloring of this African bug allow it to hide in mud and debris, where it ambushes insect prey.

$^3/_8$–$^1/_2$ in
1–1.5 cm

$^3/_8$–$^7/_{10}$ in
1–1.3 cm

hair-fringed hind leg

$^3/_4$–4 in
8–10 cm

SAUCER BUG
Ilyocoris cimicoides
F: Naucoridae

The saucer bug traps air from the surface under its folded wings and hunts prey in the shallow margins of lakes and slow rivers. This species is from Europe.

$^1/_4$ in
6 mm

pair of appendages acts as breathing siphon

GIANT WATER BUG
Lethocerus sp.
F: Belostomatidae

Using its strong front legs and toxic saliva, this tropical bug can overpower larger vertebrate prey such as frogs and fishes.

WATER MEASURER
Hydrometra stagnorum
F: Hydrometridae

This slow-moving European bug lives around the margins of ponds, lakes, and rivers and feeds on small insects and crustaceans.

COMMON GREEN CAPSID
Lygocoris pabulinus
F: Miridae

This widespread bug can be a serious pest to a range of plants, including fruit crops such as raspberries, pears, and apples.

»

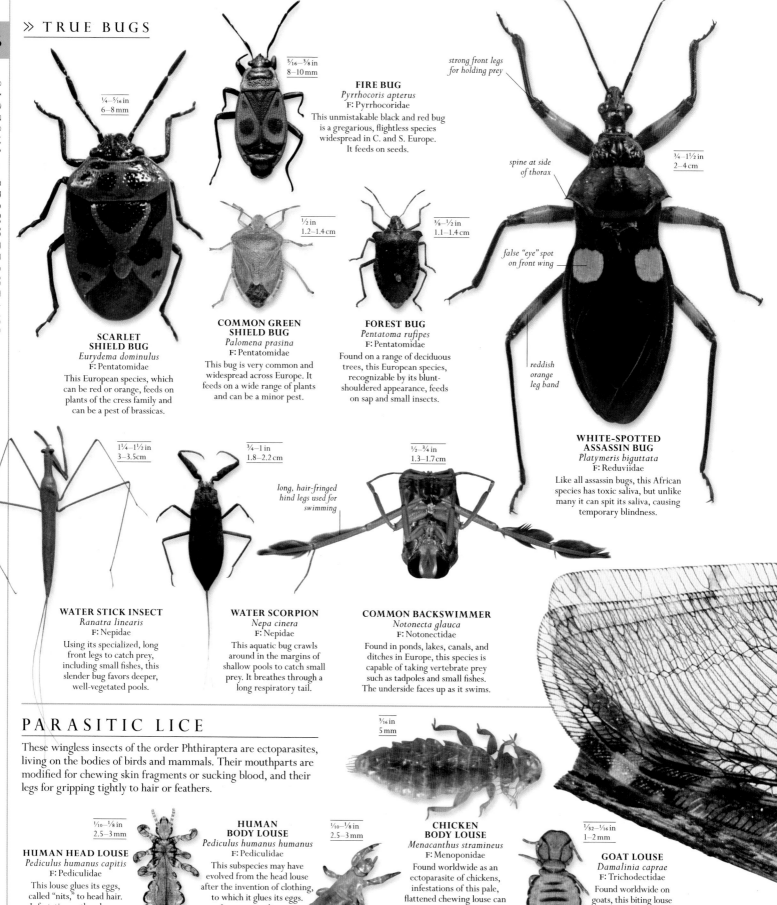

» TRUE BUGS

¼–⁵⁄₁₆ in
6–8 mm

⁵⁄₁₆–³⁄₈ in
8–10 mm

FIRE BUG
Pyrrhocoris apterus
F: Pyrrhocoridae
This unmistakable black and red bug
is a gregarious, flightless species
widespread in C. and S. Europe.
It feeds on seeds.

strong front legs
for holding prey

spine at side
of thorax

¾–1½ in
2–4 cm

false "eye" spot
on front wing

½ in
1.2–1.4 cm

³⁄₈–½ in
1.1–1.4 cm

reddish
orange
leg band

**SCARLET
SHIELD BUG**
Eurydema dominulus
F: Pentatomidae
This European species, which
can be red or orange, feeds on
plants of the cress family and
can be a pest of brassicas.

**COMMON GREEN
SHIELD BUG**
Palomena prasina
F: Pentatomidae
This bug is very common and
widespread across Europe. It
feeds on a wide range of plants
and can be a minor pest.

FOREST BUG
Pentatoma rufipes
F: Pentatomidae
Found on a range of deciduous
trees, this European species,
recognizable by its blunt-
shouldered appearance, feeds
on sap and small insects.

**WHITE-SPOTTED
ASSASSIN BUG**
Platymeris biguttata
F: Reduviidae
Like all assassin bugs, this African
species has toxic saliva, but unlike
many it can spit its saliva, causing
temporary blindness.

1¼–1½ in
3–3.5 cm

¾–1 in
1.8–2.2 cm

½–¾ in
1.3–1.7 cm

long, hair-fringed
hind legs used for
swimming

WATER STICK INSECT
Ranatra linearis
F: Nepidae
Using its specialized, long
front legs to catch prey,
including small fishes, this
slender bug favors deeper,
well-vegetated pools.

WATER SCORPION
Nepa cinera
F: Nepidae
This aquatic bug crawls
around in the margins of
shallow pools to catch small
prey. It breathes through a
long respiratory tail.

COMMON BACKSWIMMER
Notonecta glauca
F: Notonectidae
Found in ponds, lakes, canals, and
ditches in Europe, this species is
capable of taking vertebrate prey
such as tadpoles and small fishes.
The underside faces up as it swims.

PARASITIC LICE

These wingless insects of the order Phthiraptera are ectoparasites,
living on the bodies of birds and mammals. Their mouthparts are
modified for chewing skin fragments or sucking blood, and their
legs for gripping tightly to hair or feathers.

³⁄₁₆ in
5 mm

¹⁄₁₀–⅛ in
2.5–3 mm

HUMAN HEAD LOUSE
Pediculus humanus capitis
F: Pediculidae
This louse glues its eggs,
called "nits," to head hair.
Infestation outbreaks are
common among school
children. A related species
attacks chimpanzees.

**HUMAN
BODY LOUSE**
Pediculus humanus humanus
F: Pediculidae
This subspecies may have
evolved from the head louse
after the invention of clothing,
to which it glues its eggs.
It transmits disease.

¹⁄₁₀–⅛ in
2.5–3 mm

**CHICKEN
BODY LOUSE**
Menacanthus stramineus
F: Menoponidae
Found worldwide as an
ectoparasite of chickens,
infestations of this pale,
flattened chewing louse can
cause feather loss and infection.

¹⁄₃₂–¹⁄₁₆ in
1–2 mm

GOAT LOUSE
Damalinia caprae
F: Trichodectidae
Found worldwide on
goats, this biting louse
can also survive for
a few days on sheep
but is not able to
breed on them.

BARKLICE AND BOOKLICE

Common on vegetation and in litter, insects of the order Psocoptera are small, squat, and soft-bodied. The head has threadlike antennae and bulging eyes. They eat microflora and some species are pests of stored products.

5/32–1/4 in
4–6 mm

PSOCOCERASTIS GIBBOSA
F: Psocidae

Native to Europe and parts of Asia, this relatively large bark louse can be found on a wide range of deciduous and coniferous trees.

1/64–1/24 in
0.6–1.5 mm

LIPOSCELIS LIPARIUS
F: Liposcelididae

This very widespread species favors dark, damp microhabitats and can be a pest in libraries and granaries if the humidity is too high.

THRIPS

Members of the order Thysanoptera are tiny insects, typically with two pairs of narrow, hair-fringed wings. They have large compound eyes and distinctive mouthparts for piercing and sucking.

1/32–1/24 in
1–1.5 mm

FLOWER THRIP
Frankliniella sp.
F: Thripidae

Flower thrips are found all over the world and can be pests on crops such as peanuts, cotton, sweet potatoes, and coffee.

SNAKEFLIES

Snakeflies, in the order Raphidioptera, are woodland insects with a long prothorax, a broad head, and two pairs of wings. Adults and larvae eat aphids and other soft prey.

RAPHIDIA NOTATA
F: Raphidiidae

Found in deciduous or coniferous woodlands in Europe, this snakefly is usually associated with oak trees, where it eats aphids.

1/2–3/4 in
1.6–1.8 cm

ALDERFLIES AND DOBSONFLIES

The Megaloptera have two pairs of wings, which are held rooflike over the body when they are at rest. The aquatic larvae, which are predatory and have abdominal gills, pupate on land in soil, moss, or rotting wood.

EASTERN DOBSONFLY
Corydalus cornutus
F: Corydalidae

Found in N. America, the male of this species has very long mandibles used for combat and for gripping the female.

4–5½ in
10–14 cm

ALDERFLY
Sialis lutaria
F: Sialidae

The female of this widespread species lays a mass of up to 2,000 eggs on twigs or leaves near water.

1/2–1 in
1.4–2.6 cm

LACEWINGS AND RELATIVES

Insects of the order Neuroptera have conspicuous eyes and biting mouthparts. The pairs of net-veined wings are held rooflike over the body at rest. The larvae have sickle-shaped mouthparts, which form sharp, sucking tubes.

1¼ in
3 cm

MANTISPA STYRIACA
F: Mantispidae

Like a miniature praying mantid, this species, found in S. and C.Europe, lives in lightly wooded areas, where it hunts small flies.

½ in
1.4 cm

SPOON-WINGED LACEWING
Nemoptera sinuata
F: Nemopteridae

Common to areas of S.E. Europe, this delicate species feeds on nectar and pollen at flowers in woodland and open grassland.

1½ in
4 cm

OWLFLY
Libelloides macaronius
F: Ascalaphidae

Capturing insect prey in midair, this species will only fly on warm sunny days. It can be found in C. and S. Europe and parts of Asia.

spotted wings

GIANT LACEWING
Osmylus fulvicephalus
F: Osmylidae

This European species can be found among shady woodland vegetation by streams, where it eats small insects and pollen.

3/8–1/2 in
1–1.2 cm

GREEN LACEWING
Chrysopa perla
F: Chrysopidae

This widespread European species has a characteristic bluish-green tinge and black markings and is commonly found in deciduous woodland.

ANTLION
Palpares libelluloides
F: Myrmeleontidae

This large, day-flying, Mediterranean species with distinctively mottled wings can be found in rough grassland, warm scrubby habitats, and sand dunes.

2–2¼ in
5–5.5 cm

males have organs for clasping females

½ in
1.3 cm

BEETLES

Members of the order Coleoptera—the largest insect order—range from minute to very large species. A distinguishing feature are the toughened front wings, called elytra, which meet down the body's midline and protect the larger membranous hindwings. Beetles occupy every aquatic and terrestrial habitat, where they are scavengers, herbivores, or predators.

elongated head

VIOLET GROUND BEETLE
Carabus violaceus
F: Carabidae
A nocturnal hunter, this beetle is common in many habitats, including gardens. It is native to Europe and parts of Asia.

1–1⅖ in
2.8–3.4 cm

1½ in
4 cm

⅝6–⅜ in
8–10 mm

⅛–³⁄₁₆ in
3–5 mm

WOODWORM BEETLE
Anobium punctatum
F: Anobiidae
Adapted to breeding in wood in buildings and furniture, this species is now widespread and can be a serious pest.

JEWEL BEETLE
Chrysochroa chinensis
F: Buprestidae
This metallic species is native to India and S.E. Asia, where its larvae burrow into the wood of deciduous trees.

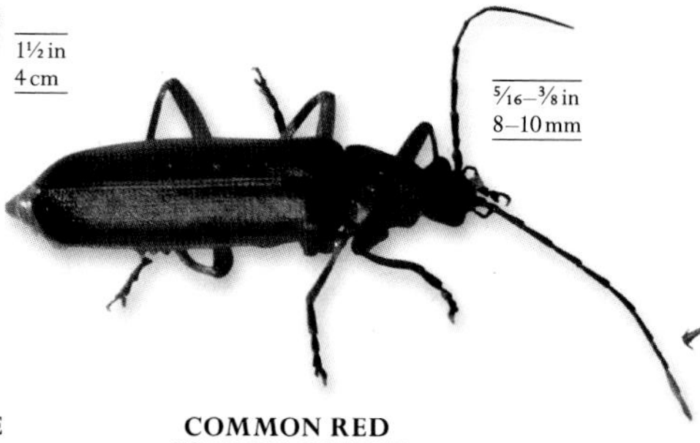

COMMON RED SOLDIER BEETLE
Rhagonycha fulva
F: Cantharidae
A common sight on flowers in the summer, this European beetle can be found in meadows and woodland margins.

VIOLIN BEETLE
Mormolyce phyllodes
F: Carabidae
The shape of this S.E. Asian beetle allows it to squeeze under bracket fungi and tree bark, where it feeds on insect larvae and snails.

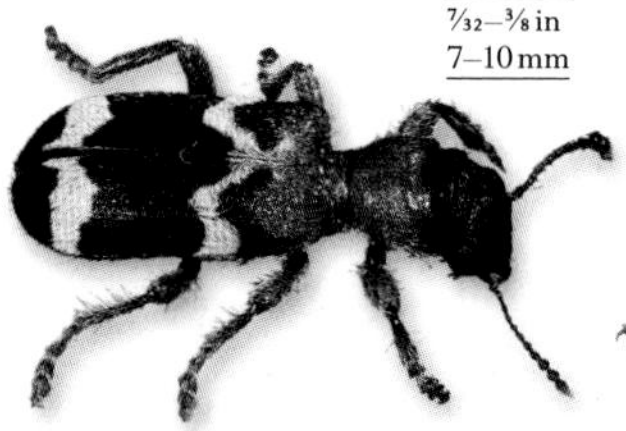

2¼–4 in
6–10 cm

wide, flat elytron

⁷⁄₃₂–⅜ in
7–10 mm

1–1½ in
2.5–3.8 cm

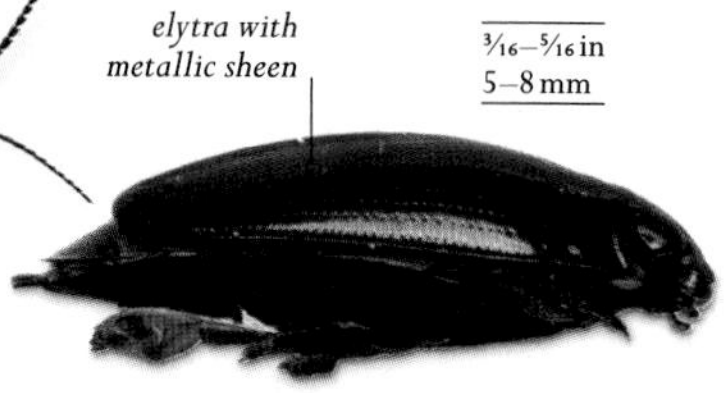

elytra with metallic sheen

³⁄₁₆–⁵⁄₁₆ in
5–8 mm

ANT BEETLE
Thanasimus formicarius
F: Cleridae
Associated with coniferous trees in Europe and N. Asia, the larvae and adults of this beetle prey on bark beetle larvae.

⁵⁄₁₆–⅜ in
8–10 mm

1¼ in
3 cm

LARDER BEETLE
Dermestes lardarius
F: Dermestidae
Found in Europe and parts of Asia, this beetle feeds on animal remains but also lives in buildings, where it eats stored produce.

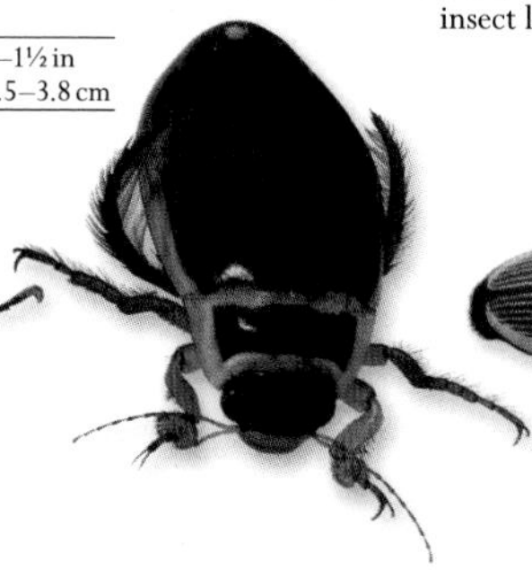

GREAT DIVING BEETLE
Dytiscus marginalis
F: Dytiscidae
This large beetle lives in weedy ponds and lakes in Europe and N. Asia. It eats insects, frogs, newts, and small fishes.

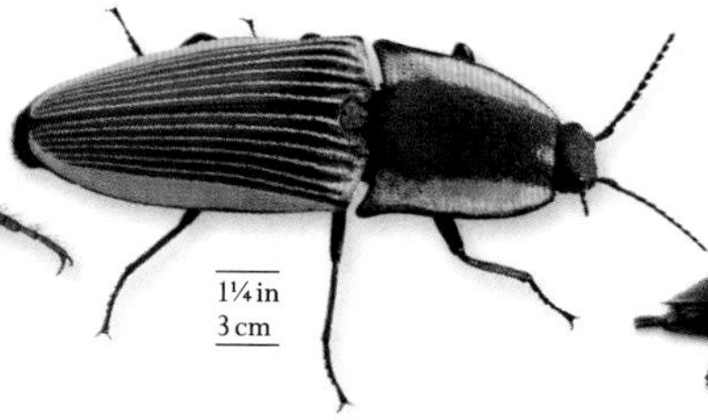

CLICK BEETLE
Chalcolepidius limbatus
F: Elateridae
This beetle is found in woodland and grassland in the warmer parts of S. America. Its larvae are predators in rotten wood and soil.

WHIRLIGIG BEETLE
Gyrinus marinus
F: Gyrinidae
A common European species, this beetle is a surface-dweller on ponds and lakes. It uses its paddle-shaped legs to skim over the water.

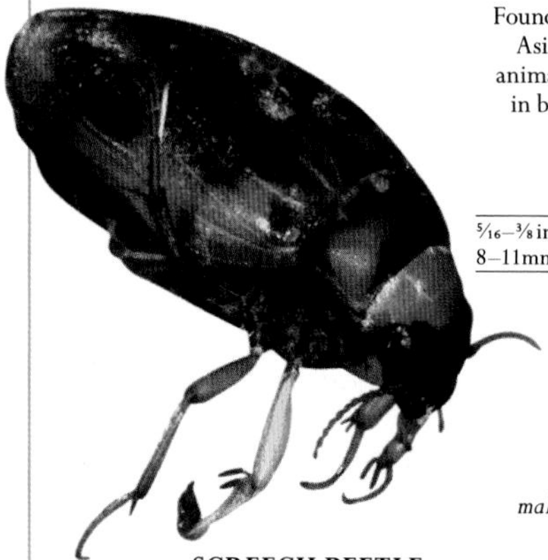

⁵⁄₁₆–⅜ in
8–11 mm

1 in
2.5 cm

male has wings ♂

¼–⅜ in
6–10 mm

½–¾ in
1.5–2 cm

curved horns

SCREECH BEETLE
Hygrobia hermanni
F: Hygrobiidae
This European beetle feeds on small invertebrate prey in slow-moving rivers and muddy ponds. It makes a high-pitched grating noise if handled.

COMMON GLOWWORM
Lampyris noctiluca
F: Lampyridae
This beetle is found across Europe and Asia, where it favors rough grassland. The wingless female—known as a glowworm—emits a greenish light to attract a mate.

FOUR-SPOTTED HISTER BEETLE
Hister quadrimaculatus
F: Histeridae
Widespread across Europe, this beetle is found in dung and sometimes carrion, where it feeds on small insects and their larvae.

MINOTAUR BEETLE
Typhaeus typhoeus
F: Geotrupidae
This species is found in sandy areas in W. Europe and buries sheep and rabbit droppings, on which the larvae feed, in burrows.

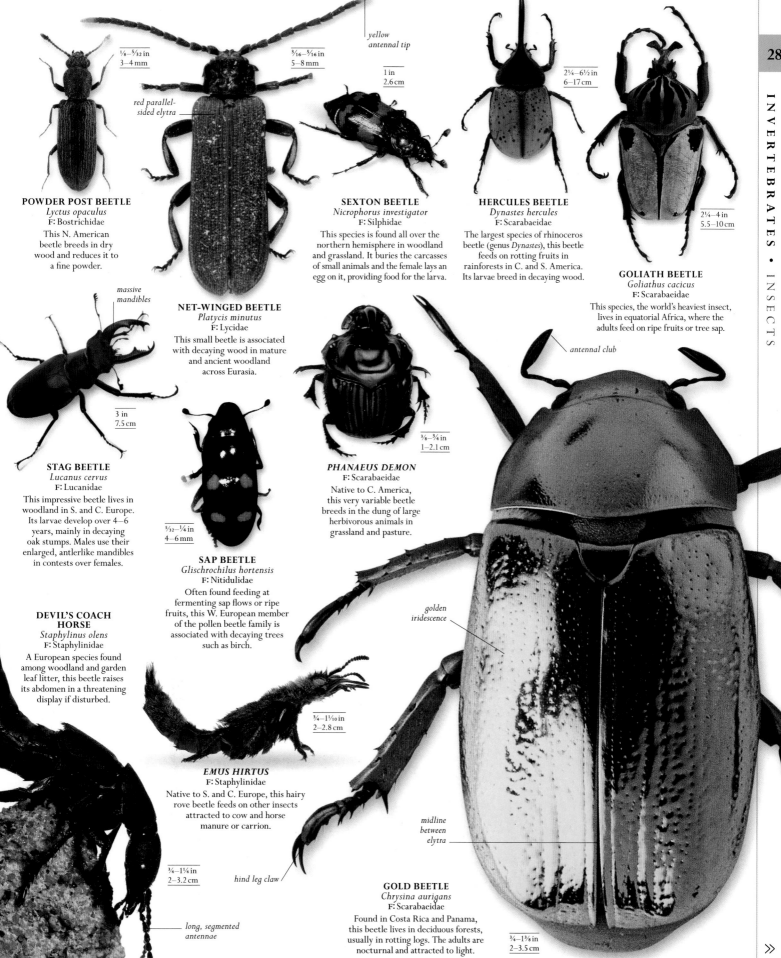

POWDER POST BEETLE
Lyctus opaculus
F: Bostrichidae
This N. American beetle breeds in dry wood and reduces it to a fine powder.

⅛–5/32 in
3–4 mm

red parallel-sided elytra

3/16–5/16 in
5–8 mm

yellow antennal tip

1 in
2.6 cm

SEXTON BEETLE
Nicrophorus investigator
F: Silphidae
This species is found all over the northern hemisphere in woodland and grassland. It buries the carcasses of small animals and the female lays an egg on it, providing food for the larva.

2¼–6½ in
6–17 cm

HERCULES BEETLE
Dynastes hercules
F: Scarabaeidae
The largest species of rhinoceros beetle (genus *Dynastes*), this beetle feeds on rotting fruits in rainforests in C. and S. America. Its larvae breed in decaying wood.

2¼–4 in
5.5–10 cm

GOLIATH BEETLE
Goliathus cacicus
F: Scarabaeidae
This species, the world's heaviest insect, lives in equatorial Africa, where the adults feed on ripe fruits or tree sap.

antennal club

massive mandibles

NET-WINGED BEETLE
Platycis minutus
F: Lycidae
This small beetle is associated with decaying wood in mature and ancient woodland across Eurasia.

3 in
7.5 cm

STAG BEETLE
Lucanus cervus
F: Lucanidae
This impressive beetle lives in woodland in S. and C. Europe. Its larvae develop over 4–6 years, mainly in decaying oak stumps. Males use their enlarged, antlerlike mandibles in contests over females.

5/32–¼ in
4–6 mm

SAP BEETLE
Glischrochilus hortensis
F: Nitidulidae
Often found feeding at fermenting sap flows or ripe fruits, this W. European member of the pollen beetle family is associated with decaying trees such as birch.

⅜–¾ in
1–2.1 cm

PHANAEUS DEMON
F: Scarabaeidae
Native to C. America, this very variable beetle breeds in the dung of large herbivorous animals in grassland and pasture.

golden iridescence

DEVIL'S COACH HORSE
Staphylinus olens
F: Staphylinidae
A European species found among woodland and garden leaf litter, this beetle raises its abdomen in a threatening display if disturbed.

¾–1 1/10 in
2–2.8 cm

EMUS HIRTUS
F: Staphylinidae
Native to S. and C. Europe, this hairy rove beetle feeds on other insects attracted to cow and horse manure or carrion.

midline between elytra

¾–1¼ in
2–3.2 cm

hind leg claw

long, segmented antennae

GOLD BEETLE
Chrysina aurigans
F: Scarabaeidae
Found in Costa Rica and Panama, this beetle lives in deciduous forests, usually in rotting logs. The adults are nocturnal and attracted to light.

¾–1⅜ in
2–3.5 cm

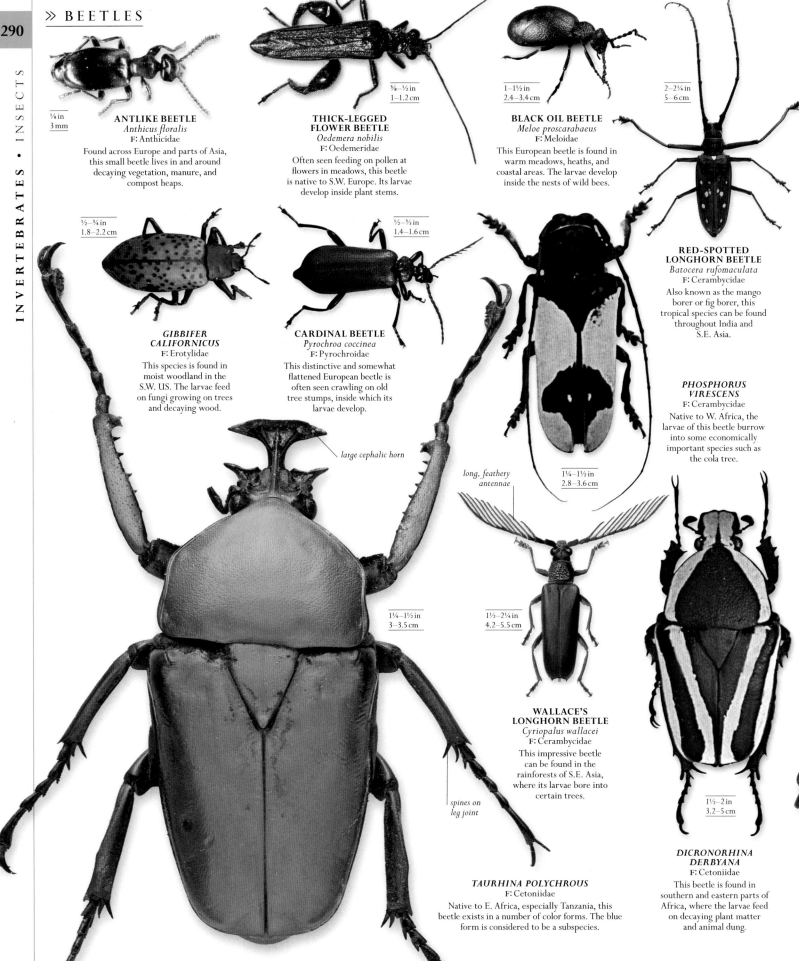

⅛ in
3 mm

ANTLIKE BEETLE
Anthicus floralis
F: Anthicidae
Found across Europe and parts of Asia,
this small beetle lives in and around
decaying vegetation, manure, and
compost heaps.

⅜–½ in
1–1.2 cm

**THICK-LEGGED
FLOWER BEETLE**
Oedemera nobilis
F: Oedemeridae
Often seen feeding on pollen at
flowers in meadows, this beetle
is native to S.W. Europe. Its larvae
develop inside plant stems.

1–1½ in
2.4–3.4 cm

BLACK OIL BEETLE
Meloe proscarabaeus
F: Meloidae
This European beetle is found in
warm meadows, heaths, and
coastal areas. The larvae develop
inside the nests of wild bees.

2–2¼ in
5–6 cm

**RED-SPOTTED
LONGHORN BEETLE**
Batocera rufomaculata
F: Cerambycidae
Also known as the mango
borer or fig borer, this
tropical species can be found
throughout India and
S.E. Asia.

½–¾ in
1.8–2.2 cm

**GIBBIFER
CALIFORNICUS**
F: Erotylidae
This species is found in
moist woodland in the
S.W. US. The larvae feed
on fungi growing on trees
and decaying wood.

½–⅝ in
1.4–1.6 cm

CARDINAL BEETLE
Pyrochroa coccinea
F: Pyrochroidae
This distinctive and somewhat
flattened European beetle is
often seen crawling on old
tree stumps, inside which its
larvae develop.

1¼–1½ in
2.8–3.6 cm

**PHOSPHORUS
VIRESCENS**
F: Cerambycidae
Native to W. Africa, the
larvae of this beetle burrow
into some economically
important species such as
the cola tree.

large cephalic horn

long, feathery
antennae

1¼–1½ in
3–3.5 cm

1½–2¼ in
4.2–5.5 cm

spines on
leg joint

**WALLACE'S
LONGHORN BEETLE**
Cyriopalus wallacei
F: Cerambycidae
This impressive beetle
can be found in the
rainforests of S.E. Asia,
where its larvae bore into
certain trees.

TAURHINA POLYCHROUS
F: Cetoniidae
Native to E. Africa, especially Tanzania, this
beetle exists in a number of color forms. The blue
form is considered to be a subspecies.

1½–2 in
3.2–5 cm

**DICRONORHINA
DERBYANA**
F: Cetoniidae
This beetle is found in
southern and eastern parts of
Africa, where the larvae feed
on decaying plant matter
and animal dung.

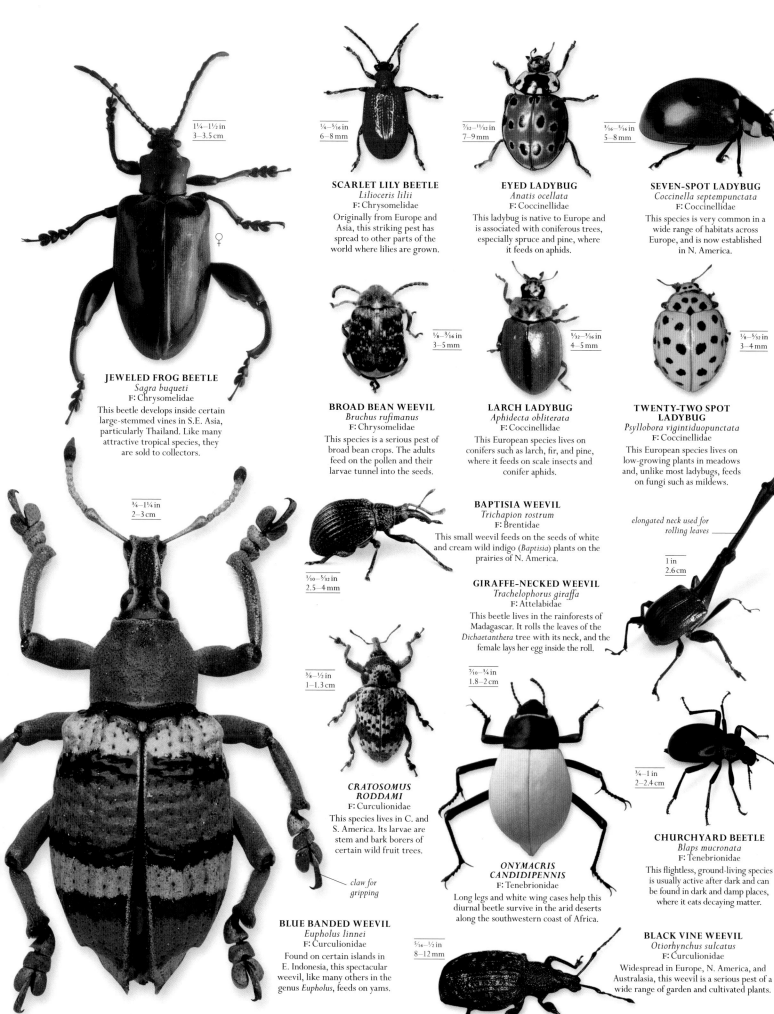

¼–1½ in
3–3.5 cm

♀

JEWELED FROG BEETLE
Sagra buqueti
F: Chrysomelidae
This beetle develops inside certain large-stemmed vines in S.E. Asia, particularly Thailand. Like many attractive tropical species, they are sold to collectors.

¾–1¼ in
2–3 cm

claw for gripping

BLUE BANDED WEEVIL
Eupholus linnei
F: Curculionidae
Found on certain islands in E. Indonesia, this spectacular weevil, like many others in the genus *Eupholus*, feeds on yams.

¼–⁵⁄₁₆ in
6–8 mm

SCARLET LILY BEETLE
Lilioceris lilii
F: Chrysomelidae
Originally from Europe and Asia, this striking pest has spread to other parts of the world where lilies are grown.

⅛–³⁄₁₆ in
3–5 mm

BROAD BEAN WEEVIL
Bruchus rufimanus
F: Chrysomelidae
This species is a serious pest of broad bean crops. The adults feed on the pollen and their larvae tunnel into the seeds.

¹⁄₁₀–⁵⁄₃₂ in
2.5–4 mm

BAPTISIA WEEVIL
Trichapion rostrum
F: Brentidae
This small weevil feeds on the seeds of white and cream wild indigo (*Baptisia*) plants on the prairies of N. America.

GIRAFFE-NECKED WEEVIL
Trachelophorus giraffa
F: Attelabidae
This beetle lives in the rainforests of Madagascar. It rolls the leaves of the *Dichaetanthera* tree with its neck, and the female lays her egg inside the roll.

⅜–½ in
1–1.3 cm

CRATOSOMUS RODDAMI
F: Curculionidae
This species lives in C. and S. America. Its larvae are stem and bark borers of certain wild fruit trees.

⁵⁄₁₆–½ in
8–12 mm

ONYMACRIS CANDIDIPENNIS
F: Tenebrionidae
Long legs and white wing cases help this diurnal beetle survive in the arid deserts along the southwestern coast of Africa.

⁷⁄₃₂–¹¹⁄₃₂ in
7–9 mm

EYED LADYBUG
Anatis ocellata
F: Coccinellidae
This ladybug is native to Europe and is associated with coniferous trees, especially spruce and pine, where it feeds on aphids.

⁵⁄₃₂–³⁄₁₆ in
4–5 mm

LARCH LADYBUG
Aphidecta obliterata
F: Coccinellidae
This European species lives on conifers such as larch, fir, and pine, where it feeds on scale insects and conifer aphids.

³⁄₁₆–⁵⁄₁₆ in
5–8 mm

SEVEN-SPOT LADYBUG
Coccinella septempunctata
F: Coccinellidae
This species is very common in a wide range of habitats across Europe, and is now established in N. America.

⅛–⁵⁄₃₂ in
3–4 mm

TWENTY-TWO SPOT LADYBUG
Psyllobora vigintiduopunctata
F: Coccinellidae
This European species lives on low-growing plants in meadows and, unlike most ladybugs, feeds on fungi such as mildews.

elongated neck used for rolling leaves

1 in
2.6 cm

¾–1 in
2–2.4 cm

CHURCHYARD BEETLE
Blaps mucronata
F: Tenebrionidae
This flightless, ground-living species is usually active after dark and can be found in dark and damp places, where it eats decaying matter.

BLACK VINE WEEVIL
Otiorhynchus sulcatus
F: Curculionidae
Widespread in Europe, N. America, and Australasia, this weevil is a serious pest of a wide range of garden and cultivated plants.

SCORPIONFLIES

These predatory insects of the order Mecoptera generally have an elongated and cylindrical body. The large, narrow wings may be clear or marked with spots or bands of dark color. Some species are short-winged or completely wingless. The head, which has large eyes and threadlike antennae, is characteristically extended downward in the form of a beak, at the end of which are biting mouthparts.

SNOW SCORPIONFLY
Boreus hyemalis
F: Boreidae
The adult of this small, wingless European insect can only be seen in the fall and winter. It breeds among moss.

COMMON SCORPIONFLY
Panorpa communis
F: Panorpidae
Native to W. Europe, this insect lives in shady hedgerows and woodland margins and can often be seen resting on nettles.

¾–1 in
1.8–3 cm

FLEAS

Members of the order Siphonaptera, fleas are laterally flattened, wingless, blood-sucking ectoparasites of mammals and some birds. Their head has short, piercing mouthparts and pairs of simple lateral eyes. The hind legs are highly specialized for jumping.

CAT FLEA
Ctenocephalides felis
F: Pulicidae
This species is very common worldwide. Also found on dogs, it can feed on the blood of several other animals, too—including humans.

1/32–1/16 in
1–2 m

TRUE FLIES

True flies belong to the order Diptera, and are recognized by a single pair of membranous front wings. The hindwings are reduced, forming a pair of balancing organs called halteres. Most flies are beneficial to humans as pollinators, predators, and recyclers, but wild and domestic animals and many people are affected by fly-borne diseases. Some flies are also crop pests.

proboscis for stabbing and sucking

MOSQUITO
Culex sp.
F: Culicidae
There are more than 1,000 species in this genus worldwide. While some are disease vectors, many are essential in food chains.

5/32–3/8 in
4–10 m

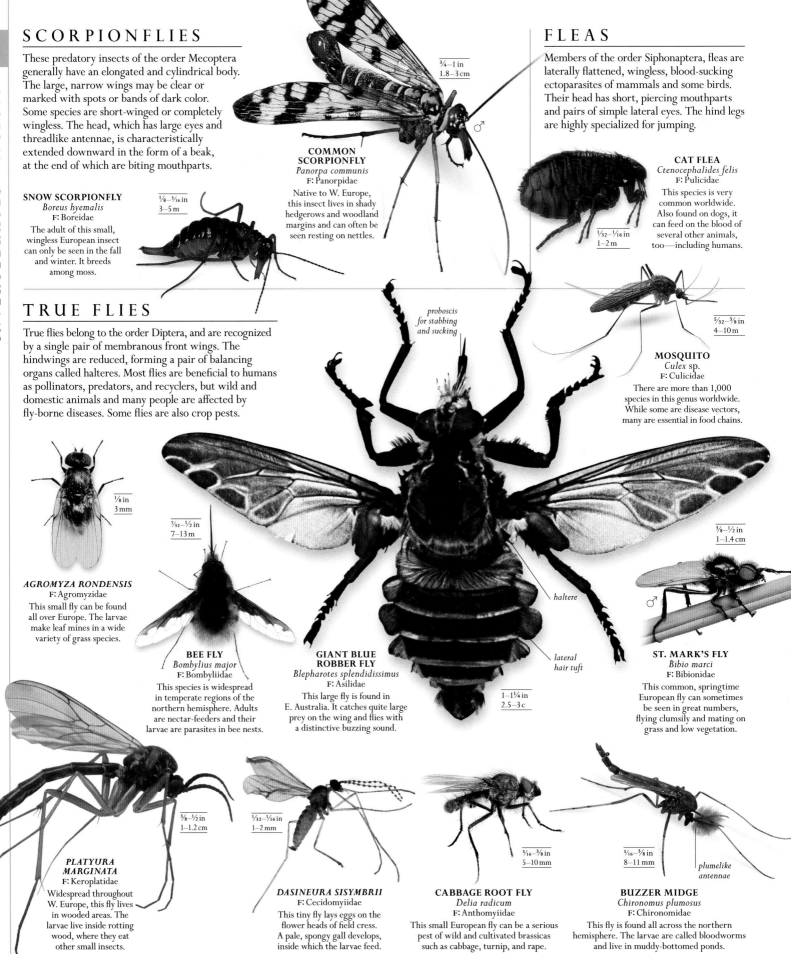

AGROMYZA RONDENSIS
F: Agromyzidae
This small fly can be found all over Europe. The larvae make leaf mines in a wide variety of grass species.

⅛ in
3 mm

BEE FLY
Bombylius major
F: Bombyliidae
This species is widespread in temperate regions of the northern hemisphere. Adults are nectar-feeders and their larvae are parasites in bee nests.

7/32–½ in
7–13 m

GIANT BLUE ROBBER FLY
Blepharotes splendidissimus
F: Asilidae
This large fly is found in E. Australia. It catches quite large prey on the wing and flies with a distinctive buzzing sound.

haltere

lateral hair tuft

1–1¼ in
2.5–3 c

3/8–½ in
1–1.4 cm

ST. MARK'S FLY
Bibio marci
F: Bibionidae
This common, springtime European fly can sometimes be seen in great numbers, flying clumsily and mating on grass and low vegetation.

PLATYURA MARGINATA
F: Keroplatidae
Widespread throughout W. Europe, this fly lives in wooded areas. The larvae live inside rotting wood, where they eat other small insects.

3/8–½ in
1–1.2 cm

DASINEURA SISYMBRII
F: Cecidomyiidae
This tiny fly lays eggs on the flower heads of field cress. A pale, spongy gall develops, inside which the larvae feed.

1/32–1/16 in
1–2 mm

CABBAGE ROOT FLY
Delia radicum
F: Anthomyiidae
This small European fly can be a serious pest of wild and cultivated brassicas such as cabbage, turnip, and rape.

3/16–3/8 in
5–10 mm

BUZZER MIDGE
Chironomus plumosus
F: Chironomidae
This fly is found all across the northern hemisphere. The larvae are called bloodworms and live in muddy-bottomed ponds.

5/16–3/8 in
8–11 mm

plumelike antennae

HOUSE FLY
Musca domestica
F: Muscidae

This species is found worldwide and is the most common fly around dwellings, where it transmits a number of disease-causing organisms to foodstuffs.

prominent red eyes

long legs can be shed if caught

¾–1 in
2–3 cm

MARSH CRANE FLY
Tipula oleracea
F: Tipulidae

Originally from Europe, this crane fly has been introduced to N. America and now some upland parts of S. America. It is often found by water.

GRASS FLY
Meromyza pratorum
F: Chloropidae

This fly occurs across the northern hemisphere, especially in sandy coastal regions. The larvae burrow in stems of marram grass and reed grass.

¼–⁵⁄₁₆ in
6–8 mm

red-orange wing base

⁵⁄₃₂–¼ in
4–6 mm

¼–⁷⁄₃₂ in
6–7 mm

⁵⁄₃₂–¼ in
4–6 mm

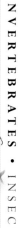

⁵⁄₃₂ in
4 mm

POECILOBOTHRUS NOBILITATUS
F: Dolichopodidae

This European species can be found in damp habitats near water. Males display in sunny patches by waving their wings.

LESSER HOUSE FLY
Fannia canicularis
F: Fanniidae

Able to breed in almost any decaying, semiliquid matter, this species is particularly associated with human habitation.

SIMULIUM ORNATUM
F: Simuliidae

This small fly lives in Europe and Asia but has been introduced elsewhere. The adults feed on animal blood and can transmit bovine onchocerciasis, a disease commonly known as river blindness.

⅛–³⁄₁₆ in
3–5 mm

½–⅔ in
1.4–1.8 cm

MOTH FLY
Clogmia albipunctata
F: Psychodidae

This very widespread fly species resembles a tiny moth. The larvae breed in dark, wet places such as drains, tree holes, and sewage.

FLESH FLY
Sarcophaga carnaria
F: Sarcophagidae

Widely distributed through Europe and Asia, this fly feeds on nectar and liquids from rotting matter. The female lays small larvae on carrion.

males with longer eyestalks win fights for territory

⁵⁄₁₆–½ in
8–14 mm

FERRUGINOUS BEE-GRABBER
Sicus ferrugineus
F: Conopidae

This European fly lays eggs in the abdomen of certain bumble bees. The larvae develop as internal parasites, killing the host.

ROTHSCHILD'S ACHIAS
Achias rothschildi
F: Platystomatidae

Males of this Papua New Guinean fly have very elongate eyestalks, which are used in territorial and mating displays.

½–⅔ in
1.5–1.8 cm

pale face with large, compound eyes

¹⁄₃₂–¹⁄₁₆ in
1–2 mm

FARMYARD MIDGE
Culicoides nubeculosus
F: Ceratopogonidae

This widespread European fly breeds in mud contaminated by dung or sewage. Adults suck the blood of horses and cattle.

⅜–½ in
1–1.2 cm

BLUEBOTTLE
Calliphora vicina
F: Calliphoridae

Found in Europe and N. America, this species is especially common in urban areas and breeds in dead pigeons and rodents.

» TRUE FLIES

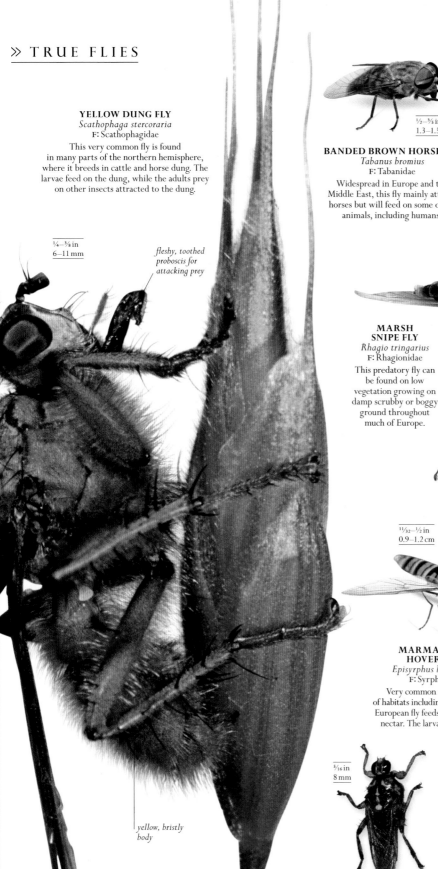

YELLOW DUNG FLY
Scathophaga stercoraria
F: Scathophagidae
This very common fly is found
in many parts of the northern hemisphere,
where it breeds in cattle and horse dung. The
larvae feed on the dung, while the adults prey
on other insects attracted to the dung.

¼–⅜ in
6–11 mm

fleshy, toothed
proboscis for
attacking prey

yellow, bristly
body

½–⅗ in
1.3–1.5 cm

BANDED BROWN HORSE FLY
Tabanus bromius
F: Tabanidae
Widespread in Europe and the
Middle East, this fly mainly attacks
horses but will feed on some other
animals, including humans.

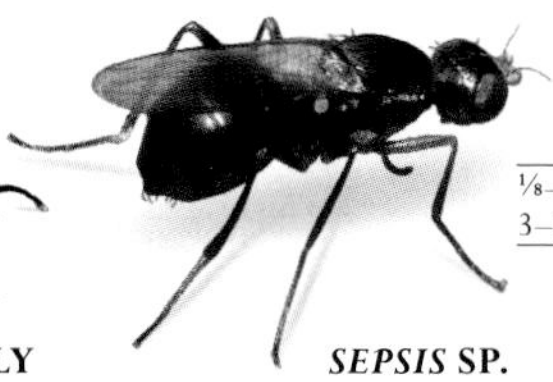

⅛–³⁄₁₆ in
3–5 mm

SEPSIS SP.
F: Sepsidae
These flies are common and
widespread in a wide variety
of habitats. The larvae
develop in animal dung
and rotting matter.

¹⁄₁₆–⅛ in
2–3 mm

COMMON FRUIT FLY
Drosophila melanogaster
F: Drosophilidae
A common laboratory
animal, this widespread
species has a distinctive dark
patch on the abdomen and
breeds in rotting fruits.

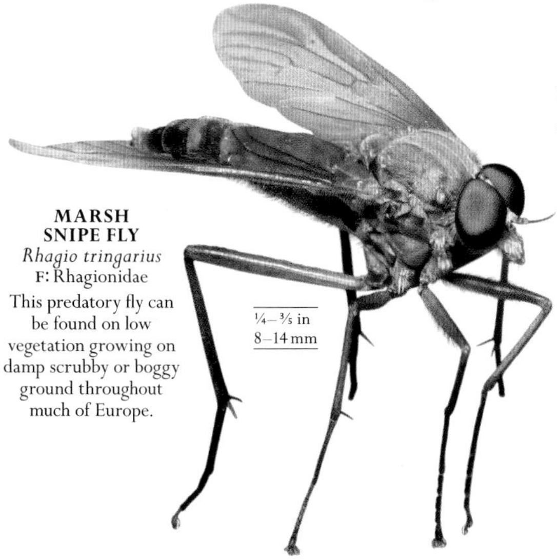

MARSH
SNIPE FLY
Rhagio tringarius
F: Rhagionidae
This predatory fly can
be found on low
vegetation growing on
damp scrubby or boggy
ground throughout
much of Europe.

¼–⅗ in
8–14 mm

⅜–½ in
1–1.2 cm

SYRPHUS RIBESII
F: Syrphidae
The larvae of this nectar-
feeding fly are major predators
of aphids in the northern
hemisphere. Its coloration
mimics that of wasps and
bees to deter predators.

¹¹⁄₃₂–½ in
0.9–1.2 cm

MARMALADE
HOVERFLY
Episyrphus balteatus
F: Syrphidae
Very common in a variety
of habitats including gardens, this
European fly feeds on pollen and
nectar. The larvae eat aphids.

⅜–½ in
1–1.2 cm

LEUCOZONA
LEUCORUM
F: Syrphidae
This northern hemisphere
species can be found visiting
flowers in damp woodlands in
the spring and early summer.
Its larvae prey on aphids.

⅗–¾ in
1.5–2 cm

DRONE FLY
Eristalis tenax
F: Syrphidae
This European fly, which
has been introduced to
N. America, is a good mimic
of the honey bee. The larvae
breed in stagnant water.

⁵⁄₁₆ in
8 mm

FOREST FLY
Hippobosca equina
F: Hippoboscidae
Occurring mainly in wooded
areas in Europe and parts of Asia,
this bloodsucking fly attacks
horses, deer, and sometimes cattle.

⁵⁄₃₂–³⁄₁₆ in
4–5 mm

PENICILLIDIA FULVIDA
F: Nycteribiidae
Widespread in sub-Saharan
Africa, this bloodsucking,
wingless, ectoparasitic fly
can be found on a wide
range of bat species.

1½–2¾ in
3.5–7 cm

GAUROMYDAS HEROS
F: Mydidae
This S. American insect, the
world's largest fly, breeds inside
the nests of leaf-cutter ants,
where the young are thought to
eat scarab beetle larvae.

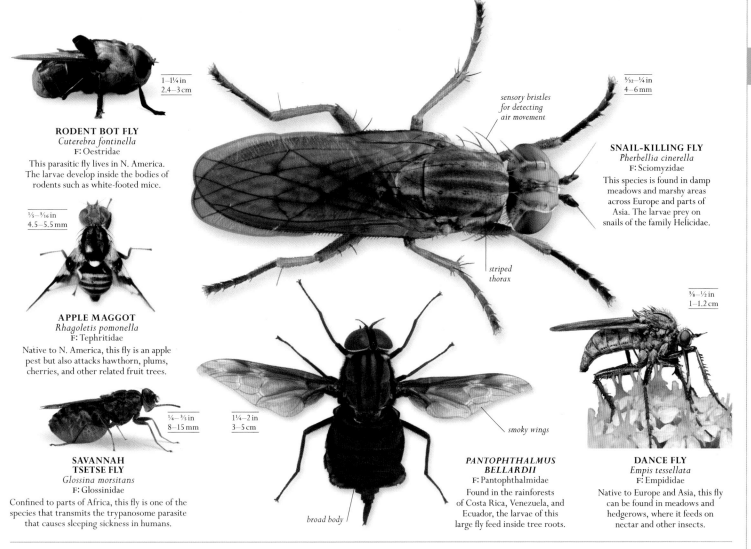

RODENT BOT FLY
Cuterebra fontinella
F: Oestridae
This parasitic fly lives in N. America.
The larvae develop inside the bodies of
rodents such as white-footed mice.

1–1¼ in
2.4–3 cm

APPLE MAGGOT
Rhagoletis pomonella
F: Tephritidae
Native to N. America, this fly is an apple
pest but also attacks hawthorn, plums,
cherries, and other related fruit trees.

⅕–³⁄₁₆ in
4.5–5.5 mm

**SAVANNAH
TSETSE FLY**
Glossina morsitans
F: Glossinidae
Confined to parts of Africa, this fly is one of the
species that transmits the trypanosome parasite
that causes sleeping sickness in humans.

¼–⅗ in
8–15 mm

*sensory bristles
for detecting
air movement*

SNAIL-KILLING FLY
Pherbellia cinerella
F: Sciomyzidae
This species is found in damp
meadows and marshy areas
across Europe and parts of
Asia. The larvae prey on
snails of the family Helicidae.

⁵⁄₃₂–¼ in
4–6 mm

*striped
thorax*

smoky wings

**PANTOPHTHALMUS
BELLARDII**
F: Pantophthalmidae
Found in the rainforests
of Costa Rica, Venezuela, and
Ecuador, the larvae of this
large fly feed inside tree roots.

1¼–2 in
3–5 cm

broad body

DANCE FLY
Empis tessellata
F: Empididae
Native to Europe and Asia, this fly
can be found in meadows and
hedgerows, where it feeds on
nectar and other insects.

⅜–½ in
1–1.2 cm

CADDISFLIES

Closely related to the Lepidoptera, these slender,
rather mothlike insects of the order Trichoptera
are covered with hairs, not scales. The head has
long, threadlike antennae and weakly developed
mouthparts. The two pairs of wings are held tent-
like over the body at rest. Caddisfly larvae are aquatic
and often construct species-specific portable
shelters out of small stones or plant fragments.

SALT AND PEPPER MICROCADDIS
Agraylea multipunctata
F: Hydroptilidae
This small caddisfly is widespread and common
in N. America. It has narrow wings and
breeds in algae-rich ponds and lakes.

⅛–⅕ in
3–4.5 mm

GREAT RED SEDGE
Phryganea grandis
F: Phryganeidae
This European caddisfly breeds in weedy
lakes and slow rivers. The larvae make a case
from cut leaf sections arranged in a spiral.

1⅛–1¼ in
2.8–3.2 cm

DARK-SPOTTED SEDGE
Philopotamus montanus
F: Philopotamidae
This European caddisfly has shortish
antennae and breeds in fast-flowing rocky
streams. The larvae make tubelike nets on
the undersides of rocks.

¼–½ in
8–13 mm

MARBLED SEDGE
Hydropsyche contubernalis
F: Hydropsychidae
This caddisfly flies in the evening and
breeds in rivers and streams. The aquatic
larvae weave a net to catch food.

½–⅗ in
1.2–1.5 cm

MOTTLED SEDGE
Glyphotaelius pellucidus
F: Limnephilidae
This European caddisfly breeds in
lakes and small ponds. The larvae
make a case out of fragments of
withered tree leaves.

½–⅗ in
1.2–1.7 cm

MOTHS AND BUTTERFLIES

Members of the order Lepidoptera are covered with minute scales. They have large compound eyes and a feeding tube called a proboscis. Larvae, known as caterpillars, pupate, metamorphosing into their adult form. Most moths are nocturnal and rest with their wings open; the majority of butterflies are diurnal and rest with their wings closed.

HERCULES MOTH
Coscinocera hercules
F: Saturniidae
Found in New Guinea and Australia, this moth is one of the largest in the world. Only the male has the long hindwing tails.

8–10½ in
20–27 cm

hindwing tail

4–6½ in
10–16 cm

AMERICAN MOON MOTH
Actias luna
F: Saturniidae
This lime-green N. American moth has distinctive long hindwing tails. The caterpillars feed on a range of deciduous trees.

4–6½ in
10–16 cm

POLYPHEMUS MOTH
Antheraea polyphemus
F: Saturniidae
Common and widespread in the US and S. Canada, this moth has large eyespots on the wings, to startle predators.

GIANT LEOPARD MOTH
Hypercompe scribonia
F: Erebidae
The distribution of this striking moth extends from S.E. Canada as far south as Mexico. The caterpillars feed on a variety of plants.

2¼–3½ in
6–9 cm

OAK EGGAR
Lasiocampa quercus
F: Lasiocampidae
Ranging from Europe to N. Africa, the caterpillars of this moth feed on the foliage of bramble, oak, heather, and other plants.

1¾–3 in
4.5–7.5 cm

LAPPET MOTH
Gastropacha quercifolia
F: Lasiocampidae
Found in Europe and Asia, the specific name of this large moth refers to its resting appearance, which resembles a cluster of dead oak leaves.

2–3¼ in
5–8.5 cm

PINE-TREE LAPPET
Dendrolimus pini
F: Lasiocampidae
This moth is widespread in coniferous forests across Europe and Asia, where the caterpillars feed on pine, spruce, and fir.

2–3¼ in
5–8 cm

1½–1¾ in
3.5–4.5 cm

AUSTRALIAN MAGPIE MOTH
Nyctemera amicus
F: Erebidae
Widespread in Australia and also found in New Zealand, the caterpillars of this day-flying moth feed on plants such as groundsel and ragwort.

white chevrons on forewing

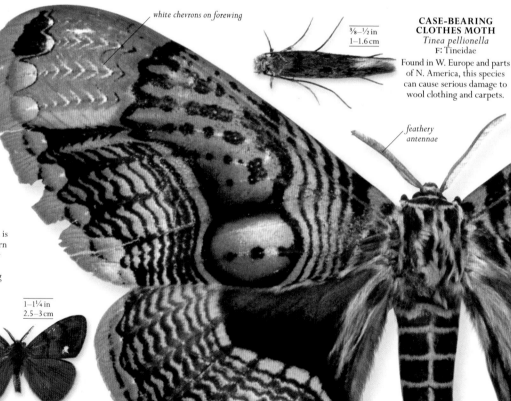

⅜–½ in
1–1.6 cm

CASE-BEARING CLOTHES MOTH
Tinea pellionella
F: Tineidae
Found in W. Europe and parts of N. America, this species can cause serious damage to wool clothing and carpets.

feathery antennae

1¾–2¾ in
4.5–7 cm

GARDEN TIGER
Arctia caja
F: Erebidae
This unmistakable moth is found across the northern hemisphere. The hairy caterpillars feed on a range of low-growing plants and shrubs.

2¾–3¼ in
7–8 cm

ILIA UNDERWING
Catocala ilia
F: Erebidae
This widespread species from N. America has distinctive, red-banded hindwings. The caterpillars feed on the foliage of oak trees.

1–1¼ in
2.5–3 cm

VAPOURER MOTH
Orgyia antiqua
F: Erebidae
Now found across the northern hemisphere, the female of this European species has tiny wings and is unable to fly.

abdomen

GIANT AGRIPPA
Thysania agrippina
F: Noctuidae

9½–12 in
24–30 cm

Found in C. and parts of S. America, this species has one of the largest wingspans of any moth in the world.

SNOUT MOTH
Vitessa suradeva
F: Pyralidae

This moth occurs in India, parts of S.E. Asia, and New Guinea. The caterpillars feed in a web on young leaves of poisonous shrubs.

1½–2 in
4–5 cm

SMALL MAGPIE
Anania hortulata
F: Pyralidae

1–1¼ in
2.4–2.8 cm

Found in hedgerows and waste ground, the caterpillars of this very common European species feed on rolled nettle leaves.

SILVER-SPOTTED GHOST MOTH
Sthenopis argenteomaculatus
F: Hepialidae

2¼–4 in
6–10 cm

This species occurs in S. Canada and parts of the US. The caterpillars mainly feed inside the roots of alder trees.

ACACIA CARPENTER MOTH
Endoxyla encalypti
F: Cossidae

The robust, white caterpillars of this large and distinctive Australian moth burrow into the woody tissue of certain species of the acacia tree.

3½–4¾ in
9–12 cm

2½–3¾ in
6.5–9.5 cm

DIVA MOTH
Divana diva
F: Castniidae

Found in tropical forests in S. America, this day-flying species is well camouflaged at rest despite its brightly colored hindwings.

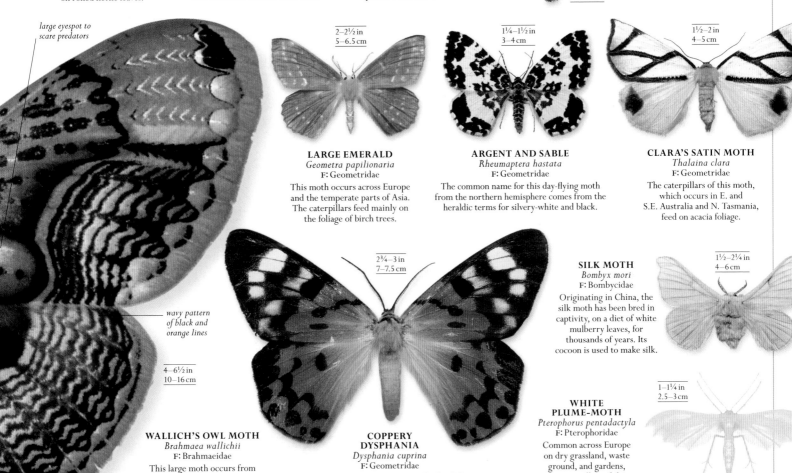

large eyespot to scare predators

2–2½ in
5–6.5 cm

LARGE EMERALD
Geometra papilionaria
F: Geometridae

This moth occurs across Europe and the temperate parts of Asia. The caterpillars feed mainly on the foliage of birch trees.

1¼–1½ in
3–4 cm

ARGENT AND SABLE
Rheumaptera hastata
F: Geometridae

The common name for this day-flying moth from the northern hemisphere comes from the heraldic terms for silvery-white and black.

1½–2 in
4–5 cm

CLARA'S SATIN MOTH
Thalaina clara
F: Geometridae

The caterpillars of this moth, which occurs in E. and S.E. Australia and N. Tasmania, feed on acacia foliage.

wavy pattern of black and orange lines

4–6½ in
10–16 cm

2¾–3 in
7–7.5 cm

WALLICH'S OWL MOTH
Brahmaea wallichii
F: Brahmaeidae

This large moth occurs from N. India to China and Japan. The caterpillars eat the foliage of ash, privet, and lilac.

COPPERY DYSPHANIA
Dysphania cuprina
F: Geometridae

Widespread across S.E. Asia, this brightly colored moth flies during the day and is thought to be unpalatable to birds.

SILK MOTH
Bombyx mori
F: Bombycidae

1½–2¼ in
4–6 cm

Originating in China, the silk moth has been bred in captivity, on a diet of white mulberry leaves, for thousands of years. Its cocoon is used to make silk.

WHITE PLUME-MOTH
Pterophorus pentadactyla
F: Pterophoridae

1–1¼ in
2.5–3 cm

Common across Europe on dry grassland, waste ground, and gardens, the caterpillars of this distinctive species feed on hedge bindweed.

long legs

2–2¼ in
5–6 cm

REGENT SKIPPER
Euschemon rafflesia
F: Hesperiidae
This brightly colored species is
native to tropical and subtropical
forests in E. Australia, where
it can be seen feeding at flowers.

1¾–2½ in
4.5–6.2 cm

GUAVA SKIPPER
Phocides polybius
F: Hesperiidae
This species can be found from
S. Texas to as far south as Argentina.
The caterpillars feed inside
rolled-up guava leaves.

1½–2 in
3.5–5 cm

HORNET MOTH
Sesia apiformis
F: Sesiidae
To deter predators, the adults of this
moth look very similar to hornets but are
harmless. The caterpillars bore into the
trunk and roots of poplars and willows in
Europe and the Middle East.

2¼–2½ in
5.5–6.5 cm

BUFF-TIP
Phalera bucephala
F: Notodontidae
This moth is found in Europe
and eastward to Siberia. At rest
its folded wings wrap around its
body for camouflage, closely
resembling a broken-off twig.

2¼–2¾ in
6–7 cm

ELEPHANT HAWK MOTH
Deilephila elpenor
F: Sphingidae
This pretty, pinkish hawk moth is
widespread in the temperate parts
of Europe and Asia. The caterpillars
feed on bedstraw and willowherb.

3½–4¾ in
9–12 cm

VERDANT SPHINX
Euchloron megaera
F: Sphingidae
This distinctive moth is widespread
throughout sub-Saharan Africa. The
caterpillars feed on the foliage of
creepers of the grape family.

COMMON MORPHO
Morpho peleides
F: Nymphalidae
This butterfly is widespread
in tropical forests in
C. and S. America.
The adults feed on
the juices of
rotting fruits.

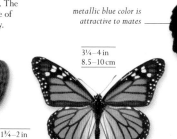

3¾–6 in
9.5–15 cm

*metallic blue color is
attractive to mates*

1¼–1½ in
3–4 cm

SIX-SPOT BURNET
Zygaena filipendulae
F: Zygaenidae
Brightly colored and distasteful
to birds, this day-flying moth can
be found in meadows and woodland
clearings across Europe.

1¾–2 in
4.5–5 cm

LITTLE WOOD SATYR
Euptychia cymela
F: Nymphalidae
This woodland butterfly is found
from S. Canada to N. Mexico. The
caterpillars feed on grasses in
clearings close to water.

3¼–4 in
8.5–10 cm

MONARCH BUTTERFLY
Danaus plexippus
F: Nymphalidae
A well-known migrant, this butterfly has
spread from the Americas to many other
parts of the world. The caterpillars feed
on the milkweed plant.

2–2¼ in
5–6 cm

QUEEN CRACKER
Hamadryas arethusa
F: Nymphalidae
The common name for this
butterfly, found in forests
from Mexico to Bolivia,
refers to the clicking noise
it makes when it flies.

WHITE ADMIRAL
Ladoga camilla
F: Nymphalidae
This butterfly can be found
across temperate Europe and
Asia to Japan. The caterpillars
feed on the foliage of the
honeysuckle.

hindwing tail

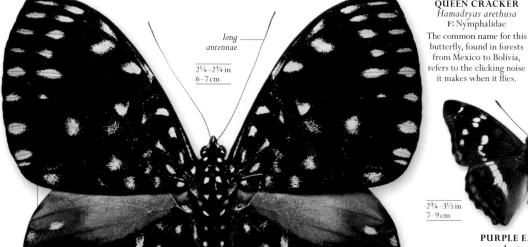

*long
antennae*

2¼–2¾ in
6–7 cm

2¾–3½ in
7–9 cm

PURPLE EMPEROR
Apatura iris
F: Nymphalidae
Found in dense oak woodland throughout
Europe and Asia as far east as Japan, the
male of this butterfly is iridescent purple
whereas the female is dull brown.

3½–4¾ in
9–12 cm

INDIAN LEAF BUTTERFLY
Kallima inachus
F: Nymphalidae
With undersides that resemble a
brown leaf, this butterfly is perfectly
camouflaged at rest with its wings
folded. It occurs from India to S. China.

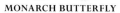

SMALL WHITE
Pieris rapae
F: Pieridae

This butterfly can now
be found worldwide. The
caterpillars feed on wild and
cultivated cabbages and
mustards, and can be pests.

1½–2 in
3.5–5 cm

pointed front
wing

3–3¾ in
7.5–9.5 cm

1½–2½ in
4–6.5 cm

CALIFORNIA DOG-FACE
Zerene eurydice
F: Pieridae

This butterfly is restricted to parts
of California and, sometimes,
western Arizona. It breeds on
a shrub called Napa False Indigo.

1½–2 in
4–5 cm

ORANGE TIP
Anthocharis cardamines
F: Pieridae

This butterfly can be found
in meadows across temperate
Europe and Asia as far as Japan.
The caterpillars feed on lady's
smock and garlic mustard.

2¼–3¼ in
7–8.5 cm

ORANGE-BARRED
SULPHUR
Phoebis philea
F: Pieridae

The distribution of this
butterfly extends from S. Brazil
to C. America, S. US, and
sporadically farther north. The
caterpillars feed on sennas.

MADAGASCAN SUNSET MOTH
Chrysiridia rhipheus
F: Uraniidae

This colorful, day-flying moth,
with iridescent wing scales, is endemic
to Madagascar. It feeds on certain
poisonous shrubs of the spurge family.

iridescent red
wing marking

dark spots on
hindwing

2–2¾ in
5–7 cm

TIGER
PIERID
Dismorphia amphione
F: Pieridae

This colorful butterfly, which
mimics unpalatable butterfly species
for protection from predators, is
widespread and common from
Mexico southward to S. America.

2¼–3¼ in
6–8 cm

SMALL POSTMAN
Heliconius erato
F: Nymphalidae

This butterfly is common along
forest edges and open ground
from C. America to S. Brazil. The
caterpillars eat passionflower foliage.

4–6 in
10–15 cm

UNDERSIDE

OWL BUTTERFLY
Caligo idomeneus
F: Nymphalidae

The underside of this large species, native to
S. America, has prominent owllike eyespots
to deter predators when the butterfly is at rest.

2¼–3 in
5.5–7.5 cm

BLACK-VEINED
WHITE
Aporia crataegi
F: Pieridae

Occurring across Europe, N. Africa,
and Asia as far as Japan, this
distinctive butterfly breeds on
hawthorn and blackthorn.

2–2¾ in
5–7 cm

CLEOPATRA
Gonepteryx cleopatra
F: Pieridae

This butterfly is common in
countries around the Mediterranean,
particularly in lightly wooded coastal
areas and scrubland. The caterpillars
feed on species of buckthorn.

» MOTHS AND BUTTERFLIES

2⅖–3¼ in
7–8.5 cm

BIG GREASY BUTTERFLY
Cressida cressida
F: Papilionidae
This species is found in grassland and drier forests in Australia and Papua New Guinea, where its pipevine foodplants grow.

2¼–4 in
6–10 cm

ZEBRA SWALLOWTAIL
Photographium marcellus
F: Papilionidae
Found in damp woodlands in eastern N. America, this butterfly has distinctive black and white markings. The caterpillars feed on pawpaw.

10–12 in
25–31 cm

QUEEN ALEXANDRA'S BIRDWING
Ornithoptera alexandrae
F: Papilionidae
Found only in S.E. Papua New Guinea east of the Owen Stanley Ranges, this endangered species, the largest butterfly in the world, is now protected.

♂

6–7 in
15–18 cm

♂

RAJAH BROOKE'S BIRDWING
Troides brookiana
F: Papilionidae
This butterfly lives in the tropical forests of Borneo and Malaysia, where the adults sip fruit juices and nectar. The caterpillars feed on pipevines.

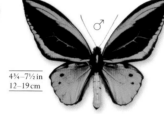

♂

4¾–7½ in
12–19 cm

COMMON GREEN BIRDWING
Ornithoptera priamus
F: Papilionidae
This large butterfly occurs from Papua New Guinea and the Solomon Islands to tropical N. Australia. The caterpillars eat pipevines.

3–3½ in
7.5–9 cm

SWALLOWTAIL
Papilio machaon
F: Papilionidae
This species is found in wet meadow, fenland, and other habitats across the northern hemisphere. The caterpillars feed on various umbellifers.

1½–2¼ in
4–5.5 cm

GREEN DRAGONTAIL
Lamproptera meges
F: Papilionidae
This unmistakable butterfly, which hovers while it feeds at flowers, can be found from India to China and through S. and S.E. Asia.

2¼–3½ in
6–9 cm

APOLLO
Parnassius apollo
F: Papilionidae
This butterfly can be found in flower-rich meadows in mountainous regions of Europe and Asia. The caterpillars feed on stonecrop.

red spot

short, robust antenna

1¾–2 in
4.5–5 cm

BLUE TRIANGLE
Graphium sarpedon
F: Papilionidae
Common and widespread from India to China, Papua New Guinea, and Australia, this butterfly feeds on nectar and drinks from puddles.

zig-zag pattern on hindwing

2¾–3¼ in
7–8 cm

SCARCE SWALLOWTAIL
Iphiclides podalirius
F: Papilionidae
Despite its name, this species is quite widespread in Europe and across temperate Asia to China. The caterpillars feed on blackthorn.

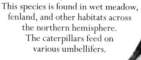

3¼–3½ in
8–9 cm

SPANISH FESTOON
Zerynthia rumina
F: Papilionidae
This species lives in scrub, meadows, and rocky hillsides in S.E. France, Spain, Portugal, and parts of N. Africa. The caterpillars feed on species of birthwort.

TIGER SWALLOWTAIL
Papilio glaucus
F: Papilionidae

3¼–5½ in
8–14 cm

This species is widespread in N. America. The young caterpillars, which resemble bird droppings, feed on a variety of trees and shrubs.

SMALL COPPER
Lycaena phlaeas
F: Lycaenidae

Common in Europe, N. Africa, and Asia as far as Japan, this species can also be found in N. America. The caterpillars feed on sorrels and docks.

1–1¼ in
2.5–3 cm

BLUE THAROPS
Menander menander
F: Lycaenidae

This fast-flying butterfly occurs in tropical forests from Panama to northern S. America. Little is known about its life cycle or caterpillars.

1¼–1½ in
3–4 cm

BROWN HAIRSTREAK
Thecla betulae
F: Lycaenidae

1½–1¾ in
3.5–4.5 cm

This species can be found in hedgerows, scrubland, and woodland across Europe and temperate Asia. The caterpillars feed at night on blackthorn.

ADONIS BLUE
Lysandra bellargus
F: Lycaenidae

1–1½ in
2.5–3.5 cm

This European species can be found in chalky grassland, where the caterpillars feed on horseshoe vetch. The male is sky blue while the female is brown.

SONORAN BLUE
Philotes sonorensis
F: Lycaenidae

¾–1 in
2–2.5 cm

This rare butterfly is restricted to rocky washes and desert cliffs in California. The caterpillars feed on succulents known as stonecrop or "live-forever" plants.

DUKE OF BURGUNDY FRITILLARY
Hamearis lucina
F: Lycaenidae

1¼–1½ in
3–4 cm

This species ranges across C. Europe as far as the Urals and favors flower meadows, where its food plants, cowslip and primrose, grow.

SAWFLIES, WASPS, BEES, AND ANTS

Hymenoptera typically have two pairs of wings, joined in flight by tiny hooks. Apart from sawflies, they have a "wasp-waist" and females have an ovipositor that may be modified to sting. Many wasps are predatory or parasitic. Bees are vital pollinators and ants play a role in many ecosystems.

STEM SAWFLY
Cephus nigrinus
F: Cephidae

7/32–11/32 in
7–9 mm

This slender, all-black sawfly is widespread in W. Europe. The larvae bore downward inside the stems of wild and some cultivated grasses.

fine tufts on wing margin

7/32–11/32 in
7–9 mm

TENTHREDO ARCUATA
F: Tenthredinidae

Widespread in meadows, this distinctive black- and yellow-marked European sawfly lays its eggs on clovers.

¾–9/10 in
1.8–2.2 cm

CIMBICID SAWFLY
Trichiosoma lucorum
F: Cimbicidae

This stout-bodied European sawfly is found in woodland, hedgerows, and scrubby areas. The larvae feed on birch and willow.

PERGID SAWFLY
Unknown sp.
F: Pergidae

These herbivorous insects are found in Australia and S. America. Many species attack eucalyptus trees, and young larvae feed gregariously.

¾ in
2 cm

ROSE SAWFLY
Arge ochropus
F: Argidae

Recognizable by the black along the wing edge, this European sawfly visits flat-topped flowers to feed. The larvae feed on wild roses.

7/32–⅜ in
7–10 mm

RED-HEADED SAWFLY
Acantholyda erythrocephala
F: Pamphiliidae

7/32–11/32 in
7–9 mm

Originally found in Europe and Asia, this sawfly has spread to North America. The larvae feed communally under silken webs among pine needles.

slender antennae

¾–1½ in
2–4 cm

large head

HORNTAIL
Urocerus gigas
F: Siricidae

This impressive species can be found across the northern hemisphere. The female drills deep into softwood trees to lay her eggs.

ovipositor

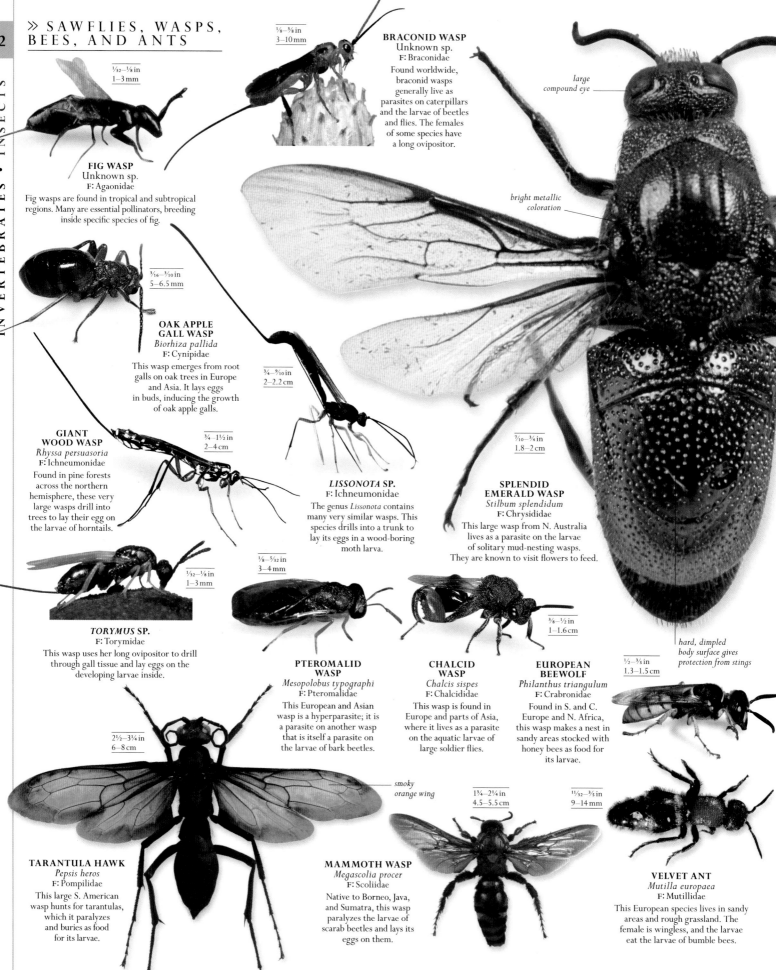

1/32–1/8 in
1–3 mm

FIG WASP
Unknown sp.
F: Agaonidae
Fig wasps are found in tropical and subtropical regions. Many are essential pollinators, breeding inside specific species of fig.

1/8–3/8 in
3–10 mm

BRACONID WASP
Unknown sp.
F: Braconidae
Found worldwide, braconid wasps generally live as parasites on caterpillars and the larvae of beetles and flies. The females of some species have a long ovipositor.

large compound eye

bright metallic coloration

3/16–3/10 in
5–6.5 mm

OAK APPLE GALL WASP
Biorhiza pallida
F: Cynipidae
This wasp emerges from root galls on oak trees in Europe and Asia. It lays eggs in buds, inducing the growth of oak apple galls.

3/4–9/10 in
2–2.2 cm

GIANT WOOD WASP
Rhyssa persuasoria
F: Ichneumonidae
Found in pine forests across the northern hemisphere, these very large wasps drill into trees to lay their egg on the larvae of horntails.

3/4–1 1/2 in
2–4 cm

LISSONOTA SP.
F: Ichneumonidae
The genus *Lissonota* contains many very similar wasps. This species drills into a trunk to lay its eggs in a wood-boring moth larva.

7/10–3/4 in
1.8–2 cm

SPLENDID EMERALD WASP
Stilbum splendidum
F: Chrysididae
This large wasp from N. Australia lives as a parasite on the larvae of solitary mud-nesting wasps. They are known to visit flowers to feed.

1/32–1/8 in
1–3 mm

TORYMUS SP.
F: Torymidae
This wasp uses her long ovipositor to drill through gall tissue and lay eggs on the developing larvae inside.

1/8–5/32 in
3–4 mm

PTEROMALID WASP
Mesopolobus typographi
F: Pteromalidae
This European and Asian wasp is a hyperparasite; it is a parasite on another wasp that is itself a parasite on the larvae of bark beetles.

CHALCID WASP
Chalcis sispes
F: Chalcididae
This wasp is found in Europe and parts of Asia, where it lives as a parasite on the aquatic larvae of large soldier flies.

3/8–1/2 in
1–1.6 cm

EUROPEAN BEEWOLF
Philanthus triangulum
F: Crabronidae
Found in S. and C. Europe and N. Africa, this wasp makes a nest in sandy areas stocked with honey bees as food for its larvae.

hard, dimpled body surface gives protection from stings

1/2–3/5 in
1.3–1.5 cm

2 1/2–3 1/4 in
6–8 cm

smoky orange wing

TARANTULA HAWK
Pepsis heros
F: Pompilidae
This large S. American wasp hunts for tarantulas, which it paralyzes and buries as food for its larvae.

1 3/4–2 1/4 in
4.5–5.5 cm

MAMMOTH WASP
Megascolia procer
F: Scoliidae
Native to Borneo, Java, and Sumatra, this wasp paralyzes the larvae of scarab beetles and lays its eggs on them.

11/32–3/5 in
9–14 mm

VELVET ANT
Mutilla europaea
F: Mutillidae
This European species lives in sandy areas and rough grassland. The female is wingless, and the larvae eat the larvae of bumble bees.

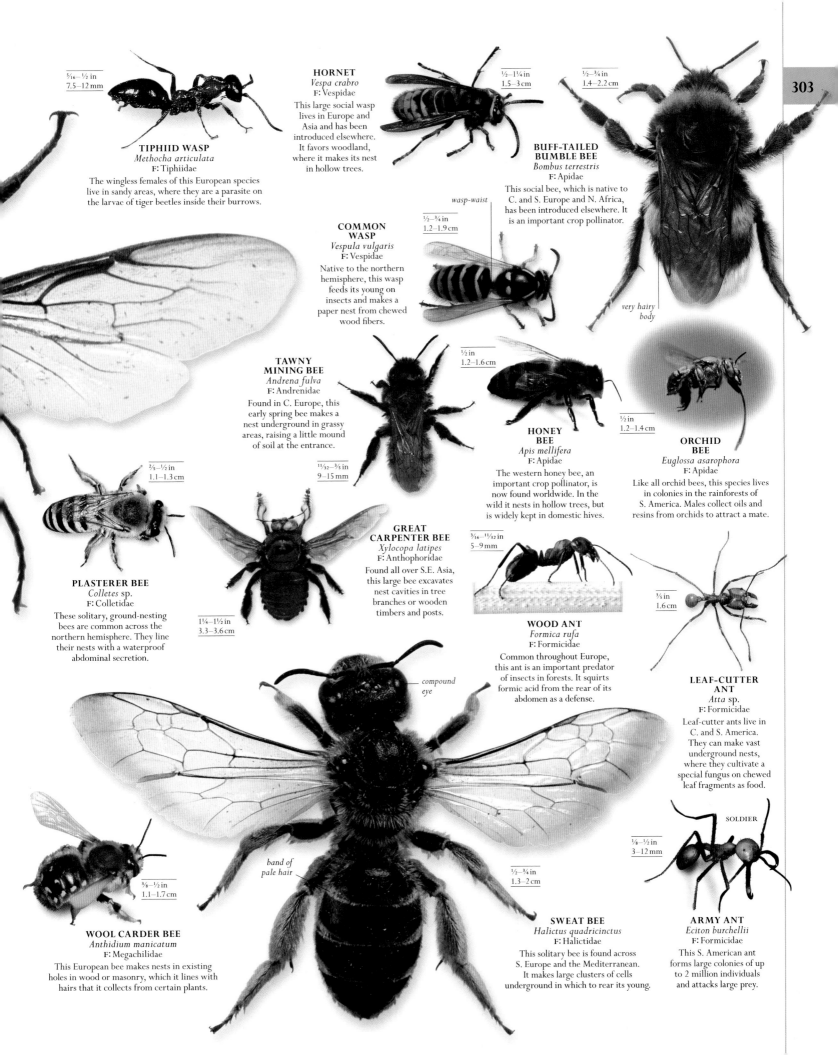

TIPHIID WASP
Methocha articulata
F: Tiphiidae
The wingless females of this European species live in sandy areas, where they are a parasite on the larvae of tiger beetles inside their burrows.

5/16–1/2 in
7.5–12 mm

HORNET
Vespa crabro
F: Vespidae
This large social wasp lives in Europe and Asia and has been introduced elsewhere. It favors woodland, where it makes its nest in hollow trees.

1/2–1 1/4 in
1.5–3 cm

BUFF-TAILED BUMBLE BEE
Bombus terrestris
F: Apidae
This social bee, which is native to C. and S. Europe and N. Africa, has been introduced elsewhere. It is an important crop pollinator.

1/2–3/4 in
1.4–2.2 cm

very hairy body

COMMON WASP
Vespula vulgaris
F: Vespidae
Native to the northern hemisphere, this wasp feeds its young on insects and makes a paper nest from chewed wood fibers.

wasp-waist

1/2–3/4 in
1.2–1.9 cm

TAWNY MINING BEE
Andrena fulva
F: Andrenidae
Found in C. Europe, this early spring bee makes a nest underground in grassy areas, raising a little mound of soil at the entrance.

HONEY BEE
Apis mellifera
F: Apidae
The western honey bee, an important crop pollinator, is now found worldwide. In the wild it nests in hollow trees, but is widely kept in domestic hives.

1/2 in
1.2–1.6 cm

ORCHID BEE
Euglossa asarophora
F: Apidae
Like all orchid bees, this species lives in colonies in the rainforests of S. America. Males collect oils and resins from orchids to attract a mate.

1/2 in
1.2–1.4 cm

PLASTERER BEE
Colletes sp.
F: Colletidae
These solitary, ground-nesting bees are common across the northern hemisphere. They line their nests with a waterproof abdominal secretion.

2/5–1/2 in
1.1–1.3 cm

GREAT CARPENTER BEE
Xylocopa latipes
F: Anthophoridae
Found all over S.E. Asia, this large bee excavates nest cavities in tree branches or wooden timbers and posts.

11/32–3/5 in
9–15 mm

1 1/4–1 1/2 in
3.3–3.6 cm

WOOD ANT
Formica rufa
F: Formicidae
Common throughout Europe, this ant is an important predator of insects in forests. It squirts formic acid from the rear of its abdomen as a defense.

3/16–11/32 in
5–9 mm

LEAF-CUTTER ANT
Atta sp.
F: Formicidae
Leaf-cutter ants live in C. and S. America. They can make vast underground nests, where they cultivate a special fungus on chewed leaf fragments as food.

3/5 in
1.6 cm

compound eye

band of pale hair

WOOL CARDER BEE
Anthidium manicatum
F: Megachilidae
This European bee makes nests in existing holes in wood or masonry, which it lines with hairs that it collects from certain plants.

3/8–1/2 in
1.1–1.7 cm

SOLDIER

1/8–1/2 in
3–12 mm

SWEAT BEE
Halictus quadricinctus
F: Halictidae
This solitary bee is found across S. Europe and the Mediterranean. It makes large clusters of cells underground in which to rear its young.

1/2–3/4 in
1.3–2 cm

ARMY ANT
Eciton burchellii
F: Formicidae
This S. American ant forms large colonies of up to 2 million individuals and attacks large prey.

RIBBON WORMS

Marine worms of the phylum Nemertea are voracious predators, seizing their prey with a proboscis and swallowing it whole or sucking its fluids.

Ribbon worms are soft slimy animals, cylindrical in shape or somewhat flattened. Many simply wriggle across the seabed, but some can also swim. One species, *Lineus longissimus*, the bootlace worm, has been recorded growing to 100 ft (30 m) in length. Like others, it has a fragile body that is easily broken.

The proboscis of a ribbon worm is not part of its gut, but emerges from a sac on its head just above the mouth. In some species the proboscis has sharp spikes for gripping or piercing prey, and even venom for immobilizing it. These well-armed species prey on other invertebrates, such as crustaceans, polychaete worms, and mollusks. Other ribbon worms feed on dead material.

PHYLUM	NEMERTEA
CLASSES	4
ORDERS	4
FAMILIES	48
SPECIES	1,350

BRYOZOANS

These minute animals form corallike colonies, filter-feeding with microscopic tentacles and, in turn, providing food for grazing invertebrates.

Many bryozoans resemble coral, but they are more advanced animals than cnidarians. A colony— genetically a single individual— is made up of thousands of tiny bodies called zooids, each consisting of a retractable fan of tentacles surrounding the mouth. Microscopic, beating hairs on the tentacles waft particles of food down to the mouth, which leads into a U-shaped gut. Waste exits by the anus in the body wall. Although a magnifying glass is needed to see individual zooids, colonies assume a variety of forms, depending upon species. Some are encrustations on rock or seaweed, others grow into bristly thickets or fleshy lobes. Many are heavily calcified like stony corals; others are soft.

PHYLUM	BRYOZOA
CLASSES	3
ORDERS	7
FAMILIES	About 160
SPECIES	6,409

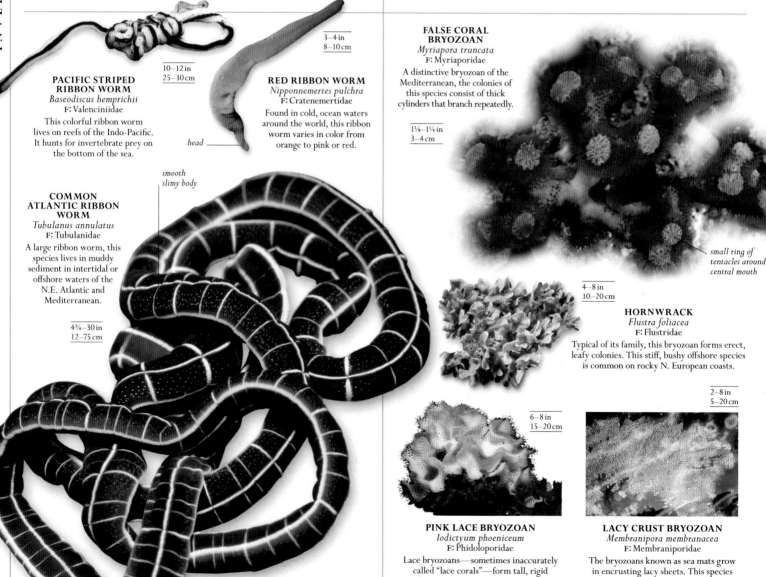

PACIFIC STRIPED RIBBON WORM
Baseodiscus hemprichii
F: Valenciniidae
This colorful ribbon worm lives on reefs of the Indo-Pacific. It hunts for invertebrate prey on the bottom of the sea.

10–12 in
25–30 cm

3–4 in
8–10 cm

RED RIBBON WORM
Nipponnemertes pulchra
F: Cratenemertidae
Found in cold, ocean waters around the world, this ribbon worm varies in color from orange to pink or red.

head

smooth slimy body

COMMON ATLANTIC RIBBON WORM
Tubulanus annulatus
F: Tubulanidae
A large ribbon worm, this species lives in muddy sediment in intertidal or offshore waters of the N.E. Atlantic and Mediterranean.

4¾–30 in
12–75 cm

FALSE CORAL BRYOZOAN
Myriapora truncata
F: Myriaporidae
A distinctive bryozoan of the Mediterranean, the colonies of this species consist of thick cylinders that branch repeatedly.

1⅛–1¼ in
3–4 cm

small ring of tentacles around central mouth

4–8 in
10–20 cm

HORNWRACK
Flustra foliacea
F: Flustridae
Typical of its family, this bryozoan forms erect, leafy colonies. This stiff, bushy offshore species is common on rocky N. European coasts.

6–8 in
15–20 cm

2–8 in
5–20 cm

PINK LACE BRYOZOAN
Iodictyum phoeniceum
F: Phidoloporidae
Lace bryozoans—sometimes inaccurately called "lace corals"—form tall, rigid colonies. This colorful species lives on coastlines around S. and E. Australia.

LACY CRUST BRYOZOAN
Membranipora membranacea
F: Membraniporidae
The bryozoans known as sea mats grow in encrusting lacy sheets. This species forms fast-growing colonies on kelp fronds of the N.E. Atlantic.

LAMPSHELLS

Lampshells look very like bivalve mollusks, but feed instead with tentacles. They belong to a different, very ancient phylum.

The soft body of a lampshell is enclosed within a two-valved shell that has one valve larger than the other. The shell is either attached to the seabed with a rubbery stalk or is directly cemented to rock. The shell valves are positioned top-and-bottom around the animal—not around the sides, as in bivalve mollusks. Like mollusks, lampshells have a fleshy mantle attached to the inner surface of the shell, which encloses a mantle cavity. In the mantle cavity there is a ring of tentacles with microscopic beating hairs that drive particles to a central mouth—just like the feeding structure in bryozoans.

PHYLUM	BRACHIOPODA
CLASSES	3
ORDERS	5
FAMILIES	About 30
SPECIES	414

The fossil record shows that lampshells were much more common and demonstrated much greater diversity in the warm, shallow seas of the Paleozoic Era. They began to decline dramatically during the time of the dinosaurs—perhaps because bivalve mollusks were more successful.

¾–1¼ in
2–3 cm

EUROPEAN LAMPSHELL
Terebratulina retusa
F: Cancellothyrididae
Found from the N.E. Atlantic to the Mediterranean, this lampshell has pear-shaped valves and attaches to vertical rocks using a short stalk.

1¼–2¼ in
3–5.5 cm

hinged, two-valved shell

PACIFIC LAMPSHELL
Terebratalia transversa
F: Terebrataliidae
An abundant lampshell of the northern Pacific, this short-stalked species has a variable smooth or ribbed shell.

⅜–½ in
1–1.5 cm

INARTICULATED LAMP SHELL
Novocrania anomala
F: Craniidae
A brachiopod of North Atlantic waters, this species cements its shell to rock, and superficially resembles a limpet.

MOLLUSKS »

Mollusks are a large, highly diverse group, ranging from blind filter-feeding bivalves cemented to rock, through voraciously grazing snails and slugs, to lively, intelligent octopus and squid.

A typical mollusk has a soft body carried on a large muscular foot, and a head that senses the world using eyes and tentacles. The viscera (internal organs) are contained in a visceral hump, which is covered by a fleshy mantle. This mantle overhangs the edge of the hump, creating a groove called the mantle cavity, which is used for breathing. In most species, this mantle also secretes substances used to make the shell. Most mollusks feed with a tongue called a radula. Coated in teeth made of chitin, the radula moves forward and backward through the mouth to rasp at food. Bivalves lack a radula. They feed instead by siphoning water through their shell. Most species trap particles of food in the mucus on their gills.

PHYLUM	MOLLUSKA
CLASSES	9
ORDERS	53
FAMILIES	609
SPECIES	71,719

These limpets have grazed the rock around them bare, but they cannot reach the green algae growing on top of their own shell.

WITH AND WITHOUT SHELLS

A shell is not only a refuge from predators: it is also protection from drying out. Some snails even have a trapdoor called an operculum to seal the opening of the shell. The shell is hardened with minerals obtained from the diet and surrounding water. It is coated in tough protein; and smooth on the inside, to allow the body to slide in and out—some groups have a lining of mother-of-pearl. Many shell-less mollusks defend themselves with chemicals that make them distasteful, or even poisonous, and advertise this with striking colors.

Most cephalopods—tentacled mollusks, such as octopus and squid—lack an obvious shell. Members of this group are entirely predatory and have a horny beak for chewing on flesh. Their muscular foot is modified into their characteristic tentacles, which are used for grasping and swimming.

GETTING OXYGEN

Most mollusks are aquatic and so they have gills for breathing, usually projecting into the mantle cavity and irrigated by a current of water. In most land snails and slugs the mantle cavity is filled with air and functions as a lung. Because many freshwater snails probably evolved from this group, they have a lung, too, and must frequently surface to breathe.

APLACOPHORANS

More wormlike than mollusklike, the Aplacophora are small cylindrical burrowers that feed on detritus or other invertebrates in the sediment of deep oceans. They lack a shell, but their clearly molluskan features include a rasping, tonguelike radula. The mantle cavity is reduced to an opening at the rear of the body, into which the animal discharges waste from the gut.

⅛–3¼ in
3 mm–8 cm

COMMON GLISTENWORM
Chaetoderma sp.
F: Chaetodermatidae
This wormlike mollusk is covered by a tough cuticle. It burrows through muddy sediment in the depths of the North Atlantic.

BIVALVES

Highly specialized aquatic mollusks, bivalves are instantly recognizable, with a hinged shell that opens up to access food and oxygen-rich water.

Bivalves are identifed from their shell, which consist of two plates, called valves, joined together by a hinge. To protect against predators and the dangers of drying out, the shell clamps tightly shut by contracting powerful muscles, sealing most or all of the body inside. Seashore species are regularly exposed by receding tides.

Within the class there is a huge variation in size, ranging from the tiny fingernail clam at about ¼ in (6 mm) to the giant clam, which can grow to 4½ ft (1.4 m) across. Some bivalves attach to rocks and hard surfaces by a bundle of tough threads,

called a byssus. Others use a well-developed muscular foot for burrowing into muddy sediment. A few bivalves, such as scallops, swim freely, using a method of jet propulsion.

Water is pumped in and out of the shell through tubes, called siphons. The water supplies oxygen and food, which are collected as the water passes over modified gills. Food particles caught in the sticky mucus around the gills are wafted toward the mouth by microscopic hairs.

BIVALVES AND PEOPLE
Mussels, clams, and oysters are important sources of food, and some oysters create fine-quality pearls by coating a foreign object with layers of nacre. Bivalves are also useful indicators of water quality, as many cannot survive high levels of pollution.

PHYLUM	MOLLUSKA
CLASS	BIVALVIA
ORDERS	19
FAMILIES	105
SPECIES	9,733

A queen scallop claps the two valves of its shell together to accelerate away from the unwelcome attention of a predatory starfish.

OYSTERS AND SCALLOPS

The Ostreoida feed on tiny food particles filtered from the seawater. Many oyster species live permanently submerged in coastal waters, fixed to the rock by the gland that produces byssus thread in other bivalves. Scallops, by contrast, can swim freely by clapping their valves open and shut.

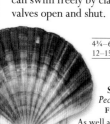

4¾–6 in
12–15 cm

GREAT SCALLOP
Pecten maximus
F: Pectinidae

As well as swimming freely, scallops have jet-propelled escape responses. This commercial species inhabits fine sands of European coasts.

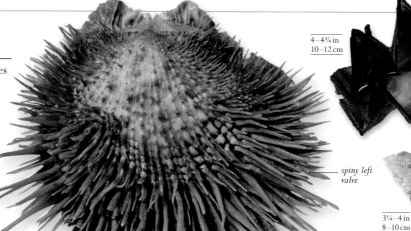

spiny left valve

4–4¾ in
10–12 cm

CAT'S TONGUE OYSTER
Spondylus linguafelis
F: Spondylidae
One of a family of thorny oysters with a colorful mantle, the cat's tongue oyster of the Pacific is named for its spiny exterior.

4–4¾ in
10–12 cm

COCK'S COMB OYSTER
Lopha cristagalli
F: Ostreidae
Close relatives of scallops, many species of oyster are prized as food and for their pearls. This is an Indo-Pacific species.

3¼–4 in
8–10 cm

EDIBLE OYSTER
Ostrea edulis
F: Ostreidae
Formerly abundant in Europe, this commercially important species is now overfished in some areas. The species is inedible during summer months when breeding.

ARC CLAMS AND RELATIVES

The shell of these bivalves is closed by two strong muscles, and hinge along a straight line bearing a continuous series of teeth. Like their relatives, the Arcoida have a reduced foot and large gills that help trap food particles.

2–2¾ in
5–7 cm

NOAH'S ARK
Arca noae
F: Arcidae
Arc shells are square-edged, thick-shelled bivalves. Noah's Ark is a byssus-attached intertidal species on rocky coasts of the E. Atlantic.

2–2¼ in
5–6 cm

EUROPEAN BITTERSWEET
Glycymeris glycymeris
F: Glycymerididae
Dog shells are rounded relatives of arc shells. This N.E. Atlantic species, fished in Europe, has a sweet flavor but is tough when overcooked.

MARINE MUSSELS

The Mytiloida have a distinctively shaped, elongate, asymmetrical shell that attaches to rocks by byssus threads. Only one of their two shell-closing muscles is well developed.

3¼–4 in
8–10 cm

COMMON MUSSEL
Mytilus edulis
F: Mytilidae
The most commercially important mussel in Europe, this long-lived species occurs in dense beds. It can tolerate the low salinity of estuarine waters.

FAN MUSSELS AND RELATIVES

Members of the Pterioida include long-hinged, winged oysters and T-shaped hammer-oysters as well as fan mussels. The group also contains commercially important marine pearl oysters.

10–16 in
25–40 cm

FLAG FAN MUSSEL
Atrina vexillum
F: Pinnidae
Fan mussels, such as this species from western European coastlines, have a triangular shell attached by a byssus in soft sediment.

FRESHWATER MUSSELS

The Unionoida are the only exclusively freshwater order of bivalves. The tiny larvae use their valves to clamp onto fishes, forming cysts on the gill fins and feeding on blood or mucus before dropping off as juvenile mussels.

4–6 in
10–15 cm

3½–4 in
9–10 cm

SWAN MUSSEL
Anodonta sp.
F: Unionidae

Bitterling fishes lay eggs in living freshwater mussels, such as this Eurasian species. The mollusk becomes a nursery for fish fry.

FRESHWATER PEARL MUSSEL
Pinctada margaritifera
F: Margaritiferidae

Well-known for its good-quality pearls, this mussel lives buried in sand or gravel beneath fast-flowing rivers in Eurasia and N. America.

TRUE CLAMS AND BORING BIVALVES

Long-siphoned bivalves in the order Myoida typically burrow in mud or bore through wood or rock. In piddocks the front of the shell acts like a file to bore burrows in soft rock. Shipworms use their shell valves to drill through wood.

4¾–6 in
12–15 cm

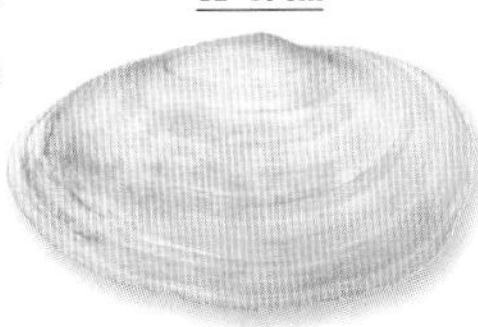

4¾–6 in
12–15 cm

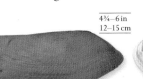

COMMON PIDDOCK
Pholas dactylus
F: Pholadidae

This common piddock of the N.E. Atlantic is phosphorescent and, like other species, lives in burrows bored in wood or clay.

SOFT SHELL CLAM
Mya arenaria
F: Myidae

An edible clam with a thin shell, this N. Atlantic species is especially abundant in muddy estuaries, where it burrows in soft sediment.

SHIPWORM
Teredo navalis
F: Teredinidae

½–¾ in
1.5–2 cm

This highly modified, widespread bivalve uses its ridged shell valves as a drill to bore deep, chalk-lined burrows in wood, damaging ships.

WATERING POTS AND RELATIVES

The order Pholadomyoida includes clams and tropical watering pots. Scarcely recognizable as bivalves, watering pots live encased in chalky tubes, and draw detritus and water into the tube through a perforated plate at the anterior end.

6–6½ in
15–17 cm

PHILIPPINE WATERING POT
Verpa philippinensis
F: Penicillidae

This bizarre Indo-Pacific watering pot belongs to a family named for their wide, perforated ends, fringed with tubules. It lives partially buried in sediment.

COCKLES AND RELATIVES

The largest order of bivalves, the Veneroida include a large range of marine animals. Most have short siphons that are often fused together. Some, notably the cockles, are agile, capable of burrowing or even leaping with their foot. Others are attached to rocks by a byssus.

1¼–1½ in
3–4 cm

ZEBRA MUSSEL
Dreissena polymorpha
F: Dreissenidae

This freshwater mussel is anchored by byssus threads, but can detach and crawl on a slender foot. Native to E. European waters, it has spread elsewhere.

COMMON EDIBLE COCKLE
Cerastoderma edule
F: Cardiidae

Sand-burrowing cockles have a shell with radiating ribs. This N.E. Atlantic species, often found in vast numbers, is fished in N. Europe.

1½–2 in
4–5 cm

1–1¼ in
2.5–3 cm

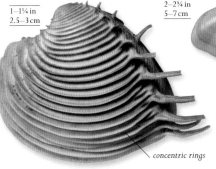

2–2¾ in
5–7 cm

concentric rings

ROYAL COMB VENUS
Hysteroconcha dione
F: Veneridae

Named after the comblike spines on its shell, this Venus clam occurs on tropical American coastlines.

2–3¼ in
5–8 cm

RED CALLISTA
Callista erycina
F: Veneridae

Many Venus clams are prized by collectors, including this Indo-Pacific species, also known as the sunset clam.

algae within mantle exposed to sunlight

GIANT CLAM
Tridacna gigas
F: Cardiidae

The world's largest bivalve, this long-lived and now endangered coral-dwelling animal lives on sandy beds in the Indo-Pacific.

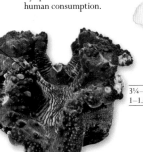

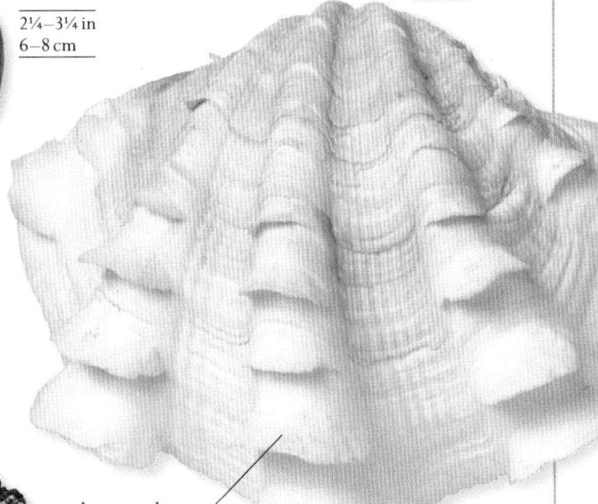

3¼–4½ ft
1–1.4 m

RINGED DOSINIA
Dosinia anus
F: Veneridae

This Venus clam is found on the coastlines of New Zealand, and is one of many species harvested for human consumption.

2¼–3¼ in
6–8 cm

6–8 in
15–20 cm

SWORD RAZOR CLAM
Ensis siliqua
F: Pharidae

Razor shells, such as this N.E. Atlantic species, live in burrows, feeding and respiring through their emergent siphon, but drop quickly using their muscular foot when disturbed.

3¼–3½ in
8–9 cm

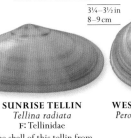

SUNRISE TELLIN
Tellina radiata
F: Tellinidae

The shell of this tellin from the Caribbean has a variable banding pattern and is often found on beaches.

STRIPED TELLIN
Tellinella virgata
F: Tellinidae

This Indo-Pacific species has a decoratively patterned shell, as do many tellins.

¾–1½ in
2–4 cm

CRADLE DONAX
Donax cuneatus
F: Donacidae

One of a family of triangular wedge shells that are fast burrowers in surf, this is a tropical Indo-Pacific species.

2–2¼ in
5–6 cm

OBLONG TRAPEZIUM
Trapezium oblongum
F: Trapeziidae

An Indo-Pacific bivalve, this animal lives attached to rocks by byssus threads, usually in crevices or beneath coral rubble.

WEST AFRICAN TELLIN
Peronaea madagascariensis
F: Tellinidae

Tellins, such as this rose-colored tropical species, feed on particles in sediment, drawn in through their long, extendable siphon.

1¼–1½ in
3–4 cm

large, scale-like flutes

12–16 in
30–40 cm

FLUTED GIANT CLAM
Tridacna squamosa
F: Cardiidae

Like other Indo-Pacific giant clams, this species opens its shell during the day so that algae in its colored mantle can photosynthesize to make food.

GASTROPODS

Gastropods are by far the largest class of mollusks. The word gastropod means "stomach-foot," because the animal appears to crawl on its belly.

Most snails and slugs glide along a stream of slime on a single muscular foot, rasping at their food with a radula, a tonguelike sandpaper. Most gastropods use their radula to rasp at vegetation, algae, or the fine film of microbes that coat underwater rocks, but some are predatory. They usually have a distinct head with well-developed sensory tentacles. Snails have a coiled or helical shell into which they can withdraw; slugs lost their shell during the course of evolution. The class also includes a number of animals that deviate from this basic plan, such as swimming sea slugs. Gastropods originated in the oceans and they still have their greatest diversity there; other species live in fresh water or on land.

TORSION

The juvenile gastropod undergoes a process called torsion: the entire body within the shell twists around 180 degrees, so that the respiratory mantle cavity comes to lie above the animal's head. This allows the vulnerable head to withdraw into the safety of the shell. In marine snails, such as periwinkles and limpets, this body form persists into adulthood. Because of this, marine snails are described as prosobranchs, meaning "forward gills." In sea slugs, however, the body twists back again, and they are called opisthobranchs, or "hind gills."

PHYLUM	MOLLUSKA
CLASS	GASTROPODA
ORDERS	21
FAMILIES	409
SPECIES	About 67,000

TRUE LIMPETS

Primitive algae-grazing limpets, the Patellogastropoda have a slightly coiled, conical shell. Their powerful muscles clamp down firmly on intertidal rocks, protecting them from predators, desiccation, and waves.

1¼–2 in
3–5 cm

COMMON LIMPET
Patella vulgata
F: Patellidae

This species grazes algae from rocks on shores of the N.E. Atlantic at high tide, then returns to a depression in the rock that matches the shape of its shell.

NERITES AND RELATIVES

A small but diverse group, Cycloneritimorpha are well known in the fossil record. They include marine, freshwater, and land-living forms, some with a coiled shell, and a few like limpets. There are also species with a trapdoor to their shell.

ZIG-ZAG NERITE
Neritina communis
F: Neritidae

An inhabitant of Indo-Pacific mangroves, this is a highly variable species, with white, black, red, or yellow shells even in the same population.

½–¾ in
1.2–2 cm

¾–2 in
2–5 cm

BLEEDING TOOTH NERITE
Nerita peloronta
F: Neritidae

Named for the blood-red mark on its shell opening, this Caribbean intertidal mollusk can survive out of water for prolonged periods.

TOP SHELLS AND RELATIVES

The Vetigastropoda are marine gastropods that graze algae and microbes with a brushlike radula. The shells range from those of keyhole limpets, with the barest suggestion of a spire, to the coiled globes and pyramids of top shells, which can seal themselves inside using an operculum, or trapdoor.

3¼–4¾ in
8–12 cm

GIANT TOP SHELL
Rochia nilotica
F: Tegulidae
Many top shells have a thick mother-of-pearl lining. This large Indo-Pacific species is used to make jewelry.

shell shaped like spinning top

banded pattern

LISTER'S KEYHOLE LIMPET
Diodora listeri
F: Fissurellidae
Keyhole limpets are close relatives of abalones and top shells. Like others, this W. Atlantic species ejects oxygen-depleted water through its "keyhole."

2–2¾ in
5–7 cm

½–1¾ in
1.5–4.5 cm

SILVER MOUTH TURBAN SHELL
Turbo argyrostomus
F: Turbinidae
Turban shells are close relatives of top shells, but have a calcified operculum. This is an Indo-Pacific species.

¾–1 in
2–2.5 cm

CHECKERED TOP SHELL
Phorcus turbinatus
F: Trochidae
Top shells have a "spinning-top" shell that is closed with a circular operculum. This is a Mediterranean species.

8–12 in
20–30 cm

RED ABALONE
Haliotis rufescens
F: Haliotidae
Abalones have an ear-shaped shell lined with thick mother-of-pearl and punctuated with holes for exhaling water. This N.E. Pacific kelp-grazing species is the largest.

TOWER SHELLS AND RELATIVES

These tall-spired snails typically live in muddy or sandy sediment. They feed on particles in the water that they circulate through their mantle cavity. The Cerithioidea are found in marine, freshwater, and estuarine habitats. They are slow-moving and often group together in very large numbers.

GREAT SCREW SHELL
Turritella terebra
F: Turritellidae

1–2¼ in
2.5–5.5 cm

A filter feeder of muddy sediments, this Indo-Pacific snail belongs to a group variously known as tower shells or auger shells.

2⅜–6½ in
6–17 cm

ROUGH CERITH
Rhinoclavis aspera
F: Cerithiidae

Ceriths are snails often abundant in shallow-water tropical marine sediments. Like others, this Indo-Pacific species lays strings of eggs attached to solid material.

WEST INDIAN WORM SHELL
Vermicularia spirata
F: Turritellidae

Free-floating males of this Caribbean unwound screw shell attach to solid material, often embedding themselves in sponges. They then develop into larger sedentary females.

1–6½ in
2.5–16 cm

PERIWINKLES, WHELKS, AND RELATIVES

The largest and most diverse order of marine snails, the Caenogastropoda are divided into several groups. Epitoniids, including wentletraps (bottom-living) and violet snails (floating), are specialist predators of cnidarians. Littorinids graze on algae, and include periwinkles, cowries, and conches. Whelks and their relatives are predators that project a long siphon out through a groove in the shell.

PRECIOUS WENTLETRAP
Epitonium scalare
F: Epitoniidae

1–2¾ in
2.5–7 cm

Wentletraps (from the German for spiral staircase) prey on anemones and corals and have cutting jaws. This is an Indo-Pacific species.

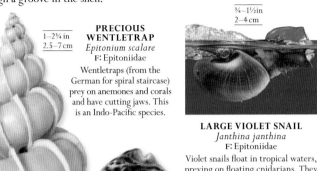

¾–1½ in
2–4 cm

LARGE VIOLET SNAIL
Janthina janthina
F: Epitoniidae

Violet snails float in tropical waters, preying on floating cnidarians. They secrete mucus to create rafts of bubbles that keep them buoyed up.

TIGER COWRIE
Cypraea tigris
F: Cypraeidae

4–6 in
10–15 cm

The fleshy lobes of a cowrie's mantle wrap right around the smooth shell when it crawls. This Indo-Pacific species preys on other invertebrates.

ATLANTIC SLIPPER LIMPET
Crepidula fornicata
F: Calyptraeidae

¾–2 in
2–5 cm

Unrelated to true limpets, filter-feeding slipper limpets form mating towers, with smaller males on top changing sex to replace dead females beneath.

FOOL'S CAP
Capulus ungaricus
F: Capulidae

½–2¼ in
1.5–6 cm

This North Atlantic gastropod resembles unrelated limpets, attaching itself to stones or even the shell of other mollusks, such as scallops.

older spikes often broken

SUNBURST CARRIER
Stellaria solaris
F: Xenophoridae

Carrier snails cement objects such as pebbles or shells of other animals to their shell for camouflage. This species inhabits Indo-Pacific waters.

2¾–5 in
6–13 cm

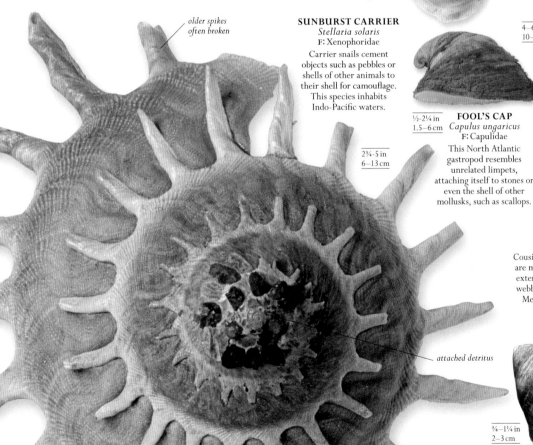

attached detritus

COMMON PELICAN'S FOOT
Aporrhais pespelecani
F: Aporrhaididae

Cousins of conches, pelican's foot snails are mud-dwelling detritus-eaters with extensions to their shell that look like webbed feet. This species inhabits the Mediterranean and the North Sea.

12–16½ in
30–42 cm

COMMON PERIWINKLE
Littorina littorea
F: Littorinidae

¾–1¼ in
2–3 cm

Periwinkles, such as this European species, form a family of intertidal snails with a globular spired shell closed by an operculum.

PINK CONCH
Aliger gigas
F: Strombidae

6–12 in
15–31 cm

Conches are mostly tropical, medium to large marine grazing snails. *Strombus* conches, such as this giant W. Atlantic species, have a flare-lipped shell.

≫

» PERIWINKLES, WHELKS, AND RELATIVES

GIANT FROG SHELL
Tutufa bubo
F: Bursidae

4–12½ in
10–32 cm

Tropical marine frog shells are warty. This Indo-Pacific species eats bristleworms using a proboscis, first anesthetizing them with saliva.

warty shell

canal for siphon

TRUMPET TRITON
Charonia tritonis
F: Charoniidae

4–20 in
10–50 cm

Tropical intertidal tritons and related frog and helmet shells are predators of other invertebrates. This Indo-Pacific species eats the coral-feeding crown-of-thorns starfish.

POLI'S NECKLACE SHELL
Euspira nitida
F: Naticidae

³⁄₁₆–1¼ in
5 mm–3 cm

This is a European member of a family of sand-burrowing predators of bivalves. Necklace shells are named for their necklacelike ribbons of spawn.

COMMON NORTHERN WHELK
Buccinum undatum
F: Buccinidae

3¼–4¼ in
8–11 cm

A large predator of other mollusks and tube worms, this North Atlantic snail also consumes carrion, and is a popular seafood.

PRICKLY PACIFIC DRUPE
Drupa ricinus
F: Muricidae

¾–1¼ in
2–3 cm

A member of the dog whelk family, this snail lives on coral reefs of the Indo-Pacific, where it preys on bristle worms.

DOG WHELK
Nucella lapillus
F: Muricidae

1¼–1½ in
3–4 cm

This North Atlantic snail belongs to a whelk family that catch prey by using the radula to drill holes in the shells of barnacles and mollusks, helped by secretions of enzymes.

TEXTILE CONE
Conus textile
F: Conidae

3½–6 in
9–15 cm

Cone shells use their radula to harpoon their prey and inject venom. Some, such as this Indo-Pacific species, can be harmful to humans.

TENT OLIVE
Oliva porphyria
F: Olividae

1⅛–5 in
3–13 cm

The largest of the olive shells, this species from the Pacific coasts of Mexico and S. America has a colorful, glossy shell.

BANDED TULIP
Cinctura lilium
F: Fasciolariidae

3–4½ in
7.5–11.5 cm

This is a Caribbean snail related to *Buccinum* whelks. Other sand-dwelling species with a longer siphonal canal are called spindle shells.

IMPERIAL HARP
Harpa costata
F: Harpidae

2¼–4¼ in
6–11 cm

Harp snails are sand-living predators of crabs, trapping them with their broad foot and digesting them with saliva. This is an Indian Ocean species.

SUBULATE AUGER
Terebra subulata
F: Terebridae

2¾–8 in
7–20 cm

Typical of many augers, this Indo-Pacific species has a patterned shell. Augers burrow through surface layers of sand to prey on worms.

FRESHWATER GILLED SNAILS

The Architaenioglossa are the only order of gilled snails containing no marine species. Most live in fresh water, but some live on land. Apple snails have gills in the mantle cavity that can function as a lung, allowing them to survive periods of drought. All species have an operculum, or trap door, for closing their shell.

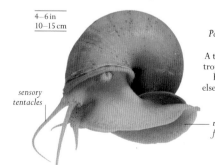

CHANNELED APPLE SNAIL
Pomacea canaliculata
F: Ampullariidae

4–6 in
10–15 cm

A typical apple snail, this tropical American species has been introduced elsewhere and has become an invasive pest.

sensory tentacles

muscular foot

COMMON RIVER SNAIL
Viviparus viviparus
F: Viviparidae

A European freshwater snail with gills, this is a relative of the apple snails and, like them, has a trapdoor (operculum) to close its shell.

1¼–1½ in
3–4 cm

SEA HARES

Sensory water-tasting head stalks, a common
feature of sea slugs, are so big in the large sea hares
that they resemble ears. The Anaspidea have a
small internal shell and, like sea angels, can swim
with their foot flaps.

foot flap

2¾–8 in
7–20 cm

SPOTTED SEA HARE
Aplysia punctata
F: Aplysiidae

Like other seaweed-eating sea hares, this
European species gathers in large breeding groups. As
a defense, it releases ink when disturbed by predators.

SWIMMING SEA SLUGS

The Gymnosomata are also called sea angels, because
they have evolved flaps from their muscular foot and use
them to fly through water. The related sea butterflies or
Thecosomata also have flaps, but retain a fragile shell.

*winglike flaps
for swimming*

1½–2 in
4–5 cm

COMMON SEA ANGEL
Clione limacina
F: Clionidae

Sea angels swim freely with soft, transparent "wings."
This cold-water species preys on a related gastropod
called the Arctic sea butterfly (*Limacina helicina*).

TRUE SEA SLUGS

The Nudibranchia are the largest group of sea slugs.
Nudibranch means "naked gills" and refers to the fact
that these animals' gills are exposed on their back,
rather than enclosed in a mantle cavity. Many species
have vibrant color patterns to advertise their toxicity.

2–3¼ in
5–8 cm

BLACK-MARGINED
SEA SLUG
Doriprismatica atromarginata
F: Chromodorididae

This common Indo-Pacific sea slug lives in
shallow waters, feeding on sponges. It varies
in color from off-white to pale yellow.

1½–2 in
4–5 cm

OPALESCENT SEA SLUG
Hermissenda crassicornis
F: Myrrhinidae

This intertidal northern Pacific species belongs to a
group of sea slugs with body outgrowths that store
the still-potent stinging organs of their jellyfish prey.

4–5 in
10–13 cm

VARICOSE SEA SLUG
Phyllidia varicosa
F: Phyllidiidae

This Indo-Pacific sea slug is a
common inhabitant of rocks,
rubble, and sand in coastal waters,
where it preys on sponges.

¾–2 in
2–5 cm

ANNA'S SEA SLUG
Chromodoris annae
F: Chromodorididae

Like other *Chromodoris* sea slugs,
this variable species from the
western Pacific Ocean specializes
in feeding on sponges.

*external
gills*

*rhinophores
(scent detectors)*

2¾–3¼ in
7–8 cm

head end

ELEGANT SEA SLUG
Okenia elegans
F: Goniodorididae

Typical of its family, this sea slug feeds
on sea squirts. It occurs in waters around
Europe, including the Mediterranean.

*gills obtain
oxygen*

4–4¾ in
10–12 cm

VARIABLE NEON SEA SLUG
Nembrotha kubaryana
F: Polyceridae

This Indo-Pacific sea slug preys on sea squirts and
incorporates their defensive chemicals into its
mucus. Many of its relatives prey on bryozoans.

12–16 in
30–40 cm

mantle

SPANISH DANCER
Hexabranchus sanguineus
F: Hexabranchidae

A giant Indo-Pacific sea slug, this species gets its name
from its free-swimming movements; its red mantle is said
to resemble the ruffled skirt of a flamenco dancer.

COMMON OCTOPUS
Octopus vulgaris

Whether it is extricating a lobster from a lobster pot or hunting a scuttling crab, the common octopus is one of the most intelligent of all the invertebrates. This sea-living mollusk has excellent vision and eight grasping arms, or tentacles, which are also used for crawling. The octopus can change color in an instant and squeeze through very narrow crevices. And its horny beak can bite into prey, whose scattered shell fragments frequently litter the sand around its lair. But in spite of its versatility, the common octopus is short-lived. Upon hatching, an infant octopus enters the oceanic plankton with half a million of its siblings, before descending to the seabed two months later. If it survives being eaten, it takes little more than a year to mature and spawn its own offspring before dying. The common octopus is widespread in tropical and warm temperate oceanic waters, and may consist of several similar species.

SIZE Tentacle span 5–10 ft (1.5–3 m)
HABITAT Rocky coastal waters
DISTRIBUTION Widespread in tropical and warm temperate waters
DIET Crustaceans, shelled mollusks

fleshy mantle encloses a cavity that contains vital organs, including gills for breathing

> INTELLIGENT MOLLUSK
An octopus is an agile predator with an advanced nervous system. Two-thirds of its neurons are located in its arms, which operate with a remarkable degree of independence from its brain. Experiments show that it also has considerable problem-solving intelligence, and excellent long- and short-term memory.

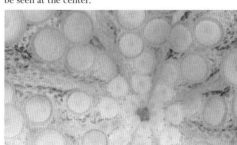

∨ UNDERSIDE
The octopus is a cephalopod, which literally means "head-footed." Its eight mobile arms radiate from the underside of the head, where the mouth can be seen at the center.

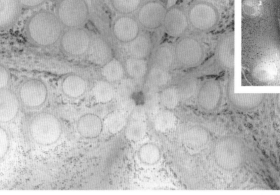

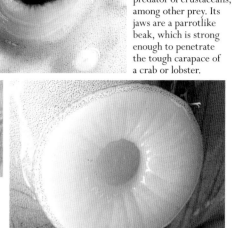

< BEAK
This octopus is a predator of crustaceans, among other prey. Its jaws are a parrotlike beak, which is strong enough to penetrate the tough carapace of a crab or lobster.

< SKIN
The skin contains special cells called chromatophores. These contain pigment which can change the octopus's color—to blend in with its surroundings or to signal its mood if angry or afraid.

∧ FUNNEL, OPEN AND CLOSED
On one side of the mantle, just behind the head, is a funnel, which the octopus uses in three ways: to expel water after gills have extracted oxygen; to rapidly discharge water as jet propulsion for a quick getaway; and to squirt a confusing cloud of ink at the enemy before fleeing.

∧ SUCKER
Each arm has two series of suckers. These give the octopus an excellent grip, allowing it to negotiate the sea bottom and coral reefs with ease. The suckers also have receptors that enable the octopus to taste what it is touching.

prominent eye with a horizontal, slit-shaped pupil

tough, warty skin can change color, texture, and shape for camouflage

long, muscular arms for moving over the seabed and grasping objects

cuplike suckers

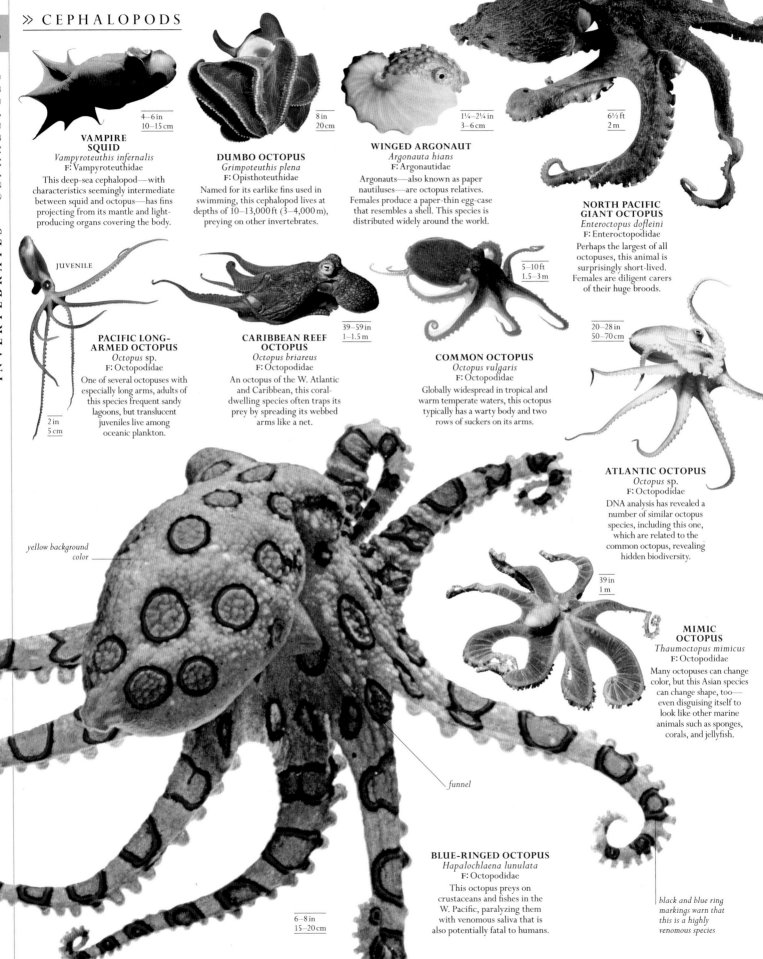

4–6 in
10–15 cm

VAMPIRE SQUID
Vampyroteuthis infernalis
F: Vampyroteuthidae

This deep-sea cephalopod—with characteristics seemingly intermediate between squid and octopus—has fins projecting from its mantle and light-producing organs covering the body.

8 in
20 cm

DUMBO OCTOPUS
Grimpoteuthis plena
F: Opisthoteuthidae

Named for its earlike fins used in swimming, this cephalopod lives at depths of 10–13,000 ft (3–4,000 m), preying on other invertebrates.

1¼–2¼ in
3–6 cm

WINGED ARGONAUT
Argonauta hians
F: Argonautidae

Argonauts—also known as paper nautiluses—are octopus relatives. Females produce a paper-thin egg-case that resembles a shell. This species is distributed widely around the world.

6½ ft
2 m

NORTH PACIFIC GIANT OCTOPUS
Enteroctopus dofleini
F: Enteroctopodidae

Perhaps the largest of all octopuses, this animal is surprisingly short-lived. Females are diligent carers of their huge broods.

JUVENILE

2 in
5 cm

PACIFIC LONG-ARMED OCTOPUS
Octopus sp.
F: Octopodidae

One of several octopuses with especially long arms, adults of this species frequent sandy lagoons, but translucent juveniles live among oceanic plankton.

39–59 in
1–1.5 m

CARIBBEAN REEF OCTOPUS
Octopus briareus
F: Octopodidae

An octopus of the W. Atlantic and Caribbean, this coral-dwelling species often traps its prey by spreading its webbed arms like a net.

5–10 ft
1.5–3 m

COMMON OCTOPUS
Octopus vulgaris
F: Octopodidae

Globally widespread in tropical and warm temperate waters, this octopus typically has a warty body and two rows of suckers on its arms.

20–28 in
50–70 cm

ATLANTIC OCTOPUS
Octopus sp.
F: Octopodidae

DNA analysis has revealed a number of similar octopus species, including this one, which are related to the common octopus, revealing hidden biodiversity.

39 in
1 m

MIMIC OCTOPUS
Thaumoctopus mimicus
F: Octopodidae

Many octopuses can change color, but this Asian species can change shape, too—even disguising itself to look like other marine animals such as sponges, corals, and jellyfish.

yellow background color

funnel

6–8 in
15–20 cm

BLUE-RINGED OCTOPUS
Hapalochlaena lunulata
F: Octopodidae

This octopus preys on crustaceans and fishes in the W. Pacific, paralyzing them with venomous saliva that is also potentially fatal to humans.

black and blue ring markings warn that this is a highly venomous species

CHITONS

Flattened, rock-hugging chitons are among the most primitive of mollusks. Most are grazers of algae and microbe films, and occur in coastal waters.

A chiton's shell, popularly known as a "coat-of-mail" shell, is made up of eight interlocking plates that are sufficiently flexible to allow the animal to bend when gliding over uneven rocks—or even to roll up when disturbed. Chitons have no eyes or tentacles,

although the shell itself contains cells that enable them to react to light. The plates are fringed by the edge of the mantle, known as the girdle. The mantle forms a fleshy skirt around the chiton. This overhangs grooves running down each side of the animal, which channel water, delivering oxygen to gills that project into the grooves. The radula, or rasping tongue, is coated with minute teeth reinforced with iron and silica, so chitons can graze on the toughest encrusting algae.

PHYLUM	MOLLUSKA
CLASS	POLYPLACOPHORA
ORDERS	2
FAMILIES	20
SPECIES	1,026

girdle fringe

eight-plated shell

LINED CHITON
Tonicella lineata
F: Ischnochitonidae
This brightly colored chiton occurs on North Pacific coasts, where it may be camouflaged when feeding on red encrusting algae.

½–2 in
4–5 cm

3¼ in/8 cm

MARBLED CHITON
Chiton marmoratus
F: Chitonidae
Like other chitons, the shell plates of this Caribbean mollusk are composed entirely of a chalky mineral called aragonite.

1½–2 in/4–5 cm

GREEN CHITON
Chiton glaucus
F: Chitonidae
A variable-colored chiton, this species occurs along the coastlines of New Zealand and Tasmania and, like most chitons, is active at night.

¾–3¼ in/2–8 cm

WEST INDIAN FUZZY CHITON
Acanthopleura granulata
F: Chitonidae
This spiky-fringed Caribbean chiton can tolerate exposure to sun, and can live high up in the intertidal zone.

1 in/2.5 cm

DECKED CHITON
Ischnochiton comptus
F: Ischnochitonidae
Ischnochiton chitons have a spiny or scaly girdle fringe. This is a common intertidal species of W. Pacific shores.

12–13 in
30–33 cm

GUMBOOT CHITON
Cryptochiton stelleri
F: Acanthochitonidae
In one chiton family the fleshy girdle overlaps the chain mail plates. In this North Pacific species, the largest chiton, it forms a leathery skin.

2–3¼ in
5–8 cm

BRISTLED CHITON
Mopalia ciliata
F: Mopaliidae
Named for its prominent hairy girdle, this chiton of the Pacific coast of N. America is sometimes found beneath long-moored boats.

1½–2¾ in
4–7 cm

HAIRY CHITON
Chaetopleura papilio
F: Chaetopleuridae
A bristly-girdled species with brown-striped chainmail plates, the hairy chiton is found beneath rocks on South African coasts.

TUSK SHELLS

Curious, mud-shoveling tusk shells live in marine sediments and have a tubular curved shell—open at both ends and unlike the shell of any other mollusk.

Tusk shells are common, but are rarely seen because they usually occur far offshore, with the wide base of the hollow tusk buried in mud. The eyeless head and foot reach deeper into the sediment, probing for food with their tentaclelike structures called captacula, and tasting the mud with chemical detectors. When food is found—small invertebrates or detritus—the tentacles bring it to the mouth, where it is ground up by the radula, a rasping tongue typical of mollusks. Tusk shells lack gills; the animal's fleshy mantle encloses a water-filled, oxygen-extracting tube that runs the length of the shell. When levels of oxygen run low, the tusk shell contracts its foot and squirts the stale water out through the top end of the shell. Fresh water enters by the same route.

PHYLUM	MOLLUSKA
CLASS	SCAPHOPODA
ORDERS	2
FAMILIES	13
SPECIES	576

1¼–1½ in
3–4 cm

EUROPEAN TUSK
Antalis dentalis
F: Dentaliidae
A northeast Atlantic species, the European tusk is widespread in sandy areas of coastal waters. In places its empty shells are found in huge numbers.

2–3¼ in
5–8 cm

BEAUTIFUL TUSK
Pictodentalium formosum
F: Dentaliidae
This colorful tropical tusk shell has been recorded in marine sediments in Japan, the Philippines, Australasia, and New Caledonia.

ECHINODERMS

Echinoderms include an astonishing array of sea creatures, ranging from filter-feeding feather stars to grazing urchins and predatory starfish.

This is the only large invertebrate phylum entirely restricted to salt water. Echinoderms are sluggish, slow-moving inhabitants of the sea floor and most have a body form built around a five-part radial symmetry. Echinoderm means "spiny-skinned"— a reference to the hard, calcified structures, called ossicles, that make up the internal skeleton of these animals. In starfish these are spaced out within the soft tissue so that the animal is reasonably flexible, but in urchins they are fused together to form a solid internal shell. In sea cucumbers the ossicles are so minute and sparsely distributed—or even absent—that the entire animal is soft-bodied.

HYDRAULIC FEET
Echinoderms are the only animals with a water transport system. Sea water is drawn into the center of the body through a perforated, sievelike plate, usually on the upper surface of the body.

The water circulates through tubes that run through the animal and is forced into tiny, soft projections near the surface, called tube feet, which can move back and forth and stick to surfaces. In feather stars these tube feet point upward on the feathery arms, capturing particles of food, which are then moved toward the central mouth. In other echinoderms thousands of downward-pointing tube feet are used in locomotion—together they pull the animal over sediment and rock.

DEFENSE
The tough, prickly skin of most echinoderms offers good protection from potential predators, but these animals defend themselves in other ways, too. Many urchins are also covered with formidable spines; some can inflict serious wounds in humans. Many species have tiny pincerlike projections, sometimes venomous, that can remove cluttering debris as well as deter potential predators. Soft-bodied sea cucumbers rely on noxious chemicals to protect them—and are often brightly colored as a warning. As a last resort, some sea cucumbers can even spew out sticky tangles or eject their gut.

PHYLUM	ECHINODERMATA
CLASSES	5
ORDERS	35
FAMILIES	173
SPECIES	7,447

The tiny tube feet on which echinoderms move around the seabed are seen here on a crown-of-thorns starfish, *Acanthaster planci*.

FEATHER STARS

A feather star carries five basic filter-feeding arms, which in some species branch into a dense cluster. A mouth and anus point upward from the center of the star. Some members of the class Crinoidea crawl along on their featherlike arms. Others, known as sea lilies, are attached by a stalk.

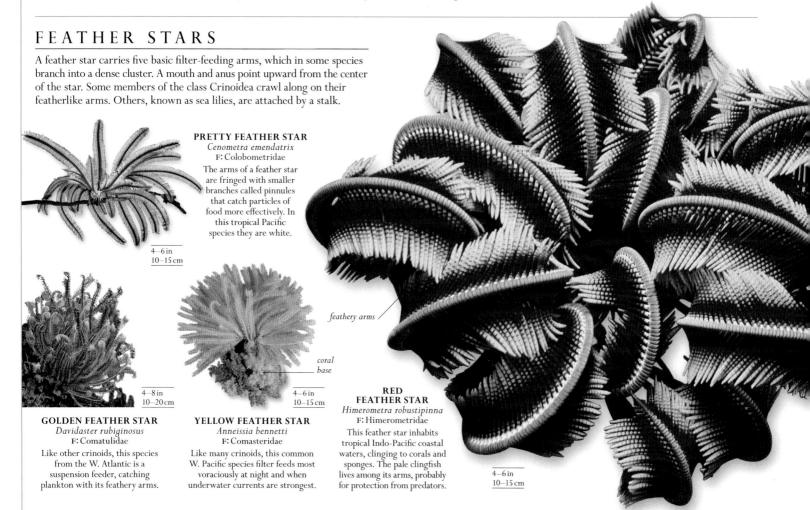

PRETTY FEATHER STAR
Cenometra emendatrix
F: Colobometridae
The arms of a feather star are fringed with smaller branches called pinnules that catch particles of food more effectively. In this tropical Pacific species they are white.

4–6 in
10–15 cm

GOLDEN FEATHER STAR
Davidaster rubiginosus
F: Comatulidae
Like other crinoids, this species from the W. Atlantic is a suspension feeder, catching plankton with its feathery arms.

4–8 in
10–20 cm

YELLOW FEATHER STAR
Anneissia bennetti
F: Comasteridae
Like many crinoids, this common W. Pacific species filter feeds most voraciously at night and when underwater currents are strongest.

4–6 in
10–15 cm

feathery arms

coral base

RED FEATHER STAR
Himerometra robustipinna
F: Himerometridae
This feather star inhabits tropical Indo-Pacific coastal waters, clinging to corals and sponges. The pale clingfish lives among its arms, probably for protection from predators.

4–6 in
10–15 cm

URCHINS AND RELATIVES

Sea urchins, which belong to the class Echinoidea, gave their name to the whole phylum. The hard ossicles of an urchin are plates that interlock to form a shell called a test. The animal's spines help in locomotion and defense. The mouth generally faces downward and the anus upward—with rows of tube feet running between the two.

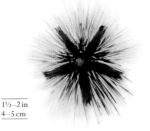

4–4¼ in
10–11 cm

INDO-PACIFIC SAND DOLLAR
Sculpsitechinus auritus
F: Astriclypeidae
This Indo-Pacific urchin of coastal sands is a typical member of a group of burrowing sea dollars, named for their flattened appearance.

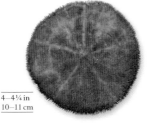

1½–2 in
4–5 cm

BANDED SEA URCHIN
Echinothrix calamaris
F: Diadematidae
The shorter spines of this Indo-Pacific reef-dwelling urchin are painfully venomous. Cardinalfishes (Apogonidae) often hide among them for protection.

8 in
20 cm

RED URCHIN
Astropyga radiata
F: Diadematidae
This Indo-Pacific lagoon species belongs to a family of tropical urchins with long hollow spines. It is often carried by a crab, *Dorippe frascone*.

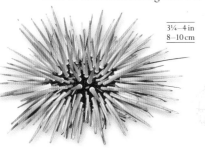

3¼–4 in
8–10 cm

MATHA'S URCHIN
Echinometra mathaei
F: Echinometridae
This Indo-Pacific species belongs to a group of urchins characterized by thick spines. It lives in crevices on coral reefs.

6½–7 in
16–18 cm

EDIBLE URCHIN
Echinus esculentus
F: Echinidae
This large, globular urchin common on N.E. Atlantic coasts has a variable color. It grazes on invertebrates and algae, especially kelp.

FIRE URCHIN
Asthenosoma varium
F: Echinothuriidae
This somewhat flat Indo-Pacific species of sandy lagoons has a shell sufficiently flexible for the urchin to enter crevices. It can deliver a painful sting.

flexible test

8–10 in
20–25 cm

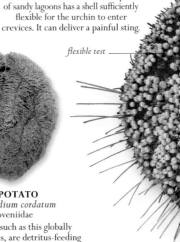

3¼–4 in
8–10 cm

PURPLE URCHIN
Strongylocentrotus purpuratus
F: Strongylocentrotidae
An inhabitant of the underwater kelp forests along the Pacific coast of N. America, this urchin is often used in biomedical research.

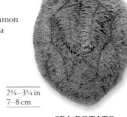

2¾–3¼ in
7–8 cm

SEA POTATO
Echinocardium cordatum
F: Loveniidae
Heart urchins, such as this globally widespread species, are detritus-feeding burrowers that lack the obvious radial symmetry of other urchins.

short, venomous spines

SEA CUCUMBERS

Sea cucumbers are soft, tubular animals, with a mouth surrounded by a ring of food-collecting tentacles at one end, and an anus at the other. Some Holothuroidea shovel or burrow in sediments using their tentacles; others use rows of tube feet to crawl along the seabed.

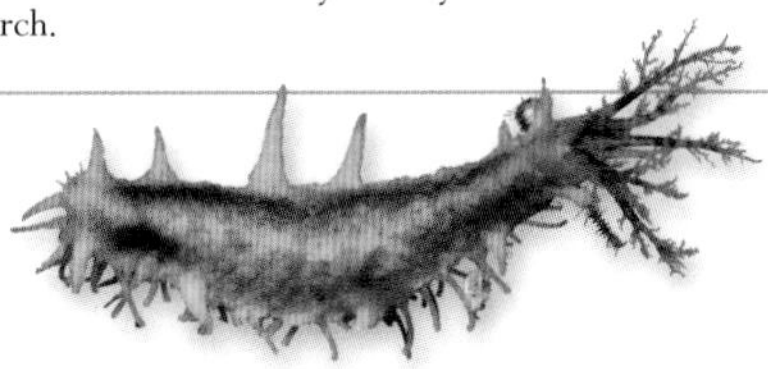

YELLOW SEA CUCUMBER
Colochirus robustus
F: Cucumariidae
This echinoderm belongs to a family of thick-skinned sea cucumbers. Like this Indo-Pacific species, many are covered in knobbly projections.

2–3¼ in
5–8 cm

6–7 in
15–18 cm

SEA APPLE CUCUMBER
Pseudocolochirus violaceus
F: Cucumariidae
Sea apples are brightly colored, highly poisonous, reef-dwelling sea cucumbers. This species has yellow or orange tube feet, but variable body color.

warning coloration

sticky mass ejected to repel predators

OCELLATED SEA CUCUMBER
Bohadschia argus
F: Holothuriidae
Distributed from the W. Indian Ocean around Madagascar to the South Pacific, this large species is variable in color, but is always spotted.

15–23½ in
38–60 cm

10–12 in
25–30 cm

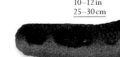

EDIBLE SEA CUCUMBER
Holothuria (Halodeima) edulis
F: Holothuriidae
Large sea cucumbers are especially abundant in tropical waters. This Indo-Pacific species is harvested and dried for human consumption.

23½ in
60 cm

VERMIFORM SEA CUCUMBER
Synapta maculata
F: Synaptidae
Typical of its family, this Indo-Pacific species is a soft-bodied, wormlike sea cucumber that burrows in soft sediment. It lacks tube feet.

spikes provide grip on ocean floor

14–16 in
35–40 cm

GIANT CALIFORNIA SEA CUCUMBER
Apostichopus californicus
F: Stichopodidae
One of the largest sea cucumbers of N. America's Pacific coast, this fleshy, spiked species is fished as a local delicacy.

PRICKLY REDFISH
Thelenota ananas
F: Stichopodidae
This echinoderm belongs to a family of sea cucumbers covered in fleshy spikes. It occurs on sandy seabeds around Indo-Pacific coral reefs.

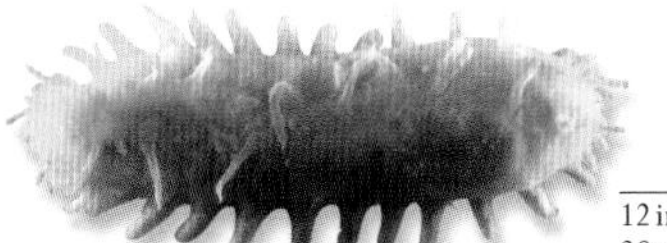

23½–30 in
60–75 cm

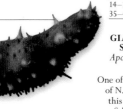

12 in
30 cm

DEEP SEA CUCUMBER
Kolga hyalina
F: Elpidiidae
Virtually worldwide in distribution at depths of up to 5,000 ft (1,500 m), this is one of many little-known sea cucumbers found on ocean floors.

BRITTLE STARS

These animals are named for the fact that their long, thin, sometimes branching, arms easily break off. There is one opening to the gut in the central disk—a downward-facing mouth. Some Ophiuroidea use their arms to trap food particles; others are predatory.

long, flexible arms

8–12 in
20–30 cm

SHORT-SPINED BRITTLE STAR
Ophioderma sp.
F: Ophiodermatidae

A brittle star of W. Atlantic coastal waters, this species lives in submerged grass beds, where it preys on other invertebrates, such as shrimp.

COMMON BRITTLE STAR
Ophiothrix fragilis
F: Ophiotrichidae

This N.E. Atlantic brittle star—often found in dense populations—has spiny arms, some of which it holds up to trap food particles.

4¾–6 in
12–15 cm

8–12 in
20–30 cm

BLACK BRITTLE STAR
Ophiocomina nigra
F: Ophiocomidae

This large brittle star of European waters can either filter feed on particles or scavenge on detritus. It can be common on rocky coasts with strong currents.

8–10 in
20–25 cm

GORGON'S HEAD BRITTLE STAR
Gorgonocephalus caputmedusae
F: Gorgonocephalidae

This species, named for its coiled, snakelike branching arms, is common around European coasts. Larger animals are found in places of strong currents, where they can collect more food.

STARFISH

Like most other echinoderms, starfish crawl on the seabed using their rows of tube feet. These run along grooves in the starfish's arms. Most species of Asteroidea are five-armed, but a few have more than five arms, and some are virtually spherical in shape. Many prey on other slow-moving invertebrates; others scavenge or feed on detritus. Although their skin is studded with hard ossicles, starfish are sufficiently flexible to catch prey.

14–16 in
35–40 cm

PURPLE SUNSTAR
Solaster endeca
F: Solasteridae

Sunstars are large, spiny starfish with many arms. Living on offshore mud in the cold waters of the northern hemisphere, this species has from 7 to 13 arms.

20–23½ in
50–60 cm

SEVEN ARM STARFISH
Luidia ciliaris
F: Luidiidae

Unlike most other starfish, this large Atlantic species has seven arms. It has long tube feet and is fast-moving in its pursuit of other echinoderms.

8–9½ in
20–24 cm

MOSAIC STARFISH
Plectaster decanus
F: Echinasteridae

Like a number of other members of its family, this striking species of S.W. Pacific rocky coasts varies greatly in color.

8–10 in
20–25 cm

WARTY STARFISH
Echinaster callosus
F: Echinasteridae

Belonging to a family of stiff-bodied starfish with conical arms, this W. Pacific species is unusual for its pink and white warts.

tube feet on underside of arms

broad base to arms

coarse, spiny upper surface

RED CUSHION STAR
Porania (Porania) pulvillus
F: Poraniidae

This distinctive, smooth-topped starfish with short arms and fringe of spines occurs on European rocky coasts, often on kelp holdfasts.

4–4¾ in
10–12 cm

4¾–5 in
10–12 cm

BLOOD HENRY STARFISH
Henricia oculata
F: Echinasteridae

This relative of the warty starfish lives in tidal pools and kelp forests of the N.E. Atlantic. Its body secretes mucus to trap food particles.

4–10 in
10–25 cm

OCHER STARFISH
Pisaster ochraceus
F: Asteriidae

This N. American Pacific relative of the common Atlantic starfish—like others of its family—hunts invertebrates. It is an important predator of mussels.

16–20 in
40–50 cm

COMMON STARFISH
Asterias rubens
F: Asteriidae

This common N.E. Atlantic predator of other invertebrates is unusual among echinoderms in tolerating estuarine conditions. It often occurs in huge local populations.

32–39 in
80–100 cm

GIANT SUNFLOWER STARFISH
Pycnopodia helianthoides
F: Asteriidae

One of the world's largest starfish, this many-armed species preys on mollusks and other echinoderms. It lives among offshore seaweed on the N.E. Pacific coast.

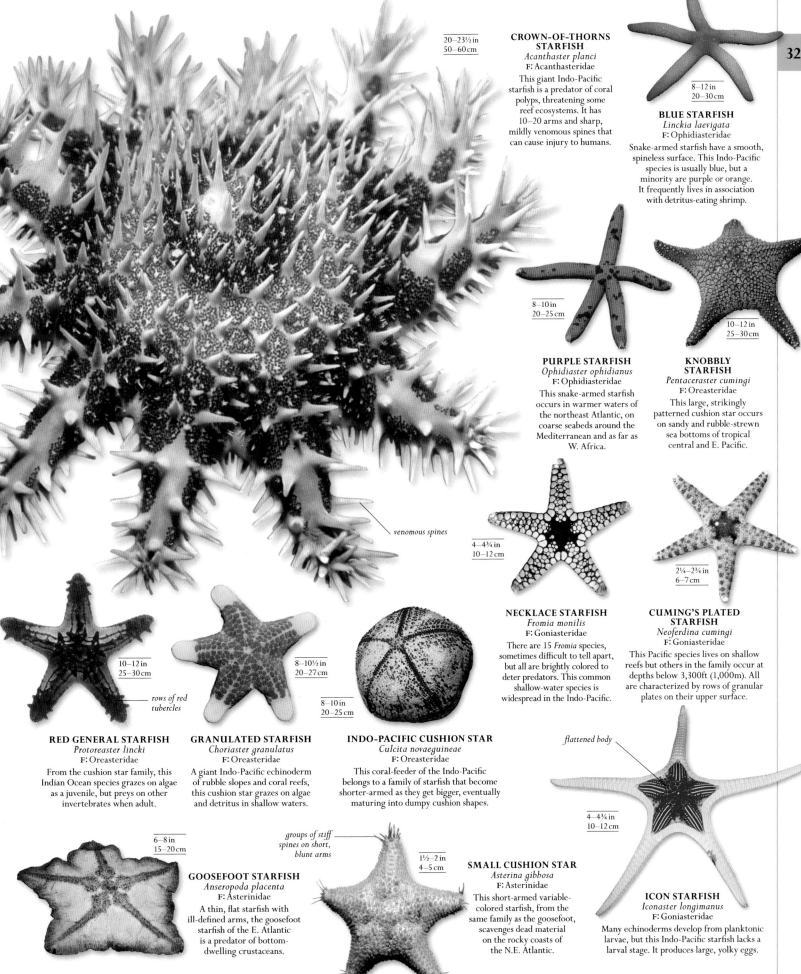

CROWN-OF-THORNS STARFISH
Acanthaster planci
F: Acanthasteridae

20–23½ in
50–60 cm

This giant Indo-Pacific starfish is a predator of coral polyps, threatening some reef ecosystems. It has 10–20 arms and sharp, mildly venomous spines that can cause injury to humans.

venomous spines

BLUE STARFISH
Linckia laevigata
F: Ophidiasteridae

8–12 in
20–30 cm

Snake-armed starfish have a smooth, spineless surface. This Indo-Pacific species is usually blue, but a minority are purple or orange. It frequently lives in association with detritus-eating shrimp.

PURPLE STARFISH
Ophidiaster ophidianus
F: Ophidiasteridae

8–10 in
20–25 cm

This snake-armed starfish occurs in warmer waters of the northeast Atlantic, on coarse seabeds around the Mediterranean and as far as W. Africa.

KNOBBLY STARFISH
Pentaceraster cumingi
F: Oreasteridae

10–12 in
25–30 cm

This large, strikingly patterned cushion star occurs on sandy and rubble-strewn sea bottoms of tropical central and E. Pacific.

NECKLACE STARFISH
Fromia monilis
F: Goniasteridae

4–4¾ in
10–12 cm

There are 15 *Fromia* species, sometimes difficult to tell apart, but all are brightly colored to deter predators. This common shallow-water species is widespread in the Indo-Pacific.

CUMING'S PLATED STARFISH
Neoferdina cumingi
F: Goniasteridae

2¼–2¾ in
6–7 cm

This Pacific species lives on shallow reefs but others in the family occur at depths below 3,300ft (1,000m). All are characterized by rows of granular plates on their upper surface.

rows of red tubercles

RED GENERAL STARFISH
Protoreaster lincki
F: Oreasteridae

10–12 in
25–30 cm

From the cushion star family, this Indian Ocean species grazes on algae as a juvenile, but preys on other invertebrates when adult.

GRANULATED STARFISH
Choriaster granulatus
F: Oreasteridae

8–10½ in
20–27 cm

A giant Indo-Pacific echinoderm of rubble slopes and coral reefs, this cushion star grazes on algae and detritus in shallow waters.

INDO-PACIFIC CUSHION STAR
Culcita novaeguineae
F: Oreasteridae

8–10 in
20–25 cm

This coral-feeder of the Indo-Pacific belongs to a family of starfish that become shorter-armed as they get bigger, eventually maturing into dumpy cushion shapes.

flattened body

GOOSEFOOT STARFISH
Anseropoda placenta
F: Asterinidae

6–8 in
15–20 cm

A thin, flat starfish with ill-defined arms, the goosefoot starfish of the E. Atlantic is a predator of bottom-dwelling crustaceans.

groups of stiff spines on short, blunt arms

SMALL CUSHION STAR
Asterina gibbosa
F: Asterinidae

1½–2 in
4–5 cm

This short-armed variable-colored starfish, from the same family as the goosefoot, scavenges dead material on the rocky coasts of the N.E. Atlantic.

ICON STARFISH
Iconaster longimanus
F: Goniasteridae

4–4¾ in
10–12 cm

Many echinoderms develop from planktonic larvae, but this Indo-Pacific starfish lacks a larval stage. It produces large, yolky eggs.

CHORDATES

Only three to four percent of known animal species are chordates, yet they include the largest, fastest, and most intelligent animals alive today. Most chordates have a skeleton made of bone or cartilage, but their defining feature is the rodlike notochord—the evolutionary forerunner of the spine.

Fossil records indicate that the earliest known true chordates were small, streamlined animals, just an inch long. Living at least 500 million years ago, they had no hard parts, except for a stiff but flexible cartilaginous notochord. The notochord ran the length of their body, creating a framework for their muscles to pull against. Today's chordates all have this feature, and a small number keep it throughout life. However, in the vast majority of chordates—from fishes and amphibians to reptiles, birds, and mammals—the notochord is present only in the early embryo. As the embryo develops, the notochord disappears, replaced by an internal skeleton made of cartilage or bone. These animals are known as vertebrates, from the column of vertebrae that makes up their spine.

Unlike shells or body cases, bony skeletons work on an amazing spectrum of scales. The smallest vertebrate, a freshwater fish called *Paedocypris progenetica*, is less than ⅜ in (1 cm) long and weighs several billion times less than a blue whale, which is the largest chordate, and the biggest animal that has ever lived.

LIFESTYLES

It is perhaps surprising that one of the simplest chordate groups, the sea squirts (tunicates), do not have a skeleton or backbone, and spend their adult lives fastened in one place. However, in their planktonic larval stage they do have a notochord, indicating their shared ancestry with vertebrate chordates such as us. In contrast to these simple chordates, vertebrate chordates are often fast-moving animals, with rapid reactions, a well-developed nervous system, and a sizeable brain. Birds, mammals, and a few fishes use energy from food to keep their body at a constant optimum temperature.

Chordates vary in the way they reproduce and raise their young. All mammals bear live young, except for the duck-billed platypus and echidnas. Almost all other chordates lay eggs, though there are examples of live-bearing species in all vertebrate groups except birds. The number of offspring produced is related to the amount of parental care. Some fishes produce millions of eggs, and play no part in raising their young, whereas mammals and birds have much smaller families.

TUNICATES
The subphylum Tunicata contains just over 3,000 species. Larval tunicates have a notochord, and resemble tadpoles. The adults are filter feeders, the most common type being sea squirts.

AMPHIBIANS
Numbering more than 8,200 species, amphibians include frogs and toads, salamanders, newts, and caecilians. Most start their lives as aquatic larvae, but as adults they take up life on land.

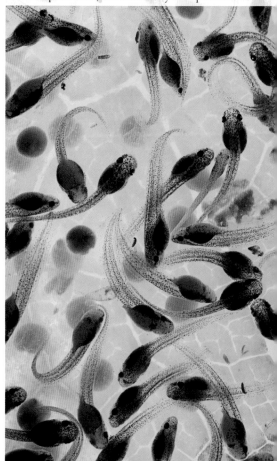

CHORDATE TREE

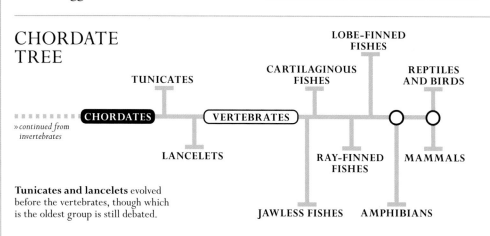

LOBE-FINNED FISHES

TUNICATES   CARTILAGINOUS FISHES   REPTILES AND BIRDS

CHORDATES — VERTEBRATES

» continued from invertebrates

LANCELETS   RAY-FINNED FISHES   MAMMALS

JAWLESS FISHES   AMPHIBIANS

Tunicates and lancelets evolved before the vertebrates, though which is the oldest group is still debated.

LANCELETS
Small marine chordates with slender bodies and a notochord throughout life, lancelets live half buried in the seabed. There are 20 species in the subphylum Cephalochordata.

FISHES
Fishes are the most diverse and numerous of all the vertebrate chordates. They are classified into a number of different classes, within the subphylum Vertebrata. These classes of fishes are often grouped together as a superclass, Pisces, and their variety reflects their evolutionary past. Almost 34,000 living species are known.

REPTILES
With over 11,000 species, reptiles inhabit every continent on Earth, with the exception of Antarctica. Unlike birds and mammals, they are cold-blooded, and covered in scales.

BIRDS
Birds are the only living animals that have feathers. There are about 10,000 species of birds, though recent studies suggest there may be up to 18,000. All lay eggs, and many show highly developed parental care.

MAMMALS
Readily identified by their fur or hair (even whales have some), mammals are the only chordates that raise their young on milk. There are more than 6,300 known mammal species.

FISHES

Fishes are the most diverse chordates and can be found in all types of watery habitat—from tiny, freshwater pools to the deep ocean. Almost without exception, they breathe using feathery gills to absorb oxygen from the water, and almost all swim using fins.

PHYLUM	CHORDATA
CLASSES	PETROMYZONTI
	MYXINI
	ELASMOBRANCHII
	HOLOCEPHALI
	ACTINOPTERYGII
	COELACANTHI
	DIPNEUSTI
ORDERS	66–81
FAMILIES	560–588
SPECIES	About 33,900

Overlapping scales cover the body of a bicolor parrotfish, forming a flexible layer that protects against injury.

Predatory sharks lunge into a huge shoal of sardines, which swim ever closer together for protection.

The huge mouth of a male gold-specs jawfish provides an unlikely nest for its eggs. He will not feed during incubation.

Fishes are not a natural, single group but are, in fact, seven classes of vertebrate chordates, of which the familiar ray-finned (bony) fishes are by far the most populous. The majority of fishes are cold-blooded and so their body temperature matches that of the surrounding water. A few top predators, such as white sharks, can maintain a flow of warm blood to the brain, eyes, and main muscles, which allows them to hunt actively even in very cold water. Most fishes are protected by scales or bony plates embedded in the skin. In fast swimmers these are lightweight, and provide streamlining as well as protection from abrasion and disease. Although a few fishes slither or shuffle over the seabed, most swim using fins. With the notable exception of rays, the tail fin usually provides the main propulsive force. Paired pectoral and pelvic fins (on the sides and underside respectively) provide stability and maneuverability, helped by up to three dorsal fins on the fish's upper side and one or two anal fins. Fishes avoid collisions—especially in large shoals—by using special sensory organs that detect the vibrations made as they or other animals move through the water. Most fishes have a row of these organs running along both sides of the body, called the lateral line. While they also use the senses of hearing, touch, sight, taste, and smell found in other chordates, only fishes have a lateral line system.

REPRODUCTIVE STRATEGIES

The seven classes of fishes have very different ways of reproducing. Most ray-finned fishes have external fertilization and shed masses of eggs and sperm directly into the water to compensate for the huge numbers that are eaten before they can develop into juvenile fishes; a few instead guard their eggs until hatching. In contrast, cartilaginous fishes (sharks, rays, and chimaeras) all have internal fertilization and produce eggs or young at an advanced developmental stage. This requires a high input of energy—only a few young are produced at one time, but these have a good chance of survival. The eggs of lampreys (jawless fishes) hatch into larvae that live and feed for many months before undergoing metamorphosis into the adult form.

BRIGHT SPECTACLE >
The fabulous colors of the Banggai cardinalfish, seen here sheltering among giant anemones, attract aquarium owners.

JAWLESS FISHES

Unlike other vertebrates, jawless fishes have no biting jaws, although they do have teeth. Once abundant and diverse, few living representatives remain.

There are two living groups of jawless fishes—lampreys and hagfishes—and the relationship between them is still debated. Instead of jaws, lampreys have a round, sucker mouth encircled by rasping teeth, while hagfishes have a slitlike mouth with teeth inside on the tongue. Circular gill openings run along the sides, seven in lampreys and a variable number (1–16) in hagfishes. For body support and muscle attachment, both groups have a simple, rodlike structure called a notochord. In addition, lampreys may have a few cartilage supports around the notochord.

DIFFERING LIFECYCLES

While hagfishes live permanently in the deep ocean, lampreys inhabit temperate coastal waters and fresh waters worldwide. All lampreys breed in fresh water. Coastal species are anadromous—that is, like salmon, they swim up rivers into fresh water to spawn, after which they die. The eggs hatch into wormlike, burrowing larvae, known as ammocoetes. These live in mud and feed on detritus. After about three years the larvae of the anadromous species change into adults and swim out to sea where they feed for several years. The freshwater species live and breed in rivers and lakes. Hagfishes lay eggs on the seabed.

PARASITES AND SCAVENGERS

Lampreys are notorious for feeding parasitically on larger fishes when they are adult. Clinging on with its sucker mouth, a lamprey grinds its way through the skin of its victim eating either, or both, flesh and blood. They can be a nuisance to fishermen, as they damage and kill fishes in nets and fish farms. However, many lampreys, especially freshwater species, just eat small invertebrates. The sucker mouth is also useful when the fishes are swimming upstream against the current—they use it to cling onto rocks so they can have a rest and to move pebbles when excavating a nest site in the river bed.

Hagfishes live in mud burrows in the seabed, emerging at night to feed on both live and dead invertebrates. If the opportunity arises, they will feed on rotting whale or fish carcasses, from which they rasp off bits of flesh. If attacked, hagfishes produce copious, gill-clogging slime to deter the predator.

PHYLUM	CHORDATA
CLASSES	PETROMYZONTI
	MYXINI
ORDERS	2
FAMILIES	4
SPECIES	128

LAMPREYS AND HAGFISHES

The orders Petromyzontiformes and Myxiniformes are the only two groups of living jawless fishes—often called agnathans. They share an eellike shape, lack of jaws, and possession of a notochord, but also have numerous anatomical differences. Lampreys have a long larval stage spent in fresh water, while hagfishes lay eggs on the seabed that hatch into miniature adults.

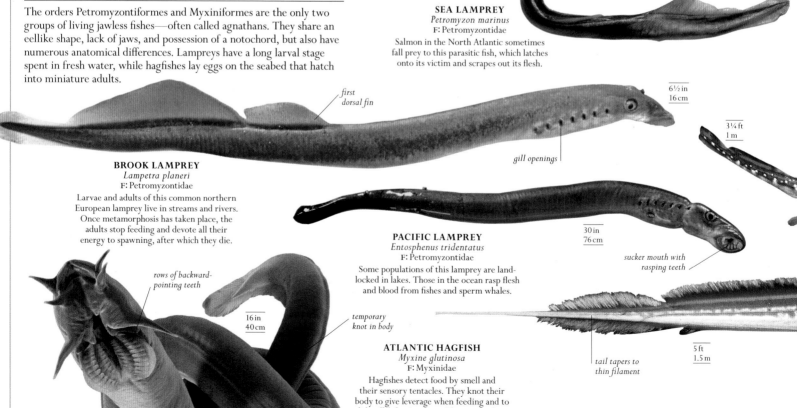

4 ft / 1.2 m

SEA LAMPREY
Petromyzon marinus
F: Petromyzontidae
Salmon in the North Atlantic sometimes fall prey to this parasitic fish, which latches onto its victim and scrapes out its flesh.

6½ in
16 cm

3¼ ft
1 m

first dorsal fin

gill openings

BROOK LAMPREY
Lampetra planeri
F: Petromyzontidae
Larvae and adults of this common northern European lamprey live in streams and rivers. Once metamorphosis has taken place, the adults stop feeding and devote all their energy to spawning, after which they die.

PACIFIC LAMPREY
Entosphenus tridentatus
F: Petromyzontidae
Some populations of this lamprey are land-locked in lakes. Those in the ocean rasp flesh and blood from fishes and sperm whales.

30 in
76 cm

sucker mouth with rasping teeth

rows of backward-pointing teeth

16 in
40 cm

temporary knot in body

ATLANTIC HAGFISH
Myxine glutinosa
F: Myxinidae
Hagfishes detect food by smell and their sensory tentacles. They knot their body to give leverage when feeding and to slide off defensive slime. This species lives in the N. Atlantic and Mediterranean.

tail tapers to thin filament

5 ft
1.5 m

CARTILAGINOUS FISHES

These fishes have a skeleton of pliable cartilage rather than the hard bone found in most other vertebrate groups. Most are predators with acute senses.

Sharks, rays, and chimaeras are all cartilaginous fishes, although chimaeras have distinct anatomical differences from the other two groups. In chimaeras, the upper jaw is fused to the braincase and so cannot be moved independently. They also have teeth that grow continuously. In contrast, sharks regularly lose their hard, enamel-covered teeth but replace them from extra rows lying flat behind the active teeth—a characteristic that helps make sharks some of the most formidable predators on Earth. The skin of all three groups—sharks, rays, and chimaeras—is protected by toothlike scales called dermal denticles.

SEARCHING FOR PREY

The ocean is home to most cartilaginous fishes and while many live on the seabed, the larger predatory shark species and plankton feeders roam in open water. The bullshark and over 100 other species can enter estuaries and swim up rivers and a few live entirely in rivers. Most open-water cartilaginous fishes are continually on the move because, unlike bony fishes, they do not have a gas-filled swim bladder to maintain neutral buoyancy and therefore may sink if they stop swimming. The whale shark and other surface-feeding species have a large, oily liver to prevent this. Predatory sharks are well known for their amazing ability to smell blood and hone in on wounded fishes and mammals. Cartilaginous fishes also have the ability to detect the weak electrical fields that surround living creatures and, while not unique to them, this sense is developed to an extraordinary extent in this group.

REPRODUCTION

All cartilaginous fishes mate and have internal fertilization. Chimaeras and many smaller sharks and rays lay eggs, each protected by a tough egg capsule and commonly known as a "mermaid's purse." However, about sixty percent of cartilaginous fishes give birth to well-developed live young that are nourished in the uterus by egg yolk or by a placental connection to the mother. Unlike mammals, the young are independent from birth and neither parent takes any interest in their offspring.

PHYLUM	CHORDATA
CLASSES	ELASMOBRANCHII
	HOLOCEPHALI
ORDERS	14–17
FAMILIES	57
SPECIES	1,338

A white shark, *Carcharodon carcharias*, shows the razor-sharp, replaceable rows of teeth that make it such a fearsome predator.

CHIMAERAS

Also known as rabbitfishes because of their fused, platelike teeth, the Holocephali are a small class of cartilaginous fishes with one order, Chimaeriformes, and about 56 species. A strong, venomous spine in front of the first of two dorsal fins provides protection in their deepwater habitat.

large eye provides better vision in deep, dark water

PACIFIC SPOOKFISH
Rhinochimaera pacifica
F: Rhinochimaeridae

4¼ ft
1.3 m

The long, conical snout of this chimaera is covered in sensory pores that detect the electrical fields of its prey.

SPOTTED RATFISH
Hydrolagus colliei
F: Chimaeridae

Like most chimaeras, this N.E. Pacific species uses its large pectoral fins to glide and flap along while searching for food.

4¼ ft
1.3 m

ELEPHANT FISH CHIMAERA
Callorhinchus milii
F: Callorhinchidae

Using its long fleshy snout as a plow, this chimaera unearths shellfish from muddy seabeds in southern Australia and New Zealand.

RATFISH
Chimaera monstrosa
F: Chimaeridae

Typically living below 1,000 ft (300 m) in the Mediterranean and E. Atlantic, ratfishes swim in small groups and search for seabed invertebrates.

pectoral fins flap for swimming

SIX- AND SEVEN-GILL SHARKS

Most sharks have five pairs of gill slits, but those in the order Hexanchiformes have either six or seven pairs. All six known species live in deep water. Two of them, called frilled sharks, have a long, soft, eellike body and have recently been placed in their own order, Chlamydoselachiformes.

18 ft
5.5 m

BLUNTNOSE SIXGILL SHARK
Hexanchus griseus
F: Hexanchidae

Rocky seamounts around the world are the haunt of this huge, green-eyed shark that weighs up to 1,325 lb (600 kg).

6½ ft
2 m

FRILLED SHARK
Chlamydoselachus anguineus
F: Chlamydoselachidae

Scattered records indicate this shark has a worldwide distribution. It has brilliant white teeth that may attract its prey of fishes and squid.

DOGFISH SHARKS AND RELATIVES

A large and varied order, the Squaliformes contains at least 143 species and includes gulper, lantern, sleeper, and kitefin sharks. These all have two dorsal (back) fins and no anal fin. All species so far studied bear live young.

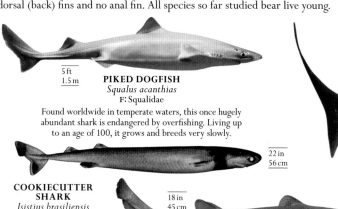

5 ft
1.5 m

PIKED DOGFISH
Squalus acanthias
F: Squalidae

Found worldwide in temperate waters, this once hugely abundant shark is endangered by overfishing. Living up to an age of 100, it grows and breeds very slowly.

22 in
56 cm

COOKIECUTTER SHARK
Isistius brasiliensis
F: Dalatiidae

This widespread, tropical shark is an ectoparasite of dolphins and large fishes. Sucking on with thick lips, it twists around and bites out a round plug of flesh.

18 in
45 cm

VELVET BELLY LANTERN SHARK
Etmopterus spinax
F: Etmopteridae

This lantern shark lives in the deep E. Atlantic. It has tiny light organs on its belly, which help in finding a mate.

8–14 ft
2.4–4.3 m

GREENLAND SHARK
Somniosus microcephalus
F: Somniosidae

One of only a few sharks that live in Arctic waters, this huge, sluggish shark often scavenges for drowned land animals.

rough skin

spines in sail-like dorsal fins

5 ft
1.5 m

ANGULAR ROUGHSHARK
Oxynotus centrina
F: Oxynotidae

Recently placed in its own order Echinorhiniformes, this deep-water shark has two saillike dorsal fins and rough skin. It lives in the E. Atlantic.

CARPET SHARKS

The 46 or so sharks of the Orectolobiformes have two dorsal fins, one anal fin, and sensory barbels hanging down from the nostrils. With the exception of the whale shark, they live quietly on the seabed, feeding on fishes and invertebrates.

white spots

mouth at end of snout

WHALE SHARK
Rhincodon typus
F: Rhincodontidae

This, the largest known fish, cruises tropical oceans, feeding on plankton and tiny fishes. Each shark has its own unique pattern of spots.

40–65 ft
12–20 m

3½ ft
1.1 m

EPAULETTE CATSHARK
Hemiscyllium ocellatum
F: Hemiscylliidae

The bold eyespot mark of this long-tailed shark may deter predators. It clambers among coral in the South Pacific using its fins.

TASSELLED WOBBEGONG
Eucrossorhinus dasypogon
F: Orectolobidae

With its fringe of skin tassels, flattened body, and camouflage pattern, this S.W. Pacific coral reef resident is hard to spot.

4 ft
1.2 m

"beard" of branching tassels

NURSE SHARK
Ginglymostoma cirratum
F: Ginglymostomatidae

By day this shark lies hidden in rock crevices, emerging at night to hunt in the warm coastal waters of the Atlantic and the E. Pacific.

10 ft / 3 m

SAWSHARKS

Sawsharks have a flat head with gills on the sides and a long snout (rostrum) with teeth on the underside and edges. Two long sensory barbels hang from the rostrum and help find buried food. Most of the nine species of Pristiophoriformes live in the tropics.

LONGNOSE SAWSHARK
Pristiophorus cirratus
F: Pristiophoridae

Sandy seabeds off southern Australia are home to this sawshark. It may use its rostrum to deter predators as well as kill prey.

4½ ft
1.4 m

BULLHEAD SHARKS

The Heterodontiformes are small bottom-living sharks with a blunt, sloping head, crushing teeth, and two dorsal (back) fins each preceded by a sharp spine. They lay unique spiral-shaped eggs.

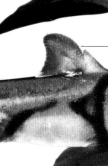

spine

5½ ft
1.7 m

PORT JACKSON SHARK
Heterodontus portusjacksoni
F: Heterodontidae

Using its paddlelike front fins, this shark crawls over the seabed off southern Australia in search of sea urchins to eat.

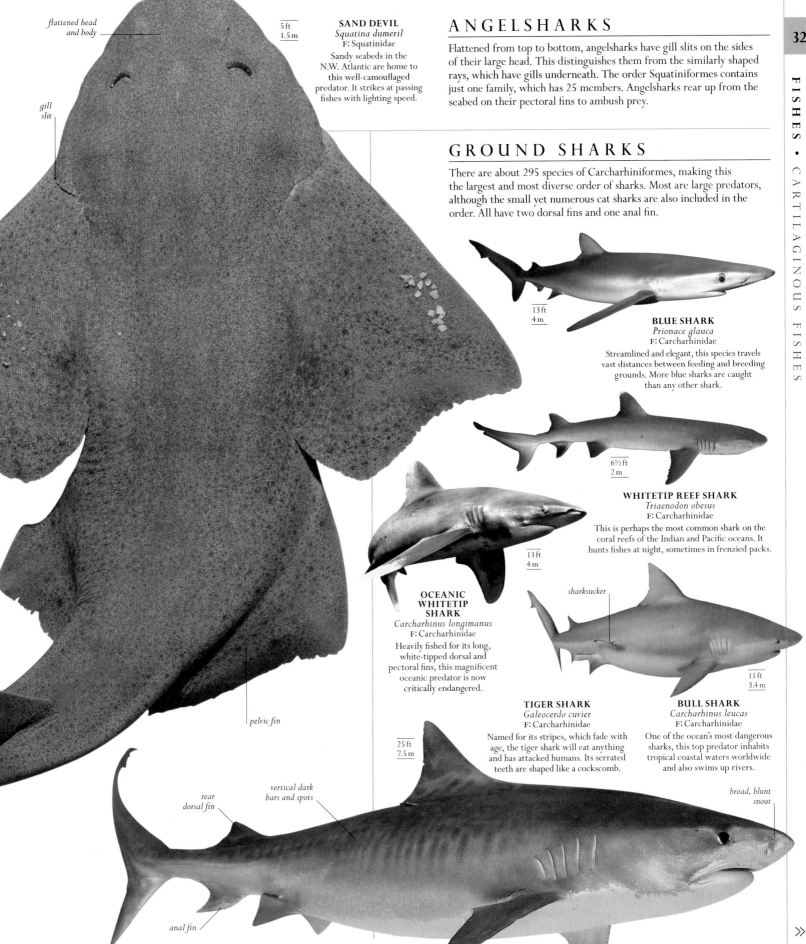

flattened head and body

gill slit

5 ft
1.5 m

SAND DEVIL
Squatina dumeril
F: Squatinidae
Sandy seabeds in the
N.W. Atlantic are home to
this well-camouflaged
predator. It strikes at passing
fishes with lighting speed.

ANGELSHARKS

Flattened from top to bottom, angelsharks have gill slits on the sides
of their large head. This distinguishes them from the similarly shaped
rays, which have gills underneath. The order Squatiniformes contains
just one family, which has 25 members. Angelsharks rear up from the
seabed on their pectoral fins to ambush prey.

GROUND SHARKS

There are about 295 species of Carcharhiniformes, making this
the largest and most diverse order of sharks. Most are large predators,
although the small yet numerous cat sharks are also included in the
order. All have two dorsal fins and one anal fin.

13 ft
4 m

BLUE SHARK
Prionace glauca
F: Carcharhinidae
Streamlined and elegant, this species travels
vast distances between feeding and breeding
grounds. More blue sharks are caught
than any other shark.

6½ ft
2 m

WHITETIP REEF SHARK
Triaenodon obesus
F: Carcharhinidae
This is perhaps the most common shark on the
coral reefs of the Indian and Pacific oceans. It
hunts fishes at night, sometimes in frenzied packs.

13 ft
4 m

**OCEANIC
WHITETIP
SHARK**
Carcharhinus longimanus
F: Carcharhinidae
Heavily fished for its long,
white-tipped dorsal and
pectoral fins, this magnificent
oceanic predator is now
critically endangered.

sharksucker

11 ft
3.4 m

TIGER SHARK
Galeocerdo cuvier
F: Carcharhinidae
Named for its stripes, which fade with
age, the tiger shark will eat anything
and has attacked humans. Its serrated
teeth are shaped like a cockscomb.

BULL SHARK
Carcharhinus leucas
F: Carcharhinidae
One of the ocean's most dangerous
sharks, this top predator inhabits
tropical coastal waters worldwide
and also swims up rivers.

pelvic fin

25 ft
7.5 m

*rear
dorsal fin*

*vertical dark
bars and spots*

*broad, blunt
snout*

anal fin

BLUE-SPOTTED RIBBONTAIL RAY
Taeniura lymma

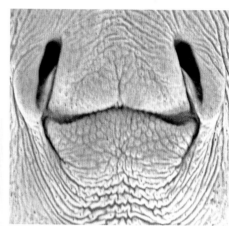

Stingrays are notorious for the painful wounds they can inflict with their barbed tail, which can, in exceptional circumstances, be lethal. However, the tropical blue-spotted ribbontail ray, like other stingrays, uses its sting just to defend itself. Much of its time is spent resting motionless on sandy patches among coral, hidden under overhangs. Often only its blue-edged tail gives it away to divers—if disturbed, it will swim away, flapping its two winglike pectoral fins. The best time to see one is on a rising tide when the stingrays swim inshore to feed on invertebrates in shallow water.

SIZE 28–35 in (70–90 cm), including tail
HABITAT Sandy patches in coral reefs
DISTRIBUTION Indian Ocean, W. Pacific
DIET Mollusks, crabs, shrimp, worms

PELVIC FINS >
In this female the urogenital opening is visible between the pelvic fins on the underside. After mating, females produce up to seven live young following a few months' to a year's gestation.

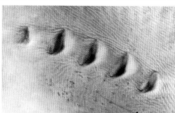

< MOUTH
The mouth is on the underside of the stingray, allowing it to extract mollusks and crabs hidden beneath the surface of the sand. Two plates of small teeth within the mouth are used to crush the shells of its prey.

< GILL SLITS
After passing over the gills, water leaves the body through five pairs of gill slits on the underside.

∨ TOP HOLE
Water is drawn in through two spiracles on top of the head behind the eyes, and leaves through gill slits on the underside. The elevated position of the spiracles helps prevent sand from getting in.

< BACK SPINES
This species has relatively smooth skin, but has two parallel rows of tiny spines running down its back, as well as other scattered spines.

spiracle behind eye

∧ ARMED TAIL
The tail is armed with one or two sharp, barbed spines that cause physical wounds and inject venom if the ray is attacked or stepped on.

barbed tail

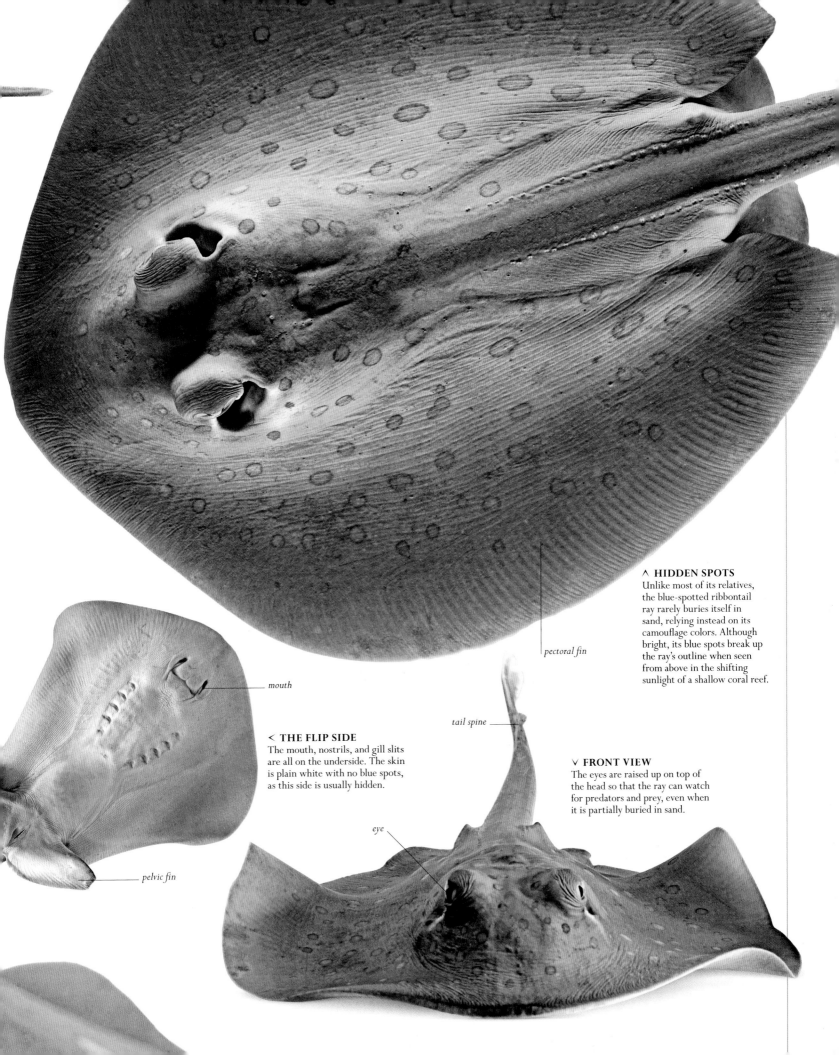

∧ **HIDDEN SPOTS**
Unlike most of its relatives, the blue-spotted ribbontail ray rarely buries itself in sand, relying instead on its camouflage colors. Although bright, its blue spots break up the ray's outline when seen from above in the shifting sunlight of a shallow coral reef.

pectoral fin

mouth

< **THE FLIP SIDE**
The mouth, nostrils, and gill slits are all on the underside. The skin is plain white with no blue spots, as this side is usually hidden.

tail spine

∨ **FRONT VIEW**
The eyes are raised up on top of the head so that the ray can watch for predators and prey, even when it is partially buried in sand.

eye

pelvic fin

RAY-FINNED FISHES

These fishes are bony, with a hard, calcified skeleton. Their fins are supported by a fan of jointed rods called rays, made of bone or cartilage.

Ray-finned fishes are able to swim with much greater precision than cartilaginous fishes. Using their highly mobile and versatile fins, they can execute maneuvers such as hovering, braking, and even swimming backward. The fins themselves can be delicate and flexible or strong and spiny and they often have important secondary uses, such as in defense, display, and camouflage.

With the exception of many bottom-living and deep-sea species, most ray-finned fishes have a buoyancy aid in the form of a gas-filled swim bladder. This allows them to adjust their buoyancy at different depths by adding or removing gas to and from the bladder via the bloodstream.

MYRIAD ADAPTATIONS

The vast majority of fishes are ray-finned and the group is hugely diverse—ranging from tiny gobies to the gigantic ocean sunfishes. Species have evolved to inhabit every conceivable aquatic niche from tropical coral reefs to the waters beneath the ice-shelves of Antarctica, from the depths of the ocean to shallow desert pools. Herbivores, carnivores, and scavengers are all represented in the group and its members exhibit many ingenious hunting and defense strategies, as well as cooperation between species.

SAFETY IN NUMBERS

Most ray-finned fishes shed eggs and sperm into the water and fertilization is external. In some cases, fewer eggs are laid and parental care is provided. For example, jawfishes and some cichlids protect their eggs and young in their mouth, while sticklebacks and many wrasses build nests of weed and debris. Some species protect their eggs with such vigor that even scuba divers are warned off.

Most species, however, lay eggs in vast numbers. The millions of floating eggs and fish larvae are an important food source for other aquatic creatures, but those that survive drift and disperse the species. Populations of fishes reproducing in this way are less vulnerable to overfishing, as numbers can recover when fishing stops, but stocks—even of such prolific breeders as Atlantic cod—will eventually succumb if intense levels of fishing are continued.

PHYLUM	CHORDATA
CLASS	ACTINOPTERYGII
ORDERS	47–59
FAMILIES	495–523
SPECIES	About 32,400

A pair of bluecheek butterflyfishes (*Chaetodon semilarvatus*) swim in unison as they patrol their patch of coral reef.

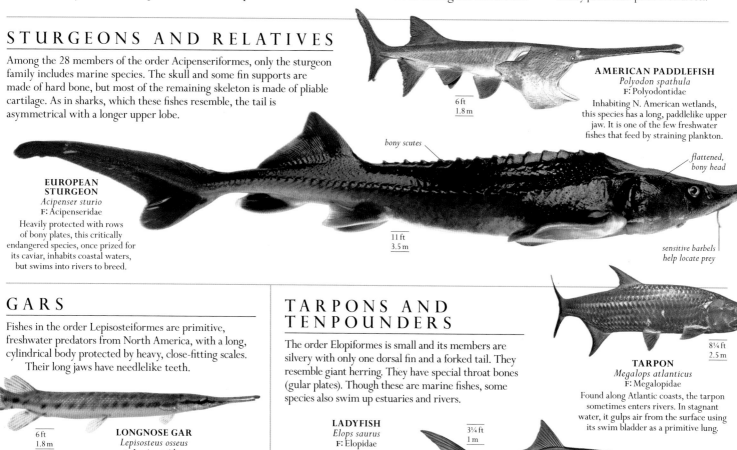

STURGEONS AND RELATIVES

Among the 28 members of the order Acipenseriformes, only the sturgeon family includes marine species. The skull and some fin supports are made of hard bone, but most of the remaining skeleton is made of pliable cartilage. As in sharks, which these fishes resemble, the tail is asymmetrical with a longer upper lobe.

6 ft
1.8 m

AMERICAN PADDLEFISH
Polyodon spathula
F: Polyodontidae
Inhabiting N. American wetlands, this species has a long, paddlelike upper jaw. It is one of the few freshwater fishes that feed by straining plankton.

bony scutes

flattened, bony head

EUROPEAN STURGEON
Acipenser sturio
F: Acipenseridae
Heavily protected with rows of bony plates, this critically endangered species, once prized for its caviar, inhabits coastal waters, but swims into rivers to breed.

11 ft
3.5 m

sensitive barbels help locate prey

GARS

Fishes in the order Lepisosteiformes are primitive, freshwater predators from North America, with a long, cylindrical body protected by heavy, close-fitting scales. Their long jaws have needlelike teeth.

6 ft
1.8 m

LONGNOSE GAR
Lepisosteus osseus
F: Lepisosteidae
A skilled predator, this long, thin fish hangs motionless in the water, hidden by vegetation, before thrusting forward to capture its prey.

TARPONS AND TENPOUNDERS

The order Elopiformes is small and its members are silvery with only one dorsal fin and a forked tail. They resemble giant herring. They have special throat bones (gular plates). Though these are marine fishes, some species also swim up estuaries and rivers.

LADYFISH
Elops saurus
F: Elopidae
The ladyfish moves in large schools, close to W. Atlantic shores. When alarmed, it skips over the water's surface.

3¼ ft
1 m

8¼ ft
2.5 m

TARPON
Megalops atlanticus
F: Megalopidae
Found along Atlantic coasts, the tarpon sometimes enters rivers. In stagnant water, it gulps air from the surface using its swim bladder as a primitive lung.

BONYTONGUES AND RELATIVES

As their name suggests, the fishes of the order Osteoglossiformes have many sharp teeth on the tongue and roof of the mouth, which help them to seize and hold prey. These fishes live in fresh water, mainly in the tropics. The group has many fishes with unusual shapes.

ARAPAIMA
Arapaima gigas
F: Arapaimidae

Weighing up to 440 lb (200 kg), this S. American species is one of the largest freshwater fishes. Its swim bladder acts as a lung and it must gulp air regularly.

dorsal fin set far back on body

15 ft
4.5 m

gray to green body

9 in
23 cm

ELEPHANTNOSE FISH
Gnathonemus petersii
F: Mormyridae

This African fish finds its way in its murky water habitat by generating weak electrical pulses. Its long lower jaw probes the mud for food.

34 in
87 cm

CLOWN KNIFEFISH
Chitala chitala
F: Notopteridae

This slender, humpbacked fish lives in wetlands of S.E. Asia. In stagnant water, it gulps air and absorbs oxygen from its swim bladder.

EELS

The Anguilliformes have a long, thin, snakelike body with smooth skin, in which the scales are either absent or deeply embedded. The fins are often limited to one long fin running down the back, around the tail, and along the belly. Eels are found in marine and freshwater habitats.

5 ft
1.5 m

ZEBRA MORAY
Gymnomuraena zebra
F: Muraenidae

This tropical, boldly striped eel has dense pebblelike teeth to eat tough-shelled crabs, mollusks, and sea urchins.

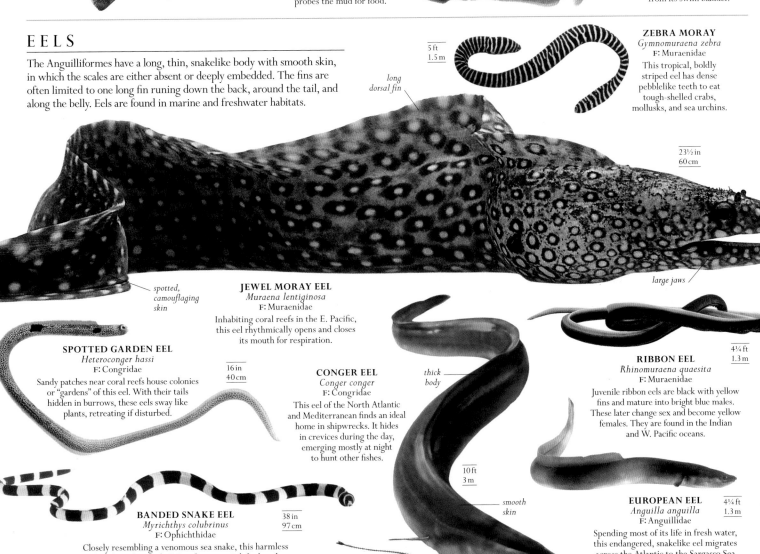

long dorsal fin

23½ in
60 cm

large jaws

spotted, camouflaging skin

JEWEL MORAY EEL
Muraena lentiginosa
F: Muraenidae

Inhabiting coral reefs in the E. Pacific, this eel rhythmically opens and closes its mouth for respiration.

SPOTTED GARDEN EEL
Heteroconger hassi
F: Congridae

Sandy patches near coral reefs house colonies or "gardens" of this eel. With their tails hidden in burrows, these eels sway like plants, retreating if disturbed.

16 in
40 cm

CONGER EEL
Conger conger
F: Congridae

This eel of the North Atlantic and Mediterranean finds an ideal home in shipwrecks. It hides in crevices during the day, emerging mostly at night to hunt other fishes.

thick body

4¼ ft
1.3 m

RIBBON EEL
Rhinomuraena quaesita
F: Muraenidae

Juvenile ribbon eels are black with yellow fins and mature into bright blue males. These later change sex and become yellow females. They are found in the Indian and W. Pacific oceans.

10 ft
3 m

smooth skin

BANDED SNAKE EEL
Myrichthys colubrinus
F: Ophichthidae

38 in
97 cm

Closely resembling a venomous sea snake, this harmless eel of the Indian and W. Pacific oceans is left alone by predators. It searches in sand burrows for small fishes.

EUROPEAN EEL
Anguilla anguilla
F: Anguillidae

4¼ ft
1.3 m

Spending most of its life in fresh water, this endangered, snakelike eel migrates across the Atlantic to the Sargasso Sea to spawn and then die.

SWALLOWERS AND GULPERS

Living in the deep ocean, these bizarre, eellike fishes, belonging to the order Saccopharyngiformes, have no tail fin or pelvic fins, and no scales. They do not have ribs and their large jaws are modified for a wide gape. Like true eels, they are believed to spawn once and then die.

3¼ ft
1 m

long, whiplike tail

huge, loosely hinged mouth

PELICAN GULPER EEL
Eurypharynx pelecanoides
F: Eurypharyngidae

This deep-sea, eellike fish has huge jaws and an expandable stomach, which enable it to swallow prey almost as large as itself.

MILKFISHES AND RELATIVES

With only two exceptions, including the milkfish itself, the Gonorynchiformes live in fresh water. They have a pair of pelvic fins that are set well back on the belly.

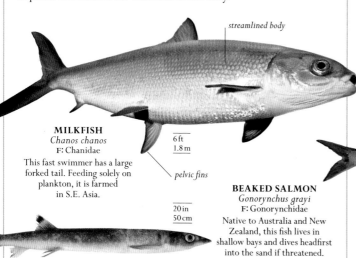

streamlined body

MILKFISH
Chanos chanos
F: Chanidae

6 ft
1.8 m

This fast swimmer has a large forked tail. Feeding solely on plankton, it is farmed in S.E. Asia.

pelvic fins

20 in
50 cm

BEAKED SALMON
Gonorynchus grayi
F: Gonorynchidae

Native to Australia and New Zealand, this fish lives in shallow bays and dives headfirst into the sand if threatened.

SARDINES AND RELATIVES

This predominantly marine order includes many commercially important species. The Clupeiformes are silvery, with loose scales, one dorsal fin, a forked tail, and a keel-shaped belly. Most live in large schools and are preyed on by sharks, tuna, and other large fishes.

PERUVIAN ANCHOVETA
Engraulis ringens
F: Engraulidae

This tiny plankton-eater lives in enormous shoals along the western coast of S. America, where it is a major source of food for humans, pelicans, and larger fishes.

8 in
20 cm

ALLIS SHAD
Alosa alosa
F: Clupeidae

33 in
83 cm

In spring, adults of this migratory species swim from the sea into European rivers to spawn, often swimming very long distances.

18 in
45 cm

ATLANTIC HERRING
Clupea harengus
F: Clupeidae

Huge shoals of these silvery fishes swim in unison, eating planktonic copepods. Highly prized, many N.E. Atlantic stocks are very heavily fished.

CARP AND RELATIVES

This is one of the largest freshwater fish orders, with more than 4,000 species worldwide. The Cypriniformes have the standard "fish shape," with a single dorsal fin. They typically have large scales. The teeth are in the throat instead of the jaws. Many are familiar aquarium fishes, including loaches, minnows, and carp.

red-edged dorsal fin

tigerlike black stripes

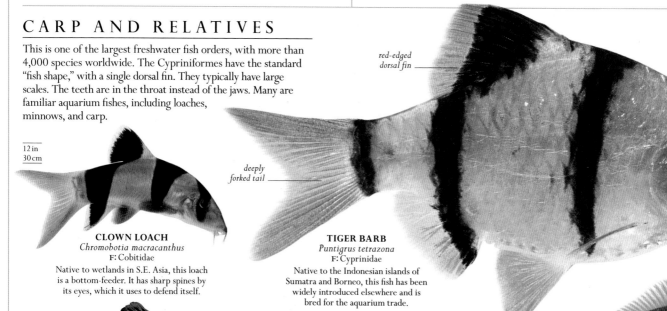

12 in
30 cm

CLOWN LOACH
Chromobotia macracanthus
F: Cobitidae

Native to wetlands in S.E. Asia, this loach is a bottom-feeder. It has sharp spines by its eyes, which it uses to defend itself.

deeply forked tail

TIGER BARB
Puntigrus tetrazona
F: Cyprinidae

Native to the Indonesian islands of Sumatra and Borneo, this fish has been widely introduced elsewhere and is bred for the aquarium trade.

2¾ in
7 cm

BITTERLING
Rhodeus amarus
F: Cyprinidae

4¼ in
11 cm

This European fish lays its eggs inside the mantle cavity of a freshwater mussel, where they eventually hatch and the fry escape.

5 ft
1.5 m

GRASS CARP
Ctenopharyngodon idella
F: Cyprinidae

A native of Asia, this species feeds on aquatic plants. For this reason, it has been introduced in Europe and the US to keep drainage channels clear of weeds.

4 in
10 cm

GOLDFISH
Carassius auratus
F: Cyprinidae

Originally native to C. Asia and China, the goldfish has been introduced all over the world and there are now many varieties.

large, silvery scales

4 ft
1.2 m

protrusible mouth

COMMON CARP
Cyprinus carpio
F: Cyprinidae

With a mouth that can be thrust out and sensory barbels, a carp finds food by grubbing through the bottom mud. Now introduced worldwide, it was originally from China and C. Europe.

2¼ in
6 cm

ZEBRAFISH
Danio rerio
F: Cyprinidae

This active little fish spawns frequently and is common in ponds and lakes in S. Asia. It is bred in aquaria and laboratories.

CHARACINS AND RELATIVES

These freshwater fishes are mostly carnivores, with well-developed teeth. As well as a normal dorsal fin, most also have a small, fatty adipose fin near the tail. Of the 23 families in the order Characiformes, the piranhas are the most notorious predators.

MEXICAN TETRA
Astyanax mexicanus
F: Characidae
While the normal form of the Mexican tetra lives in streams and can see well, this variant lives in cave pools and is blind.

4¾ in
12 cm

dark gray head

adipose fin

13 in
33 cm

spots on body

RED PIRANHA
Pygocentrus nattereri
F: Serrasalmidae
Native to S. American rivers, this fish usually eats invertebrates and fishes, but schools of them, frenzied by blood, can kill large mammals with their razor-sharp teeth.

red belly when fully grown

2½ in
6.5 cm

RIVER HATCHETFISH
Gasteropelecus sternicla
F: Gasteropelecidae
This insectivorous S. American fish can leap out of the water to escape predators and to catch aerial prey.

3¼ ft
1 m

TIGERFISH
Hydrocynus vittatus
F: Alestidae
Native to African rivers, this large predator has fanglike teeth and can eat fishes half as long as itself.

16 in
40 cm

LONGSNOUT DISTICHODUS
Distichodus lusosso
F: Distichodontidae
Unlike some of its piranha relatives, this is a peaceful herbivorous fish. It lives in streams in equatorial Africa.

CATFISHES

The predominantly freshwater catfishes have a long body and many whiskery mouth barbels used to help find food. There is a sharp spine in front of the dorsal fin. Most species in this order (Siluriformes) grub around on the bottom, searching for algae and small invertebrates.

12½ in
32 cm

STRIPED EEL CATFISH
Plotosus lineatus
F: Plotosidae
Juveniles of this tropical marine species form dense, ball-shaped shoals for protection. Solitary adults defend themselves with their venomous fin spines.

3¼ ft
1 m

TIGER SHOVELNOSE CATFISH
Pseudoplatystoma fasciatum
F: Pimelodidae
The long barbels of this S. American fish help it find food as it hunts over riverbeds at night searching out small fishes.

16 ft
5 m

SHEATFISH
Silurus glanis
F: Siluridae
From wetlands in C. Europe and Asia, this huge fish has been known to weigh more than 660 lb (300 kg), but due to heavy fishing, none this size exists today.

long anal fin

spine visible through transparent body

6 in
15 cm

BROWN BULLHEAD
Ameiurus nebulosus
F: Ictaluridae
This N. American catfish has venomous fin spines that help ward off predators when it is guarding its nest.

sensory mouth barbels help find food

GLASS CATFISH
Kryptopterus bicirrhis
F: Siluridae
A native of S.E. Asia, this small fish hangs motionless in the water; its transparent body makes it hard for predators to see.

20½ in
52 cm

SALMON AND RELATIVES

This order of marine and freshwater fishes includes many anadromous members (moving from sea to fresh water to breed). Powerful predators, the Salmoniformes have a large tail, a single dorsal fin, and a much smaller adipose fin.

adipose fin

SOCKEYE SALMON
Oncorhynchus nerka
F: Salmonidae
In red breeding color, this N. Pacific fish migrates into N. American and Asian rivers to spawn and die. The salmon are hunted by brown bears while on their difficult journey upriver.

33 in
84 cm

hooked jaw on breeding male

CISCO
Coregonus artedi
F: Salmonidae
Widespread in N. American lakes and large rivers, this fish forms shoals that feed on plankton and invertebrates.

22½ in
57 cm

ARCTIC CHAR
Salvelinus alpinus
F: Salmonidae
Clean, cold water is a must for this fish. Some individuals live in high-altitude lakes, while others migrate from sea to river.

3¼ ft
1 m

RAINBOW TROUT
Oncorhynchus mykiss
F: Salmonidae
Although native to N. America, this fish has been introduced into fresh waters for food and sport throughout the world.

4 ft
1.2 m

PIKES AND RELATIVES

Found in cool, fresh waters across the northern hemisphere, the Esociformes are fast and agile. The dorsal and anal fins are set far back, near the tail, to give these predatory fishes an instant forward thrust.

single dorsal fin set far back

distinctive markings

MUD MINNOW
Umbra krameri
F: Umbridae
The European mud minnow is now rare, as the small ditches and canals of its home—the Danube and Dniester river systems—disappear.

6½ in
17 cm

LIZARDFISHES AND RELATIVES

This diverse order of marine fishes inhabits shallow coastal waters as well as the deep sea. The Aulopiformes have a large mouth, with many small teeth, and can catch large prey. They have a single dorsal fin and a much smaller adipose fin.

triangular head

elongated pectoral fins

long pelvic fin used as prop

VARIEGATED LIZARDFISH
Synodus variegatus
F: Synodontidae
This resident of tropical reefs of the Indian and Pacific oceans perches on a coral head, remaining completely still, and then darts out to catch fishes.

16 in
40 cm

16 in
40 cm

BOMBAY DUCK
Harpadon nehereus
F: Synodontidae
During the monsoon season in the Indo-Pacific, schools of this small fish gather near river deltas to feed on material that is washed down.

FEELER FISH
Bathypterois longifilis
F: Ipnopidae
Perching on its pelvic fins and tail above the deep, muddy ocean floor, this fish uses its filamentlike pectoral fins to detect prey.

14½ in
37 cm

elongated pelvic fins

LANTERNFISHES AND RELATIVES

Lanternfishes are small, slim fishes with many photophores (light organs) that help them communicate in their deep, dark ocean habitat. Members of the order Myctophiformes, these species have large eyes and many migrate toward the surface at night to feed.

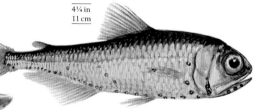

4¼ in
11 cm

SPOTTED LANTERNFISH
Myctophum punctatum
F: Myctophidae
Living in the dark depths of the Atlantic, this lanternfish uses its impressive array of photophores for camouflage as well as communication.

DRAGONFISHES AND RELATIVES

A majority of these deep-sea fishes have photophores to help them hunt, hide, and find mates. Most members of the order Stomiiformes are fearsome-looking predators, with large teeth and sometimes a long chin barbel.

row of photophores
large eyes

2¾ in
7 cm

PACIFIC HATCHETFISH
Argyropelecus affinis
F: Sternoptychidae
The silvery, thin body of this fish helps to camouflage and hide it from predators. It is found in temperate, tropical, and subtropical waters.

long, slender body

14 in
35 cm

9½ in
24 cm

NORTHERN STOPLIGHT LOOSEJAW
Malacosteus niger
F: Stomiidae
Found globally in temperate, tropical, and subtropical oceans, this fish pinpoints its shrimp prey with a beam of red bioluminescent light invisible to the shrimp.

SLOANE'S VIPERFISH
Chauliodus sloani
F: Stomiidae
Viperfishes have long, transparent fangs that protrude when the mouth is closed. Light from photophores confuses their prey in the depths of tropical and subtropical oceans.

KNIFEFISHES

Knifefishes have a sideways flattened, knifelike body, and a single long anal fin, which they use to glide backward and forward. The electric eel is atypical, with a long, round body. The Gymnotiformes live in fresh water and can produce electrical impulses.

duckbill-shaped snout

23½ in
60 cm

3¼ ft
99 cm

CHAIN PICKEREL
Esox niger
F: Esocidae
When hunting, this N. American pike uses delicate movements of its fins to hover motionless before striking with lightning speed.

ELECTRIC EEL
Electrophorus electricus
F: Gymnotidae
This large S. American fish can produce an electric shock of up to 600 volts—enough to kill other fishes and stun a human.

8¼ ft
2.5 m

BANDED KNIFEFISH
Gymnotus carapo
F: Gymnotidae
Found in murky wetlands in C. and S. America, this fish produces mild electric currents, which it uses to sense its surroundings.

SMELTS AND RELATIVES

Smelts resemble small, slim salmon and, like them, most have an adipose fin on the back near the tail. Some of the Osmeriformes have a distinctive smell, like the European smelt, which smells like fresh cucumber.

18 in
45 cm

EUROPEAN SMELT
Osmerus eperlanus
F: Opisthoproctidae
Resembling a cross between a herring and a trout, this fish is common in North Sea estuaries. In spring, smelt swim up rivers from the sea to spawn.

CAPELIN
Mallotus villosus
F: Osmeridae
Found in cold Arctic and nearby waters, this small fish forms large shoals that are a crucial food source for many seabirds; its abundance or scarcity determines the breeding success of these birds.

10 in
25 cm

OARFISHES AND RELATIVES

The 23 members of this marine order, called Lampriformes, are colorful fishes of the open ocean, the adults having crimson fins. In many species, the rays of the dorsal fin extend as long streamers. Most fishes in this order are ocean wanderers that are rarely seen.

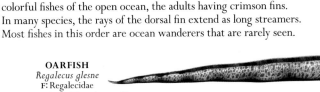

crest of elongated dorsal fin rays

OARFISH
Regalecus glesne
F: Regalecidae
The longest bony fish in the world, this huge fish has been the subject of many stories about sea serpents. It is found in tropical to temperate waters worldwide.

BLUE JUVENILE

36 ft
11 m

pectoral fin

6½ ft
2 m

OPAH
Lampris guttatus
F: Lampridae
This rarely seen fish beats its long pectoral fins like wings. Inhabiting tropical to temperate oceans, it eats squid and small fishes.

moderately forked tail fin

ANGLERFISHES AND RELATIVES

The 376 or so Lophiiformes include some of the most bizarre of all marine fishes. A modified fin ray on top of the head acts as a fishing lure (bioluminescent in deep-water species), which attracts prey toward the fish's cavernous mouth.

8 in
20 cm

RED-LIPPED BATFISH
Ogcocephalus darwini
F: Ogcocephalidae
The oddly shaped batfish scrambles over the seabed using its paired pectoral and pelvic fins as props. Its red lips are a mystery.

COFFINFISH
Chaunax endeavouri
F: Chaunacidae
Coffinfishes lie on muddy seabeds in the S.W. Pacific, waiting for small fishes to stray within reach.

9 in
22 cm

ANGLER
Lophius piscatorius
F: Lophiidae
A fringe of seaweed-shaped flaps around its mouth helps disguise this N.E. Atlantic angler. It can strike with lightning speed.

6½ ft
2 m

8 in
20 cm

FANFIN ANGLER
Caulophryne jordani
F: Caulophrynidae
In the dark ocean depths it is difficult to find a mate. Once successful, the tiny male fanfin angler latches onto the female permanently.

4½ in
11.5 cm

WARTY FROGFISH
Antennarius maculatus
F: Antennariidae
The well-camouflaged warty frogfish clambers over coral reefs using its limblike pectoral fins.

SARGASSUMFISH
Histrio histrio
F: Antennariidae
While most frogfishes live on the seabed, this one hides among rafts of floating *Sargassum* seaweed, which it closely resembles.

8 in
20 cm

large pectoral fins used for scrambling

skin flaps for camouflage

COD AND RELATIVES

The Gadiformes include many important marine commercial species. Most have two or three soft dorsal fins on their back and many have a chin barbel. Grenadiers live in deep water and have a long thin tail.

6½ ft
2 m

ATLANTIC COD
Gadus morhua
F: Gadidae
Overfishing has reduced the average weight of Atlantic cod to 24 lb (11 kg) from an historical maximum weight of over 200 lb (90 kg).

chin barbel

3 ft
91 cm

ALASKA POLLOCK
Gadus chalcogrammus
F: Gadidae
The lack of a chin barbel and a protruding lower jaw help distinguish this fish from cod. It lives in cold Arctic waters.

20 in
50 cm

SHORE ROCKLING
Gaidropsarus mediterraneus
F: Lotidae
Equipped with three sensory mouth barbels, this eellike rockling searches for food in N.E. Atlantic rock pools.

colorful tail fin with rounded edge

4 ft
1.2 m

BURBOT
Lota lota
F: Lotidae
Unlike all other Gadiformes, the burbot lives in fresh water. It is found in deep lakes and rivers across the northern hemisphere.

3¼ ft
1 m

PACIFIC GRENADIER
Coryphaenoides acrolepis
F: Macrouridae
A long, scaly tail and bulbous head give this abundant deep-water cod relative its alternative name of Pacific rattail.

CUSK EELS

Most members of the order Ophidiiformes live in the ocean and are long, slim, eellike fishes. They have thin pelvic fins and long dorsal and anal fins, which in many species join onto the tail fin.

PEARLFISH
Carapus acus
F: Carapidae
The adult pearlfish finds refuge inside a sea cucumber, entering tail-first through the anus, exiting at night to feed.

8½ in
21 cm

GRAY MULLETS

Gray mullets are silvery striped fishes with two widely separated dorsal fins; the first has sharp spines, the second has soft rays. Members of the Mugiliformes are distributed worldwide. They are vegetarian, feeding on fine algae and detritus.

30 in
75 cm

GOLDEN GRAY MULLET
Liza aurata
F: Mugilidae
Harbors, estuaries, and coastal waters in the N.E. Atlantic are all likely haunts of the this mullet, which often lives in shoals.

341

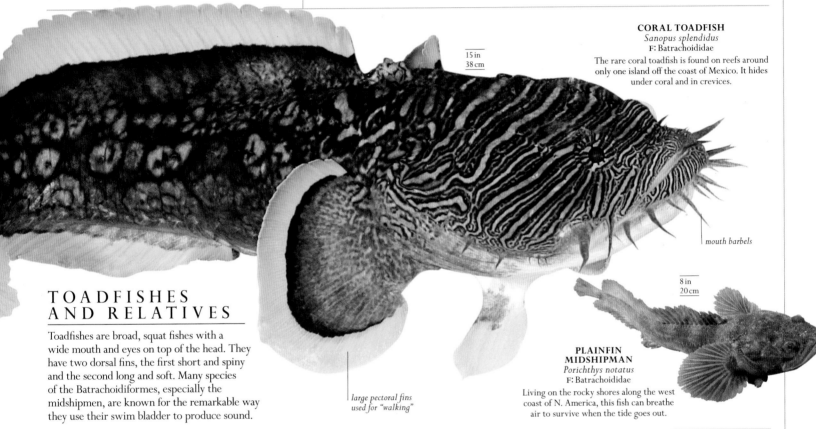

15 in
38 cm

CORAL TOADFISH
Sanopus splendidus
F: Batrachoididae
The rare coral toadfish is found on reefs around only one island off the coast of Mexico. It hides under coral and in crevices.

mouth barbels

TOADFISHES AND RELATIVES

Toadfishes are broad, squat fishes with a wide mouth and eyes on top of the head. They have two dorsal fins, the first short and spiny and the second long and soft. Many species of the Batrachoidiformes, especially the midshipmen, are known for the remarkable way they use their swim bladder to produce sound.

large pectoral fins used for "walking"

8 in
20 cm

PLAINFIN MIDSHIPMAN
Porichthys notatus
F: Batrachoididae
Living on the rocky shores along the west coast of N. America, this fish can breathe air to survive when the tide goes out.

SILVERSIDES AND RELATIVES

These small, slim, silvery fishes often live in large shoals. There are more than 330 Atheriniformes, occuring both in marine and freshwater habitats. Most have two dorsal fins, the first of which has flexible spines, and a single anal fin.

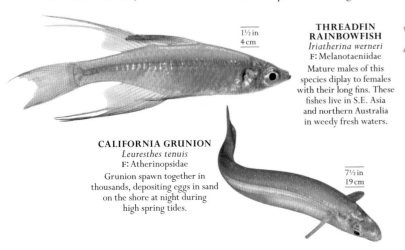

1½ in
4 cm

THREADFIN RAINBOWFISH
Iriatherina werneri
F: Melanotaeniidae
Mature males of this species diplay to females with their long fins. These fishes live in S.E. Asia and northern Australia in weedy fresh waters.

CALIFORNIA GRUNION
Leuresthes tenuis
F: Atherinopsidae
Grunion spawn together in thousands, depositing eggs in sand on the shore at night during high spring tides.

7½ in
19 cm

NEEDLEFISHES AND RELATIVES

With a long, thin, rodlike body and extended beaklike jaws, these silvery fishes are well camouflaged in the open ocean. Flying fishes—with their large, paired pectoral (side) and pelvic (belly) fins—also belong to this order, the Beloniformes.

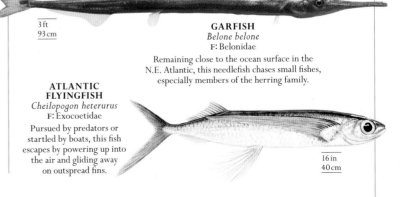

3 ft
93 cm

GARFISH
Belone belone
F: Belonidae
Remaining close to the ocean surface in the N.E. Atlantic, this needlefish chases small fishes, especially members of the herring family.

ATLANTIC FLYINGFISH
Cheilopogon heterurus
F: Exocoetidae
Pursued by predators or startled by boats, this fish escapes by powering up into the air and gliding away on outspread fins.

16 in
40 cm

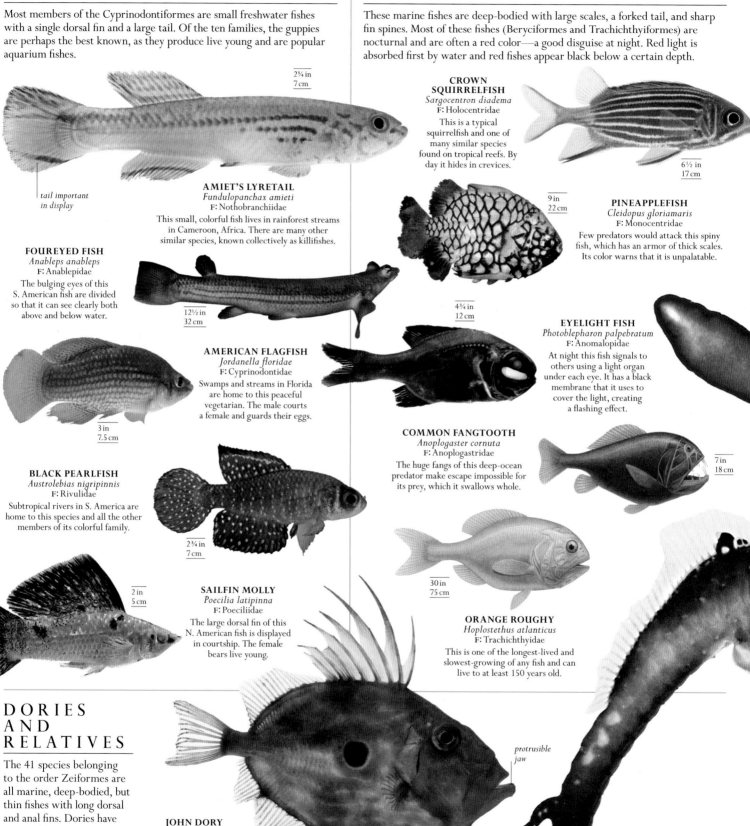

KILLIFISHES AND RELATIVES

Most members of the Cyprinodontiformes are small freshwater fishes with a single dorsal fin and a large tail. Of the ten families, the guppies are perhaps the best known, as they produce live young and are popular aquarium fishes.

2¾ in
7 cm

tail important
in display

AMIET'S LYRETAIL
Fundulopanchax amieti
F: Nothobranchiidae
This small, colorful fish lives in rainforest streams in Cameroon, Africa. There are many other similar species, known collectively as killifishes.

FOUREYED FISH
Anableps anableps
F: Anablepidae
The bulging eyes of this S. American fish are divided so that it can see clearly both above and below water.

12½ in
32 cm

AMERICAN FLAGFISH
Jordanella floridae
F: Cyprinodontidae
Swamps and streams in Florida are home to this peaceful vegetarian. The male courts a female and guards their eggs.

3 in
7.5 cm

BLACK PEARLFISH
Austrolebias nigripinnis
F: Rivulidae
Subtropical rivers in S. America are home to this species and all the other members of its colorful family.

2¾ in
7 cm

2 in
5 cm

SAILFIN MOLLY
Poecilia latipinna
F: Poeciliidae
The large dorsal fin of this N. American fish is displayed in courtship. The female bears live young.

DORIES AND RELATIVES

The 41 species belonging to the order Zeiformes are all marine, deep-bodied, but thin fishes with long dorsal and anal fins. Dories have protrusible jaws—they can be shot out to capture prey. They feed on a wide range of small fishes that they stalk head on.

JOHN DORY
Zeus faber
F: Zeidae
Viewed face on, this ultra-thin fish is hard to see and can stalk and strike its prey very effectively.

3 ft
90 cm

SQUIRRELFISHES AND ROUGHIES

These marine fishes are deep-bodied with large scales, a forked tail, and sharp fin spines. Most of these fishes (Beryciformes and Trachichthyiformes) are nocturnal and are often a red color—a good disguise at night. Red light is absorbed first by water and red fishes appear black below a certain depth.

CROWN SQUIRRELFISH
Sargocentron diadema
F: Holocentridae
This is a typical squirrelfish and one of many similar species found on tropical reefs. By day it hides in crevices.

6½ in
17 cm

9 in
22 cm

PINEAPPLEFISH
Cleidopus gloriamaris
F: Monocentridae
Few predators would attack this spiny fish, which has an armor of thick scales. Its color warns that it is unpalatable.

4¾ in
12 cm

EYELIGHT FISH
Photoblepharon palpebratum
F: Anomalopidae
At night this fish signals to others using a light organ under each eye. It has a black membrane that it uses to cover the light, creating a flashing effect.

COMMON FANGTOOTH
Anoplogaster cornuta
F: Anoplogastridae
The huge fangs of this deep-ocean predator make escape impossible for its prey, which it swallows whole.

7 in
18 cm

30 in
75 cm

ORANGE ROUGHY
Hoplostethus atlanticus
F: Trachichthyidae
This is one of the longest-lived and slowest-growing of any fish and can live to at least 150 years old.

protrusible
jaw

STICKLEBACKS AND SEAMOTHS

Most sticklebacks inhabit still and slow-moving fresh water, but some live in the sea, as do seamoths. The Gasterosteiformes have a long, thin, stiff body protected by bony scutes (plates) along the sides and sharp spines along the back.

spines

4¼ in
11 cm

THREE-SPINED STICKLEBACK
Gasterosteus aculeatus
F: Gasterosteidae
This little fish is widespread in fresh water and shallow seas in the northern hemisphere. The male performs an elaborate courtship dance.

bony scutes

males have red belly when breeding

SEAMOTH
Eurypegasus draconis
F: Pegasidae
Unlike the closely related sticklebacks, the tropical marine seamoth is flat and has large winglike pectoral (side) fins.

2¾ in
7 cm

PIPEFISHES AND SEAHORSES

Seahorses and other similar fishes in the order Syngnathiformes are encased in a body armor of bony plates, making their bodies very stiff. The group includes both marine and freshwater species. Seahorses have a small mouth at the end of a tubular snout, which they use to feed on tiny planktonic crustaceans.

tiny pectoral fin helps maintain position

18 in
46 cm

WEEDY SEADRAGON
Phyllopteryx taeniolatus
F: Syngnathidae
This large, bizarrely shaped Australian seadragon hides among seaweeds on rocky reefs, camouflaged by its many leaflike skin flaps.

YELLOW SEAHORSE
Hippocampus kuda
F: Syngnathidae
As in all seahorses, the male of this species has a belly pouch in which he broods eggs laid there by the female.

12 in
30 cm

long, tubular snout

6 in
15 cm

RAZORFISH
Aeoliscus strigatus
F: Centriscidae
Hanging head down amongst urchin spines disguises the razorfish, which habitually swims in this position. This is an Indo-Pacific species.

RINGED PIPEFISH
Dunckerocampus dactyliophorus
F: Syngnathidae
The long, thin body of this coral-reef dweller is typical of pipefishes. It hovers under and between corals and rocks.

7 in
18 cm

skin flaps provide camouflage in seaweed

TRUMPETFISH
Aulostomus chinensis
F: Aulostomidae
Trumpetfishes often shadow moray eels, which hunt over coral reefs in order to snap up small fishes that the eels have flushed out.

32 in
80 cm

CLINGFISHES AND RELATIVES

Clingfishes are small, usually marine, bottom-dwelling fishes. Most of the Gobiesociformes possess a sucker formed from modified pelvic fins with which they cling to rocks. Their eyes are set high on the head and they have one dorsal fin.

3¼ in
8 cm

CONNEMARA CLINGFISH
Lepadogaster candolii
F: Gobiesocidae
Living in rocky shallows in the N.E. Atlantic, this little fish is exposed to strong waves, but can cling on tightly.

6½ in
16 cm

ROBUST GHOST PIPEFISH
Solenostomus cyanopterus
F: Solenostomidae
Large pelvic fins enable the ghost pipefish to drift and swim slowly among weeds and seagrasses as it hunts minute invertebrates.

SWAMPEELS AND RELATIVES

These tropical and subtropical freshwater fishes have an eellike body and most Synbranchiformes have tiny or no fins. This makes it easy for them to slither through swamps and marshes and burrow in mud.

3¼ ft
1 m

FIRE EEL
Mastacembelus erythrotaenia
F: Mastacembelidae
Living in flooded lowland plains and slow rivers in S.E. Asia, this edible spiny eel feeds on insect larvae and worms.

MARBLED SWAMPEEL
Synbranchus marmoratus
F: Synbranchidae
Able to breathe air if necessary, this almost finless fish survives in very small bodies of water in C. and S. America.

5 ft
1.5 m

FLATFISHES

Pleuronectiformes start life in the plankton as normal upright juveniles. As they grow, they become flattened sideways and lie on one side on the seabed. The underneath eye migrates around to join the other on the upper side.

1 m
3¼ ft

EUROPEAN PLAICE
Pleuronectes platessa
F: Pleuronectidae
This North Atlantic commercial flatfish lies camouflaged on the seabed with its right side uppermost. It emerges at night to feed.

ATLANTIC HALIBUT
Hippoglossus hippoglossus
F: Pleuronectidae
One of the largest flatfishes, the Atlantic halibut lies on its left side with both eyes on the upward-facing right side.

8¼ ft
2.5 m

COMMON SOLE
Solea solea
F: Soleidae
Although they can live to be 30 years old, most sole do not survive this long, as they are a valuable commercial fish.

28 in
70 cm

right eye has moved to top of the fish

TURBOT
Scophthalmus maximus
F: Scophthalmidae
The ability to alter its color to match the seabed helps the turbot escape predators' attention. It is a valuable North Atlantic species.

3¼ ft
1 m

TRIGGERFISHES, PUFFERFISHES, AND RELATIVES

This diverse group of marine and freshwater fishes includes the huge ocean sunfish and the poisonous pufferfish. Instead of normal teeth, the Tetraodontiformes have fused tooth plates or just a few large teeth. Their scales are modified to form protective spines or plates.

20 in
50 cm

CLOWN TRIGGERFISH
Balistoides conspicillum
F: Balistidae
This brightly colored coral-reef fish can wedge itself in a crevice by erecting its dorsal spines, which lock in position.

large tail

10 in
25 cm

SPOTTED BOXFISH
Ostracion meleagris
F: Ostraciidae
Encased in a rigid box of fused bony plates, and with poisonous skin, this Indo-Pacific reef fish is left alone by predators.

males have violet blue sides

top is the left side of the fish

6 in
15 cm

WHITE-SPOTTED PUFFER
Arothron hispidus
F: Tetraodontidae
The neurotoxin contained within the skin and organs of this fish can easily kill a human and deters natural predators.

20 in
50 cm

LONG-SPINE PORCUPINEFISH
Diodon holocanthus
F: Diodontidae
Found in all tropical seas, this spiny fish can suck in water to inflate its body and turn itself into a prickly ball—a very effective deterrent against predators.

MIMIC FILEFISH
Paraluteres prionurus
F: Monacanthidae

The mimic filefish is avoided by predators because it closely resembles the saddled pufferfish, which has toxic flesh.

4¼ in
11 cm

flat back forms top of bony protective box

SCORPIONFISHES AND RELATIVES

Mostly bottom-living and marine, fishes in this large order have a large spiny head with a unique bony strut across the cheek. Most Scorpaeniformes have sharp, sometimes venomous spines in their dorsal fins and many are camouflage experts.

SMALLSCALED SCORPIONFISH
Scorpaena porcus
F: Scorpaenidae

Well concealed by skin flaps on the head and its ability to change color, the scorpionfish is hard to spot.

14½ in
37 cm

TUB GURNARD
Chelidonichthys lucerna
F: Triglidae

Walking over the seabed on three mobile rays on each of its pectoral fins, the tub gurnard can probe for hidden invertebrates.

30 in
75 cm

JUVENILE

18 in
45 cm

RED LIONFISH
Pterois volitans
F: Scorpaenidae

The striped colors warn predators that even juveniles of this coral-reef fish have venomous dorsal fin spines. Adults darken with age and can be almost black.

8½ in
21 cm

LUMPSUCKER
Cyclopterus lumpus
F: Cyclopteridae

A strong sucker on the belly allows this rotund North Atlantic fish to cling onto wave-battered rocks and guard its eggs.

23½ in
60 cm

BIG BAIKAL OILFISH
Comephorus baikalensis
F: Comephoridae

About a quarter of the body of this fish is oil, which provides it with buoyancy. It is endemic to Lake Baikal, Russia.

small mouth has stout teeth to tear off sponges

venomous dorsal fin spines

large, upturned mouth

OCEAN SUNFISH
Mola mola
F: Molidae

Often lying on its side at the surface, this fish is nearly the heaviest bony fish. The record is held by the bump-head sunfish *M.alexandrini* at 5,070 lb (2,300 kg).

11ft
3.3 m

LONGSPINED BULLHEAD
Taurulus bubalis
F: Cottidae

The color of this coastal and shore fish varies widely, according to its background. For example, fishes living among red seaweeds are red.

10 in
25 cm

STONEFISH
Synanceia verrucosa
F: Synanceiidae

It is extremely difficult to spot this well camouflaged tropical reef fish. A sting from its venomous spines can be fatal to humans.

16 in
40 cm

BULLHEAD
Cottus gobio
F: Cottidae

The bullhead lives among stones and vegetation in freshwater streams and rivers throughout much of Europe. The male guards the eggs.

7 in
18 cm

20 in
50 cm

FLYING GURNARD
Dactylopterus volitans
F: Dactylopteridae

This species uses its huge fanlike pectoral fins to "fly" through the water. It "takes off" from the seabed if disturbed.

RED LIONFISH
Pterois volitans

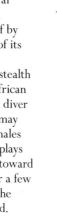

The red lionfish is a night hunter, patrolling tropical coral and rocky reefs in the western Pacific in search of small fishes and crustaceans. It corrals its prey against the reef by spreading out its wide pectoral fins—one on each side of its body—before gulping in the prey with lightning speed. Sometimes it stalks its prey out in the open, relying on stealth and a final quick rush, similar to a lion hunting on an African plain. Protected by venomous spines, it will often face a diver or potential predator head-on if approached. One male may collect a small harem of females and will charge other males that stray too close. When ready to spawn, the male displays to the female, circling around her before they swim up toward the surface to shed eggs and sperm into the water. After a few days the eggs hatch into planktonic larvae that drift in the plankton for about a month before settling on the seabed.

SIZE 18 in (45 cm) in length
HABITAT Coral and rock reefs
DISTRIBUTION Pacific Ocean, introduced into W. Atlantic
DIET Fishes and crustaceans

> VIBRANT WARNING
The contrasting striped pattern of the red lionfish acts as a warning to predators that the fish is venomous and they should keep away. On land, stinging wasps use a similar method to avoid being eaten.

eye is disguised by a dark stripe that confuses potential predators

< MENACING PREDATOR
This efficient predator has been released from aquariums into the Caribbean, where it is a threat to native reef fishes. It hunts at dusk using its large eyes and acute sense of smell.

head tentacle

∧ VARIED STRIPES
The striped pattern differs between individuals, and becomes much less obvious in breeding males, which can be very dark.

< PECTORAL FINS
The soft rays of the pectoral fins are joined together for part of their length by a fine membrane, which is marked with circular splotches of color.

VENOMOUS SPINES >
Sharp spines in this fish's dorsal, anal, and pelvic fins can be used to inject venom that causes extreme pain in humans, but is rarely fatal. The spines are purely defensive.

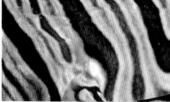

fleshy "whiskers" may help disguise the fish's large, open mouth when approaching prey

< TAIL UP
Lionfishes often hang slightly head down and tail up, ready to pounce on passing prey. The tail helps the fishes to maintain this position, rather than to swim fast.

dorsal fin made up
of sharp spines

wide rays of pectoral
fins spread out fully
when the lionfish
ambushes prey

tail

anal fin

SPOTTED CTENOPOMA
Ctenopoma acutirostre
F: Anabantidae

4¾ in
12 cm

This tropical freshwater fish is found in the Congo basin of Africa. It often stalks its prey with its head held down.

SIAMESE FIGHTING FISH
Betta splendens
F: Osphronemidae

2½ in
6.5 cm

The original range of this Asian freshwater fish is unclear, but it has been bred for centuries for its fighting ability, particularly strong in males.

POWDERBLUE SURGEONFISH
Acanthurus leucosternon
F: Acanthuridae

9 in
23 cm

This Indian Ocean fish has a sharp, bladelike structure sheathed on either side of its tail base, which it can flick out to slash a would-be attacker.

BUTTERFLY BLENNY
Blennius ocellaris
F: Blenniidae

8 in
20 cm

Like many of its relatives, this fish of the N.E. Atlantic lives on the seabed and guards its eggs, which are often laid in an empty shell.

broad tail fin

FIREFISH
Nemateleotris magnifica
F: Microdesmidae

3½ in
9 cm

Hovering above its coral reef burrow, this fish gathers plankton from the water and darts back into safety if it senses danger.

elongated first ray on dorsal fin

MANDARINFISH
Synchiropus splendidus
F: Callionymidae

2¼ in
6 cm

Native to the Pacific, this is one of the most colorful of all tropical reef fishes. Its vivid colors warn predators of its foul taste.

huge, saillike dorsal fin

ATLANTIC SAILFISH
Istiophorus albicans
F: Istiophoridae

10 ft
3.2 m

This ocean predator uses its upper jaw, which extends as a long spear, to slash into shoals of fishes and stun them.

spear-shaped snout

torpedo-shaped body
projecting lower jaw

GREAT BARRACUDA
Sphyraena barracuda
F: Sphyraenidae

6½ ft
2 m

Found globally in tropical and subtropical waters, this solitary predator stalks its prey and then accelerates rapidly as it strikes.

ATLANTIC MACKEREL
Scomber scombrus
F: Scombridae

23½ in
60 cm

Living in large shoals in the North Atlantic, this fish feeds voraciously on small fishes and plankton. Its streamlined body makes it a fast swimmer.

BUMBLEBEE FISH
Brachygobius doriae
F: Gobiidae

1½ in
4 cm

Found in S.E. Asia, this bottom-dwelling goby is tolerant of brackish water and lives in estuaries and mangroves.

PAINTED GOBY
Pomatoschistus pictus
F: Gobiidae

3½ in
9 cm

This and other similar gobies are common in shallow sediment areas of the N.E. Atlantic. DNA analysis suggests gobies might be placed in a separate order, Gobiiformes.

NORTHERN BLUEFIN TUNA
Thunnus thynnus
F: Scombridae

15 ft
4.5 m

Found globally, this tuna is one of the world's most valuable commercial fishes. A fast, dynamic predator, it roams widely, hunting small fishes.

long tail

10 in
25 cm

high-set, bulbous eyes

ATLANTIC MUDSKIPPER
Periophthalmus barbarus
F: Gobiidae

As long as it stays moist, this mudskipper can stay out of water for hours by absorbing oxygen through its skin.

LOBE-FINNED FISHES

Once considered to be the ancestors of land vertebrates, lobe-finned fishes have fins that resemble primitive limbs, with a fleshy base preceding the fin membrane.

Like the ray-finned fishes, these are bony fishes with a hard skeleton but their fins have a different structure. The fin membrane is supported on a muscular lobe that projects out from the body and is strong enough to let some of these fishes shuffle along on their paired pectoral and pelvic fins. Bones and cartilage inside the lobes provide attachment for muscles. There are many fossil groups of lobe-finned fishes, but living representatives are confined to the marine coelacanths and the freshwater lungfishes.

A LIVING FOSSIL

Coelacanths are nocturnal and secretive. The first living specimen was discovered in 1938—until then only fossil species over 65 million years old were known. The historic find belonged to a species living in deep rocky areas in the western Indian Ocean. A second species was found in Indonesian waters in 1998. They have several unique structural features including an incomplete backbone, residual notochord, and a tail fin with a characteristic extra middle lobe. Their scales are heavy, bony plates and they are not long-distance swimmers. Unlike the egg-laying lungfishes, coelacanths bear live young as the eggs hatch internally. Gestation may be as long as three years, and this makes their survival precarious if individuals continue to be caught in the nets of deep-sea trawlers.

BREATHING OUT OF WATER

Although most of their fossil ancestors lived in the ocean, modern lungfishes are limited to freshwater habitats in South America, Africa, and Australia. All can, to some extent, breathe air via a connection to the swim bladder—useful when pools dry out seasonally. Some species can survive buried in mud for many months and would die if kept permanently submerged in water, while others still rely mainly on gills for breathing. Their shape and the fact that the larvae of some species have external gills led early zoologists to believe that lungfishes were amphibians.

PHYLUM	CHORDATA
CLASSES	COELACANTHI
	DIPNEUSTI
ORDERS	3
FAMILIES	4
SPECIES	8

DEBATE

FISHES ON LAND

While it is generally accepted that land-living vertebrates evolved from a primitive fish or fishlike ancestor in the sea, finding that ancestor is more difficult. Recent work suggests that lungfishes are more closely related to tetrapods (four-limbed vertebrates such as mammals) than are coelacanths. In 2002 a fossil lobe-finned fish, *Styloichthys*, was found in China. It appears to show close links between lungfishes and tetrapods. The debate continues, but currently coelacanths are not considered direct ancestors of tetrapods.

AFRICAN LUNGFISHES

All four species of African lungfishes have a long body and threadlike pectoral and pelvic fins, collectively called paired fins. The Lepidosireniformes breathe through a pair of lungs derived from their swim bladder.

6½ ft
2 m

WEST AFRICAN LUNGFISH
Protopterus annectens
F: Protopteridae
When the lakes in which it lives dry out, this lungfish survives by burying itself in mud, forming a cocoon with an inlet for air.

COELACANTHS

The two primitive fishes from the order Coelacanthiformes have limblike, fleshy bases to their pectoral and pelvic fins, and large bony scales. In life, these fishes are a metallic blue with pale spots, but the color fades on death. Interesting 'head-standing' behavior has been observed from a submersible.

triple-lobed tail fin

4½ ft
1.4 m

INDONESIAN COELACANTH
Latimeria menadoensis
F: Latimeriidae
Molecular studies have shown this to be a separate species from the coelacanth, although they are physically similar. It lives in the Celebes Sea.

AUSTRALIAN LUNGFISHES

The single species in the order Ceratodontiformes has a long body, large scales, paddlelike paired fins, and a tapering tail. It can breathe through its lungs for short periods, but cannot survive if its habitat dries out.

body flecked with white speckles

AUSTRALIAN LUNGFISH
Neoceratodus forsteri
F: Neoceratodontidae
Living in deep pools and rivers, this lungfish can survive in stagnant water by gulping air into its swim bladder.

6 ft
1.8 m

limblike fins

6½ ft
2 m

fin with fleshy stalk

COELACANTH
Latimeria chalumnae
F: Latimeriidae
This fish lives off the coasts of southern Africa and Madagascar in steep, rocky, underwater terrain and hides in deep ocean caves during the day.

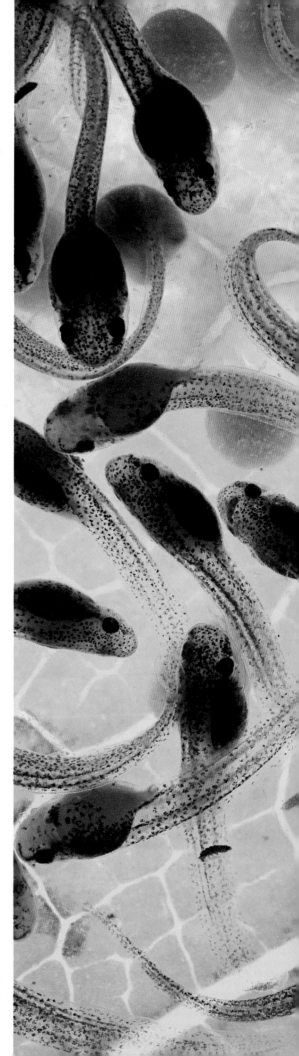

AMPHIBIANS

Amphibians are cold-blooded vertebrates that thrive in freshwater habitats. Some spend their whole lives in water; others only require water to breed. On land, they must find damp places, as their skin is permeable and does not protect them from drying out.

PHYLUM	CHORDATA
CLASS	AMPHIBIA
ORDERS	3
FAMILIES	74
SPECIES	8,212

Large numbers of male and female Costa Rican golden toads gather in a pool to mate. The species is now extinct.

A female of the North American Jefferson's salamander attaches a mass of eggs to a submerged twig in early spring.

When the eggs hatch, the male red-backed poison frog carries the tadpoles to bromeliad pools in the canopy.

DEBATE
OUR BIGGEST CHALLENGE?

A third of the world's amphibians face extinction in the near future and this poses a huge conservation challenge. This danger of extinction is largely due to the destruction and pollution of freshwater habitats, but amphibians are also threatened by the global spread of the disease chytridiomycosis, caused by a fungus that invades their soft skin.

The three orders of living amphibians are thought to share a common ancestor, but their origins are still uncertain. There is an enormous gap in the fossil record between the first land animals to evolve from fishes—the tetrapods, some 375 million years ago— and a froglike creature that lived 230 million years ago.

Amphibians have a unique and complex life cycle and occupy very different ecological niches at different stages of their lives. For most amphibians, their eggs hatch into larvae—called tadpoles in frogs and toads— that live in water, often in very high densities, and feed mostly on algae and other plant material. As larvae, they grow rapidly and then undergo a complete change of form—metamorphosis—to become land-living adults. As adults, all amphibians are carnivores, most feeding on insects and other small invertebrates. They typically lead secretive and solitary lives, except when they return to ponds and streams to breed. Most amphibians thus require two very different habitats during their lives: water and land. During metamorphosis, they undergo a wide variety of anatomical and physiological changes—from aquatic creatures that swim with a tail and breathe through gills, to terrestrial animals that move with four limbs and breathe through lungs.

DIVERSE PARENTAL CARE
Some amphibians produce huge numbers of eggs that are left to fend for themselves, so that only a very few will survive. Many others have evolved various forms of parental care. This is generally associated with the production of far fewer offspring, so that reproductive success is achieved by caring for a manageable number of young, rather than by producing as many eggs as possible. Parental care takes many forms, including defending eggs or larvae against predators, feeding tadpoles with unfertilized eggs, and carrying tadpoles from one place to another. In some species, such as the midwife toads and some poison frogs, parental care is the duty of the father. In many salamanders and caecilians, the mother is the sole protector of the young. In a few frog species, mother and father establish a durable pair-bond and share parental duties.

READY TO HATCH >
Tadpoles of the Tanzanian Mitchell's reed frog wriggle within their egg membranes, just prior to hatching.

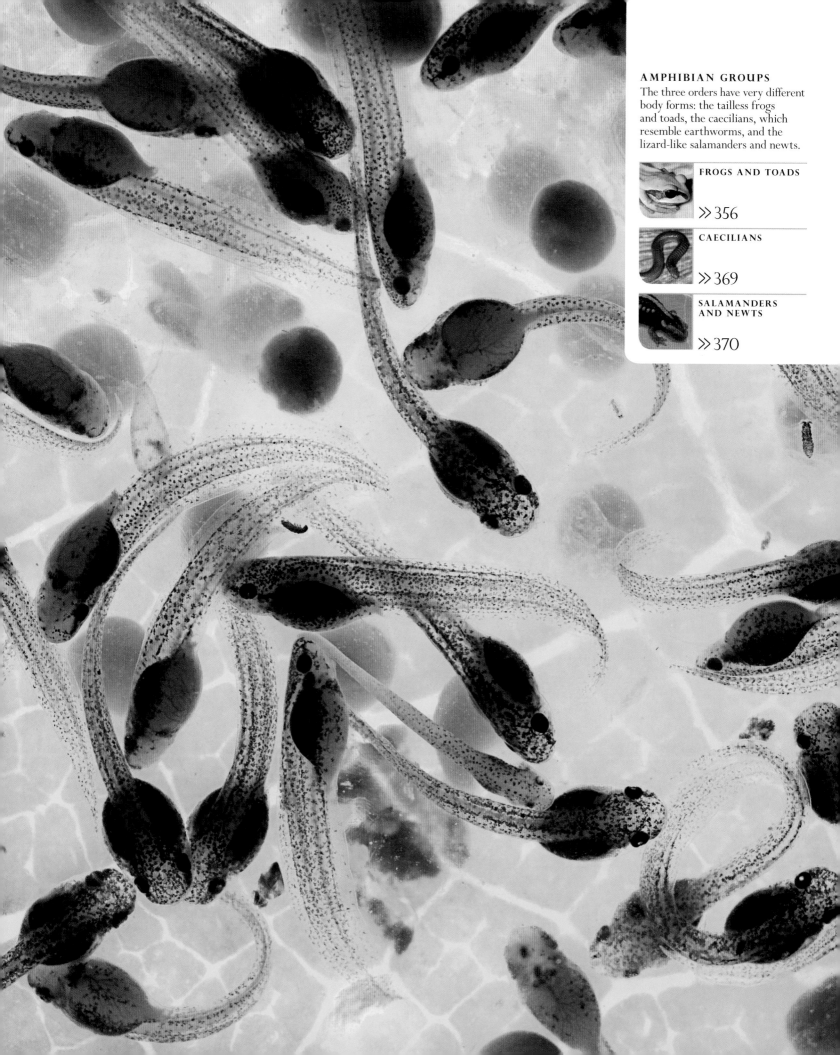

AMPHIBIAN GROUPS
The three orders have very different
body forms: the tailless frogs
and toads, the caecilians, which
resemble earthworms, and the
lizard-like salamanders and newts.

FROGS AND TOADS

With powerful hind limbs tucked under its body, a wide mouth, and protruding eyes, the typical frog has a unique and unmistakable body shape.

Anura, the name of the order of frogs and toads, means "animals without a tail." As adults, all other amphibians have a tail, but in the anurans, the tail gradually disappears during metamorphosis from the larval to the adult stage of the life cycle. The larvae, known as tadpoles, feed chiefly on plant material and have a spherical body that contains the long, coiled gut required by such a diet. The adults, in contrast, are totally carnivorous, feeding on a wide range of insects and other invertebrates, with larger species also taking small reptiles and mammals, as well as other frogs.

INGENIOUS ADAPTATIONS

Frogs and toads catch their prey by ambush, most spectacularly by jumping. In many frogs and toads, the hind limbs are modified for jumping, being much longer than the forelimbs and very muscular. Jumping is also an effective method of escaping from predators, of which frogs and toads have many.

Not all anurans jump, however. Many have hind limbs adapted for other kinds of locomotion, including swimming, burrowing, climbing, and, in a few species, gliding through the air. Most frogs and toads live in damp habitats, close to the pools and streams in which they breed, but there are several species adapted to life in very arid areas. The greatest diversity of anurans is found in the tropics, particularly in rainforests. Many are active by day, while others are nocturnal. Some species are cleverly camouflaged; in contrast, others are brightly colored, advertising that they are poisonous or unpleasant to taste.

COURTSHIP AND BREEDING

Anurans differ from other amphibians in having a voice and very good hearing. Males of most species call to attract females, making sounds characteristic of their species. In all but a very few species, fertilization is external, the male shedding sperm onto the eggs as they emerge from the female's body. To do this, the male clasps the female from above—a position known as amplexus. The duration of amplexus varies between species—from a few minutes to several days.

PHYLUM	CHORDATA
CLASS	AMPHIBIA
ORDER	ANURA
FAMILIES	54
SPECIES	7,244

FROG OR TOAD

The distinction between frogs and "toads" is biologically meaningless and the two words are used differently in different parts of the world. In Europe and North America, for example, the term "toad" refers to species in the family Bufonidae, but this also includes the South and Central American harlequin frogs. In general, toads have rough skins, are slow-moving, and often burrow in the ground, whereas frogs are smooth-skinned, agile, and fast-moving, and spend much of their time in water. A group of smooth-skinned, aquatic frogs native to Africa were once known as clawed toads, but are now called clawed frogs.

ALYTIDAE

Males of the Alytidae, a small family of terrestrial frogs, call at night to attract females. When mating, the male midwife toad attaches fertilized eggs to his back. He carries them until they are ready to hatch, when he releases the tadpoles into water. Occasionally, a male may carry egg strings from more than one female. With painted frogs, a female mates with several males, producing up to 1,000 eggs, which are dropped into water.

SQUEAKERS AND RELATIVES

The Arthroleptidae are a large, diverse family of frogs occurring across sub-Saharan Africa—in forest, woodland, and grassland, some at high altitudes. Its members range from tiny "squeakers," named for their high-pitched call, to large treefrogs.

¾–1¼ in
2–3 cm

eggs attached to male's back

vertical, slitlike pupil

1¼–2 in
3–5 cm

COMMON MIDWIFE TOAD
Alytes obstetricans
Found in W. and C. Europe, this midwife toad has a plump body and powerful forelimbs adapted for digging. It hides in a burrow by day.

1¼–1½ in
3–4 cm

RIO BENITO LONG-FINGERED FROG
Cardioglossa gracilis
An inhabitant of lowland forest, this frog breeds in streams. Males call from nearby slopes.

very long third finger

WEST AFRICAN SCREECHING FROG
Arthroleptis poecilonotus
The female of this small species lays large eggs in soil cavities. The male is noted for his loud call.

1–1½ in
2.5–4 cm

1½–2¼ in
4–5.5 cm

WEST CAMEROON FOREST TREEFROG
Leptopelis nordequatorialis
This large treefrog lives in montane grasslands of W. Africa. Males call to females near water when breeding, and eggs are deposited in ponds or marshes.

AFRICAN TREEFROG
Leptopelis modestus
This species is found near streams in forests of W. and C. Africa. Females are larger than males.

GLASS FROGS

Found in Central and South America, frogs of the family Centrolenidae are known as "glass frogs" because many species have transparent skin on the underside through which their internal organs are visible.

silver eyes with black reticulations

LIMON GIANT GLASS FROG
Sachatamia ilex
This arboreal frog lives in wet vegetation near streams. It has dark green bones, visible through its skin.

adhesive pad at tip of digits

¾–1¼ in
2–3 cm

1–1½ in
2.5–3.5 cm

FLEISCHMANN'S GLASS FROG
Hyalinobatrachium fleischmanni
Males of this species are territorial, using calls to defend their territory and to attract females. The females lay eggs on leaves over water.

¾–1¼ in
2–3 cm

WHITE-SPOTTED COCHRAN FROG
Sachatamia albomaculata
This species is found in wet lowland forests, and breeds near streams. Males call to females from low vegetation nearby.

¾–1¼ in
2–3 cm

EMERALD GLASS FROG
Espadarana prosoblepon
Males of this arboreal frog are fiercely territorial, defending their space by calling, and occasionally fighting rivals while hanging upside down.

CERATOPHRYIDAE

These South American horned frogs have a very large head and wide mouth that enable them to eat animals nearly as large as themselves. They are "sit-and-wait" predators, remaining still and well-camouflaged before their prey comes within range.

"horn" above eye

ORNATE HORNED FROG
Ceratophrys ornata
A voracious predator living in grasslands in Argentina, this species breeds after heavy rain, in temporary pools and ditches.

3½–5½ in
9–14 cm

wide mouth

3¼–5 in
8–13 cm

CRANWELL'S HORNED FROG
Ceratophrys cranwelli
Spending much of its life underground, this large frog emerges after heavy rain to mate and lay its eggs in pools.

1½–4 in
4–10 cm

BUDGETT'S FROG
Lepidobatrachus laevis
This frog has a flattened body, a wide mouth, and fangs. It spends dry periods in a cocoon below ground, emerging to breed after rain.

ROBBER FROGS

Found in North, Central, and South America, the Craugastoridae produce eggs that develop directly into small adults, without a tadpole stage. Eggs may be deposited on the ground or in vegetation. Many species show parental care of the eggs.

BROAD-HEADED RAIN FROG
Craugastor megacephalus
This C. American frog hides in a burrow by day and emerges at night. It lays its eggs in leaf litter.

1¼–2¾ in
3–7 cm

¾–2 in
2–5 cm

ISLA BONITA ROBBER FROG
Craugastor crassidigitus
Native to the humid forests of C. America, this ground-living frog is also found in coffee plantations and pastures.

1–2¼ in
2.5–5.5 cm

FITZINGER'S ROBBER FROG
Craugastor fitzingeri
In this forest species, the male calls to the larger female from an elevated perch. Eggs are laid in the ground and are guarded by the female.

TRUE TOADS

Distributed worldwide, the family Bufonidae is large and diverse. Its members are characterized by shortened forelimbs, hind limbs that are used for walking or hopping, dry warty skin, and parotoid glands behind the eyes. However, this family also includes the more slender and long-limbed South and Central American harlequin frogs and the stubfoot toads.

horizontal pupil

green blotches on back

warty skin

3½–4¾ in
9–12 cm

COMMON AFRICAN TOAD
Sclerophorus regularis
Common throughout Africa except the driest parts of Sahara and Namib deserts, this thickset toad breeds in dams and ponds. Males produce rasping, ducklike calls to attract females.

2–4½ in
5–11.5 cm

2–4 in
5–10 cm

MALAYAN TREE TOAD
Rentapia hosii
Found in E. Asia, this toad is unusual in living a largely arboreal life. It has adhesive pads on its toes that enable it to climb trees.

GREEN TOAD
Bufotes viridis
A native of sandy habitats, this colorful toad is found in Europe and W. Asia. It emerges from its burrow in spring to breed in ponds.

2–4 in
5–10 cm

NATTERJACK
Epidalea calamita
Compared with other true toads, this species has short legs and runs like a mouse. Occurring across Europe, it breeds from spring to summer.

CERATOBATRACHIDAE

Found in Southeast Asia, China, and several Pacific islands, frogs of this family produce large eggs that hatch directly into small frogs. In many species, the tips of the fingers and toes are enlarged.

FIJI GROUND FROG
Cornufer vitianus
The populations of this species on several of the Fijian Islands have been wiped out following the introduction of mongooses.

1–4¼ in
2.5–11 cm

hornlike projection above eye

flat, triangular head

2–3¼ in
5–8 cm

SOLOMON ISLANDS HORNED FROG
Cornufer guentheri
This species has a pointed snout and hornlike projections above its eyes. It hides among dead leaves.

BOMBINATORIDAE

These small aquatic toads are found in Europe and Asia. They have a flattened body and many are brightly colored. Fire-bellied toads are active by day, but the dull-colored barbourulas from the Philippines and Borneo are nocturnal.

prominent eyes

bright red underside

bright green coloration

1¼–2 in
3–5 cm

ORIENTAL FIRE-BELLIED TOAD
Bombina orientalis
Found in China and Korea, this small, squat frog can produce a toxic skin secretion. If attacked, it displays its bright belly colors.

RAIN FROGS

Members of the family Brevicipitidae are found in eastern and southern Africa. During mating, the much smaller male is glued onto a female's back by a special skin secretion.

1¼–2 in
3–5 cm

DESERT RAIN FROG
Breviceps macrops
Living away from standing water, this burrowing frog lives and breeds among Namibian sand dunes, occasionally moistened by sea fog.

AMERICAN TOAD
Anaxyrus americanus
Found in eastern
N. America, this toad is
quite variable in color.
Breeding occurs in ponds
where males produce
long trilling calls.

large parotoid glands

short, warty legs

2–3½ in
5–9 cm

VARIABLE HARLEQUIN FROG
Atelopus varius
This aggressive species from Panama
and Costa Rica has vivid and variable
coloration. It lives near streams and
is active during the day.

1–2¼ in
2.5–6 cm

TRUANDO TOAD
Rhaebo haematiticus
Living among leaf litter in
forests of C. and S. America, this
broad-headed toad lays its eggs
in long strings in rocky pools.

1½–3¼ in
4–8 cm

GUYANAN STUBFOOT TOAD
Atelopus barbotini
This small toad from Guyana has a
flattened body. It breeds throughout
the year in forest streams.

1–1½ in
2.5–4 cm

PANAMANIAN GOLDEN FROG
Atelopus zeteki
This brightly colored frog from Panama
breeds in pools after heavy rain. It may
now be extinct in the wild.

2–4 in
5–10 cm

EUROPEAN COMMON TOAD
Bufo bufo
Found throughout Europe and
N. Africa, this species breeds in spring,
with males outnumbering the much
larger females by about three to one.

3¼–8 in
8–20 cm

**GREEN
CLIMBING FROG**
Incilius coniferus
Native to C. and S. America,
this nocturnal toad is
often found climbing
among vegetation.

2¼–3¾ in
5.5–9.5 cm

parotoid glands
secrete toxin

olive-
brown,
warty skin

4–9½ in
10–24 cm

CANE TOAD
Rhinella marina
One of the world's
largest toads, this
American species was
introduced in Australia,
where it is now a serious
threat to native wildlife.

RHINODERMATIDAE

This South American family includes two species
with a pointed snout and cryptic coloration, and
the males carry the eggs and tadpoles in their
mouth. The remaining species is rare
and little is known about it.

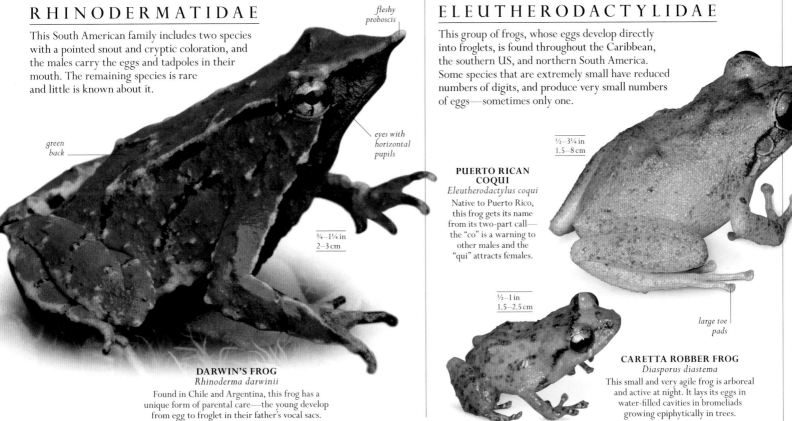

fleshy
proboscis

eyes with
horizontal
pupils

green
back

¾–1¼ in
2–3 cm

DARWIN'S FROG
Rhinoderma darwinii
Found in Chile and Argentina, this frog has a
unique form of parental care—the young develop
from egg to froglet in their father's vocal sacs.

ELEUTHERODACTYLIDAE

This group of frogs, whose eggs develop directly
into froglets, is found throughout the Caribbean,
the southern US, and northern South America.
Some species that are extremely small have reduced
numbers of digits, and produce very small numbers
of eggs—sometimes only one.

**PUERTO RICAN
COQUI**
Eleutherodactylus coqui
Native to Puerto Rico,
this frog gets its name
from its two-part call—
the "co" is a warning to
other males and the
"qui" attracts females.

½–3¼ in
1.5–8 cm

large toe
pads

½–1 in
1.5–2.5 cm

CARETTA ROBBER FROG
Diasporus diastema
This small and very agile frog is arboreal
and active at night. It lays its eggs in
water-filled cavities in bromeliads
growing epiphytically in trees.

CANE TOAD
Rhinella marina

One of the largest toads in the world, the cane toad is a hardy creature with a huge appetite. Also known as the marine toad, it is mainly a resident of dry environments, scrub, and savanna. It commonly lives around human settlements and is often seen under streetlights, waiting for insects to fall. The female is larger than the male, and the largest females can lay more than 20,000 eggs in a single clutch. Males attract females with a slow, low-pitched trill. They have few enemies, as at all stages of their life cycle they are distasteful or toxic to potential predators. In Australia they have become a major pest—they are poisonous to native and domestic animals, injurious to humans, and breed so prolifically as to be out of control.

SIZE 4–9½ in (10–24 cm)
HABITAT Non-forested habitats
DISTRIBUTION C. and S. America; introduced to Australia and elsewhere
DIET Terrestrial invertebrates

parotoid gland

adult coloration is yellow, olive, or reddish brown

warts on male's skin develop dark, sharp spines in the breeding season

NIGHT HUNTER >
Protected by their poisonous skin and unafraid of predators, cane toads emerge at night from their daytime hiding places to hop around in search of prey.

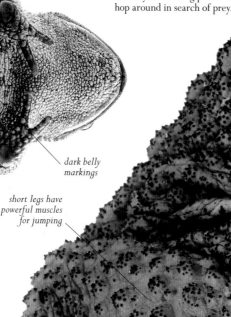

dark belly markings

short legs have powerful muscles for jumping

∧ PALE BELLY
The cane toad's belly and throat are relatively smooth, and mostly pale in color. Toads have permeable skin and must hide by day to conserve water.

DEBATE
PEST CONTROL

The cane toad gets its name from Australia, where it was introduced into Queensland in 1935 to control insect pests on sugar cane farms. The toads flourished in Australia, feasting on native fauna and building up much denser populations than in their native habitat. Still spreading at an alarming rate, they now occur throughout eastern and northern Australia, and are likely to migrate further. Scientists are working on methods to limit their numbers and contain their territorial expansion.

nostril

wide mouth used to
take prey of any size
that will fit inside it

male's throat is distended
when he produces his loud,
trill-like mating call

∨ IRIDESCENT IRIS
Like most toads, cane toads have large,
protruding eyes. They have very good eyesight,
enabling them to detect small, moving objects
and to lunge accurately at their insect prey.

GLAND >
The enormous parotoid
glands—located on each side of
the head—secrete a powerful
toxin that is distasteful to some
predators and lethal to most.

EAR >
Toads rely on their hearing
to identify potential
enemies. At night, it is
especially important for
females, who locate males
by their calls.

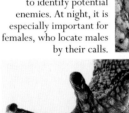

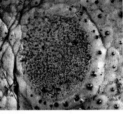

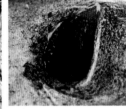

∧ NOSTRIL
The cane toad relies more
on its sense of smell to find
food than other toads—and
breathes more with its lungs
than with its skin.

< HIND FOOT
The long toes on the hind foot,
each with a horny tip, provide
a firm grip on the ground
when a toad pushes off to
hop or jump.

FRONT FOOT ∧
In the breeding season,
males develop dark, horny
nuptial pads on their first
three fingers. These enable
them to clasp females
firmly during mating.

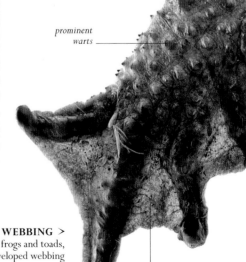

prominent
warts

WEBBING >
Compared with many other frogs and toads,
cane toads have poorly developed webbing
between their toes. This reflects the relatively
small part of their lives spent in water.

web of skin

POISON-DART FROGS

Called poison or poison-dart frogs, the Dendrobatidae are noted for their bright coloration. This warns predators that the skin of these frogs contains powerful toxins, which are derived from their insect prey. These frogs are found in the forests of Central and South America. They are active by day.

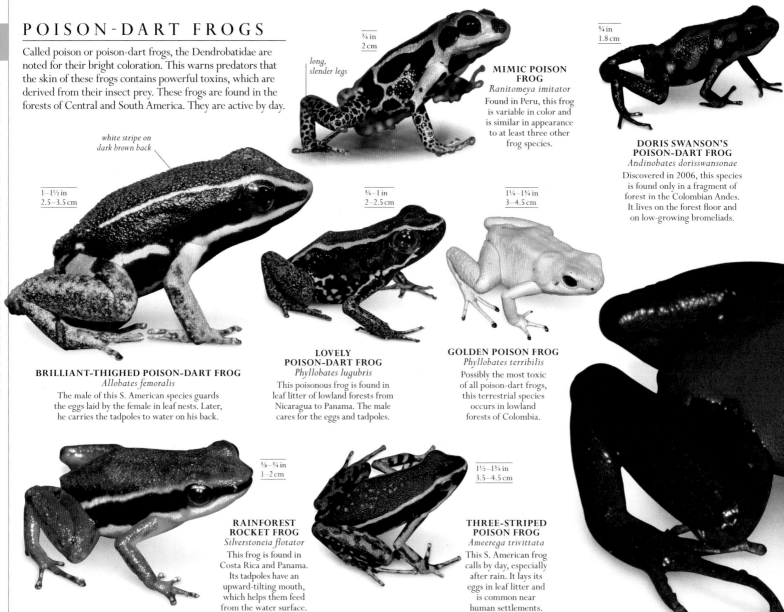

¾ in
2 cm

long, slender legs

MIMIC POISON FROG
Ranitomeya imitator
Found in Peru, this frog is variable in color and is similar in appearance to at least three other frog species.

¾ in
1.8 cm

DORIS SWANSON'S POISON-DART FROG
Andinobates dorisswansonae
Discovered in 2006, this species is found only in a fragment of forest in the Colombian Andes. It lives on the forest floor and on low-growing bromeliads.

white stripe on dark brown back

1–1½ in
2.5–3.5 cm

¾–1 in
2–2.5 cm

1¼–1¾ in
3–4.5 cm

BRILLIANT-THIGHED POISON-DART FROG
Allobates femoralis
The male of this S. American species guards the eggs laid by the female in leaf nests. Later, he carries the tadpoles to water on his back.

LOVELY POISON-DART FROG
Phyllobates lugubris
This poisonous frog is found in leaf litter of lowland forests from Nicaragua to Panama. The male cares for the eggs and tadpoles.

GOLDEN POISON FROG
Phyllobates terribilis
Possibly the most toxic of all poison-dart frogs, this terrestrial species occurs in lowland forests of Colombia.

⅜–¾ in
1–2 cm

1½–1¾ in
3.5–4.5 cm

RAINFOREST ROCKET FROG
Silverstoneia flotator
This frog is found in Costa Rica and Panama. Its tadpoles have an upward-tilting mouth, which helps them feed from the water surface.

THREE-STRIPED POISON FROG
Ameerega trivittata
This S. American frog calls by day, especially after rain. It lays its eggs in leaf litter and is common near human settlements.

MARSUPIAL FROGS

Found in South and Central America, the Hemiphractidae carry their eggs on their back, where they hatch directly into froglets. Some also carry the eggs in a pouch, hence the name "marsupial frogs."

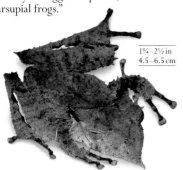

1¾–2½ in
4.5–6.5 cm

SUMACO HORNED TREEFROG
Hemiphractus proboscideus
This species is found in Colombia, Ecuador, and Peru. The female carries her eggs on her back, but does not have a pouch.

REED AND SEDGE FROGS

Also called African treefrogs, this large family, the Hyperoliidae, includes many agile climbers that gather in trees, bushes, or reeds near water to find mates and lay eggs. Some of the species are brightly colored, with marked differences between the sexes.

RED-LEGGED KASSINA
Kassina maculosa
This is an aquatic E. African frog with adhesive disks on its toes. Its eggs, laid on submerged vegetation, hatch into large tadpoles.

prominent eyes

red patch on leg

2¼–2½ in
5.5–6.5 cm

1–1½ in
2.5–3.5 cm

FOULASSI BANANA FROG
Afrixalus paradorsalis
Found in W. Africa, this frog lays its eggs in a folded leaf above water. The male attracts females with a clicklike call.

GREEN AND BLACK POISON-DART FROG
Dendrobates auratus
The males of this species fight to secure a territory. The male defends the eggs and, when they hatch, carries tadpoles to small pools in bromeliads.

1–2¼ in
2.5–6 cm

bright red body

toes with adhesive pads

1¼–1½ in
3–4 cm

YELLOW-HEADED POISON FROG
Dendrobates leucomelas
Found in wet forests of northern S. America, this frog derives its skin toxin from the ants on which it feeds.

rounded snout

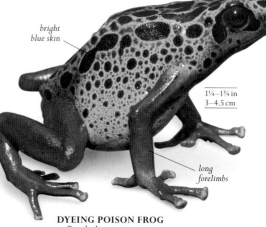

bright blue skin

1¼–1¾ in
3–4.5 cm

long forelimbs

DYEING POISON FROG
Dendrobates tinctorius
This is the blue, or *azureus*, form of this variable S. American species. Both sexes protect the eggs and are very aggressive while defending their territory.

STRAWBERRY POISON FROG
Oophaga pumilio
The female of this species cares for the young by carrying tadpoles to water-filled bromeliads and feeding them with unfertilized eggs.

¾–1 in
2–2.5 cm

SPLASHBACK POISON FROG
Adelphobates galactonotus
Living in leaf litter in Brazilian forests, this frog lays eggs on the ground and carries the tadpoles to water.

½–¾ in
1.5–2 cm

RIO MADEIRA POISON FROG
Adelphobates quinquevittatus
Found in Brazil and Peru, this tiny frog carries its tadpoles to water-filled holes, where the female feeds them with unfertilized eggs.

¾ in
2 cm

GRANULAR POISON FROG
Oophaga granulifera
This species is found in Costa Rica and Panama. Females care for the young by feeding them with unfertilized eggs.

¾–1 in
2–2.5 cm

BRAZIL-NUT POISON FROG
Adelphobates castaneoticus
This species is found in Brazil. The males of this frog place tadpoles individually into small, water-filled tree-holes. The tadpoles are voracious feeders.

SMITH'S REED FROG
Hyperolius tuberilinguis
This agile frog is noted for its loud call. In the mating season, thousands of males may gather around a pond to produce a deafening chorus.

large adhesive disk on toes

¾–1½ in
2–3.5 cm

large eyes

1¼–1¾ in
3–4.5 cm

BOLIFAMBA REED FROG
Hyperolius bolifambae
This small W. African frog occurs in bushland, and breeds in pools. The male's call is a high-pitched buzz.

AUSTRALIAN GROUND FROGS

The Limnodynastidae, found in Australia and New Guinea, include many terrestrial and burrowing frog species. Two recently extinct species uniquely brooded their eggs in the stomach.

1¼–2¼ in
3–6 cm

BROWN-STRIPED MARSH FROG
Limnodynastes peronii
This Australian frog survives dry periods by burying itself in the ground. It emerges to breed after heavy rain, and deposits its eggs in floating foam nests.

TREEFROGS

The Hylidae are a large family widespread worldwide. They are especially well represented in the New World. These frogs have long, slender limbs and adhesive disks on their fingers and toes. Most are arboreal and nocturnal. Many gather to breed in noisy choruses.

SPLENDID LEAF FROG
Cruziohyla calcarifer
This frog occurs in C. and northern S. America and lives high up in trees. It glides from one tree to another, using its extended webbed feet as parachutes.

2–3½ in
5–9 cm

1–1½ in
2.5–4 cm

RUFOUS-EYED STREAM FROG
Duellmanohyla rufioculis
Found in forests of Costa Rica, this frog breeds in fast-flowing streams. Its tadpoles have a modified mouth with which they attach themselves to rocks.

brown
upperparts with
dark patches

¾–1¼ in
2–3 cm

SPRING PEEPER
Pseudacris crucifer
Found in moist woodland in the E. US and Canada, the spring peeper's distinctive, high-pitched call indicates that spring is underway.

2–2¾ in
5–7 cm

PARADOX FROG
Pseudis paradoxa
This aquatic frog is so named because its tadpoles are four times longer than the adult. It is found in S. America and Trinidad.

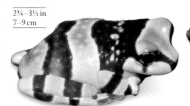

2¾–3½ in
7–9 cm

MISSION GOLDEN-EYED TREEFROG
Trachycephalus resinifictrix
Living high in the canopy of S. American forests, this frog deposits its eggs in water-filled tree-holes, where the tadpoles then develop into froglets.

ROSENBERG'S TREEFROG
Boana rosenbergi
Males of this C. and S. American frog dig pools in damp ground in which the eggs are laid. They defend them against rivals in fights that can be fatal.

2¼–3 in
5.5–7.5 cm

1½–2 in
4–5 cm

ORANGE-LEGGED LEAF FROG
Phyllomedusa hypochondrialis
A native of dry habitats in northern S. America, this climbing frog reduces water loss by rubbing a waxy secretion over its skin.

adhesive
disks on toes

2¼–4¼ in
5.5–11 cm

COPE'S BROWN TREEFROG
Ecnomiohyla miliaria
This large C. American treefrog has fringes of skin on the limbs that may enable it to glide from one tree to another.

1½–2¼ in
3.5–5.5 cm

BOULENGER'S SNOUTED TREEFROG
Scinax boulengeri
Found in C. America and Colombia, this frog breeds in temporary pools created after rain. Males return night after night to the same calling perch to attract mates.

1–4 in
2.5–10 cm

CUBAN TREEFROG
Osteopilus septentrionalis
Native to Cuba, the Cayman Islands, and the Bahamas, this treefrog has been introduced to Florida, where it preys on native frogs, reducing their population.

1¼–2 in
3–5 cm

EUROPEAN TREEFROG
Hyla arborea
Males of this European treefrog gather in spring, calling in loud choruses to attract females. Mating pairs descend to lay eggs in nearby ponds.

large, prominent,
red eyes

1½–2¾ in
4–7 cm

pale
underparts

RED-EYED TREEFROG
Agalychnis callidryas
An excellent climber, this treefrog mates in trees overhanging water. It lays its eggs on leaves and, upon hatching, the tadpoles fall into the water.

1¼–2 in
3–5 cm

LEMUR FROG
Agalychnis lemur
This C. American treefrog is nocturnal, sleeping on the underside of leaves during the day. It lays its eggs on leaves over water.

2¾–4 in
7–10 cm

MANAUS SLENDER-LEGGED TREEFROG
Osteocephalus taurinus
This arboreal frog lives in S. American forests. It mates after rain, and lays its eggs on the surface of pools.

SMALL-HEADED TREEFROG
Dendropsophus microcephalus
Found in C. and S. America and Trinidad, this frog breeds in pools. It is pale yellow by day, and red-brown by night.

¾–1¼ in
2–3 cm

horizontal pupils

2–4 in
5–10 cm

AUSTRALIAN GREEN TREEFROG
Litoria caerulea
Found in N.E. Australia and New Guinea, this agile climber is often found close to human habitations.

NEW ZEALAND FROGS

This family, the Leiopelmatidae, is composed of four species of frogs that are all confined to New Zealand. They are characterized by extra vertebrae and, unlike most frogs, they kick their legs alternately when swimming. They live in humid forests and are nocturnal.

1–1½ in
2.5–3.5 cm

COROMANDEL NEW ZEALAND FROG
Leiopelma archeyi
Found only in New Zealand's North Island, this terrestrial frog lays its eggs under logs. It is critically endangered, as a result of habitat loss and disease.

TROPICAL GRASS FROGS

The Leptodactylidae are a large, varied family of mostly ground-dwelling frogs from North, Central, and South America, and the West Indies, with a pointed snout and long, powerful hind limbs. Their breeding habits are varied and many lay their eggs in nests of foam.

warty skin

1¼–1½ in
3–4 cm

TÚNGARA FROG
Engystomops pustulosus
During mating, the female of this C. American species produces a secretion that the male whips into a floating foam nest in which the eggs are laid.

MANTELLAS

Found only on the islands of Madagascar and Mayotte, the Mantellidae are active by day. Many are brightly colored, warning predators of powerful toxins in their skin. Most species are threatened by the loss of habitat and by the international pet trade.

SPINY-HEADED TREEFROG
Anotheca spinosa
Native to Mexico and C. America, this large frog lives in bromeliads and banana plants, laying its eggs in water-filled cavities.

2¼–3¼ in
6–8 cm

1½–3¼ in
4–8 cm

MASKED TREEFROG
Smilisca phaeota
Active only at night, this inhabitant of humid forests in C. and S. America lays its eggs in small pools.

1¾–3¼ in
4.5–8 cm

WHITE-LIPPED BRIGHT-EYED FROG
Boophis albilabris
This large arboreal frog, found only in Madagascar, lives near the streams in which it breeds. It has fully webbed hind feet.

2–2¼ in
5–6 cm

ELEGANT MADAGASCAN FROG
Spinomantis elegans
An inhabitant of rocky outcrops, this frog is found at high altitudes, even above the tree line. It breeds in streams.

¾–1 in
2–2.5 cm

bony projection

rough, moist skin

2–3 in
5–7.5 cm

YUCATECAN SHOVEL-HEADED TREEFROG
Triprion petasatus
An inhabitant of lowland forests in Mexico and C. America, this frog retreats into tree-holes, using the bony projection on its head to seal the opening.

¾–1¼ in
2–3 cm

MADAGASCAN GOLDEN MANTELLA
Mantella aurantiaca
The bright color of this tiny frog from Madagascan rainforests warns potential predators that its skin secretes a powerful toxin.

PAINTED MANTELLA
Mantella madagascariensis
This Madagascan frog breeds in forest streams. It is under threat due to loss of habitat. The male's call consists of short chirps.

NARROW-MOUTHED FROGS

Frogs of the large and diverse family Microhylidae are found in the Americas, Asia, Australia, and Africa. Most are ground-dwelling and some live in burrows. Most have a short snout, a plump, often teardrop-shaped, body, and stout hind legs.

2–3 in
5–7.5 cm

PAINTED TOAD
Kaloula pulchra
Widespread in Asia, this species has adapted well to human settlements. It protects itself with a noxious, sticky skin secretion.

1¼–2¼ in
3–6 cm

TOMATO FROG
Dyscophus antongilii
Native to Madagascar, this frog stays buried in soil by day, emerging to feed at night. A sticky skin secretion protects it against predators.

3¼–4¾ in
8–12 cm

BLACK-SPOTTED NARROW-MOUTHED FROG
Kalophrynus pleurostigma
This frog from the Philippines protects itself by producing a sticky secretion. It breeds in small pools after rain.

EASTERN NARROW-MOUTHED TOAD
Gastrophryne carolinensis
Found in the S.E. US, this burrowing toad breeds in water bodies of all sizes. The male's call sounds like a bleating lamb.

¾–1½ in
2–3.5 cm

GIANT STUMP-TOED FROG
Stumpffia grandis
This small terrestrial frog is found in leaf litter in high-altitude forests in Madagascar.

¾–1 in
2–2.5 cm

MEGOPHRYIDAE

Found across Asia, this is a small family of frogs whose body shape and color patterns enable them to be camouflaged among leaves. They walk rather than jump, and most are ground-living.

hornlike projections on eyelids

cryptic coloration with black markings

ASIAN HORNED FROG
Pelobatrachus nasuta
This frog conceals itself among dead leaves, while it waits for prey. Females lay their eggs under rocks and logs in streams.

2¾–5½ in
7–14 cm

PARSLEY FROGS

Made up of only four species, the family Pelodytidae is confined to Europe and the Caucasus. Named for the green markings on their skin, parsley frogs breed after rain, laying their eggs in broad strips.

COMMON PARSLEY FROG
Pelodytes punctatus
When climbing smooth, vertical surfaces, this European frog uses its underside as a suction cup. Both sexes call during breeding.

1¼–2 in
3–5 cm

TONGUELESS FROGS

These aquatic frogs are well adapted to life in water: they have a flattened body, fully webbed hind feet, and eyes that protrude upward, enabling them to see above the surface. As their common name suggests, the Pipidae lack a tongue. They feed on a wide range of prey and scavenge on dead animals.

eggs on female's back

muscular hind legs

DWARF SURINAM TOAD
Pipa parva
In this wholly aquatic species found in Venezuela and Colombia, the eggs develop on the female's back.

1–1¾ in
2.5–4.5 cm

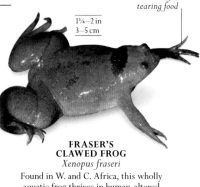

claws used for tearing food

1¼–2 in
3–5 cm

FRASER'S CLAWED FROG
Xenopus fraseri
Found in W. and C. Africa, this wholly aquatic frog thrives in human-altered habitats and is harvested by people for food.

SPADEFOOT TOADS

A small family found in Eurasia and North Africa, the Pelobatidae are characterized by horny projections on their hind feet. They use them to burrow into the ground, where they wait for rain.

COMMON SPADEFOOT TOAD
Pelobates fuscus
Found in Europe and Asia, this species is variable in color. It has a plump body, which it inflates when attacked.

1½–3¼ in
4–8 cm

DICROGLOSSIDAE

This diverse family of frogs is found across Africa, Asia, and several Pacific islands. Most are ground-living, but are found close to water. Many lay their eggs in water and have free-living tadpoles.

INDIAN BULLFROG
Hoplobatrachus tigerinus
This large, voracious feeder from S. Asia breeds during the monsoon. The male has a particularly loud call.

2½–6½ in
6.5–17 cm

RAJAMALLY WART FROG
Fejervarya kirtisinghei
Found only in Sri Lanka, this frog lives in leaf litter near streams and thrives in plantations and gardens.

1–1¾ in
2.5–4.5 cm

1½–2½ in
4–6.5 cm

COMMON SKITTERING FROG
Euphlyctis cyanophlyctis
Widespread in S. Asia, this aquatic frog is noted for its ability to skitter over the surface of water.

MARTEN'S PUDDLE FROG
Occidozyga martensii
Found in China and S.E. Asia, this small frog occurs in pools near forest streams and rivers.

½–¾ in
1.5–2 cm

GOLIATH FROGS

The family Conrauidae comprises six species of large, semiaquatic frogs from Africa that live and breed in fast-flowing rivers and streams. Males are thought not to call, and females lay their eggs among gravel and rocks on the riverbed.

4–16 in
10–40 cm

long, webbed toes

powerful hind legs for swimming

GOLIATH BULLFROG
Conraua goliath
The world's largest frog, this W. African species has an aquatic lifestyle; its powerful legs and webbed feet make it a strong swimmer.

TRUE FROGS

Known as true frogs, this large family is found in most parts of the world. Most of the Ranidae have powerful hind limbs that enable them to jump athletically on land and to swim powerfully in water. They typically breed in early spring, with many laying their eggs communally.

EDIBLE FROG
Pelophylax esculentus
This frog is a hybrid between the widespread European *Pelophylax lessonae* and other more local species. It lives in and close to water.

3¼–4¾ in
8–12 cm

PICKEREL FROG
Rana palustris
Found over much of N. America, this frog breeds in spring, with females laying their eggs in clumps containing two to three thousand eggs.

2¼–2¾ in
6–7 cm

1½–3¼ in
3.5–8 cm

WOOD FROG
Rana sylvaticus
The only American frog found north of the Arctic Circle, it breeds in early spring in temporary pools free of fishes.

AMERICAN BULLFROG
Rana catesbeianus
This voracious predator has tadpoles that may take four years to develop, reaching a large size. It is the largest N. American frog.

3½–8 in
9–20 cm

black spots on green to brown body

white vocal sac

male has thick forelimbs

EUROPEAN COMMON FROG
Rana temporaria
Also known as the grass frog, this species lives mostly on land, migrating to ponds in spring to breed. It lays eggs in clumps.

2–4 in
5–10 cm

PHRYNOBATRACHIDAE

This family of small, terrestrial or semixaquatic frogs is confined to sub-Saharan Africa. Most breed throughout the year, laying their eggs in water. They reach maturity in five months.

warty skin

½–¾ in
1.5–2 cm

GOLDEN PUDDLE FROG
Phrynobatrachus auritus
This frog is so named because it breeds in very small pools. A ground-dwelling species, it is found in C. African rainforests.

ORNATE FROGS AND GRASS FROGS

Found in open country in Africa, Madagascar, and the Seychelles, the Ptychadenidae include many brightly colored frogs. A streamlined body and strong hind legs make them prodigious jumpers.

1¾–2¾ in
4.5–7 cm

MASCARENE RIDGED FROG
Ptychadena mascareniensis
Common on agricultural land, this frog has long legs and a pointed snout. It breeds in puddles, wheel ruts, and ditches.

AFRO-ASIAN TREEFROGS

The Rhacophoridae range across Africa and much of Asia, and are mostly arboreal frogs. The family also includes flying frogs that glide from one tree to another. Many lay their eggs in foam nests, where the eggs and tadpoles stay protected against predators.

1¾–2¼ in
4.5–6 cm

AFRICAN FOAM-NEST TREEFROG
Chiromantis rufescens
Found in the forests of W. and C. Africa, this frog lays its eggs in a foam nest attached to a branch overhanging water.

1½–2¼ in
4–6 cm

SOUTHERN WHIPPING FROG
Taruga longinasus
Endangered by the loss of much of its habitat, this arboreal frog occurs in Sri Lanka's remaining patches of rainforest.

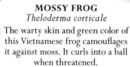

MOSSY FROG
Theloderma corticale
The warty skin and green color of this Vietnamese frog camouflages it against moss. It curls into a ball when threatened.

2¾–3½ in
7–9 cm

shiny green coloration

3½–4 in
9–10 cm

WALLACE'S FLYING FROG
Rhacophorus nigropalmatus
This arboreal species from the rainforests of S.E. Asia has webbed feet that enable it to glide between trees.

long, fully webbed fore- and hind feet

PYXICEPHALIDAE

Inhabiting a variety of habitats in sub-Saharan Africa, members of this family range in size from huge bullfrogs to typical pond frogs and tiny moss frogs. Most lay eggs in water, although some small species lay eggs on land.

AFRICAN BULLFROG
Pyxicephalus adspersus
Males of this large African species aggressively defend their eggs and tadpoles. They dig channels in soil to help the tadpoles reach open water.

olive-green body with dark markings

3¼–9 in
8–23 cm

very wide mouth

powerful limbs

MEXICAN BURROWING TOAD

The only member of the family Rhinophrynidae, the burrowing toad specializes in digging in soil and eating ants. It has a long, thin tongue that it sticks out of its narrow mouth.

2¼–3¼ in
6–8 cm

MEXICAN BURROWING TOAD
Rhinophrynus dorsalis
This unusually shaped toad spends most of its life burrowing underground, emerging only after rain to breed in temporary pools.

AMERICAN SPADEFOOT TOADS

This family consists of toads that live on dry land and stay inactive underground for long periods. The Scaphiopodidae emerge after rain to breed in temporary pools. These may evaporate quickly, so the tadpoles develop very rapidly.

2¼–3½ in
5.5–9 cm

1½–2¼ in
4–6 cm

PLAINS SPADEFOOT
Spea bombifrons
Found in arid plains of N. America and Mexico, this frog burrows by day. It gathers to breed in large numbers after heavy rain.

mottled greenish brown skin

COUCH'S SPADEFOOT
Scaphiopus couchii
This N. American frog lives in arid areas, spending much of its time underground. It emerges at night to feed and, after heavy rain, to breed.

STRABOMANTIDAE

Many small frogs native to South America and the Caribbean make up this family. They are all direct developers: there is no tadpole stage and the egg hatches straight into a miniature adult.

LIMON ROBBER FROG
Pristimantis cerasinus
Hiding in leaf litter by day, but arboreal by night, this small frog is found in humid lowland forests of C. America.

½–1½ in
1.5–3.5 cm

¾–1½ in
2–4 cm

CHIRIQUI ROBBER FROG
Pristimantis cruentus
This small, terrestrial frog, found in C. and S. America, lays its eggs in crevices on tree trunks.

½–1 in
1.5–2.5 cm

PYGMY RAIN FROG
Pristimantis ridens
Found in the forests of C. and S. America, this tiny nocturnal frog thrives in gardens and lays its eggs in leaf litter.

CAECILIANS

Caecilians are long-bodied, limbless amphibians with little or no tail. Ring-shaped folds (annuli) in the skin give them a segmented appearance.

All caecilians live in the tropics. They vary in length from 4¾ in (12 cm) to 5¼ ft (1.6 m). Most live underground, burrowing in soft soil, using their pointed, bony head as a shovel. They emerge at night, especially after rain, to feed on earthworms, termites, and other insects. Others live in water and resemble eels, rarely moving onto the land. These have a fin on the tail. With only rudimentary eyes, all rely on smell to find food and mates. A pair of retractable tentacles between the eyes and the nostrils transmit chemical signals to the nose.

In all caecilians, the eggs are fertilized internally. Some species lay eggs, but in others the eggs are retained inside the female's body. The young emerge either as gilled larvae or as small adults.

PHYLUM	CHORDATA
CLASS	AMPHIBIA
ORDER	GYMNOPHIONA
FAMILIES	10
SPECIES	214

20 in
50 cm

PURPLE CAECILIAN
Gymnopis multiplicata
This terrestrial caecilian from C. America lives in a wide range of habitats. The eggs hatch inside the female.

ICHTHYOPHIIDAE

Found in Asia, these caecilians lay eggs in soil near water. Females remain with their clutches, defending them until the larvae have made their way to open water.

13 in
33 cm

yellow stripe along body

KOH TAO CAECILIAN
Ichthyophis kohtaoensis
Found in a variety of habitats in S.E. Asia, this caecilian lays its eggs on land, but its larvae live in water.

DERMOPHIIDAE

These thick-set, cylindrical caecilians from Africa live beneath the surface and are mostly brown, gray, or dark purple in color but may be yellow. Some species lack eyes. They give birth to live young.

CAECILIIDAE

Most species in this family are burrowers. Found in most tropical regions of the world, they vary greatly in length, some growing to more than 5 ft (1.5 m). In some the eggs hatch into larvae; in others the larvae develop inside the female.

14½–36 in
37–91 cm

SANTA ROSA CAECILIAN
Caecilia attenuata
A rarely seen, grayish blue, burrowing species from the moist lowland forests of Amazonian Ecuador, this caecilian also occurs in degraded habitats including plantations and gardens.

SALAMANDERS AND NEWTS

Unlike their fellow amphibians, the frogs, salamanders and newts normally have a slender, lizardlike body, a long tail, and four legs similar in size.

Newts and salamanders, also known as urodeles, are generally found in damp habitats and are largely confined to the northern hemisphere. They are numerous in the Americas, ranging from Canada to northern South America. They vary considerably in size, from species over 3¼ ft (1 m) in length, to tiny creatures about ¾ in (2 cm) long.

AMPHIBIOUS LIFESTYLES

Some species, notably the newts, spend part of their life in water, part on land. Some salamander species live their entire lives in water, while others are wholly terrestrial. Most have smooth, moist skin through which they breathe to a greater or lesser extent.

The salamanders of one family, the plethodontids, have no lungs and breathe entirely through their skin and the roof of their mouth. Urodeles have a relatively small head, compared to frogs and toads; they also have smaller eyes, smell being the most important sense used in finding food and in social interactions. Most species, especially the terrestrial ones, are nocturnal, hiding under a log or rock during the day.

REPRODUCTION

In the majority of species, the eggs are fertilized inside the female. Males, however, do not have a penis but package sperm in capsules called spermatophores which are passed to the female during mating. In many species sperm transfer is preceded by elaborate courtship in which the male induces the female to cooperate with him. She may, of course, reject his advances. In many newts, males develop dorsal crests and bright colors in the breeding season.

Many species lay their eggs in water. When the larvae hatch, they have a long, slender body, a deep, finlike tail, and large, feathery external gills. The larvae are carnivorous, feeding on tiny water creatures. The exceptions to this rule are the wholly terrestrial species of salamander. These lay their eggs on land and the larval stage is completed within the egg, which hatches to produce a miniature adult.

PHYLUM	CHORDATA
CLASS	AMPHIBIA
ORDER	CAUDATA
FAMILIES	10
SPECIES	754

A male alpine newt sniffs a female before courting her. Odor helps to identify the gender and species of potential partners.

SIRENS

The wholly aquatic salamanders of the family Sirenidae are found in the southern US and Mexico. They retain larval features in the adult stage and resemble eels. They have external gills and no hind limbs.

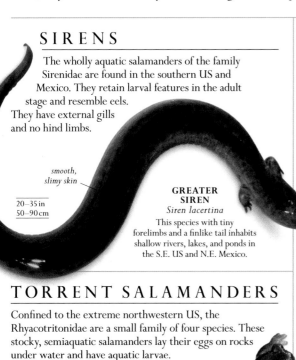

smooth, slimy skin

20–35 in
50–90 cm

GREATER SIREN
Siren lacertina
This species with tiny forelimbs and a finlike tail inhabits shallow rivers, lakes, and ponds in the S.E. US and N.E. Mexico.

TORRENT SALAMANDERS

Confined to the extreme northwestern US, the Rhyacotritonidae are a small family of four species. These stocky, semiaquatic salamanders lay their eggs on rocks under water and have aquatic larvae.

COLUMBIA TORRENT SALAMANDER
Rhyacotriton kezeri
3–4½ in
7.5–11.5 cm
Confined to the forests of Oregon and Washington, this salamander lays its eggs in springs. Intensive logging has caused its numbers to decline considerably.

NEWTS AND EUROPEAN SALAMANDERS

The Salamandridae are small- and medium-sized salamanders and newts found in Europe, North Africa, Asia, and North America. Eggs are fertilized inside the female, with sperm being transferred from the male in a spermatophore (sperm capsule) during an elaborate courtship.

orange poison glands

4¾–8 in
12–20 cm

CALIFORNIA NEWT
Taricha torosa
This nocturnal newt enters ponds in spring to mate and lay eggs. It secretes a lethal nerve toxin to discourage predators.

4¾–7 in
12–18 cm

CROCODILE NEWT
Tylototriton verrucosus
Found in C. Asia, this robust newt breeds in ponds after the monsoon. Its orange markings signal distasteful secretions.

broad head with rounded jaws

6–12 in
15–30 cm

flattened tail

SHARP-RIBBED SALAMANDER
Pleurodeles waltl
This large Spanish and Moroccan species has a unique mode of defense. When grasped, it pushes out the sharp tips of its ribs through its skin.

cylindrical tail

blue
markings
on tail

2¼–4¾ in
6–12 cm

2¾–4 in
7–10 cm

orange
underside

ALPINE NEWT
Ichthyosaura alpestris
Found across much of
N. Europe, this newt breeds
in early spring. The female
wraps eggs individually in
the leaves of water plants.

7–11 in
18–28 cm

SPECTACLED SALAMANDER
Salamandrina terdigitata
This secretive salamander is found only
in Italy and lives in streams in hilly areas.
It has a long, slender, and flattened body.

5–6½ in
13–17 cm

SPOTLESS STOUT NEWT
Paramesotriton labiatus
Found in mountain streams in China,
this newt is equipped with a large tail for
swimming. It attaches its eggs to rocks.

large,
prominent eyes

4–5½ in
10–14 cm

**SARDINIAN BROOK
SALAMANDER**
Euproctus platycephalus
Found only in Sardinia, this slender
salamander lives in streams, laying its
eggs under rocks. It is endangered
due to degradation of its habitat.

FIRE SALAMANDER
Salamandra salamandra
This European salamander spends most
of its time on land, entering water only
to lay eggs. The glands on its head can
spray a toxic secretion at enemies.

large glands
secrete toxin

3½–5 in
9–13 cm

EASTERN NEWT
Notophthalmus viridescens
This pond-breeding newt is from eastern
N. America. Juveniles, or "efts," are
terrestrial, bright red in color, and very toxic.

2½–5½ in
6.5–14 cm

LORESTAN NEWT
Neurergus kaiseri
Found only in Iran, this species lives
in streams. It is critically endangered
due to loss of its habitat and its
popularity as a pet.

orange and
black limbs

JAPANESE NEWT
Cynops pyrrhogaster
This newt spends much of its life in water. Its
brightly colored belly warns potential predators
that the glands in its skin secrete a toxin.

3½–4¾ in
9–12 cm

2¾–4 in
7–10 cm

SMOOTH NEWT
Lissotriton vulgaris
Common across Europe and W. Asia,
this small amphibian breeds in ponds.
The male mates with the female
after elaborate courtship.

GREAT CRESTED NEWT
Triturus cristatus
Found across Europe and C. Asia, males of this large,
pond-breeding species develop a spectacular dorsal crest
in spring, and display it vigorously to potential mates.

4–7 in
10–18 cm

MARBLED NEWT
Triturus marmoratus
Found in France and Spain, this newt lives
in woodland, heathland, and hedgerows.
It enters ponds to breed in spring.

4–5½ in
10–14 cm

LUNGLESS SALAMANDERS

Containing more than 390 species, the Plethodontidae are the largest salamander family. Members of this family have no lungs, but breathe through their mouth and skin. Apart from six European species, all live in North, Central, and South America, inhabiting a wide variety of habitats and feeding mainly on small invertebrates.

small legs

2¾–4¼ in
7–11 cm

CUKRA CLIMBING SALAMANDER
Bolitoglossa striatula
A small salamander with webbed digits, this species is nocturnal, hiding by day among banana leaves. It is found in Costa Rica, Honduras, and Nicaragua.

SEAL SALAMANDER
Desmognathus monticola
This stout-bodied species lives in a burrow by day and is active at night. It often perches on a rock.

long, slender
tail and body

3¼–5 in
8–13 cm

ALLEGHANY MOUNTAIN DUSKY SALAMANDER
Desmognathus ochrophaeus
Mostly terrestrial, this salamander is often found foraging in forests in large numbers after heavy rain, and sometimes climbing trees and shrubs.

ALLEN'S WORM SALAMANDER
Oedipina alleni
Found in leaf litter in lowland forests of Costa Rica, this salamander coils up its elongated body and tail when attacked.

4–6½ in
10–16 cm

THREE-LINED SALAMANDER
Eurycea guttolineata
Found in and around water, the slender three-lined salamander is a very good swimmer, but spends much of its time in a burrow.

4¼–6 in
11–15 cm

black stripe along
the side of the belly

2¾–4¼ in
7–11 cm

BLUE RIDGE TWO-LINED SALAMANDER
Eurycea wilderae
Found around springs and streams, this small species is common in the wooded mountains of the S. Appalachians. It mates in the fall and lays its eggs in winter.

2¾–4¾ in
7–12 cm

MISSISSIPPI SLIMY SALAMANDER
Plethodon mississippi
Found in hardwood forests, this terrestrial salamander protects itself from predators by producing a sticky skin secretion. It lays its eggs on land.

4½–8½ in
11.5–21 cm

REDBACK SALAMANDER
Plethodon cinereus
A terrestrial species, this salamander hides under bark by day and hunts for insects and other prey in foliage after dark.

GIANT SALAMANDERS

The Cryptobranchidae are three large, wholly aquatic species—one each from Japan, China, and North America. They feed on a wide variety of prey, from worms to small mammals. The family includes the world's largest salamander—the Chinese giant salamander, which is about 6 ft (1.8 m) long.

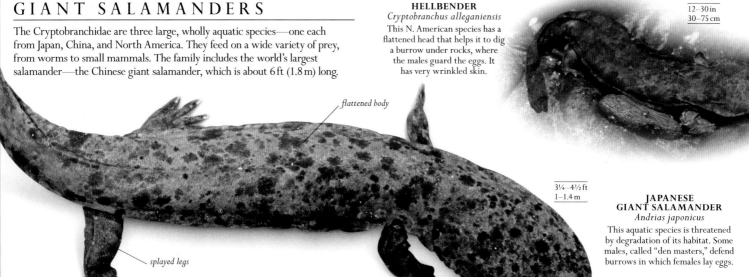

HELLBENDER
Cryptobranchus alleganiensis
This N. American species has a flattened head that helps it to dig a burrow under rocks, where the males guard the eggs. It has very wrinkled skin.

12–30 in
30–75 cm

flattened body

3¼–4½ ft
1–1.4 m

JAPANESE GIANT SALAMANDER
Andrias japonicus
This aquatic species is threatened by degradation of its habitat. Some males, called "den masters," defend burrows in which females lay eggs.

splayed legs

SPRING SALAMANDER
Gyrinophilus porphyriticus

4¾–7½ in
12–19 cm

An agile inhabitant of mountain streams and springs, this colorful salamander is most commonly found hiding under a log or rock.

2¾–4¾ in
7–12 cm

ITALIAN CAVE SALAMANDER
Hydromantes italicus

Found near streams and springs in the mountains of N. Italy, this salamander lives in caves and rock crevices.

3–6 in
7.5–15.5 cm

ENSATINA SALAMANDER
Ensatina eschscholtzii

This smooth-skinned, N. American salamander has a plump tail, which is narrow at the base. It waves its tail at enemies in a defensive posture.

2–3½ in
5–9 cm

FOUR-TOED SALAMANDER
Hemidactylium scutatum

A terrestrial salamander, the adults of this species live in mosses, but the larvae grow in water. There is a marked narrowing at the base of its tail.

ASIATIC SALAMANDERS

About 50 small- to medium-sized species form the Hynobiidae. They are confined to Asia, with some found in mountain streams. They lay their eggs in ponds or streams, and their larvae have external gills. Some species have claws to grasp rocks.

4–6½ in
10–16 cm

OITA SALAMANDER
Hynobius dunni

Females of this endangered Japanese species produce eggs in sacs, and males then compete to fertilize them externally.

MOLE SALAMANDERS

These large creatures mostly live in burrows and emerge to forage by night. The Ambystomatidae comprises 33 species, all found in North America. Some species, notably the Mexican axolotl, are aquatic as adults and retain larval features, such as external gills.

TIGER SALAMANDER
Ambystoma tigrinum

A thickset salamander found throughout much of N. America, this species migrates in spring to ponds to mate and lay eggs.

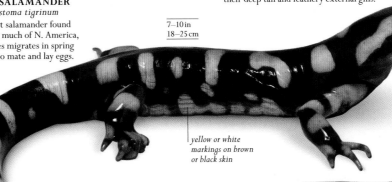

7–10 in
18–25 cm

yellow or white markings on brown or black skin

MARBLED SALAMANDER
Ambystoma opacum

This short-tailed, stocky salamander breeds in the fall and lays its eggs in dried-up ponds that fill with rain in winter.

AXOLOTL
Ambystoma mexicanum

4–12 in
10–30 cm

Adult axolotls never leave the water and resemble very large salamander larvae, with their deep tail and feathery external gills.

broad head with small eyes

3½–4¼ in
9–11 cm

AMERICAN GIANT SALAMANDERS

The four large and aggressive species of the *Dicamptodontidae* are found in the damp conifer forests of western North America. They breed in clear, unpolluted, and permanent streams, where their larvae live and develop for two to four years before they metamorphose.

6½–12 in
17–30 cm

CALIFORNIA GIANT SALAMANDER
Dicamptodon ensatus

This large, nocturnal salamander is threatened by loss and degradation of its forest habitat. This species has an aquatic larval stage.

massive head

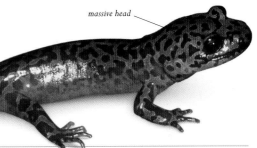

marbled coloration

MUDPUPPIES AND RELATIVE

Five of the six species in the Proteidae are found in North America; the sixth one—the cave-living olm—lives in Europe. Members of this family retain the larval form into their adult life, with a long, slender body, external gills, and small eyes.

8–20 in
20–50 cm

MUDPUPPY
Necturus maculosus

A voracious predator, the mudpuppy, or waterdog, feeds on a variety of invertebrates, fishes, and amphibians. The female aggressively defends her developing eggs.

8–12 in
20–30 cm

EUROPEAN OLM
Proteus anguinus

The wholly aquatic olm lives in dark, flooded caves in Slovenia and Montenegro. It is blind and has white, pink, or gray skin.

AMPHIUMAS

Occurring in eastern North America, the Amphiumidae are a family of three large, wholly aquatic salamanders with an eellike body and tiny limbs. Females guard the eggs in a nest on land. Amphiumas can survive drought by burrowing into mud and forming a cocoon. They feed on worms, mollusks, fishes, snakes, and small amphibians.

16–43 in
40–110 cm

THREE-TOED AMPHIUMA
Amphiuma tridactylum

This large salamander, with slimy skin and a long tail, can give a painful bite. Males breed every year, while females breed in alternate years.

REPTILES

Reptiles are a sophisticated, diverse, and successful group of ectothermic (cold-blooded) vertebrates. Although commonly associated with hot, dry environments, they are found in a wide range of habitats and climates around the world.

PHYLUM	CHORDATA
CLASS	REPTILIA
ORDERS	4
FAMILIES	92
SPECIES	11,050

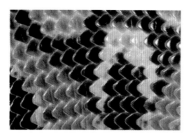

Reptile scales are a protective layer of skin with overlapping patches thickened with keratin and sometimes bone.

The shelled egg allows reptiles to reproduce out of water. The outer layer protects the embryo from dehydration.

Crocodiles can walk with their body raised well off the ground, whereas turtles manage only a sprawling gait.

Reptiles modify their behavior to regulate body temperature. They bask in the morning sun to absorb heat energy.

The first reptiles evolved from amphibians more than 295 million years ago. They are the ancestors not only of modern-day reptiles but also of mammals and birds. During the Mesozoic Era, reptiles such as dinosaurs, aquatic ichthyosaurs and plesiosaurs, and aerial pterosaurs dominated Earth. Reptile groups that still exist today evolved in this period and survived the mass extinction event that killed off the dinosaurs 66 million years ago.

All reptiles share certain characteristics, such as scaly skin and behavioral thermoregulation—using outside sources to maintain a constant body temperature, by basking in the sun, for example. But the various groups exhibit obvious differences. Turtles and tortoises are protected by a unique, heavily armored shell. Lizards, crocodilians, and tuataras have four limbs and a long tail. Many squamates, or scaled reptiles—lizards, snakes, and amphisbaenians—have evolved limbless forms.

Hot desert ecosystems are typically dominated by lizards and snakes but reptiles of all types are found in all habitats in the tropics and subtropics. Fewer species are found in cooler temperate climates. Within each ecosystem reptiles are important both as predators and as prey. Most are carnivores, feeding on a wide range of other animals. A few large lizards and tortoises are exclusively herbivores, but many species are opportunistic omnivores.

BEHAVIOR AND SURVIVAL

Some species are solitary whereas others are highly social. Although sluggish and inactive when cool, once their preferred body temperature is reached by basking or pressing against a warm rock, they can be very active.

Reproductive behavior can be complex, with males actively defending territories and courting females. Although some squamates give birth to live young, in most cases the male reptile fertilizes the female, who then lays eggs in an underground nest. All reptiles are independent and self-feeding from birth, although some crocodilians give parental care for two or more years.

Reptiles are exploited by humans for their skin or as food. This, along with habitat loss, pollution, and climate change, threatens the survival of many species.

SEA TURTLE IN ITS ELEMENT >
The green turtle may appear primitive but, with efficient flippers and a flattened shell, is well adapted to its aquatic life.

TURTLES AND TORTOISES

With their bony shell, stout limbs, and toothless, beaklike mouth, turtles and tortoises have changed little from species that lived 200 million years ago.

The order includes sea turtles, freshwater turtles, and terrestrial tortoises. Some extinct turtles were gigantic, but most modern ones are more modestly sized. The exceptions are the large sea turtles and a few terrestrial tortoises inhabiting isolated islands.

PROTECTION AND LOCOMOTION

A turtle's shell is formed of numerous fused bones, overlaid by scutes. The domed upperpart is known as the carapace and the lower part, the plastron. Scutes are horny plates and new ones form beneath the old ones each year. The degree of protection provided by the shell varies between species and is much reduced in some aquatic turtles. Not all species can retract the head into the shell—some tuck their head along one shoulder under the edge of the shell.

Tortoises are slow animals, but some sea turtles can reach speeds of 19 mph (30 kph) when swimming, as a result of the modification of their forelimbs into flippers. Although they need to breathe air, many turtles are tolerant of low oxygen levels and can remain submerged for hours. Turtles have a slow metabolism and are generally long-lived.

Some species are carnivores, others herbivores, but most turtles are omnivorous. Animal prey is either slow-moving or ambushed as the turtle lies hidden. Food is broken up by the "beak," a sharp, keratinous jaw covering.

NESTING AND BREEDING

Turtles do not defend territories, but they can have extensive home ranges and may develop social hierarchies. Otherwise turtles often congregate on river banks or lakesides to bask or nest.

Both in water and on land, males of some species engage in elaborate courtship of females before copulation. Fertilization is internal and the females lay shelled eggs, like those of other reptiles and birds. These are spherical or elongated, with rigid or flexible shells. Clutches are laid in nests dug up by the female in terrestrial sites. Nearly all marine species come to land only to nest. In many species, but not all, the gender of the hatchlings depends on incubation temperature.

PHYLUM	CHORDATA
CLASS	REPTILIA
ORDER	TESTUDINES
FAMILIES	14
SPECIES	353

A green sea turtle hatchling enters the sea. Many of its breeding beaches are protected, but it remains endangered.

AUSTRO-AMERICAN SIDE-NECKED TURTLES

From South America and Australasia, the Chelidae include carnivorous and omnivorous species. Their characteristically long neck cannot be retracted, so it is turned sideways under the edge of the shell. They lay elongated eggs with leathery shells.

13½ in
34 cm

MACQUARIE TURTLE
Emydura macquarii
Widely distributed through the Murray River basin of Australia, this species eats amphibians, fishes, and algae. Males are smaller than females.

REIMANN'S SNAKE-NECKED TURTLE
Chelodina reimanni
This turtle from New Guinea eats crustaceans and mollusks. When threatened, it tucks its large head under the side of its carapace.

30 in
75 cm

scalloped carapace with keel (central ridge)

COMMON SNAKE-NECKED TURTLE
Chelodina longicollis
A shy, freshwater turtle of Australia, this species has a long neck, which enables it to raise its head out of water and capture prey.

10 in
25 cm

20 in
50 cm

MATAMATA
Chelus fimbriatus
This S. American turtle uses its unusual appearance as camouflage when ambushing prey, which it sucks into its mouth.

long snout

AFRICAN SIDE-NECKED TURTLES

Most of the Pelomedusidae are carnivorous and occupy freshwater habitats. When threatened, they can hide their head and neck beneath the edge of the shell. Found throughout Africa and Madagascar, they survive dry conditions by burying themselves in mud.

brown carapace

8 in
20 cm

scales on head resemble helmet

AFRICAN HELMETED TURTLE
Pelomedusa subrufa
Widespread in sub-Saharan Africa, this carnivorous turtle is highly sociable and often hunts in packs to bring down large prey.

BIG-HEADED TURTLE

The sole member of the Platysternidae, the big-headed turtle is an endangered species from shallow forest streams in southern China and Southeast Asia. It forages in streams, bottom-walking rather than swimming.

7 in
18 cm

BIG-HEADED TURTLE
Platysternon megacephalum
This small, carnivorous turtle has a flattened body, a very large head with powerful jaws, and a long tail.

AMERICAN SIDE-NECKED RIVER TURTLES

Closely related to African side-necked turtles, most species of the family Podocnemididae are found in tropical South America, barring one in Madagascar. These are herbivorous turtles found in various freshwater habitats. They cannot retract their neck into their shell.

12½ in
32 cm

RED-HEADED AMAZON RIVER TURTLE
Podocnemis erythrocephala
This species is found in swamps of the Rio Negro region of the Amazon basin in S. America.

SNAPPING TURTLES

Native to North and Central America, these large, aquatic turtles are noted for their aggression. The Chelydridae have a rough shell and a powerful head, with heavy, crushing jaws. They are effective predators, ambushing a variety of animals, but they also eat plants.

22 in
55 cm

COMMON SNAPPING TURTLE
Chelydra serpentina
A robust turtle, it often lies in wait for prey half-buried in mud. It occupies freshwater habitats from eastern N. America as far south as Ecuador.

massive head with strong jaws and sharp, pointed beak

tongue with wormlike lure

32 in
80 cm

ALLIGATOR SNAPPING TURTLE
Macrochelys temminckii
One of the largest freshwater turtles in the world, this N. American species has a worm-shaped lure on its tongue to attract prey as it lies in wait.

heavily built carapace with three rows of conical scutes

SOFTSHELL TURTLES

These aquatic predators inhabit freshwater habitats in North America, Africa, and southern Asia. The Trionychidae have a flattened shell covered with leathery skin rather than keratinous scutes. The adult shell length ranges in size from 10 in (25 cm) to more than 3¼ ft (1 m).

grayish-green carapace with raised ridges

22 in
55 cm

SPINY SOFTSHELL TURTLE
Apalone spinifera
A native of eastern N. America, this species eats mainly insects and aquatic invertebrates.

10½ in
27 cm

INDIAN FLAPSHELL TURTLE
Lissemys punctata
This Indian turtle has a rear flap on either side of the plastron that protects the hind limbs when they are withdrawn into the body.

14 in
35 cm

CHINESE SOFT-SHELLED TURTLE
Pelodiscus sinensis
Hunted for its meat, this turtle of E. Asia is rare in its native habitat, but tens of thousands are bred annually on farms.

PIG-NOSED TURTLE

The only species in the family Carettochelyidae, this turtle is an omnivore. Its carapace lacks hard scutes, but is still rigid. Its snout is adapted for breathing air while submerged.

PIG-NOSED RIVER TURTLE
Carettochelys insculpta
This nocturnal turtle is found in New Guinea and N. Australia. Like sea turtles, it has forelimbs modified as flippers for aquatic flight.

clawed flippers

28 in
70 cm

AMERICAN MUD AND MUSK TURTLES

These turtles from the New World emit a strong scent when threatened. The Kinosternidae tend to walk along the bottom of lakes and rivers, rather than swim, and are opportunistic omnivores. They lay elongated, hard-shelled eggs.

EASTERN MUD TURTLE
Kinosternon subrubrum
An omnivore, this freshwater turtle feeds on the bottom of slow, shallow watercourses in the S.E. US.

5 in
13 cm

COMMON MUSK TURTLE
Sternotherus odoratus
This freshwater turtle from eastern N. America is an omnivore. When threatened, it not only exudes a nauseating musk, but also bites.

5 in
13 cm

LEATHERBACK SEA TURTLE

There is only one species in the Dermochelyidae. This turtle is capable of maintaining an elevated body temperature, which allows it to swim in cold waters. The shell has no scutes; the leathery skin covers a layer of insulating oily tissue.

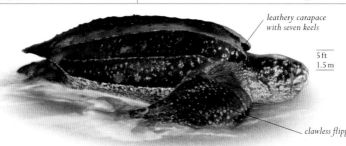

leathery carapace with seven keels

clawless flippers

LEATHERBACK SEA TURTLE
Dermochelys coriacea
Feeding mainly on jellyfish, this oceanic sea turtle is the world's largest turtle. It has a worldwide distribution that includes subarctic waters.

5 ft
1.5 m

SEA TURTLES

Sea turtles occur throughout the world's oceans, mainly in coastal waters. They are highly adapted to the marine environment, with a streamlined body and broad, paddlelike limbs. The Cheloniidae come to land to nest on beaches. Most species in this family are endangered.

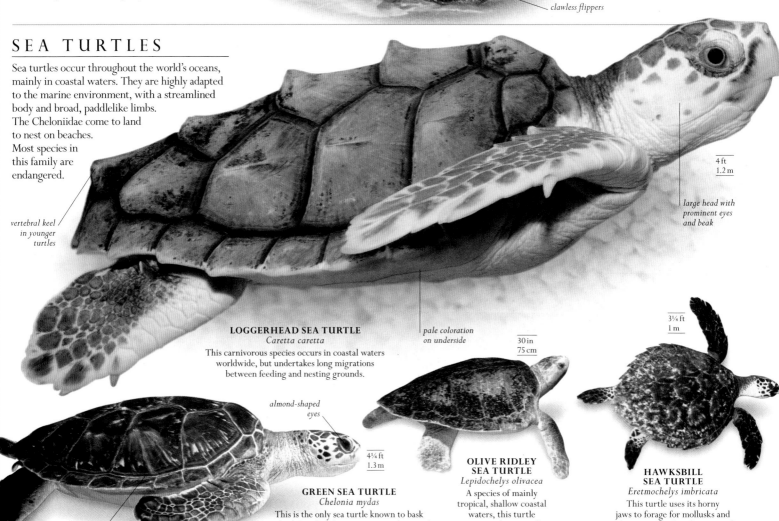

vertebral keel in younger turtles

large head with prominent eyes and beak

4 ft
1.2 m

LOGGERHEAD SEA TURTLE
Caretta caretta
This carnivorous species occurs in coastal waters worldwide, but undertakes long migrations between feeding and nesting grounds.

pale coloration on underside

30 in
75 cm

almond-shaped eyes

flattened carapace with large scutes

4¼ ft
1.3 m

GREEN SEA TURTLE
Chelonia mydas
This is the only sea turtle known to bask on land. It is found in temperate and tropical oceans worldwide and is a strict herbivore.

OLIVE RIDLEY SEA TURTLE
Lepidochelys olivacea
A species of mainly tropical, shallow coastal waters, this turtle eats a wide range of invertebrates and algae.

3¼ ft
1 m

HAWKSBILL SEA TURTLE
Eretmochelys imbricata
This turtle uses its horny jaws to forage for mollusks and other prey. It is found in tropical oceans throughout the world.

POND TURTLES

These turtles range from being fully aquatic to fully terrestrial. Most of them are found in North America, but one species in this family comes from Europe. Diet varies between species, although many are herbivores. The Emydidae are often brightly colored and intricately marked.

row of spines on carapace

FALSE MAP TURTLE
Graptemys pseudogeographica
This N. American turtle occurs in freshwater habitats with abundant vegetation. Females are almost twice as large as males.

10½ in
27 cm

pattern of yellow lines on neck and head

strong, clawed front legs

11 in
28 cm

RED-EARED SLIDER
Trachemys scripta elegans
This N. American species is mainly herbivorous and is common in the pet trade. Its ability to colonize new habitats has allowed it to become widely established in Europe and Asia.

10½ in
27 cm

YELLOW-BELLIED SLIDER
Trachemys scripta scripta
Named for its habit of sliding into water when disturbed, this turtle of the S. US is a diurnal omnivore.

9 in
23 cm

DIAMONDBACK TERRAPIN
Malaclemys terrapin
This is a diurnal species of brackish waters in eastern N. America. It has powerful jaws adapted for eating crustaceans and mollusks.

10 in
25 cm

PAINTED TURTLE
Chrysemys picta
Widely distributed in N. America, this small, freshwater turtle is active during summer. During winter, it lies torpid under water.

8 in
20 cm

CAROLINA BOX TURTLE
Terrapene carolina
Males of this N. American species have evolved highly curved claws that help grip the female's domed shell while mating.

distinctively patterned shell

ORNATE BOX TURTLE
Terrapene ornata
This is an omnivorous, terrestrial turtle of central N. America. It digs burrows to escape extreme heat or cold.

5½ in
14 cm

strong claws used for burrowing

5 in
13 cm

SPOTTED TURTLE
Clemmys guttata
Identifiable by its spots, this small turtle feeds on aquatic invertebrates and plants in the marshes of eastern N. America.

10 in
26 cm

CHICKEN TURTLE
Deirochelys reticularia
This shy turtle of swamps in eastern N. America extends its long neck to strike at crayfish and other prey.

8½ in
21 cm

EUROPEAN POND TURTLE
Emys orbicularis
Widespread in Europe, this is a highly aquatic turtle. It basks on logs or rocks, but rapidly dives into the water if disturbed.

5 in
13 cm

WOOD TURTLE
Glyptemys insculpta
This turtle occupies the damp forests in northeastern N. America. Unusually, males and females perform an elegant dance together during courtship.

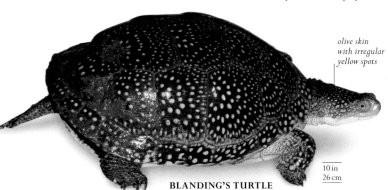

olive skin with irregular yellow spots

10 in
26 cm

BLANDING'S TURTLE
Emydoidea blandingii
Found mainly across the Great Lakes region of N. America, this turtle is an omnivore and especially skilled at capturing crayfish.

15 in
38 cm

FLORIDA REDBELLY TURTLE
Pseudemys nelsoni
Restricted to Florida, this turtle lives in lakes and slow-moving streams. During courtship, the male caresses the female's head with his forelimbs.

16 in
40 cm

RED-BELLIED TURTLE
Pseudemys rubriventris
Restricted to the N.E. USA, this omnivorous, diurnal species prefers large, deep bodies of water. Females are larger than males.

ALDABRA GIANT TORTOISE
Aldabrachelys gigantea

The last remaining species of giant tortoise on islands in the Indian Ocean, the Aldabra giant tortoise can weigh more than 660 lb (300 kg). Though resident on three islets of the Aldabra atoll, over 90 percent of the tortoises live on Grande-Terre—the largest islet by far—despite the scarcity of water and inadequate supply of vegetation. Poor conditions there have inhibited the growth of individual tortoises, with many not reaching sexual maturity. However, they are highly sociable, more so than the larger tortoises on other islets. Males are larger than females, but courtship is a gentle affair. Their eggs are buried underground, and hatch in the rainy season. The whole population is vulnerable to natural disasters and rising sea levels.

SIZE 4 ft (1.2 m)
HABITAT Grassy areas
DISTRIBUTION Aldabra, Indian Ocean
DIET Vegetation

< HORNY BEAK
The tortoise's large mouth contains no teeth. It cuts vegetation with its sharp, horny beak, takes it into the mouth with its tongue, and swallows it whole.

^ LEATHERY VISAGE
The skin is tough, leathery, and—depending on islet—either gray or brown. The skin is creased into folds around the neck. If it senses danger, the tortoise pulls its head back inside its shell.

< EAR
Tortoises have no external ear flap; so the eardrum is situated in a hollow.

< EYE
The eye is relatively large and has a well-developed eyelid. Tortoises can see in color, particularly in the red and yellow parts of the spectrum; this probably helps them to find colorful fruit.

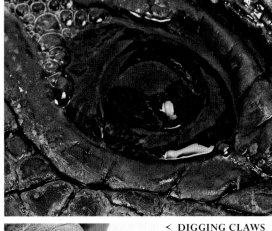

< DIGGING CLAWS
Tortoises have sturdy, elephantlike hind limbs, with five claws on each foot. Females have larger claws than males, which they use when excavating a nesting site.

horny scales cover the front and hind limbs

hind limbs are elephantine, with strong claws

^ FORELEG
The front limbs are cylindrical in shape, and longer than the hind limbs. This allows the tortoise to raise its body well above the ground as it walks. The legs are covered in large, tough scales.

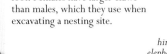
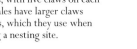

< TAIL
The giant tortoise has a short tail, and can tuck it to one side under the back of the carapace. Males have longer tails than females.

scutes on the upper shell show growth rings

MADE TO LAST ⌄
Compared to giant tortoises from the Galapagos Islands, this species has a more rounded head, a pointed snout, and a smiling mouth. It can live for more than a hundred years.

POND AND RIVER TURTLES

Found both in the Old and New Worlds, the Geoemydidae are freshwater and terrestrial turtles, ranging in adult shell length from 5½–20 in (14 to 50 cm). They vary in their dietary preferences from being herbivorous to carnivorous. Many exhibit sexual dimorphism, with females being larger than males.

9 in
23 cm

BROWN WOOD TURTLE
Rhinoclemmys annulata
A herbivorous tortoise, this species lives in tropical forests of C. America. It is mainly active in the mornings and after rains.

12 in
30 cm

GOLDEN COIN TURTLE
Cuora trifasciata
This is a carnivorous turtle from S. China. Its name derives from its value in the illegal wildlife market. Its use in traditional medicine threatens its survival.

pronounced
spinal keel

6½ in
17 cm

yellow
stripes
behind
eyes

YELLOW-MARGINATED BOX TURTLE
Cuora flavomarginata
An omnivore of China's rice paddies, this species avoids deep water and spends hours basking on land.

5 in
13 cm

ASIAN LEAF TURTLE
Cyclemys dentata
This omnivorous turtle is found in slow-moving waters in S.E. Asia. It defends itself by emitting foul-smelling liquid.

5 in
13 cm

BLACK-BREASTED LEAF TURTLE
Geoemyda spengleri
An inhabitant of wooded mountains in S. China, this turtle eats small invertebrates and fruits. Its carapace is rectangular, keeled, and spiky.

TORTOISES

Found in the Americas, Africa, and southern Eurasia, members of the family Testudinidae can reach great sizes. They have a dome shell into which they can retract their head. Fully terrestrial, tortoises are characterized by elephantine limbs. They lay hard-shelled eggs.

13 in
33 cm

ELONGATED TORTOISE
Indotestudo elongata
Found in tropical S.E. Asia, this tortoise eats fruits and carrion. It hides in moist leaf-litter when the weather is dry.

16 in
40 cm

SERRATED HINGE-BACK TORTOISE
Kinixys erosa
This is an omnivorous species found in swamps in tropical W. Africa. Older tortoises develop hinges in the rear of their carapaces.

28 in
70 cm

RED-FOOTED TORTOISE
Chelonoidis carbonaria
Although known to eat carrion, this tortoise is mainly vegetarian. It is found in a variety of habitats of northeastern S. America.

12 in
30 cm

DESERT TORTOISE
Gopherus agassizii
This tortoise lives in small burrows in deserts of southwestern N. America. Mainly vegetarian, it sometimes preys on animals.

16 in
40 cm

RADIATED TORTOISE
Astrochelys radiata
Restricted to S. Madagascar, this endangered species eats mainly plants and is active during the early part of the day.

highly
domed scutes

INDIAN STAR TORTOISE
Geochelone elegans
Found in dry areas of India and Sri Lanka, this herbivorous species mates and breeds in the monsoon season.

15 in
38 cm

PANCAKE TORTOISE
Malacochersus tornieri
Found in E. Africa, this
omnivorous species inhabits
rocky areas. Its flat shape
allows it to hide in cracks.

7 in
18 cm

*growth rings on the
carapace
of juveniles*

*flattened
scute*

ALDABRA GIANT TORTOISE
Aldabrachelys gigantea
Restricted to the Aldabra atoll in the
Indian Ocean, this large herbivore is able
to drink through its nostrils.

4 ft
1.2 m

large forelimbs

GALAPAGOS TORTOISE
Chelonoidis nigra
One of the world's largest tortoises,
this mainly vegetarian species has
11 subspecies, from different islands
in the Galapagos archipelago.

7½ in
19 cm

HERMANN'S TORTOISE
Testudo hermanni
This herbivorous species inhabits dry forests
of coastal Italy and S. France. It hibernates
during the cold winter months.

*jaw with sharp
cutting edge for
eating vegetation*

STEPPES TORTOISE
Agrionemys horsfieldii
A herbivore from the dry
deserts and steppes of
C. Asia, this species escapes
the daytime heat by
sheltering in burrows.

4 ft
1.2 m

11 in
28 cm

TUATARA

**Superficially like a lizard, New Zealand's
tuatara belongs to a much more ancient
order of reptiles. Its closest relatives
became extinct 100 million years ago.**

The tuatara has many anatomical features that set it
apart from lizards. Most striking are its wedge-shaped
"teeth"—actually serrations of the jawbone.
The upper jaw has a double row that fits
over a single row on the lower jaw.

Tuataras are very long lived, but are vulnerable to
introduced ground predators. Inhabiting coastal
forests, they are active at low body temperatures
and emerge from their burrows at night to hunt for
invertebrates, and birds' eggs and chicks. Males are
territorial and nesting is communal. Eggs take up
to four years to form and are incubated for 11 to 16
months before hatching. Incubation
temperature determines the
sex of the hatchlings.

PHYLUM	CHORDATA
CLASS	REPTILIA
ORDER	RHYNCHOCEPHALIA
FAMILIES	1
SPECIES	1

TUATARA

Often portrayed as "living fossils,"
the Sphenodontidae are modern
representatives of reptiles that
coexisted with dinosaurs. This
primitive species is found only on
offshore islands of New Zealand.

stout tail

TUATARA
Sphenodon punctatus
This reptile has clawed limbs to dig burrows.
It can shed its tail to escape predators. The
male uses its dorsal crest for display.

23½ in
60 cm

*powerful,
clawed limbs*

LIZARDS

Typically, lizards have four legs and a long, thin tail, but there are also many legless species. They have scaly skin and firm jaw articulation.

All lizards are ectothermic, obtaining heat energy from the environment. Although seen as predominately tropical or desert animals, they have a global distribution; lizards are found from beyond the Arctic Circle in Europe to the tip of South America. Highly adaptable, they occupy a wide range of terrestrial habitats with many being arboreal or rock-dwelling. Many legless species are adapted for burrowing, a few lizards are effective gliders from trees, and others are semiaquatic, including a marine species on the Galapagos Islands.

STRATEGIES FOR SURVIVAL

Lizards range in length from about ½ in (1.5 cm) to 10 ft (3 m) in the case of the Komodo dragon, but the majority are between 4 and 12 in (10–30 cm). Although most are carnivores, about two percent of species are primarily herbivorous. Many lizards are prey for other carnivores. Their defense

mechanisms depend on agility, camouflage, and bluff. Many species are able to shed their tail to distract the attention of predators as they escape. The tail grows back. In some families of lizards, the skin can change color. This ability can be used for camouflage or for sexual or social signaling. Unusually, in many lizards the pineal gland on the top of the head acts as a light-sensitive "third eye."

VARIED LIFESTYLES

Although some species are solitary, many lizards have complex social structures, with the males maintaining their territories through visual signals. Many species lay eggs in underground nests, while others retain the eggs in the oviduct until hatching. There are also truly viviparous species, in which the mother provides nutrition via a placenta. Males have paired hemipenes—sexual organs used for internal fertilization—although some species are parthenogenetic, with females reproducing without the participation of a male. Some lizards tend their eggs during incubation but very few exhibit any maternal care of the hatchlings.

PHYLUM	CHORDATA
CLASS	REPTILIA
ORDER	SQUAMATA
FAMILIES (LIZARDS)	38
SPECIES	6,687

SQUAMATA: AN ALL-INCLUSIVE ORDER

Traditionally, lizards and snakes were considered two distinct groups, but modern genetic research makes it clear that they are not. Primitive squamates (scaled reptiles) were lizardlike animals that arose in the mid-Jurassic. Available evidence suggests that snakes evolved from lizards during the mid-Cretaceous. By this time various families of lizards had branched off on their own evolutionary course. Some lizard families are thus closer to snakes than they are to other lizards. Of living lizards, the monitors are thought to be those most closely related to snakes.

CHAMELEONS

Restricted to the Old World, the members of the family Chamaeleonidae have long limbs with grasping feet and a prehensile tail for gripping branches—an adaptation for life in trees. Their eyes are capable of moving independently in any direction to locate insects or small vertebrates. They catch their prey with a long, sticky tongue that shoots out from the mouth. Their ability to change color is for display and camouflage. Many species are threatened with extinction.

3¼ in
8 cm

BEARDED PYGMY-CHAMELEON
Rieppeleon brevicaudatus
This unusual, small chameleon is from E. Africa. Drab coloration and patterning along its body allow it to resemble a dead leaf.

PARSON'S CHAMELEON
Calumma parsonii
The largest chameleon in the world, this species is restricted to Madagascar. It hunts invertebrates in the canopy of montane forests.

28 in
70 cm

12 in
30 cm

prehensile tail

MEDITERRANEAN CHAMELEON
Chamaeleo chamaeleon
This chameleon spends its time in bushes searching for insects. It is found in N. Africa and around the Mediterranean.

JACKSON'S CHAMELEON
Trioceros jacksonii
This diurnal, arboreal species is found in E. Africa. The male is identifiable by the three horns on its snout, which are for display.

green skin coloration sometimes changes to brown

12 in
30 cm

dorsal spines

20 in
51 cm

GIANT SPINY CHAMELEON
Furcifer verrucosus
This large chameleon is a native of the humid, coastal region of Madagascar. A shy species, it relies on camouflage to ambush its insect prey.

22 in
56 cm

grasping, four-toed feet

PANTHER CHAMELEON
Furcifer pardalis
Found only in dry forests in Madagascar, this lizard hunts insects in trees by stealth. Males are highly territorial.

23½ in
60 cm

VEILED CHAMELEON
Chamaeleo calyptratus
This species is from the southern coast of the Arabian Peninsula. Males have a large casque on the head, while females have a smaller one.

CHISEL-TEETH LIZARDS

Common in Africa, southern Asia, and Australasia, the small lizards of the family Agamidae—which also includes flying lizards and water dragons—are the Old World equivalents of the iguanas. Spines, flaps, and crests often adorn the head and back of these lizards, which lay soft-shelled eggs. Males are brightly colored while the females are duller.

AUSTRALIAN WATER DRAGON
Intellagama lesueurii
3¼ ft
1 m
The largest water dragon in Australia, this species lives beside water into which it dives to escape predators. Adults eat invertebrates and small vertebrates.

ORIENTAL GARDEN LIZARD
Calotes versicolor
16 in
40 cm
This agile, diurnal lizard is often found hunting insects in trees around human habitation. It is common in S. Asia.

3¼ ft
1 m

long limbs

long tail helps with balance when climbing

long, laterally flattened tail used in swimming

ASIAN WATER DRAGON
Physignathus cocincinus
A powerful swimmer, this lizard seeks refuge in water when threatened. It is found in riverside trees in S. Asia.

BOULENGER'S PRICKLENAPE
Acanthosaura crucigera
10 in
26 cm
This slow-moving lizard from Asia preys on insects while perched on a branch. Males use the long spines on their neck when fighting each other.

16 in
40 cm

NORTH AFRICAN MASTIGURA
Uromastyx acanthinura
This herbivorous lizard inhabits the harsh desert environments of N. Africa. It uses its clublike, spiny tail in defense.

chisel-shaped teeth on outer rim of mouth

CENTRAL BEARDED DRAGON
Pogona vitticeps
20 in
50 cm
Found in arid woodland in Australia, this lizard has a beard of spines. It lays eggs in an underground nest.

8 in
20 cm

GREEN-STRIPED TREE DRAGON
Diploderma splendidum
This colorful lizard is from humid montane forests of China, where it hunts insects. It lays small clutches of five to seven eggs.

5–7 in
15–18 cm

THORNY DEVIL
Moloch horridus
This Australian lizard feeds exclusively on ants. The body is covered with large, thorny spines, which make it difficult for a predator to swallow. The spines also channel rainwater or morning dew into its mouth.

large skin frill makes head look larger

35 in
90 cm

strong hind legs allow this lizard to run away bipedally

FRILLED LIZARD
Chlamydosaurus kingii
Common in subtropical woodland in Australia, this species feeds on insects and other lizards in trees and on the ground. When threatened, it unfurls its frill and opens its mouth wide.

16 in
40 cm

RAINBOW LIZARD
Agama agama
This common, insectivorous lizard from Africa is gray at night, but becomes brightly colored in sunlight. Males use their bright skin colors during territorial disputes.

PANTHER CHAMELEON
Furcifer pardalis

This large chameleon is endemic to Madagascar off the east coast of southern Africa, and has also been introduced to Mauritius and Réunion. This species lives in trees in humid scrubland, and its feet are so well adapted for grasping branches that it is difficult for the lizard to walk across a flat surface. Active during the day, it moves slowly through the branches to hunt its insect food by stealth. When it spots its next meal, the chameleon focuses on its victim with both eyes, then shoots out its long tongue to grasp the insect and pull it back into its large mouth. The chameleon's uncanny ability to change color is an indicator of mood and social status only, and is not used for camouflage. When faced with a rival, it rapidly inflates its body and changes color, putting on a display of dominance that is usually enough to decide a dispute.

SIZE 16–22 in (40–56 cm)
HABITAT Trees in humid scrub
DISTRIBUTION Madagascar
DIET Arthropods, crustaceans

a ridge of protective spines runs along the midline of the back

casque, or bony shield, at the back of the head

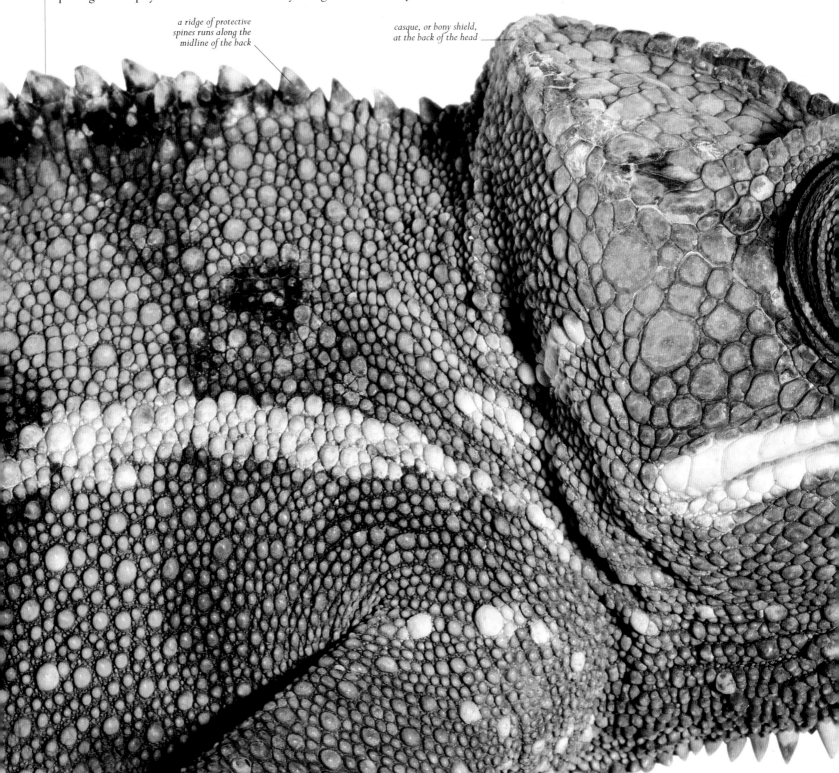

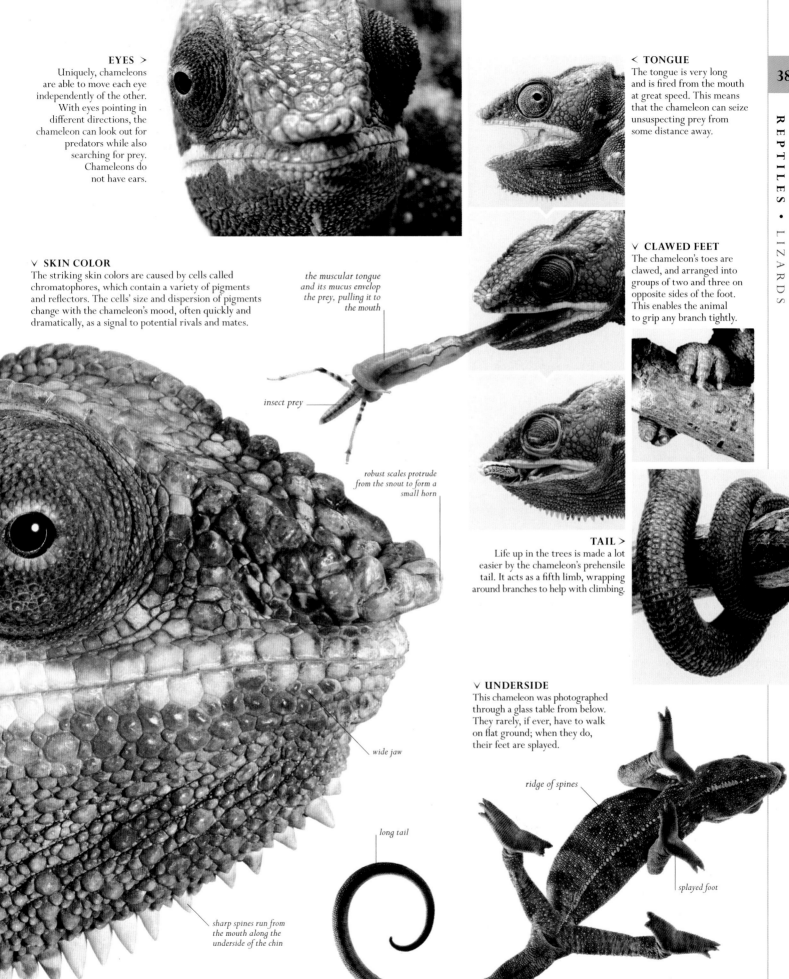

EYES >
Uniquely, chameleons are able to move each eye independently of the other. With eyes pointing in different directions, the chameleon can look out for predators while also searching for prey. Chameleons do not have ears.

< TONGUE
The tongue is very long and is fired from the mouth at great speed. This means that the chameleon can seize unsuspecting prey from some distance away.

v SKIN COLOR
The striking skin colors are caused by cells called chromatophores, which contain a variety of pigments and reflectors. The cells' size and dispersion of pigments change with the chameleon's mood, often quickly and dramatically, as a signal to potential rivals and mates.

the muscular tongue and its mucus envelop the prey, pulling it to the mouth

insect prey

robust scales protrude from the snout to form a small horn

v CLAWED FEET
The chameleon's toes are clawed, and arranged into groups of two and three on opposite sides of the foot. This enables the animal to grip any branch tightly.

TAIL >
Life up in the trees is made a lot easier by the chameleon's prehensile tail. It acts as a fifth limb, wrapping around branches to help with climbing.

wide jaw

v UNDERSIDE
This chameleon was photographed through a glass table from below. They rarely, if ever, have to walk on flat ground; when they do, their feet are splayed.

ridge of spines

long tail

splayed foot

sharp spines run from the mouth along the underside of the chin

GECKOS

Widely distributed in the tropics and subtropics, the superfamily Gekkonidae number more than 1,000 species in 100 genera. They are noted for being very vocal and for their ability to climb smooth surfaces. Many lay hard-shelled eggs, while others lay soft-shelled eggs. Some species are viviparous.

WESTERN BANDED GECKO
Coleonyx variegatus

4¾ in
12 cm

This terrestrial gecko hunts invertebrates in the deserts of the W. US. Unlike many geckos, it has moveable eyelids.

10 in
25 cm

AFRICAN FAT-TAILED GECKO
Hemitheconyx caudicinctus

Common in the W. Sahara, this species has a tail that stores fat. It lacks the adhesive foot pads of many other geckos.

8½ in
21 cm

COMMON LEOPARD GECKO
Eublepharis macularius

Originally from southern C. Asia, this gecko is a popular pet, which has been bred for different markings and colors.

6 in
15 cm

MOORISH GECKO
Tarentola mauritanica

Common around the Mediterranean, this agile gecko preys on insects. It lives on rock faces but often enters houses.

8 in
20 cm

WONDER GECKO
Teratoscincus scincus

To defend itself, this C. Asian species waves its tail slowly, causing a row of large scales on the tail to rub together, making a hissing sound.

bright coloration

MADAGASCAR DAY GECKO
Phelsuma madagascariensis

This colorful diurnal species is typically found on trees. Like many geckos, it lays hard-shelled eggs that stick to a branch.

adhesive pads on toes

10 in
25 cm

8 in
20 cm

KUHL'S FLYING GECKO
Ptychozoon kuhli

This gecko's webbed toes and skin flaps help break its fall as it parachutes from trees in the rainforests of S.E. Asia.

16 in
40 cm

TOKAY GECKO
Gekko gecko

Named after its harsh "to-kay" call, this large gecko from S.E. Asia often lives in houses.

13½ in
34 cm

RING-TAILED GECKO
Cyrtodactylus louisiadensis

Found in New Guinea, this large, nocturnal gecko feeds aggressively on invertebrates and small frogs found on the ground.

6 in
15 cm

BROOK'S HOUSE GECKO
Hemidactylus brookii

This gecko lives in close proximity to humans. Native to N. India, it now also occurs in Hong Kong, Shanghai, and the Philippines.

4 in
10 cm

MEDITERRANEAN GECKO
Hemidactylus turcicus

Often betrayed by its mewing cry, this small European gecko is common in houses, where it hunts insects attracted to lights.

FLAP-FOOTED LIZARDS

All lizards of this group have an elongated body, no front limbs, and much-reduced hind limbs. Restricted to Australasia, the 36 species in the family Pygopodidae hunt insects by burrowing or on the surface. They are related to geckos and lay soft-shelled eggs.

4¾ in
12 cm

FRASER'S DELMA
Delma fraseri

This insectivorous, Australian lizard inhabits spinifex grassland. It is well adapted to moving through the stiff blades of grass.

8½ in
21 cm

COMMON SCALY FOOT
Pygopus lepidopodus

Snakelike in appearance, this lizard is widespread in Australia. It is a diurnal predator of insects and tunnel-inhabiting spiders.

23½ in
60 cm

BURTON'S SNAKE-LIZARD
Lialis burtonis

Native to Australia, this lizard has an elongated, wedge-shaped snout that it uses to grasp skinks, which it consumes whole.

ANOLES

Most anoles come from around the Caribbean. A diverse group, the Dactyloidae are typically small, arboreal, and insectivorous. Although often green or brown, they change skin color according to mood and environment. Both sexes aggressively defend their territories.

KNIGHT ANOLE
Anolis equestris

The largest anole, this species is restricted to Cuba. Adhesive pads on its toes enable it to scale smooth walls.

20 in
50 cm

CAROLINA ANOLE
Anolis carolinensis

Males show their dominance by bobbing their heads and flaring their brightly colored throat fan.

8 in
20 cm

HELMETED LIZARDS

Nine species of arboreal lizards from Central and
South America constitute the family Corytophanidae.
They are closely related to the iguanas. All species have
a well-developed head crest. Long legs and a long
tail allow these lizards to run at high speeds
to evade predators.

saillike dorsal crest

*bright
orange iris*

HELMETED IGUANA
Corytophanes cristatus
This C. American iguana lives
on a diet of large arthropods. By
feeding infrequently, it minimizes
the time it spends in the open or
is conspicuous to predators.

*hard, conelike casque at
rear of head*

GREEN BASILISK
Basiliscus plumifrons
Native to C. American
rainforests, this species inhabits
riverbanks. The crests on its
head, back, and tail are
supported by bony spines.

26 in
65 cm

long legs

**EASTERN
CASQUE-HEADED IGUANA**
Laemanctus longipes
This large, insectivorous species lives in
small groups of one male and two to three
females. It is native to C. America.

*slender, green
feet and legs*

28 in
70 cm

*long tail aids balance when
climbing or running*

NORTH AMERICAN
SPINY LIZARDS

Restricted to North and Central America, the Phrynosomatidae are a
diverse family of lizards that prefer arid environments and hunt insects.
They are generally small, dull in color, and spiny.
Most lay eggs, but those living at high
altitudes are viviparous.

**COLORADO DESERT
FRINGE-TOED LIZARD**
Uma notata
Closable nostrils and ear flaps, overlapping
jaws, and interlocking eyelids help protect this
desert-living species as it burrows in sand.

3¼ in
8 cm

GREEN SPINY LIZARD
Sceloporus malachiticus
This diurnal, arboreal lizard from
C. America has stiff, keeled scales,
which give it a spiny appearance.

8 in
20 cm

DESERT HORNED LIZARD
Phrynosoma platyrhinos
This lizard of N. American deserts
primarily eats ants. Its flattened body
is an adaptation to maximize heat
absorption while basking in the sun.

6 in
15 cm

IGUANAS

Largely confined to the Americas, the Iguanidae
number nearly 30 species in 8 genera. They are
diverse in color. Most are diurnal, carnivorous
predators, although larger species are
herbivorous. They all lay eggs.

MARINE IGUANA
Amblyrhynchus cristatus
Native to the Galapagos
Islands, this aquatic lizard is
well adapted to feeding on
underwater algae. Nasal
glands help remove salt.

2¼–5 ft
0.7–1.5 m

GREEN IGUANA
Iguana iguana
Widespread in C. and
S. America, this large iguana
is herbivorous. Males defend
their territory by vigorously
nodding their heads to
display dominance.

dorsal crest

6½ ft
2 m

**BLACK
IGUANA**
Ctenosaura similis
This gregarious species
from C. America lives in
colonies with a dominant
male. Although usually
herbivorous, it sometimes
eats small lizards.

35 in
90 cm

*long,
whiplike tail*

SKINKS

Distributed worldwide, the Scincidae are a diverse group of 1,400 species. While many are active diurnal predators, several others are nocturnal, limbless, burrowing lizards. They use chemical as well as visual communication. Although typically oviparous, many species are viviparous.

14 in
35 cm

bright coloration on flanks

weak legs

FIRE SKINK
Mochlus fernandi
This insectivorous skink is native to humid, forested areas of W. Africa. Its attractive coloration makes it a popular lizard to keep in captivity.

EMERALD TREE SKINK
Lamprolepis smaragdina
Native to islands of the W. Pacific, this arboreal lizard preys on insects on bare tree trunks.

10 in
25 cm

8½ in
21 cm

FIVE-LINED SKINK
Plestiodon fasciatus
This N. American skink coils around its eggs to protect them during incubation. It prefers woodland and feeds on ground-living insects.

PERCIVAL'S LANCE SKINK
Acontias percivali
This legless African skink burrows through leaf litter to hunt its invertebrate prey. It gives birth to up to three live young.

12 in
30 cm

WALL AND SAND LIZARDS

Lizards of the family Lacertidae are found in a wide variety of habitats throughout the Old World. They are active predators and have complex social systems, with males defending their territory. Almost all species lay eggs. These lizards typically have a large head.

LARGE PSAMMODROMUS
Psammodromus algirus
Found in the W. Mediterranean region, this small lizard inhabits dense, bushy areas. During breeding, males develop a red patch on their throat.

3 in
7.5 cm

8 in
20 cm

OCELLATED LIZARD
Timon lepidus
The largest European member of the family, this lizard lives in dry, bushy places. It eats insects, eggs, and small mammals.

VIVIPAROUS LIZARD
Zootoca vivipara
Found throughout Europe and up to 9,900 ft (3,000 m) in the Alps, this ground-dwelling lizard lives in a variety of habitats. It gives birth to live young.

6 in
15 cm

FRINGE-TOED LIZARD
Acanthodactylus erythrurus
Found in the Iberian Peninsula and N. Africa, this species has fringes of spiny scales on its toes that enable it to cross over loose sand.

3½ in
9 cm

GRAN CANARIA GIANT LIZARD
Gallotia stehlini
This large species is restricted to the shrubland of Grand Canary Island. It is diurnal and herbivorous.

32 in
80 cm

3 in
7.5 cm

ITALIAN WALL LIZARD
Podarcis siculus
A ground-living species, found around the N. Mediterranean, this lizard inhabits grassy places and often lives in close proximity to humans.

WHIPTAILS AND RACERUNNERS

These fast-running American lizards occupy a range of habitats. Smaller species of the Teiidae are insectivorous but larger ones are carnivorous. All 120 species are oviparous, although many whiptails are all-female and reproduce parthenogenetically, laying viable, fertile eggs without mating.

AMAZON RACERUNNER
Ameiva ameiva
Powerful jaws allow this S. American species to feed on small vertebrates and insects that it catches in open terrestrial habitats.

18 in
45 cm

long tail used in defense

RED TEGU
Salvator rufescens
This large species of the arid regions in central S. America is an active predator and scavenger, but is also known to eat plants.

4 ft
1.2 m

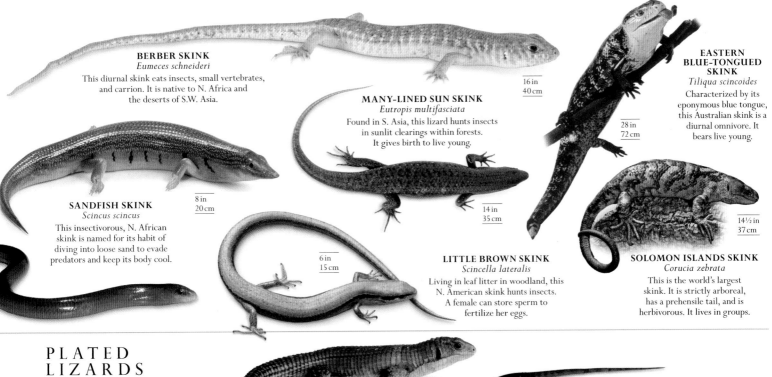

BERBER SKINK
Eumeces schneideri
This diurnal skink eats insects, small vertebrates, and carrion. It is native to N. Africa and the deserts of S.W. Asia.

16 in
40 cm

MANY-LINED SUN SKINK
Eutropis multifasciata
Found in S. Asia, this lizard hunts insects in sunlit clearings within forests. It gives birth to live young.

14 in
35 cm

EASTERN BLUE-TONGUED SKINK
Tiliqua scincoides
Characterized by its eponymous blue tongue, this Australian skink is a diurnal omnivore. It bears live young.

28 in
72 cm

SANDFISH SKINK
Scincus scincus
This insectivorous, N. African skink is named for its habit of diving into loose sand to evade predators and keep its body cool.

8 in
20 cm

6 in
15 cm

LITTLE BROWN SKINK
Scincella lateralis
Living in leaf litter in woodland, this N. American skink hunts insects. A female can store sperm to fertilize her eggs.

SOLOMON ISLANDS SKINK
Corucia zebrata
This is the world's largest skink. It is strictly arboreal, has a prehensile tail, and is herbivorous. It lives in groups.

14½ in
37 cm

PLATED LIZARDS

There are 32 species of plated lizards. All are from sub-Saharan Africa and lay eggs. They have a cylindrical body with well-developed legs for hunting insects in rocky areas and the savanna. Solitary in nature, the Gerrhosauridae are often aggressive to other members of their species.

19 in
48 cm

ROUGH-SCALED PLATED LIZARD
Broadleysaurus major
This omnivorous species is widely distributed in the savanna of E. Africa. It occupies crevices in rocky outcrops or termite nests.

14 in
36 cm

MADAGASCAN GIRDLED LIZARD
Zonosaurus madagascariensis
This insectivorous lizard is native to Madagascar. It prefers to be solitary, and feeds on the ground in open, dry habitats.

GIRDLED LIZARDS

Limited to southern and eastern Africa, the Cordylidae are named for the rings of spiny scales that encircle their tail. Their flattened body has limited space for live young or eggs. In viviparous species, this restricts litter size; oviparous species lay only two eggs.

CAPE GIRDLED LIZARD
Cordylus cordylus
Endemic to southern Africa, this lizard lives in dense colonies. Adults are aggressive, and form social hierarchies under a dominant male.

8½ in
21 cm

spiny scales

MICROTEIID LIZARDS

There are 165 species of the Gymnophthalmidae in tropical South America. Generally small, they are characterized by large scales on their back. These are diurnal and secretive insectivorous lizards and their dull coloration provides camouflage in leaf litter. Most species lay eggs.

BROMELIAD LIZARD
Anadia ocellata
An arboreal species found in C. America, the bromeliad lizard hunts insects and takes refuge in foliage.

3¼ in
8 cm

muscular body

glossy head scales

CROCODILE LIZARD

The crocodile lizard is the only representative of the Shinisauridae. There are two recognized sub-species found in China and northern Vietnam. This little-studied lizard is endangered because of the popularity of the lizards as pets.

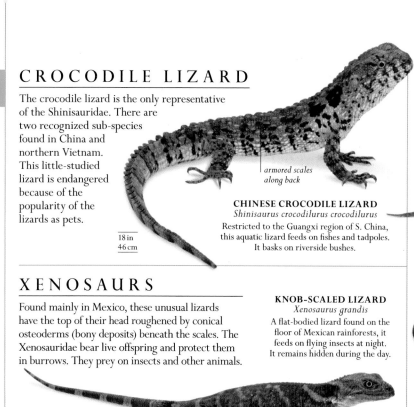

armored scales along back

18 in
46 cm

CHINESE CROCODILE LIZARD
Shinisaurus crocodilurus crocodilurus
Restricted to the Guangxi region of S. China, this aquatic lizard feeds on fishes and tadpoles. It basks on riverside bushes.

XENOSAURS

Found mainly in Mexico, these unusual lizards have the top of their head roughened by conical osteoderms (bony deposits) beneath the scales. The Xenosauridae bear live offspring and protect them in burrows. They prey on insects and other animals.

KNOB-SCALED LIZARD
Xenosaurus grandis
A flat-bodied lizard found on the floor of Mexican rainforests, it feeds on flying insects at night. It remains hidden during the day.

10 in
25 cm

NIGHT LIZARDS

The 30 or so Xantusiidae species are restricted to North and Central America. These secretive, viviparous lizards are crepuscular (active at twilight) or diurnal rather than nocturnal. They have a large headshield and rectangular belly scales.

YELLOW-SPOTTED NIGHT LIZARD
Lepidophyma flavimaculatum
Found among rotting logs in the wet forests of C. America, this species feeds on insects.

5 in
13 cm

AMERICAN LEGLESS LIZARDS

Found only in the deserts of western North America, the Anniellidae comprises two species. They have a long, cylindrical body and a small head, and burrow to hunt invertebrates. They give birth to one or two live young.

CALIFORNIA LEGLESS LIZARD
Anniella pulchra
An insectivorous, wormlike lizard, this species burrows in sand or loose soil. It can detach its tail to trick predators.

5½ in
14 cm

BEADED LIZARDS

Native to the arid areas of western North America, the Helodermatidae have salivary glands modified to produce venom that is delivered along grooved teeth. Active at night, they hunt a wide range of invertebrates and also eat carrion.

pink and black beadlike scales

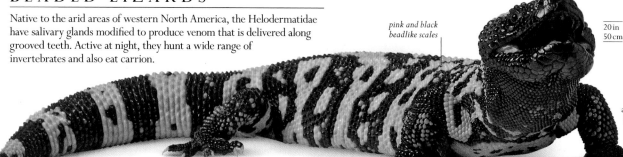

20 in
50 cm

GILA MONSTER
Heloderma suspectum
This lizard of the deserts of southwestern N. America is an egg-laying species. It hunts insects and ground-dwelling vertebrates, including small mammals.

MONITOR LIZARDS

Widely distributed throughout tropical Africa, Asia, and Australia, the Varanidae are the largest of all living lizards. They have an elongated body and strong legs, and many produce toxic saliva. Their choice of prey depends on their size.

5 ft
1.5 m

HEATH MONITOR
Varanus rosenbergi
Found around the coasts of southern Australia, this monitor has a varied diet. It is a strong digger and often forages underground for food.

6½ ft
2 m

ASIAN WATER MONITOR
Varanus salvator
This monitor is found in rainforests and wet habitats of S. Asia. Its large size allows it to tackle a variety of prey.

4¼ ft
1.3 m

SAVANNA MONITOR
Varanus exanthematicus
Found in the savanna of sub-Saharan Africa, this species forages for invertebrates and other animals small enough to swallow.

long, sinuous tail

ANGUID LIZARDS

The Anguidae include both legless forms and species with normal limbs. Most lizards are surface dwellers and occur in a range of habitats in the Old and New Worlds. Though often insectivorous, these lizards will feed on a range of prey. They lay eggs.

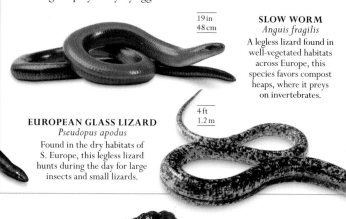

19 in
48 cm

SLOW WORM
Anguis fragilis
A legless lizard found in well-vegetated habitats across Europe, this species favors compost heaps, where it preys on invertebrates.

EUROPEAN GLASS LIZARD
Pseudopus apodus
Found in the dry habitats of S. Europe, this legless lizard hunts during the day for large insects and small lizards.

4 ft
1.2 m

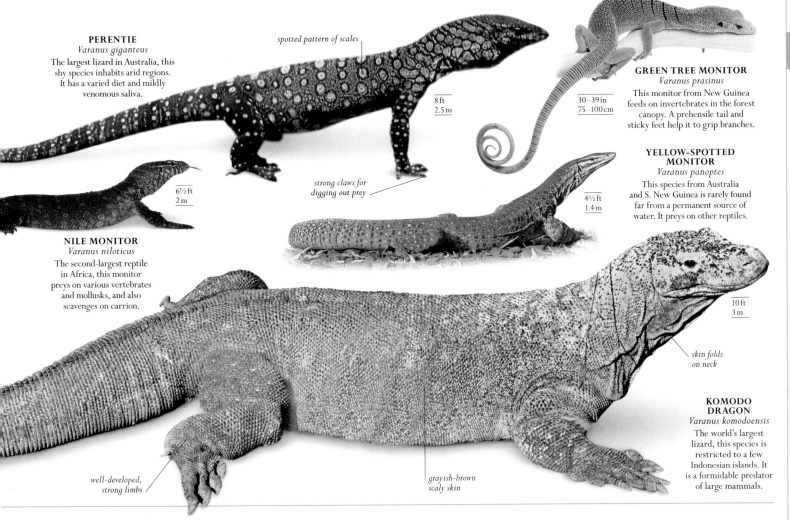

PERENTIE
Varanus giganteus
The largest lizard in Australia, this shy species inhabits arid regions. It has a varied diet and mildly venomous saliva.

spotted pattern of scales

8 ft
2.5 m

GREEN TREE MONITOR
Varanus prasinus
This monitor from New Guinea feeds on invertebrates in the forest canopy. A prehensile tail and sticky feet help it to grip branches.

30–39 in
75–100 cm

strong claws for digging out prey

6½ ft
2 m

NILE MONITOR
Varanus niloticus
The second-largest reptile in Africa, this monitor preys on various vertebrates and mollusks, and also scavenges on carrion.

YELLOW-SPOTTED MONITOR
Varanus panoptes
This species from Australia and S. New Guinea is rarely found far from a permanent source of water. It preys on other reptiles.

4½ ft
1.4 m

10 ft
3 m

skin folds on neck

KOMODO DRAGON
Varanus komodoensis
The world's largest lizard, this species is restricted to a few Indonesian islands. It is a formidable predator of large mammals.

well-developed, strong limbs

grayish-brown scaly skin

AMPHISBAENIANS

They may resemble legless lizards, but the burrowing reptiles in the suborder Amphisbaenia are quite distinct in their anatomy and behavior.

The amphisbaenians, or worm-lizards, are reptiles wholly adapted to a subterranean lifestyle, spending almost their entire lives underground. Most species have lost all trace of their limbs and have elongated bodies with smooth scales.

They hunt for soil invertebrates by means of scent and sound, killing them in their crushing jaws. Burrowing involves the worm-lizard propelling itself forward by contracting and extending its body while the specially adapted head pushes through the soil. The eyes are protected by tough, translucent skin. Fertilization is internal, with some species laying eggs, others bearing live young. They are found in South America, Florida, Africa, the Middle East, and southern Europe.

PHYLUM	CHORDATA
CLASS	REPTILIA
ORDER	SQUAMATA
FAMILIES (AMPHISBAENIANS)	6
SPECIES	195

WORM-LIZARDS

Amphisbaenians are burrowing reptiles with a head specially adapted for digging. They are found in Europe, sub-Saharan Africa, and South America. They are formidable predators of soil invertebrates, hunting by scent and sound.

EUROPEAN WORM-LIZARD
Blanus cinereus
Rarely seen above ground, this burrowing reptile from Spain and Morocco forages in leaf litter for invertebrates, particularly ants.

12 in
30 cm

rudimentary eye

black and white markings

18 in
45 cm

SPECKLED WORM-LIZARD
Amphisbaena fuliginosa
This worm-lizard hunts invertebrates while burrowing through leaf litter in the rainforests of S. America. It surfaces during heavy rains.

6½ in
17 cm

LANG'S ROUND-HEADED WORM-LIZARD
Chirindia langi
Found in southern Africa, this species burrows through sandy soils to forage for termites. If caught, it can shed its tail in a bid to escape.

SNAKES

Snakes are predators with an elongated body and skin made up of overlapping scales. In a number of species modified teeth, or fangs, are used to deliver venom.

Although most numerous in tropical regions, snakes have adapted to live at colder latitudes and at altitude and are found on all continents except Antarctica. They are usually terrestrial with many living in trees, but some species burrow, some are semiaquatic, and others live entirely in the sea. The smallest are tiny, threadlike creatures, but the largest can reach 33 ft (10 m) in length. Most are between 1–6½ ft (30 cm and 2 m).

A UNIQUE ARRAY OF SENSES

Some snakes have external vestiges of hind limbs, but they move by muscular contractions creating traction between the scales on their undersides and the surface beneath. A large number of flexibly linked vertebrae allows bending and coiling in any direction. The scaly skin is regularly shed to allow growth. Many snakes rely on just one elongated lung for breathing, while the gut is a simple tube with a large, muscular stomach. Instead of eyelids that can be opened and closed, each eye is covered by a clear, protective scale. Eyesight can be good, but varies according to lifestyle. Lacking external ears, snakes detect sounds as vibrations from the ground as well as in the air. Their key sense is smell, not through the nostrils, but from airborne chemicals collected by the forked tongue. Some species can also detect the body heat of mammals and birds.

KILLING TO LIVE

All snakes are carnivorous. Their sharp teeth curve backward to grasp and hold prey. Whether they eat their victim alive or kill it by injection of venom or by constriction, they consume their prey whole. Their uniquely elastic jaws allow them to swallow animals much larger than a snake's own head. Snakes defend themselves by camouflage, warning coloration, or mimicry. When threatened, they may strike out and bite.

Species in cold places tend to bear live offspring, whereas those in warmer climates lay eggs. Fertilization is internal. The male deposits sperm by means of one of his paired sexual organs, known as hemipenes.

PHYLUM	CHORDATA
CLASS	REPTILIA
ORDER	SQUAMATA
FAMILIES (SNAKES)	30
SPECIES	3,789

If threatened, the Mojave rattlesnake uses its characteristic rattle, bares its fangs, and coils up, ready to strike at the aggressor.

BOAS

The majority of Boidae are found in Central and South America, but a few species of boas are also present in Madagascar and New Guinea. These snakes are found from forests to marshes, where they prey on vertebrates, killing them by constriction. This family includes the anaconda, the largest living snake, and other species of varying sizes. Most give birth to live young.

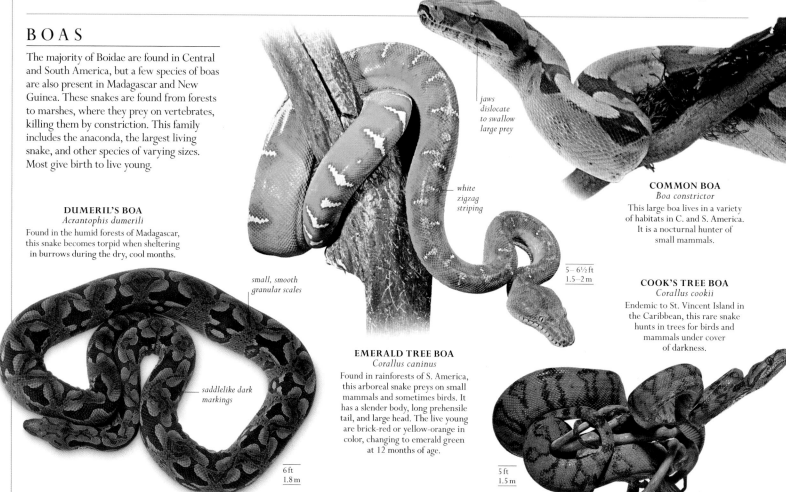

DUMERIL'S BOA
Acrantophis dumerili
Found in the humid forests of Madagascar, this snake becomes torpid when sheltering in burrows during the dry, cool months.

saddlelike dark markings

6 ft
1.8 m

small, smooth granular scales

white zigzag striping

jaws dislocate to swallow large prey

EMERALD TREE BOA
Corallus caninus
Found in rainforests of S. America, this arboreal snake preys on small mammals and sometimes birds. It has a slender body, long prehensile tail, and large head. The live young are brick-red or yellow-orange in color, changing to emerald green at 12 months of age.

COMMON BOA
Boa constrictor
This large boa lives in a variety of habitats in C. and S. America. It is a nocturnal hunter of small mammals.

5–6½ ft
1.5–2 m

COOK'S TREE BOA
Corallus cookii
Endemic to St. Vincent Island in the Caribbean, this rare snake hunts in trees for birds and mammals under cover of darkness.

5 ft
1.5 m

EAST AFRICAN SAND BOA
Gongylophis colubrinus
This African snake, with a stout body and
short tail, waits in a burrow with only its head
exposed to catch unsuspecting prey.

3½ ft
1.1 m

head with
small eyes

small, smooth scales
assist in burrowing

35 in
90 cm

headlike tail
used in defense

CALABAR GROUND BOA
Calabaria reinhardtii
Native to W. Africa, the Calabar ground
boa hunts in burrows for small mammals.
It is the only egg-laying boa.

mottled brown
coloration for
camouflage

RAINBOW BOA
Epicrates cenchria
Microscopic ridges
on the scales of this
S. American snake
refract light to give it an
iridescent sheen. It hunts
for mammals in forests.

6½ ft
2 m

long,
muscular body

smooth, glossy scales

3¼ ft
1 m

8 ft
2.5 m

**NEW GUINEA
GROUND BOA**
Candoia aspera
A sharply angled snout
helps to identify this
New Guinea snake. It is a
slow-moving terrestrial
hunter of small vertebrates.

32 in
80 cm

RUBBER BOA
Charina bottae
Preferring cool, humid conditions,
this burrowing species is found
at high altitudes and as
far north as British Columbia.

33 ft
10 m

ROSY BOA
Lichanura trivirgata
This is a slow-moving
snake from the deserts
of the W. US. It ambushes
mammals and kills them
by constriction.

dull yellow color
with prominent
black spots

GREEN ANACONDA
Eunectes murinus
This aquatic S. American
species is the largest of the
New World snakes. It hides
in water and kills its prey
by constriction.

3¼ ft
1 m

thick, muscular body
used to constrict prey

COMMON BOA
Boa constrictor

The common boa is a large terrestrial snake from the tropics of Central and South America. Although they are often found in woodland and scrub, boas easily adapt to a variety of habitats. Lying motionless on the ground, the snake ambushes its mammalian prey—grasping its victim with its jaws, it coils around the body and kills by constriction, tightening its coils with every exhalation of its prey. Once dead, the prey is swallowed head first. The left and right sides of the boa's lower jaw separate in front, and can open wide to take astonishingly large prey. Usually a solitary species, males will actively seek out a partner at breeding times, attracted by the scent emitted by the female. Larger than males, female boas give birth to 30–50 live offspring, each measuring about 12 in (30 cm) in length. This species is regularly seen in captivity but its propensity to bite makes it an unpredictable pet.

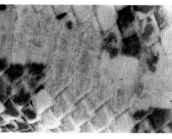

SIZE Up to 8 ft (2.5 m)
HABITAT Open woodland and scrubland
DISTRIBUTION C. and S. America
DIET Mammals, birds, reptiles

∧ COLOR VARIANTS
There are numerous subspecies of common boa, which are defined on the basis of size and color patterns. Color also varies considerably between localities.

> CAMOUFLAGE
Pigmentation of some scales forms patches of color that disrupt the outline of the snake. This helps with camouflage when hunting prey or evading predators.

∧ BELLY SCALES
The robust belly scales grip most surfaces, allowing the common boa to pull itself along and climb trees. It periodically sheds its skin.

> SIZEABLE FOE
The common boa hunts by sight and smell, so the eyes and nostrils are prominent on the triangular head. Feeding can be infrequent—a small prey item may last a snake two to three weeks. A large snake that kills a deer will not need to feed for at least six months.

distinctive, saddlelike markings

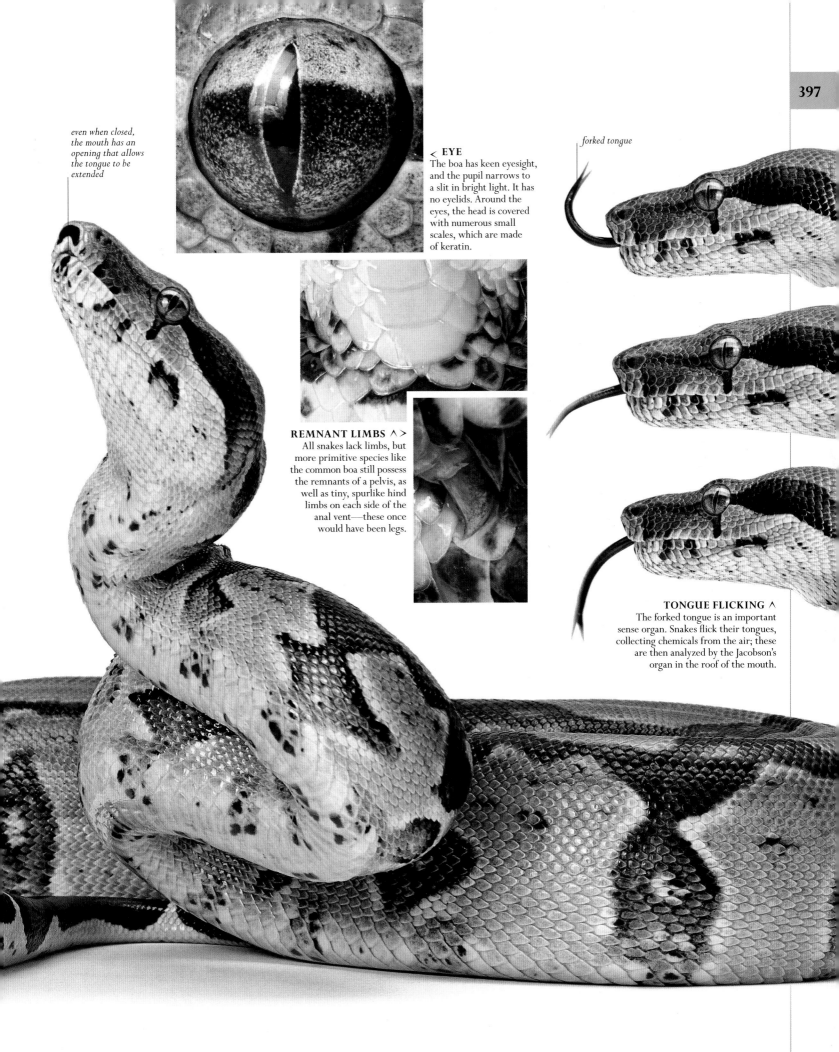

even when closed, the mouth has an opening that allows the tongue to be extended

< **EYE**
The boa has keen eyesight, and the pupil narrows to a slit in bright light. It has no eyelids. Around the eyes, the head is covered with numerous small scales, which are made of keratin.

forked tongue

REMNANT LIMBS ∧ >
All snakes lack limbs, but more primitive species like the common boa still possess the remnants of a pelvis, as well as tiny, spurlike hind limbs on each side of the anal vent—these once would have been legs.

TONGUE FLICKING ∧
The forked tongue is an important sense organ. Snakes flick their tongues, collecting chemicals from the air; these are then analyzed by the Jacobson's organ in the roof of the mouth.

COLUBRIDS

With about 2,000 species worldwide, colubrids form the largest family of snakes. The members of the family Colubridae inhabit a wide range of habitats, from deserts to wetlands. They have a varied diet. Many species lay eggs.

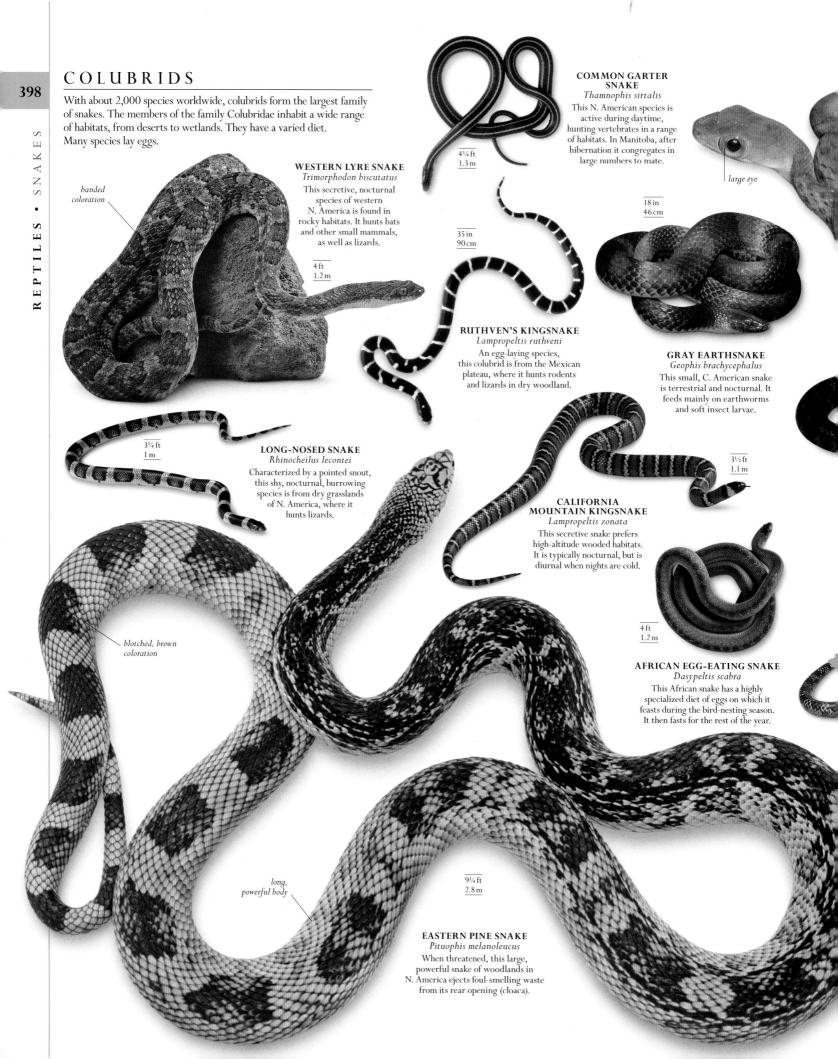

banded coloration

WESTERN LYRE SNAKE
Trimorphodon biscutatus
This secretive, nocturnal species of western N. America is found in rocky habitats. It hunts bats and other small mammals, as well as lizards.

4 ft
1.2 m

COMMON GARTER SNAKE
Thamnophis sirtalis
This N. American species is active during daytime, hunting vertebrates in a range of habitats. In Manitoba, after hibernation it congregates in large numbers to mate.

4¼ ft
1.3 m

large eye

35 in
90 cm

RUTHVEN'S KINGSNAKE
Lampropeltis ruthveni
An egg-laying species, this colubrid is from the Mexican plateau, where it hunts rodents and lizards in dry woodland.

18 in
46 cm

GRAY EARTHSNAKE
Geophis brachycephalus
This small, C. American snake is terrestrial and nocturnal. It feeds mainly on earthworms and soft insect larvae.

3¼ ft
1 m

LONG-NOSED SNAKE
Rhinocheilus lecontei
Characterized by a pointed snout, this shy, nocturnal, burrowing species is from dry grasslands of N. America, where it hunts lizards.

CALIFORNIA MOUNTAIN KINGSNAKE
Lampropeltis zonata
This secretive snake prefers high-altitude wooded habitats. It is typically nocturnal, but is diurnal when nights are cold.

3½ ft
1.1 m

blotched, brown coloration

4 ft
1.2 m

AFRICAN EGG-EATING SNAKE
Dasypeltis scabra
This African snake has a highly specialized diet of eggs on which it feasts during the bird-nesting season. It then fasts for the rest of the year.

long, powerful body

9¼ ft
2.8 m

EASTERN PINE SNAKE
Pituophis melanoleucus
When threatened, this large, powerful snake of woodlands in N. America ejects foul-smelling waste from its rear opening (cloaca).

GREEN HIGHLAND RACER
Drymobius chloroticus
This fast, agile snake is from the rainforests of C. America. It is usually found near water, where it feeds on frogs.

3¼ ft / 1 m — long, slender body

WESTERN INDIGO SNAKE
Drymarchon corais
This is one of the longest snakes in N. America. It often shares a burrow with the gopher tortoise.

10 ft / 3 m

BANDED FLYING SNAKE
Chrysopelea pelias
Despite its name, this S. Asian snake does not actively fly. Instead, it glides down from high branches by making its underside concave.

4 ft / 1.2 m

NORTHERN WATER SNAKE
Nerodia sipedon
This aquatic, viviparous species from eastern N. America is active throughout the day and the night. It feeds on amphibians and fishes.

4½ ft / 1.4 m

short snout

MUD SNAKE
Farancia abacura
This N. American snake preys on aquatic salamanders, grasping them with strongly curved teeth. Females remain coiled around their eggs until they hatch.

7 ft / 2.1 m

FALSE CORAL SNAKE
Erythrolamprus mimus
The bright coloration of the harmless false coral snake from S. America is similar to that of highly poisonous coral snakes.

26 in / 65 cm

striking red, white, and black bands

MOELLENDORFF'S RATSNAKE
Elaphe moellendorffi
This species is from dry limestone regions of China and Vietnam. It has an elongated snout and a relatively long tail.

6 ft / 1.8 m

REDBACK COFFEE SNAKE
Ninia sebae
This harmless snake from C. America has an ability to extend its neck as a form of intimidation.

16 in / 40 cm

NORTHERN CAT-EYED SNAKE
Leptodeira septentrionalis
This nocturnal, arboreal species from C. America has large eyes to help it hunt for vertebrates and eggs of tree frogs.

3¼ ft / 99 cm

BROWN BLUNT-HEADED VINE SNAKE
Imantodes cenchoa
The large eyes of this slender snake help it hunt lizards in the dark. It is found in tropical American rainforests.

4¼ ft / 1.3 m

BROWN TREESNAKE
Boiga irregularis
Originally from Australia and New Guinea, this snake was accidentally introduced on the island of Guam in the W. Pacific, where it has decimated the native fauna.

3¼ ft / 1 m

FALSE WATER COBRA
Hydrodynastes gigas
Native to S. American rainforests, this is a semiaquatic snake. Like cobras, it can flatten its neck to appear more intimidating.

10 ft / 3 m

GREEN VINE SNAKE
Oxybelis fulgidus
Long and delicate, this arboreal snake is from C. and S. American rainforests. It holds its prey in the air until its venom immobilizes the victims.

6½ ft / 2 m

characteristic yellow collar

CALICO SNAKE
Oxyrhopus petolarius
This species feeds on lizards and other small vertebrates. It is a terrestrial and diurnal inhabitant of S. American rainforests.

3½ ft / 1.1 m

4 ft / 1.2 m

GRASS SNAKE
Natrix natrix
Widely distributed throughout Europe, this species often feigns death when threatened. At home in water, it regularly feeds on amphibians.

olive-gray body color

» COLUBRIDS

COMMON TRINKET SNAKE
Coelognathus helena
This Indian species enlarges its neck and rises up to intimidate its enemies. It hunts mammals, usually at night.

4½ ft
1.4 m

SMOOTH SNAKE
Coronella austriaca
This secretive European snake is found on heaths and kills its prey by constriction. Females produce eggs that hatch within their bodies.

BROWN-HEADED CENTIPEDE SNAKE
Tantilla ruficeps
This burrowing snake is found in the tropical forests of C. America. It is a mainly diurnal insect-eater.

8 in
20 cm

32 in
80 cm

ROUGH GREEN SNAKE
Opheodrys aestivus
Native to woodlands in southeastern N. America, this arboreal snake is a diurnal predator of insects. It is oviparous.

5¼ ft
1.6 m

23½ in
60 cm

WESTERN HOGNOSED SNAKE
Heterodon nasicus
This snake of the N. American prairies uses its specialized, enlarged teeth to puncture the lungs of toads, making them easier to swallow.

head with prominent eyes
pale underside

6 ft
1.8 m

DIADEM SNAKE
Spalaerosophis diadema cliffordi
Native to deserts in N. Africa, this subspecies is diurnal during cooler months, but turns nocturnal during summer months.

BALKAN WHIPSNAKE
Hierophis gemonensis
Restricted to the Balkans, this snake is found in dry scrubland and olive groves. It hunts during the day, preying on lizards.

3¼ in
1 m

brown markings on muscular body

DAHL'S WHIPSNAKE
Platyceps najadum
From the Mediterranean region, this snake is found in dry, stony habitats. It hunts small lizards and grasshoppers during the day.

4½ ft
1.4 m

6 ft
1.8 m

RED CORNSNAKE
Pantherophis guttatus
This snake is common in southeastern N. America, but is rarely seen. It hunts small mammals in woodland.

RED-TAILED GREEN RATSNAKE
Gonyosoma oxycephalum
This fast-moving snake hunts for birds and mammals in trees. It is found in the rainforests of S.E. Asia.

7¾ ft
2.4 m

long, slender body

SOUTHERN WATER SNAKE
Nerodia fasciata
Living in wetlands, this colubrid preys on amphibians and fishes. It is found in the S. US.

5¼ ft
1.6 m

6½ ft
2 m

YELLOW RATSNAKE
Spilotes pullatus
A predator of small vertebrates, this large snake is rarely found far from water. It is widely distributed in S. and C. America.

LAMPROPHIIDAE

Found mainly in Africa, the diverse lamprophiid snakes occupy varied habitats—terrestrial, arboreal, and semiaquatic. Predating a range of vertebrates and invertebrates, they subdue their prey by either constriction or venom.

MOLE VIPER
Atractaspis fallax
This poisonous snake from E. Africa has large front fangs. It hunts other burrowing vertebrates underground.

30 in
75 cm

MONTPELLIER SNAKE
Malpolon monspessulanus
Inhabiting dry scrub and rocky hillsides around the Mediterranean, this long, slender snake hunts small vertebrates during the day.

6½ ft
2 m

GIANT MALAGASY HOGNOSE SNAKE
Leioheterodon madagascariensis
This large, diurnal snake, with an upturned snout, hunts lizards and amphibians in grasslands and forests in Madagascar.

6 ft
1.8 m

ASIAN PIPE SNAKES

Found in Sri Lanka and S.E. Asia, these small, burrowing snakes of the Cylindrophiidae have a cylindrical body and smooth, shiny scales. They live in wet habitats, sheltering in tunnels, but emerge at night to hunt other snakes and eels. They give birth to live young.

CEYLONESE PIPE SNAKE
Cylindrophis maculatus
Endemic to Sri Lanka, this harmless burrowing snake eats invertebrates. It uses its flattened tail to mimic the movements of the venomous cobra.

26 in
65 cm

SEA SNAKES

These highly venomous snakes from the tropical coastal waters of the Indian and Pacific oceans belong to the subfamily Hydrophiinae, part of the larger Elapidae. They are well adapted for swimming, with a paddle-like tail. All but a few breed in water, bearing live young. They eat eels and other fishes.

YELLOW-LIPPED SEAKRAIT
Laticauda colubrina
This snake hunts fishes at night in tropical Indo-Pacific seas. It can move on land as well.

4½ ft
1.4 m

COBRAS AND RELATIVES

Widely distributed in the tropics, the venomous snakes of the Elapidae have short, permanently erect fangs at the front of the mouth. Varied in body forms, they occupy a wide variety of habitats. Some species lay eggs, while others bear live young.

small, slender head

32 in
80 cm

CENTRAL AMERICAN CORAL SNAKE
Micrurus nigrocinctus
This venomous snake of C. America forages in leaf litter of tropical forests. Its bright coloration serves as a warning to potential aggressors.

4 ft
1.2 m

YELLOW-FACED WHIP SNAKE
Demansia psammophis
Widely distributed across Australia, this slender snake is an active diurnal hunter of lizards. It prefers dry, open habitats.

30 in
75 cm

EASTERN SHIELD-NOSE SNAKE
Aspidelaps scutatus fulafulus
This nocturnal predator of small lizards and mammals is found in southern African savanna. It burrows in sandy soil.

26 in
65 cm

ROSEN'S SNAKE
Suta fasciata
Found in arid areas of Western Australia, this venomous snake hunts for lizards.

spread hood acts as warning

20 in
50 cm

14 in
35 cm

28 in
70 cm

RINGED BROWN SNAKE
Pseudonaja modesta
Found in dry, rocky areas in Australia, this venomous snake feeds on small skinks. Its survival is threatened by the loss of its habitat.

SOUTHERN DESERT BANDED SNAKE
Simoselaps bertholdi
Widespread in Western Australia, this small snake burrows in search of its lizard prey. The banding may confuse predators.

DESERT DEATH ADDER
Acanthophis pyrrhus
This predator of the western deserts of Australia lures its prey of small lizards and mammals by wiggling its tail. It then ambushes them.

coloration can vary depending on region

6½ ft
2 m

MONOCLED COBRA
Naja kaouthia
Common in S.E. Asia, this large snake hunts rats and other snakes in woodland and paddy fields, often close to human habitation.

7¾ ft
2.4 m

EGYPTIAN COBRA
Naja haje
This large cobra hunts small vertebrates in the deserts of N. and C. Africa. When threatened, it spreads its hood and raises its head.

16 ft
5 m

KING COBRA
Ophiophagus hannah
A forest species of tropical Asia, the massive king cobra primarily preys on other snakes. Unusually for snakes, its nest of eggs is defended by both sexes.

smooth scales on brown body

30 in
75 cm

RED SPITTING COBRA
Naja pallida
This African species not only raises its hood when threatened, but also sprays streams of venom into the face of its aggressor.

PYTHONS

Members of the family Pythonidae can be found across Africa, Asia, and Australia. Pythons detect warm-blooded prey with their facial pits. They use their mouth to grasp prey but kill by constriction. Some species coil around their eggs to help incubate them using the heat generated by their shivering.

GREEN TREE PYTHON
Morelia viridis
Found in the tropical forests of Australasia, this arboreal snake spends its time looped over a branch, ready to ambush lizards and small mammals.

5 ft
1.5 m

bright green color aids in camouflage

BLOOD PYTHON
Python curtus
This relatively short python of S.E. Asian rainforests coils around its clutch of eggs to incubate and protect them.

6 ft
1.8 m

10 ft
3 m

BLACK-HEADED PYTHON
Aspidites melanocephalus
Endemic to Australia, this species is found in a variety of habitats. It preys upon other snakes and reptiles.

BURMESE PYTHON
Python molurus
Rare in its native homeland, this Asian snake is becoming superabundant in the Florida Everglades. Its broad diet of birds, mammals, and reptiles is a threat to many local animals.

23 ft
7 m

BLIND SNAKES

The Typhlopidae have eyes covered with scales, rendering them effectively blind. Found in leaf litter of tropical forests, these small, burrowing snakes mainly eat soil invertebrates and have teeth only on their upper jaw. Most of them lay eggs.

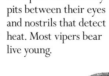

30 in
75 cm

BLACKISH BLIND SNAKE
Anilios nigrescens
Tough scales allow this burrowing snake from E. Australia to resist attack from ants while eating their eggs and larvae.

14 in
35 cm

EURASIAN BLIND SNAKE
Xerotyphlops vermicularis
This wormlike blind snake from Europe inhabits dry, open areas. It burrows to feed on invertebrates, mainly ant larvae.

THREAD SNAKES

These small, slender snakes are rarely seen because of their habit of burrowing to hunt invertebrates. Belonging to the Leptotyphlopidae, they are found in the tropics of the Americas, Africa, and southwest Asia.

12 in
30 cm

SENEGAL BLIND SNAKE
Myriopholis rouxestevae
This snake inhabits the soil of tropical forests in western Africa. First studied in 2004, it is believed to eat invertebrates.

VIPERS

The Viperidae are characterized by a thick body, keeled scales, and a triangular head. Long, tubular, hinged fangs at the front of the mouth efficiently inject venom into their vertebrate prey. Pit vipers have sensory pits between their eyes and nostrils that detect heat. Most vipers bear live young.

5 ft
1.5 m

FER-DE-LANCE
Bothrops atrox
Found in the forests of tropical America, this venomous snake is characterized by a sharply pointed head. It hunts birds and mammals by night.

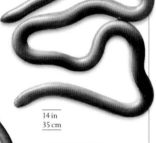

34 in
85 cm

23½ in
60 cm

DESERT HORNED VIPER
Cerastes cerastes
Occurring in the deserts of N. Africa and the Sinai Peninsula, this snake lies buried in sand to ambush small mammals and lizards.

HOGNOSE PIT VIPER
Porthidium nasutum
This diurnal hunter of C. America prefers humid, open forests. Pit vipers, which have heat-detecting sensory pits between their eyes and nostrils, bear live young.

blotched markings

SPOTTED PYTHON
Antaresia maculosa
This python is found on the rocky hillsides of N. Australia. It feeds on bats by catching them at the entrance of their roosting caves.

4½ ft
1.4 m

SUNBEAM SNAKES

Characterized by iridescent scales, the Xenopeltidae include only two species, both restricted to Southeast Asia. They live in burrows in forest habitats and hunt amphibians, other reptiles, and small mammals. They reproduce by laying eggs.

33 ft
10 m

4¼ ft
1.3 m

netlike pattern

powerful, muscular body useful for constricting prey

RETICULATED PYTHON
Python reticulatus
One of the longest snakes in the world, this species inhabits Asian rainforests. It kills large mammals by constriction.

SUNBEAM SNAKE
Xenopeltis unicolor
A flat head helps this snake burrow into decaying vegetation. It emerges at dusk to feed on amphibians and small mammals.

23½ in
60 cm

4 ft
1.2 m

4¼ ft
1.3 m

COPPERHEAD
Agkistrodon contortrix
Banded markings help this snake to hide in leaf litter in rocky woodland. It is found in eastern N. America.

35 in
90 cm

COMMON ADDER
Vipera berus
This diurnal species hunts small mammals and lizards. It is widely distributed across a variety of habitats in Eurasia.

short, dark-edged blotches

ASP VIPER
Vipera aspis
This European viper prefers warm, dry habitats, where it feeds on small mammals. It bears up to 20 live young.

WEST AFRICAN GABOON VIPER
Bitis rhinoceros
Occurring in tropical rainforests in Africa, this large, heavily built snake preys on mammals by ambushing them.

7 ft
2.1 m

WESTERN DIAMONDBACK RATTLESNAKE
Crotalus atrox
A nocturnal hunter of mammals, this large snake is widespread in arid areas of western N. America.

PRAIRIE RATTLESNAKE
Crotalus viridis
This snake from the midwestern US hunts mammals at dawn and dusk. It hides in crevices during the day.

6½ ft
2 m

large head

6 ft
1.8 m

PUFF ADDER
Bitis arietans
Most active at night, this venomous snake usually lies still to ambush its vertebrate prey. It is found in rocky grassland in Africa.

3¼ ft
1 m

MALAYAN PIT VIPER
Calloselasma rhodostoma
Occurring in S.E. Asia, this species forages by night in open areas adjacent to forests for rodents and lizards.

CROCODILES AND ALLIGATORS

Crocodilians are large, carnivorous, aquatic reptiles. Their armored skin and powerful jaws make them formidable predators, yet they are sociable animals and caring parents.

All crocodilians—crocodiles, alligators, and gharials—have a similar body plan: an elongated snout armed with numerous, sharp, unspecialized teeth, a streamlined body, a long, muscular tail, and skin armored with bony plates. The saltwater crocodile is the largest living reptile.

SWIMMING AND FEEDING

Crocodilians are found in tropical regions worldwide, inhabiting a variety of freshwater and marine habitats. The eyes, ears, and nostrils are on top of the head, which enables the animal to be almost completely submerged while hunting. The powerful tail is used for swimming and strong legs allow crocodilians to move readily on land with the body raised well above the ground.

Although generalist carnivores, feeding on fishes, reptiles, birds, and mammals, crocodilians exhibit sophisticated feeding behaviors to locate and ambush specific prey, such as migrating mammals and fishes. Small prey is swallowed whole but large prey is first drowned before the crocodilian rolls over while grasping the carcass to twist off pieces of meat. Digestion is aided by stones in the stomach and highly acidic gastric secretions.

SOCIAL BEHAVIOR AND BREEDING

Behaviorally, crocodilians have more in common with birds, their closest living relatives, than with other reptiles. Adults form loose groupings—particularly at good feeding sites—and use a wide range of vocalizations and body language to interact with each other.

In the breeding season, dominant males control territories and actively court females. Fertilization is internal and clutches of rigid-shelled eggs are laid in a nest constructed and guarded by the female. The sex of individuals is determined by the incubation temperature. Hatchlings call to stimulate the female to excavate the nest and carry them in her mouth to water. Mortality is high in young crocodilians but once they are longer than 3¼ ft (1 m), they have few natural enemies.

PHYLUM	CHORDATA
CLASS	REPTILIA
ORDER	CROCODYLIA
FAMILIES	3
SPECIES	25

Exploiting a fast-flowing stream, these Yacare caimans lie in wait with their mouth held open, waiting for a fish to come within reach.

GHARIALS

The endangered Gavialidae are restricted to India. The slender snout has rows of numerous sharp teeth that help to catch fishes. Males develop a wartlike growth on the tip of the snout.

olive-green body

23 ft
7 m

GHARIAL
Gavialis gangeticus
Among the largest of the Crocodylia, this Asian species has a fearsome array of teeth, which help it to catch fishes, but it is not known to attack humans.

CROCODILES

Relatively unspecialized in lifestyle, habitat, and diet, the Crocodylidae are most easily recognized by the exposure of the fourth tooth of the lower jaw, when the jaws are closed. These tropical reptiles occupy many habitats around rivers and coasts.

6½ ft
2 m

DWARF CROCODILE
Osteolaemus tetraspis
From the forests of tropical Africa, this small crocodile has a heavy armor of scales on its neck and back. It is a nocturnal hunter of fishes and frogs.

CUBAN CROCODILE
Crocodylus rhombifer
Endemic to Cuba, this medium-sized crocodile inhabits swamps, and hunts fishes and small mammals. It lays its eggs in holes in the ground.

11 ft
3.5 m

13 ft
4 m

SIAMESE CROCODILE
Crocodylus siamensis
Restricted to S.E. Asia, this large crocodile is critically endangered in the wild. Found in freshwater marshes, it feeds on a variety of prey.

powerful tail propels crocodile through water

ALLIGATORS AND CAIMANS

Members of the family Alligatoridae feed on fishes, birds, and mammals that share their aquatic habitat. These reptiles are distributed throughout freshwater swamps and rivers of the tropical and subtropical Americas. The only species living outside the Americas is the rare Chinese alligator.

coloration darkens with age

16 ft
5 m

AMERICAN ALLIGATOR
Alligator mississippiensis
Conservation efforts have now made this N. American species quite common. It preys on birds, small mammals, and turtles.

strong legs allow smooth movement on land

6½ ft
2 m

CHINESE ALLIGATOR
Alligator sinensis
This critically endangered species of the Yangtze valley of China lies dormant in burrows during the cold winter months.

8¼ ft
2.5 m

SPECTACLED CAIMAN
Caiman crocodilus
This caiman feeds on a wide range of prey. Widespread in C. and S. America, it appears to be the only crocodilian that readily occupies man-made aquatic habitats.

rounded head with broad snout

10 ft
3 m

mottled pattern

BROAD-SNOUTED CAIMAN
Caiman latirostris
A mound-nesting species found across much of central S. America, this caiman is characterized by a broad snout. It hunts mammals and birds.

5½ ft
1.7 m

bony armor

CUVIER'S DWARF CAIMAN
Paleosuchus palpebrosus
This S. American species is the smallest of the New World members of this order. It has a doglike skull and bony-armored skin.

5 ft
1.5 m

SCHNEIDER'S DWARF CAIMAN
Paleosuchus trigonatus
This small crocodile of S. American rainforests is semiterrestrial. It builds its nest against a termite mound, which keeps the eggs warm.

SALTWATER CROCODILE
Crocodylus porosus
The largest of all living reptiles, this common Indo-Pacific species readily crosses the open ocean. Its diet is wide-ranging and unspecialized.

23 ft
7 m

greenish brown body, often covered in algae

eyes set high on head

16 ft
5 m

NILE CROCODILE
Crocodylus niloticus
Widely distributed in Africa, this large crocodile typically lives in fresh water but has been seen around coasts as well. Its diet varies with age, with adults taking larger prey.

CUBAN CROCODILE
Crocodylus rhombifer

This strikingly colored crocodile is native to Cuba.
It is medium-sized, robust, and armored with bony scales. More terrestrial than other crocodilian species, it raises its belly off the ground and adopts a "high walk" posture when moving on land. Its preferred prey are turtles, which are crushed by the strong teeth at the rear of its mouth. Hunting pressure and habitat loss meant that, in the 1960s, the few remaining wild Cuban crocodiles were captured and released into Zapata Swamp, a reserve in the south of the island. Although protected, the population remains small, and hybridization with American crocodiles on the reserve threatens the purity of this species.

SIZE 9¾–11 ft (3–3.5 m)
HABITAT Freshwater swamps
DISTRIBUTION Cuba
DIET Fishes, turtles, small mammals

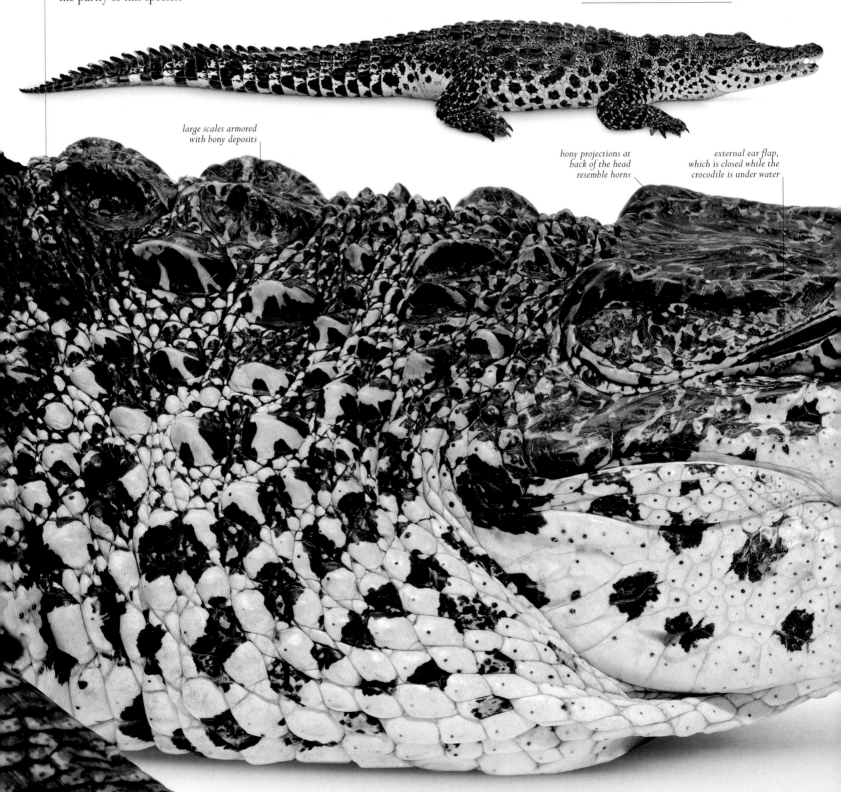

large scales armored
with bony deposits

bony projections at
back of the head
resemble horns

external ear flap,
which is closed while the
crocodile is under water

REPTILES • CROCODILES AND ALLIGATORS

MUSCULAR LIMBS ∨ >
This crocodile can race over short distances, powered by its strong rear legs. Because they are used for walking rather than swimming, the feet have no webbing between the toes.

∨ NOSTRILS
Paired nostrils are located on the nasal disk, an elevated pad of tissue at the end of the snout. Under water, the nostrils can be closed by valves.

∧ TAIL RIDGE
Scales with bony projections increase the depth of the tail—useful when the crocodile is swimming. Rich in blood vessels, they also help to absorb heat when basking.

∧ FORMIDABLE JAWS
Crocodiles cannot chew their food, but they use their powerful jaws to grasp their prey. Sensitive taste buds on the tongue allow the crocodile to reject unpleasant tasting objects. Evaporation from the mouth cools the body.

< BELLY SCALES
The scales on the crocodile's underside are small and uniform in size and pattern. These scales are prized by the leather trade.

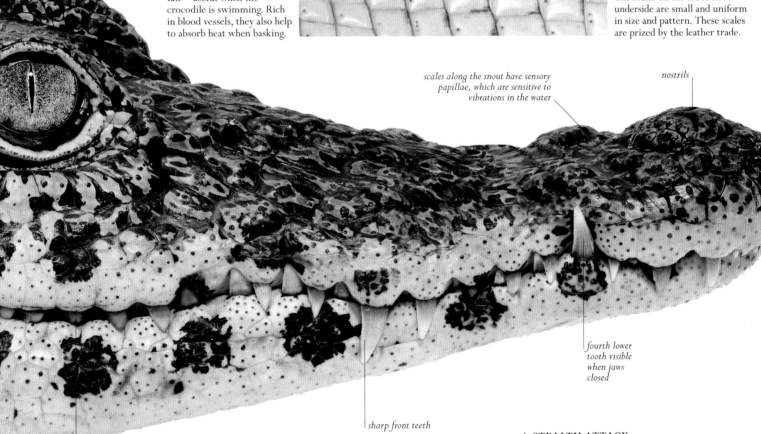

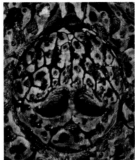

scales along the snout have sensory papillae, which are sensitive to vibrations in the water

nostrils

fourth lower tooth visible when jaws closed

sharp front teeth

blunt back teeth are more robust than the front teeth

∧ STEALTH ATTACK
Sharp eyes and sensitive ears are located on the top of the head, so that prey can be detected while the body is submerged. This enables the crocodile to take its victims by surprise. The long lower jaw can be closed with tremendous force to grasp struggling prey securely.

BIRDS

Birds have busy, active lives. Many are spectacularly beautiful; others sing complex—even musical—songs. They have the intelligence and parental devotion of a mammal combined with characteristics more reminiscent of a reptilian ancestor.

PHYLUM	CHORDATA
CLASS	AVES
ORDERS	40
FAMILIES	250
SPECIES	10,770

Asymmetrical flight feathers, with a narrow outer blade and broader inner blade, give birds better control in the air.

Many chicks are altricial—born in a helpless state. Blind and naked, they require prolonged parental care.

Elaborate nests are built by the male weaver bird to impress prospective mates, testament to the brain power of birds.

DEBATE

ARE BIRDS DINOSAURS?

Traditionally birds are in a class separate from reptiles, including dinosaurs. But today scientists believe that all descendants of a common ancestor should be classified together to reflect evolutionary relationships—a method known as cladistics. This makes birds as much a part of the dinosaur group as *Tyrannosaurus*.

Birds are the only living animals with feathers. They are warm-blooded vertebrates that stand on two legs and their forelimbs are considerably modified as wings. As well as facilitating flight, feathers insulate the body, so many birds stay active in icy-cold conditions, just like furry mammals. Feathers can also be colorful, showing that birds rely on flaunting visual cues to communicate and find a mate. Birds evolved from a group of two-legged, carnivorous dinosaurs that included *Tyrannosaurus*. It is possible—even likely—that these ancestors were already feathered.

TAKING TO THE AIR

No one knows for sure how—or why—the first birds took to the air, but the fact that they did had effects that would be forever imprinted upon the bird's body. Their wristbones became fused together as arms evolved into wings. The breastbone evolved a huge keel for the attachment of larger muscles for powering the flapping wings. Already their dinosaur ancestors had air-filled, but strong bones. But birds coupled this with a strong cardiovascular system powered by a mammallike four-chambered heart and a rapid metabolism to generate lots of energy. A system of air sacs—linked to the bone chambers—expanded their breathing capacity to flush stale air out of lungs more efficiently than in mammals.

Many reptilian features still remain. The bare parts of a bird's legs and toes have horny reptilian scales. Birds excrete a semisolid waste product called uric acid (not a solution of urea, as in mammals). Waste from the kidneys and gut mix and exit via a common reptilian opening called a cloaca. But a bird's brain is better developed than that of a reptile, because it is supported by a warm-blooded body and fast metabolism. This makes birds not only skilled at flying, but remarkably sophisticated in the ways they acquire food or raise a family. Young birds hatch from reptilian hard-shelled eggs, but they are reared by intelligent parents who will invest much time and energy into raising them to maturity. Sixty million years of evolution mean that a bird is no longer just a reptile with feathers.

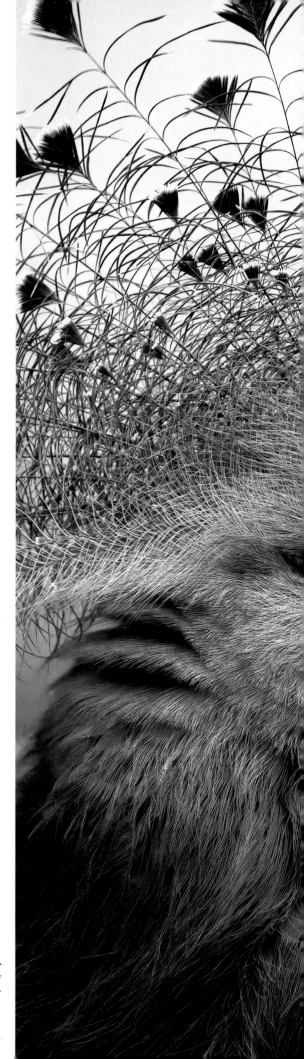

CROWNING GLORY >
Birds' remarkable colors and structures often appear overly extravagant, such as this pigeon's elaborate head crest.

BIRD GROUPS

The 40 orders range from heavy-bodied ground-dwellers—a few of which are incapable of flight—to masters of the air, such as soaring albatrosses, dive-bombing falcons, and hovering hummingbirds. They exploit habitats from deep sea to mountain peaks, from deserts to rainforests and wetlands.

TINAMOUS
» 410

RATITES
» 410

FOWL, GAMEBIRDS, AND RELATIVES
» 412

WATERFOWL
» 416

PENGUINS
» 420

LOONS
» 424

ALBATROSSES, PETRELS, AND SHEARWATERS
» 425

GREBES
» 427

FLAMINGOS
» 428

STORKS
» 429

IBISES, BITTERNS, HERONS, AND PELICANS
» 430

CORMORANTS, GANNETS, AND RELATIVES
» 433

TROPICBIRDS
» 433

BIRDS OF PREY
» 434

FALCONS AND CARACARAS
» 441

BUSTARDS
» 442

CRANES, RAILS, AND RELATIVES
» 443

KAGU AND SUNBITTERN
» 445

WADERS, GULLS, AND AUKS
» 448

SANDGROUSE
» 456

PIGEONS AND DOVES
» 457

PARROTS AND COCKATOOS
» 460

TURACOS
» 464

HOATZIN
» 465

CUCKOOS
» 465

OWLS
» 467

MESITES
» 470

NIGHTJARS
» 471

HUMMINGBIRDS AND SWIFTS
» 473

TROGONS
» 476

MOUSEBIRDS
» 476

KINGFISHERS AND RELATIVES
» 477

CUCKOO-ROLLER
» 479

HORNBILLS, HOOPOES, AND WOOD HOOPOES
» 482

WOODPECKERS AND TOUCANS
» 483

SERIEMAS
» 487

PASSERINES
» 488

TINAMOUS

The ground-dwelling tinamous from Central and South America resemble Old World partridges, but they are most closely related to ratites.

Tinamous, the sole family in the order Tinamiformes, are small, ground-dwelling birds with a rounded body and short legs. All have a very short tail, which gives them a dumpy appearance, and some species are crested.

PHYLUM	CHORDATA
CLASS	AVES
ORDER	TINAMIFORMES
FAMILIES	1
SPECIES	47

Unlike the ratites, tinamous have a ridged breastbone for the attachment of flight muscles, a feature that is seen in all other birds. Their wings are much better developed than those of the ratites, and they can fly, although only over short distances. Usually they are reluctant to take to the air and prefer instead to run away from predators. Tinamous have a comparatively small heart and lungs, which perhaps explains why these birds become exhausted so easily.

Some groups of tinamous inhabit woodlands and forest, while others are found in open grassland. All feed on seeds, fruit, insects, and sometimes small vertebrates. The cryptic plumage patterns of most species makes them difficult to spot in the wild. They are more likely to be recognized by their distinctive call.

BEAUTIFUL EGGS

Male tinamous mate with many different females, and it is the males that tend the eggs and young. The nests are made in leaf litter on the ground. Tinamou eggs—brightly colored turquoise, red, or purple—have a porcelainlike gloss.

DEBATE

ORIGINS OF TINAMOUS

Traditionally, tinamous are grouped separately from the flightless ratites. But the two groups share a pattern of skull bones not found in other birds, suggesting that they have a common ancestor. Scientists are not certain whether this ancestor was flightless or able to fly; the latter is more likely.

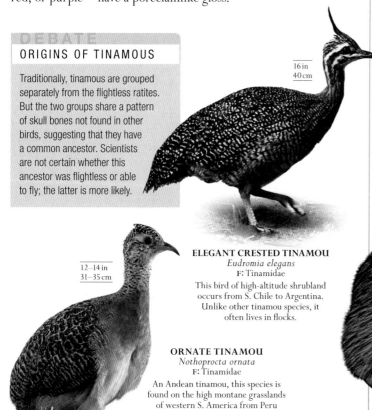

16 in
40 cm

ELEGANT CRESTED TINAMOU
Eudromia elegans
F: Tinamidae
This bird of high-altitude shrubland occurs from S. Chile to Argentina. Unlike other tinamou species, it often lives in flocks.

12–14 in
31–35 cm

ORNATE TINAMOU
Nothoprocta ornata
F: Tinamidae
An Andean tinamou, this species is found on the high montane grasslands of western S. America from Peru to N. Argentina.

RATITES

A group that includes the largest living birds, the ratites are flightless and strong-footed. Some flock in open habitats; others live alone in forests.

Ratites evolved in the southern hemisphere and there is evidence to suggest they have flying ancestors. They lack some of the features associated with flight, but retain others.

PHYLUM	CHORDATA
CLASS	AVES
ORDERS	4
FAMILIES	4
SPECIES	13

Ratites do not have the bony ridge on their breastbone that anchors powerful flight muscles in other birds, but they still have wings, and the part of the brain that controls flight is well developed.

The ostriches, which make up the order Struthioniformes, are found in dry open areas in Africa. They are powerful-limbed birds that can run at high speed on flat ground. Somewhat similar in appearance, though less massively built, are the rheas—the order Rheiformes—from the grasslands of South America. Like most ratites, both these groups of birds have a reduced number of toes: the rhea has three toes, the ostrich only two. Both have large wings that they use in displays or to assist their balance when sprinting.

AUSTRALASIAN SPECIES

The Australasian ratites include the cassowaries and emus of the order Casuariiformes and the kiwis, the Apterygiformes. All are forest birds, save the emu, which inhabits scrubland and grassland. The tiny wings of these ratites are invisible under their plumage, which resembles shaggy hair rather than feathers. The nocturnal kiwis, familiar as the national symbol of New Zealand, are the only ratites that do not have a reduced number of toes.

The male ostrich scrapes out a shallow nest in the ground, in which his harem of females lay up to 50 eggs.

20–26 in
50–65 cm

furlike plumage

26–28 in
65–70 cm

NORTH ISLAND KIWI
Apteryx mantelli
F: Apterygidae
One of several New Zealand kiwis, this nocturnal ratite detects buried invertebrate prey by smell, using nostrils at its probing bill tip.

TOKOEKA
Apteryx australis
F: Apterygidae
This is the South Island cousin of *A. mantelli*. Lighter in color than its relative, it has been recognized as a separate species after DNA analysis.

coarse, shaggy plumage

helmetlike casque

red wattles

5–6 ft
1.5–1.8 m

SOUTHERN CASSOWARY
Casuarius casuarius
F: Casuariidae

Cassowaries are fruit-eating rainforest birds. With a range that extends through New Guinea and N. Australia, this is the most widespread species of its family.

brown plumage

5½–7 ft
1.7–2.1 m

EMU
Dromaius novaehollandiae
F: Dromaiidae

Australia's largest living bird is found across the continent in open grassy areas. Like the related cassowary, its chicks are striped for camouflage.

almost bare neck

black body plumage

3–3¼ ft
92–100 cm

LESSER RHEA
Rhea pennata
F: Rheidae

Smaller than the greater rhea, its northern cousin, this ratite inhabits the S. Andes and Patagonia, where it lives in small flocks.

4¼–4½ ft
1.3–1.4 m

GREATER RHEA
Rhea americana
F: Rheidae

This central S. American ratite has a breeding system that does not involve pair-bonding. Males tend a large nest of eggs produced by many females.

white primary feathers

♂

5½–8¾ ft
1.7–2.7 m

large eye

OSTRICH
Struthio camelus
F: Struthionidae

The largest birds in the world, ostriches live in savanna and semidesert areas of Africa. Breeding males mate with several females.

gray neck

scaly legs

♀

grayish-brown plumage

5½–8¾ ft
1.7–2.7 m

26–28 in
65–70 cm

GREAT SPOTTED KIWI
Apteryx haastii
F: Apterygidae

One of two species of kiwis with gray mottled plumage, this bird is confined to mountain areas of western South Island.

LITTLE SPOTTED KIWI
Apteryx owenii
F: Apterygidae

Introduced mammals have driven this small kiwi to the edge of extinction. It survives on predator-free satellite islands around New Zealand.

14–18 in
35–45 cm

featherless thighs

two-toed feet

SOMALI OSTRICH
Struthio molybdophanes
F: Struthionidae

Separated from other populations of ostrich by the Rift Valley, this E. African gray-necked form is now regarded as a separate species.

FOWL, GAME BIRDS, AND RELATIVES

Strong footed game birds are adapted to a wide range of habitats. They are primarily ground-dwellers and feed mostly on plants.

Although they may be good fliers, the majority of game birds, the order Galliformes, take to the air only when threatened. None, except the common and Japanese quails, sustains a long migratory flight. Most game birds spend their lives on the ground. Only the members of the curassow family of tropical America are habitual tree-dwellers; these birds make their nests in trees, too. Most primitive of the game bird group are the big-footed megapodes of the Indo-Pacific forests. They bury their eggs in heaps of compost or in volcanic sand and rely on heat generated by decomposition or volcanic action to incubate them. All other game birds take a more active role in parenthood, although only the female raises the brood. The precocious offspring can run and feed soon after hatching.

Male game birds often have flamboyant plumage, which they use in courtship displays to attract a harem of females. Some species are armed with leg spurs to battle with competitors.

DOMESTICATED SPECIES

The relative ease with which many game bird species, including the turkey, can be kept in captivity has meant that they are economically important in many countries. Domesticated chickens around the world are all descendants of the red jungle fowl of Southeast Asia.

PHYLUM	CHORDATA
CLASS	AVES
ORDER	GALLIFORMES
FAMILIES	5
SPECIES	300

DEBATE

WHAT IS A PARTRIDGE?

Traditional categories of birds may not reflect biological relationships. Many game birds are called partridges, but the European gray partridge differs subtly from others in that, for example, males are more distinctively patterned than females—perhaps making it a closer ally of pheasants.

22 in
55 cm

28 in
70 cm

31–36 in
78–92 cm

BRUSH TURKEY
Alectura lathami
F: Megapodiidae
Males of this E. Australian megapode control the temperature of their incubation mounds by adding or removing rotting vegetation. More females hatch at high temperatures.

black markings down the front of the throat

GREAT CURASSOW
Crax rubra
F: Cracidae
Ranging from Mexico to Ecuador, this large curassow has a curly crest like other curassows. Males are black, while females are brown.

MALEO
Macrocephalon maleo
F: Megapodiidae
This megapode (large-footed bird) from Sulawesi island in Indonesia lays its eggs in sand and relies on heat from the sun or underground volcanic activity for incubation.

23½ in
60 cm

33 in
84 cm

MALLEEFOWL
Leipoa ocellata
F: Megapodiidae
Like many megapodes, this southern Australian species makes a compost mound to lay its eggs. It feeds primarily on various seeds, but is thought to be omnivorous.

brown, white, and black wings

BARE-FACED CURASSOW
Crax fasciolata
F: Cracidae
This S. American curassow has a sparsely feathered face. Unlike other related species, it has no fleshy wattles or bill knob.

gray-brown upperparts

19–21 in
48–53 cm

olive-brown breast

18 in
46 cm

27 in
69 cm

23–26 in
59–65 cm

white-tipped tail

PLAIN CHACHALACA
Ortalis vetula
F: Cracidae
Ranging from Texas to Costa Rica, this northernmost member of the American curassow family is the only one found in the US.

GRAY-HEADED CHACHALACA
Ortalis cinereiceps
F: Cracidae
Named after their calls, chachalacas are brown-colored relatives of curassows. This is a species of scrubby habitat from Honduras to Colombia.

BLUE-THROATED PIPING GUAN
Pipile cumanensis
F: Cracidae
This S. American bird has glossy black plumage typical of tree-dwelling piping guans. They are named for their shrill breeding call.

HIGHLAND GUAN
Penelopina nigra
F: Cracidae
This C. American bird is more ground-dwelling than other guans. It is perhaps the only member of its family to nest on the ground.

26–30 in
66–76 cm

22 in
55 cm

26–30 in
67–75 cm

BEARDED GUAN
Penelope barbata
F: Cracidae
Named for its streaked neck, the bearded guan, like related species in this genus, is a rainforest bird. It is found in Ecuador and Peru.

SPIX'S GUAN
Penelope jacquacu
F: Cracidae
This S. American species belongs to a group of brown guans that resemble chachalacas, but have shorter legs. It nests in trees and has a loud mating call.

30 in
76 cm

red throat wattle

CAUCA GUAN
Penelope perspicax
F: Cracidae
A bronze-sheened relative of the spix's guan, this bird is found only in the Cauca Valley region in the western and northern parts of Colombia.

bare, bluish head

27–30 in
68–75 cm

24–28 in
61–71 cm

DUSKY-LEGGED GUAN
Penelope obscura
F: Cracidae
The only brown guan with dark rather than reddish legs, this species occurs in central S. America, from Brazil to northern Argentina.

CHESTNUT-BELLIED GUAN
Penelope ochrogaster
F: Cracidae
The throat wattle that is typical of brown guans is especially well-developed in this species from south-central Brazil.

cape of long feathers

10–12 in
26–31 cm

9½–10½ in
24–27 cm

10 in
25 cm

VULTURINE GUINEA FOWL
Acryllium vulturinum
F: Numididae
Guinea fowl are African game birds that live in groups, but are monogamous. They all have a bare head. This is the largest species of guinea fowl and is found in E. Africa.

MOUNTAIN QUAIL
Oreortyx pictus
F: Odontophoridae
This native of the Rocky Mountains, like other American quails, is a ground-nesting, monogamous species.

CALIFORNIAN QUAIL
Callipepla californica
F: Odontophoridae
Found from Oregon to California, this species is characterized by its forward-drooping crest, comprised of a cluster of six feathers.

GAMBEL'S QUAIL
Callipepla gambelii
F: Odontophoridae
This relative of the Californian quail has a longer crest. It occurs in the desert in the south of California and the ranges of the two do not overlap.

»

» FOWL, GAME BIRDS, AND RELATIVES

12½–14 in
32–35 cm

CHUKAR PARTRIDGE
Alectoris chukar
F: Phasianidae

With a range extending across central Eurasia, this is the most widespread of the red-legged partridges—typically birds of dry country, with striking black markings.

black stripes on flanks

15 in
38 cm

ARABIAN PARTRIDGE
Alectoris melanocephala
F: Phasianidae

This large, red-legged partridge occurs in semidesert regions of the Arabian Peninsula and Yemen. Its preferred habitat is juniper woodland.

8½ in
21 cm

CHESTNUT-BELLIED HILL PARTRIDGE
Arborophila javanica
F: Phasianidae

Found in S.E. Asian rainforests, hill partridges are small, cryptically colored game birds with stumpy tails. Many, such as this Javan species, have distinctive head markings.

11½–12½ in
29–32 cm

GRAY PARTRIDGE
Perdix perdix
F: Phasianidae

This is the most widespread of a small genus of dark-bellied Eurasian game birds that are more related to pheasants than other partridges.

11–12 in
28–30 cm

GRAY FRANCOLIN
Francolinus pondicerianus
F: Phasianidae

As with related jungle fowl, males of this S. Asian francolin have leg spurs that they use in fighting.

10–15 in
25–38 cm

RED-NECKED FRANCOLIN
Francolinus afer
F: Phasianidae

A ground-nesting bird of forest and grassland, this species is widespread through Africa, from the Democratic Republic of Congo to the Cape.

10 in
26 cm

ROULROUL
Rollulus rouloul
F: Phasianidae

Related to hill partridges, the S.E. Asian roulroul has a more colorful plumage. Males are dark blue and have a reddish crest; females are green and lack the crest.

12 in
31 cm

CHINESE BAMBOO PARTRIDGE
Bambusicola thoracicus
F: Phasianidae

There are two species of bamboo partridge, birds of E. Asia that are related to francolins and jungle fowl. This bird is native to China.

6½–7 in
16–18 cm

COMMON QUAIL
Coturnix coturnix
F: Phasianidae

Unusually for a game bird, this tiny species of W. Eurasian grasslands and semidesert is migratory in the northern parts of its range.

15½–16 in
39–40 cm

black throat

SPRUCE GROUSE
Canachites canadensis
F: Phasianidae

Like other woodland grouse, this N. American species can digest the needles of conifer and spruce trees, rejected by most animals.

16–20 in
40–50 cm

SOOTY GROUSE
Dendragapus fuliginosus
F: Phasianidae

One of several N. American grouse that have an inflatable neck sac, this dark-colored species inhabits the pine forests of the Pacific coast.

16–18½ in
41–47 cm

SHARP-TAILED GROUSE
Tympanuchus phasianellus
F: Phasianidae

This N. American grouse is a more widespread, more northerly cousin of prairie chickens. Males have purple sacs for display.

17 in
43 cm

GREATER PRAIRIE CHICKEN
Tympanuchus cupido
F: Phasianidae

This species is from the C. US. As with other species of prairie chicken, males engage in group courtship displays by inflating their colored neck sac.

15–16 in
38–41 cm

LESSER PRAIRIE CHICKEN
Tympanuchus pallidicinctus
F: Phasianidae

Found in southern N. America, this small species of prairie chicken has a colored neck sac and drumming call, typical of *Tympanuchus* grouse.

WESTERN CAPERCAILLIE
Tetrao urogallus
F: Phasianidae

Found in western Eurasian coniferous forests, this is the largest member of the grouse family. Males display to females with rattling, popping calls.

23½–34 in
60–87 cm

ROCK PTARMIGAN
Lagopus muta
F: Phasianidae

One of three related grouse that turn white in winter, the rock ptarmigan is a tundra and mountain bird with a circumpolar distribution.

13½–14 in
34–36 cm

RED GROUSE
Lagopus lagopus scotica
F: Phasianidae

Unlike most ptarmigan species, this British race of the widespread circumpolar willow grouse does not turn white in winter.

15–16 in
38–41 cm

pale gray tail feathers

KALIJ PHEASANT
Lophura leucomelanos
F: Phasianidae

This pheasant occurs in forests from the Himalayas to Myanmar. There is also an introduced population in the Hawaiian Islands.

22–30 in
55–75 cm

SATYR TRAGOPAN
Tragopan satyra
F: Phasianidae

This is a Himalayan species of *Tragopan*, a genus of Asian tree-nesting pheasants. Male tragopans have an expandable fleshy lappet and horns that are used for display.

23½–28 in
60–70 cm

SIAMESE FIREBACK
Lophura diardi
F: Phasianidae

Like many pheasants, this S.E. Asian bird has bare, red facial skin. As with other species of *Lophura*, this feature is common to both sexes.

23½–32 in
60–80 cm

21–35 in
53–89 cm

COMMON PHEASANT
Phasianus colchicus
F: Phasianidae

Native to the woodlands of C. and E. Eurasia, this species has been introduced into western Europe, where it is now common in farmland.

23½–47 in
60–120 cm

LADY AMHERST'S PHEASANT
Chrysolophus amherstiae
F: Phasianidae

Like many other pheasants, only males of this species have spectacular plumage, and only females care for the young. This bird is from China and Myanmar.

16–20 in
40–50 cm

white-tipped tail fanned in display

bald head

PALAWAN PEACOCK-PHEASANT
Polyplectron napoleonis
F: Phasianidae

Male peacock-pheasants lack the long plumes of their peafowl relatives, but display tail feathers marked with iridescent eyespots.

2½–7¼ ft
0.8–2.2 m

INDIAN PEAFOWL
Pavo cristatus
F: Phasianidae

This tropical Asian species occurs in India and Sri Lanka. Males court females by raising their long plumes, which are just above the tail.

16–31 in
41–78 cm

RED JUNGLE FOWL
Gallus gallus
F: Phasianidae

Unlike related francolins, males of this Asian ancestor of domestic chickens mate with many females. They have a comb and wattles, which are absent in females.

3½–4 ft
1.1–1.2 m

WILD TURKEY
Meleagris gallopavo
F: Phasianidae

Turkeys are large, wattled game birds native to N. America. This species from the S. US is the ancestor of domesticated turkeys.

WATERFOWL

Species in this group are web-footed birds adapted for surface swimming. The majority feed on plants, although some eat small aquatic animals.

Most waterfowl have short legs set well back on the body. These birds, the Anseriformes, swim by propelling themselves with webbed feet, and all have plumage kept waterproofed with oil taken from a gland near the tail. South American screamers and the magpie-goose from Australasia have partially webbed feet, spending much of their time on land or wading in marshes. These birds belong to two ancient families. All other waterfowl are classified in a single family, the most primitive group being the whistling ducks.

SWANS, GEESE, AND DUCKS
In swans and geese, the sexes have similar plumage patterns. These long-necked, long-winged birds occur mostly outside the tropics. Many northern species breed close to the Arctic and migrate southward to overwinter. Ducks are generally smaller and have a shorter neck than others in the Anatidae family.

Unlike geese and swans, male ducks are usually more brightly plumaged than the females, especially when breeding. Both sexes of many duck species have a vividly colored wing-patch called a speculum.

The typical "duckbills" of waterfowl have internal plates that have evolved to strain aquatic food drawn in by a muscular tongue. All species retain this feature, even though some have adopted different feeding methods. Geese graze on grasslands, but swans are more aquatic and use their long neck to plunge their head deep in water. Many ducks dabble—upending to feed on or just below the water surface. Others, such as pochards, dive under water to gather food. The saw-billed ducks have a narrow bill with serrated edges to catch fishes. One group of ducks, including eiders, scoters, and mergansers, includes accomplished sea divers.

NESTING
Most waterfowl are monogamous, some pairing for life. They nest largely on the ground, but a few nest in trees. The marine ducks come inland to breed. All waterfowl produce downy chicks that can walk and swim soon after hatching.

PHYLUM	CHORDATA
CLASS	AVES
ORDER	ANSERIFORMES
FAMILIES	3
SPECIES	177

Flying in formation reduces air drag, enabling birds like these snow geese to conserve energy on long migrations.

21–22 in
53–56 cm

RED-BREASTED GOOSE
Branta ruficollis
F: Anatidae
This is the most brightly colored member of the *Branta* group of dark geese. It breeds in northwestern Siberia, where it nests close to raptors, possibly for protection from foxes.

striking red, black, and white plumage

PINK-FOOTED GOOSE
Anser brachyrhynchus
F: Anatidae
This small, gray goose breeds on rocky outcrops of tundra in Greenland and Iceland, and overwinters in western Europe.

23½–30 in
60–75 cm

20–43 in
50–110 cm

CANADA GOOSE
Branta canadensis
F: Anatidae
The largest of the *Branta* geese, this species is native to N. America, but has been introduced to northern Europe.

23–28 in
58–71 cm

BARNACLE GOOSE
Branta leucopsis
F: Anatidae
This goose breeds on Arctic tundra in Greenland and Russia, where it avoids predators by nesting on cliff tops.

22–28 in
56–71 cm

HAWAIIAN GOOSE
Branta sandvicensis
F: Anatidae
Confined to the Hawaiian Islands, this goose has reduced foot-webbing and strong claws, and is adapted to climbing on to rocky lava flows.

BAR-HEADED GOOSE
Anser indicus
F: Anatidae
This bird is adapted to the thin air of montane C. Asia. Migrating high over the Himalayas, it overwinters in India and Myanmar.

28–30 in
71–76 cm

pale gray body

30–35 in
76–89 cm

GRAYLAG GOOSE
Anser anser
F: Anatidae
Widespread across the grasslands and wetlands in Eurasia, this typical gray *Anser* goose is the wild ancestor of domesticated geese.

EMPEROR GOOSE
Anser canagicus
F: Anatidae

Found in northeastern Siberia and Alaska, this goose grazes on coastal grass and seaweed. It is less gregarious than most other geese.

finely barred, gray body

26–35 in
66–89 cm

ASHY-HEADED GOOSE
Chloephaga poliocephala
F: Anatidae

S. American *Chloephaga* geese are perhaps more related to ducks than to other geese. This species occurs in Chile and Argentina. Like others in its genus, it may perch in trees.

20–23½ in
50–60 cm

EGYPTIAN GOOSE
Alopochen aegyptiaca
F: Anatidae

Belonging to a group of southern hemisphere geese that are related to ducks, the Egyptian goose is widespread through Africa.

28–29 in
71–73 cm

PLUMED WHISTLING DUCK
Dendrocygna eytoni
F: Anatidae

Distinct from geese and true ducks, whistling ducks are named for their unique call. This is an Australian species.

15½–17½ in
39–44 cm

MAGPIE-GOOSE
Anseranas semipalmata
F: Anseranatidae

From Australian wetlands, this species has long legs and partially webbed feet. It is not closely related to any other waterfowl.

28–35 in
70–90 cm

CAPE BARREN GOOSE
Cereopsis novaehollandiae
F: Anatidae

This distinctive goose is restricted to southern Australia and offshore islands, where it gathers in small flocks and grazes on grassland.

30–39 in
75–100 cm

23½–30 in
60–75 cm

BLUE-WINGED GOOSE
Cyanochen cyanoptera
F: Anatidae

This goose has a thick plumage as an adaptation to the cool highlands of its native Eritrea and Ethiopia.

COSCOROBA SWAN
Coscoroba coscoroba
F: Anatidae

Resembling a goose, this is the smallest of swans. It is confined to swamps in the southern parts of Chile and Argentina.

3–4 ft
0.9–1.2 m

blackish plumage

red bill

3½–4½ ft
1.1–1.4 m

BLACK SWAN
Cygnus atratus
F: Anatidae

This sooty-black swan with white wing tips may nest in huge colonies. Native to Australia and Tasmania, it has been introduced to New Zealand, Europe, and N. America.

TRUMPETER SWAN
Cygnus buccinator
F: Anatidae

Related to other vocal swans, such as the Eurasian whooper swan (*C. cygnus*), this N. American species has a loud, honking call.

straight neck

MUTE SWAN
Cygnus olor
F: Anatidae

This species breeds in Europe and C. Asia. Like other swans, it grazes on underwater vegetation by submerging its head.

4¼–5¼ ft
1.3–1.6 m

3½–4 ft
1–1.2 m

all-white body

5–6 ft
1.5–1.8 m

BLACK-NECKED SWAN
Cygnus melancoryphus
F: Anatidae

Found in southern S. America, this species spends more time on water than other swans and nests on floating vegetation.

WHITE-BACKED DUCK
Thalassornis leuconotos
F: Anatidae

This bird from Africa and Madagascar is related to whistling ducks, but spends more time on water, nesting on islands of vegetation.

15–16 in
38–40 cm

AFRICAN PYGMY-GOOSE
Nettapus auritus
F: Anatidae

Like other pygmy-geese, this African species nests in tree hollows. It is usually found in wetlands with water lilies, on which it feeds.

12–13 in
30–33 cm

ORINOCO GOOSE
Neochen jubata
F: Anatidae

A S. American relative of ducks, this goose occurs on tropical wet savanna and forest edges along rivers.

24–26 in
61–66 cm

SOUTHERN SCREAMER
Chauna torquata
F: Anhimidae

Screamers are large, bulky birds that live in S. American marshlands. Like other screamers, this species has bony wing spurs that may be used in fighting.

33–37 in
83–95 cm

>>

cream, black, and green head pattern

15½–17 in
39–43 cm

17–22 in
43–56 cm

18–22 in
45–56 cm

AMERICAN WIGEON
Mareca americana
F: Anatidae

This species feeds by dabbling on the surface of shallow waters and occasionally upends. Huge flocks winter in the Caribbean after breeding in N. America.

BAIKAL TEAL
Sibirionetta formosa
F: Anatidae

This distinctive duck breeds in cold, open forests in Siberia on the edge of the tundra, and migrates to E. Asia in the winter.

NORTHERN SHOVELER
Spatula clypeata
F: Anatidae

Both sexes of dabbling (surface-feeding) ducks typically have a colored wing-patch called a speculum. This widespread wetland species of the northern hemisphere has a green speculum.

22–26 in
55–65 cm

20–26 in
50–65 cm

15–20 in
38–51 cm

♀
20–26 in
50–65 cm

♂

MALLARD
Anas platyrhynchos
F: Anatidae

Widespread in the northern hemisphere, the surface-feeding mallard can interbreed with related species, perhaps indicating that this group has only recently evolved.

13–16 in
33–40 cm

orange cheek plumes

INDIAN RUNNER
Anas platyrhynchos
F: Anatidae

A domestic descendant of the wild mallard, this long-necked breed originated in the Malay Peninsula and India in the 19th century.

DOMESTIC DUCK
Anas platyrhynchos
F: Anatidae

Most domesticated ducks have descended from the mallard. They are raised for their meat, eggs, or down, or kept for ornamental purposes.

WHITE-CHEEKED PINTAIL
Anas bahamensis
F: Anatidae

This saltwater dabbling duck is found in S. American estuaries and mangrove swamps. Unlike temperate northern pintails, both sexes look alike.

BUFFLEHEAD
Bucephala albeola
F: Anatidae

The smallest sea duck in N. America, the bufflehead nests in tree-holes and sometimes uses those vacated by woodpeckers.

15–20 in
38–51 cm

17–20 in
43–51 cm

HARLEQUIN DUCK
Histrionicus histrionicus
F: Anatidae

A highly buoyant sea duck, this species rides choppy waters and nests beside fast-flowing streams in eastern N. America, Iceland, and western Russia.

WOOD DUCK
Aix sponsa
F: Anatidae

The newly hatched chicks of this N. American tree-perching duck jump down from their nests in high tree-holes to reach the water below.

MANDARIN DUCK
Aix galericulata
F: Anatidae

Wrongly thought to be monogamous, this tree-nesting duck is a symbol of love in its native northeastern Asia. It has been introduced in Europe and California.

16–20 in
41–51 cm

♂

TORRENT DUCK
Merganetta armata
F: Anatidae

A powerful swimmer, this high-altitude S. American duck lives in fast-flowing rivers of the Andes and nests beneath riverside rocks.

♀

17–18 in
43–46 cm

RINGED TEAL
Callonetta leucophrys
F: Anatidae

Like other tropical ducks, this S. American teal is not migratory and retains its plumage colors throughout the year.

white flank patch

14–15 in
35–38 cm

SURF SCOTER
Melanitta perspicillata
F: Anatidae
The N. American surf scoter, like other scoter species, breeds near fresh water and winters at sea. The male's body is entirely black.

18–22 in
46–55 cm

HOODED MERGANSER
Lophodytes cucullatus
F: Anatidae
Found in N. America, this bird has a saw-edged bill for catching fishes. It dives with powerful kicks of the feet.

16½–20 in
42–50 cm

BRONZE-WINGED DUCK
Speculanas specularis
F: Anatidae
This duck is found along the rivers of S. America. It is locally called "dog-duck" after the barking call of the female.

18–22 in
46–54 cm

white cheek patch

PINK-EARED DUCK
Malacorhynchus membranaceus
F: Anatidae
The pink head spot of this widespread Australian species is less distinctive than its zebra pattern. It consumes plankton, strained from water by beak flaps.

14–18 in
36–45 cm

RUDDY DUCK
Oxyura jamaicensis
F: Anatidae
This N. American duck, now introduced in Europe, has a stiff tail which it uses as a rudder when diving.

14–17 in
35–43 cm

TUFTED DUCK
Aythya fuligula
F: Anatidae
A Eurasian member of the pochard group, the tufted duck feeds mostly, though not entirely, on invertebrates in contrast to its largely vegetarian relatives.

16–18½ in
40–47 cm

CANVASBACK
Aythya valisineria
F: Anatidae
This N. American bird is the largest species of pochard, a group of ducks with typically stocky bodies and large heads.

19–24 in
48–61 cm

LONG-TAILED DUCK
Clangula hyemalis
F: Anatidae
Unlike most other Arctic sea ducks, the long-tailed duck breeds in saltwater as well as freshwater habitats. Males have a distinctive long tail.

15–23 in
38–58 cm

SMEW
Mergellus albellus
F: Anatidae
A member of the group of cavity-nesting mergansers, this species is the only small white duck found in northern Eurasia.

14–17½ in
35–44 cm

COMMON SHELDUCK
Tadorna tadorna
F: Anatidae
This gooselike duck is largely a coastal resident in Europe, but in Asia it migrates southward from inland regions during winter. It nests in burrows.

24–25 in
61–63 cm

RED-BREASTED MERGANSER
Mergus serrator
F: Anatidae
Widespread across the northern hemisphere, the red-breasted merganser breeds on coasts and spends more time at sea than other mergansers.

20½–23 in
52–58 cm

orange head lobe

KING EIDER
Somateria spectabilis
F: Anatidae
This bird breeds along Arctic tundra coastlines. Its large body size may help it to dive deeply for invertebrate prey.

17–25 in
43–63 cm

rose blush on breast

ROSYBILL
Netta peposaca
F: Anatidae
This S. American duck is related to the diving pochards, but spends more time feeding at the water surface. Only the males have a red bill.

22 in
55–56 cm

CRESTED DUCK
Lophonetta specularioides
F: Anatidae
This Andean species is a possible relic of a S. American lineage that was ancestral to more widespread dabbling ducks, such as the mallard.

20–24 in
51–61 cm

STELLER'S EIDER
Polysticta stelleri
F: Anatidae
Like other related sea ducks of Arctic and subarctic regions, this eider overwinters farther south in huge flocks, sometimes numbering 20,000 birds.

17–19 in
43–48 cm

PENGUINS

With their two-toned plumage, erect posture, and waddling walk, penguins are instantly recognizable as the classic symbol of the southern oceans.

Residents of the coastal regions of the southern hemisphere, all penguins are adapted to life in cold water. Most species live around islands encircling Antarctica, although some are found on southern coastlines of South America, Africa, and Australasia.

These flightless birds of the order Sphenisciformes probably descended from a common ancestor shared with the albatrosses. They may also be distant cousins of the divers from the northern hemisphere.

SPECIAL ADAPTATIONS

Penguins have legs set far back toward the tail, a feature that provides excellent propulsion in water and is seen also in birds such as divers and grebes. On land penguins walk upright, but with webbed feet placed flat on the ground their gait is ungainly. In common with other flightless birds, penguins have reduced wings, but these are modified in such a way that they can be used as flippers. In effect, penguins "fly" under water.

A penguin's short, densely packed feathers have downy bases that trap warm air, and a layer of fat beneath the skin provides further insulation. The feather tips are greasy and waterproof, lubricated by oil produced from a large gland on the rump. A complex system of blood-flow through the legs and feet ensures that the penguin's body is not chilled by standing on snow or ice. The largest penguin species incubate their eggs on their feet. All penguins have counter-shaded plumage (dark above and pale below), which in the sea provides camouflage against ocean-going predators, such as leopard seals.

FORAGING AND NESTING

Penguins can dive for fishes, shrimp, and krill more than 200 times each day. When they have eggs to incubate and young to tend, the parents take turns to forage. In the case of emperor penguins, one of the few species to breed on the Antarctic pack ice, the males incubate the eggs alone throughout the winter while the females remain feeding at sea. Most penguins are colonial breeders and often return to the same nest site each season.

PHYLUM	CHORDATA
CLASS	AVES
ORDER	SPHENISCIFORMES
FAMILIES	1
SPECIES	18

FLYING ANCESTORS?

In the early 20th century, flightless birds were popularly believed to be primitive. It was hoped that study of their embryos would reveal evidence of a direct link to a dinosaurian past. Penguin eggs became sought after, and members of Robert Falcon Scott's last expedition (1910–13) made a courageous journey to an emperor rookery in the bleak Antarctic winter. By the time the eggs they collected received scientific attention, the embryo theory had been disproved. Modern studies of anatomy, fossils, and DNA show that the ancestors of penguins, and other flightless birds, could fly.

KING PENGUIN
Aptenodytes patagonicus
F: Spheniscidae
Resembling the emperor penguin, this sub-Antarctic species has an orange-yellow neck and breast markings, and incubates a single egg on its feet.

3–3¼ ft
90–100 cm

LITTLE PENGUIN
Eudyptula minor
F: Spheniscidae
A burrow-nester, this bird is the smallest of all penguins. It occurs along the coasts of S. Australia and New Zealand.

14–16 in
35–40 cm

EMPEROR PENGUIN
Aptenodytes forsteri
F: Spheniscidae
The largest penguin, this species breeds in colonies on Antarctic ice. Males incubate eggs during the bitter polar winter.

white feathers contrast with black head and wings

3½–4 ft
1.1–1.2 m

ROCKHOPPER PENGUIN
Eudyptes chrysocome
F: Spheniscidae
The smallest of the sub-Antarctic crested penguins, this bird gets its name from its habit of clambering over rocks and boulders.

18–23 in
45–58 cm

yellow plume

28 in
70 cm

FIORDLAND PENGUIN
Eudyptes pachyrhynchus
F: Spheniscidae
Nesting in the cool coastal forests of S. New Zealand, this species has hairlike crest feathers and a red bill typical of *Eudyptes* penguins.

22–23½ in
55–60 cm

MACARONI PENGUIN
Eudyptes chrysolophus
F: Spheniscidae
This bird inhabits islands in the far southern Atlantic and Indian oceans, but is the only crested penguin to breed on the Antarctic Peninsula.

stubby bill

white
eye ring

28—32 in
71—80 cm

18—30 in
46—75 cm

CHINSTRAP
PENGUIN
Pygoscelis antarcticus
F: Spheniscidae
The chinstrap penguin dives
for krill and fishes. It breeds
on Antarctic coasts and
islands of the South Atlantic.

thin
black streak
across face

26—28 in
67—72 cm

GENTOO PENGUIN
Pygoscelis papua
F: Spheniscidae
This penguin breeds on
the Antarctic Peninsula
and Southern Ocean
islands. Its nest is a
simple cluster of sticks,
stones, and feathers.

blue-black
upperparts

30 in
75 cm

ADELIE PENGUIN
Pygoscelis adeliae
F: Spheniscidae
One of three "brush-tailed"
Pygoscelis penguins from Antarctica
and adjacent islands, the Adelie
penguin breeds in colonies of more
than 200,000 pairs.

YELLOW-EYED
PENGUIN
Megadyptes antipodes
F: Spheniscidae
A cousin of *Eudyptes* crested
penguins, this rare New Zealand
bird nests in scrub, but not in
dense colonies like the other
Eudyptes penguins.

GALAPAGOS PENGUIN
Spheniscus mendiculus
F: Spheniscidae
This is the only penguin to
breed in tropical waters
cooled by the Humboldt
Current, which flows along
the western coast of
S. America. It nests
in rock crevices.

black face

black
breast band

19—20 in
48—51 cm

HUMBOLDT PENGUIN
Spheniscus humboldti
F: Spheniscidae
Found along the Pacific coast of
southern S. America, this species
belongs to a group of burrow-nesters
characterized by bold banded
patterns from flanks to thighs.

27—28 in
68—70 cm

26—28 in
65—70 cm

24—30 in
61—76 cm

MAGELLANIC
PENGUIN
Spheniscus magellanicus
F: Spheniscidae
Closely related to the Humboldt
penguin, this banded penguin
lives in colonies around the
southern tip of S. America
and the Falkland Islands.

JACKASS PENGUIN
Spheniscus demersus
F: Spheniscidae
Named for its
donkeylike braying call,
this bird is the only
penguin to breed in
Africa, in colonies on
southwestern coasts.

KING PENGUIN
Aptenodytes patagonicus

The king penguin is the second-largest penguin species. Only its close cousin, the emperor penguin, is bigger. Unlike the emperor, the king inhabits subantarctic islands. It hunts fishes, ignoring the krill taken by its rivals, and dives to extraordinary depths to get them—sometimes to more than 660 ft (200 m). It produces just one egg at a time and takes more than a year to raise the single young. This means adults cannot breed annually and the enormous breeding colonies—containing juveniles of different ages—maintain a permanent presence on favored islands of the southern oceans.

SIZE 35–39 in (90–100 cm)
HABITAT Flat coastal plains and waters around subantarctic islands
DISTRIBUTION Islands of S. Atlantic and S. Indian Oceans
DIET Mostly lanternfishes, occasionally squid

black upper mandible

penguins drink sea water and excrete excess salt as brine through the nostrils

< SPINY TONGUE
A penguin's tongue is muscular and spiny: projections on the tongue surface called papillae have evolved into backward-facing barbs that help grip fishes caught in dives.

∨ KEEN EYE
Penguins hunt by sight and have good underwater vision. Bioluminescent lanternfishes, caught on night dives, predominate in the diet of this species.

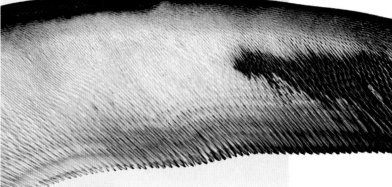

∧ WING PROPULSION
Penguins are flightless birds, but they are propelled by feet and wings during dives. Their flipperlike wings effectively make them "fly" under water.

< DENSE FEATHERS
The feathers, arranged in an outer oily waterproof layer and inner downy insulating layers, are adapted for diving in cold water.

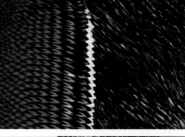

∧ SCALY SKIN
Scales on the legs and feet are reminiscent of the reptilian roots of all birds; the dark skin may help to spread heat to eggs and young.

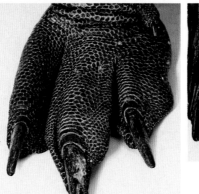

∧ EGG PROTECTOR
The single egg is incubated above the feet and beneath a fold of warm skin called a brood pouch. Once the chick has hatched, it uses the brood pouch for shelter.

WEBBED FOOT >
The kicking action of webbed feet helps propel the bird forward under water—and on land, when "tobogganing" over snow on its belly.

< FIRM TAIL
The short tail consists of stiffened feathers, used as a rudder under water. In smaller species, it is used as a prop on land.

∨ **COUNTERSHADED DIVER**
Below its distinctive yellow markings, the king penguin has the tuxedo-style plumage pattern typical of penguins: white beneath and dark above. When diving, this pattern offers camouflage from aquatic predators. Seen from below, the pale belly is harder to see against the sunlit surface, and when seen from above, the bird's dark back blends in with the dark of the water below.

yellow markings are due to pigments called carotenoids, which are absent from some penguin species

yellow stripe on lower mandible

yellow breast

LOONS

These web-footed, fish-eating birds of Arctic waters have rear-set legs that limit their movement on land but provide excellent thrust for swimming.

Loons are the only family in the order Gaviiformes. The alternative name for these birds—loon—may originate from their haunting "lunatic" wailing cries uttered during the breeding season, or possibly from their ungainly motion when out of water. A loon's legs are set so far back on the body that the bird can only manage an awkward shuffle on land. In water, loons swim and dive with ease. Their streamlined body and spearlike bill closely resemble those of penguins, which have similar habits—it is possible, though not certain, that these birds may share a common ancestor.

Although their pointed wings are relatively small for their body size, loons are fast fliers. To get themselves airborne, the larger species have to patter across open water; only the red-throated loon is able to take off from land. All species fly southward to overwinter.

SHARED PARENTING
Territorial male loons choose nest sites in vegetation on the shores of clear Arctic lakes, where both sexes incubate and tend the young. Young loon chicks ride on the backs of their parents, but they can swim and even dive soon after hatching. After breeding, loons lose their striking head and neck patterns, and their duller body plumage makes it more difficult to distinguish the different species.

PHYLUM	CHORDATA
CLASS	AVES
ORDER	GAVIIFORMES
FAMILIES	1
SPECIES	5

Loons fly with their outstretched head held slightly lower than the body, which gives them a hunchbacked appearance.

COMMON LOON
Gavia immer
F: Gaviidae
One of the largest loons, this bird breeds on lakes in subarctic regions of N. America and Iceland. It overwinters on coasts further south, including those around Britain.

27–36 in
69–91 cm

striped neck patch

black head and neck

30–36 in
76–91 cm

WHITE-BILLED LOON
Gavia adamsii
F: Gaviidae
A large species of Arctic waters, this white-billed bird can be distinguished from other loons by its yellowish white bill.

grayish head and neck

21–27 in
53–69 cm

23–29 in
58–74 cm

23–29 in
58–73 cm

BLACK-THROATED LOON
Gavia arctica
F: Gaviidae
Largely confined to Eurasia when breeding, this species sometimes reaches Alaska, but winters further south, including the Pacific coast of N. America.

RED-THROATED LOON
Gavia stellata
F: Gaviidae
The smallest loon, this bird breeds on small tundra pools in circumpolar regions and migrates south to Europe, China, and the southeastern US to overwinter.

PACIFIC LOON
Gavia pacifica
F: Gaviidae
This species has similar striped plumage to that of the black-throated loon. Both species have a white throat when not breeding.

white spots in summer

ALBATROSSES, PETRELS, AND SHEARWATERS

Long-winged albatrosses and their relatives spend much of their lives in the air, traveling great distances as they scan the surface of the oceans for fishes.

These birds, the order Procellariiformes, are master aeronauts, rarely returning to land except to breed. Albatrosses, petrels, and shearwaters—called "tubenoses" because of the tubular nasal protuberances on their bill—are globally widespread, ocean-going birds, but show the greatest diversity in the southern hemisphere.

Unusually for birds, they locate their sparsely scattered sea prey by smell. All but the smallest diving species in this group have elongated wings, and almost all have their webbed feet positioned so far back that some cannot easily walk. These

birds deter predators by regurgitating a noxious oil that they produce in their stomach and sometimes forcefully eject. This oil is also nutritious enough to feed their chicks.

SLOW REPRODUCTION

Tubenoses form pair-bonds, sometimes for life, which in the larger species can be several decades. Many species breed in colonies on remote islands, often returning to the same site year after year. Smaller species nest in cavities and burrows. These birds have low breeding rates, but the parents invest heavily in rearing their young. Typically, tubenoses make only one breeding attempt per season and produce a single egg. Despite a lengthy incubation period, chicks are helpless upon hatching and mature very slowly.

PHYLUM	CHORDATA
CLASS	AVES
ORDER	PROCELLARIIFORMES
FAMILIES	4
SPECIES	147

Long-lived wandering albatrosses are monogamous. They reinforce their pair-bonding with an elaborate courtship dance.

WANDERING ALBATROSS
Diomedea exulans
F: Diomedeidae
The largest of the giant Southern Ocean *Diomedea* albatrosses, this bird forms a life-long pair bond and produces a single young every two years.

dark wings turn white with age

3½–4½ ft
1.1–1.4 m

mostly white body

pale pink bill

LAYSAN ALBATROSS
Phoebastria immutabilis
F: Diomedeidae
Albatrosses from the northern Pacific include those that breed in the tropics, like this small species that nests on islands—including Hawaii—when breeding.

30–32 in
77–80 cm

27–29 in
68–74 cm

BLACK-FOOTED ALBATROSS
Phoebastria nigripes
F: Diomedeidae
Like other northern Pacific albatrosses, this small, dark species frequently interrupts its gliding flight with bursts of wing-flapping.

black "brow"

18–20 in
45–50 cm

NORTHERN FULMAR
Fulmarus glacialis
F: Procellariidae
This gull-like petrel is common throughout the northern hemisphere. It nests on cliffs and can eject a foul-smelling stomach oil to repel predators.

BLACK-BROWED ALBATROSS
Thalassarche melanophrys
F: Diomedeidae
This is one of several dark-backed albatrosses called mollymawks, from the southern hemisphere. It nests in dense colonies, producing one egg each year.

32–37 in
80–95 cm

»

» ALBATROSSES, PETRELS, AND SHEARWATERS

sooty brown head

12 in
30 cm

AUDUBON'S SHEARWATER
Puffinus lherminieri
F: Procellariidae

This small bird breeds on tropical oceanic islands. Different populations are considered by some to be separate species.

PINK-FOOTED SHEARWATER
Ardenna creatopus
F: Procellariidae

A variable species existing in dark and light forms, this shearwater nests on islands off Chile. It migrates to the eastern Pacific during summer.

19 in
48 cm

BULLER'S SHEARWATER
Ardenna bulleri
F: Procellariidae

This shearwater nests on islands off N. New Zealand but roams across the Pacific outside the breeding season.

18–18½ in
45–47 cm

6½–8 in
17–20 cm

18–22 in
45–56 cm

CORY'S SHEARWATER
Calonectris diomedea
F: Procellariidae

A large shearwater that glides with bowed wings, this species breeds on islands in the eastern Atlantic and disperses more widely in winter.

ANTARCTIC PRION
Pachyptila desolata
F: Procellariidae

Prions are small, gray Southern Ocean petrels that skim the sea to strain plankton with their flattened bill.

12 in
31 cm

JOUANIN'S PETREL
Bulweria fallax
F: Procellariidae

A tropical petrel of the northwestern Indian Ocean, this bird has a weaving flight. It is closely related to the gliding shearwaters.

14–16 in
36–41 cm

16 in
41 cm

17 in
43 cm

ANTARCTIC PETREL
Thalassoica antarctica
F: Procellariidae

A bird of sub-Antarctic waters, this large species dives for fishes and squid. It breeds on islands around the Antarctic.

SNOW PETREL
Pagodroma nivea
F: Procellariidae

One of the few birds that breed on the Antarctic, this petrel breeds farther south than any other bird, even wandering to the South Pole.

BLACK-CAPPED PETREL
Pterodroma hasitata
F: Procellariidae

Like many of the small, fast species known as gadfly petrels, this one has a tropical distribution. It breeds on islands of the West Indies.

scattered dark feathers on white body

massive yellowish bill

15½–16 in
39–40 cm

34–39 in
86–99 cm

7½–8½ in
19–21 cm

CAPE PETREL
Daption capense
F: Procellariidae

One of several members of this family, mostly circumpolar in the southern hemisphere, the cape petrel breeds on islands around Antarctica and winters farther north.

SOUTHERN GIANT PETREL
Macronectes giganteus
F: Procellariidae

This carrion-eating bird breeds in the South Atlantic. Unlike many other petrels, it has strong enough legs to walk well on land.

BAND-RUMPED STORM PETREL
Oceanodroma castro
F: Hydrobatidae

Typical of northern hemisphere storm petrels, this tiny bird has a white-banded rump and forked tail. It is seen in both the Atlantic and Pacific.

GREBES

Birds of ponds and lakes, grebes characteristically swim low in the water and dive propelled by their feet. They feed on small aquatic animals.

Like many other diving birds, grebes have legs that are set far back on the body, making them clumsy on land but agile in water. The toes of their feet are lobed, providing good power strokes during diving, but minimizing drag in the water between strokes. The feet may be used for steering, too; in other diving birds, the tail is used for steering.

A grebe's tail is nothing more than a tuft of feathers. This serves more as a social signal than as a rudder, and is often cocked to expose white feathers beneath. The plumage is dense and well waterproofed by an oil gland. Uniquely in birds,

the gland produces a secretion that is 50 percent paraffin. Grebes have small wings and many species are often reluctant to fly, although northern species are migratory, traveling from inland habitats to the coast to overwinter.

Traditionally, grebes, the order Podicipediformes, are classified as close relatives of divers, penguins, and albatrosses. However, new research suggests that they may be related to flamingos.

RITUAL AND REPRODUCTION
In the breeding season, some grebes perform elaborate courtship rituals. They nest on floating mats of vegetation in freshwater habitats. The young are fully mobile upon hatching and can swim, but seek refuge on a parent's back for the first few weeks.

PHYLUM	CHORDATA
CLASS	AVES
ORDER	PODICIPEDIFORMES
FAMILIES	1
SPECIES	23

The courtship ritual of the great crested grebe culminates with both birds rising up from the water clutching a clump of weeds.

9–11½ in
23–29 cm

LITTLE GREBE
Tachybaptus ruficollis
F: Podicipedidae
Found across the Old World, this bird is the most widespread of the small, dumpy grebes known as dabchicks. It has a reddish neck when breeding.

9½–14 in
24–36 cm

WHITE-TUFTED GREBE
Rollandia rolland
F: Podicipedidae
A native of southern S. America, this is a bird of open lakes with abundant water weeds. A related short-winged Andean species is flightless.

12–15 in
30–38 cm

PIED-BILLED GREBE
Podilymbus podiceps
F: Podicipedidae
This American bird is more stocky and has a stubbier bill than other grebes. Northern birds overwinter in the Caribbean, but tropical populations are sedentary.

16–20 in
40–50 cm

gray flanks

RED-NECKED GREBE
Podiceps grisegena
F: Podicipedidae
This species breeds in Eurasia and N. America and overwinters farther south in coastal waters. Like other grebes, it migrates at night.

11–13½ in
28–34 cm

BLACK-NECKED GREBE
Podiceps nigricollis
F: Podicipedidae
Like other *Podiceps* grebes, this sharp-billed diver has a colorful head plumage when breeding. It is found across the northern hemisphere.

black head

18–20 in
46–51 cm

GREAT CRESTED GREBE
Podiceps cristatus
F: Podicipedidae
This Old World grebe is known for its courtship display. Like other grebes, both sexes are colorful.

10–11½ in
25–29 cm

dark gray back

SILVERY GREBE
Podiceps occipitalis
F: Podicipedidae
This S. American grebe is a colonial breeder, congregating on alkaline or salty lakes. It is found from the Andes to the Falkland Islands.

22–30 in
55–75 cm

WESTERN GREBE
Aechmophorus occidentalis
F: Podicipedidae
One of two similar western N. American grebes, this species has a range from Canada to Mexico; northern populations winter off the Pacific coast.

white throat, breast, and belly

FLAMINGOS

These remarkable birds live on salty lagoons and alkaline lakes. Once classified with storks, flamingos are now thought to be related to grebes.

Extreme gregariousness defines flamingo life. These birds of the order Phoenicopteriformes congregate in huge flocks, sometimes numbering hundreds of thousands. These vast groups are so tightly packed that individual birds cannot easily take flight and must initially walk or run if disturbed. But the open habitat that flamingos favor, coupled with the vigilance of many birds, ensures that predators are easily spotted.

The large flocks are necessary to stimulate breeding, and flamingo courtship involves group displays. Pairs build mud nests and territory is simply determined by how far a bird's neck will stretch from the nest. A few days after hatching, flamingo chicks gather together in large crèches. The parent birds feed their young with a liquid food called crop milk.

FILTER FEEDERS

Flamingos have a unique style of filter-feeding using a specially adapted bill. Holding its head upside-down, a flamingo strains planktonic algae and shrimps from the water through hairlike structures lining the bill. Pigments absorbed from the food give flamingos their characteristic pink color. There are few competitors for food, as the organisms are taken from otherwise barren inland waters that are either very salty or caustic.

PHYLUM	CHORDATA
CLASS	AVES
ORDER	PHOENICOPTERIFORMES
FAMILIES	1
SPECIES	6

Some of the biggest flocks of flamingos are seen in the African Rift Valley, where lesser flamingos gather to feed.

CHILEAN FLAMINGO
Phoenicopterus chilensis
F: Phoenicopteridae
The most widespread flamingo in S. America, the Chilean flamingo has a range from Peru to Tierra del Fuego, and is distinguished by gray legs with pink "knees."

3¼–4¼ ft
1–1.3 m

pinkish white plumage

pink "knee" is the ankle

slender gray leg

black-tipped bill

4–4½ ft
1.2–1.4 m

CARIBBEAN FLAMINGO
Phoenicopterus ruber
F: Phoenicopteridae
This Caribbean species differs from the greater flamingo in being somewhat smaller and having a pinker plumage.

extremely long neck

pale pink bill

bright red wing feathers

3½–5 ft
1.1–1.5 m

GREATER FLAMINGO
Phoenicopterus roseus
F: Phoenicopteridae
With a range spanning Africa, S. Europe, and C. Asia, this is the largest and most widespread of the flamingos.

3¼–3½ ft
1–1.1 m

ANDEAN FLAMINGO
Phoenicoparrus andinus
F: Phoenicopteridae
One of two flamingo species restricted to the higher altitudes of the Andes, this distinctively yellow-legged flamingo may wander nomadically between lakes for food.

LESSER FLAMINGO
Phoeniconaias minor
F: Phoenicopteridae
The smallest flamingo, this species from Africa and southern Asia occurs in huge numbers on highly alkaline lakes.

32–39 in
80–100 cm

STORKS

Most species in this group are wetland birds with long legs for walking through marshy or grassy habitats. They snatch prey with their long bill.

Storks, order Ciconiformes, are tall, upright birds, most often found in open spaces. While some are heronlike waterside species, others forage on dry ground. They have a long-striding walk but can perch on open branches and many nest in trees. Their varied diet includes fishes, amphibians, small reptiles and mammals, and large insects. In Africa, storks feed on small animals fleeing grassland fires, and exploit swarms of locusts and grasshoppers.

Storks have a strong and more or less pointed bill. The marabou uses its heavy, powerful bill for everything from displacing vultures at carcasses of dead mammals to scavenging at refuse tips. Other storks have slimmer bills, some more or less tapered to an upswept or downcurved tip.

All storks have broad wings with "fingered" tips. They soar on rising thermals of warm air to gain height before gliding off over great distances to a new feeding area or on long migratory journeys. Large flocks of white storks cross the Mediterranean Sea at its narrower points, in the same way as migratory birds of prey.

COLONIAL NESTING

Many birds of this group are gregarious when breeding, and their nesting colonies may include several species. All species produce helpless young that must be reared in the nest for some weeks.

CLASS	CHORDATA
AVES	AVES
ORDER	CICONIIFORMES
FAMILIES	1
SPECIES	19

Storks typically nest in trees, but in western Europe white storks make good use of level platforms high on buildings.

3–4 ft
0.9–1.2 m

WOOD STORK
Mycteria americana
F: Ciconiidae
This N. American bird belongs to a group of storks with ibislike bills. It submerges its open bill in shallow water, snapping it shut on moving prey.

long, heavy bill

4–5 ft
1.2–1.5 m

MARABOU
Leptoptilos crumeniferus
F: Ciconiidae
Like other *Leptoptilos* storks, this African species is bareheaded, which enables it to scavenge on carrion without soiling its feathers. It flies with its head retracted.

4½–5 ft
1.4–1.5 m

red and black bill

black and white plumage

SADDLE-BILL STORK
Ephippiorhynchus senegalensis
F: Ciconiidae
Related to the jabiru, this African stork has a slightly upturned bill with a yellow "saddle." It occurs in solitary pairs, even when nesting.

4–4½ ft
1.2–1.4 m

JABIRU
Jabiru mycteria
F: Ciconiidae
A large American stork, the jabiru is the tallest flying bird in S. America. It expands its featherless neck sac when excited.

3¼–4 ft
1–1.2 m

EUROPEAN WHITE STORK
Ciconia ciconia
F: Ciconiidae
Three *Ciconia* species breed outside the tropics. This European bird uses thermal updrafts on migration routes over land to winter in Africa.

30–36 in
75–91 cm

WOOLLY-NECKED STORK
Ciconia episcopus
F: Ciconiidae
The most widespread tropical stork, this bird occurs in both Africa and Asia. It prefers wetland habitats but may stray into pasture.

black plumage

gap in bill

32–37 in
81–94 cm

AFRICAN OPENBILL
Anastomus lamelligerus
F: Ciconiidae
Small, tropical wetland storks, openbills use their distinctive bill to catch and manipulate mollusks before eating them. This species occurs in mainland Africa and Madagascar.

IBISES, BITTERNS, HERONS, AND PELICANS

Most species in the order Pelicaniformes are wetland birds with long legs for walking through marshy or grassy habitats. Their long toes help them to stretch forward when hunting or grasp perches in mangrove swamps or wet thickets. These birds have a long bill for snatching prey.

While most birds in this group eat amphibians and fishes, they sometimes eat small mammals and insects as well. Herons, egrets, and bitterns have modified vertebrae, forming an S-shaped neck that gives them lightning-fast thrust to snatch, but not spear, prey. Ibises probe for prey with their thin, curved bill, whereas spoonbills sweep their partly opened bill—with a flat, rounded tip—sideways through shallow water in search of food. The bill snaps shut when prey is detected by touch. Herons and egrets retract their necks in flight, unlike spoonbills and ibises.

Pelicans are massively built, but broad-winged and expert fliers. They have a long bill with a deep, flexible pouch. When feeding (mostly when swimming, but brown pelicans also plunge-dive) they scoop up a large volume of water, filtering out food as the pouch empties.

PHYLUM	CHORDATA
CLASS	AVES
ORDER	PELICANIFORMES
FAMILIES	5
SPECIES	118

This pelican chick reaches deep into its parent's throat to feed on regurgitated, partly digested fishes.

gray-green back

greenish-black cap

yellow legs and feet

GREEN HERON
Butorides virescens
F: Ardeidae
A small heron of N. American wetlands, the green heron may sometimes bait fishes with food at the water's edge.

16–22 in
40–55 cm

32–39 in
80–100 cm

GREAT EGRET
Ardea alba
F: Ardeidae
Not a true egret but a large white heron, this marsh bird is widespread throughout much of the world.

long, narrow crest

35–39 in
90–98 cm

black tinge to white foreneck

GRAY HERON
Ardea cinerea
F: Ardeidae
Familiar across Eurasia and Africa, the gray heron, like other larger herons, breeds in colonies. It builds stick nests in trees.

28–32 in
70–80 cm

LITTLE BITTERN
Ixobrychus minutus
F: Ardeidae
Among the smallest of the heron-bittern family, this skulking Old World species clambers around reeds—sometimes freezing upright in typical bittern manner.

10½–15 in
27–38 cm

EURASIAN BITTERN
Botaurus stellaris
F: Ardeidae
Like its relatives, this bittern is more frequently heard than seen. It has a loud, booming call.

32–39 in
80–100 cm

WHITE-NECKED HERON
Ardea pacifica
F: Ardeidae
This large heron of wet regions in Australia and New Guinea hunts for insects and small vertebrates in marshes and grassland.

LITTLE EGRET
Egretta garzetta
F: Ardeidae
This Old World species has started to colonize America. It has a distinct black bill and legs, and one race has yellow feet.

22–26 in
55–65 cm

23½–28 in
60–70 cm

22–22½ in
55–57 cm

23–25 in
58–63 cm

WHITE-FACED HERON
Egretta novaehollandiae
F: Ardeidae
A member of the egret group from Indonesia, Australia, and New Zealand, this species has a varied diet that includes insects and frogs.

TRICOLORED HERON
Egretta tricolor
F: Ardeidae
This is an inhabitant of American swamps, where, like related species, it uses its daggerlike bill to snatch small animals.

LITTLE BLUE HERON
Egretta caerulea
F: Ardeidae
This American species belongs to the egret group. The purplish head and neck turn gray-blue outside the breeding season.

CATTLE EGRET
Bubulcus ibis
F: Ardeidae

Actually a small white heron, this globally widespread species forages on grassland and often follows livestock to catch disturbed prey.

19–21 in
48–53 cm

27–32 in
68–82 cm

black-tipped bill

BOAT-BILLED HERON
Cochlearius cochlearius
F: Ardeidae

This heron has a broad bill, adapted for scooping animal prey. It inhabits C. and S. American mangrove swamps.

18–20 in
45–50 cm

16½–18 in
42–45 cm

INDIAN POND HERON
Ardeola grayii
F: Ardeidae

Common in S. Asia, this heron stalks aquatic prey, but it may also flush fishes by flying low over water.

long neck plumes when breeding

BLACK-CROWNED NIGHT HERON
Nycticorax nycticorax
F: Ardeidae

Night herons have good night vision. Found in most warm regions except Australia, this is the most widespread species.

23–26 in
58–65 cm

bare head and neck

black wing plumes

26–30 in
65–75 cm

23–30 in
59–76 cm

gray body

STRAW-NECKED IBIS
Threskiornis spinicollis
F: Threskiornithidae

So called because of the strawlike feathers at the base of its neck, this nomadic ibis occurs in New Guinea and Australia.

SACRED IBIS
Threskiornis aethiopicus
F: Threskiornithidae

This is a common bird of the wetlands and grasslands of Africa and Madagascar. It has been introduced into America and Europe.

27–30 in
69–76 cm

22–24 in
56–61 cm

SCARLET IBIS
Eudocimus ruber
F: Threskiornithidae

The national bird of Trinidad, the scarlet ibis occurs in tropical America and acquires its scarlet pigment from eating crustaceans.

REDDISH EGRET
Egretta rufescens
F: Ardeidae

This American species has white and reddish gray color forms. When fishing, it sometimes holds out its wings to reduce sun glare.

AUSTRALIAN IBIS
Threskiornis molucca
F: Threskiornithidae

This is a common species throughout Australia. It often invades urban areas, where it is sometimes considered a pest.

30 in
75–77 cm

22–26 in
55–65 cm

30–35 in
76–89 cm

HADADA IBIS
Bostrychia hagedash
F: Threskiornithidae

A common African species found in grasslands, forests, parks, and gardens, the hadada ibis is named after its distinctive flight call.

BLACK-FACED IBIS
Theristicus melanopis
F: Threskiornithidae

Found in S. America, this species occurs in temperate grassland habitat from the Andes to Patagonia.

GLOSSY IBIS
Plegadis falcinellus
F: Threskiornithidae

Found throughout warm parts of the world, this is the most widespread ibis species. Its tree-nesting colonies may be accompanied by herons.

≫ IBISES, BITTERNS, HERONS, AND PELICANS

AFRICAN SPOONBILL
Platalea alba
F: Threskiornithidae
The only spoonbill confined
to wetlands of Africa, this
species is characterized
by its red face
and legs.

35–36 in
90–92 cm

EURASIAN SPOONBILL
Platalea leucorodia
F: Threskiornithidae
This species breeds in Eurasia
and overwinters in Africa.
Mature birds have a
broad yellow tip to
the black bill.

32–35 in
80–90 cm

gray
plumage

large
bill

4–5 ft
1.2–1.5 m

SHOEBILL
Balaeniceps rex
F: Balaenicipitidae
Restricted to swamps from Sudan
to Zambia, this wading bird uses
its huge bill to scoop vertebrate
prey from muddy waters.

pinkish red
patch on wing

pink breast
tuft when
breeding

ROSEATE SPOONBILL
Platalea ajaja
F: Threskiornithidae
Like other spoonbills, this
distinctive American species feeds
on small aquatic animals by swinging
its bill from side to side in water.

28–34 in
71–86 cm

22 in
56 cm

HAMMERKOP
Scopus umbretta
F: Scopidae
This African wetland bird
builds huge, heavy-walled
nests using sticks and mud to
protect its young, who are
often left for long periods.

bold head and
neck pattern

3¼–4½ ft
1–1.4 m

orange throat pouch

all-white plumage

4¼–5¼ ft
1.3–1.6 m

**AMERICAN
WHITE PELICAN**
Pelecanus erythrorhynchos
F: Pelecanidae
This pelican breeds on inland lakes
of N. America and overwinters
on the coasts. In the breeding season,
it develops a flat "horn" on its bill.

4¼–5 ft
1.3–1.5 m

SPOT-BILLED PELICAN
Pelecanus philippensis
F: Pelecanidae
A species from S. Asia, the spot-billed pelican,
like most other pelicans, fishes by scooping
prey while swimming on the water surface.

BROWN PELICAN
Pelecanus occidentalis
F: Pelecanidae
This gray and brown pelican from the
S. US to S. America is a coastal-
breeding species and, unlike other
pelicans, plunge-dives for fishes.

CORMORANTS, GANNETS, AND RELATIVES

The order Suliformes includes marine species, though some cormorants and anhingas also visit freshwater habitats. The birds in this group have webbing between all four toes.

Frigatebirds cross vast areas of sea but rarely settle on water, taking flying fishes or stealing food from other seabirds in the air. Cormorants and anhingas feed on a variety of aquatic species by diving from the surface, while gannets and boobies plunge from a height and are more exclusively fish-eaters.

PHYLUM	CHORDATA
CLASS	AVES
ORDER	SULIFORMES
FAMILIES	4
SPECIES	61

32 in
81 cm

BLUE-FOOTED BOOBY
Sula nebouxii
F: Sulidae
Found on rocky coasts from California to Peru and the Galapagos Islands, this booby flaunts its blue feet as part of its courtship display.

MASKED BOOBY
Sula dactylatra
F: Sulidae
This is the largest of the tropical boobies, with a black and white plumage and long, yellow bill. It most closely resembles the related cold-water gannets.

31–36 in
80–92 cm

yellow tinge to back of head

white upperparts

NORTHERN GANNET
Morus bassanus
F: Sulidae
The three similar species of cold-water gannets breed in large colonies on rocky coasts. This one is a native of the North Atlantic.

snakelike neck

spearlike bill

3–3¼ ft
90–100 cm

28 in
71 cm

30–37 in
75–95 cm

20–22 in
50–55 cm

LITTLE PIED CORMORANT
Microcarbo melanoleucos
F: Phalacrocoracidae
This is an Australasian member of a group of primitive, short-billed "micro-cormorants" that are largely associated with freshwater or estuarine habitats.

3¼–3½ ft
1–1.1 m

ANHINGA
Anhinga anhinga
F: Anhingidae
This American species has a long, cormorantlike neck and a straight bill for spearing fishes. Anhingas are also called snake birds or darters.

RED-FACED SHAG
Phalacrocorax urile
F: Phalacrocoracidae
A deep-diving seabird found from Japan to the Bering Sea, this species belongs to a group of North Pacific cormorants that are highly adapted to the marine environment.

MAGNIFICENT FRIGATEBIRD
Fregata magnificens
F: Fregatidae
Infrequent feeding opportunities and long hours in flight cause frigatebirds, such as this American species, to have a low breeding rate and the longest period of parental care of any bird.

TROPICBIRDS

Large, ternlike seabirds with a long, spikelike tail projection, tropicbirds nest on tropical islands and forage widely at sea.

The three species in this order, the Phaethontiformes, have all four toes joined by webs, like gannets and boobies, but are unable to stand. They locate fishes by sight while hovering before plunging from the air. At their breeding colonies, groups of males display in swirling, circular flights, swaying their tail spike to impress watching females.

PHYLUM	CHORDATA
CLASS	AVES
ORDER	PHAETHONTIFORMES
FAMILIES	1
SPECIES	3

black "mask"

WHITE-TAILED TROPICBIRD
Phaethon lepturus
F: Phaethontidae
Tropicbirds are streamer-tailed seabirds with feet so feeble that they use their bellies to shuffle on land. This species frequents most tropical coastlines.

30–32 in
76–80 cm

RED-BILLED TROPICBIRD
Phaethon aethereus
F: Phaethontidae
Found from the E. Pacific to the Atlantic, this typical, hole-nesting tropicbird rears a single, slow-growing chick, an adaptation to scattered ocean food resources.

mostly white with black wingtips

3–3½ ft
0.9–1.1 m

long, whiplike tail projection

BIRDS OF PREY

The largest and most important group of day-flying hunters, almost all the birds in the order Accipitriformes are exclusively meat-eaters. In some habitats they are the top predators.

Large birds of prey, such as the harpy eagle and Philippine eagle of tropical rainforests, are so powerful they can kill large monkeys and small deer. The largest species, which include Old World vultures and American condors, feed on dead animals, and many, such as kites and buzzards, eat carrion as well as catch live prey. Exceptions include the largely vegetarian palm-nut vulture of Africa.

Birds of prey typically have keen vision and hunt by sight. The turkey vulture detects prey by smell, and other species watch and follow it down to hidden carcasses. Birds in this group have a strong, hooked bill for dismembering prey. Larger vultures are well-equipped to rip hide and flesh but have relatively weak feet. Most have a bare head, which prevents an unhygienic, sticky mess when probing inside animal carcasses. New World vultures may be more closely related to storks than to the similar-looking Old World vultures. Eagles, hawks, buzzards, and kites have stronger feet with long, curved, sharp claws, used for catching and killing prey. The unique secretary bird of African plains hunts on foot.

Some kites have a forked tail for greater control in flight. Bird-eating hawks have rather short wings and a long tail, for extra maneuverability through woodland, while other species have longer, broader wings that allow them to soar on rising air currents and glide long distances with minimal effort. In this way, heavy vultures can move very long distances to find food each day, and some species perform very long annual migrations. Their "fingered" wingtips reduce turbulence at the wingtip, giving greater efficiency.

PHYLUM	CHORDATA
CLASS	AVES
ORDER	ACCIPITRIFORMES
FAMILIES	4
SPECIES	266

The pale underwings of a turkey vulture catch the sunlight as it tilts, twists, and turns, searching for food below.

25–32 in
64–81 cm

TURKEY VULTURE
Cathartes aura
F: Cathartidae
Unusually for birds, this widespread American vulture locates rotting carcasses by smell. It often nests in dark recesses, such as under large rocks or stumps.

26–32 in
67–81 cm

contrasting black and white wings

22–26 in
56–66 cm

white streaks on massive wings

ANDEAN CONDOR
Vultur gryphus
F: Cathartidae
S. America's largest flying land bird, this species soars on updrafts in the Andes, locating carrion by sight or by following other scavengers, such as turkey vultures.

KING VULTURE
Sarcoramphus papa
F: Cathartidae
This large bird soars high above forests of tropical America looking for carrion. It is distinguished by its colorful head and fleshy bill wattle.

BLACK VULTURE
Coragyps atratus
F: Cathartidae
More gregarious than the related turkey vulture, the black vulture is an opportunistic scavenger found from the C. US to Chile.

EUROPEAN HONEY-BUZZARD
Pernis apivorus
F: Accipitridae

Belonging to a group of tropical raptors that eat bee and wasp larvae, this species breeds in Eurasia and overwinters in Africa.

20½–23½ in
52–60 cm

EURASIAN BUZZARD
Buteo buteo
F: Accipitridae

A common raptor, this species has light and dark color forms. Its northern populations overwinter in tropical Africa and Asia.

20–22½ in
51–57 cm

dark brown to white coloration

LONG-LEGGED BUZZARD
Buteo rufinus
F: Accipitridae

This buzzard of semideserts and mountains breeds in C. Europe and C. Asia, with some populations migrating to N. Africa in winter.

20–26 in
50–65 cm

WHITE-TAILED KITE
Elanus leucurus
F: Accipitridae

Typical of its genus, this sharp-browed kite habitually hovers when hunting. It occurs from the US to S. America, outside the Amazon basin.

12½–15 in
32–38 cm

WHITE-EYED BUZZARD
Butastur teesa
F: Accipitridae

This small buzzard-hawk from S. Asia is more terrestrial than related species, hunting small animals and insects on the ground.

15–17 in
38–43 cm

20–25 in
50–64 cm

SWALLOW-TAILED KITE
Elanoides forficatus
F: Accipitridae

This insect-eating bird of prey is a graceful and agile flier. It breeds in the S.E. US and C. America, and overwinters in S. America.

BRAHMINY KITE
Haliastur indus
F: Accipitridae

With a range from India to Australasia, this riverside and coastal scavenger also hunts for live prey such as fishes and small mammals.

white neck ruff

sharply hooked bill

red-brown tail

17–20 in
43–51 cm

head plumes

3¼–4½ ft
1–1.4 m

OSPREY
Pandion haliaetus
F: Accipitridae

Found almost worldwide, the fish-eating osprey plunges for its prey and has a reversible outer toe for a better grip of its slippery catch.

20½–26 in
52–66 cm

elongated central tail feathers

PALM-NUT VULTURE
Gypohierax angolensis
F: Accipitridae

Unusually for a vulture, this African species has a largely vegetarian diet, consisting of oil palm fruit. However, it also feeds on fishes and carrion.

23½ in
60 cm

SECRETARY BIRD
Sagittarius serpentarius
F: Accipitridae

One of the few raptors to hunt on the ground, this long-legged bird of African savanna chases small animals, often stamping to disable them.

long legs

4¼–5 ft
1.3–1.5 m

28–38 in
71–96 cm

BALD EAGLE
Haliaeetus leucocephalus
F: Accipitridae
This N. American sea eagle, the
US's national emblem, captures
or scavenges fishes, sometimes
hunting cooperatively. It breeds
in woodland near water.

WHITE-BELLIED SEA EAGLE
Haliaeetus leucogaster
F: Accipitridae
With a distribution from India
to Australasia along lakes and
rivers, like other large eagles,
this fishing raptor builds
huge stick nests.

28–34 in
72–85 cm

22–28 in
55–72 cm

28–35 in
70–90 cm

BONELLI'S EAGLE
Aquila fasciatus
F: Accipitridae
A woodland- and mountain-dwelling species,
this long-winged buzzardlike eagle has a range
that extends from S. Eurasia to N. Africa.

WHITE-HEADED VULTURE
Trigonoceps occipitalis
F: Accipitridae
Found in N., E., and S. Africa,
this vulture is often seen in pairs,
and is usually outnumbered by
other vulture species when
feeding on carcasses.

EGYPTIAN VULTURE
Neophron percnopterus
F: Accipitridae
A relative of the palm-nut vulture, this
bird of S. Eurasia and Africa routinely
uses rocks to crack open ostrich eggs.

28–33 in
70–83 cm

EASTERN IMPERIAL EAGLE
Aquila heliaca
F: Accipitridae
True eagles, including those of
the *Aquila* genus, such as this
Eurasian species, are described
as "booted" because of their
fully feathered legs.

22–26 in
55–65 cm

23½–28 in
60–70 cm

AFRICAN HAWK-EAGLE
Aquila spilogaster
F: Accipitridae
A small raptor from
sub-Saharan Africa,
this hawk eagle hunts in
wooded savanna and
hilly country.

pale head

24–30 in
61–75 cm

23½–39 in
60–100 cm

*dark brown
flight feathers*

feathered legs

GOLDEN EAGLE
Aquila chrysaetos
F: Accipitridae
A graceful soaring bird, this large,
long-tailed eagle occurs in open country
across the northern hemisphere. It
frequents forests in some regions.

CHANGEABLE HAWK-EAGLE
Spizaetus cirrhatus
F: Accipitridae
Often crested, Asian hawk-eagles are
forest raptors. This variable species
has dark and pale forms and is found
from the Himalayas to Indonesia.

AFRICAN WHITE-BACKED VULTURE
Gyps africanus
F: Accipitridae
One of the commonest vultures of the sub-Saharan savanna, this species gathers in large numbers near carcasses. These birds are seen in towns and villages.

35–39 in
90–98 cm

3¼–4 ft
1–1.2 m

LAPPET-FACED VULTURE
Torgos tracheliotus
F: Accipitridae
Like the related *Gyps* vultures, this carrion-eater of arid Africa has a long neck and a bare head to prevent carcasses from soiling its plumage.

34–38 in
85–97 cm

3¼–4¼ ft
1–1.3 m

EURASIAN GRIFFON
Gyps fulvus
F: Accipitridae
This vulture occurs in mountainous regions of S.W. Eurasia and N.E. Africa. It breeds and roosts among rocks and on ledges.

neck ruff turns white with age

RÜPPELL'S VULTURE
Gyps rueppelli
F: Accipitridae
An African relative of the Eurasian griffon, this darker species inhabits arid areas. It has been recorded as flying at higher altitudes than any other bird.

bulbous bill

BEARDED VULTURE
Gypaetus barbatus
F: Accipitridae
Native to the mountains of Africa and Eurasia, this diamond-tailed, solitary vulture subsists largely on marrow obtained by dropping bones on rocks to split them open.

18–20 in
46–51 cm

BLACK-COLLARED HAWK
Busarellus nigricollis
F: Accipitridae
A relative of the snail kite, this species occurs in C. and S. American wetlands, where it fishes by dropping feet first into floating vegetation.

18–22 in
46–56 cm

WHITE HAWK
Pseudastur albicollis
F: Accipitridae
This forest hawk of C. and S. America hunts for reptiles, especially snakes. It has been described as lethargic and easily approached.

14–16 in
36–40 cm

23½–26 in
60–66 cm

DARK CHANTING GOSHAWK
Melierax metabates
F: Accipitridae
An African hawk of dry, open country, this bird resembles a harrier when in flight. It has a musical piping call.

3–3½ ft
0.9–1.1 m

17–22 in
43–56 cm

SNAIL KITE
Rostrhamus sociabilis
F: Accipitridae
This bird from Florida and C. and S. American marshes has a strongly curved bill, adapted for feeding on aquatic snails.

AFRICAN HARRIER-HAWK
Polyboroides typus
F: Accipitridae
This raptor from sub-Saharan Africa eats oil-palm fruit and hunts small vertebrates. Its flexible "double-jointed" legs help it take prey from tree-holes.

»

RÜPPELL'S VULTURE
Gyps rueppelli

One of the iconic scavengers of the African plains, from Senegal east to Sudan and Tanzania, Rüppell's vulture flies so high in roaming for food that its blood is specially adapted to capture oxygen in the thin air. This vulture patrols dry mountainous terrain, leaving its cliff-top roost sites in early morning to rise on updrafts instead of thermals. It uses acute vision to find carcasses and will wait patiently—days, if necessary—for predators to leave a kill. Like most vultures, it eats soft, rotting flesh and offal. However, its longer neck means it can reach deeper into corpses than many of its competitors—gorging its fill until, with some difficulty, it returns to the skies.

SIZE 34–38 in (85–97 cm)
HABITAT Gorges in dry, open country
DISTRIBUTION North and East Africa
DIET Carrion

> **THIRD EYELID**
A typical feature of birds, this membrane cleans the surface of the eye, and may help protect against flying debris during frenzied feeding.

∨ **WHITE COLLAR**
White fluffy feathers encircling the base of the neck form a ruff. This may be discolored with dust and blood from carcasses.

< **SCALLOPING**
This vulture's dark wing feathers are broadly tipped with lighter marking, giving the bird a scalloped appearance from a distance.

nostril

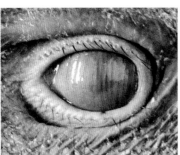

< **PLUMAGE**
Beneath the patterned feathers that mark the contours of the body are fluffy down feathers that trap body heat—vital at high altitude.

hooked bill

< **LEG**
Although the bird's upper legs are feathered, its lower legs are bare and so remain relatively clean when the bird is feeding on carcasses.

> **A HEAD FOR GORE**
Although the sparse down of the vulture's head and neck is frequently blood-stained, a fully feathered head would get clogged with sticky debris as the bird reached deep inside the carcasses of large mammals to feed. The vulture's hooked bill is used to tear semi-rotten flesh, and is long enough to probe carcasses.

rows of covert feathers smooth passage of air flow

longer, stiffer flight feathers provide thrust and lift for flying

∧ **WING**
Long, broad wings help the vulture soar or glide, saving energy. Takeoff can be a struggle after a heavy meal.

< **FOOT**
Because vultures use their feet for walking rather than killing, they lack the big talons that are typical of predatory birds of prey.

*pinkish gray skin
of head and neck
has a light coating
of down*

» BIRDS OF PREY

17–18½ in
43–47 cm

MONTAGU'S HARRIER
Circus pygargus
F: Accipitridae

Eurasian harriers, such as the Montagu's harrier, migrate to Africa and S. Asia. This species inhabits grassland and reed beds.

♀

19–22 in
48–56 cm

WESTERN MARSH HARRIER
Circus aeruginosus
F: Accipitridae

The males of this species are brown, like the females, with gray on their wings and tail. Males of other harrier species are all gray.

17½–20½ in
44–52 cm

NORTHERN HARRIER
Circus hudsonius
F: Accipitridae

Harriers are characterized by a narrow tail; narrow, pointed wings; and long legs. This species is widespread in N. America.

12–14½ in
30–37 cm

LIZARD BUZZARD
Kaupifalco monogrammicus
F: Accipitridae

A native of the African savanna, this bird mostly preys on large insects, such as grasshoppers, but also eats small vertebrates.

25–27 in
63–68 cm

BLACK-BREASTED SNAKE EAGLE
Circaetus pectoralis
F: Accipitridae

A bird of African grasslands, this eagle feeds on lizards and small mammals as well as snakes.

19–24 in
48–62 cm

NORTHERN GOSHAWK
Accipiter gentilis
F: Accipitridae

This large hawk from N. America and Eurasia is capable of maneuvering through tall trees to catch squirrels and grouse.

11–16 in
28–40 cm

EURASIAN SPARROWHAWK
Accipiter nisus
F: Accipitridae

One of nearly 50 species of *Accipiter* hawks, this hunter of small birds is found in woodland habitats from Europe to Japan.

10–14 in
25–35 cm

SHIKRA
Accipiter badius
F: Accipitridae

A typical *Accipiter* hawk with a long tail and short wings, the Old World shikra has a dashing flight for catching small animals, including birds.

red facial skin

long, broad wings

23½–26 in
60–66 cm

RED KITE
Milvus milvus
F: Accipitridae

Like other *Milvus* kites, this species from Europe and the Middle East is slightly weak-footed, but accomplished at soaring. It frequently feeds on carrion.

22–23½ in
55–60 cm

BLACK KITE
Milvus migrans
F: Accipitridae

Found in open country across Eurasia, Africa, and Australasia, the black kite has a varied diet that includes live fish and small mammals, carrion, and sometimes human refuse.

22–30 in
55–75 cm

CRESTED SERPENT EAGLE
Spilornis cheela
F: Accipitridae

Belonging to a group of Asian snake eagles, this species inhabits a range from India to the Philippines, and is often found near fresh water.

22–28 in
55–70 cm

BATELEUR
Terathopius ecaudatus
F: Accipitridae

This African savanna bird is the only snake eagle that regularly eats carrion. *Bateleur* means "acrobat" in French, a reference to this bird's acrobatic flight.

red feet

FALCONS AND CARACARAS

Falcons are no longer considered to be closely related to the main order of birds of prey, the Accipitriformes, but instead have their own order, the Falconiformes.

Falcons have a hooked bill, similar to that of the Accipitriformes, but with an added "notch" used to kill prey that is caught and held in the feet. They also have strong feet with sharp, curved claws. Most have long wings and are marvelous fliers. Some falcons hunt by catching large insects and birds in the air, while others dive onto prey on the ground. Some even take on larger birds in dramatic dives, stooping to strike at speed. Several "hover" at a fixed point above the ground, to search for prey below, often detecting them by the ultraviolet light emitted by their urine. Caracaras also hunt and scavenge on the ground. Unlike most true birds of prey, falcons make no nest but lay eggs on a ledge or in an old nest of some other species.

PHYLUM	CHORDATA
CLASS	AVES
ORDER	FALCONIFORMES
FAMILIES	1
SPECIES	67

slate-gray
upperparts

dark
"moustache"

yellow
feet

9½–13 in
24–33 cm

13½–23 in
34–58 cm

PEREGRINE FALCON
Falco peregrinus
F: Falconidae
As the fastest bird of prey, this falcon dives steeply on prey in midair. It is found throughout the world in open country, including tundra and semidesert.

10–12 in
26–30 cm

AMUR FALCON
Falco amurensis
F: Falconidae
Unusually for falcons, this raptor habitually gathers in flocks; it breeds in marshy woodland across Siberia and China and migrates to southern Africa to overwinter.

8–12 in
20–31 cm

AMERICAN KESTREL
Falco sparverius
F: Falconidae
Kestrels are small falcons that habitually hover when hunting. This species is distributed throughout the Americas, including the Caribbean islands.

BAT FALCON
Falco rufigularis
F: Falconidae
This swift-flying American bird hunts birds, bats, and large insects at twilight. It is found from Mexico to Argentina.

9–12 in
23–30 cm

COMMON KESTREL
Falco tinnunculus
F: Falconidae
Like other kestrels, this species of open country across Eurasia and Asia uses updrafts to maintain its hovering, as it scans the ground for prey.

12½–15½ in
32–39 cm

MERLIN
Falco columbarius
F: Falconidae
A nimble predator with a dashing flight, this falcon captures birds in midair over hills and moorland across the northern hemisphere.

7–8½ in
18–21 cm

AFRICAN PYGMY FALCON
Polihierax semitorquatus
F: Falconidae
This African bird dives for insects and lizards on the ground. It breeds in the nests of weaver birds and may raise a brood cooperatively with others.

16–18 in
40–46 cm

19–21 in
48–53 cm

yellowish-red
facial skin

black crown
and crest

CRESTED CARACARA
Caracara cheriway
F: Falconidae
This common species of caracara is found in open country, from the S. US to northern S. America. It nests in trees or on the ground.

21–24 in
53–62 cm

STRIATED CARACARA
Phalcoboenus australis
F: Falconidae
The tendency of this fearless caracara to attack newborn lambs has led to its persecution in its native Falkland Islands.

YELLOW-HEADED CARACARA
Milvago chimachima
F: Falconidae
A hawklike scavenger that also eats oil palm fruit, this bird of southern S. America frequents savanna and forest edges.

MOUNTAIN CARACARA
Phalcoboenus megalopterus
F: Falconidae
Relatives of falcons, caracaras are more sluggish and longer legged. Like other caracaras, this species of the high Andes is a scavenger, but also hunts small animals.

19½–23 in
49–58 cm

BUSTARDS

Bustards are medium to very large ground-living birds (including some of the heaviest flying birds) with long, stout legs but a short bill.

Birds in this order, the Otidiformes, have a steady, long-striding walk and pick insects, reptiles, small mammals, and vegetable matter from the ground in semideserts, open grasslands, and crops. In flight, they show large white areas on their long, broad wings. Larger species are powerful, steady fliers, while smaller ones have rapid wingbeats, looking more like small wildfowl or gamebirds with square-tipped wings. Social birds, they form flocks where numbers remain sufficiently high.

PHYLUM	CHORDATA
CLASS	AVES
ORDER	OTIDIFORMES
FAMILIES	1
SPECIES	26

KORI BUSTARD
Ardeotis kori
F: Otidae
Among the heaviest flying birds, weighing up to 42 lb (19 kg), this bustard occurs across eastern and southern Africa. It feeds on small vertebrates, carrion, and seeds.

3¼–4½ ft
1–1.4 m

AUSTRALIAN BUSTARD
Ardeotis australis
F: Otidae
This bustard inhabits grasslands and open woodland of Australia and southern New Guinea. Males have a throat sac, which they inflate when displaying.

2½–5 ft
0.8–1.5 m

22–26 in
55–65 cm

orange-brown wings

28–43 in
70–110 cm

HOUBARA BUSTARD
Chlamydotis undulata
F: Otidae
A native of arid country, this bird is found in open plains and barren desert areas of N. Africa and the Canary Islands.

rufous breast band

GREAT BUSTARD
Otis tarda
F: Otidae
Male bustards are bigger than females, especially so in this species of the Eurasian steppes. Adult birds take six years to mature and develop plumes.

16–18 in
40–45 cm

LITTLE BUSTARD
Tetrax tetrax
F: Otidae
This diminutive bustard breeds in open country across Eurasia and migrates southward in winter. In flight, it resembles a shelduck.

21 in
53 cm

RED-CRESTED BUSTARD
Lophotis ruficrista
F: Otidae
Like other bustards, this southern African species has a spectacular courtship display. Males perform tumbling aerial flights, and pairs call in duet.

CRANES, RAILS, AND RELATIVES

From graceful dancing cranes to small skulking rails, this order contains a wide variety of ground-dwelling birds of both dry and wetland habitats.

Mostly long-legged and long-billed, birds in the order Gruiformes are behaviorally very diverse. Cranes walk on the ground, do not perch in trees, and have short toes with the hind toe reduced or absent. The finfoots and coots have lobed, not webbed, feet. Rails and crakes have longer toes and a body that is deep but very slim, helping movement through dense vegetation, including reedbeds, in marshes, or drier areas. Some on remote islands are flightless.

PHYLUM	CHORDATA
CLASS	AVES
ORDER	GRUIFORMES
FAMILIES	6
SPECIES	188

olive-brown wings and body

26–28 in
65–70 cm

YELLOW-LEGGED BUTTONQUAIL
Turnix tanki
F: Turnicidae
Like other buttonquails, this east Asian species inhabits tropical grasslands. The females are more colorful, and compete for the drab males, who take care of the young.

LIMPKIN
Aramus guarauna
F: Aramidae
Mostly nocturnal, this small cousin of cranes inhabits tropical American wetlands and uses its tweezerlike bill to extract snails from shells.

6 in
15 cm

7½–9 in
19–23 cm

9–12 in
22–30 cm

CORNCRAKE
Crex crex
F: Rallidae
This shy grassland crake is best identified by its rasping "crex crex" call. It breeds in Eurasia and overwinters in Africa.

BLACK CRAKE
Amaurornis flavirostra
F: Rallidae
This African crake is widespread south of the Sahara. Unlike many of its secretive relatives, it is often seen out in the open.

long, gray legs

BUFF-BANDED RAIL
Gallirallus philippensis
F: Rallidae
Unlike some of the other Indo-Pacific *Gallirallus* species that are flightless or nearly so, this rail has managed to disperse to many oceanic islands, from the Philippines to New Zealand.

11–13 in
28–33 cm

15–19 in
38–48 cm

5–7 in
13–18 cm

YELLOW RAIL
Coturnicops noveboracensis
F: Rallidae
One of several barred-backed rails, this tiny, secretive N. American species is most easily detected by its nocturnal clicking call.

KING RAIL
Rallus elegans
F: Rallidae
This species is found in eastern N. America, Mexico, and Cuba. It forages in concealed locations for insects, spiders, shrimp, and snails.

9–11 in
23–28 cm

WATER RAIL
Rallus aquaticus
F: Rallidae
Rallus rails are long-billed marsh birds with a narrow body for passing through reedbeds. Like other members of the genus, this Eurasian bird is rarely seen away from dense cover.

8–10½ in
20–27 cm

VIRGINIA RAIL
Rallus limicola
F: Rallidae
This long-distance migrant has a range from N. America to northern S. America. It is a secretive bird that is difficult to spot.

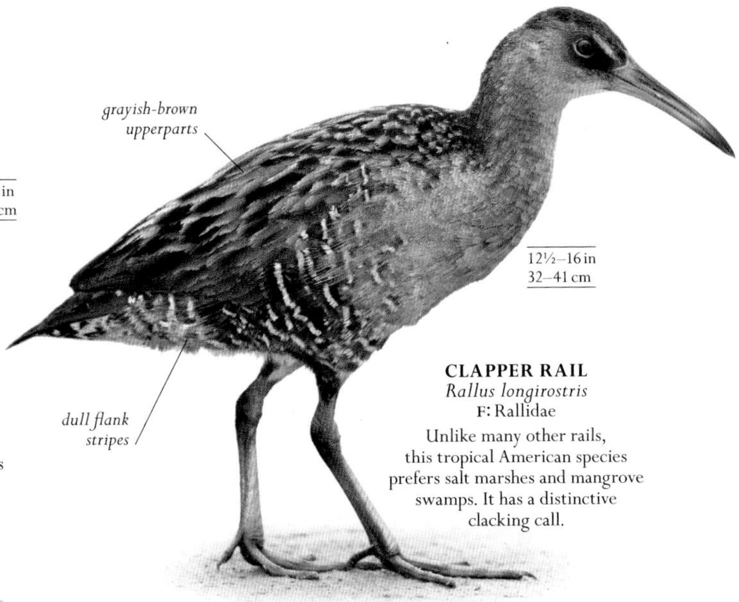

grayish-brown upperparts

12½–16 in
32–41 cm

CLAPPER RAIL
Rallus longirostris
F: Rallidae
Unlike many other rails, this tropical American species prefers salt marshes and mangrove swamps. It has a distinctive clacking call.

dull flank stripes

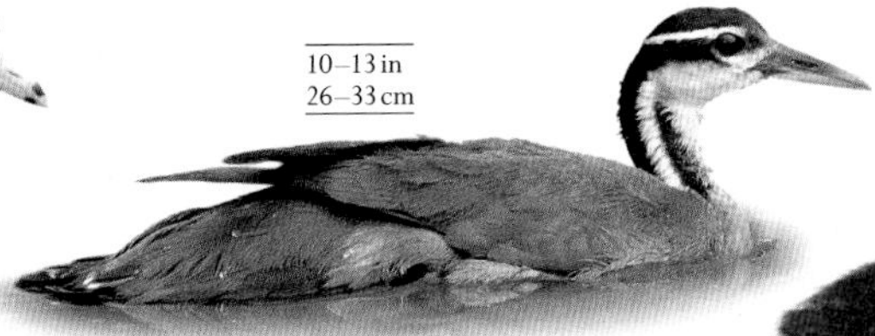

10–13 in
26–33 cm

SUNGREBE
Heliornis fulica
F: Heliornithidae
This species is a tropical American finfoot. Like all finfoots, it is a secretive bird of slow-moving water, where it feeds on small animals.

red bill with yellow tip

15½–16 in
39–40 cm

AMERICAN COOT
Fulica americana
F: Rallidae
Found from N. America to northern S. America, this bird, unlike true rails, lives on water. It forages in shallow water and feeds on land.

8½–10½ in
21–27 cm

RUDDY-BREASTED CRAKE
Porzana fusca
F: Rallidae
This crake occurs in wetlands of eastern Asia, but is also seen in mangroves or drier habitats. It has distinctive chestnut-colored underparts.

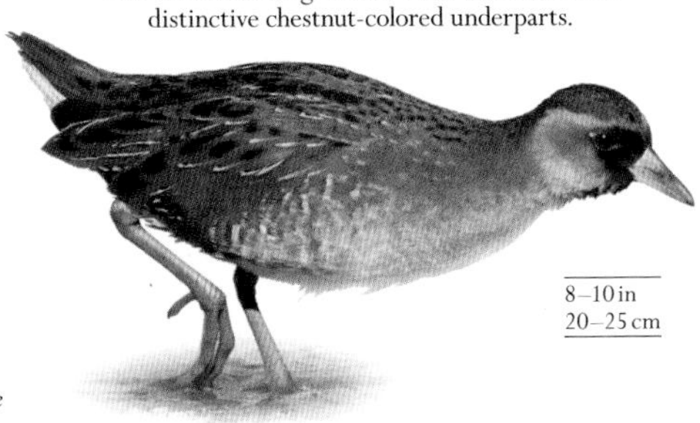

12–14 in
30–36 cm

purplish-blue underparts

green wings

8–10 in
20–25 cm

SORA RAIL
Porzana carolina
F: Rallidae
This crake is the most common N. American member of the rail family. It breeds in shallow wetland and overwinters in the Caribbean.

WHITE-BROWED CRAKE
Porzana cinerea
F: Rallidae
Shorter bills and distinctive calls distinguish crakes from rails. This gray-fronted bird is a typical *Porzana* crake. Its range extends from the Malay Peninsula to Polynesia.

7–9 in
18–22 cm

yellow legs

PURPLE GALLINULE
Porphyrio martinica
F: Rallidae
Occurring in tropical American marshes, this species belongs to a group of rails called swamphens. It is characterized by blue-purple plumage and a blue forehead shield.

COMMON MOORHEN
Gallinula chloropus
F: Rallidae
Moorhens are noisy birds with dark plumage and jerky movements. One of several *Gallinula* species, this one has an almost global distribution.

12½–14 in
32–35 cm

GRAY-WINGED TRUMPETER
Psophia crepitans
F: Psophiidae
Named for their loud calls,
trumpeters are gregarious Amazonian
ground birds that are weak fliers.
This is a black-bodied species,
which—like other trumpeters—
has a "hunchback" appearance.

19–22 in
48–56 cm

gray-white plumage

bare red head

BROLGA
Grus rubicunda
F: Gruidae
This Australian crane
has a bare red head and
black dewlap (flap under
the chin). It performs a
spectacular prancing
courtship display.

3½–4 ft
1–1.2 m

3¼–3½ ft
1–1.1 m

BLUE CRANE
Anthropoides paradiseus
F: Gruidae
The long wing plumes of this African
crane resemble a tail. When not
breeding, it is nomadic, frequenting
lakeside, grassland, and farmland.

slate-gray
plumage

**GRAY CROWNED
CRANE**
Balearica regulorum
F: Gruidae
African crowned cranes are
the only cranes that can grip
branches, enabling them to roost
in trees. This is the most
southerly species.

red
wattle

4½–5 ft
1.4–1.5 m

**RED-CROWNED
CRANE**
Grus japonensis
F: Gruidae
Declining in numbers, this
bird breeds in Siberia and
overwinters in Korea and
China. The heaviest of all
cranes, like related species,
it has a bare, red
crown patch.

3½ ft
1.1 m

3½–4 ft
1.1–1.2 m

SANDHILL CRANE
Grus canadensis
F: Gruidae
A N. American crane, this
species also reaches westward
into Siberia. It migrates south in
family groups, as far as Mexico.

3½–4 ft
1.1–1.2 m

COMMON CRANE
Grus grus
F: Gruidae
Frequenting marsh,
heathland, and tundra,
this bird breeds in Eurasia
and migrates to N. Africa
and S. Asia, often in a
V-formation.

KAGU AND SUNBITTERN

**Recently separated from the cranes
and rails, the tropical kagu and
sunbittern are unique birds of damp
forest, limited to very restricted
geographical ranges.**

A handful of species worldwide are difficult to
place in a relationship with others: these two
appear to be closely related, but look very
different. Their position in the evolutionary
tree is still controversial. The kagu reveals
broad gray and white bands across its wings,
when they are spread in courtship displays.
The sunbittern likewise shows unexpected patches
of color on its upper wings and tail, which are
fanned horizontally to create a single, broad
pattern when it is displaying.

PHYLUM	CHORDATA
CLASS	AVES
ORDER	EURYPYGIFORMES
FAMILIES	2
SPECIES	2

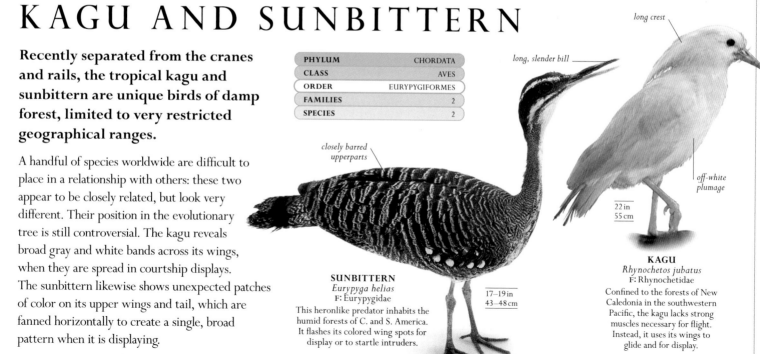

long crest

long, slender bill

closely barred
upperparts

off-white
plumage

22 in
55 cm

KAGU
Rhynochetos jubatus
F: Rhynochetidae
Confined to the forests of New
Caledonia in the southwestern
Pacific, the kagu lacks strong
muscles necessary for flight.
Instead, it uses its wings to
glide and for display.

17–19 in
43–48 cm

SUNBITTERN
Eurypyga helias
F: Eurypygidae
This heronlike predator inhabits the
humid forests of C. and S. America.
It flashes its colored wing spots for
display or to startle intruders.

neck feathers lack interlocking barbs, giving a loose, "hairy" appearance

juveniles lose their buff face feathers as they mature, leaving a patch of white cheek skin

red throat sac

⌃ CROWN AND COLOR

This species sports a bristly golden crown, a neat black forehead, and bare white cheeks, edged with a patch of red, which is broader in East African birds. Both males and females have a red throat sac that can be inflated with air, then rapidly deflated to give a booming call.

GRAY CROWNED CRANE
Balearica regulorum

Gray crowned cranes belong to the Gruidae family of birds, famous for their choreography; for these birds, dance is an important part of life. On the open savanna they display by jumping, flapping their wings, and bowing—sometimes, it seems, just to relieve aggression or reinforce pair-bonding—but mainly to court their mate by flaunting their elaborate head ornamentation. They lack the long, coiled windpipe of the longer-billed crane species, so instead of calling like a bugle they honk like a goose. They also boom during courtship by puffing air from an inflated red throat sac. During the breeding season pairs retire to wetter habitats, where thicker vegetation conceals their nest—a circular platform made from grasses and sedges. Here, the chicks are hidden from predators, allowing the parents to roost, uniquely for cranes, perched high in trees.

SIZE 3½ ft (1.1 m)
HABITAT Open country
DISTRIBUTION E. and S. Africa
DIET Grasses, seeds, invertebrates, small vertebrates

black head feathers give the appearance of a bulging forehead

nostril

bill is shorter and stouter than in other crane species

THIRD EYELID >
Also known as a nictitating membrane, from the Latin *nictare*, meaning "to blink," this translucent eyelid—the same as in other birds—moves across the eye to clean its surface.

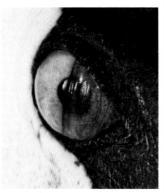

GOLDEN PLUMES >
When the wings are folded, long, golden feathers on the upper wings—just above the flight feathers—hang down over the side of the body.

∨ **RUFFLED NECK**
Long, tapered contour feathers give the crowned crane a shaggy appearance around the upper part of the body and the lower part of the neck. Most of the bird's plumage is gray.

FEET AND CLAWS >
Crowned cranes have long hind toes, setting them apart from other crane species. This enables them to perch in trees—perhaps a vestige of tree-dwelling ancestors.

∧ **LONG LEGS**
Although long legs help in dancing displays, and are good for wading, crowned cranes have shorter legs than those of other crane species.

white wing covert feathers

∧ **WING**
When crowned cranes fly overhead, their white underwing patches are clearly visible. Although strong-winged, these tropical birds—unlike other cranes—do not sustain long-distance migrations.

black primary flight feathers

brown secondary flight feathers

WADERS, GULLS, AND AUKS

Largely birds of coastal areas, waders are diverse in form and habit. Many are adapted for feeding in mud and water, with long legs and a probing bill.

The birds collectively known as waders, or Charadriiformes, fall into three main groups. Of these, two groups are made up of shorebirds. Plovers and allied species are mostly short-legged, short-billed birds that feed on small invertebrates found close to the surface of the ground. Some birds in this group, such as lapwings, favor drier habitats inland. Other species in the plover group are more adapted to wetland. Stilts and avocets sweep shallow water with their needlelike bill, while oystercatchers use their long, stout bill for opening mollusk shells. In the group consisting of sandpipers, snipe, and allies, the birds use their long bill to probe deep mud. Some of their relatives, such as shanks and curlews, have long legs, too, and so can wade into deeper water to feed.

OCEANIC BIRDS

The final group in this order is made up of gulls, terns, jaegers, and auks—web-footed birds that are the most ocean-going of the order. They may spend much of their lives at sea and some travel immense distances on migration. Gulls are opportunistic predators that are also seen foraging far inland. Auks, distributed around the Arctic, are specialized for diving for ocean-swimming prey. Their black-and-white coloration gives them a superficial resemblance to penguins, but they are not close relatives.

PHYLUM	CHORDATA
CLASS	AVES
ORDER	CHARADRIIFORMES
FAMILIES	19
SPECIES	383

EURASIAN STONE CURLEW
Burhinus oedicnemus
F: Burhinidae
Named after their shrill, curlewlike calls, the mostly nocturnal stone curlews are related to plovers. This widespread Eurasian species occurs in dry mudflats inland.

16–17½ in
40–44 cm

GREAT STONE CURLEW
Esacus recurvirostris
F: Burhinidae
Large *Esacus* stone curlews have a chisellike bill for hunting prey, such as crabs, near water. This is a S. Asian species.

19½–22 in
49–55 cm

SNOWY SHEATHBILL
Chionis albus
F: Chionidae
This is one of two white sheathbills, Antarctic cousins of plovers, which scavenge and consume carrion and the food and young of other birds.

13½–16 in
34–41 cm

8–9 in
20–22 cm

MAGELLANIC PLOVER
Pluvianellus socialis
F: Chionidae
This unusual S. American wader—the only one to regurgitate food for chicks—is more closely related to sheathbills than other plovers.

IBISBILL
Ibidorhyncha struthersii
F: Ibidorhynchidae
The only plover relative with a long, down-curved bill, this bird plucks invertebrates from stony riverbeds in mountains of C. Asia.

15–16 in
38–41 cm

dark brown to black body

long, red bill

16½–18½ in
42–47 cm

pink legs

AMERICAN BLACK OYSTERCATCHER
Haematopus bachmani
F: Haematopodidae
Black oystercatchers generally have more restricted ranges than pied (having two or more colors) species. This species is confined to the west coast of N. America.

EURASIAN OYSTERCATCHER
Haematopus ostralegus
F: Haematopodidae
The most widespread oystercatcher, this pied species breeds in N. Eurasia. Like others, it uses its long bill to open bivalve mollusks.

16–18 in
40–45 cm

CRAB PLOVER
Dromas ardeola
F: Dromadidae
The crab plover may be more closely related to gulls than to true plovers. Found around Indian Ocean coastlines, it uses its stout bill for eating crabs.

13–16 in
33–40 cm

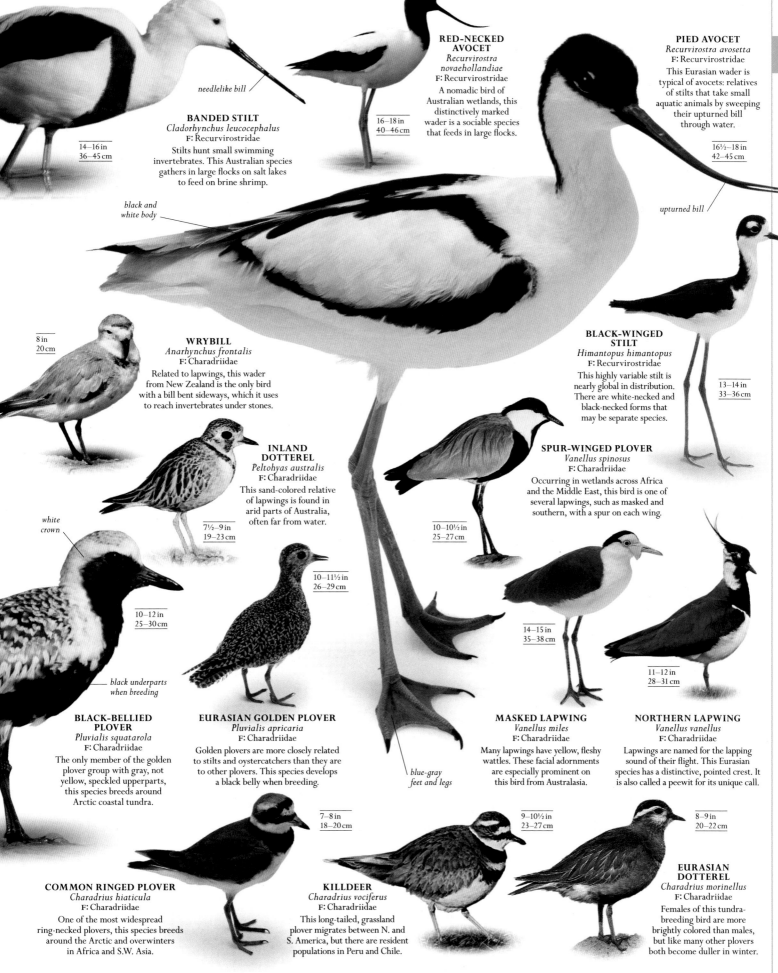

needlelike bill

BANDED STILT
Cladorhynchus leucocephalus
F: Recurvirostridae
Stilts hunt small swimming
invertebrates. This Australian species
gathers in large flocks on salt lakes
to feed on brine shrimp.

14–16 in
36–45 cm

**RED-NECKED
AVOCET**
*Recurvirostra
novaehollandiae*
F: Recurvirostridae
A nomadic bird of
Australian wetlands, this
distinctively marked
wader is a sociable species
that feeds in large flocks.

16–18 in
40–46 cm

PIED AVOCET
Recurvirostra avosetta
F: Recurvirostridae
This Eurasian wader is
typical of avocets: relatives
of stilts that take small
aquatic animals by sweeping
their upturned bill
through water.

16½–18 in
42–45 cm

upturned bill

*black and
white body*

WRYBILL
Anarhynchus frontalis
F: Charadriidae
Related to lapwings, this wader
from New Zealand is the only bird
with a bill bent sideways, which it uses
to reach invertebrates under stones.

8 in
20 cm

**BLACK-WINGED
STILT**
Himantopus himantopus
F: Recurvirostridae
This highly variable stilt is
nearly global in distribution.
There are white-necked and
black-necked forms that
may be separate species.

13–14 in
33–36 cm

**INLAND
DOTTEREL**
Peltohyas australis
F: Charadriidae
This sand-colored relative
of lapwings is found in
arid parts of Australia,
often far from water.

7½–9 in
19–23 cm

SPUR-WINGED PLOVER
Vanellus spinosus
F: Charadriidae
Occurring in wetlands across Africa
and the Middle East, this bird is one of
several lapwings, such as masked and
southern, with a spur on each wing.

10–10½ in
25–27 cm

*white
crown*

**BLACK-BELLIED
PLOVER**
Pluvialis squatarola
F: Charadriidae
The only member of the golden
plover group with gray, not
yellow, speckled upperparts,
this species breeds around
Arctic coastal tundra.

10–12 in
25–30 cm

*black underparts
when breeding*

EURASIAN GOLDEN PLOVER
Pluvialis apricaria
F: Charadriidae
Golden plovers are more closely related
to stilts and oystercatchers than they are
to other plovers. This species develops
a black belly when breeding.

10–11½ in
26–29 cm

MASKED LAPWING
Vanellus miles
F: Charadriidae
Many lapwings have yellow, fleshy
wattles. These facial adornments
are especially prominent on
this bird from Australasia.

14–15 in
35–38 cm

NORTHERN LAPWING
Vanellus vanellus
F: Charadriidae
Lapwings are named for the lapping
sound of their flight. This Eurasian
species has a distinctive, pointed crest. It
is also called a peewit for its unique call.

11–12 in
28–31 cm

*blue-gray
feet and legs*

COMMON RINGED PLOVER
Charadrius hiaticula
F: Charadriidae
One of the most widespread
ring-necked plovers, this species breeds
around the Arctic and overwinters
in Africa and S.W. Asia.

7–8 in
18–20 cm

KILLDEER
Charadrius vociferus
F: Charadriidae
This long-tailed, grassland
plover migrates between N. and
S. America, but there are resident
populations in Peru and Chile.

9–10½ in
23–27 cm

**EURASIAN
DOTTEREL**
Charadrius morinellus
F: Charadriidae
Females of this tundra-
breeding bird are more
brightly colored than males,
but like many other plovers
both become duller in winter.

8–9 in
20–22 cm

»

11–12 in
28–31 cm

9–12 in
23–31 cm

long tail when breeding

8–10½ in
20–27 cm

golden neck patch

12–23 in
31–58 cm

6½–9 in
17–23 cm

AFRICAN JACANA
Actophilornis africanus
F: Jacanidae
Typical of the jacana family, this African wetland bird has enlarged feet for walking on floating vegetation.

BRONZE-WINGED JACANA
Metopidius indicus
F: Jacanidae
This species is widespread across India and S.E. Asia. Males have a flattened "forearm" bone that helps them lift chicks with their wings.

♀

6–7½ in
15–19 cm

COMB-CRESTED JACANA
Irediparra gallinacea
F: Jacanidae
Found from Asia to Australia, this species is named for its fleshy head wattle. As with other jacanas, males incubate the eggs and care for the young.

9–10 in
23–25 cm

PHEASANT-TAILED JACANA
Hydrophasianus chirurgus
F: Jacanidae
This southern Asian spur-winged jacana is the only species that changes plumage—losing its long tail—outside the breeding season.

WATTLED JACANA
Jacana jacana
F: Jacanidae
The dominant female of this S. American species breeds with many males, perhaps to compensate for crocodile predation of eggs.

PLAINS WANDERER
Pedionomus torquatus
F: Pedionomidae
This quaillike bird of Australian grassland— the only member of its family—is actually related to the wetland jacanas.

9–10 in
23–26 cm

GREATER PAINTED-SNIPE
Rostratula benghalensis
F: Rostratulidae
Related to the jacanas, the short-legged painted-snipes share their female-dominant mating system. This bird is from Old World tropical wetlands.

SHORT-BILLED DOWITCHER
Limnodromus griseus
F: Scolopacidae
Dowitchers are related to snipes and are reddish colored in their breeding plumage. The short-billed dowitcher is confined to the Americas.

9–10 in
23–25 cm

RED KNOT
Calidris canutus
F: Scolopacidae
Like many migratory waders, the Arctic-breeding red knot molts its distinctive summer plumage to more drab colors in winter.

6½–8 in
16–20 cm

DUNLIN
Calidris alpina
F: Scolopacidae
Globally common, the dunlin is typical of waders that breed around the Arctic and flock to the warmer south in the winter.

6½–7½ in
17–19 cm

JACK SNIPE
Lymnocryptes minimus
F: Scolopacidae
Birds of the snipe-woodcock group have a long bill, short legs, and are camouflaged by their plumage. This Old World bird is the smallest in this group.

10–10½ in
25–27 cm

COMMON SNIPE
Gallinago gallinago
F: Scolopacidae
Typical of waders, chicks of this globally widespread species are active soon after hatching. Unlike other waders, snipes feed their chicks.

long, slightly upturned bill

rufous (reddish) underparts during breeding

8–8½ in
20–21 cm

SANDERLING
Calidris alba
F: Scolopacidae
This sandpiper breeds within the Arctic Circle, farther north than other waders. In winter, it flocks on sandy shores farther south.

black and white upperparts

14½–16½ in
37–42 cm

HUDSONIAN GODWIT
Limosa haemastica
F: Scolopacidae
This is one of two American godwits, and is named after its breeding habitat, which includes the shores of Hudson Bay.

16–17½ in
40–44 cm

BLACK-TAILED GODWIT
Limosa limosa
F: Scolopacidae
Typical of godwits, this Old World species has a slightly upturned bill. It ranges far inland to forage on grassland, heath, and pasture.

7–7½ in
18–19 cm

♀

RED-NECKED PHALAROPE
Phalaropus lobatus
F: Scolopacidae
This species has a breeding range that includes much of the Arctic region. Females are brightly colored and attract males with displays during breeding; males care for offspring.

LONG-BILLED CURLEW
Numenius americanus
F: Scolopacidae

18–26 in
45–66 cm

A typical curlew, this American species is a large wader with a long, down-curved bill for probing deeply in mud for invertebrate prey.

WHIMBREL
Numenius phaeopus
F: Scolopacidae

16–16½ in
40–42 cm

A mid-sized curlew, the whimbrel has a distinctive trilling call and breeds in circumpolar regions. Some individuals winter as far away as Australia.

BUFF-BREASTED SANDPIPER
Tryngites subruficollis
F: Scolopacidae

7–8 in
18–20 cm

This wader breeds in the tundra of N. America and extreme E. Siberia. It winters in the grasslands of S. America.

large neck ruff in breeding male

COMMON REDSHANK
Tringa totanus
F: Scolopacidae

10½–11½ in
27–29 cm

Most waders breed near fresh water, but the Old World common redshank—named after its colored legs—may sometimes breed on salt marshes.

LESSER YELLOWLEGS
Tringa flavipes
F: Scolopacidae

9–10 in
23–25 cm

Yellowlegs belong to the shank group. This species breeds in forests in Alaska and Canada, and overwinters in the Caribbean.

AMERICAN WOODCOCK
Scolopax minor
F: Scolopacidae

10–11 in
26–28 cm

Like other woodcocks and snipes, this American bird is superbly camouflaged and its eyes are placed high, enabling all-around vision for predators.

WANDERING TATTLER
Tringa incana
F: Scolopacidae

10–12 in
26–30 cm

This shank species breeds in Alaska, during which its breast is barred. It winters farther south along the Pacific coastline of America.

RUFF
Calidris pugnax
F: Scolopacidae

8–12 in
20–30 cm

Males of this Old World marsh and meadow species transform during breeding: they lose their gray winter plumage and take on a rufous (reddish brown) and black coloration, with a showy neck ruff.

black-spotted chest and belly in summer

reddish legs

SPOTTED SANDPIPER
Actitis macularius
F: Scolopacidae

7–8 in
18–20 cm

Like the related common sandpiper (*A. hypoleucos*) from Eurasia, this American sandpiper has a short bill for pecking food from dry ground.

RUDDY TURNSTONE
Arenaria interpres
F: Scolopacidae

9–9½ in
22–24 cm

This bird, which breeds across the northern hemisphere, is named for the way it flips objects to search for prey.

SPOON-BILLED SANDPIPER
Calidris pygmaea
F: Scolopacidae

5½–6½ in
14–16 cm

Like spoonbills, this wader from E. Asia uses its distinctive spoonlike bill to sweep shallow water for invertebrate prey.

WHITE-BELLIED SEEDSNIPE
Attagis malouinus
F: Thinocoridae

10½–11½ in
27–29 cm

Four species of short-billed, herbivorous seedsnipe live in the open habitats of S. America. This species is confined to the southern tip.

>>

11–13 in
28–33 cm

WHITE TERN
Gygis alba
F: Laridae

This small, all-white tern from the islands of the tropical Atlantic and Indian oceans is notable for laying its eggs on bare branches in trees.

16–16½ in
40–42 cm

white cheek stripe

INCA TERN
Larosterna inca
F: Laridae

A distinctively patterned bird, the Inca tern is found in Peru and Chile, where it breeds on rocky coastlines.

bright red legs

9–9½ in
22–24 cm

BLACK TERN
Chlidonias niger
F: Laridae

This small freshwater tern breeds in the wetlands of the northern hemisphere and winters in S. America and Africa.

LESSER CRESTED TERN
Thalasseus bengalensis
F: Laridae

In this relative of the greater crested tern, the bill changes from yellow to orange during breeding.

14–14½ in
35–37 cm

GREATER CRESTED TERN
Thalasseus bergii
F: Laridae

This Old World tern belongs to a group of related crested terns, characterized by a black nape tuft.

18–19½ in
46–49 cm

black cap

18½–21½ in
47–54 cm

9–9½ in
22–24 cm

LITTLE TERN
Sternula albifrons
F: Laridae

The Old World little tern belongs to a group of coastal species with a white "blaze" above the eye.

black markings on tips of underwings

CASPIAN TERN
Hydroprogne caspia
F: Laridae

The largest tern species, this bird is found in most continents. It is a colonial ground-nester, like most sea terns.

13–14 in
33–36 cm

SAUNDERS'S TERN
Sternula saundersi
F: Laridae

A small tern from the Red Sea and Indian Ocean, Saunders's tern used to be considered a race of the little tern.

9–9½ in
23–24 cm

SOOTY TERN
Onychoprion fuscatus
F: Laridae

This white-blazed sea tern breeds on tropical islands. It is also known as the wideawake tern because of its noisy colonies.

long, black legs

13–15 in
33–38 cm

12½–13½ in
32–34 cm

12–12½ in
30–32 cm

BRIDLED TERN
Onychoprion anaethetus
F: Laridae

This tern of tropical and subtropical regions has a white eye-blaze, similar to little and sooty terns. It spends much of its time at sea.

ROSEATE TERN
Sterna dougallii
F: Laridae

As with related sea terns, the roseate tern's black cap fades in winter. This bird is found mostly in the southern hemisphere and is migratory.

WHITE-CHEEKED TERN
Sterna repressa
F: Laridae

This species from the Red Sea and Indian Ocean can be identified by its plumage, which is darker than that of other gray terns.

12–12½ in
30–32 cm

BLACK-NAPED TERN
Sterna sumatrana
F: Laridae

Found on the Indian and Pacific oceans, the black-naped tern nests in small colonies, usually apart from other terns.

13–14 in
33–35 cm

ARCTIC TERN
Sterna paradisaea
F: Laridae

This tern migrates the greatest distance of any animal: from its Arctic breeding grounds to Antarctica. It feeds on fishes and crustaceans.

16–20 in
40–50 cm

BLACK SKIMMER
Rynchops niger
F: Laridae

Skimmers are the only birds with projecting lower bill mandibles, used to skim water for fishes. This species is found in N. and S. America.

16–18 in
40–45 cm

BROWN NODDY
Anous stolidus
F: Laridae

Noddies are dark or white tropical members of the tern family. The largest species, the brown noddy is widespread around the world.

stout, hooked bill

20½–21 in
52–54 cm

brownish-gray body

SOUTH POLAR SKUA
Stercorarius maccormicki
F: Stercorariidae

This large bird, with a reputation for attacking other seabirds, is one of the few waders to breed on the Antarctic coast.

18–20 in
46–51 cm

POMARINE JAEGER
Stercorarius pomarinus
F: Stercorariidae

Jaegers (also called skuas in Europe) are aggressive gull-like birds. This Arctic species will kill and eat other seabirds and even attack humans that threaten its nest.

PARASITIC JAEGER
Stercorarius parasiticus
F: Stercorariidae

This is the most common Arctic species of jaeger. Like most of its relatives, it mobs other seabirds to steal their prey.

19–21 in
48–53 cm

16–18 in
41–46 cm

LONG-TAILED JAEGER
Stercorarius longicaudus
F: Stercorariidae

Like other species, this smallest member of the jaeger family is migratory. It breeds in circumpolar regions and winters farther south.

9½–10 in
24–25 cm

MARBLED MURRELET
Brachyramphus marmoratus
F: Alcidae

This small American auk nests in trees in coniferous woods. Fledged chicks leave the nest at night and head for the sea.

9½–10½ in
24–27 cm

6½–7½ in
17–19 cm

DOVEKIE
Alle alle
F: Alcidae

This tiny auk breeds on Arctic islands and winters at sea farther south. It feeds on tiny fishes and crustaceans.

14½–15½ in
37–39 cm

RAZORBILL
Alca torda
F: Alcidae

The North Atlantic razorbill has a flattened, white-barred bill. As with other auks, its pointed eggs cannot roll off cliff-top breeding sites.

CRESTED AUKLET
Aethia cristatella
F: Alcidae

Like other auklets, this North Pacific species feeds on planktonic crustaceans. Pairs anoint themselves with secretions from their backs during courtship.

dark brown to black head

12–14 in
30–36 cm

PIGEON GUILLEMOT
Cepphus columba
F: Alcidae

This North Pacific auk, fully adapted to cold conditions, cannot migrate south through warmer waters, just as southern hemisphere penguins cannot move northward.

white wing patch

red feet

12–12½ in
30–32 cm

BLACK GUILLEMOT
Cepphus grylle
F: Alcidae

This species of northern N. American and Eurasian coasts breeds in sparser colonies than other auks and winters mainly inshore.

11–11½ in
28–29 cm

RHINOCEROS AUKLET
Cerorhinca monocerata
F: Alcidae

This North Pacific relative of puffins shares their burrow-nesting habits, and is named for the hornlike bill projection in breeding adults.

10–11½ in
26–29 cm

ATLANTIC PUFFIN
Fratercula arctica
F: Alcidae

A small, North Atlantic member of the auk family, this species—like other puffins—nests in colonies, in burrows, usually below turf.

13½–14 in
34–36 cm

TUFTED PUFFIN
Fratercula cirrhata
F: Alcidae

Like its Atlantic relative, this larger Pacific puffin catches small fishes and can hold many at a time, crosswise, in its bill.

15–16 in
38–41 cm

COMMON MURRE
Uria aalge
F: Alcidae

A typical diving member of the auk family, this bird breeds on North Atlantic and Pacific coasts, and winters at sea.

SANDGROUSE

Sand-colored plumage camouflages these birds in their desert habitats, where they are well adapted for living in an extremely dry environment.

With their rounded body and short legs, sandgrouse might be mistaken for partridges until they take flight, darting quickly and acrobatically in the air. These birds, the Pterocliformes, are found in arid areas of Asia, Africa, Madagascar, and southern Europe. They are not related to the subarctic grouse species but are more closely related to pigeons.

Sandgrouse have long pointed wings and all species have cryptic plumage—mottled on the back, but often boldly marked with brown or white stripes or blotches on the head and underparts. They are

sociable birds that gather in flocks to make early morning, or sometimes evening, forays to drinking holes, often traveling considerable distances. They feed exclusively on seeds.

WATER CARRIERS

Sandgrouse breed during the rainy season to take advantage of seed harvests. Their nest is nothing more than a shallow depression on the ground. Both parents incubate and care for their young. Remarkably, the males supply water to their chicks, which are normally raised far from water sources. The male's belly feathers soak up and retain moisture when the bird visits a drinking hole. Back at the nest, the chicks drink from the parent's soaked feathers.

PHYLUM	CHORDATA
CLASS	AVES
ORDER	PTEROCLIFORMES
FAMILIES	1
SPECIES	16

Namaqua sandgrouse drink at a watering hole. Like all sandgrouse, they gather in large flocks to confuse predators.

PALLAS'S SANDGROUSE
Syrrhaptes paradoxus
F: Pteroclidae
One of two C. Asian sandgrouses, this large species has feathered toes, a long tail, and a long flight feather on each wing.

12–16 in
30–41 cm

barred buff plumage

long, pointed tail

CHESTNUT-BELLIED SANDGROUSE
Pterocles exustus
F: Pteroclidae
With a range that extends from Senegal to Kenya, and east as far as India, this bird of open desert gathers in huge flocks.

12–13 in
31–33 cm

10–11 in
25–28 cm

black and white bands on forecrown (male only)

DOUBLE-BANDED SANDGROUSE
Pterocles bicinctus
F: Pteroclidae
A bird of southern African savanna and open woodland, this is one of several sandgrouse species in which the males have distinctive belly stripes.

white-barred wings

black breast band

10½–12 in
27–30 cm

CROWNED SANDGROUSE
Pterocles coronatus
F: Pteroclidae
This yellow-throated sandgrouse is found from the stony desert of the Sahara to Pakistan. It can tolerate high temperatures and even brackish water.

LICHTENSTEIN'S SANDGROUSE
Pterocles lichtensteinii
F: Pteroclidae
This small bird is less gregarious than other species of sandgrouse. It inhabits bushy scrubland, and semidesert regions from N. and E. Africa to Pakistan.

9½–10 in
24–26 cm

PIGEONS AND DOVES

This is a highly successful group of seed-eating and fruit-eating birds that are distributed almost worldwide outside the coldest regions.

After the parrots, pigeons and doves form the largest group of vegetarian tree-perching birds: the order Columbiformes. But, whereas parrots have a heavy, hooked bill for cracking large nuts, pigeons have a less robust bill for feeding on smaller seeds and grain. Some tropical groups, such as the Indo-Pacific fruit doves, are specialized fruit-eaters of the rainforest canopy.

Most birds in this order are short-legged and a few spend almost all of their time on the ground. Pigeons and doves are unusual among birds in being able to suck up water when drinking without tilting back their head, because of a pumping action in their food-pipe. This allows them to drink continuously, which is a particular benefit in arid environments. They can store food in their bill, and they feed their chicks on a secretion from the bill. This secretion is similar to mammalian milk.

THREATS AND EXTINCTIONS

The success of many pigeon species is due to high rates of reproduction. However, some species are threatened by humans and others have already been lost. The 17th-century demise of the dodo, a vulnerable flightless species, was the result of human impact. And the passenger pigeon, once one of North America's commonest birds, was hunted to extinction in the 1900s.

PHYLUM	CHORDATA
CLASS	AVES
ORDER	COLUMBIFORMES
FAMILIES	1
SPECIES	344

DEBATE
A MODIFIED PIGEON?

Mid-19th-century scientists placed the extinct flightless dodo from Mauritius in the pigeon order, and recent analysis indicates that it was related to the Nicobar pigeon. So the dodo was really a modified pigeon, with Indo-Pacific ancestors that might already have been flightless.

EUROPEAN TURTLE DOVE
Streptopelia turtur
F: Columbidae
With a range from Africa to Eurasia, turtle doves are small, slim relatives of *Columba* pigeons. Many, such as this western species, have distinctive neck markings.

black and white striped neck patch

10–11 in
26–28 cm

10–10½ in
25–27 cm

red patch around eyes

white-speckled wings

LAUGHING DOVE
Streptopelia senegalensis
F: Columbidae
Common in villages and oases in Africa and southern Asia, this turtle dove is named after its distinctive chuckling song.

15–17 in
38–43 cm

BROWN CUCKOO-DOVE
Macropygia amboinensis
F: Columbidae
Long-tailed *Macropygia* doves are cuckoolike birds of Indo-Pacific rainforests. Like others, this species from the Moluccas (Indonesia), New Guinea, and Australia has many races.

NAMAQUA DOVE
Oena capensis
F: Columbidae
This long-tailed, ground-feeding species belongs to a group of small African wood doves. Its range extends into Madagascar and Saudi Arabia.

10–11 in
26–28 cm

13–15 in
33–38 cm

SPECKLED PIGEON
Columba guinea
F: Columbidae
This large African pigeon is common in open country south of the Sahara, and it often gathers around towns and villages.

12½ in
32 cm

15–17 in
38–43 cm

12–14 in
31–35 cm

12–14 in
31–35 cm

PINK PIGEON
Nesoenas mayeri
F: Columbidae
Perhaps related to turtle doves, this rare bird is restricted to Mauritius. Its once threatened population recovered due to a successful captive-breeding program.

WOODPIGEON
Columba palumbus
F: Columbidae
This large W. Eurasian pigeon is commonly found in woods and farmland, and often enters parks and gardens.

ROCK PIGEON
Columba livia
F: Columbidae
True wild pigeons inhabit cliffs and hedges. In its natural environment, this species lives in mountainous regions of Europe and Asia.

DOMESTIC PIGEON
Columba livia
F: Columbidae
Domesticated and feral descendants of the rock pigeon are found in urban areas worldwide. These birds have variable plumage patterns.

WOMPOO FRUIT DOVE
Ptilinopus magnificus
F: Columbidae
Fruit doves are multicolored relatives of imperial pigeons (*Ducula* sp.). This large species from New Guinea and Australia lives in the rainforest canopy.

yellow markings on wings

11½–22 in
29–55 cm

13–16 in
33–40 cm

6½–9 in
17–23 cm

INCA DOVE
Columbina inca
F: Columbidae
From arid regions of the S. US and C. America, this species belongs to a group of mostly drab-colored tropical American ground-doves.

8 in
20 cm

NICOBAR PIGEON
Caloenas nicobarica
F: Columbidae
Possibly related to the extinct Mauritian dodo, this pigeon inhabits coastal regions and island forests from Malaysia to New Guinea.

DIAMOND DOVE
Geopelia cuneata
F: Columbidae
A small, nomadic bird from the dry interiors of Australia, this dove occurs in groups, flocking in large numbers at watering holes.

10–12 in
25–31 cm

WHITE-TIPPED DOVE
Leptotila verreauxi
F: Columbidae
A tropical relative of the mourning dove, this species is widespread in C. and S. America, and reaches as far north as Texas.

dark green tail

12 in
30 cm

14 in
35 cm

9–13½ in
23–34 cm

dark pink head and breast

MINDANAO BLEEDING-HEART PIGEON
Gallicolumba criniger
F: Columbidae
The five species of Philippine bleeding-heart pigeons are named for their blood-red breast patch. This bird inhabits the southern islands of the archipelago.

SULAWESI GROUND DOVE
Gallicolumba tristigmata
F: Columbidae
From the forests of Sulawesi, Indonesia, this terrestrial bird is related to Philippine bleeding-heart pigeons, and perhaps to various doves of arid Australia.

MOURNING DOVE
Zenaida macroura
F: Columbidae
So called because of its mournful call, this long-tailed dove occurs in open country throughout N. and C. America, and the Caribbean.

emerald-green wings and back

18 in
45 cm

15½–17½ in
39–44 cm

PIED IMPERIAL PIGEON
Ducula bicolor
F: Columbidae
Imperial pigeons are large, fruit-eating rainforest birds. This species from S.E. Asia and Australasia has a white plumage that is often stained due to its diet.

16–18 in
40–46 cm

9–11 in
23–28 cm

EMERALD DOVE
Chalcophaps indica
F: Columbidae
An iridescent green bird, this ground-feeding species occurs in rainforests from India to islands of the southwest Pacific. It has a diet of fruits and seeds.

TOPKNOT PIGEON
Lopholaimus antarcticus
F: Columbidae
A large, hawklike pigeon from E. Australia, this species is named for the double crest of feathers on its forecrown and crown.

GREEN IMPERIAL PIGEON
Ducula aenea
F: Columbidae
Found in rainforest canopy, from India to S.E. Asia, this large bird has a deep, booming call and has a diet consisting mainly of fruits.

VICTORIA CROWNED PIGEON
Goura victoria
F: Columbidae
Crowned pigeons are the largest of the pigeons. This bird from N. New Guinea is distinguished from the southern crowned pigeon by its white-tipped crest.

29–30 in
74–75 cm

side-flattened tail

14–15 in
36–38 cm

WONGA PIGEON
Leucosarcia melanoleuca
F: Columbidae
Confined to eastern Australia, this distinctively patterned pigeon inhabits woodland and scrub from S. Queensland to Victoria.

PHEASANT PIGEON
Otidiphaps nobilis
F: Columbidae
Recent research suggests this New Guinean ground pigeon belongs to a group that includes crowned pigeons and perhaps even the extinct dodo.

18–20 in
45–50 cm

fanlike crown

SOUTHERN CROWNED PIGEON
Goura scheepmakeri
F: Columbidae
This species inhabits the forests of the southern part of New Guinea. It has bluish-gray upperparts, a maroon chest, and a lacy crest.

bluish-gray plumage

30 in
75 cm

maroon chest

KEY WEST QUAIL-DOVE
Geotrygon chrysia
F: Columbidae
Quail-doves are birds of tropical American forests. This iridescent species occurs in the Caribbean, including the Bahamas.

10½–12 in
27–31 cm

10–11 in
25–28 cm

AFRICAN GREEN PIGEON
Treron calvus
F: Columbidae
More than 20 species of *Treron* pigeons occur in the tropical regions of Africa and Asia. This species is widespread south of the Sahara.

13–14 in
33–36 cm

COMMON BRONZEWING
Phaps chalcoptera
F: Columbidae
Bronzewings are fast-flying Australian pigeons that feed on the ground. They have iridescent wing patches, which are particularly extensive in this widespread woodland species.

grayish body

white wing patch

8–9 in
20–22 cm

SPINIFEX PIGEON
Geophaps plumifera
F: Columbidae
This Australian bronzewing pigeon is found in arid, rocky habitats with an abundance of spinifex grass, in which it nests.

iridescent wing patch

12–14 in
31–35 cm

CRESTED PIGEON
Ocyphaps lophotes
F: Columbidae
One of several species of Australian bronzewing pigeons, this species is widespread in open country throughout the continent.

PARROTS AND COCKATOOS

Most parrots are birds of tropical forests, although a few favor open habitats. This order, the Psittaciformes, is diverse and often colorful.

The most recognizable feature of a parrot is its decurved bill. Both mandibles of the bill articulate with its skull: the upper mandible can move up, just as the lower can move down. This allows the bird not only to crack open hard seeds and nuts, but also to use its bill for gripping when clambering through trees. The legs are strong and the feet, with two forward-facing and two backward-facing toes, can grasp and manipulate foodstuffs. Bright colors are common, often in both sexes, with green plumage predominating. This color is the result of feather structure, which scatters light on yellow pigment. The Australasian cockatoos—parrots with a distinctive erectile crest—lack this feather texture and so lack green and blue colors.

Australasia's high parrot diversity, which includes the brush-tongued, nectar-feeding lorikeets, suggests that the group originated in this part of the world. The most primitive parrots, the kea and the nocturnal, flightless kakapo, are both found in New Zealand.

SOCIAL SPECIES
Parrots are sociable, often living in large flocks, and almost all species form strong pair-bonds. Their engaging personalities make them popular as pets, but the international cagebird trade has driven many species to the edge of extinction.

PHYLUM	CHORDATA
CLASS	AVES
ORDER	PSITTACIFORMES
FAMILIES	4
SPECIES	398

Green-winged macaws socialize at a clay-lick, which provides sodium in an area where the mineral is very limited.

barred green plumage

23½ in
60 cm

14 in
36 cm

KEA
Nestor notabilis
F: Psittacidae
An opportunistic alpine omnivore, the kea eats carrion and live petrel chicks. It belongs to an ancient group of parrots in New Zealand that includes the kakapo.

19 in
48 cm

short tail

KAKAPO
Strigops habroptila
F: Psittacidae
The only flightless parrot, this large, nocturnal bird occurs in the small islands off New Zealand. Males make booming calls to attract females.

GALAH
Eolophus roseicapilla
F: Psittacidae
The only cockatoo with a deep pink neck and underparts, this small relative of the white cockatoos is widespread throughout Australia in areas with scattered trees.

5–6 in
13–15 cm

♀

4¾–6 in
12–15 cm

♂

19½ in
49 cm

20–24 in
50–61 cm

VERNAL HANGING PARROT
Loriculus vernalis
F: Psittacidae
Hanging parrots are Indo-Pacific birds, even though recent genetic evidence suggests affinities with African lovebirds. This species is found from India to Thailand.

BLUE-CROWNED HANGING PARROT
Loriculus galgulus
F: Psittacidae
Hanging parrots, unusually among birds, sleep upside down. Like other such parrots, this small forest bird from S.E. Asia has a short tail and mainly green plumage.

SULPHUR-CRESTED COCKATOO
Cacatua galerita
F: Psittacidae
White cockatoos are screeching parrots with a range that extends from Indonesia to the Pacific Rim. This species occurs in New Guinea and Australia.

RED-TAILED BLACK COCKATOO
Calyptorhynchus banksii
F: Psittacidae
One of several Australian black cockatoos with colored tail panels, this glossy species has a typical wailing call and ponderous wing beats.

COCKATIEL
Nymphicus hollandicus
F: Psittacidae
A bird of Australia's dry interior, this species resembles a parakeet, but its DNA shows it is a miniature cockatoo.

♂ ♀

12½ in
32 cm

CHATTERING LORY
Lorius garrulus
F: Psittacidae
One of a group of green-winged red lories from New Guinea and surrounding islands, this species is from the Moluccas.

12 in
30 cm

10 in
25 cm

DUSKY LORY
Pseudeos fuscata
F: Psittacidae
The patchy brown coloration of this species is unique among lories. This bird occurs in New Guinea and nearby islands.

12 in
30 cm

BLACK-WINGED LORY
Eos cyanogenia
F: Psittacidae
The *Eos* lories are vividly red and violet Indonesian birds. This is the easternmost species, restricted to the area around Geelvink Bay, New Guinea.

9½ in
24 cm

OLIVE-HEADED LORIKEET
Trichoglossus euteles
F: Psittacidae
Found on the island of Timor, this long-tailed relative of the rainbow lorikeet has bright green plumage, typical of many parrots.

7 in
18 cm

VARIED LORIKEET
Psitteuteles versicolor
F: Psittacidae
A small lorikeet from northern Australian woodlands, this species nests in eucalyptus tree cavities, like many other parrots in its habitat.

bright scarlet head

♀

blue band across back and chest

♂

RAINBOW LORIKEET
Trichoglossus haematodus
F: Psittacidae
A variable species, this bird is found in many habitats with nectar-giving blossom. Its range includes most of Australasia and the islands of the southwest Pacific.

10–12 in
25–30 cm

pale green streak on green back

17 in
43 cm

purple rump

AUSTRALIAN KING PARROT
Alisterus scapularis
F: Psittacidae
King parrots are rainforest birds from tropical Australasia and represent an evolutionary link to Asian parakeets. This species occurs in E. Australia.

mainly green body

13–15½ in
33–39 cm

ECLECTUS PARROT
Eclectus roratus
F: Psittacidae
Males and females of this parrot are so different they were originally classified as separate species. They are found in tropical Australasian rainforests.

7 in
18 cm

BUDGERIGAR
Melopsittacus undulatus
F: Psittacidae
A nomadic Australian bird, this small parrot from arid regions flocks to watering holes. Despite being a seed-eater, it is related to nectar-feeding lorikeets.

10½ in
27 cm

RED-FRONTED PARAKEET
Cyanoramphus novaezelandiae
F: Psittacidae
Small parakeets with colored foreheads have diversified in the southwest Pacific, and include the only parakeets—like this species—from New Zealand.

13½–15 in
34–38 cm

AUSTRALIAN RINGNECK
Barnardius zonarius
F: Psittacidae
Named for its yellow neck ring, this Australian bird has black- and green-headed forms. It is widespread in woodland.

black mask

18½ in
47 cm

yellow belly

12 in
30 cm

14 in
36 cm

♂

8 in
20 cm

♀

EASTERN ROSELLA
Platycercus eximius
F: Psittacidae
Rosellas are broad-tailed parrots
with a range from Australia to
nearby Pacific islands. A typical
rosella, this variable E. Australian
bird has white cheeks.

TURQUOISE PARROT
Neophema pulchella
F: Psittacidae
A member of a group of small,
green grass parrots from
Australia, the turquoise parrot
is an open woodland bird from
the southeast of the continent.

RED-FAN PARROT
Deroptyus accipitrinus
F: Psittacidae
Possibly related more closely
to macaws than to other
short-tailed parrots, this
S. American parrot raises a
red neck ruff when excited.

**ROSE-RINGED
PARAKEET**
Psittacula krameri
F: Psittacidae
This is the most widespread Asian
parakeet that reaches westward
into N. Africa. This species has also
been introduced in Europe.

MASKED SHINING PARROT
Prosopeia personata
F: Psittacidae
One of the three species of
shining parrots restricted to Fiji,
this parrot is declining in numbers
due to loss of its native forest.

blue plumage
and upperparts

15–16½ in
38–42 cm

16–18½ in
40–47 cm

♀

♂

gray-black
undertail

white face

13 in
33 cm

16 in
40 cm

SUPERB PARROT
Polytelis swainsonii
F: Psittacidae
This S.E. Australian bird
belongs to a group of long-tailed
Australian parrots and breeds in
red gum forests. It is rapidly
declining in number.

PRINCESS PARROT
Polytelis alexandrae
F: Psittacidae
A nomad from central
Australia, this bird follows
watercourses and gathers
near spinifex grass. It often
breeds in small colonies
in eucalyptus trees.

yellow
undertail

pale gray
underparts

6 in
15 cm

6½–7 in
17–18 cm

14–14½ in
35–37 cm

red tail

GRAY PARROT
Psittacus erithacus
F: Psittacidae
A native of African rainforests,
this highly intelligent species is an
accomplished mimic. This has led to
its exploitation in the bird trade.

**YELLOW-COLLARED
LOVEBIRD**
Agapornis personatus
F: Psittacidae
Lovebirds are small African parrots
that live in small flocks. Unlike
most parrots, this Tanzanian
species builds a nest—a domed
structure in a tree cavity.

**ROSY-FACED
LOVEBIRD**
Agapornis roseicollis
F: Psittacidae
This social lovebird from dry
woodland and semidesert
habitats of S.W. Africa
habitually gathers near
watering holes.

BROWN-NECKED PARROT
Poicephalus robustus
F: Psittacidae
The largest of a group of mainly green-gray
parrots from Africa, this bird occurs in forests
from Gambia to the Cape of Good Hope.

white facial patch with lines of black feathers

34 in
85 cm

BLUE-AND-YELLOW MACAW
Ara ararauna
F: Psittacidae

Macaws are large, long-tailed parrots with sparsely feathered facial patches. One of two blue and yellow species, this bird comes from northern S. America.

powerful bill

15 in
38 cm

16 in
40 cm

BLUE-FRONTED PARROT
Amazona aestiva
F: Psittacidae

A member of a large group of mostly green parrots, this species occurs in open forest habitats of central-eastern S. America.

ST. VINCENT PARROT
Amazona guildingii
F: Psittacidae

Several Caribbean *Amazona* parrots are faced with extinction. This large species is being conserved by an ongoing breeding program in its native island of St. Vincent.

22–23½ in
55–60 cm

RED-FRONTED MACAW
Ara rubrogenys
F: Psittacidae

This small macaw is restricted to the arid scrub of central Bolivia. Its small population is threatened by habitat loss and wildlife trafficking.

31–35 in
79–89 cm

bright yellow feathers on blue wings

SCARLET MACAW
Ara macao
F: Psittacidae

Like other macaws, this noisy, flocking bird has a heavy bill for breaking nuts and palm fruit. Its range extends from S. Mexico to C. Brazil.

long, red tail

4¾–5½ in
12–14 cm

9½–11 in
24–28 cm

BLUE-HEADED PARROT
Pionus menstruus
F: Psittacidae

A relative of the *Amazona* parrots, this small parrot is common in lowland forests from Costa Rica to Bolivia.

PACIFIC PARROTLET
Forpus coelestis
F: Psittacidae

Parrotlets are tiny green American parrots; only New Guinean pygmy parrots are smaller. This species occurs in western Ecuador and Peru.

3¼ ft
1 m

HYACINTH MACAW
Anodorhynchus hyacinthinus
F: Psittacidae

Unlike other macaws, this giant Brazilian parrot has its facial patch reduced to an eye-ring—similar to that of the related *Aratinga* parakeets.

8–10 in
20–25 cm

12 in
30 cm

white eye-ring

JANDAYA PARAKEET
Aratinga jandaya
F: Psittacidae

Aratinga is a genus of predominantly green "mini-macaws" found in northeastern Brazil. Some have extensive splashes of golden coloration.

10 in
25 cm

17½–18 in
44–46 cm

BURROWING PARAKEET
Cyanoliseus patagonus
F: Psittacidae

A Patagonian relative of macaws, this colonial breeder nests in tunnels in earth banks. It forms faithful pair-bonds, unlike most birds.

YELLOW-CHEVRONED PARAKEET
Brotogeris chiriri
F: Psittacidae

Native to central S. America, escaped cagebirds of this species have established feral populations in warmer parts of the US.

11½ in
29 cm

MONK PARAKEET
Myiopsitta monachus
F: Psittacidae

This colonial breeder from temperate S. America is the only parrot to build a stick nest, often a large communal structure.

red belly

MAROON-BELLIED PARAKEET
Pyrrhura frontalis
F: Psittacidae

Relatives of *Aratinga*, most *Pyrrhura* parakeets have prominent splashes of maroon or red, such as this species from eastern S. America.

TURACOS

Confined to Africa, this group includes the strikingly colorful turacos, which are largely green, blue or purple, the dull, grayish go-away birds (named for their distinctive nasal calls), and the plantain-eaters.

In the order Musophagiformes, these birds share certain features with cuckoos, such as foot structure, which appears to be coincidental rather than indicative of any relationship. They live in forests, subsisting largely on seasonal fruit. Although their bill is remarkably short, a long tail helps them maintain balance when reaching for food in the tree canopy. They move heavily through foliage and fly short distances across forest clearings. Turacos have vividly colored plumage that derives from green and red pigments, which are uniquely copper-based. Most species in this group have a short, thick, upstanding crest.

PHYLUM	CHORDATA
CLASS	AVES
ORDER	MUSOPHAGIFORMES
FAMILIES	1
SPECIES	23

red crest

16–17 in
40–43 cm

white face

HARTLAUB'S TURACO
Tauraco hartlaubi
F: Musophagidae
A blue-crested member of the green turaco group, this species is found in highland forest in E. Africa.

17 in
43 cm

RED-CRESTED TURACO
Tauraco erythrolophus
F: Musophagidae
One of few red-crested turacos, this species has a less bright body plumage than others in the green group. It is commonly found in the evergreen forests and woodlands of Angola.

long tail

bright crimson flight feathers

KNYSNA TURACO
Tauraco corythaix
F: Musophagidae
A South African relative of the green turaco, this species shares the white eye-stripes, but differs in having a white-tipped crest.

18–18½ in
45–47 cm

16 in
40 cm

RUSPOLI'S TURACO
Tauraco ruspolii
F: Musophagidae
A distinctively white-crested green turaco, this species has one of the most restricted ranges of any turaco, being confined to forests in S. Ethiopia.

GREEN TURACO
Tauraco persa
F: Musophagidae
With a range from Senegal to Angola, this is the most widespread of a group of green turacos with crimson wing-feathers visible in flight.

20–21½ in
51–54 cm

18–20 in
45–50 cm

VIOLET TURACO
Musophaga violacea
F: Musophagidae
One of two glossy, violet-colored turacos, this species ranges through forests of W. and C. Africa.

16–17 in
40–43 cm

ROSS'S TURACO
Musophaga rossae
F: Musophagidae
The second-largest turaco, this species is distinguished by its upright, red crest. It is an E. African counterpart of the violet turaco.

18½–20 in
47–50 cm

fanlike crest

yellow bill with red tip

20 in
50 cm

19 in
48 cm

blue body

GRAY GO-AWAY BIRD
Corythaixoides concolor
F: Musophagidae
Named after its call, "kay-waaay," this southern African species is a typical go-away bird with a pointed, shaggy crest.

BARE-FACED GO-AWAY BIRD
Corythaixoides personatus
F: Musophagidae
Like other go-away birds, this E. African species is found in savanna woodland. It raises its crest when excited.

long, broad tail

GREAT BLUE TURACO
Corythaeola cristata
F: Musophagidae
This fan-crested turaco from W. and C. Africa is the largest and most distinctive member of the turaco family.

28–30 in
70–75 cm

EASTERN GRAY PLANTAIN-EATER
Crinifer zonurus
F: Musophagidae
Like the related go-away birds, plantain-eaters are drably colored. Despite their name, they favor figs, not plantains. This species is from E. Africa.

HOATZIN

The hoatzin is an enigmatic South American bird that has an entirely vegetarian diet of leaves.

Some features of the hoatzin, such as strong feet and a short, curved bill, are similar to those of gamebirds. However, the hoatzin also shares features with cuckoos, a relationship that appeared to be confirmed by early DNA (deoxyribonucleic acid) studies. Later research has thrown doubt on this—leaving the hoatzin in a group of its own, the only representative of the order Opisthocomiformes.

PHYLUM	CHORDATA
CLASS	AVES
ORDER	OPISTHOCOMIFORMES
FAMILIES	1
SPECIES	1

HOATZIN
Opisthocomus hoazin
F: Opisthocomidae
Of uncertain taxonomic affiliation, the hoatzin is a plant-eater of S. American riverine forests. Its chicks have claws on their wings for clambering through branches.

spiky crest

24–26 in
61–66 cm

long neck

CUCKOOS

Cuckoos are rather short-legged, long-winged, mostly softly feathered birds that include dull, grayish species as well as a small group with patches of vivid emerald green.

All members of the Cuculiformes have feet with two toes pointing forward and two backward. The most primitive cuckoos are heavy-bodied American birds that forage on or near the ground—a lifestyle taken to the extreme by the sprinting roadrunner. Inspite of their reputation, not all cuckoos lay their eggs in other birds' nests.

Almost all these ground cuckoos build nests and raise their own young, as do their Old World counterparts, such as the tropical coucals and couas. Cuckoos that lay eggs in nests of other bird species are called "brood parasites." Remarkably, this habit has evolved independently at least twice in Old World cuckoos. The chicks of crested cuckoos (*Clamator* sp.) outgrow their host nest so quickly that host chicks die from starvation. Another line of cuckoos, which includes the Eurasian common cuckoo, evolved chicks with more murderous intent: they actively eject host eggs and chicks.

PHYLUM	CHORDATA
CLASS	AVES
ORDER	CUCULIFORMES
FAMILIES	1
SPECIES	149

thick, black bars on belly

long wings

14–16 in
35–40 cm

GREAT SPOTTED CUCKOO
Clamator glandarius
F: Cuculidae
With a range from Europe to Africa, this cuckoo lays eggs in magpie nests, but, like other *Clamator* species, does not evict the host's young.

JACOBIN CUCKOO
Clamator jacobinus
F: Cuculidae
Clamator cuckoos are large crested Old World cuckoos. Like others, this tropical species from Africa and Asia eats hairy caterpillars that are avoided by competitors.

12–13½ in
32–34 cm

13½ in
34 cm

HIMALAYAN CUCKOO
Cuculus saturatus
F: Cuculidae
This species from Asia and Australia is typical of a group of Old World barred cuckoos that resemble sparrowhawks—a fact that perhaps helps them to flush hosts from nests.

11–13½ in
28–34 cm

FAN-TAILED CUCKOO
Cacomantis flabelliformis
F: Cuculidae
This brown-breasted cuckoo of Australasian woodlands is one of very few cuckoos that are found in the Pacific—and the only one in Fiji.

9½ in
24 cm

9½–11 in
24–28 cm

PALLID CUCKOO
Cacomantis pallidus
F: Cuculidae
Many Old World cuckoos have barred plumage, but in some, such as this Australian species, the barring occurs only in immature birds.

12–13 in
30–33 cm

COMMON CUCKOO
Cuculus canorus
F: Cuculidae
This widespread Eurasian species overwinters in Africa and S. Asia. It has the well-known "cooc-coo" call that gives cuckoos their name.

DIDERIC CUCKOO
Chrysococcyx caprius
F: Cuculidae
An African member of a group of small bronze-sheened tropical cuckoos, this species lays its eggs in the nest of the weaver bird.

GRAY-BELLIED CUCKOO
Cacomantis passerinus
F: Cuculidae
Found from the Malay Peninsula to Australia, this bird is one of many Indo-Pacific, brown-breasted species that are closely related to *Cuculus* cuckoos.

9 in
23 cm

BRUSH CUCKOO
Cacomantis variolosus
F: Cuculidae
Related to Asian brown-breasted cuckoos, this plain-colored species breeds in southern Asia. The montane populations migrate to warm lowlands in winter.

6½–7½ in
17–19 cm

KLAAS'S CUCKOO
Chrysococcyx klaas
F: Cuculidae
This small African cuckoo is a close relative of the dideric cuckoo, differing in being greener and lacking white wing spots.

6½–7 in
16–18 cm

>>

GIANT COUA
Coua gigas
F: Cuculidae
Couas are Madagascan ground-cuckoos with blue facial skin and long eyelashes. This bird lives in dry coastal forests.

GUIRA CUCKOO
Guira guira
F: Cuculidae
This shaggy-looking bird belongs to the ani group of S. American cuckoos. It occurs in noisy flocks and nests communally in trees.

YELLOW-BILLED CUCKOO
Coccyzus americanus
F: Cuculidae
One of several American tree cuckoos with brown plumage and white-spotted tails, this species migrates between N. and S. America.

BLACK-BELLIED CUCKOO
Piaya melanogaster
F: Cuculidae
Like other American tree cuckoos, this species builds its own nest and cares for its young. It is found from Colombia to Bolivia.

GREATER COUCAL
Centropus sinensis
F: Cuculidae
Like other coucals, this S. Asian species is a strong-legged bird with spurred feet and long hind claws.

streaked, reddish-brown wings

PHEASANT-COUCAL
Centropus phasianinus
F: Cuculidae
Parental care in coucals is provided mostly by the males. This Australasian species has a black body when breeding, and builds a cup-shaped nest in grass.

rufous head crest

PHEASANT-CUCKOO
Dromococcyx phasianellus
F: Cuculidae
This American ground cuckoo inhabits the forest floor in tropical S. America, where it lays its eggs in the nests of smaller passerines.

PAVONINE CUCKOO
Dromococcyx pavoninus
F: Cuculidae
A smaller relative of the pheasant-cuckoo, this tropical S. American bird is a ground predator. It feeds on invertebrates.

long, graduated tail

long tail

GROOVE-BILLED ANI
Crotophaga sulcirostris
F: Cuculidae
A heavy-billed American cuckoo occurring from California to Argentina, the communal-nesting ani is a clumsy flier but runs well.

white cheek stripe

gray-brown upperparts with white spots

CORAL-BILLED GROUND CUCKOO
Carpococcyx renauldi
F: Cuculidae
One of three species of Asian ground cuckoos related to Madagascan couas, this bird inhabits rainforests of S.E. Asia.

GREATER ROADRUNNER
Geococcyx californianus
F: Cuculidae
This fast-running predator is an American ground cuckoo. The roadrunner inhabits deserts and nests in cacti. and occurs in a range from the US to Mexico.

COMMON KOEL
Eudynamys scolopaceus
F: Cuculidae
Unusually for cuckoos, this tropical brood parasite, found from Asia to Australasia, is a fruit-eater. Males are black and females are gray-brown.

OWLS

With acute senses, powerful weaponry, and noiseless flight, owls are superbly adapted as nighttime hunters. Only a few species are active by day.

Although not related to birds of prey, owls, which belong to the order Strigiformes, are similarly equipped with a hooked bill and strong talons. Most species have cryptic plumage for camouflage when roosting during the day. Barn owls are distinct in having a heart-shaped facial disk; owls with a round facial disk are more typical, although they are a family of great diversity, ranging from hawklike owlets to massive eagle owls.

SIGHT AND HEARING

All owls have large, forward-facing eyes that are efficient at letting in light and allow the birds to see well in dim conditions. Their binocular vision is good for judging distance during attacks, but the eyes are fixed in their sockets. To track moving prey, an owl must turn its whole head. The bird can do this through a wide range of movement by means of a long, flexible neck concealed beneath the plumage. Owls also have highly sensitive hearing, helped by the facial disk, which serves to guide sounds to the large ear openings. A downward-pointing bill minimizes interference with sound. An owl can determine the direction of prey with great accuracy from sounds. Most of the nocturnal species have one ear slightly higher than the other, an arrangement that helps them to pick up sounds in a vertical direction, too.

SILENT HUNTERS

Hunting owls usually swoop on their prey and extend their feet to make the grab. The approach is almost silent on large rounded wings that require minimal flapping. A soft fringe of serrated flight feathers breaks up air turbulence, muffling the sound of wing-beats. Some day-flying species lack these fringed feathers. Most owls prey on small mammals such as mice and voles, but some of the smaller species take large insects, and a few owls are specialized fish-eaters. An owl tears larger carcasses apart with its hooked bill, but swallows smaller prey whole. The indigestible parts of a meal, such as bones and fur, are compressed in one part of the stomach and later regurgitated in the form of a pellet.

PHYLUM	CHORDATA
CLASS	AVES
ORDER	STRIGIFORMES
FAMILIES	2
SPECIES	243

DEBATE
SCREECH OWLS

There are more than 60 species of owls in the genus *Otus*. In their native forests, these small, cryptically patterned birds are most reliably identified by their call. Although individual species' calls differ in frequency and duration of notes, *Otus* owls fall into two well-defined groups: scops owls, from the Old World, which call with slow notes; and American screech owls, which have faster, piercing trills. The distinction was enough for some ornithologists to advocate placing the screech owls in their own separate genus—*Megascops*. Recent research shows that this division is underpinned by differences in DNA.

BARN OWL
Tyto alba
F: Tytonidae
Found virtually worldwide outside deserts and polar regions, this is the most widespread species of owl. It is famous for its blood-curdling screech.

heart-shaped facial disk

10–18 in
25–45 cm

golden upperparts

10–17 in
26–43 cm

ASHY-FACED OWL
Tyto glaucops
F: Tytonidae
Confined to the Caribbean island of Hispaniola, this owl of dry forest is threatened by competition from the stronger barn owl.

7½–8 in
19–20 cm

EURASIAN SCOPS OWL
Otus scops
F: Strigidae
This small, agile owl from W. Eurasia is a typical tufted scops owl of the Old World. It nests in cavities of trees or buildings.

rufous body

9–9½ in
22–24 cm

RAINFOREST SCOPS OWL
Otus rutilus
F: Strigidae
Restricted to Madagascar, this owl is common in wooded areas. Most are gray; the rarer reddish-brown birds are confined to rainforests.

7½–10 in
19–25 cm

WESTERN SCREECH OWL
Megascops kennicottii
F: Strigidae
A common owl of western N. America, this species favors wooded areas near rivers but will also enter parkland and towns.

6½–10 in
16–25 cm

EASTERN SCREECH OWL
Megascops asio
F: Strigidae
Widespread throughout eastern N. America, this owl occurs in gray or rufous (reddish brown) color varieties, with more rufous birds farther east in its range.

concentric dark rings on facial disc

18½–21 in
47–53 cm

23½–24 in
60–62 cm

dark brown eyes

pale gray plumage with brown markings

14½–15½ in
37–39 cm

26–28 in
65–70 cm

BROWN WOOD OWL
Strix leptogrammica
F: Strigidae
This owl occurs in lowland tropical forests throughout India and S.E. Asia. Hard to see, this wood owl can be recognized by its distinctive call.

TAWNY OWL
Strix aluco
F: Strigidae
Occurring throughout Eurasia, the tawny owl frequents farmland, towns, and gardens. It belongs to a group of related wood owls.

17–20 in
43–50 cm

18½–19 in
47–48 cm

BARRED OWL
Strix varia
F: Strigidae
A large wood owl native to eastern N. America, this aggressive species is spreading westward and displacing the smaller spotted owl.

URAL OWL
Strix uralensis
F: Strigidae
A relative of the great gray owl, this N. Eurasian species frequents coniferous and broadleaf forests, and may also enter towns.

SPOTTED OWL
Strix occidentalis
F: Strigidae
Native to western N. America, this wood owl occurs in mature coniferous forest, where it hunts flying squirrels and similar-sized prey.

GREAT GRAY OWL
Strix nebulosa
F: Strigidae
A large owl of the circumpolar regions, this species occurs in coniferous forest. It sometimes hunts during the day and preys on large rodents and birds.

8½–10 in
21–25 cm

18–20 in
45–50 cm

23½–30 in
60–75 cm

10–11 in
25–28 cm

RUFESCENT SCREECH OWL
Megascops ingens
F: Strigidae
This little-known bird from wet montane forests in northern S. America is larger than many other screech owls, and has smaller ears.

TROPICAL SCREECH OWL
Megascops choliba
F: Strigidae
The most widespread tropical American screech owl, this species occurs in a range from Costa Rica to Argentina. It exists in brown and gray forms.

heavily barred underparts

DESERT EAGLE-OWL
Bubo ascalaphus
F: Strigidae
A bird of the Sahara Desert, this species is a smaller, paler, and longer-legged relative of the Eurasian eagle-owl.

EURASIAN EAGLE-OWL
Bubo bubo
F: Strigidae
One of the largest species of owl, this widespread eagle-owl occurs throughout Eurasia, and hunts animals as large as deer.

8½–9 in
22–23 cm

26–30 in
66–75 cm

18–27 in
46–68 cm

BLACK-CAPPED SCREECH OWL
Megascops atricapilla
F: Strigidae
Screech owls have diversified in forests throughout the Americas. Many have conspicuous ear tufts. This species is confined to C. and S. Brazil.

GREAT HORNED OWL
Bubo virginianus
F: Strigidae
The most widespread American owl, this eagle-owl is found from Alaska to Argentina, in habitats as diverse as forests and deserts.

VERREAUX'S EAGLE-OWL
Bubo lacteus
F: Strigidae
This is the largest African owl. It is widespread south of the Sahara Desert, where it feeds on small game animals.

ear tuft

white facial disc

14 in
36 cm

14–18 in
36–45 cm

ELF OWL
Micrathene whitneyi
F: Strigidae
Like other insect-eaters,
this tiny Mexican desert owl
does not need to fly silently,
so it lacks the fringed,
sound-muffling wing
feathers of larger species.

5–6 in
13–15 cm

9–9½ in
22–24 cm

*brilliant
yellow eyes*

NORTHERN HAWK-OWL
Surnia ulula
F: Strigidae
This relative of pygmy owls
is found in subarctic forests.
With its small head, long tail,
and daytime activity, it is the
most hawklike owl.

**SOUTHERN
WHITE-FACED OWL**
Ptilopsis granti
F: Strigidae
Like many other owl
species, this bird of
sub-Saharan Africa
breeds in the stick nests
of other birds.

STRIPED OWL
Pseudoscops clamator
F: Strigidae
A S. American
eared owl, this species
lives in open, marshy
habitats and nests in
ground vegetation
or low tree-holes.

*broad
ear tufts*

18 in
46 cm

SNOWY OWL
Nyctea scandiaca
F: Strigidae
An Arctic species of eagle-owl,
this large, ground-nesting predator
breeds on open tundra, where it
feeds on lemmings and ptarmigan.

20½–28 in
52–71 cm

SPECTACLED OWL
Pulsatrix perspicillata
F: Strigidae
This is a distinctive species
from C. and S. American forests.
Its fledglings are white with
a contrasting black face.

5–6 in
13–15 cm

**LEAST
PYGMY OWL**
*Glaucidium
minutissimum*
F: Strigidae
Like other related small
owls, this woodland
bird of Paraguay and
S.E. Brazil is active
during the day and night.

**NORTHERN
PYGMY OWL**
Glaucidium gnoma
F: Strigidae
Pygmy owls often fearlessly
tackle prey bigger than
themselves. This little
predator from western
N. America has been
seen attacking grouse.

6–6½ in
15–17 cm

6½–7 in
17–18 cm

**FERRUGINOUS
PYGMY OWL**
Glaucidium brasilianum
F: Strigidae
Typical of pygmy owls,
this American species has
eyespots on the back of the
head to confuse predators. It
flicks its tail when excited.

6–7 in
15–18 cm

CUBAN PYGMY OWL
Glaucidium siju
F: Strigidae
Many owls nest in decayed
tree cavities, but pygmy
owls habitually choose old
woodpecker holes. This
species is restricted to Cuba.

18–18½ in
46–47 cm

long claws

BUFFY FISH OWL
Ketupa ketupu
F: Strigidae
This S.E. Asian relative of
eagle-owls catches fishes and
other aquatic prey with the
help of its long-clawed feet.

» OWLS

8½–11 in
21–28 cm

TENGMALM'S OWL
Aegolius funereus
F: Strigidae

This owl of northern circumpolar forests is adept at locating small mammals under snow. It hunts during daylight.

7½–10 in
19–25 cm

BURROWING OWL
Athene cunicularia
F: Strigidae

A long-legged owl, this partly diurnal species inhabits grassland and deserts across the Americas, nesting in burrows of prairie dogs or other animals.

LONG-EARED OWL
Asio otus
F: Strigidae

Found in forests and heathland through much of the northern hemisphere, this owl can flatten its long ear tufts, rendering them invisible.

12–14½ in
31–37 cm

SHORT-EARED OWL
Asio flammeus
F: Strigidae

yellow eye

This owl is found in open country in a range that includes the Americas, Eurasia, and N. Africa, and is even found on many Pacific islands.

13½–17 in
34–43 cm

PEL'S FISHING OWL
Scotopelia peli
F: Strigidae

The largest African fishing owl, this bird frequents riverside forests, where it hunts from low perches in slow-running waters. It eats mainly fishes, but also crabs and frogs.

25–26 in
63–65 cm

pale brown, heart-shaped face

yellow eye

heavily streaked upperparts

barred underparts

BARKING OWL
Ninox connivens
F: Strigidae

Found from the Moluccas, Indonesia, to Australia, this woodland hawk-owl is named for its doglike barking call.

15–17 in
38–43 cm

MOREPORK
Ninox novaeseelandiae
F: Strigidae

Typical of Australasian hawk-owls, this bird has large, yellow eyes. Unusually for owls, the male of the species is larger than the female.

12–14 in
30–35 cm

BLACK-AND-WHITE OWL
Ciccaba nigrolineata
F: Strigidae

This barred owl of dense tropical forests from Mexico to Ecuador is a close relative of the *Strix* wood owls.

15 in
38 cm

MESITES

These small, insectivorous birds of forest and scrub in Madagascar are flightless, or almost so.

In the family Mesitornithidae, mesites resemble passerines in their use of song for territorial defense, but they are social birds. Small groups often feed, rest, and preen together. Two species of the family are monogamous and their sexes look alike, while one is polygamous, with marked differences between the male and the female bird.

PHYLUM	CHORDATA
CLASS	AVES
ORDER	MESITORNITHIFORMES
FAMILIES	1
SPECIES	3

WHITE-BREASTED MESITE
Mesitornis variegatus
F: Mesitornithidae

This bird is found in three isolated areas of undisturbed deciduous forest where small groups forage among leaf litter on the ground for insects, spiders, and seeds.

12 in
31 cm

NIGHTJARS

All nightjars and their relatives are nocturnal. Most are insect-eaters, snatching their prey on the wing using an exceptionally wide bill gape.

Voracious predators by night, concealed by cryptic plumage during the day, nightjars have similarities to owls, and some ornithologists have thought the two groups to be related. Recent studies, however, link nightjars with swifts and hummingbirds. The connection is supported by features such as the weakness of their feet and the ability of some species to enter a torpid state.

Nightjars belong to the order Caprimulgiformes. Typically, they have a big head and long wings for quick aerial maneuvers. Their short bill opens enormously wide to catch flying insects. The gape is particularly wide in the aptly named frogmouths from Australasia, which are predators of small vertebrates. Of the group, only the oilbird is vegetarian: a fruit-eater that roosts by day in caves.

All these birds are masters of camouflage. Nightjars blend perfectly with forest litter when on the ground; in trees they perch lengthwise on a branch to minimize detection. Frogmouths and potoos freeze when disturbed to mimic tree stumps, closing their eyes to complete the effect.

MINIMAL NESTS

All in this group are minimalist nesters. Nightjars simply lay their eggs in leaf litter on the ground. For oilbirds, a nest is a pile of droppings on a cave ledge, while potoos use nothing more than a depression on a branch.

PHYLUM	CHORDATA
CLASS	AVES
ORDER	CAPRIMULGIFORMES
FAMILIES	4
SPECIES	122

A great potoo opens its enormous gape, displaying a key feature of a group of birds that can scoop up flying insects on the wing.

OILBIRD
Steatornis caripensis
F: Steatornithidae
A fruit-eating relative of nightjars, this cave-nesting species from northern S. America is the only nocturnal bird that navigates by echolocation.

16–19 in
41–48 cm

orange-yellow eye

long bristles at base of bill

9–9½ in
22–24 cm

12½–18 in
32–46 cm

TAWNY FROGMOUTH
Podargus strigoides
F: Podargidae
This nocturnal Australian bird hunts from perches. Like other frogmouths, it roosts motionless during the day, resembling a broken branch.

COMMON NIGHTHAWK
Chordeiles minor
F: Caprimulgidae
Nighthawks lack the bill bristles of nightjars, and this absence perhaps helps in their aerial pursuit of insects. This insect-eater migrates between N. and S. America.

erect, stumplike posture

14–16 in
36–41 cm

gray-brown, barklike plumage

8 in
20 cm

OCELLATED POORWILL
Nyctiphrynus ocellatus
F: Caprimulgidae
Poorwills are small American nightjars, so called because of their mournful, monotonous calls. This dark-colored species occurs in tropical forests.

gray mottled plumage

COMMON POTOO
Nyctibius griseus
F: Nyctibiidae
Potoos are nocturnal insectivores of C. and S. America. This species has a haunting call. Its plumage provides excellent camouflage against tree bark.

7½–8½ in
19–21 cm

COMMON POORWILL
Phalaenoptilus nuttallii
F: Caprimulgidae
From arid parts of the US and Mexico, this small nightjar is one of the few birds that enters a hibernationlike torpor in winter.

9½–11 in
24–28 cm

COMMON PAURAQUE
Nyctidromus albicollis
F: Caprimulgidae
This C. and S. American nighthawk frequents scrubby habitats. It is often seen resting on dirt roads at night.

reddish brown collar

10–11 in
26–28 cm

EUROPEAN NIGHTJAR
Caprimulgus europaeus
F: Caprimulgidae
Like many nightjars of northern temperate regions, this species is migratory. Found in W. Eurasian heath and woodland, it overwinters in Africa.

SPOT-TAILED NIGHTJAR
Caprimulgus maculicaudus
F: Caprimulgidae
Found from Mexico to Paraguay, this species may be more closely related to tropical American nighthawks than Old World nightjars.

8 in
20 cm

10–10½ in
25–27 cm

LARGE-TAILED NIGHTJAR
Caprimulgus macrurus
F: Caprimulgidae
Among the most widespread of the Old World nightjars, this species ranges from Pakistan to Australia and New Guinea.

PLAIN NIGHTJAR
Caprimulgus inornatus
F: Caprimulgidae
This African bird occurs from Mauritania to Saudi Arabia, and migrates south to Liberia, Congo, and Tanzania.

9 in
22 cm

9 in
23 cm

DUSKY NIGHTJAR
Caprimulgus saturatus
F: Caprimulgidae
Perhaps genetically closer to American poorwills than Old World nightjars, this species occurs in montane forests of Costa Rica and Panama.

LONG-TAILED NIGHTJAR
Caprimulgus climacurus
F: Caprimulgidae
The long tail of some nightjars may be used in a courtship display. This African species has a range from Senegal to Ethiopia.

9–11 in
23–28 cm

LADDER-TAILED NIGHTJAR
Hydropsalis climacocerca
F: Caprimulgidae
This Amazonian species belongs to a group of S. American nightjars, in which males have a long, forked tail with white markings. These give the bird its name and are possibly used to attract females.

10–14 in
25–35 cm

MADAGASCAR NIGHTJAR
Caprimulgus madagascariensis
F: Caprimulgidae
This nightjar from Madagascar and Aldabra Island (Seychelles) has a call that resembles the sound of a marble bouncing on a hard floor.

8½ in
21 cm

9 in
22 cm

mottled plumage for camouflage

elongated tail feather

SHORT-TAILED NIGHTHAWK
Lurocalis semitorquatus
F: Caprimulgidae
One of several tropical American nighthawks, this bird has a short tail and long wings, which give it a batlike appearance on its insect-chasing flights.

LONG-TRAINED NIGHTJAR
Macropsalis creagra
F: Caprimulgidae
Males of many nightjars have an elaborate tail. This species evolved from tropical American nighthawks and is limited to eastern S. America.

13½–30 in
34–76 cm

STANDARD-WINGED NIGHTJAR
Macrodipteryx longipennis
F: Caprimulgidae
This African species eats insects, including moths and beetles. Breeding males have flaglike flight feathers, longer than the body, which they raise in courtship displays.

8½–9 in
21–23 cm

flaglike flight feathers

HUMMINGBIRDS AND SWIFTS

The dashing flight of swifts—fastest of all birds—and the audible hovering of hummingbirds characterize the master aeronauts united in this order.

Both swifts and hummingbirds, members of the order Apodiformes, have tiny feet that are useful only for perching, but in compensation the birds have exceptional maneuverability on the wing. Swifts skim through the skies to catch flying insects.

Hummingbirds have the wing control to fly backward, some of them whirring at more than 70 wing-beats per second—sustained by a high metabolism fueled by energy-rich nectar. Insects and spiders provide hummingbirds with protein to feed to their chicks. Hummingbirds use cobwebs to glue together their thimble-sized nests, while swifts use saliva. The swifts' range is almost worldwide, reaching oceanic islands. Hummingbirds are found only in the Americas.

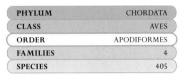

PHYLUM	CHORDATA
CLASS	AVES
ORDER	APODIFORMES
FAMILIES	4
SPECIES	405

COMMON SWIFT
Apus apus
F: Apodidae
This Eurasian swift nests in holes in cliffs or buildings, making it a common bird in towns. It overwinters in Africa.

6½ in
16–17 cm

WHITE-THROATED SWIFT
Aeronautes saxatalis
F: Apodidae
Found in canyons and mountains of western N. America and C. America, this bird gathers in flocks in the evenings to roost.

CHIMNEY SWIFT
Chaetura pelagica
F: Apodidae
Originally nesting in caves and tree-holes, this eastern N. American species now nests primarily in chimneys in urban centers. It overwinters in S. America.

6–7 in
15–18 cm

4¾–6 in
12–15 cm

8–9 in
20–22 cm

ALPINE SWIFT
Tachymarptis melba
F: Apodidae
Distinctively white-bellied, with a dark undertail, the Alpine swift occurs in southern Eurasia, Africa, and Madagascar. It feeds on large insects.

SCALE-THROATED HERMIT
Phaethornis eurynome
F: Trochilidae
Hermits are dull-colored hummingbirds with a long, curving bill for probing *Heliconia* flowers. This species is from eastern S. America.

decurved bill

4¾ in
12 cm

deeply forked tail

WHITE-TIPPED SICKLEBILL
Eutoxeres aquila
F: Trochilidae
This bird has a range that extends from Costa Rica to Peru. It uses its decurved bill to feed from similarly curved *Heliconia* flowers.

5 in
13 cm

violet "ear" patch

4¼ in
11 cm

metallic blue throat

violet-blue underparts

5½–6 in
14–15 cm

4 in
10 cm

GREEN-FRONTED LANCEBILL
Doryfera ludovicae
F: Trochilidae
This inhabitant of the forests of the Andes takes nectar from five species of epiphytes (plants that grow on other plants), one of which is a mistletoe.

WHITE-VENTED VIOLET-EAR
Colibri serrirostris
F: Trochilidae
From a group of violet-eared hummingbirds, most of which occur in highland forests, this is a species of S. American savanna.

VIOLET SABERWING
Campylopterus hemileucurus
F: Trochilidae
Saberwings are named for the thickened shafts of their flight feathers. This C. American bird is the largest hummingbird outside S. America.

»

3½ in
9 cm

2¾ in
7 cm

2¾–3½ in
7–9 cm

brown-black
plumage

HOODED VISORBEARER
Augastes lumachella
F: Trochilidae
Visorbearers, with their
green facial mask, are
hummingbirds of arid
habitats, such as highland
savanna. This species
occurs in E. Brazil.

BRAZILIAN RUBY
Clytolaema rubricauda
F: Trochilidae
Named for the male's
brilliant ruby-red throat,
this hummingbird is found
in S.E. Brazil, but may be
closely related to species
found in the Andes.

RUBY-THROATED HUMMINGBIRD
Archilochus colubris
F: Trochilidae
This tiny bee hummingbird
is the eastern US's only
breeding hummingbird. It
flies nonstop to the Gulf
of Mexico to overwinter.

3¼ in
8 cm

4–4¼ in
10–11 cm

RUBY TOPAZ
Chrysolampis mosquitus
F: Trochilidae
This S. American, open-country
species has a ruby-red crown and
yellow throat, but in poor light the
bird can appear all black.

4–7 in
10–18 cm

4¼ in
11 cm

bright orange
tail

LONG-TAILED SYLPH
Aglaiocercus kingi
F: Trochilidae
Belonging to a diverse group
of Andean hummingbirds
called coquettes, this bird
inhabits woodland borders
and gardens. Males have a
long, deeply forked tail.

BUFF-TAILED CORONET
Boissonneaua flavescens
F: Trochilidae
Found in Colombia, Venezuela,
and Ecuador, this relative of
Andean hummingbirds forages
near flowering trees from
mid-level to forest canopy,
often with other birds.

BUFF-BELLIED HUMMINGBIRD
Amazilia yucatanensis
F: Trochilidae
Belonging to a large
group of emerald
hummingbirds found
in C. America, this bird
is a Mexican species
of open woodland.

3½ in
9 cm

BLUE-CHINNED SAPPHIRE
Chlorostilbon notatus
F: Trochilidae
One of many emerald
hummingbirds with metallic
green plumage, this is a
forest and farmland bird
from northern S. America.

dark
ear patch

short,
straight bill

BLUE-THROATED HUMMINGBIRD
Lampornis clemenciae
F: Trochilidae
This large Mexican species belongs to a
group of hummingbirds called mountain
gems that originated in C. America. The
northerly populations are migratory.

4¼ in
11 cm

3½ in
9 cm

4¾ in
12 cm

4 in
10 cm

COLLARED INCA
Coeligena torquata
F: Trochilidae
Incas form a group of forest
hummingbirds native to the
Andes. The collared inca is
one of the most widespread
of the group, found from
Colombia to Bolivia.

SPECKLED HUMMINGBIRD
Adelomyia melanogenys
F: Trochilidae
A common hummingbird from the
Andes, this species has duller plumage
than many of its relatives in the
coquette group. Both sexes are similar.

ANNA'S HUMMINGBIRD
Calypte anna
F: Trochilidae
A bee hummingbird of western N. America,
this species overwinters farther north than
other hummingbirds. The bird feeds on
nectar from a wide range of flowers.

WHITE-EARED HUMMINGBIRD
Basilinna leucotis
F: Trochilidae
This emerald hummingbird frequents pine and oak montane forests, often near streams, from S. Arizona to Nicaragua.

3½–4 in
9–10 cm

ANDEAN HILLSTAR
Oreotrochilus estella
F: Trochilidae
An Andean coquette hummingbird, this species lives higher than other species of the group, and feeds in air so thin that it cannot hover.

5 in
13 cm

4 in
10 cm

3½ in
9 cm

LUCIFER HUMMINGBIRD
Calothorax lucifer
F: Trochilidae
This bee hummingbird is found in a range from the southern US to Mexico. It lives in semideserts, especially those with *Agave* plants.

purple throat patch

straight, black bill

FORK-TAILED WOODNYMPH
Thalurania furcata
F: Trochilidae
A relative of the C. American emerald hummingbirds, this S. American species occurs in lowland forests as far south as N. Argentina.

iridescent green throat

BROAD-BILLED HUMMINGBIRD
Cynanthus latirostris
F: Trochilidae
Males of this Mexican scrub hummingbird fly back and forth in a pendulum display to attract females. The species belongs to the "emerald" group.

2–2¼ in
5–6 cm

4¾ in
12 cm

4 in
10 cm

9–10 in
23–26 cm

SWORD-BILLED HUMMINGBIRD
Ensifera ensifera
F: Trochilidae
The only bird with a bill longer than its body, this member of the Andean inca group favors the trumpet-shaped flowers of passion vines.

BEE HUMMINGBIRD
Mellisuga helenae
F: Trochilidae
This tiny hummingbird is confined to Cuba, but is related to migratory N. American hummingbirds. Males of the species are the smallest of all birds.

RACKET-TAILED PUFFLEG
Ocreatus underwoodii
F: Trochilidae
Belonging to the inca group, this bee-sized hummingbird of Andean forests is attracted to brush blossoms, such as those of legumes.

4 in
10 cm

STRIPE-BREASTED STARTHROAT
Heliomaster squamosus
F: Trochilidae
Male starthroats have a vividly colored band around their throat. Although starthroats are related to C. American mountain gems, this species is from eastern Brazil.

white spot near the eye

2¾–3½ in
7–9 cm

rufous (reddish brown) upperparts

RUFOUS HUMMINGBIRD
Selasphorus rufus
F: Trochilidae
Sustaining the longest migration of comparably small birds, this aggressively territorial bee hummingbird takes a route from Alaska to Mexico.

3½ in
9 cm

3½ in
9 cm

WHITE-NECKED JACOBIN
Florisuga mellivora
F: Trochilidae
Genetic research suggests that this large, canopy-dwelling bird of tropical American lowlands belongs to a small lineage independent of other groups in its family.

4¾ in
12 cm

GORGETED SUNANGEL
Heliangelus strophianus
F: Trochilidae
Sunangels belong to the coquette group of Andean hummingbirds. This species is found from Colombia to Ecuador and lives in wet forest thickets.

3½ in
9 cm

CALLIOPE HUMMINGBIRD
Stellula calliope
F: Trochilidae
One of the migratory bee hummingbirds, this bird breeds in open forests of western N. America and overwinters in the Mexican semidesert.

PLOVERCREST
Stephanoxis lalandi
F: Trochilidae
This distinctive bird may be related to larger saberwings of C. America, but is found in montane forests of eastern S. America.

TROGONS

Gaudily colored birds of tropical forests, the fruit-eating trogons have a wide bill, delicate plumage, and a foot structure unique to this group.

Crow-sized birds belonging to the order Trogoniformes are found in tropical America, Africa, and Asia. The males have brighter plumage than the females. Long-tailed and short-winged, trogons are capable fliers, though reluctant to break cover. The first and second toes of a trogon's foot face backward, whereas the third and fourth toes face forward. This arrangement is seen in no other birds. These unusual feet are so weak that a trogon can barely manage more than a shuffle when perched.

Trogons use their wide bill for tackling large fruit and taking invertebrates such as caterpillars. They also use their bill for excavating nesting holes in rotten wood and termite mounds.

PHYLUM	CHORDATA
CLASS	AVES
ORDER	TROGONIFORMES
FAMILIES	1
SPECIES	43

dark violet-blue crown

iridescent green-blue plumage

14–39 in
35–100 cm

long tail (only in male)

12–14 in
31–36 cm

RED-HEADED TROGON
Harpactes erythrocephalus
F: Trogonidae
An Asian trogon found from the Himalayas to Sumatra, this elusive bird, like many of its cousins, perches motionless for long periods.

10½–12½ in
27–32 cm

ORANGE-BREASTED TROGON
Harpactes oreskios
F: Trogonidae
The color pattern of this bird, dull upperparts, and bright underparts, is typical of many Asian trogons. This species is a native of S.E. Asia.

10–11 in
26–28 cm

RESPLENDENT QUETZAL
Pharomachrus mocinno
F: Trogonidae
Quetzals are iridescent American trogons. The male's long tail in this C. American species is almost half its overall length.

11–12 in
28–30 cm

ELEGANT TROGON
Trogon elegans
F: Trogonidae
The only trogon with a range that extends to the US, this bird occurs in montane forests from southern Arizona to C. America.

10–10½ in
25–27 cm

MASKED TROGON
Trogon personatus
F: Trogonidae
Males of this species from S. American montane forests have plumage that resembles that of the elegant trogon.

CUBAN TROGON
Priotelus temnurus
F: Trogonidae
One of two species of *Priotelus* trogons confined to the Caribbean, this is the national bird of Cuba.

jagged-tipped tail

MOUSEBIRDS

Named for their scurrying, rodentlike behavior, the mousebirds are a small group of drab-colored, long-tailed birds confined to sub-Saharan Africa.

PHYLUM	CHORDATA
CLASS	AVES
ORDER	COLIIFORMES
FAMILIES	1
SPECIES	6

Mousebirds, the sole family of the order Coliiformes, are softly feathered in shades of brown or gray and have an erectile crest and long tail. Gregarious and agile, they resemble parakeets in behavior. These tree-dwelling birds build twiggy, cup-shaped nests. The chicks hatch at an advanced stage of development and quickly master flight. Ornithologists think that mousebirds are the remnant of a more diverse group that, in prehistoric times, ranged beyond Africa. Fossil mousebirds have been found in Europe. Their relationship to other birds is uncertain: mousebirds could be allied with trogons, kingfishers, or woodpeckers.

SPECKLED MOUSEBIRD
Colius striatus
F: Coliidae
The largest and one of the most widespread of mousebirds, this species occurs from Nigeria to South Africa in savanna and open woodland.

12–14 in
30–35 cm

13–14 in
33–35 cm

BLUE-NAPED MOUSEBIRD
Urocolius macrourus
F: Coliidae
Urocolius mousebirds are stronger fliers and less "mouselike" than *Colius* mousebirds. This scrubland dweller is found from Senegal to Tanzania.

KINGFISHERS AND RELATIVES

This order, Coraciiformes, has a worldwide distribution. All members perch with three toes pointing forward, as do the Passerines, but their two outer toes are fused at the base.

Kingfishers often have bright green or blue colors, created not by pigments but by the scattering of light from the microstructure of their feathers. They follow diverse lifestyles, not all reflected in the popular English name. Fish-eating species locate prey by sight and dive onto it, either from a fixed perch or after a hover in midair. Others also hunt land animals, including lizards, rodents, and insects. A large head and a strong neck helps these birds to dive

efficiently, and their sharp-edged, long bill helps grip slippery prey. Fish-eaters have a slender bill, while a broader, scooplike bill is a feature of woodland kingfishers that catch their prey directly from dry ground. Their short legs are ideal for perching, but kingfishers cannot walk. Their feet are, however, also used for excavating elongated nesting cavities in earth or sand banks.

OTHER RELATIVES

Other members of this order, the American motmots and the Old World rollers, are all land hunters. All kingfisher relatives display a wide variety of bill shapes. In the aerial-hunting bee-eaters, the long bill holds stinging insects away from the head, and is used like forceps to squeeze out insect venom.

PHYLUM	CHORDATA
CLASS	AVES
ORDER	CORACIIFORMES
FAMILIES	6
SPECIES	177

Despite their daggerlike bill, kingfishers grasp prey, rather than stab it, like almost all other fish-eating birds.

LILAC-BREASTED ROLLER
Coracias caudatus
F: Coraciidae
Like other *Coracias* rollers, this African bird is a predator of ground animals, swooping to take lizards, rodents, and large invertebrates.
12½–14 in
32–36 cm

11–12 in
28–30 cm

14–15 in
36–38 cm

11½–12½ in
29–32 cm

EUROPEAN ROLLER
Coracias garrulus
F: Coraciidae
The most widespread of the *Coracias* rollers, named for their somersaulting courtship displays, this western Eurasian species overwinters in Africa.

RUFOUS-CROWNED ROLLER
Coracias naevius
F: Coraciidae
Somewhat dull in color compared to other species, this large roller inhabits arid areas across sub-Saharan Africa.
14–16 in
36–41 cm

BLUE-BELLIED ROLLER
Coracias cyanogaster
F: Coraciidae
This C. African roller, like many of its relatives, nests in tree cavities. It is common throughout its extensive range.

RACQUET-TAILED ROLLER
Coracias spatulatus
F: Coraciidae
This distinctive fork-tailed roller from E. Africa has outer tail feathers with projecting "flag-tipped" vanes.

PITTA-LIKE GROUND ROLLER
Atelornis pittoides
F: Brachypteraciidae
As the name suggests, ground rollers are mostly terrestrial. This rainforest species is the most brightly colored of the family.

black stripe through eye

white throat

10 in
26 cm

DOLLARBIRD
Eurystomus orientalis
F: Coraciidae
Named for its pale, dollar-shaped underwing markings, this roller occurs in woodlands from the Himalayas to Australia.
10½–12 in
27–30 cm

long, brown tail with dark barring

18½–20½ in
47–52 cm

yellow bill

BROAD-BILLED ROLLER
Eurystomus glaucurus
F: Coraciidae
Typical of its genus, this African species is longer winged and more agile in pursuit of flying prey than *Coracias* rollers.
10½–12 in
27–30 cm

LONG-TAILED GROUND ROLLER
Uratelornis chimaera
F: Brachypteraciidae
Like other ground rollers, this species from the dry forests of S.W. Madagascar nests in holes excavated in the ground.

>>

» KINGFISHERS AND RELATIVES

4½ in
11 cm

JAMAICAN TODY
Todus todus
F: Todidae

Todies are tiny, green kingfisherlike birds from the Caribbean, with a flattened bill for catching insects in flight. This species is found only in the woodlands of Jamaica.

bright blue back and wings

WHITE-THROATED KINGFISHER
Halcyon smyrnensis
F: Alcedinidae

Like many other tree kingfishers, this southern Asian species is a noisy bird. It has whinnying and laughterlike calls.

white chest and throat

11½ in
29 cm

GRAY-HEADED KINGFISHER
Halcyon leucocephala
F: Alcedinidae

Tree kingfishers are a group of mainly insect-eating forest birds. The gray-headed kingfisher is an African species that may also eat fishes.

8 in
20 cm

dark eye patch contrasts with the white head

heavy bill

16–18½ in
41–47 cm

AFRICAN PYGMY KINGFISHER
Ispidina picta
F: Alcedinidae

This African bird belongs to a group of river kingfishers, but mainly feeds on insects so occurs away from water in woodland and savanna.

4¾ 5 in
12–13 cm

GREEN KINGFISHER
Chloroceryle americana
F: Alcedinidae

A tropical American bird, this water kingfisher plunge-dives for fishes and aquatic insects and often perches on rocks in streambeds.

7–8 in
18–20 cm

LAUGHING KOOKABURRA
Dacelo novaeguineae
F: Alcedinidae

Widespread in Australian woodland, this large tree kingfisher preys on reptiles as well as invertebrates, and has a loud, chuckling call.

YELLOW-BILLED KINGFISHER
Syma torotoro
F: Alcedinidae

This tree kingfisher occurs in rainforests of New Guinea and N. Australia. It feeds mainly on insects, but also eats worms and small lizards.

7–8½ in
18–21 cm

azure-blue upperparts

orange underparts

AZURE KINGFISHER
Ceyx azureus
F: Alcedinidae

Like other river kingfishers, the azure kingfisher nests in holes along riverbanks. It is found in creeks and mangroves in Australia and New Guinea.

6½–7½ in
17–19 cm

LITTLE KINGFISHER
Ceyx pusillus
F: Alcedinidae

This tiny Australasian species of mangroves is a river kingfisher. It eats tiny fishes, insect larvae, small crustaceans, and shrimps.

4¾–5 in
12–13 cm

COMMON KINGFISHER
Alcedo atthis
F: Alcedinidae

Found in Eurasia and N. Africa, this bird is typical of a group of diminutive, short-tailed river kingfishers from the Old World.

6½ in
16–17 cm

PIED KINGFISHER
Ceryle rudis
F: Alcedinidae

From Africa and S. Asia, this water kingfisher of rivers and lakes often gathers in small groups. It can hover in one spot for long periods of time.

11–11½ in
28–29 cm

11–14 in
28–35 cm

single blue
chest band

BELTED KINGFISHER
Megaceryle alcyon
F: Alcedinidae
A water kingfisher from
N. America, the belted
kingfisher is a fishing specialist
that often hovers over water
before diving for prey.

12–14 in
30–35 cm

**BUFF-BREASTED
PARADISE KINGFISHER**
Tanysiptera sylvia
F: Alcedinidae
Many tree kingfishers, such as this
species from New Guinea and
N. Australia, nest in holes in
treetop termite mounds.

14½ in
37 cm

**BROWN-WINGED
KINGFISHER**
Pelargopsis amauroptera
F: Alcedinidae
A tree kingfisher of mangrove
forests, this species occupies
a range from India to the
Malay Peninsula.

14½–16 in
37–41 cm

STORK-BILLED KINGFISHER
Pelargopsis capensis
F: Alcedinidae
Occurring in southern Asia, this bird is found
near water in forests, especially outside the
range of the related brown-winged kingfisher.

10–11 in
25–28 cm

**COLLARED
KINGFISHER**
Todiramphus chloris
F: Alcedinidae
Favoring mangroves,
this tree kingfisher is widespread
in coastal areas across southern
Asia to the Pacific.

white central
tail feathers

18 in
46 cm

RUFOUS MOTMOT
Baryphthengus martii
F: Momotidae
Like other motmots,
this stout-billed bird from
C. and northern S. America
sallies for large insects
and small vertebrates.

turquoise-blue
crown

**BLUE-CROWNED
MOTMOT**
Momotus momota
F: Momotidae
Like the related bee-eaters
and many kingfishers, this
typical brightly colored
motmot from tropical
America nests in
tunnels in riverbanks.

16 in
41 cm

**TURQUOISE-
BROWED MOTMOT**
Eumomota superciliosa
F: Momotidae
As with other motmots,
the racket-tipped tail
of this C. American
motmot is formed by
weakened feather barbs
that break away.

13 in
33 cm

black
eye-stripe

long, slightly
curved bill

10–11½ in
25–29 cm

EUROPEAN BEE-EATER
Merops apiaster
F: Meropidae
Like other bee-eaters, this
bird from southwestern
Eurasia and Africa catches
insects in flight. Before eating
bees, it removes the sting
from its prey.

9 in
23 cm

**RAINBOW
BEE-EATER**
Merops ornatus
F: Meropidae
The only bee-eater in Australia,
southernmost populations of
this species migrate northward
to overwinter in northern
Australia and Indonesia.

8½–9½ in
22–24 cm

**WHITE-FRONTED
BEE-EATER**
Merops bullockoides
F: Meropidae
This African bee-eater breeds
in large colonies, which have
a complex social system with
non-breeding birds assisting
in raising the young.

8½–10 in
22–25 cm

GREEN BEE-EATER
Merops orientalis
F: Meropidae
Found in dry, open habitats
in Africa and S. Asia, the
green bee-eater nests in
looser colonies than other
species of bee-eater.

CUCKOO-ROLLER

**Comprising a single rare species from
the Madagascar forest, the order
Leptosomiformes is placed close to the
quetzals of South America, though its
affinity to other orders is unclear.**

The cuckoo-roller, which survives in fragmented
remnant forests, seems less endangered than many
other Madagascan species. It feeds like a flycatcher
on large insects, but also eats small reptiles. Similar
to cuckoos, it is able to digest hairy caterpillars,
which most species avoid. Again similar to cuckoos,
its two toes face forward and two backward.

PHYLUM	CHORDATA
CLASS	AVES
ORDER	LEPTOSOMIFORMES
FAMILIES	1
SPECIES	1

Small eye in
bulky head

Male has
iridescent
upperparts

Short,
thick bill

CUCKOO-ROLLER
Leptosomus discolor
F: Leptosomatidae
Superficially resembling a cuckoo,
this Madagascan predator of small
animals is not closely related to true
rollers. It nests in tree cavities.

15½–19½ in
40–50 cm

SHOVEL-BILLED ASSASSIN ∨

A bold, dry-land, woodland-edge kingfisher, the laughing kookaburra sits on an exposed perch looking for prey. It uses its shovel-shaped bill and muscular neck to grab small snakes and rodents, hitting larger prey against the ground to subdue them.

spiky crown feathers form short crest

broad, dark cheek-patch below large eye

lower mandible used to scoop up small prey

white underside helps hide the bird against the sky from prey below

LAUGHING KOOKABURRA
Dacelo novaeguineae

Laughing kookaburras are famously vocal. A first loud, chuckling laugh spreads into a loud chorus from several birds, the staccato notes echoing through the Australian bush. "Laughing jackass" was a popular nickname for this large-headed, short-legged bird; another, "bushman's clock," reflects the regularity of calls at dawn and dusk. Its nest is an unlined hollow in a tree stump or in a hole excavated with its bill in a termite mound. Like other kingfishers, its three to five eggs are pure white. If food is short, the final egg is small and the resultant weak chick is likely to be killed by its stronger siblings. Adult laughing kookaburras form lifelong pair-bonds, but are helped at the nest by up to five immature birds from previous years, and all may incubate the eggs and then brood and feed the chicks. This species is widespread and relatively common, but recent wildfires in Australian woodland have seriously affected many populations.

SIZE 16–18½in (41–47cm)
HABITAT Open, bushy woodland and eucalypt forest
DISTRIBUTION Eastern Australia, introduced in other parts of Australia and New Zealand
DIET Lizards, snakes, small mammals, insects, worms, snails, and other invertebrates

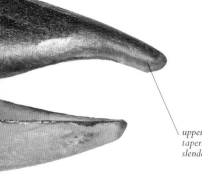

upper mandible tapers to a slender point

EYE >
A sensitive area of the retina in each eye allows exceptionally sharp vision. The laughing kookaburra moves its head to align a second, sensitive area in each eye to give binocular vision when searching for prey below.

∨ TONGUE
The broad, pointed tongue sits neatly inside the deep lower mandible. Sticky with mucus, it has a small "hook" at the rear, and helps to manipulate and swallow wriggling or muscular prey items.

∨ FOOT
As with other kingfishers, the long outer toe and middle toe are fused together. The very short leg gives a secure, upright stance on a perch, but means the laughing kookaburra can only shuffle when on the ground.

< NOSTRIL
An open nostril high at the base of the bill minimizes the risk of clogging by sand or mud. The long, bristly feathers behind it help protect the eye.

TAIL ∨
The laughing kookaburra's tail is long for a kingfisher. In flight, the tail is spread wide, revealing all 12 feathers.

outer tail feathers have more white bands than inner feathers

WING ∧
Short and broad, the wing is curved into an airfoil: the thick, muscular inner "arm" tapers to flat, bladelike flight feathers behind and at the tip.

HORNBILLS, HOOPOES, AND WOOD HOOPOES

Hornbills (including some giants of the bird world), hoopoes, wood hoopoes, and the similar scimitarbills form an Old World order, the Bucerotiformes, some boldly colored and all strikingly patterned and distinctive in form.

Hoopoes walk on the ground and probe for invertebrates and small reptiles with their slender, curved bill, while wood hoopoes hop on trees and probe inside loose bark. Hornbills have a larger, longer bill, some with a large, horny adornment on top. They range from medium-sized to very large forest species. Ground hornbills are extremely large, heavy birds that feed exclusively on the ground. Some hornbills nest in tree holes and males seal the females inside with mud, feeding them through a narrow slit until the young hatch.

PHYLUM	CHORDATA
CLASS	AVES
ORDER	BUCEROTIFORMES
FAMILIES	4
SPECIES	74

reddish bill

16½–18 in
42–45 cm

gray, white, and black plumage

white underparts

RED-BILLED HORNBILL
Tockus erythrorhynchus
F: Bucerotidae
Tockus hornbills are small African predators, most with a red or yellow bill. This species inhabits savanna and open woodland from Senegal to Namibia.

TRUMPETER HORNBILL
Bycanistes bucinator
F: Bucerotidae
This hornbill inhabits the forests of eastern to southern Africa. It is similar to the related silvery-cheeked hornbill, but has red facial skin.

23–26 in
58–65 cm

30–32 in
75–80 cm

SILVERY-CHEEKED HORNBILL
Bycanistes brevis
F: Bucerotidae
This fruit-eating pied hornbill is a forest species from E. Africa. As with other hornbills, the male has a larger bill than the female, with a more prominent casque.

ORIENTAL PIED HORNBILL
Anthracoceros albirostris
F: Bucerotidae
The most widespread of the Asian pied hornbills, this bird occurs from the Himalayas to Bali (Indonesia), in cultivated land as well as forest.

28 in
70 cm

3¼ in
1 m

black-tipped crest, raised when landing

28 in
70 cm

10–12½ in
25–32 cm

black and white bars on wings

12–14 in
30–36 cm

MALABAR PIED HORNBILL
Anthracoceros coronatus
F: Bucerotidae
Like most other Asian hornbills, this pied species from India and Sri Lanka is omnivorous, and eats a great deal of fruit.

NORTHERN GROUND HORNBILL
Bucorvus abyssinicus
F: Bucorvidae
One of two predatory ground hornbills of African grasslands, this species from Senegal to Kenya tolerates drier conditions than its southern counterpart.

HOOPOE
Upupa epops
F: Upupidae
This species nests in tree-holes in Africa and Eurasia. It has a flickering, butterflylike flight and a strong bill to dig for ground-living invertebrates.

GREEN WOOD HOOPOE
Phoeniculus purpureus
F: Phoeniculidae
This African wood hoopoe climbs trees like a woodpecker, but has a hoopoelike bill to probe rotting wood for invertebrates.

WOODPECKERS AND TOUCANS

Mostly tree-dwelling cavity nesters, birds of this order, the Piciformes, have a common foot structure. More than half of the species are woodpeckers.

The woodpeckers are almost worldwide in distribution. These birds cling to trunks with zygodactyl feet typical of the order (two toes pointing forward, two backward) and, propped by their stiffened tail, drum through the wood with their strong bill. They extract prey with a long barbed tongue. Other birds in the order are largely confined to the tropics. Honeyguides feed on the wax from raided bee nests. Jacamars hunt large insects, while puffbirds are flycatchers. Barbets have a serrated bill for eating fruits; the tropical American barbets are cousins of the toucans. Honeyguides are brood parasites, laying eggs in the nests of other birds, including woodpeckers. Others in this group nest in tree-holes, or may dig burrows in the ground or in termite mounds.

PHYLUM	CHORDATA
CLASS	AVES
ORDER	PICIFORMES
FAMILIES	9
SPECIES	445

21–23½ in
53–60 cm

WHITE-THROATED TOUCAN
Ramphastos tucanus
F: Ramphastidae
Like other toucans, this species from northern S. America breeds in tree cavities, often making use of abandoned woodpecker nests.

17 in
43 cm

RED-BREASTED TOUCAN
Ramphastos dicolorus
F: Ramphastidae
Among the smallest of the *Ramphastos* toucans, this species from eastern S. America is the only one with extensive red underparts.

CUVIER'S TOUCAN
Ramphastos tucanus cuvieri
F: Ramphastidae
This Amazonian bird is often regarded as a subspecies of the white-throated toucan, from which it differs by having a darker bill.

22–23½ in
55–60 cm

19 in
48 cm

CHANNEL-BILLED TOUCAN
Ramphastos vitellinus
F: Ramphastidae
Ramphastos toucans are black with a white or yellow chest. In this variable S. American species, the chest color depends on race.

12–14 in
30–35 cm

EMERALD TOUCANET
Aulacorhynchus prasinus
F: Ramphastidae
The most widespread of a group of small green toucans, this species occurs from Mexico to Bolivia. The species consists of several races.

14 in
35 cm

♀

SPOT-BILLED TOUCANET
Selenidera maculirostris
F: Ramphastidae
This S. Brazilian species is one of several dichromatic toucanets, which are the only toucans with different-colored sexes. The females are marked with brown.

pale orange around eye

22–26 in
55–65 cm

TOCO TOUCAN
Ramphastos toco
F: Ramphastidae
This bird from northern S. America is the largest of the toucans. Unlike other forest species, it also inhabits more open woodland habitats.

massive, black-tipped orange bill

blue eye-skin

SAFFRON TOUCANET
Pteroglossus bailloni
F: Ramphastidae
Found in S.E. Brazil, this toucan has a yellow-olive plumage. It is related to the more brightly colored aracaris.

14–16 in
35–40 cm

16 in
41 cm

COLLARED ARACARI
Pteroglossus torquatus
F: Ramphastidae
With the most northerly range of the aracaris, this species occurs in humid forests from S. Mexico to northern S. America.

14½ in
37 cm

ornate black bill with yellow bands

CHESTNUT-EARED ARACARI
Pteroglossus castanotis
F: Ramphastidae
Aracaris are long-tailed, gregarious toucans. Most have a red rump and prominent bands on the belly. This species is from northwestern S. America.

7½ in
19 cm

6½ in
17 cm

9 in
23 cm

BLACK-BILLED BARBET
Lybius guifsobalito
F: Ramphastidae
Many African barbets occur in more open country than the forest-dwelling species of the Americas and Asia. This species occurs in E. Africa.

BLACK-SPOTTED BARBET
Capito niger
F: Ramphastidae
This barbet from northern S. America is rarely seen, but its froglike call is often heard.

RED-HEADED BARBET
Eubucco bourcierii
F: Ramphastidae
This species is found from Costa Rica to Peru. It is unusual among barbets in being largely silent.

BLUE-THROATED BARBET
Psilopogon asiaticus
F: Ramphastidae
Asian barbets often forage with other fruit-eaters in the forest canopy. This species occurs from the Himalayas to Thailand.

9 in
23 cm

red forehead

green upperparts

10 in
26 cm

BEARDED BARBET
Lybius dubius
F: Ramphastidae
Barbets have sensory bristles at the base of their bill, which are much enlarged in this species from W. and C. Africa.

8 in
20 cm

6½ in
17 cm

large, yellowish bill

11 in
28 cm

BROWN-HEADED BARBET
Psilopogon zeylanicus
F: Ramphastidae
A typical Asian barbet, this species from the Himalayas, India, and Sri Lanka is a fruit-eater and especially favors figs.

COPPERSMITH BARBET
Psilopogon haemacephalus
F: Ramphastidae
This is a widespread barbet found in forest edges and scrub. Its incessant "tonk-tonk" hammering call is a common sound in S. Asia.

CRIMSON-FRONTED BARBET
Psilopogon rubricapillus
F: Ramphastidae
A small Asian barbet, this species is restricted to Sri Lanka and southwestern India. It is a common species that is seen in towns.

12½–13 in
32–33 cm

GREAT BARBET
Psilopogon virens
F: Ramphastidae
The largest Asian barbet, this species is a noisy bird of highland forests from the E. Himalayas to Thailand.

red vent

RED-FRONTED TINKERBIRD
Pogoniulus pusillus
F: Ramphastidae
Restricted to riverside forests of the E. African coast, this species feeds on insects and fruits, including mistletoe berries.

4–4¼ in
10–11 cm

tail narrows to point

4–4¼ in
10–11 cm

YELLOW-RUMPED TINKERBIRD
Pogoniulus bilineatus
F: Ramphastidae
Tinkerbirds are small, black and white African barbets with a repetitive call that continues through the day. This species is widespread south of the Sahara.

YELLOW-FRONTED TINKERBIRD
Pogoniulus chrysoconus
F: Ramphastidae
A more widespread relative of the red-fronted tinkerbird, this species occurs in dry, open woodland and savanna in sub-Saharan Africa.

4¼ in
11 cm

beardlike bristles

RED-FRONTED BARBET
Tricholaema diademata
F: Ramphastidae
An E. African barbet related to
the spot-flanked barbet, this species
occurs in drier habitats. Like other
barbets, it nests in tree-holes.

9 in
22 cm

**SPOT-FLANKED
BARBET**
Tricholaema lacrymosa
F: Ramphastidae
A barbet of damp
woodlands from
C. to E. Africa, the
spot-flanked barbet mainly
feeds on figs and berries.

9 in
22 cm

*white-spotted
upperparts*

6–6½ in
15–16 cm

**D'ARNAUD'S
BARBET**
Trachyphonus darnaudii
F: Ramphastidae
Trachyphonus barbets
from Africa live in
open country and spend
much time on the ground.
D'Arnaud's barbet occurs
throughout E. Africa.

TOUCAN-BARBET
Semnornis ramphastinus
F: Ramphastidae
A bird of the rainforests of
Colombia and Ecuador, this species
is intermediate between barbets and
toucans. It is exclusively a fruit-eater.

8 in
20 cm

*striking head
pattern*

9 in
23 cm

**RED-AND-YELLOW
BARBET**
Trachyphonus erythrocephalus
F: Ramphastidae
A typical African ground
barbet, this bird eats insects,
fruits, seeds, and even small
lizards. It often burrows into
termite mounds to nest.

FIRE-TUFTED BARBET
Psilopogon pyrolophus
F: Ramphastidae
The only Asian barbet with a
graduated tail, this S.E. Asian
species also has tufts of
facial bristles and a
cicadalike call.

11 in
28 cm

4¾–5 in
12–13 cm

6½ in
17 cm

**GREEN-BACKED
HONEYBIRD**
Prodotiscus zambesiae
F: Indicatoridae
Honeybirds eat insects, fruits,
and even beeswax. This African
species lays eggs in the nests of the
white-eye, a small woodland bird,
and its young kill host nestlings.

NORTHERN WRYNECK
Jynx torquilla
F: Picidae
This Eurasian woodland
ant-eater has a weaker bill
than true woodpeckers and
is named for its ability to
twist its neck.

**BAR-BREASTED
PICULET**
Picumnus aurifrons
F: Picidae
Piculets are tiny nuthatchlike
members of the woodpecker
family. They use their short
bill for taking insects from
decayed wood. This species is
from central S. America.

4 in
10 cm

4 in
10 cm

**GOLDEN-SPANGLED
PICULET**
Picumnus exilis
F: Picidae
This S. American piculet,
like other piculets, lacks stiff,
supporting tail feathers of bigger
woodpeckers, so it spends less
time on vertical trunks.

4 in
10 cm

4 in
10 cm

4 in
10 cm

OCHRACEOUS PICULET
Picumnus limae
F: Picidae
Confined to E. Brazil, the
ochraceous piculet—like other
piculets—reuses woodpecker
nesting holes, as its bill is too
small to excavate its own nest.

**OCHER-COLLARED
PICULET**
Picumnus temminckii
F: Picidae
Confined to the forests of
E. Paraguay, S.E. Brazil, and
N.E. Argentina, this piculet
is well camouflaged by its
pattern and coloration.

SPOTTED PICULET
Picumnus pygmaeus
F: Picidae
The spotted piculet
occurs only in the
tropical forests of
N.E. Brazil, where it
is relatively common.

4 in
10 cm

≫

**12 in
30 cm**

GROUND
WOODPECKER
Geocolaptes olivaceus
F: Picidae

Unusual in living on the
ground, this South African
ant-eating woodpecker
inhabits barren, rocky areas
and tunnels into earth banks
to make a nesting chamber.

**7 in
18 cm**

NUBIAN WOODPECKER
Campethera nubica
F: Picidae

This relative of the European
green woodpecker occurs in dry
parts of northeastern Africa,
where it is often seen in pairs.

**7–9 in
18–22 cm**

YELLOW-BELLIED
SAPSUCKER
Sphyrapicus varius
F: Picidae

Sapsuckers make holes in
trees to drink sap. This
fork-tailed species breeds
in N. America and migrates
to the Caribbean.

**11 in
28 cm**

COMMON
FLAME-BACKED
WOODPECKER
Dinopium javanense
F: Picidae

A tropical woodpecker,
this species occurs in
a variety of woodland
types—including
mangrove—and ranges
from India eastward
to Borneo and Java.

**6–6½ in
15–17 cm**

HEART-SPOTTED
WOODPECKER
Hemicircus canente
F: Picidae

One of two related small,
crested woodpeckers
from S.E. Asia, this
species is named for the
black, heart-shaped
marks on its back.

**16–19½ in
40–49 cm**

PILEATED WOODPECKER
Dryocopus pileatus
F: Picidae

This is the largest N. American
woodpecker. Unlike the Eurasian black
woodpecker, American members of
the *Dryocopus* genus are crested.

red cap

male has
red face streak

**18–22½ in
45–57 cm**

**9 in
23 cm**

GOLDEN-GREEN
WOODPECKER
Piculus chrysochloros
F: Picidae

Typical of a group of
green-backed woodpeckers
from tropical America,
this bird often follows
mixed-species flocks,
gleaning bark surfaces.

BLACK WOODPECKER
Dryocopus martius
F: Picidae

This large woodpecker
from woodlands across
N. Eurasia belongs
to a small group of
predominantly black species
that have a red crown.

white patch
on wing

♀

**7½–9 in
19–23 cm**

white
underparts

**9½ in
24 cm**

RED-BELLIED
WOODPECKER
Melanerpes carolinus
F: Picidae

Like other *Melanerpes*
woodpeckers, this common
N. American species stores food
in crevices. Despite its name,
the belly only has a tinge of red.

**12–13 in
31–33 cm**

GREEN
WOODPECKER
Picus viridis
F: Picidae

This European member
of a group of Old
World green-backed
woodpeckers is also
called the yaffle after its
distinct yelping call.

RED-HEADED
WOODPECKER
Melanerpes erythrocephalus
F: Picidae

A distinctive woodpecker
from N. America, this is
an aggressive species that
destroys nests and eggs of
other birds in its territory.

**7½ in
19 cm**

YELLOW-FRONTED
WOODPECKER
Melanerpes flavifrons
F: Picidae

Most *Melanerpes* woodpeckers
have partly barred plumage.
They may also have striking
colors, as does this
S. American species.

**12 in
31 cm**

ROBUST
WOODPECKER
Campephilus robustus
F: Picidae

The *Campephilus* woodpeckers
are black and white birds
with a red head. The robust
woodpecker is restricted
to eastern S. America.

red patch (in male)

HAIRY WOODPECKER
Picoides villosus
F: Picidae

A N. American relative of the
three-toed woodpecker, this
species becomes more abundant
according to the availability of
its prey—larvae of bark beetle.

red nape
(in male)

8–9 in
20–22 cm

11–12 in
28–31 cm

9 in
22–23 cm

NORTHERN FLICKER
Colaptes auratus
F: Picidae

American flickers are so
called because of the way
their colored underwings
flash in flight. This species
has red- and yellow-shafted
forms that hybridize.

dark bars
on wings

7–10 in
18–26 cm

**GREAT SPOTTED
WOODPECKER**
Dendrocopos major
F: Picidae

A widespread member
of a Eurasian group of
pied woodpeckers, this
bird is common in
forests and gardens from
Europe to S.E. Asia.

red patch
beneath tail

**MIDDLE-SPOTTED
WOODPECKER**
Dendrocoptes medius
F: Picidae

A pied woodpecker
confined to Europe and
S.W. Asia, this species
drums less than the related
great spotted woodpecker.

iridescent green
upperparts

11 in
28 cm

9 in
23 cm

red underparts

**RUFOUS-TAILED
JACAMAR**
Galbula ruficauda
F: Galbulidae

This C. and S. American
species is typical of its
genus, characterized by
iridescent green coloration
above and rufous below.

8 in
20 cm

7 in
18 cm

**THREE-TOED
JACAMAR**
Jacamaralcyon tridactyla
F: Galbulidae

The most dull-colored jacamar,
this dry forest species nests in
earth banks and has two toes
in front and one behind.

**CHESTNUT
JACAMAR**
*Galbalcyrhynchus
purusianus*
F: Galbulidae

Jacamars specialize in
catching large insects,
such as butterflies, from
the air. This is one of
two chestnut-colored
species from S. America.

GREAT JACAMAR
Jacamerops aureus
F: Galbulidae

This is the largest species
of jacamar. It occurs from
Costa Rica to Bolivia and
feeds mainly on insects,
supplemented by small lizards.

8–8½ in
20–22 cm

11 in
28 cm

**WHITE-EARED
PUFFBIRD**
Nystalus chacuru
F: Bucconidae

Like other puffbirds, this
species from central S.
America is a big-headed,
"puff-bodied" bird with a
heavy bill for catching
small animals.

6 in
15 cm

**BLACK-FRONTED
NUNBIRD**
Monasa nigrifrons
F: Bucconidae

Nunbirds are
black-bodied relatives
of puffbirds. This noisy
S. American species often
forages beneath monkey
troops, feeding on disturbed
small animal prey.

**RUSTY-BREASTED
NUNLET**
Nonnula rubecula
F: Bucconidae

Nunlets are small,
drab-colored members of
the puffbird family. This
bird is a typical
S. American species of
vine-bordered forest.

5½ in
14 cm

**SWALLOW-WINGED
PUFFBIRD**
Chelidoptera tenebrosa
F: Bucconidae

Described as martinlike
when perched and batlike
in flight, this bird of
northern S. America
sallies for flying insects
along rivers.

SERIEMAS

**Seriemas, order Cariamiformes, are
large, long-legged, noisy ground birds
of dry savanna woodland in South
America, with unclear relationships
to other bird orders.**

PHYLUM	CHORDATA
CLASS	AVES
ORDER	CARIAMIFORMES
FAMILIES	1
SPECIES	2

Once linked with cranes, seriemas have since been placed closer to falcons. The
two species are among the largest ground-living birds of South America and rarely
fly, except to reach the canopy of trees to nest and roost overnight. Their loud calls
make them more often heard than seen.

loose plumage on
neck and breast

barred wing feathers
rarely revealed

RED-LEGGED SERIEMA
Cariama cristata
F: Cariamidae

Related to the extinct, predatory,
giant terror birds, seriemas are natives
of the grasslands of S. America.
This species has a sicklelike claw
to dismember small prey.

30–35 in
75–90 cm

PASSERINES

This huge order makes up about 60 percent of all bird species. They are known collectively as perching birds, for their specialized feet.

Like many birds, passerines have four-toed feet with three toes pointing forward and one back. As a passerine lands, a muscle automatically tightens tendons running through the legs so that the toes lock around a perch, securing the bird even in sleep.

These birds, of the order Passeriformes, are found in almost all land habitats worldwide, from dense rainforest to arid desert and even freezing Arctic tundra. They range in size from the tiny American flycatchers, no bigger than hummingbirds, to the powerfully built Eurasian ravens.

DIVERSITY

The order includes birds adapted to a variety of feeding habits. The insect-eaters have a bill like a needle for probing foliage or with a wide gape for catching prey in flight. Other passerines have a stubby bill for cracking open seeds, while some have a long curved bill for extracting nectar.

Passerines combine a high metabolic rate with a comparatively large brain size, giving some the resilience to tolerate cold and a few the intelligence to master the use of simple tools. They produce helpless, naked young, which are reared in nests ranging from simple cups to elaborate mud-chambers or hanging purses of woven grass. A few passerines are brood parasites, laying their eggs in the nests of other species.

SONGBIRDS

Passerines are split into two groups, largely on the basis of the structure of their vocal box. The first group, the Suboscines, forms about a fifth of all passerines. It occurs in the Old World tropics but has greatest diversity in America; it includes broadbills, pittas, antbirds, and tyrant flycatchers. The second group, the Oscines, includes all remaining passerine families, popularly known as songbirds. Birds often communicate by voice, but the structure of their vocal box enables many passerines to deliver complex songs that are important in courtship and defending territory. Songs are often so distinctive that they can be used to identify species.

PHYLUM	CHORDATA
CLASS	AVES
ORDER	PASSERIFORMES
FAMILIES	141
SPECIES	6,456

BROADBILLS

The families Calyptomenidae and Eurylaimidae consist of forest-dwelling birds from tropical Africa and Asia. They mostly use their wide bill to catch insects in trees, but members of one group of Asian green broadbills are fruit-eaters.

6½–7 in
17–18 cm

GREEN BROADBILL
Calyptomena viridis
One of the three S.E. Asian green broadbills, this bird is a specialist fruit-eater and builds globular hanging nests.

blue bill with yellow at base

10 in
25 cm

BLACK-AND-RED BROADBILL
Cymbirhynchus macrorhynchos
This distinctively colored S.E. Asian broadbill frequents forests near water. It builds a pouchlike nest suspended from the tips of branches.

6 in
15 cm

BLACK-AND-YELLOW BROADBILL
Eurylaimus ochromalus
An Asian insect-eating broadbill, this species forages in the middle and upper levels of rainforests from Myanmar to Borneo and Sumatra.

ASITIES

The brush-tipped tongue of these Madagascan birds, the Philepittidae, suggests that they may have evolved from nectar-feeding ancestors: one genus is now fruit-eating; the other resembles the unrelated sunbirds that feed on nectar.

thin, decurved bill

3½ in
9 cm

COMMON SUNBIRD-ASITY
Neodrepanis coruscans
One of two long-billed birds from eastern Madagascar, this species evolved the nectar-feeding habit of true sunbirds, but is unrelated to them.

PITTAS

Birds of this Old World family forage for insects on the forest floor in tropical regions. The Pittidae have a rounded body and a short bill, and many species are brilliantly plumaged. Both adults incubate the eggs.

8 in
20 cm

7½ in
19 cm

BLUE-WINGED PITTA
Pitta moluccensis
Found from S. China to Borneo and Sumatra, the blue-winged pitta inhabits dense forest when breeding, but overwinters in coastal scrub.

INDIAN PITTA
Pitta brachyura
Like its relatives, this pitta from the S. Himalayas, India, and Sri Lanka builds domed nests on or near the ground.

MANAKINS

These birds, the Pipridae, are related to cotingas. The brightly colored males gather in groups called leks to display to females. For many species, courtship involves elaborate dancing movements. Females build the nest and rear offspring alone.

red cap

6 in
15 cm

5½–6 in
14–15 cm

ARARIPE MANAKIN
Antilophia bokermanni
Described as recently as 1998, this critically endangered bird is found only in the Araripe highlands of N.E. Brazil.

BLUE MANAKIN
Chiroxiphia caudata
One of the most colorful birds of the rainforest of S. Brazil, the blue manakin has a whining, catlike call.

4¼–5 in
11–13 cm

3½–4 in
9–10 cm

PIN-TAILED MANAKIN
Ilicura militaris
Males of this manakin from S.E. Brazil have elongated central tail feathers and display to females with raised rump feathers.

STRIPED MANAKIN
Machaeropterus regulus
This is an inconspicuous species of northern S. America. Males of the species have been observed making insectlike buzzing sounds while displaying.

3½ in
9 cm

GOLDEN-HEADED MANAKIN
Pipra erythrocephala
The male of this S. American species displays by jumping and whirring its wings—and sliding to one side or backward.

COTINGAS AND RELATIVES

A diverse group of tropical American passerines, the Cotingidae consists of fruit-eating or insect-eating forest birds. Males are brightly colored and are highly vocal when courting females; some species display in trees and others on the ground.

bright yellow throat

5 in
13 cm

10½–11 in
27–28 cm

BARE-THROATED BELLBIRD
Procnias nudicollis
Bellbirds are named after their metallic calls. Like other bellbirds, males of this eastern S. American species have white plumage.

FIERY-THROATED FRUIT-EATER
Pipreola chlorolepidota
Fruit-eaters are stocky green birds, males of which are usually black headed and red or yellow throated. This is a small Andean species.

bare red skin around eyes

glossy black body

purple-red patch

9 in
22 cm

11–12 in
28–30 cm

PURPLE-THROATED FRUITCROW
Querula purpurata
Males of this distinctive Amazonian cotinga have a prominent, iridescent purplish red throat patch. It can pluck fruits while hovering in midair.

BLACK-TAILED TITYRA
Tityra cayana
This canopy bird belongs to a group of big-headed fruit-eaters. It nests in tree-holes, often those abandoned by woodpeckers.

large crest hides the bill

brilliant red plumage

11–12½ in
28–32 cm

ANDEAN COCK-OF-THE-ROCK
Rupicola peruvianus
The colorful crested males of this large Andean cotinga display in groups to females. Females nest among rocks and rear young alone.

TYRANT FLYCATCHERS AND RELATIVES

Widespread throughout the Americas, these birds of the family Tyrannidae often account for a third of all passerines in South American bird communities. These birds are insectivorous and typically perch and wait for prey or glean it from foliage.

6½–8½ in
17–21 cm

6 in
15 cm

9 in
22 cm

gray head

brown tail with pale cinnamon edge

GREAT CRESTED FLYCATCHER
Myiarchus crinitus
A widespread tyrant flycatcher, this large, migratory species, like its relatives, catches insects on the wing. It often hovers when foraging.

EASTERN WOOD PEWEE
Contopus virens
This bird, with its characteristic "pewee" call, hunts by sallying—taking off from perches to catch insects in midair. It breeds in eastern N. America.

TROPICAL KINGBIRD
Tyrannus melancholicus
A large, sallying tyrant flycatcher, this aggressively territorial bird breeds in open habitats from southern N. America to S. America.

6 in
15 cm

dark brown upperparts

red underparts

4 in
10 cm

7½ in
19 cm

6½ in
17 cm

VERMILION FLYCATCHER
Pyrocephalus rubinus
A species of open country, this bird forages near the ground. Males are vivid red, whereas females are mainly gray and white.

COMMON TODY-FLYCATCHER
Todirostrum cinereum
This tiny C. and S. American tyrant flycatcher is typical of a group that forages by upward striking, but favors more open habitat than its relatives.

CLIFF FLYCATCHER
Hirundinea ferruginea
This flycatcher from northern and central S. America is swallowlike in its aerial foraging behavior, and perches on rocky outcrops.

BLACK PHOEBE
Sayornis nigricans
This tail-wagging, tyrant flycatcher forages close to the ground in tropical regions, often near water. It dives into ponds to capture minnows.

TYPICAL ANTBIRDS

From tropical American forests, these heavy-billed birds of the family Thamnophilidae hunt insects near the ground, with some following army ants to feed on insects fleeing the ants. Some are long-clawed for gripping vertical stems.

7 in
18 cm

WHITE-BEARDED ANTSHRIKE
Biatus nigropectus
This little-known, uncommon species is restricted to S.E. Brazil, where it feeds on insects in bamboo forests. It is threatened by deforestation.

ANTPITTAS AND ANT-THRUSHES

Short-tailed antpittas spend more time on the ground than the tree-dwelling ant-thrushes. Both are insectivores of South American forests and belong to the families Grallariidae and Formicariidae, respectively.

7 in
18 cm

MOUSTACHED ANTPITTA
Grallaria alleni
A rare species found in isolated localities in Colombia and Ecuador, the moustached antpitta inhabits undergrowth in humid montane forest.

AUSTRALASIAN WRENS

The Maluridae are small, cock-tailed insect-eaters that resemble wrens of the northern hemisphere, but are more closely related to nectar-feeding honeyeaters. Male fairy-wrens are patterned in blue and black. Emu-wrens and grasswrens are browner birds of grassy habitats.

STRIATED GRASSWREN
Amytornis striatus
Like most other grasswrens, this C. Australian species favors spinifex grass, where family groups dart beneath bushes.

6–7 in
15–18 cm

TAPACULOS AND CRESCENT-CHESTS

Strong-legged and weak-flying, the Rhinocryptidae and Melanopareiidae are among the most ground adapted of South American passerines. Some have long hind claws to scratch for food in soil and leaf litter.

5½–6 in
14–15 cm

COLLARED CRESCENT-CHEST
Melanopareia torquata
Crescent-chests have a longer tail than tapaculos and have been separated into a new family, the Melanopareiidae. This Brazilian species occurs in arid habitats.

GNATEATERS

Dumpy, short-tailed, long-legged insect-eaters of forest undergrowth, gnateaters of the family Conophagidae are secretive birds that forage near the ground using tyrantlike sallying or gleaning techniques. They are related to antbirds.

RUFOUS GNATEATER
Conopophaga lineata
More abundant than other gnateater species, this bird from eastern S. America often moves in mixed-species flocks and uses degraded habitats.

5 in
13 cm

VARIEGATED FAIRY WREN
Malurus lamberti
This most widespread Australian fairy wren, like others, builds a domed nest. Young may stay close to help raise the next brood.

6 in
15 cm

SPLENDID FAIRY WREN
Malurus splendens
Fairy wrens form strong pair-bonds, but partners may mate with other individuals. This species is found mostly in southern Australia.

5½ in
14 cm

white throat

bright yellow underparts

9 in
22 cm

GREAT KISKADEE
Pitangus sulphuratus
Widespread through tropical America and named after its call, this species is a typical sallying tyrant flycatcher, but also forages near the ground.

BOWERBIRDS AND CATBIRDS

The Australasian family Ptilonorhynchidae consists predominantly of fruit-eaters. Male bowerbirds—often brightly colored—build structures called bowers to attract females. They mate with many females and play no part in raising the young.

9 in
23 cm

GREEN CATBIRD
Ailuroedus crassirostris
Male catbirds—named for their catlike call—attract mates by laying leaves on the ground. This species is found in New Guinea and E. Australia.

olive upperparts

9–10 in
23–25 cm

GOLDEN BOWERBIRD
Prionodura newtoniana
The male of this small species from N. Australia attracts a mate by building a tower of sticks up to 10 ft (3 m) high.

LYREBIRDS

The Menuridae are large Australian birds that eat ground insects. They have complex vocal organs and excel at mimicking forest sounds. Males court females on display mounds by fanning their long tails and plumes.

32–38 in
80–96 cm

SUPERB LYREBIRD
Menura novaehollandiae
The most common lyrebird, this species is found in forests of S.E. Australia and Tasmania. Its lyre-shaped outer tail feathers are adorned with notches.

AUSTRALASIAN TREECREEPERS

The birds of the family Climacteridae have evolved to resemble the northern treecreepers, but are unrelated and, unlike them, do not use their tail for support when climbing up trees.

6½–7 in
16–18 cm

BROWN TREECREEPER
Climacteris picumnus
This common eastern Australian species has a distinct dark-backed form in the north of the range and a brown-backed form in the south.

OVENBIRDS AND RELATIVES

These American birds, the Furnariidae, are adept at hunting hidden invertebrates and are known for the immense diversity of their nest architecture. This includes stick nests, tunnel nests, and those that resemble clay ovens.

7½–8 in
19–20 cm

7–8 in
18–20 cm

WHITE-EYED FOLIAGE-GLEANER
Automolus leucophthalmus
A S. American bird with distinctive white irises, like many insect-eaters, this species forages in mixed flocks to flush out prey.

RUFOUS HORNERO
Furnarius rufus
Widespread in central and southern S. America, this species, typical of the hornero group, builds an ovenlike mud nest.

HONEYEATERS

The birds of the family Meliphagidae are from Australia and the southwest Pacific islands. They feed on nectar with a long, brush-tipped tongue and are important pollinators of plants in that region. Other nectar-feeders, such as sunbirds, have evolved similar features.

olive-green wings

10–12 in
25–30 cm

4–4¼ in
10–11 cm

SCARLET HONEYEATER
Myzomela sanguinolenta
This bird is an E. Australian member of a group of long-billed, flower-visiting honeyeaters that often carry daubs of pollen on their forehead.

BLUE-FACED HONEYEATER
Entomyzon cyanotis
A large, noisy honeyeater from Australia and New Guinea, this species is more insectivorous than many others, but it also feeds on fruits.

7½–8½ in
19–21 cm

LEWIN'S HONEYEATER
Meliphaga lewinii
A member of a short-billed group of honeyeaters, this species from eastern Australia eats insects, fruits, and berries.

5–6½ in
13–16 cm

EASTERN SPINEBILL
Acanthorhynchus tenuirostris
Spinebills are an ancient group of heathland honeyeaters and the most specialized nectar-feeders of the family. This species is from E. Australia.

11½–12½ in
29–32 cm

TUI
Prosthemadera novaeseelandiae
Although confined to New Zealand, this species is related to short-billed honeyeaters of Australia. It has an extraordinary vocal range.

6½–7½ in
16–19 cm

NEW HOLLAND HONEYEATER
Phylidonyris novaehollandiae
Like other honeyeaters, this white-whiskered species from southern Australia and Tasmania also takes honeydew—the waste sugar solution produced by certain sap-sucking insects.

WOODCREEPERS

The tropical American birds of the family Dendrocolaptidae are specialists at climbing up tree trunks. They have stiffened tails for support and strongly clawed front toes for gripping the bark.

7½ in
19 cm

SCALLOPED WOODCREEPER
Lepidocolaptes falcinellus
This typical buff and brown woodcreeper is found only in the forests of southeastern S. America.

THORNBILLS AND RELATIVES

The Acanthizidae is a small family of warbler- and wrenlike, insect-eating birds from Australia and adjacent islands. It also includes Australia's smallest bird—the weebill. These birds have short wings, a short tail, and drab-colored, longish legs.

4¼ in
11 cm

BUFF-RUMPED THORNBILL
Acanthiza reguloides
Thornbills are mostly gray, brown, or yellow. This eastern species has a freckled forehead, like many of this group.

dark facial mask between white stripes

4¼–5½ in
11–14 cm

WHITE-BROWED SCRUBWREN
Sericornis frontalis
Scrubwrens are birds of Australasian thickets. Mostly brown, some have white head markings, such as this species, which is widespread across Australia and Tasmania.

PARDALOTES

Dumpy birds from Australia, the Pardalotidae have a stubby bill for gleaning sap-sucking scale insects from trees. Brightly colored, they nest in deep tunnels in earth banks.

3¼–4 in
8–10 cm

SPOTTED PARDALOTE
Pardalotus punctatus
Three of the four species of pardalotes have white spots. This highly active bird is from dry forests in S. and E. Australia.

WOOD SWALLOWS, BUTCHER-BIRDS, AND RELATIVES

From Southeast Asia, New Guinea, and Australasia, the birds in the family Artamidae include wood swallows that catch insects in flight and are among the few small passerines that can soar. The Australasian butcher-birds, currawongs, and ground-dwelling Australian magpies are intelligent, highly vocal, omnivorous birds.

7½ in
19 cm

MASKED WOOD SWALLOW
Artamus personatus
This dark-faced, thick-billed wood swallow occurs in drier parts of inland Australia and is highly nomadic. Like other wood swallows, it often gathers in large flocks.

AUSTRALIAN MAGPIE
Gymnorhina tibicen
A widespread Australian species with a highly variable black and white plumage, this bird has a varied and melodious song and is a capable mimic.

13½–17½ in
34–44 cm

CROWS AND JAYS

The family Corvidae, occurring worldwide, includes some of the biggest passerines. They are intelligent, opportunistic birds that have complex social organization and strong pair-bonding. Crows have demonstrated tool use, play behavior, and perhaps even self-awareness.

22–27 in
56–69 cm

COMMON RAVEN
Corvus corax
Found in a wide range of open-country habitats across the northern hemisphere, this is the most widespread crow and the largest passerine.

13–15½ in
33–39 cm

EURASIAN JACKDAW
Corvus monedula
A small crow of W. Eurasia and N. Africa, this bird nests in cavities on crags, and is found on coastal cliffs and in urban areas.

PIED CROW
Corvus albus
A relative of the raven, this heavy-billed bird of open country is perhaps the most common member of the crow family in Africa and Madagascar.

18–20 in
46–50 cm

very long tail

10–12 in
25–30 cm

BLUE JAY
Cyanocitta cristata
This colorful jay of N. America lives in tightly bonded family groups. It is fond of acorns and disperses them, helping to distribute oak trees.

COMMON MAGPIE
Pica pica
This Eurasian bird is common in habitats ranging from open woodland to semidesert. It is related more closely to *Corvus* crows than Asian magpies.

18 in
46 cm

18–19 in
45–48 cm

ROOK
Corvus frugilegus
Routinely flocking throughout the year in Eurasia, this bare-faced crow nests in colonies in trees in open countryside.

18½–20½ in
47–52 cm

CARRION CROW
Corvus corone
This common Eurasian, largely solitary species has a wide-ranging diet. It feeds on small animals and vegetable matter, as well as carrion.

IORAS

These rainforest birds, the Aegithinidae, are usually active in the high canopy. Their green or yellow coloration camouflages them in foliage, where they glean for insects. Males can engage in elaborate courtship displays.

COMMON IORA
Aegithina tiphia
The smallest and most widespread iora, it is found across tropical Asia, from India to Borneo, sometimes in disturbed habitats. This bird builds a cup-shaped nest.

6 in
15 cm

ORIOLES

Old World passerines related to shrikes and crows, the birds of the family Oriolidae live in forest canopies and eat insects and fruits. Many species have striking yellow and black plumage. The females are usually greener than males.

10½–11½ in
27–29 cm

9½ in
24 cm

EURASIAN GOLDEN ORIOLE
Oriolus oriolus
Breeding in woodlands in W. and C. Eurasia, this oriole migrates southward to overwinter in Africa.

AUSTRALASIAN FIG-BIRD
Sphecotheres vieilloti
Stout-billed relatives of orioles from Australasia, fig-birds are gregarious fruit-eaters. This species occurs in N. and E. Australia.

SHRIKES

The family Laniidae contains predators of open country. Many species store their prey (insects and small vertebrates) by impaling them on thorns. Most occur in Africa and Eurasia; two species live in North America.

6½–7 in
17–18 cm

RED-BACKED SHRIKE
Lanius collurio
This bird breeds from Europe to Siberia and overwinters in Africa. Like other *Lanius* shrikes, it has a musical call.

BUSH SHRIKES AND RELATIVES

The family Malaconotidae is entirely confined to Africa. They are found mostly in scrubby, open woodland, and have a hooked bill for catching large insects.

CRIMSON-BREASTED GONOLEK
Laniarius atrococcineus
Members of the African bush shrike family, gonoleks have red and black plumage. This species occurs in southern Africa.

9 in
23 cm

orange-
red bill

26 in
67 cm

RED-BILLED BLUE MAGPIE
Urocissa erythrorhyncha
A forest bird from the Himalayas to E. Asia, this species robs chicks from nests and plucks at carcasses.

DRONGOS

Black, long-tailed birds of Old World tropics, the Dicruridae sally for insects, bolting out suddenly to catch their prey. They are aggressive birds, and sometimes attack larger species to defend their nests.

10 in
26 cm

CRESTED DRONGO
Dicrurus forficatus
Like other drongos, this Madagascan species has a long, forked tail and red eyes. It has a distinctive tuft of feathers at the base of its bill.

WATTLE-EYES AND RELATIVES

The Platysteiridae, wattle-eyes and their relatives, form a family of insect-eating birds from Africa. They have a flat, hooked bill with bristles at the base. Like flycatchers, they snatch their prey suddenly.

BROWN-THROATED WATTLE-EYE
Platysteira cyanea
Wattle-eyes get their name from the red skin around their eyes. This common species occurs in woodland throughout sub-Saharan Africa.

5 in
13 cm

GREEN JAY
Cyanocorax yncas
This bird feeds on fruits and seeds. S. American populations of the green jay differ sufficiently from those of C. America to be considered as separate species.

11½ in
29 cm

VANGAS AND RELATIVES

The predatory passerines of the family Vangidae include the helmet-shrikes of Africa and vangas from Madagascar. They feed on invertebrates, reptiles, and frogs, and have a range of bill shapes—chisel-, sickle-, and dagger-shaped—for different prey and feeding techniques.

VIREOS

Superficially resembling American warblers, but somewhat thicker-billed, the members of the family Vireonidae are more closely related to crows, and Old World orioles and shrikes. They take insects by gleaning or fly catching, and also eat some fruits.

AZURE-WINGED MAGPIE
Cyanopica cyanus
A sociable species, this bird breeds in colonies. Two separate populations (in Portugal and in E. Asia) of this gregarious woodland bird may constitute separate species.

12–14 in
31–35 cm

8 in
20 cm

WHITE HELMET-SHRIKE
Prionops plumatus
Widespread across sub-Saharan Africa, the white-crested helmet-shrike often gathers in small groups. This bird has a wide range of different calls.

BLACK-CAPPED VIREO
Vireo atricapilla
This species breeds in N. America and migrates to Mexico. Unlike other vireos, the sexes differ. The male is black-capped, while the female is gray-capped.

4¼ in
11 cm

13½ in
34 cm

EURASIAN JAY
Garrulus glandarius
More closely related to Old World crows than American jays, this colorful woodland bird habitually hoards acorns in the fall.

8 in
20 cm

RUFOUS VANGA
Schetba rufa
Common in the forests of Madagascar, this bird resembles a shrike, but is not closely related to the shrike family.

4¾–5 in
12–13 cm

RED-EYED VIREO
Vireo olivaceus
N. American populations of this very vocal vireo migrate to S. America, where they join resident races of the same species.

TYPICAL TITS

These small, acrobatic, usually hole-nesting birds of the family Paridae occur in wooded habitats across North America, Eurasia, and Africa. They frequently hang upside down to glean insects from foliage, and manipulate seeds and nuts to crack them open.

4¾–5½ in
12–14 cm

VARIED TIT
Sittaparus varius
This species is found in forest habitats, including conifers and bamboo forests, in N.E. Asia, Japan, and Taiwan.

4¾–6 in
12–15 cm

BLACK-CAPPED CHICKADEE
Poecile atricapillus
A typically inquisitive and acrobatic tit, this is a common N. American bird. Like other tits, it hoards seeds for later use.

GREAT TIT
Parus major
Widespread across Eurasia, this tit occurs in habitats from forest to heathland, and has a diverse range of vocalization.

5½ in
14 cm

5½–6½ in
14–16 cm

TUFTED TITMOUSE
Baeolophus bicolor
Like other tits, this eastern N. American species supplements its insect diet with seeds, holding them firmly to smash them with its bill.

4½–4¾ in
11–12 cm

BLUE TIT
Cyanistes caeruleus
This common bird of broadleaf woodlands in Europe, Turkey, and N. Africa is a frequent visitor to bird feeders in gardens.

PENDULINE TITS

These small, needle-billed birds from the family Remizidae occur in Africa and Eurasia, with one species in America. Most use cobwebs and other soft material to build flask-shaped nests that hang from branches, often over water.

4¼ in
11 cm

PENDULINE TIT
Remiz pendulinus
The only member of the family to have a wide range across Eurasia, this species occurs on marshes with trees, where it builds its pendulous nest.

3½–4¼ in
9–11 cm

VERDIN
Auriparus flaviceps
Unlike most other penduline tits, the verdin makes a spherical nest. It is found in desert scrub of the S. USA and Mexico.

BIRDS-OF-PARADISE

Mainly found in the rainforests of New Guinea, members of the family Paradisaeidae are mostly fruit-eating birds. The males display their bright, gaudy plumes in elaborate courtship displays, and spend most of their energy on this mating ritual, leaving the females to rear the young alone.

12½ in
32 cm

LESSER BIRD-OF-PARADISE
Paradisaea minor
This species occurs across N. and W. New Guinea. Males use their long, yellow flank plumes and distinctive cape for courtship displays.

yellow flank plumes

AUSTRALIAN ROBINS

The Petroicidae are chunky, round-headed insect-eaters. Unrelated to the robins of Europe or America, they are found from Australasia to the islands of the southwest Pacific. Some exhibit cooperative breeding, in which young birds help their parents to raise a new brood.

5 in
13 cm

JACKY WINTER
Microeca fascinans
This common robin uses its broad bill for fly-catching. It is widespread in woodland throughout Australia and New Guinea.

6 in
15 cm

EASTERN YELLOW ROBIN
Eopsaltria australis
Common in the woodlands and gardens of E. Australia, this bird sallies from low perches for invertebrates, which it snatches from the ground.

LONG-TAILED TITS

Birds of the family Aegithalidae are small, restless insect-eaters that build dome-shaped nests that are woven with cobwebs and lined with feathers. Most species are found in Eurasia, with one occurring in North America.

5½ in
14 cm

LONG-TAILED TIT
Aegithalos caudatus
The most widespread long-tailed tit, this woodland bird inhabits a range from N. to C. Eurasia, gathering in restless flocks when not breeding.

WAXWINGS

From the family Bombycillidae, these berry-eating birds are named for the waxy, red-tipped shafts on their wings. Three species occur in the cool northern forests of North America and Eurasia.

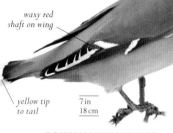

waxy red shaft on wing

yellow tip to tail

7 in
18 cm

BOHEMIAN WAXWING
Bombycilla garrulus
This sleek bird is pinkish brown, with chestnut undertail feathers. It breeds in northern taiga forest and is attracted to berried shrubs during its migration south.

SILKY FLYCATCHERS

The family Ptilogonatidae from Central America has just four species. These birds are named for their soft plumage (similar to that of related waxwings) and their feeding habits.

PHAINOPEPLA
Phainopepla nitens
This bird of the S. US and Mexico nests colonially in woodland, but is territorial when breeding in deserts.

7–8½ in
18–21 cm

MONARCHS AND RELATIVES

Generally long-tailed birds, members of the Monarchidae have a broad bill for fly-catching. Most species occur in Old World tropical forests. With the exception of the magpie-lark, they are tree-dwellers that build cup-shaped nests decorated with lichen.

black head

10–12 in
26–30 cm

reddish brown
upperparts

long tail feather

AFRICAN PARADISE-FLYCATCHER
Terpsiphone viridis
This bird has variable color forms, but all males have long tail streamers. This species occurs in savanna, south of the Sahara.

6½–15 in
17–38 cm

MAGPIE-LARK
Grallina cyanoleuca
Unlike other members of the monarch family, this Australian bird spends much time on the ground and builds a large mud nest.

AUSTRALIAN MUDNESTERS

The family Corcoracidae contains two species of social birds that feed on the ground and build large, cup-shaped nests of grass. These are held together with mud and are made in trees on horizontal branches.

APOSTLEBIRD
Struthidea cinerea
This ground-living bird associates in flocks of 6 to 20 individuals. It occurs in woodlands of northern and E. Australia.

11½–12½ in
29–32 cm

LARKS

Brown in color, with a melodious call, these birds from the family Alaudidae inhabit arid, open habitats. Most occur in Africa, with one species in North America. They usually have long hind claws that provide the stability they need to spend so much time on the ground.

7–8 in
18–20 cm

GREATER HOOPOE-LARK
Alaemon alaudipes
This long-legged lark has a curved bill and occurs in arid habitats of N. Africa and the Middle East, where it habitually runs on the ground.

HORNED LARK
Eremophila alpestris
This lark breeds on Arctic N. American and Eurasian tundra. It overwinters on coastlines farther south.

5½–6½ in
14–17 cm

7–7½ in
18–19 cm

EURASIAN SKYLARK
Alauda arvensis
Common across Eurasia, from the British Isles to Japan, this open-country bird is notable for its musical aerial song.

BULBULS

Found across the warmer parts of Eurasia and Africa, most bulbuls of the family Pycnonotidae are gregarious, noisy, fruit-eating birds. The soft plumage of many species is drably colored, marked by red or yellow feathers beneath the tail.

red cheek patch

9–10 in
23–25 cm

BLACK BULBUL
Hypsipetes leucocephalus
Common in forests and gardens in India, China, and Thailand, this species exists in dark-headed and white-headed races.

8 in
20 cm

RED-WHISKERED BULBUL
Pycnonotus jocosus
A common Asian bulbul found from India to the Malay Peninsula, this is an opportunistic woodland species that is also found near villages.

SWALLOWS AND MARTINS

From the family Hirundinidae, these swiftlike birds have long wings and a forked tail. Their short, flattened bill and wide gape help in taking insects on the wing. They build mud nests or use tree-holes or tunnels in banks.

GREATER STRIPED SWALLOW
Cecropis cucullata
This African grassland swallow breeds toward the south of the continent and migrates northward to overwinter.

8 in
20 cm

4¾–5½ in
12–14 cm

BANK SWALLOW
Riparia riparia
Like other swallows, this species migrates south to overwinter in the tropics. It nests in colonies in riverside banks across the northern hemisphere.

4¾–6 in
12–15 cm

TREE SWALLOW
Tachycineta bicolor
This N. American swallow of wooded swamps supplements its insect diet with berries. This enables it to breed farther north than other swallows.

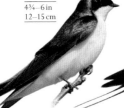

6–7½ in
15–19 cm

BARN SWALLOW
Hirundo rustica
Occurring worldwide, this is the most widespread species of swallow. It was originally a cave-nester, but now uses buildings as well.

BABBLERS, LAUGHING THRUSHES, AND RELATIVES

Generally more gregarious, noisier, and less migratory than warblers, the babblers and laughing thrushes of the families Timaliidae and Leiothrichidae have evolved into a variety of warblerlike and thrushlike forms. Some species are brightly colored.

silver-gray ear patch

9 in
23 cm

WHITE-EARED SIBIA
Heterophasia auricularis
Sibias are nectar-feeding babblers. This species is confined to Taiwan, where its distinctive call is often heard in mountain forests.

CHESTNUT-CAPPED BABBLER
Timalia pileata
A bird of low thickets in S.E. Asia, this babbler is often found near water, along with flycatchers and other babblers.

6½ in
16–17 cm

dark red wing patch

5½ in
14 cm

RED-TAILED MINLA
Minla ignotincta
This small babbler is similar to a tit. It is a noisy inhabitant of montane forest canopies in Nepal, China, and Myanmar.

7 in
18 cm

13 in
33 cm

SILVER-EARED MESIA
Leiothrix argentauris
A skulking, montane forest bird, this S.E. Asian species belongs to a group of "song babblers," which includes minlas, sibias, and laughing thrushes.

GREATER NECKLACED LAUGHING-THRUSH
Garrulax pectoralis
Laughing thrushes are large forest babblers with laughing calls. They often move in mixed flocks. This species inhabits the Himalayas and S.E. Asia.

GNATCATCHERS

These small American insect-eaters from the family Polioptilidae are related to wrens, but are more warblerlike in appearance. In common with several members of the wren family, some gnatcatchers cock their tail when foraging.

4¾ in
12 cm

BLUE-GRAY GNATCATCHER
Polioptila caerulea
This N. American gnatcatcher may flick its white-edged tail to flush insects. Unlike related gnatcatcher species, males do not have dark head markings.

TREECREEPERS

From the family Certhiidae, these are small insect-eating birds of the northern hemisphere. They forage on vertical tree trunks using their tail as a prop, habitually climbing one tree and flying to the bottom of the next.

EURASIAN TREECREEPER
Certhia familiaris
This is the most widespread *Certhia* treecreeper. It is found across Eurasia, from Britain to Japan, in broadleaf and coniferous forests.

5 in
13 cm

OLD WORLD WARBLERS AND RELATIVES

Several families, such as the large and widespread Sylviidae and those with fewer species such as Madagascar's Bernieridae, comprise a varied group of thin-billed, insectivorous birds. Some are forest species, while others inhabit low bushes, dense grasses, or taller reedbed habitats. Many are subtly patterned and difficult to tell apart.

5–6 in
13–15 cm

ICTERINE WARBLER
Hippolais icterina
This Eurasian woodland species has a more musical call than other reed warblers of the family Acrocephalidae. It migrates to southern Africa in winter.

5 in
13 cm

SEDGE WARBLER
Acrocephalus schoenobaenus
This is one of many species of reed warblers (inhabitants of wetland areas) that breeds in Eurasia and winters in Africa.

lemon-yellow underparts

7½–9 in
19–23 cm

CAPE GRASSBIRD
Sphenoeacus afer
A bird of South African shrubland, it belongs to an ancient African family, the Macrosphenidae, which evolved separately from other Old World warblers.

7–9½ in
18–24 cm

BROWN SONGLARK
Cincloramphus cruralis
This "grassbird" warbler from Australia in the family Locustellidae is a nomad of open habitats. Like larks, it sallies skyward from exposed perches.

WRENTIT
Chamaea fasciata
A dull-colored bird with a cocked tail, the wrentit is the only New World member of the family Sylviidae. It may be related to parrotbills.

6 in
15 cm

4¾ in
12 cm

VINOUS-THROATED PARROTBILL
Sinosuthura webbiana
Despite its stumpy, seed-cracking bill, this long-tailed Asian parrotbill is a member of the insectivorous Old World warblers family Sylviidae. This species occurs in China and Korea.

BLACKCAP
Sylvia atricapilla
Male *Sylvia* warblers are typically patterned with patches of black or brown. In this widespread Eurasian species, females have a brown cap.

5½ in
14 cm

SUBALPINE WARBLER
Sylvia cantillans
Like many *Sylvia* warblers, this species breeds in scrubby Mediterranean habitats and winters in Africa.

4¾–5 in
12–13 cm

BEARDED TIT

The bearded tit forms a single-species family, the Panuridae. A reedbed specialist, it eats insects in summer and hardens its stomach in winter to digest reed seeds.

6½ in
16–17 cm

BEARDED TIT
Panurus biarmicus

WHITE-EYES

Most of the Zosteropidae are characterized by a ring of white feathers around the eye. The birds of this uniform family, closely related to babblers, have a brush-tipped tongue and are nectar-feeding specialists.

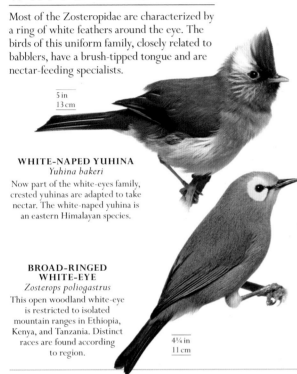

5 in
13 cm

WHITE-NAPED YUHINA
Yuhina bakeri
Now part of the white-eyes family, crested yuhinas are adapted to take nectar. The white-naped yuhina is an eastern Himalayan species.

BROAD-RINGED WHITE-EYE
Zosterops poliogastrus
This open woodland white-eye is restricted to isolated mountain ranges in Ethiopia, Kenya, and Tanzania. Distinct races are found according to region.

4¼ in
11 cm

FAIRY-BLUEBIRDS

Two species of fairy-bluebirds, from the family Irenidae, are found in Southeast Asia, where they feed on fruits—especially figs—in forest canopies. Only males have a vivid blue color; females are dull green.

bright blue upper parts

10 in
25 cm

ASIAN FAIRY-BLUEBIRD
Irena puella
The most widespread fairy-bluebird, found from India to Indonesia, this species often feeds with other fruit-eaters, such as hornbills and pigeons.

GOLDCRESTS

Among the smallest of passerines, these color-crested birds of the family Regulidae are found in cool northern forests. They have a high metabolic rate, which forces them to feed constantly when awake. They glean tiny, soft-bodied invertebrates from foliage with their needlelike bill.

3½ in
9 cm

4½ in
11 cm

GOLDCREST
Regulus regulus
All kinglets are adapted to coniferous forests. The Eurasian goldcrest has especially grooved feet and enlarged toe pads for clinging to needlelike leaves.

RUBY-CROWNED KINGLET
Regulus calendula
The red head patch of this N. American species is visible when it raises its crown feathers—a trait shared by all kinglets.

NUTHATCHES AND RELATIVES

Nuthatches, in the family Sittidae, and the wallcreeper, sole member of the family Tichodromidae, are more acrobatic than treecreepers, and do not use their tail as a support. They feed on seeds as well as insects, sometimes hoarding them in crevices.

WALLCREEPER
Tichodroma muraria
This central Eurasian mountain bird gleans insects from rocks with its pointed bill.

6½ in
16–17 cm

black stripe across eye

4¼ in
11 cm

5½ in
14 cm

orange-brown underparts

EURASIAN NUTHATCH
Sitta europaea
A widespread species, this forest bird can—like other nuthatches—crack open nuts by wedging them in cracks in tree bark.

RED-BREASTED NUTHATCH
Sitta canadensis
This N. American species has a similar body pattern to the Eurasian species, but males are more vividly colored.

WRENS

With the exception of the winter wren, the family Troglodytidae is confined to the Americas. Most species are highly vocal, but visually inconspicuous, short-winged birds that forage on insects in undergrowth; some even sleep on the ground.

CACTUS WREN
Campylorhynchus brunneicapillus
The largest wren, this species occurs in deserts of California and Mexico, where it forages in groups on the ground.

4 in
10 cm

WINTER WREN
Troglodytes troglodytes
The only wren whose range includes Eurasia, this bird is found across the northern hemisphere. It has a large number of local races.

7–9 in
18–23 cm

5½ in
14 cm

BEWICK'S WREN
Thryomanes bewickii
This long-tailed wren inhabits dry, open, wooded habitats in California and Mexico. It has a variety of songs in its repertoire.

MOCKINGBIRDS AND RELATIVES

Birds of the family Mimidae occur throughout much of the Americas, and the Caribbean and Galapagos Islands. Usually gray or brown, these strong-legged birds are highly vocal; some are accomplished mimics.

CURVE-BILLED THRASHER
Toxostoma curvirostre
A bird of arid, scrubby habitats in the S. US and Mexico, this species uses its long bill to probe the soil for invertebrates.

10½ in
27 cm

pale gray upperparts

long tail

8½–10 in
21–26 cm

NORTHERN MOCKINGBIRD
Mimus polyglottos
This N. American bird is famous for its extraordinary repertoire of varied songs, which it delivers day and night.

8½–9½ in
21–24 cm

GRAY CATBIRD
Dumetella carolinensis
Named for its catlike mewing call, this N. American bird forages on the ground. It overwinters in C. America and the Caribbean.

OXPECKERS

In the family Buphagidae, oxpeckers bound low over African savannas in undulating flight. Although their strong feet provide a tight grip on the hide of large animals as the birds feed, their short legs are unsuitable for walking on the ground.

YELLOW-BILLED OXPECKER
Buphagus africanus
Common in savanna across sub-Saharan Africa, this bird habitually perches on large mammals, eating parasites but also pecking at wounds.

7½–9 in
19–22 cm

STARLINGS AND MYNAS

From the family Sturnidae, these are mostly gregarious, noisy birds, many with a glossy plumage that has a metallic sheen. This family is divided into mynas and relatives from southern Asia and the Pacific, and true starlings from Africa and Eurasia.

10½–12 in
27–31 cm

HILL MYNA
Gracula religiosa
Found in forests of tropical Asia, this species is a popular cage bird. It is notable for its melodious call and its mimicking ability.

20 in
50 cm

WHITE-NECKED MYNA
Streptocitta albicollis
This long-tailed, magpie-like myna is restricted to rainforests in Sulawesi, Indonesia, and adjacent islands, where it usually associates in pairs.

10 in
25 cm

BALI MYNA
Leucopsar rothschildi
Found only in the rainforests of Bali, Indonesia, this striking bird is highly endangered by habitat destruction and the bird trade.

OLD WORLD FLYCATCHERS AND CHATS

Related to thrushes, the family Muscicapidae is divided into two groups: true flycatchers, with a broad bill for snatching flying insects; and chats, which include robins, nightingales, and wheatears. Some species are brightly colored, but most of these small birds have predominantly gray or brown plumage.

7 in
18 cm

BLUE-AND-WHITE FLYCATCHER
Cyanoptila cyanomelaena
Belonging to a large group of vivid blue flycatchers from tropical Asia, this species from E. Asia forages high in forests.

5 in
13 cm

COMMON STONECHAT
Saxicola torquatus
Typical of chats, this is a small, upright, perching insectivore with a harsh call. It is common in grassland in Eurasia and Africa.

reddish brown base of tail

5½ in
14 cm

5½ in
14 cm

EUROPEAN ROBIN
Erithacus rubecula
Related to chats, this bird of hedgerows and woodland is found in W. Eurasia and N. Africa. In Britain, it is a common garden visitor.

6–6½ in
15–16 cm

NORTHERN WHEATEAR
Oenanthe oenanthe
Wheatears are white-rumped chats of open country. This is the most widespread species across Eurasia. It winters in Africa.

7½–8½ in
19–21 cm

MOCKING CLIFFCHAT
Thamnolaea cinnamomeiventris
Belonging to a group of dark-colored African chats, this bird occurs in bushy, rocky habitats and can become tame near villages.

5–5½ in
13–14 cm

WHITE-BROWED SHORTWING
Brachypteryx montana
Now members of the flycatchers family, shortwings are running birds found in Asian forests. This species occurs from the Himalayas to Java, Indonesia.

5½ in
14 cm

COMMON REDSTART
Phoenicurus phoenicurus
Named for the reddish brown tail, redstarts are chatlike birds found mainly in Asia. This species occurs from W. to C. Eurasia and migrates to E. Africa.

glossy blue body

9 in
22 cm

THRUSHES AND RELATIVES

Belonging to the family Turdidae, most thrushes are woodland birds that forage on the ground for invertebrate prey, such as earthworms, snails, and insects. They are found worldwide, but the majority of species occur in the Old World. Many have melodious songs.

EUROPEAN STARLING
Sturnus vulgaris
Native to Eurasia, but introduced to N. America, this common starling roosts communally, in groups that gather after mass aerial maneuvers.

6½–8½ in
16–21 cm

9 in
22 cm

ORANGE-HEADED THRUSH
Zoothera citrina
One of many tropical Old World *Zoothera* thrushes, this species is found in forests from the Himalayas to Bali, Indonesia.

SPLENDID GLOSSY STARLING
Lamprotornis splendidus
Widespread in sub-Saharan Africa, this species belongs to a group of African starlings characterized by a metallic sheen on their plumage.

7 in
18 cm

12 in
30 cm

black band on orange breast

EASTERN BLUEBIRD
Sialia sialis
A bird of eastern N. America, this is a common species of open woodland and fields. It sometimes nests in old woodpecker holes.

7–7½ in
18–19 cm

EMERALD STARLING
Lamprotornis iris
This glossy W. African species feeds mainly on fruits, especially figs, but also takes ants.

HILDEBRANDT'S STARLING
Lamprotornis hildebrandti
This E. African glossy starling occurs in wooded savanna, where it preys on large ground insects, often flocking with other starling species.

faint spots on breast

VARIED THRUSH
Ixoreus naevius
Found in mature coniferous forests of western N. America, this bird overwinters in parks and gardens. Like other thrushes, it forages in ground litter.

7½–10 in
19–26 cm

red or white spot on blue throat

WHITE-TAILED ROBIN
Myiomela leucura
A bird of riverine forests from the Himalayas to Indochina, it usually keeps close to the ground unless disturbed.

7 in
18 cm

8–9 in
20–23 cm

SONG THRUSH
Turdus philomelos
Found from Europe to Siberia, this woodland and garden bird habitually uses a hard surface as an "anvil" to smash open snail shells.

8–11 in
20–28 cm

AMERICAN ROBIN
Turdus migratorius
Unrelated to the European robin, this N. American thrush sometimes gathers in winter roosts of up to a quarter of a million birds.

BLUETHROAT
Luscinia svecica
A relative of the nightingale, the bluethroat breeds in damp regions of N. Eurasia, and migrates to Africa and S.E. Asia.

5 in
13 cm

6½ in
17 cm

COMMON NIGHTINGALE
Luscinia megarhynchos
This brown bird of W. to C. Eurasian thickets is known for its loud, rich song, which it sings night or day.

PIED FLYCATCHER
Ficedula hypoleuca
This large genus of mainly Asian flycatchers is closely related to chats. This woodland species ranges from Europe to Siberia.

EURASIAN BLACKBIRD
Turdus merula
A common long-tailed thrush of woodland habitats ranging from Europe and N. Africa to India, the highly territorial blackbird often visits gardens.

9½–11½ in
24–29 cm

9–10½ in
22–27 cm

FIELDFARE
Turdus pilaris
This bird breeds in N. Eurasia and winters farther south, where it flocks together in fields.

LEAFBIRDS

The fruit-eating birds of the family Chloropseidae occur in forests in southeastern Asia. They use their brush-tipped tongue to take nectar, with which they supplement their diet. Male leafbirds are characteristically green, with a blue or black throat.

8 in
20 cm

ORANGE-BELLIED LEAFBIRD
Chloropsis hardwickei
This melodious bird inhabits the forest canopy at high altitudes, in a range from the Himalayas to the Malay Peninsula.

WHYDAHS AND RELATIVES

African whydah birds and related indigobirds of the family Viduidae are cuckoolike brood parasites of waxbills. The mouthparts and begging behavior of the chicks usually resemble those of the host chicks, thus fooling the host parents.

4¾–15 in
12–38 cm

EASTERN PARADISE WHYDAH
Vidua paradisaea
Typical of whydahs, breeding males of this E. African species have extraordinarily long tail feathers, which they use in display flights.

FLOWERPECKERS

These dumpy birds of the family Dicaeidae are related to the sunbirds from tropical Asia and Australasia. Largely fruit-eating, like sunbirds, they take nectar from flowers. The flowerpeckers differ in having a shorter bill.

MISTLETOEBIRD
Dicaeum hirundinaceum
This Australian flowerpecker has a small gut to process mistletoe berries rapidly, and plays an important role in dispersing seeds of this parasitic plant.

4–4¼ in
10–11 cm

WAXBILLS AND RELATIVES

Consisting of small, highly gregarious, often brightly colored seed-eaters, the family Estrildidae occurs in tropical Africa, Asia, and Australia. Many birds live in grassland or open woodland and build domelike nests. Both parents share parental duties.

5½ in
14 cm

PURPLE GRENADIER
Uraeginthus ianthinogaster
An E. African bird of dry woodland, the purple grenadier belongs to a genus of predominantly blue waxbills.

4 in
10 cm

GREEN-BACK TWINSPOT
Mandingoa nitidula
Found in western to southern Africa, this thicket-dwelling twinspot has white-spotted underparts and is more secretive than other waxbills.

COMMON WAXBILL
Estrilda astrild
Abundant throughout Africa, this tiny waxbill, like related species, is a restless, flocking bird of open land, feeding on grass seeds.

blackish bill

black and white "scaly" belly

SCALY-BREASTED MUNIA
Lonchura punctulata
This bird is common in S. Asian scrub. Male and female are alike in appearance.

4 in
10 cm

♀

♂

ZEBRA FINCH
Taeniopygia guttata
Native to the drier parts of Australia, the zebra finch is a popular cagebird around the world.

6½ in
16 cm

JAVA SPARROW
Lonchura oryzivora
This vulnerable species from Java and Bali frequents farmland and feeds on cereal crops. It has been hunted as a rice pest and for the pet trade.

4 in
10 cm

4¾ in
12 cm

♀

♂

GREEN-WINGED PYTILIA
Pytilia melba
Male *Pytilia* waxbills have red splashes on their wings. This African species is host to the paradise whydah (*Vidua paradisaea*), a brood parasite.

4¾ in
12 cm

4¾ in
12 cm

RED-THROATED PARROTFINCH
Erythrura psittacea
Parrotfinches are mostly green-bodied waxbills from S.E. Asia to the Pacific Ocean. This species occurs in grasslands of the island of New Caledonia.

OLD WORLD SPARROWS

These stubby-billed, seed-eating birds, the Passeridae, are found across Africa and Eurasia. As well as the familiar true sparrows, the group includes snowfinches, mountain-dwelling birds that occur from the Pyrenees to Tibet.

6 in
15 cm

HOUSE SPARROW
Passer domesticus
Originally native to Eurasia and N. Africa, the house sparrow has adapted to human settlements around the world.

DIPPERS

Belonging to the family Cinclidae, dippers are the only passerines that can dive and swim under water. They have adaptations for their aquatic lifestyle, such as well-oiled, waterproof feathers and oxygen-storing blood.

7 in
18 cm

WHITE-THROATED DIPPER
Cinclus cinclus
Widespread through temperate Eurasia, the white-throated dipper breeds near fast-flowing streams but may move to slower-flowing rivers in winter.

CUT-THROAT
Amadina fasciata
Named for the male's red neck patch, this is a common bird of dry African woodlands, and is often found near human habitations.

multicolored body

purple chest

GOULDIAN FINCH
Erythrura gouldiae
A brilliantly colored relative of parrotfinches, this endangered bird of northern Australia is nomadic. Males are red- or black-faced.

PIPITS AND WAGTAILS

Present on every continent, the birds of the family Motacillidae inhabit open country, where they feed on insects. Most wagtails have longer tails and are more brightly colored than more sombre pipits, and some are associated with water.

RED-THROATED PIPIT
Anthus cervinus
This pipit breeds in Arctic tundra, during which it develops a colored throat: reddish brown in males and pink in females.

olive back

GOLDEN PIPIT
Tmetothylacus tenellus
A bird of open bush and grassland, the golden pipit is confined to E. Africa, from Sudan to Tanzania.

YELLOW WAGTAIL
Motacilla flava
This widespread Eurasian species overwinters in Africa, India, and Australia. There are several races, including many with gray or black head patterns.

yellow underparts

BUFF-BELLIED PIPIT
Anthus rubescens
A typical ground-running pipit, this species breeds in Arctic tundra and winters in fields and on coasts further south.

WHITE WAGTAIL
Motacilla alba
A typical wagtail, this species is widespread throughout Eurasia— often on farmland or in towns.

SUNBIRDS

The small, fast-moving, nectar-feeding Old World tropical birds of the family Nectarinidae are similar to American hummingbirds, with their long, decurved bill and long tongue. Males are usually brightly colored with a metallic sheen. These birds are fiercely territorial.

STREAKY-BREASTED SPIDERHUNTER
Arachnothera affinis
Spiderhunters are drab, long-billed members of the sunbird family. Like other sunbirds, this S.E. Asian species eats invertebrates as well as nectar.

long, decurved bill

brilliant scarlet breast

SCARLET-CHESTED SUNBIRD
Chalcomitra senegalensis
Common throughout much of sub-Saharan Africa, this large sunbird occupies a variety of wooded habitats.

PURPLE SUNBIRD
Cinnyris asiaticus
Like other sunbirds, this species from S. Asia feeds its young mostly on insects. The male loses its brilliant plumage after breeding.

WEAVERS

Gregarious seed-eaters, the Ploceidae build elaborate nests. Males usually take sole responsibility for this duty; females use the nest as the basis for choosing mates. Most species are African, with a few found in southern Asia.

CHESTNUT WEAVER
Ploceus rubiginosus
The genus *Ploceus* includes the largest number of species of weavers. This species occurs in E. Africa.

RED-BILLED QUELEA
Quelea quelea
Widely considered the world's most abundant bird, this African species gathers in giant flocks that can cause serious damage to crops.

RED-COLLARED WIDOWBIRD
Euplectes ardens
Breeding males of widowbirds are black and some fan enlarged tails during display flights. This species is widespread in sub-Saharan Africa.

YELLOW-CROWNED BISHOP
Euplectes afer
This African bird is related to widowbirds. Breeding males are vibrantly patterned; females and nonbreeding males are red or black.

ACCENTORS

Mostly ground-dwelling passerines, the narrow-billed Prunellidae are found in Eurasia. Most species are adapted to high altitudes, but they move to lower altitudes in winter to supplement their insectivorous diet with seeds.

DUNNOCK
Prunella modularis
Unlike other accentors, this is a lowland bird, and it does not usually gather in flocks. It is widespread in temperate Eurasia.

FINCHES AND RELATIVES

Finches of the family Fringillidae have diversified in Eurasia, Africa, and the tropical regions of the Americas. From thin-billed nectar-feeders to heavy-billed, seed-cracking grosbeaks and hawfinches, these birds have evolved to cope with a wide range of foods.

EURASIAN GOLDFINCH
Carduelis carduelis
Goldfinches have a thinly pointed bill for taking seeds from the seed heads of tall plants, such as thistles. This species is widespread in Eurasia.

4¾ in
12 cm

AMERICAN GOLDFINCH
Carduelis tristis
Bright yellow goldfinches and siskins have diversified in the Americas, especially in S. America. This migratory species is from N. America.

4¾–5 in
12–13 cm

CHAFFINCH
Fringilla coelebs
The most common finch in Europe, the chaffinch is also found throughout N. Asia. It often forages with other finch species in winter.

6 in
15 cm

large, white wing patch

6½ in
17 cm

RED CROSSBILL
Loxia curvirostra
Crossbills use their uniquely crossed mandibles for extracting seeds from cones. This species is widespread in coniferous forests across the northern hemisphere.

4¾ in
12 cm

YELLOW-FRONTED CANARY
Serinus mozambicus
This bird belongs to a group of mostly yellow-colored African serins and canaries. It is common south of the Sahara.

6–6½ in
15–17 cm

GRAY-CROWNED ROSY FINCH
Leucosticte tephrocotis
A member of a group of mountain finches related to bullfinches, this N. American species is an inhabitant of high, rocky landscapes.

EVENING GROSBEAK
Hesperiphona vespertina
Grosbeaks of the finch family have a heavy, seed-crushing bill, a feature they share with the unrelated grosbeaks of the cardinal-grosbeak family. This is a N. American species.

AMERICAN BLACKBIRDS AND RELATIVES

These American species of the family Icteridae superficially resemble the common blackbird, to which they are not related. As a group, these strong-billed birds are more closely related to finches. The bill has a powerful gaping action for prying apart tough food.

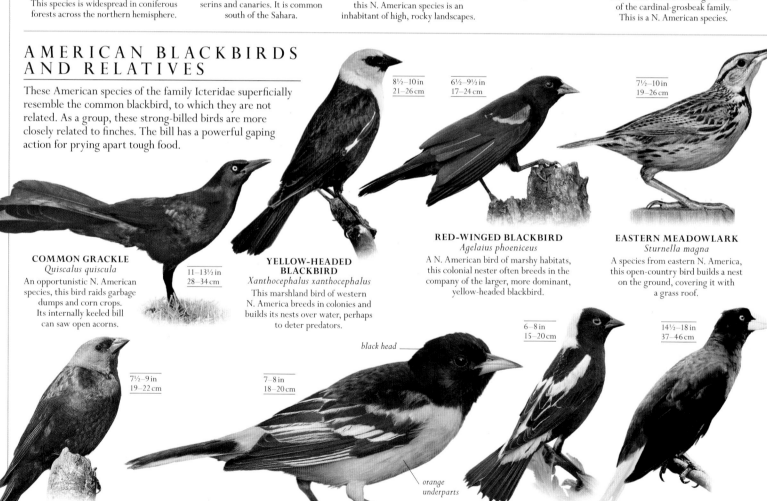

8½–10 in
21–26 cm

6½–9½ in
17–24 cm

7½–10 in
19–26 cm

COMMON GRACKLE
Quiscalus quiscula
An opportunistic N. American species, this bird raids garbage dumps and corn crops. Its internally keeled bill can saw open acorns.

11–13½ in
28–34 cm

YELLOW-HEADED BLACKBIRD
Xanthocephalus xanthocephalus
This marshland bird of western N. America breeds in colonies and builds its nests over water, perhaps to deter predators.

RED-WINGED BLACKBIRD
Agelaius phoeniceus
A N. American bird of marshy habitats, this colonial nester often breeds in the company of the larger, more dominant, yellow-headed blackbird.

EASTERN MEADOWLARK
Sturnella magna
A species from eastern N. America, this open-country bird builds a nest on the ground, covering it with a grass roof.

black head

6–8 in
15–20 cm

14½–18 in
37–46 cm

7½–9 in
19–22 cm

7–8 in
18–20 cm

orange underparts

BROWN-HEADED COWBIRD
Molothrus ater
This N. American bird produces many eggs, laying them in the nests of other passerines. Its young are raised by many different host species.

BALTIMORE ORIOLE
Icterus galbula
Feeding on insects and caterpillars in spring and summer, this species changes its diet in winter to include nectar and berries.

BOBOLINK
Dolichonyx oryzivorus
This ground-nesting N. American bird gets its name from its bubbling flight song. It migrates to central S. America in winter.

CRESTED OROPENDOLA
Psarocolius decumanus
Like other tropical American oropendolas, this colonial species weaves long nests that hang from tips of branches in open forest.

yellow forehead

4¼ in
11 cm

4 in
10 cm

PURPLE-THROATED EUPHONIA
Euphonia chlorotica
Euphonias are small birds that mainly eat fruits, especially mistletoe berries. This species is from northern S. America.

5½ in
14 cm

HOUSE FINCH
Carpodacus mexicanus
The N. American house finch is a type of rosefinch. Rosefinches are especially diverse in temperate Asia. The males are red or pale colored.

BLUE-NAPED CHLOROPHONIA
Chlorophonia cyanea
The fruit-eating chlorophonias are mainly green birds. This species is widespread in forests across S. America.

8 in
20 cm

8 in
20 cm

6–6½ in
15–16 cm

5½ in
14 cm

9 in
22 cm

PINE GROSBEAK
Pinicola enucleator
The pine grosbeak occurs in coniferous forests in the northern hemisphere. The bill of this grosbeak resembles that of the closely related bullfinches.

EURASIAN BULLFINCH
Pyrrhula pyrrhula
Bullfinches have a short, thick bill and a thickset head. This is the most widespread of the bullfinches and is found in woodlands across temperate Eurasia.

IIWI
Drepanis coccinea
This long-billed nectar-eater is found only in Hawaii. It is a bird of montane forests, the largest populations occurring at higher elevations.

JAPANESE GROSBEAK
Eophona personata
A strikingly patterned bird, the Japanese grosbeak of cold northern forests breeds in Siberia and N. Japan and winters in S. China.

AMERICAN WARBLERS

Unrelated to warblers of the Old World, these finch-related insect-eaters of the family Parulidae are found throughout North and South America. The tropical species are sedentary, but temperate ones are migratory. The males molt their bright colors in winter.

black and white streaks overall

4½–5½ in
11–14 cm

BLACK-AND-WHITE WARBLER
Mniotilta varia
This N. American warbler moves up and down tree trunks like a nuthatch, and has long hind claws for clinging to bark.

YELLOW-BREASTED CHAT
Icteria virens
This N. American bird is relatively large for a warbler. It sings at night as well as during the day, and mimics the calls of other birds.

7 in
18 cm

5½ in
14 cm

5 in
13 cm

4¼–5 in
11–13 cm

yellow body

4¾–5 in
12–13 cm

YELLOW WARBLER
Setophaga aestiva
Widespread from N. America to the Caribbean, there is a large number of local races of yellow warblers.

BAY-BREASTED WARBLER
Setophaga castanea
This bird breeds in spruce forests in eastern N. America. Its numbers fluctuate according to the abundance of its prey, the spruce budworm.

HOODED WARBLER
Setophaga citrina
Like the American redstart, this bird from broad-leaved woodland of the E. US feeds on insects by catching flies while in flight.

AMERICAN REDSTART
Setophaga ruticilla
An active forager, this N. American warbler habitually flashes its orange and black wings and tail to flush out insects. It also catches prey in the air.

dull brown feet and legs

6 in
15 cm

5 in
13 cm

4¾ in
12 cm

4¼–5½ in
11–14 cm

NORTHERN WATERTHRUSH
Parkesia noveboracensis
A large, tail-bobbing ground warbler of N. America, this bird forages in leaf litter of wet woodlands and nests in thickets near water.

PROTHONOTARY WARBLER
Protonotaria citrea
Unusually for American warblers, this species of densely wooded swamps nests in tree cavities—sometimes using old woodpecker holes.

COMMON YELLOWTHROAT
Geothlypis trichas
A bird of wet habitats, this is a migratory species like other N. American warblers. It winters in California and Mexico.

GOLDEN-WINGED WARBLER
Vermivora chrysoptera
A warbler of eastern N. America, this bird breeds in scrubby, open habitats and farmland, and benefits from deforestation.

BUNTINGS AND RELATIVES

True buntings of the Old World in the family Emberizidae are mostly ground-feeding seed-eaters. Longspurs, in both Old and New Worlds, are similar short-legged ground birds of open spaces, from plains and tundra to mountain tops in the family Calcaridae.

BLACK-HEADED BUNTING
Emberiza melanocephala
An inhabitant of scrub and olive groves, the black-headed bunting breeds in the Middle East and winters in India.

male yellow beneath and has a black cap

6½ in
17 cm

LAPLAND LONGSPUR
Calcarius lapponicus
Relatives of buntings, longspurs have been named for their long hind claw. This species has a circumpolar distribution during the breeding season.

6 in
15 cm

AMERICAN SPARROWS

Birds in the family Passerellidae resemble Old World buntings rather than true sparrows, especially in the bill structure, with a small, sharp upper mandible and often slightly kinked cutting edge. Many also have striped head patterns, with or without streaked flanks—others are more uniform in color. They search for seeds while hopping on the ground.

dark gray body

6–6½ in
15–16 cm

DARK-EYED JUNCO
Junco hyemalis
Juncos are gray-brown, ground-flocking N. American sparrows. This species, locally called the snowbird, is a common winter visitor of garden bird feeders.

7½ in
19 cm

YELLOW-THIGHED FINCH
Pselliophorus tibialis
A noisy, tropical member of the American sparrow family, this species is found only in the montane rainforests of Costa Rica and Panama.

9 in
22 cm

SPOTTED TOWHEE
Pipilo maculatus
Towhees are long-tailed American sparrows. The spotted towhee is a reddish-brown-sided bird found in thicket habitats across N. America.

TANAGERS AND RELATIVES

An American family, the Thraupidae consist of gaudily colored tanagers, found in the forests of tropical South America. These finch-related birds have evolved to exploit a range of food sources, including fruits, insects, seeds, and nectar.

VARIABLE SEED-EATER
Sporophila corvina
Belonging to a group of tropical American, stubby-billed seed-eaters, this bird has a variable plumage pattern.

4¼ in
11 cm

BLACK-MASKED FINCH
Coryphaspiza melanotis
A ground-feeding finch of grassland habitats in central S. America, this species has been assigned to the tanager family.

5½ in
14 cm

5½ in
14 cm

GREEN HONEYCREEPER
Chlorophanes spiza
This stocky species lives in forest canopy and uses its stout, decurved bill to eat fruits. It is often seen in mixed flocks of tanagers.

7–7½ in
18–19 cm

SCARLET TANAGER
Piranga olivacea
In most species of *Piranga* tanager, the males turn red during breeding. This migratory species is from N. America.

7 in
18 cm

WESTERN TANAGER
Piranga ludoviciana
This bird breeds in western N. America and overwinters in C. America. It is the only *Piranga* tanager where the breeding male is predominantly yellow.

6 in
15 cm

LARGE GROUND FINCH
Geospiza magnirostris
This seed-eater is a native of the Galapagos Islands. The species feeds less from the ground than other ground finches.

6 in
15 cm

GOLDEN-COLLARED TANAGER
Iridosornis jelskii
A member of a group of Andean forest tanagers with a yellow head streak, this species occurs in Peru and Bolivia.

7 in
18 cm

BLUE-WINGED MOUNTAIN TANAGER
Anisognathus somptuosus
Most mountain tanagers of northern S. America are colored blue and yellow, including this inhabitant of montane rainforests.

FOX SPARROW
Passerella iliaca
Widespread throughout N. America, this large sparrow has a red-streaked back and chest. It usually forages in low vegetation.

6½–7½ in
17–19 cm

pale gray band above eye

LARK-BUNTING
Calamospiza melanocorys
A bird of the N. American prairies, the lark-bunting is a ground-nesting member of the American sparrow group.

7 in
18 cm

long, rounded, plain tail

variably colored belly, streaked blackish, brown, or red

5–5½ in
13–14 cm

WHITE-CROWNED SPARROW
Zonotrichia leucophrys
This N. American bird is usually found in low vegetation or on the ground. It has distinctive black and white head markings.

6½–7½ in
17–19 cm

SONG SPARROW
Melospiza melodia
A common N. American bird from Alaska to Mexico, this species has a large number of races. It is named for its melodious song.

5½–6½ in
14–16 cm

CHIPPING SPARROW
Spizella passerina
A rufous (reddish brown)-capped bird common in open woodland across N. America, the chipping sparrow has a distinctive "chipping" trill.

RED-CRESTED CARDINAL
Paroaria coronata
Birds of the *Paroaria* genus are found in S. America. This is a common species in open woodland.

7½ in
19 cm

CARDINALS AND RELATIVES

These mostly chunky-billed birds are seed-eaters like buntings, and many have bright colors like tanagers. Related to both, the Cardinalidae belong to a large group of American passerines that are derived from finches.

5 in
13 cm

7–8½ in
18–21 cm

ROSE-BREASTED GROSBEAK
Pheucticus ludovicianus
This migratory American cardinal grosbeak has a typically heavy bill for feeding on large seeds and on insects, such as beetles.

NORTHERN CARDINAL
Cardinalis cardinalis
The males of this resident of the E. US and Mexico are red, due to chemicals called carotenoids that are acquired from their diet.

8½–9 in
21–23 cm

breeding male has blue underparts and head

green back

5½ in
14 cm

RED-LEGGED HONEYCREEPER
Cyanerpes cyaneus
Tropical American honeycreepers feed on nectar. The most widespread species, this bird has a decurved bill. Like females, males develop a dull green "eclipse" plumage after breeding.

5 in
13 cm

BLUE-NECKED TANAGER
Tangara cyanicollis
This open forest bird from northern S. America belongs to a group of tanager species with particularly colorful, iridescent plumage.

5 in
13 cm

PAINTED BUNTING
Passerina ciris
This species breeds in the S. US and overwinters in C. America and the Caribbean. Only males have the striking tricolored pattern.

INDIGO BUNTING
Passerina cyanea
The most wide ranging of the blue buntings, this species migrates between Canada and S. America. The males lose their blue plumage in winter.

MAMMALS

The mammals are very successful animals. They are able to occupy almost every terrestrial habitat, and some visit the deep oceans between breaths of air. However, this has not always been the case; at 210 million years old, mammals are a relatively recent group.

PHYLUM	CHORDATA
CLASS	MAMMALIA
ORDERS	28
FAMILIES	160
SPECIES	About 6,300

A single jawbone articulated directly to the skull gives mammals, such as this Tasmanian devil, a powerful bite.

Baleen is made of a protein called keratin, and baleen whales use it to strain sea water, trapping food in their mouth.

By sucking milk from their mother, young warthogs obtain all the nourishment they need during the first weeks of life.

DEBATE

EQUALLY WELL ADAPTED

Marsupial features are sometimes described as primitive. This implies that marsupials, whose young are born in an embryonic state and develop in a pouch, are less advanced than placental mammals, which bear conventional young. In fact, the two groups emerged at the same time about 175 million years ago.

Mammals owe their success to a unique combination of adaptations that allowed them to replace reptiles as the dominant form of animal life on Earth. Like reptiles, mammals breathe air, but unlike their scaly ancestors, they are warm-blooded, so they are able to burn fuel (food) in order to maintain a constant, warm body temperature. This creates conditions in which the internal chemical processes that sustain life can happen most efficiently, without relying directly on heat from the sun. Uniquely, mammals have hair, which reduces heat loss and allows an animal to be active in cold climates and at night. Because fur can be molted and regrown, it is seasonally adjustable, too.

VARIATIONS ON A THEME

The basic mammal skeleton is strong, with upright limbs supporting the body from below. This arrangement allows land-dwelling mammals to walk, run, and leap, but the basic structure is highly adaptable, and has been modified for swimming, as in seals and cetaceans; for flying, as in bats; and for climbing or swinging through the trees, as in some primates. The mammalian skull has powerful jaws, with a single lower jawbone articulating directly with the skull, and a diverse range of teeth adapted to a huge variety of food types. Certain other bones, which in reptiles contribute to the lower jaw, are put to a new use by mammals—they have become the three tiny bones of the inner ear, allowing for greatly enhanced hearing. The mammalian skull also serves to protect the brain, which is larger in mammals than in other groups, hinting at increased processing power. With a big brain comes intelligence, and an unrivaled ability to learn, to remember, and to engage in complex behavior.

These abilities take time to fine-tune, and young mammals do so during the extended period of parental care that begins while they are being fed on milk produced by their mother's mammary glands. These unique organs evolved from sebaceous glands, which originally provided secretions to condition the skin and perhaps to prevent eggs from drying out, and it is for them that the entire group is named.

FUR AS INSULATION >
The air trapped in a sea otter's dense fur insulates it from the cold water in which it lives.

MAMMAL GROUPS

There are three groups of mammals—the egg-laying monotremes, the pouched mammals (or marsupials), and the placental mammals. The latter group is the most diverse, and is here divided into further groups.

EGG-LAYING MAMMALS

There are just five species in this group of mammals, which are known as monotremes. All of them have a specialized snout and lay eggs.

The monotremes, which are the duck-billed platypus and the long-nosed and short-nosed echidnas, form the order Monotremata. They are found in various habitats in New Guinea, Australia, and Tasmania.

Platypuses and echidnas lay soft-shelled eggs that hatch after around ten days' incubation. The young feed on milk secreted from the female's mammary glands; monotremes have no teats. Infant echidnas live in their mother's pouch until their spines appear and then, like young platypuses, stay in a burrow for several months.

Monotremes have a specialized snout for finding and eating prey. The platypus, which is partly aquatic, has a flattened, ducklike bill covered with sensory receptors that allow the animal to locate invertebrates under water, even in murky conditions. The long, cylindrical snout and long tongue of the land-dwelling echidnas are ideal for probing the nests of ants and termites, and for catching worms. Neither the platypus nor the echidnas have teeth; instead they have grinding surfaces or horny spines on their tongue.

WHAT'S IN A NAME
The word "monotreme" means "single hole" and makes reference to the cloaca, a single posterior body opening into which the digestive, urinary, and reproductive tracts empty.

PHYLUM	CHORDATA
CLASS	MAMMALIA
ORDER	MONOTREMATA
FAMILIES	2
SPECIES	5

The large, webbed forefeet of the duck-billed platypus provide propulsion. The hind feet and tail are used to steer.

DUCK-BILLED PLATYPUS

The sole member of the family Ornithorhynchidae, the duck-billed platypus is well adapted to a semiaquatic lifestyle. It has a streamlined body, waterproof fur, webbed feet, and a flattened tail. Males have a venomous horny spur on each hind foot.

short, dense body fur

small eyes

sensitive, ducklike bill

16–25 in
40–63 cm

venomous spur of male

DUCK-BILLED PLATYPUS
Ornithorhynchus anatinus
Found in rivers and streams in E. Australia and Tasmania, this rare species uses its soft bill, which is covered in electroreceptors, to hunt for invertebrates.

ECHIDNAS

The family Tachyglossidae is made up of the long- and short-beaked echidnas. Their rounded body is covered with fur and spines, and the long snout is ideal for seeking out insects, ants, and worms.

sharp spines for defense

12–18 in
30–45 cm

19–25 in
48–63 cm

SHORT-BEAKED ECHIDNA
Tachyglossus aculeatus
Widespread in Australia, Tasmania, and New Guinea, the short-beaked echidna lays a single egg into a pouch on its abdomen.

EASTERN LONG-BEAKED ECHIDNA
Zaglossus bartoni
The largest of the monotremes, this species lives in the forested highlands of E. New Guinea.

POUCHED MAMMALS

These mammals bear live young that are born in a very immature state, and typically complete their growth in the mother's abdominal pouch.

Marsupials inhabit a vast array of habitats from desert and dry scrub to tropical rainforest. Most of them are either terrestrial or arboreal, but several glide, one is aquatic, and there are two marsupial moles that live underground. The diets of marsupials are equally wide-ranging: among the different species are carnivores, insectivores, herbivores, and omnivores. There are even marsupials that feed on nectar and pollen. Marsupials also differ markedly in size, from the tiny planigales, which are among the smallest mammals in the world, weighing less than about 1/3 oz (4.5 g), to the red kangaroo, males of which can weigh over 200 lb (90 kg).

EARLY DEVELOPMENT
Young marsupials are born blind and furless. They make their way through the mother's fur and latch on to a nipple and suckle. In around half of marsupial species, the nipples are located in a protective pouch. Some marsupials have only one offspring at a time, but others may have up to a dozen or more. The phase during which young are in the pouch is equivalent to the gestation period of placental mammals.

Kangaroos and some other marsupials are able to stop the development of an embryo before it implants in the uterus if they already have an infant occupying the pouch. The pregnancy continues when the pouch is vacated.

THE SEVEN ORDERS
Pouched mammals, known as marsupials, are now divided into seven major groups. These are: the American opossums, order Didelphimorphia; the shrew opossums, order Paucituberculata; monito del monte, the sole member of the order Microbiotheria; the Australasian carnivorous marsupials, order Dasyuromorphia; bandicoots, order Peramelemorphia; the marsupial moles, order Notoryctemorphia; and the Australasian marsupials of Diprotodontia, the largest order of pouched mammals, which includes the koala, wombats, possums, wallabies, and kangaroos.

PHYLUM	CHORDATA
CLASS	MAMMALIA
ORDERS	7
FAMILIES	18
SPECIES	OVER 350

SHREW OPOSSUMS

Fewer incisor teeth distinguish the members of the family Caenolestidae from other American marsupials. All eight species occur in the Andes in western S. America.

3½–5½ in
9–14 cm

DUSKY SHREW OPOSSUM
Caenolestes fulginosus
Living at high altitudes in Colombia, Ecuador, and Venezuela, this shrew opossum uses its large lower incisors to kill prey.

MONITO DEL MONTE

The sole member of the family Microbiotheriidae, the monito del monte is well adapted to the cold. It uses both daily torpor and seasonal hibernation to conserve energy when temperatures are low or food is in short supply.

3¼–5 in
8–13 cm

MONITO DEL MONTE
Dromiciops gliroides
Inhabiting cool bamboo forest and temperate rainforest in Chile and Argentina, this species has dense fur to limit heat loss.

MARSUPIAL MOLES

Two species of Australian marsupial moles form the family Notoryctidae. Their short limbs, large claws, and horned nose-shield help them to burrow. They lack external ears and have nonfunctional eyes.

4½–5½ in
11–14 cm

OPOSSUMS

The New World opossums of the family Didelphidae have a pointed muzzle with sensitive whiskers, and naked ears. Many species have a grasping, prehensile tail that helps when climbing. Some opossums lack pouches.

VIRGINIA OPOSSUM
Didelphis virginiana
The largest American marsupial, this species lives in grasslands and temperate and tropical forests in the US, Mexico, and C. America.

14½–20 in
37–50 cm

8–12½ in
20–32 cm

BROWN-EARED WOOLLY OPOSSUM
Caluromys lanatus
Also known as the western woolly opossum, this arboreal, solitary opossum inhabits moist forests in western and central S. America.

SOUTHERN MARSUPIAL MOLE
Notoryctes typhlops
This marsupial mole is specialized for burrowing in the sandy desert and spinifex grassland of C. Australia.

6½–11 in
16–28 cm

BARE-TAILED WOOLLY OPOSSUM
Caluromys philander
This woolly opossum's long, prehensile tail helps it move through the canopy of moist rainforests across eastern and central S. America.

≫

» OPOSSUMS

4¼–5¾ in
11–14.5 cm

dark fur
around
large eyes

4¾–9 in
12–22 cm

10–16 in
26–40 cm

WOOLLY MOUSE OPOSSUM
Marmosa sp.
Living in C. and S. America, this pouchless
marsupial with thick, woolly fur is arboreal,
nocturnal, and omnivorous.

WATER OPOSSUM
Chironectes minimus
The only aquatic marsupial, this opossum
of C. and S. America is also unique in that both sexes
have a pouch, but it is only watertight in females.

LINNAEUS'S MOUSE OPOSSUM
Murinus murina
Widespread in forests, pampas, and
plantations in S. America, this agile,
nocturnal climber has a long, prehensile tail.

prehensile
tail

3½–5½ in
9–14 cm

4¾–5¾ in
12–14.5 cm

**ELEGANT FAT-
TAILED MOUSE
OPOSSUM**
Thylamys elegans
Like several other
opossums, this marsupial
from Chile stores fat
in its tail as winter
approaches.

dependent
young

white spot
above eye

PATAGONIAN OPOSSUM
Lestodelphys halli
Living in shrub, savanna, and
grassland in Argentina, the
Patagonian opossum has the
most southerly distribution
of any opossum species.

4¾–7 in
12–18 cm

8–13 in
20–33 cm

**GRAY FOUR-
EYED OPOSSUM**
Philander opossum
This opossum has white
spots on its forehead,
making it look as if it
has four eyes. It lives
in Mexico, and
C. and S. America.

**GRAY SHORT-TAILED
OPPOSUM**
Monodelphis domestica
Found in Argentina, Brazil,
Bolivia, and Paraguay, this
short-tailed opossum
sometimes inhabits human
dwellings, as well as forest,
scrub, and grassland.

NUMBAT

The numbat is the only member of the family Myrmecobiidae.
It has a distinctive striped coat, strong claws for digging,
and a very long tongue for extracting termites
from their nests.

9–11½ in
22–29 cm

NUMBAT
Myrmecobius fasciatus
Found only in eucalyptus
forest and woodland in
Western Australia, this
diurnal marsupial is a
specialized termite-eater.

BILBIES

The family Thylacomyidae has only one species,
since the lesser bilby has been declared extinct.
Living in arid habitats, this nocturnal marsupial
does not need to drink water, as it absorbs
moisture from its food.

12–22 in
30–55 cm

GREATER BILBY
Macrotis lagotis
A burrowing species of the desert of
C. Australia, this marsupial has silky fur,
a tricolored tail, and long, rabbitlike ears.

QUOLLS, DUNNARTS, AND RELATIVES

With strong jaws and sharp canine teeth, the family Dasyuridae is made up of more than 70 large and small carnivorous marsupials. These animals also have sharp claws, except on their big toe.

white band on rump and chest

fat store in base of tail

3¾–4¼ in
9.5–10.5 cm

FAT-TAILED FALSE ANTECHINUS
Pseudantechinus macdonnellensis
This nocturnal insect-eater stores fat in the base of its tapered tail. It inhabits arid rocky habitats of C. and W. Australia.

TASMANIAN DEVIL
Sarcophilus harrisii
The largest carnivorous marsupial in the world, this nocturnal hunter lives in a variety of habitats across Tasmania.

22½–26 in
57–65 cm

2¾–5½ in
7–14 cm

BROWN ANTECHINUS
Antechinus stuartii
Antechinus males die of stress and exhaustion after their first breeding season. This species is endemic to forests in E. Australia.

THREE-STRIPED DASYURE
Myoictis sp.
This marsupial's coloration helps it to remain camouflaged on the rainforest floor in the islands of Indonesia and New Guinea.

4¾–9 in
12–23 cm

10–16 in
26–40 cm

7½–9½ in
19–24 cm

WESTERN QUOLL
Dasyurus geoffroii
Also called the chuditch, this nocturnal hunter from S.W. Australia is mainly ground-dwelling, although it can climb trees.

RED-TAILED PHASCOGALE
Phascogale calura
Possessing a brush of black hair on its red-based tail, this carnivorous marsupial inhabits woodland in S.W. Australia.

4¼–5 in
10.5–12.5 cm

CREST-TAILED MULGARA
Dasycercus cristicauda
This carnivore from W. and C. Australia inhabits arid and semiarid habitats, such as desert, heath, and grassland. It stores fat in its tail.

2¾–4 in
7–10 cm

2–3 in
5–7.5 cm

2–3 in
5–7.5 cm

2¼–3½ in
6–9 cm

KULTARR
Antechinomys laniger
This fast, agile marsupial uses its large hind feet to bound across woodland, grassland, and semidesert habitats in southern and C. Australia.

NARROW-NOSED PLANIGALE
Planigale tenuirostris
This flat-headed, nocturnal, rodentlike marsupial lives in low shrub and dry grassland in S.E. Australia.

INLAND NINGAUI
Ningaui ridei
A shrewlike marsupial with a pointed snout, this nocturnal predator hunts insects in arid spinifex grassland in C. Australia.

FAT-TAILED DUNNART
Sminthopsis crassicaudata
Living in open grassland in southern Australia, this small, nocturnal marsupial stores fat reserves in its tail.

BANDICOOTS

These omnivorous marsupials are found across Australia and New Guinea. Characterized by fused second and third toes on their hind feet and three pairs of lower incisor teeth, the Peramelidae also have short, coarse, or spiny hair.

9–15 in
22.5–38 cm

SPINY BANDICOOT
Echymipera kalubu
This forest-dwelling, nocturnal insect-eater of New Guinea has a conical snout, spiny coat, and a hairless tail.

11–14 in
28–36 cm

coarse, yellowish brown fur

melanistic (black) form of species

12–17½ in
31–44.5 cm

10½–14 in
27–35 cm

LONG-NOSED BANDICOOT
Perameles nasuta
Found in rainforest and woodland in eastern coastal Australia, this nocturnal bandicoot forages for food by digging for insects.

EASTERN BARRED BANDICOOT
Perameles gunnii
Named for the creamy bands of fur on its flanks, this bandicoot lives in grassland and grassy woodland in Australia and Tasmania.

SOUTHERN BROWN BANDICOOT
Isoodon obesulus
This short-nosed bandicoot lives in the shrubby heathland in southern Australia and several islands, including Kangaroo Island and Tasmania.

KOALA

The sole member of the family
Phascolarctidae, the koala is
adept at climbing because
of its powerful forearms,
opposable fingers and toes, and
sharp, curved claws. It sleeps
for up to 20 hours each day
because its diet of eucalyptus
leaves is low in nutrients.

large, white,
rounded ears

dense
fur

26—32 in
65—82 cm

KOALA
Phascolarctos cinereus
Feeding almost exclusively
on eucalyptus leaves,
the koala lives in the forests
and woodlands of
E. Australia. It is solitary
and nocturnal.

long, curved
claws

PYGMY POSSUMS

These small, nocturnal marsupials with
a prehensile tail are omnivorous, feeding
on insects, fruits, nectar, and pollen.
Four species of the family Burramyidae
are endemic to Australia; the fifth lives
in Australia and New Guinea.

4¼ in
10.5 cm

**LONG-TAILED
PYGMY POSSUM**
Cercartetus caudatus
This arboreal
marsupial is found
in temperate rainforest
in New Guinea and
N.E. Queensland.

4—5 in
10—13 cm

dull,
grayish brown
upperparts

**MOUNTAIN
PYGMY POSSUM**
Burramys parvus
A ground-dweller of
rocky, upland habitats in
Australia, the mountain
pygmy possum hibernates
under snow for several
months during winter.

RINGTAILS
AND RELATIVES

This family consists of ringtail possums and the
lemurlike greater glider. The Pseudocheiridae
are all arboreal and are specialized
leaf-eaters. They have an enlarged
pouch at the start of the large
intestine for fermentation of
cellulose from their diet.

COMMON RINGTAIL
Pseudocheirus peregrinus
This agile marsupial
lives in a wide range of
habitats in E. Australia and
Tasmania. It has become
a pest in New Zealand.

11½—14 in
29—35 cm

WOMBATS

Stocky marsupials with a short
tail and short limbs, wombats have
large forepaws and long claws for
burrowing. The family Vombatidae
feed on coarse grass, which they
grind using strong jaws, and digest
with an elongated gut.

COMMON WOMBAT
Vombatus ursinus
Found in forest,
heathland, and coastal
scrub in S.E. Australia,
this species is capable of
digging tunnels up to
655 ft (200 m) long.

35—45 in
90—115 cm

33—43 in
84—111 cm

silky fur mottled
brown and gray

**SOUTHERN
HAIRY-NOSED WOMBAT**
Lasiorhinus latifrons
An inhabitant of central
southern Australia, this wombat
lives colonially in warrens
but feeds alone.

14—22 in
35—55 cm

**LEMUROID RING-
TAILED POSSUM**
Hemibelideus lemuroides
Found only in a
small area of rainforest
in N.E. Queensland,
Australia, this species
of ringtail is nocturnal.

CUSCUS AND RELATIVES

The family Phalangeridae includes cuscuses, brushtail possums, and their relatives. Most members are arboreal, with an opposable thumb on the hind limbs and a prehensile tail. In cuscuses, part or all of the tail is naked, whereas in brushtail possums the tail has fur.

16½–29 in
42–74 cm

strong,
curved claws

SULAWESI BEAR CUSCUS
Ailurops ursinus
The largest of the cuscuses, this species inhabits the canopy of temperate rainforest in Sulawesi and several other Indonesian islands.

large eyes assist
sight at night

18½–22½ in
47–57 cm

COMMON SPOTTED CUSCUS
Spilocuscus maculatus
This cuscus is sexually dimorphic (males and females look different)—only the male has spots. It lives in the rainforests of New Guinea and N.E. Australia.

19½–21½ in
49–54 cm

MOUNTAIN BRUSHTAIL POSSUM
Trichosurus cunninghami
The mountain brushtail possum inhabits dense, wet forests in S.E. Australia, typically living at altitudes above 985 ft (300 m).

furry tail

12–18½ in
30–47 cm

SCALY-TAILED POSSUM
Wyulda squamicaudata
Found only in the Kimberley in N.W. Australia, this nocturnal possum is solitary and bears just one young at a time.

13–23½ in
33–60 cm

CUSCUS
Phalanger sp.
Cuscuses of this genus live on New Guinea and nearby islands. They live at different altitudes to avoid competition with one another.

grizzled,
gray fur

13½–18 in
34–45 cm

GREATER GLIDER
Petauroides volans
The largest gliding marsupial, this species can travel distances of more than 330 ft (100 m) between trees. It lives in E. Australia.

12½–16 in
32–40 cm

GREEN RINGTAIL
Pseudochirops archeri
Named for its thick, greenish fur, this solitary possum is found only in rainforests in the far north of Queensland, Australia.

prehensile tail

DAINTREE RIVER RINGTAIL POSSUM
Pseudochirulus cinereus
Resembling a lemur, this ringtail possum lives in the montane tropical rainforest in the Daintree River area in N.E. Queensland, Australia.

13½–14½ in
34–37 cm

GLIDING AND STRIPED POSSUMS

The Petauridae include striped possums and gliders, except the greater glider (left). Gliders have a thin, furred membrane between their fore and hind limbs. Striped possums emit strong smells and have an elongated fourth finger to probe for wood-boring beetles.

9½–11 in
24–28 cm

gray body with
dark stripe
along its back

6–6½ in
15–17 cm

STRIPED POSSUM
Dactylopsila trivirgata
Skunklike in appearance and odor, this possum is a nocturnal tree-dweller from N.E. Queensland, Australia, and New Guinea.

LEADBEATER'S POSSUM
Gymnobelideus leadbeateri
Found in moist, high-altitude forests in Victoria, Australia, this possum feeds on insects, and sap and gum from trees.

thick tail

6–8½ in
15–21 cm

club-shaped tail

SUGAR GLIDER
Petaurus breviceps
Partial to the sweet sap of eucalyptus trees, the sugar glider is native to N.E. Australia, New Guinea, and neighboring islands.

HONEY POSSUM

This diminutive species is the only member of the family Tarsipedidae. It has fewer teeth than other possums and a long, brush-covered tongue for probing flowers.

HONEY POSSUM
Tarsipes rostratus
Inhabiting heathland and woodland in S.W. Australia, this possum is a specialized feeder on nectar and pollen.

2½–3½ in
6.5–9 cm

long, pointed snout

FEATHER-TAILED GLIDERS

The family Acrobatidae contains three species—the feather-tailed or pygmy glider, the broad-toed feathertail glider, and the feather-tailed possum. All have tails fringed with rows of stiff hairs.

FEATHER-TAILED GLIDER
Acrobates pygmaeus
This species is the smallest gliding marsupial. It lives in E. Australian forests, where it feeds on nectar.

2–2¾ in
5–7 cm

KANGAROOS AND RELATIVES

The kangaroos and wallabies of the family Macropodidae are medium to large animals, with long hind legs for jumping. The hind feet lack a first toe, a fleshy sheath encompasses the second and third toes, forming a short grooming claw, and the strong, elongated fourth and fifth toes are weight-bearing.

26–36 in
66–92 cm

red-brown upperparts

2¼–4½ ft
0.7–1.4 m

RED KANGAROO
Osphranter rufus
The largest extant marsupial, this kangaroo is widespread in Australia, where it lives in savanna grassland and desert.

22½–42½ in
57–108 cm

COMMON WALLAROO
Osphranter robustus
Widespread throughout most of mainland Australia, the common or hill wallaroo often seeks shade in rocky outcrops.

18–21 in
45–53 cm

PARMA WALLABY
Notamacropus parma
Native to the mountains of the Great Dividing Range in Australia, the parma wallaby inhabits a range of forest habitats.

23½–34 in
60–85 cm

AGILE WALLABY
Notamacropus agilis
Unusually for wallabies, this species occurs in both Australia and New Guinea, where it inhabits grassland and open woodland.

RED-NECKED WALLABY
Notamacropus rufogriseus
This species lives in coastal forest and scrub in southeastern Australia, including the islands of Tasmania and the Bass Strait.

long tail used for support at rest, for balance when moving

joey in pouch

3–7½ ft
0.9–2.3 m

EASTERN GRAY KANGAROO
Macropus giganteus
Widespread in E. Australia, this species lives in dry woodland, scrub, and shrub. A subspecies is found in Tasmania.

2¼–7¼ ft
0.7–2.2 m

WESTERN GRAY KANGAROO
Macropus fuliginosus
The only kangaroo that does not reproduce using the method of delayed implantation of the embryo in the uterus, this species occurs in southern Australia, including Kangaroo Island.

POTOROOS

This family contains potoroos, bettongs, and rat kangaroos, small marsupials with many similarities to the larger Macropodidae. However, unlike them, the Potoroidae as adults have a single serrated upper and lower premolar in each side of the jaws.

12–14 in
30–36 cm

10–16 in
26–41 cm

LONG-NOSED POTOROO
Potorous tridactylus
Inhabiting heath and forest in S.E. Australia, this potoroo has strong, curved claws that help it to dig for underground fungi.

BRUSH-TAILED BETTONG
Bettongia penicillata
This bettong lives in forest and grassland in S.W. Australia. It has a prehensile tail that it uses for moving nesting material.

RAT-KANGAROO

The musky rat-kangaroo is the sole member of the family Hypsiprymnodontidae. It is relatively primitive; unlike other kangaroo species, it has a first toe, which is opposable and helps it grip.

6–11 in
15–28 cm

MUSKY RAT-KANGAROO
Hypsiprymnodon moschatus
This diurnal species inhabits tropical rainforest in N. Queensland, Australia, where it forages for fallen fruits, seeds, and fungi.

19½–27 in
49–69 cm

13½–43 in
34–109 cm

RED-NECKED PADEMELON
Thylogale thetis
A forest wallaby from eastern Australia, this species ventures to the forest edge at night to feed on grass, leaves, and shoots.

11½–25 in
29–63 cm

muscular thighs

long, narrow sole of foot

NORTHERN NAIL-TAILED WALLABY
Onychogalea unguifera
Also called the sandy nail-tailed wallaby, this medium-sized kangaroo occurs across N. Australia.

BRIDLED NAIL-TAIL WALLABY
Onychogalea fraenata
This nocturnal wallaby was once thought to be extinct; the only remaining wild population is in a small area of Queensland, Australia.

21½–30 in
54–77 cm

16–21½ in
40–54 cm

12–15½ in
31–39 cm

coarse, dark fur

26–34 in
66–85 cm

RUFOUS HARE-WALLABY
Lagorchestes hirsutus
Previously inhabiting the Australian mainland, the mala or rufous hare-wallaby now exists in the wild on only two islands in Western Australia.

BROWN DORCOPSIS
Dorcopsis muelleri
This forest-dwelling kangaroo is endemic to the low-altitude rainforests of western New Guinea, and three offshore islands.

QUOKKA
Setonix brachyurus
Rare in mainland Australia, this small marsupial is found on Rottnest and Bald Islands, off the southwestern coast.

22–30¾ in
55–78 cm

DORIA'S TREE KANGAROO
Dendrolagus dorianus
The heaviest of the tree-dwelling marsupials, Doria's tree kangaroo frequently descends to the ground. It lives in montane forest in New Guinea.

reddish-brown upperparts

SWAMP WALLABY
Wallabia bicolor
Darker in color than other wallabies, the swamp wallaby lives in tropical and temperate forests, and swampland in E. Australia.

20–34 in
50–85 cm

GOODFELLOW'S TREE KANGAROO
Dendrolagus goodfellowi
Also called the ornate tree-kangaroo, this species is native to the mountainous rainforests of New Guinea, where it feeds on leaves and fruit.

16½–28 in
42–71 cm

20–24 in
51–62 cm

BRUSH-TAILED ROCK WALLABY
Petrogale penicillata
This resident of S.E. Australia has roughened, padded hind feet that help it to grip when jumping in rocky habitats.

nonprehensile tail

LUMHOLTZ'S TREE KANGAROO
Dendrolagus lumholtzi
The smallest of the tree-kangaroos, this marsupial is found in the rainforests of N. Queensland, Australia.

∨ BIG BOOMER
Male red kangaroos, known as boomers, are much larger than females, sometimes weighing twice as much. Only male red kangaroos are red; the females, known as blue fliers, have bluish-gray fur all over. The males fight for access to mates, their battles taking the form of boxing matches.

velvety nose

glands in the neck and chest produce a dark, scented secretion, which male kangaroos rub on bushes to assert their dominance

kangaroos lick their forearms when hot—this cools the blood close to the skin

elastic tendon stores energy when the leg is flexed, and releases it to power the next hop

RED KANGAROO
Osphranter rufus

Kangaroos evolved in Australia, filling the niche occupied elsewhere by grazing animals such as antelope. In common with those species, kangaroos have a big stomach to take on large quantities of grass. Living out in the open is risky for all herbivores, and kangaroos share with antelope a number of adaptations for avoiding predators. They live in herds (called mobs), have sharp senses, are tall enough to scan their surroundings for danger, and possess a wonderful turn of speed. The red is the largest and swiftest kangaroo, able to hop at more than 30 miles (50 km) an hour. Only the female has a pouch, in which she carries her young—known as a joey—for the first seven months of its life. Red kangaroos are well adapted to drought—they are able to eat scrubby saltbush, which is toxic to other animals.

SIZE 2¼–4½ft (0.7–1.4 m)
HABITAT Scrubland, desert
DISTRIBUTION Australia
DIET Herbivorous

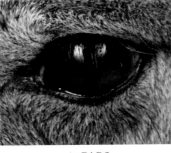

< EYE
The eyes are widely spaced toward the sides of the head, giving a very wide field of vision. This aids the kangaroo's awareness of predators.

> EARS
The ears are large and sensitive, and can be swiveled in different directions to focus on sounds that might mean danger is approaching.

∨ STRONG FIGHTER
The male kangaroo has a broad, muscular chest. The forelimbs are used to punch rivals in fights, but the most forceful blows come from double kicks of the hind feet.

∧ CLAWS
The feet have surprisingly sharp claws. These provide grip when hopping, but also serve as weapons, and as combs for grooming.

< ∨ HIND FOOT
The scientific name *Macropus* literally means "big foot." Each hind foot has four toes: an outer, weight-bearing pair, and an inner pair used for grooming.

tail acts as a counterbalance when hopping, and as a supportive fifth limb when standing still

∧ HOPPING
Red kangaroos hop on their enormous hind legs with apparently effortless grace. They are able to cover up to 30 ft (9 m) in a single bound.

SENGIS

Formerly known as elephant shrews, sengis form a single family. They are superficially similar to the true shrews, although they are not related.

Distinguished by an elongated, flexible snout, sengis are small, with long legs and tail. They move either on all fours or, especially if they need to move fast, by hopping. Sengis live in pairs within a defined area, although there is little interaction between the partners. Males and females even sometimes make separate nests, each defending the jointly held territory against intruders of its own sex. Active mostly by day, many sengis maintain a network of cleared paths around their territory, which they patrol for prey and use as quick escape routes if threatened.

Found only in Africa, sengis occupy a range of habitats, from forests and savanna to extremely arid deserts. Some take shelter in rock crevices or nests of dried leaves, while the larger species dig out shallow burrows.

Sengis are primarily insectivorous, although some species take spiders and worms. They use their sensitive nose to root on the ground and under leaf litter for food, extending their long tongue to scoop up their prey.

ACTIVE YOUNG

Depending on species, a pair of sengis may breed several times during the year. Litters are small, numbering between one and three young, which are well-developed at birth and rapidly become active.

PHYLUM	CHORDATA
CLASS	MAMMALIA
ORDER	MACROSCELIDIDEA
FAMILIES	1
SPECIES	20

DEBATE
AN ORDER OF THEIR OWN

In the past, sengis have been linked with shrews and hedgehogs, rabbits and pikas, and even ungulates. Now in their own order Macroscelididea, genetic evidence suggests their closest relatives are tenrecs and golden moles in the order Afrosoricida.

SENGIS

The Macroscelididae occur in a wide range of African habitats, from deserts to mountains to forests. Largely insectivorous, sengis (or elephant shrews) use their elongated, movable snout to search for prey, which they flick into the mouth with the help of their tongue.

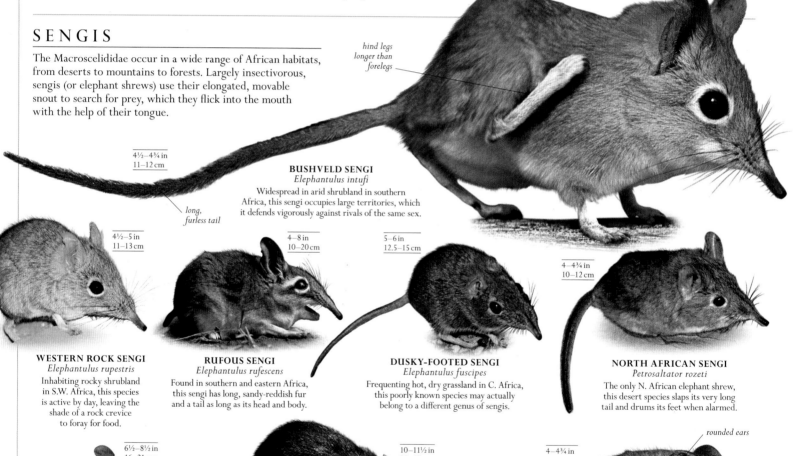

hind legs longer than forelegs

4½–4¾ in
11–12 cm

long, furless tail

BUSHVELD SENGI
Elephantulus intufi
Widespread in arid shrubland in southern Africa, this sengi occupies large territories, which it defends vigorously against rivals of the same sex.

4½–5 in
11–13 cm

WESTERN ROCK SENGI
Elephantulus rupestris
Inhabiting rocky shrubland in S.W. Africa, this species is active by day, leaving the shade of a rock crevice to foray for food.

4–8 in
10–20 cm

RUFOUS SENGI
Elephantulus rufescens
Found in southern and eastern Africa, this sengi has long, sandy-reddish fur and a tail as long as its head and body.

5–6 in
12.5–15 cm

DUSKY-FOOTED SENGI
Elephantulus fuscipes
Frequenting hot, dry grassland in C. Africa, this poorly known species may actually belong to a different genus of sengis.

4–4¾ in
10–12 cm

NORTH AFRICAN SENGI
Petrosaltator rozeti
The only N. African elephant shrew, this desert species slaps its very long tail and drums its feet when alarmed.

6½–8½ in
16–21 cm

FOUR-TOED SENGI
Petrodromus tetradactylus
Found in a range of moist habitats, this species is one of the most widespread elephant shrews. It occurs from C. to South Africa.

BLACK AND RUFOUS SENGI
Rhynchocyon petersi
This giant elephant shrew inhabits the coastal forests of E. Africa. Its orange foreparts graduate through deep red to black on the rump.

10–11½ in
25–29 cm

4–4¾ in
10–12 cm

rounded ears

ROUND-EARED SENGI
Macroscelides proboscideus
A medium-sized sengi, this southern African species occupies some of the world's driest habitats. Strictly monogamous, males strongly guard their mates.

TENRECS AND GOLDEN MOLES

Three families make up the order Afrosoricidae—the golden moles that are exclusively burrowers, the tenrecs and shrew tenrecs that have evolved into a variety of habitats, and the otter shrews that live in riverine and wetland areas.

Most species in this order are native to mainland Africa, with tenrecs also found in Madagascar. Golden moles are all very similar in appearance, with a cylindrical body and other anatomical adaptations to suit their burrowing lifestyle. In contrast, tenrecs display diverse characteristics that reflect the wide range of environments they occupy. Various species of tenrecs are found in tropical forests, where they may be terrestrial or semiarboreal. The otter shrews are aquatic, living in streams and rivers; and there are also burrowing tenrecs. Until recently, tenrecs and golden moles were classified with true moles, shrews, and hedgehogs because of their largely insectivorous diet and often similar appearance. It is now known that the order Afrosoricida evolved independently and is most closely related to the elephants, hyraxes, and sirenians.

UNUSUAL FEATURES

Tenrecs and golden moles have features once considered primitive but now recognized as adaptations to harsh environments. Such features include a low metabolic rate and body temperature. These mammals also have the ability to enter a state of torpor—for up to three days—to save energy in cold conditions, and possess highly efficient kidneys, which reduce the need to drink water.

PHYLUM	CHORDATA
CLASS	MAMMALIA
ORDER	AFROSORICIDA
FAMILIES	3
SPECIES	55

A Grant's golden mole feasts on a locust. At night it forages for food on the surface, feeding mainly on termites.

GOLDEN MOLES

These moles resemble both true moles (Talpidae) from Europe, Asia, and North America, and marsupial moles (Notoryctemorphia) from Australia, as all have similar burrowing habits. The Chrysochloridae from southern Africa have short legs with powerful digging claws, dense fur that repels moisture, and toughened skin, especially on the head. They have nonfunctional eyes covered with skin, and lack external ears.

3½–4½ in
9–11 cm

JULIANA'S GOLDEN MOLE
Neamblysomus julianae
Unique to the dry highlands of South Africa, typically in sandy soils, this species frequents well-irrigated gardens within its range.

3½–4¾ in
9–12 cm

CAPE GOLDEN MOLE
Chrysochloris asiatica
Although secretive, this is a common species in parts of South Africa. It uses the enlarged second toe on its front feet for digging.

soft, dense, glossy fur

leathery pad protects nostrils and aids burrowing

3–3¼ in
7.5–8.5 cm

webbed toes to kick soil backward

eyes covered with thick layer of skin

HOTTENTOT GOLDEN MOLE
Amblysomus hottentotus
This golden mole inhabits tunnel systems up to 655 ft (200 m) long, using the large second and third toes on its front feet for digging.

4–5½ in
10–14 cm

GRANT'S DESERT GOLDEN MOLE
Eremitalpa granti
Inhabiting the southwest African coastal dunes, one of the world's driest habitats, this species "swims" through sand rather than building tunnels.

TENRECS

From Africa and Madagascar, tenrecs have diverse body forms, resembling the unrelated shrews, mice, hedgehogs, and otters. They range in weight from ³/₁₆ oz (5g) to more than 2¼ lb (1 kg). Largely nocturnal and with poor eyesight, these insectivores use their sensitive whiskers to locate food.

4–7 in
10–18 cm

7–14 in
18–35 cm

LESSER HEDGEHOG TENREC
Echinops telfairi

Its hairs modified into spines, this species bears a remarkable resemblance to a true hedgehog, sharing its defense of rolling into a ball.

4–6½ in
10–17 cm

white-tipped spines on back

LOWLAND STREAKED TENREC
Hemicentetes semispinosus

A tailless tenrec, the coarse, spiny coat of this species is distinctively two-tone—black, with yellow stripes and crown bristles.

COMMON TENREC
Tenrec ecaudatus

A large terrestrial tenrec, with red-brown fur and spines along the body, the females have up to 29 teats, more than any other mammal.

moderately large ears

4–4¾ in
10–12 cm

RICE TENREC
Oryzorictes sp.

A burrower with well-developed forelimbs, long claws, and small eyes and ears, this tenrec can become numerous in marshes and rice paddies.

5½–9 in
14–23 cm

sharp claws

GREATER HEDGEHOG TENREC
Setifer setosus

Widespread throughout Madagascar, even in urban habitats, this species has a diverse diet, from insects, earthworms, and carrion to fruit.

AARDVARK

The only species in the Orycteropodidae, the aardvark is adapted for digging. It has an arched back, thick skin, long ears, and a tubular snout.

The aardvark's toes (four on the front foot and five on the hind) each have a flattened nail used to dig burrows and excavate insect nests, which it detects through its keen sense of smell.

It uses its long, thin tongue to catch huge numbers of insects. Its teeth are very unusual, and are one of the main reasons for it being classified in its own order. The aardvark is born with incisors and canines at the front of the jaw, but when these fall out they are not replaced. The rear teeth lack enamel, are open-rooted, and grow throughout life.

PHYLUM	CHORDATA
CLASS	MAMMALIA
ORDER	TUBULIDENTATA
FAMILIES	1
SPECIES	1

AARDVARK

Inhabiting savanna and bushland in sub-Saharan Africa, aardvarks are solitary and nocturnal. Powerful front legs with flattened claws are used to dig into the nests of ants and termites for food.

3–4½ ft
0.9–1.4 m

AARDVARK
Orycteropus afer

The yellowish skin of the aardvark is often stained red by soil, in which it burrows and forages. It also has a sparse covering of bristly hairs.

hair lighter colored on body

long, blunt, shovel-shaped claws

DUGONG AND MANATEES

A small order of fully aquatic herbivores, sirenians inhabit a range of tropical habitats, from swamps and rivers to marine wetlands and coastal waters.

Also known as sea cows, sirenians are supremely adapted to an aquatic way of life. Their front legs are paddlelike and adapted for steering; the hind legs are not visible, being restricted to two small remnant bones floating in the muscle. The body is therefore streamlined, and the flattened tail provides propulsion. A layer of blubber below their skin provides insulation, but to help counteract its natural buoyancy, the bones are dense and the lungs and diaphragm extend for the full length of the spine. The result is a hydrodynamic body, albeit rather slow-moving, capable of fine adjustments to shift its position in the water. All four species are considered to be at risk of extinction.

PHYLUM	CHORDATA
CLASS	MAMMALIA
ORDER	SIRENIA
FAMILIES	2
SPECIES	4

DUGONG

There is only one species in the Dugongidae. Slow-moving marine herbivores, dugongs graze on submerged seagrass beds, using the downturned muscular snout to uproot food.

8¼–10 ft
2.5–3 m

DUGONG
Dugong dugon
From the Indo-Pacific region, especially around Australia, the dugong's cylindrical body has no dorsal fin or hind limbs. The tail has flukes.

MANATEES

With a shorter snout than dugongs, the Trichechidae also differ in the shape of their tail, which is paddle-shaped rather than fluked. Rather sluggish, they spend much of the day sleeping under water, surfacing every 20 minutes to breathe.

AMAZONIAN MANATEE
Trichechus inunguis
A freshwater manatee from the Amazon Basin, this species usually has a characteristic white chest patch.

6½–10 ft
2–3 m

8–12¾ ft
2.5–3.9 m

FLORIDA MANATEE
Trichechus manatus latirostris
The largest living sirenian, the Florida manatee inhabits both fresh water and coastal waters of the S.E. US. It feeds on aquatic plants.

8–12¾ ft
2.5–3.9 m

ANTILLEAN MANATEE
Trichechus manatus manatus
Ranging from Mexico to Brazil, this manatee is smaller and more likely to be found farther from the shore than the Florida subspecies.

HYRAXES

Furry and rotund, modern hyraxes belong to the only extant family in the order Hyracoidea. Hyraxes were once the primary terrestrial herbivores in the Old World. Now just five species remain.

Widely represented in the fossil record of Asia, Africa, and Europe, some extinct species of hyrax were more than three feet in height.

Unlike most browsing and grazing animals, hyraxes use their cheek teeth (rather than incisors) to slice vegetation, before chewing it. With a relatively indigestible diet, they have a complex but efficient digestive system in which bacteria break down tough plant matter.

With a number of primitive mammalian features, such as poor control of their internal body temperature, hyraxes show several affinities with elephants: their upper incisor teeth are often extended into short tusks, their soles have sensitive pads, and they have high brain function.

PHYLUM	CHORDATA
CLASS	MAMMALIA
ORDER	HYRACOIDEA
FAMILIES	1
SPECIES	5

HYRAXES

The Procaviidae are rotund, short-tailed herbivores from Africa and the Middle East. Hyraxes are often seen basking in the sun, and huddling together in groups for warmth. They will also seek shade if too hot. Behaving in this way provides hyraxes with an effective means of thermoregulation.

17¼–22½ in
44–57 cm

long, silky fur

short snout

WESTERN TREE HYRAX
Dendrohyrax dorsalis
A usually dark, semi-arboreal hyrax from W. and C. Africa, this species has a distinctive white patch on its lower back and rump.

12½–22 in
32–56 cm

12½–23½ in
32–60 cm

BUSH HYRAX
Heterohyrax brucei
There are 24 subspecies of bush hyrax in rocky habitats across Africa. It feeds on grass and fruit, and also catches and eats small vertebrates.

15½–23 in
39–58 cm

SOUTHERN TREE HYRAX
Dendrohyrax arboreus
An able climber from southern Africa, usually inhabiting hollow trees, this hyrax is known for its shrieking territorial calls at night.

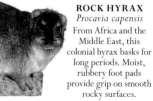

ROCK HYRAX
Procavia capensis
From Africa and the Middle East, this colonial hyrax basks for long periods. Moist, rubbery foot pads provide grip on smooth rocky surfaces.

∧ BRAIN POWER

Long acknowledged for having a good memory, the elephant is now known to solve problems, use tools, show emotions, and work with others, making it one of the most intelligent of all terrestrial animals.

elongated nose and upper lip form the mobile trunk

AFRICAN SAVANNA ELEPHANT
Loxodonta africana

A mature male African savanna elephant is larger than any other living land mammal and in exceptional cases can weigh as much as eleven tons. The head alone can exceed 880lb (400kg), including the brain, eyes, trunk, ears, tusks, and teeth. Females are smaller and form strongly bonded herds, comprising a dominant female, close relatives, and their offspring. Bull elephants form less cohesive groups led by older males, but in the breeding season, when they come into musth, they are more solitary and will fight for access to receptive females. Elephants are long-lived and may survive well into their 60s.

EYE ∨
Long eyelashes protect the small, laterally-positioned eyes, which can see clearly for only about 33ft (10m). Other senses compensate for this limitation.

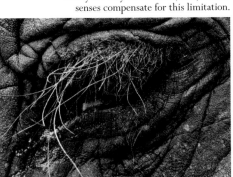

TRUNK SKIN ∨
The trunk has short sensory hairs along its length that enhance its responsiveness to touch.

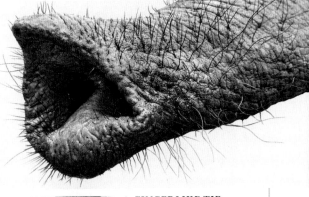

∧ FINGERLIKE TIP
The opposing fingerlike projections at the tip of the trunk are extremely sensitive and dextrous. This allows the African savanna elephant to pick up objects as small as a peanut.

EAR >
The large, thin-skinned ears help to regulate body temperature. They are also used to communicate moods, such as aggression.

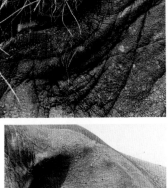

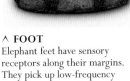

∧ FOOT
Elephant feet have sensory receptors along their margins. They pick up low-frequency underground sounds that can travel over long distances.

< TAIL
Elephants swing their tail to deter insects. The air turbulence created prevents them from landing, and any that do are swatted.

the tusks are enlarged second incisor teeth

musth glands are only found in elephants and are present in both sexes

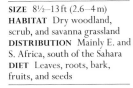

SIZE 8½–13ft (2.6–4m)
HABITAT Dry woodland, scrub, and savanna grassland
DISTRIBUTION Mainly E. and S. Africa, south of the Sahara
DIET Leaves, roots, bark, fruits, and seeds

LARGE AND HEAVY >
A bulky body supported by straight, columnar legs might suggest the elephant is a slow-moving animal. It can, however, reach speeds of 15mph (24kph) when charging.

large abdominal cavity contains about 62ft (19m) of intestines

ELEPHANTS

Largest of all land mammals, elephants are distinguished by their colossal bulk, long flexible trunk, large ears, and curving ivory tusks.

The only members of the Proboscidea, elephants inhabit grassland and forest in tropical Africa and Asia. Sometimes weighing more than five tons, elephants have a modified skeleton to bear their weight. The limb bones are extremely stout, and the toes are splayed around a pad of connective tissue. This large frame needs vast amounts of food and elephants eat for up to 16 hours a day, consuming up to 550 lb (250 kg) of vegetation.

EARS, NOSE, AND MOUTH

The elephant's muscular trunk, a fusion of the nose and upper lip, is remarkably versatile and is able to grasp morsels of food. It can also draw in water and serve as a snorkel when swimming. Sensory organs in the trunk pick up scents and vibrations, so aiding communication. An elephant's large ears are also used for communication and when spread out are a sign of aggression. However, constant flapping helps an elephant to lose heat. Tusks are incisor teeth that grow continually. They are used to dig for roots and salt, clear pathways, and mark territories.

SOCIAL STRUCTURES

Male and female elephants have different social behavior. Females form herds of related cows and their calves, led by an elderly matriarch. Bull elephants form short-term associations, but will fight with any intruding rival during the breeding season.

PHYLUM	CHORDATA
CLASS	MAMMALIA
ORDER	PROBOSCIDEA
FAMILIES	1
SPECIES	3

TWO SPECIES OR ONE?

The splitting of the African elephant into two species, the savanna elephant and the smaller forest elephant, was originally based on physical differences, but now there is sufficient genetic evidence to substantiate this division. However, some scientists are reluctant to accept this idea because the two species sometimes interbreed where their ranges overlap.

ELEPHANTS

Including elephants and the extinct mammoths, the Elephantidae form a family of large herbivores, with a long trunk, large ears, thick skin, and tusks. The trunk is used in feeding, drinking, bathing, and in social interactions. Elephants make a range of sounds, from trumpeting to subsonic rumblings, enabling them to communicate over long distances.

7¾–11 ft
2.4–3.4 m

8½–13 ft
2.6–4 m

large ears

long, curved tusks

ASIATIC ELEPHANT
Elephas maximus
Domesticated for forest operations and ceremonial use, this species has smaller ears and a more arched back than African species. Females usually lack tusks as do some males.

AFRICAN SAVANNA ELEPHANT
Loxodonta africana
The largest living land animal, this elephant has a large head and ears. Both sexes have well-developed, curved tusks.

5¼–9½ ft
1.6–2.9 m

muscular trunk projections

AFRICAN FOREST ELEPHANT
Loxodonta cyclotis
With five toenails on the forelimbs and four on the hind limbs, this small elephant with straight tusks is from the tropical forests of Central Africa.

semicircular toenails

ARMADILLOS

Recognizable by their armored shell—unique among mammals—armadillos occur in various shapes, sizes, and colors. All are native to the Americas.

Armadillos, sole members of the order Cingulata, range across a variety of habitats, feeding mostly on insects and other invertebrates. Despite having short legs, armadillos can run quickly and burrow with their strong claws to avoid predators. Their main defense consists of a bony carapace, covered with horny plates, across their upperparts. In most species, there are rigid shields over the shoulders and hips, with a varying number of bands separated by flexible skin covering the back and flanks. This allows some species to roll into a ball to protect their vulnerable, furry underparts.

Armadillos have few natural predators. Despite the threats from human hunters and habitat loss, the range of some armadillo species is expanding.

Although their armor is heavy, armadillos are effective swimmers. By inflating the stomach and intestines with air, they can increase their buoyancy and cross small bodies of water. Armadillos can hold their breath under water for several minutes, allowing them to walk over a stream bed rather than swim across it.

HABITS
Most species of armadillo are nocturnal, though they occasionally emerge during the day as well. These animals are largely solitary, associating with one another only during the mating season. Males sometimes display aggression toward rivals.

PHYLUM	CHORDATA
CLASS	MAMMALIA
ORDER	CINGULATA
FAMILIES	2
SPECIES	20

Contrary to popular belief, not all armadillos can roll into a ball in defense—only species of the genus *Tolypeutes* can.

ARMADILLOS

The only surviving family in their order, the Dasypodidae are found in the Americas. Covered with bony plates across their upperparts, they use their sharp claws to dig for invertebrates and to excavate burrows. Of the 20 or so species, some are able to roll into a ball when threatened, in order to protect their vulnerable, soft, furry underparts.

long tail in proportion to body

9½–22½ in
24–57 cm

LONG-NOSED ARMADILLO
Dasypus sp.
The seven species of this genus inhabit shaded, rocky areas. Unlike other armadillos, they have sparse, yellowish fur, mainly on their bellies.

long, pointed snout used for foraging

well-developed claws

8–12 in
20–30 cm

ANDEAN HAIRY ARMADILLO
Chaetophractus vellerosus
Unusual in having copious hair between its scales, this mammal inhabits high-altitude S. American grasslands. It is hunted for its meat and shell.

10–16 in
26–40 cm

LARGE HAIRY ARMADILLO
Chaetophractus villosus
From arid habitats in southern S. America, this species has long, coarse hairs between the 18 or so plates of protective armor across its back.

9–12 in
22–31 cm

PICHI
Zaedyus pichiy
A small, dark armadillo with thick dorsal plates, this species wedges itself into its burrow when threatened, presenting its jagged scales for defense.

16–20 in
40–50 cm

SIX-BANDED ARMADILLO
Euphractus sexcinctus
More active by day than most armadillos, this brownish-yellow species forages in grassland and forest, eating plant and animal matter.

4½–6 in
11–15 cm

PINK FAIRY ARMADILLO
Chlamyphorus truncatus
This tiny armadillo from C. Argentina is mainly subterranean, "swimming" through loose sand; its head is shielded to reduce abrasion.

2¼–3¼ ft
0.75–1 m

jointed bands on blackish upperparts

12–15 in
30–38 cm

rounded, pink ears

NORTHERN NAKED-TAILED ARMADILLO
Cabassous centralis
The protective plates of this C. and S. American armadillo do not extend to its tail. It relies mainly on dense cover to avoid predators.

GIANT ARMADILLO
Priodontes maximus
The largest armadillo, this species has tough armor and uses its long, curved third front claw to dig for food and in defense.

SIX-BANDED ARMADILLO
Euphractus sexcinctus

Despite its name, this species of armadillo may have anything from six to eight armored bands around its middle. The upper part of the animal's body is covered by a sturdy carapace. The carapace is made up of plates of bone topped with a thin covering of horny material, and it is embedded in the armadillo's skin. The bands act as joints in the carapace, giving the animal flexibility, although it is unable to curl completely into a ball like some other species. The six-banded armadillo is diurnal, spending the daylight hours trundling about its home range in search of food. It has a varied diet, ranging from roots and shoots to invertebrates and carrion.

SIZE 16–20 in (40–50 cm)
HABITAT Forest and savanna
DISTRIBUTION S. America, mainly south of Amazon Basin
DIET Omnivorous

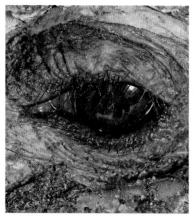

< SHIELDED EYE
The armadillo's eyes are protected by the headshield at the expense of a slightly restricted field of vision. As it has poor eyesight anyway, this makes little difference.

< NOSE
Armadillos have an acute sense of smell. They sniff and snuffle their way to most meals, and can easily detect food buried in the soil.

head is narrow and pointed and protected by a shield made of fused bony plates

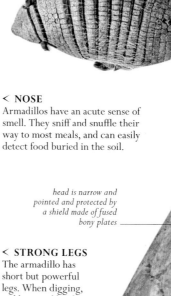

< STRONG LEGS
The armadillo has short but powerful legs. When digging, soil loosened by the front feet is kicked out of the hole by the hind feet.

∧ HAIRY SKIN
The bony carapace is interrupted by six to eight bands, which are separated by flexible skin. Long, bristly hairs sprout from between the bands – this is one of the hairy armadillos.

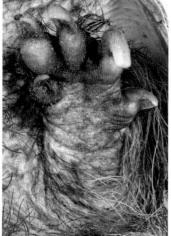

TAIL >
Glands at the base of the armadillo's tail release scent through small holes in the armor plates. The scent marks the armadillo's territory.

∧ CLAWS
Long, strong claws allow the armadillo to dig rapidly into hard ground. In a matter of minutes it is able to excavate a trench deep enough to bury itself in.

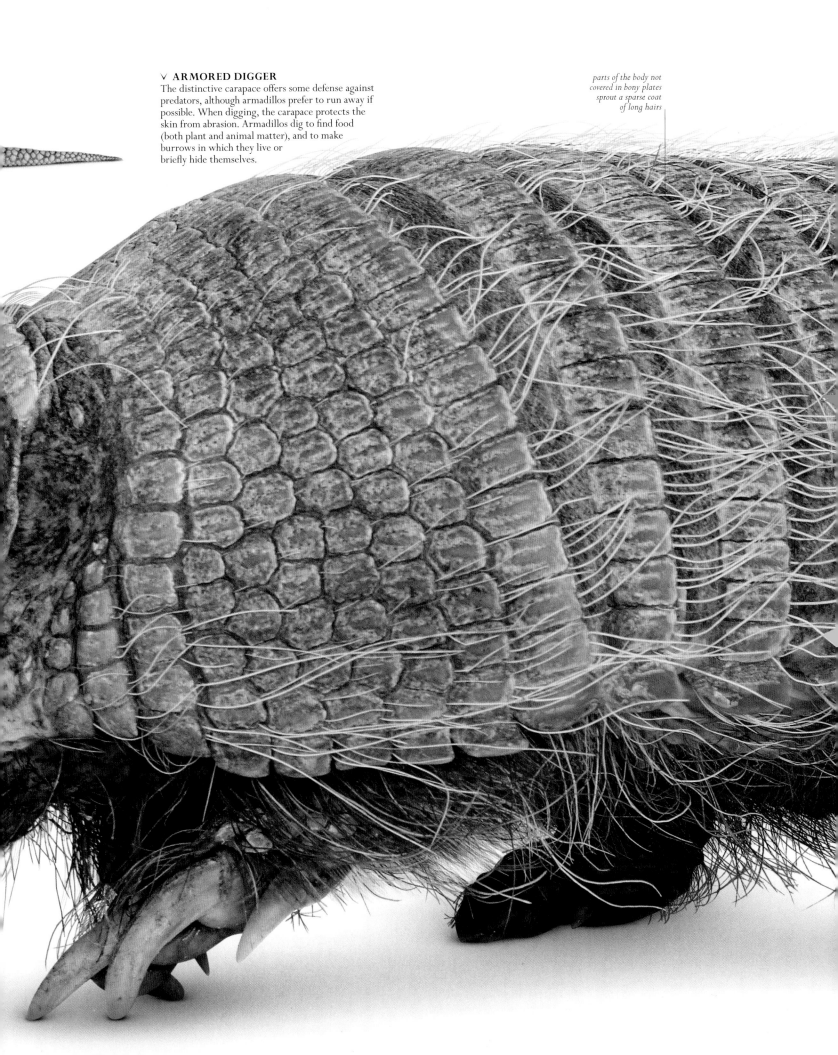

⌄ ARMORED DIGGER

The distinctive carapace offers some defense against predators, although armadillos prefer to run away if possible. When digging, the carapace protects the skin from abrasion. Armadillos dig to find food (both plant and animal matter), and to make burrows in which they live or briefly hide themselves.

parts of the body not covered in bony plates sprout a sparse coat of long hairs

SLOTHS AND ANTEATERS

Although very different in form and habits, sloths and anteaters have one characteristic in common: they share a lack of normal mammalian dentition.

With just one exception, members of this order, the Pilosa, are largely arboreal creatures: only the giant anteater is terrestrial. They are all native to Central and South America.

The two anteater families feed on ants, termites, and other insects. Anteaters have no teeth at all, relying on their long, sticky tongue to capture insects, which are then crushed in the mouth before being swallowed.

The slow-moving sloths are herbivorous. They have no incisors or canines, but instead have a number of cylindrical, rootless teeth, used to grind up their food. A sloth may take about a month to completely digest a meal, as the fibrous leaf matter passes slowly through several stomach compartments and is broken down by bacteria.

ARBOREAL LIFE

The long, prehensile tail of the arboreal anteaters allows them to lead a more active lifestyle than the sloths. On the ground, all the arboreal species move only with difficulty, although some are excellent swimmers. When walking, their mobility is restricted by their enlarged, curved claws, which the sloths use like hooks to hang from branches. Anteaters have enlarged claws only on their front feet. These claws are used to tear open insect nests in search of food, and are formidable defense weapons.

PHYLUM	CHORDATA
CLASS	MAMMALIA
ORDER	PILOSA
FAMILIES	4
SPECIES	16

To defend themselves members of the genus *Tamandua* stand up on their hind legs and lash out with their powerful front limbs.

THREE-TOED SLOTHS

Generally smaller and slower moving than two-toed sloths, the arboreal and mostly nocturnal Bradypodidae have three toes on each foot, bearing long, curved claws that help them to hang from branches. Their long, shaggy fur has a green tinge, due to algae growing on it.

23–28 in
59–72 cm

MANED SLOTH
Bradypus torquatus
This small Brazilian sloth has long, dark fur—especially around its head and neck—which often harbors algae, ticks, and moths.

BROWN-THROATED SLOTH
Bradypus variegatus
An inhabitant of the forests of C. and S. America, this sloth is the most widespread in its family. The female attracts her mate with a shrill scream.

20½–21½ in
52–54 cm

18–30 in
45–76 cm

shaggy, coarse coat

PALE-THROATED SLOTH
Bradypus tridactylus
Found in the rainforests of S. America, this species usually lives alone and may rest for more than 18 hours a day.

TWO-TOED SLOTHS

Unlike three-toed sloths, the Megalonychidae have only two toes on their forefeet, a more prominent snout, and no tail. Similarly arboreal and mainly nocturnal, they usually descend from trees head first.

two toes on front feet

21½–35 in
54–88 cm

SOUTHERN TWO-TOED SLOTH
Choloepus didactylus
This large herbivore from S. America swims well, even crossing rivers. Its main predators are large raptors, such as harpy eagles.

21½–35 in
54–88 cm

LINNAEUS'S TWO-TOED SLOTH
Choloepus hoffmanni
This solitary, S. American herbivore is found in the tropical forests of the Orinoco and Amazon river basins, where it spends most of its time sleeping or resting in the forest canopy.

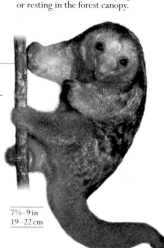

SILKY ANTEATERS

The Cyclopedidae, though well-represented in the fossil record, has seven extant species. Large, curved front claws and a prehensile tail help silky anteaters live on trees, where they nest in holes. They feed on ants and termites.

SILKY ANTEATER
Cyclopes sp.
Arboreal, nocturnal, and slow moving, silky anteaters live in forested areas of Central and South America. Their claws provide an effective defense when threatened by predators.

7½–9 in
19–22 cm

GIANT ANTEATER AND RELATIVES

Native to Central and South America, the Myrmecophagidae have an elongated snout and a long tongue. They use their powerful claws to rip open termite mounds and ant hills. Lacking teeth, they catch insects with their sticky saliva and spine-covered tongue.

18½–30 in
47–77 cm

SOUTHERN TAMANDUA
Tamandua tetradactyla
This solitary species has a prehensile tail. It forages for food at any time of day, and is active for about 7 hours in every 24.

huge, bushy tail

stiff, strawlike hair

long, tubular snout

3¼–4½ ft
1–1.4 m

GIANT ANTEATER
Myrmecophaga tridactyla
The largest anteater, with a tail almost as long as its body, this species uses its long, sticky tongue to catch up to 30,000 insects a day.

RABBITS, HARES, AND PIKAS

Two herbivorous families make up the order known as Lagomorpha. Their apparent similarities to rodents result from adaptation to a common lifestyle.

These species occupy a variety of habitats, from tropical forest to Arctic tundra. All are terrestrial herbivores or browsers. They are gnawing animals and share the diet of many rodents. Like those of rodents, the teeth of rabbits, hares, and pikas continue to grow throughout life—they are worn down by chewing. However, there are fundamental differences between the groups: rabbits and their relatives have four incisors in the upper jaw, while rodents have only two.

Relying on food that is relatively slow to break down, these animals have a modified digestive system. They produce two types of feces: moist pellets, which are eaten to gain further nutrients from the food, and dry pellets passed as waste.

ESCAPING PREDATORS

Pikas, the most rodentlike species in this order, take refuge from predators in burrows and crevices after raising the alarm with a whistle. Rabbits and hares, in contrast, have long ears to detect danger and powerful limbs to flee from predators. These animals have large eyes positioned high on each side of the head to provide almost 360-degree vision. When a predator is observed, hares drum their hind legs on the ground in warning.

PHYLUM	CHORDATA
CLASS	MAMMALIA
ORDER	LAGOMORPHA
FAMILIES	2
SPECIES	OVER 90

To survive the winter, the pika collects various plants, creates a hay pile of dried food, and stores it in its den.

RABBITS AND HARES

Native across much of the world, the Leporidae have long, movable ears and enlarged hind legs to detect and escape predators. Large eyes reflect their mainly nocturnal habits. Rabbits often inhabit permanent burrow systems, whereas the more solitary hares make only transitory shelters.

14–15 in
36–38 cm

EUROPEAN RABBIT
Oryctolagus cuniculus
Native to the Iberian penninsula, this species has been introduced for meat and fur worldwide, with devastating effects on local habitats and wildlife.

furry feet

6–12 in
15–30 cm

LOP-EARED RABBIT
Oryctolagus cuniculus
First bred in England in the 19th century, this type has long, lop-eared (floppy) ears. There is variation in both color and size.

5–7 in
13–18 cm

DWARF RABBIT
Oryctolagus cuniculus
One of the smallest breeds, the dwarf rabbit occurs in many colors and patterns. Its rounded face and small ears make it a popular pet.

10–15 in
25–38 cm

ANGORA RABBIT
Oryctolagus cuniculus
Valued for its long, soft hair, which is spun into yarn, this breed originates from Anatolia (in modern-day Turkey).

»

hairs of the coat are long, coarse, and grooved and grow away from the belly

strong, elongated fore- and hind limbs are about the same length

TOPSY-TURVY WORLD >
Adapted to living upside down in the forest canopy, Linnaeus's two-toed sloth rarely ventures to the forest floor other than to relieve itself every 3–5 days. Strong fingers with clawlike hooks for grasping branches allow it to eat, mate, give birth, and even sleep in this position.

LINNAEUS'S TWO-TOED SLOTH
Choloepus hoffmanni

Small and arboreal, Linnaeus's two-toed sloth is far removed from its giant ground-dwelling ancestors, which only became extinct about 10,000 years ago. It is one of the slowest-moving mammals and sleeps for about 13 hours a day, often curled into a ball in the fork of a tree, waking at night to forage for leaves. The sloth's fur is unusual in supporting an ecosystem that includes unique microorganisms, green algae, and various insects. In damp conditions, the coat may acquire a green hue due to the algae growing in the grooves of the hairs.

SIZE 21½–35 in (54–88 cm)
HABITAT Tropical lowland forest
DISTRIBUTION Orinoco and Amazon basins, S. America
DIET Mainly leaves

EYE >
The eyes are large and forward facing, but the eyesight is poor. Sloths are nearsighted, color blind, and have poor visual acuity.

< TEETH
The five upper and four lower teeth are peglike, lack enamel, and grow throughout life. There are no incisors.

NOSE >
A short snout with a large, snub nose at its tip provides the sloth with an excellent sense of smell.

< EAR
The sloth's hearing is poor and limited to lower frequencies. The external part of the ear is tiny and buried in the fur.

CLAWS ∨ >
The forelimb terminates in two digits, each with a 3¼–4 in (8–10 cm), hooklike claw. The hind limb has three clawed digits.

MIDDLE PARTING >
Due to their upside-down lifestyle, the fur parts in the midline of the chest and abdomen and falls down toward the back, so rain will run off.

» RABBITS AND HARES

DESERT COTTONTAIL
Sylvilagus audubonii
Found in arid regions of the S.W. US and N. and C. Mexico, the desert cottontail nests above ground rather than in a burrow.

14½–16 in
37–40 cm

MARSH RABBIT
Sylvilagus palustris
Inhabiting wetlands of N. America, this species is a strong swimmer. It walks instead of hopping, unlike most of its relatives.

16½–17½ in
42–44 cm

brown fur, sometimes turns reddish

long, black-tipped ears

20–28 in
50–70 cm

EUROPEAN HARE
Lepus europaeus
A grazer in summer and a browser of bark and buds in winter, this hare is shy and solitary, except during its springtime "boxing" courtship. Females fight off males until ready to mate.

powerful hind legs

WHITE-TAILED JACKRABBIT
Lepus townsendii
This species ranges extensively across western N. America. The northern population turns white in winter, while the southern animals develop whitish patches only on their flanks.

22–26 in
56–66 cm

14–20½ in
36–52 cm

SNOWSHOE HARE
Lepus americanus
Adapted to harsh N. American winters, this species has a white winter coat for camouflage, and large hind feet to allow movement on soft snow.

22–28 in
55–70 cm

ARCTIC HARE
Lepus arcticus
Adapted to polar and mountainous areas, this hare survives with thick fur that turns white in winter, and by digging snow holes for shelter.

BLACK-TAILED JACKRABBIT
Lepus californicus
Widespread on western N. American prairies and farmland, this species is prone to huge local fluctuations. It has a black tail and black ear-tips.

20½–24 in
52–61 cm

ANTELOPE JACKRABBIT
Lepus alleni
Very long ears and insulated, reflective fur help this large hare to keep cool in its desert grassland habitats in Mexico.

22–26 in
55–67 cm

MOUNTAIN HARE
Lepus timidus
From polar regions and mountains in Europe and Asia, this hare has a white winter coat and a tail that stays white all year round.

20–22 in
51–55 cm

18–22 in
45–55 cm

CAPE HARE
Lepus capensis
Common in open habitats of Africa and the Middle East, this species is very similar to the closely related European hare.

long legs

PIKAS

The Ochotonidae are small herbivores that live in rocky mountainsides and open steppes in North America and Asia. Alert to predators, they produce high-pitched alarm calls as they run for cover in crevices and burrows.

6½–8½ in
16–21 cm

AMERICAN PIKA
Ochotona princeps
This N. American montane scree dweller dries food piles in the sun, and stores them in its den for the winter.

RODENTS

Ranging from tiny mice to animals the size of a pig, rodents live in almost every habitat type. They make up nearly half of all mammal species.

Members of the order Rodentia are distinguished by a pair of prominent upper and lower incisor teeth (often orange or yellow in color), which continue to grow throughout life. Gnawing— a characteristic behavior of all rodents—wears the teeth away at the same rate as they grow. Rodents have no canine teeth, just a long space between the incisors and the three or four cheek teeth in each jaw. The scientific classification of rodents is also based on other features of their teeth and jaws that are not visible externally.

SPECIES DIVERSITY

To suit their various lifestyles, rodents often have special adaptations such as webbed toes, big ears, or long hind feet for a bounding gait. Some rodents burrow, others live in trees or in the water, but none live in the sea. Many species inhabit desert regions, where they may never drink, deriving all the water they need from food.

THE IMPACT OF RODENTS

Several rodent species carry fatal diseases that have killed millions of people. Others consume or contaminate huge quantities of stored food meant for humans. The house mouse has become the world's most widespread wild mammal through its close association with humans. This species inhabits every continent, except Antarctica, even surviving in mines and cold stores. Some rodents cause damage to crops or trees, or burrow in inconvenient places. Beavers can transform entire habitats, affecting hundreds of other animal and plant species.

On the positive side, rodents are a vital food source for predatory animals and, in some countries, for humans as well. Many small rodents, such as hamsters, are specially bred as pets.

PHYLUM	CHORDATA
CLASS	MAMMALIA
ORDER	RODENTIA
FAMILIES	34
SPECIES	About 2,500

DEBATE
RODENT ORGANIZATION

The huge diversity of species in the order Rodentia makes the task of classifying them rather complicated. With 34 families of modern rodent to organize, experts in zoological classification have divided them into two subgroups, according to differences in their skulls, teeth, and jaws. The Sciurognathi (squirrel-jaws) are globally widespread, and include all squirrels, beavers, and mouselike rodents. The Hystricognathi (porcupine-jaws) includes the cavies, porcupines, chinchillas, and the capybara, and these are mostly restricted to the southern hemisphere and the tropics.

MOUNTAIN BEAVER

Only one species of the once widespread family Aplodontiidae has survived into modern times. The mountain beaver, or sewellel, lives in burrows in the damp forests and plantations of western North America.

flattened head

12–16 in
30–40 cm

MOUNTAIN BEAVER
Aplodontia rufa
This most primitive type of living rodent inhabits forested coastal mountains of W. Canada and the US.

SQUIRRELS AND CHIPMUNKS

Occurring almost everywhere, except in the polar regions, Australia, and the Sahara Desert, members of the family Sciuridae range from tropical rainforests to Arctic tundra, and from the tops of trees to underground tunnels. This family includes typical bushy-tailed tree squirrels, many species of burrow-digging ground squirrels, chipmunks, and marmots. They feed mainly on nuts and seeds.

AMERICAN RED SQUIRREL
Tamiasciurus hudsonicus
The chattering calls and sharp barks of this red squirrel are common sounds of coniferous woodlands in Canada and the N. US.

9–12 in
23–30 cm

8–10 in
20–26 cm

GRAY SQUIRREL
Sciurus carolinensis
Common in the E. US, this species was introduced to parts of Europe, where it is slowly displacing the native red squirrel.

EURASIAN RED SQUIRREL
Sciurus vulgaris
When this squirrel molts into its summer coat, the long hairs on its ears are lost.

SOUTHERN FLYING SQUIRREL
Glaucomys volans
A strictly nocturnal native of the eastern US, this squirrel lives in tree holes and attics, often in groups during winter.

6½–8 in
17–20 cm

gliding membrane

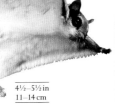

4½–5½ in
11–14 cm

12½–15½ in
32–39 cm

8–10 in
20–26 cm

12–15 in
30–38 cm

PALE GIANT SQUIRREL
Ratufa affinis
Of the four species of giant squirrel, this species inhabits the forests of the Malay Peninsula, Borneo, and Sumatra.

GRIZZLED GIANT SQUIRREL
Ratufa macroura
One of the world's largest tree squirrels, this species occurs in S. India and Sri Lanka. It eats fruits, flowers, and insects.

PREVOST'S SQUIRREL
Callosciurus prevostii
There are 17 subspecies of these tricolored squirrels. Prevost's is found in Malaysia, Borneo, Sumatra, and nearby islands, such as Sulawesi.

»

» SQUIRRELS AND CHIPMUNKS

long, bushy tail

6½–10½ in
17–27 cm

GAMBIAN SUN SQUIRREL
Heliosciurus gambianus
A common species in savanna woodlands across Africa, from Senegal to Zimbabwe, this squirrel feeds mainly on the seeds of Acacia trees.

CAPE GROUND SQUIRREL
Geosciurus inauris
This coarsely haired squirrel escapes the extreme temperatures of the semideserts of South Africa by sheltering in burrows.

COLUMBIAN GROUND SQUIRREL
Urocitellus columbianus
A rather large, bushy tailed ground squirrel, this species forms colonies in meadows and forest edges from Idaho north into W. Canada.

6–8 in
15–20 cm

GOLDEN-MANTLED GROUND SQUIRREL
Callospermophilus lateralis
Appearing like a larger version of the chipmunk, this species is a common sight in forests and mountains of the W. US.

10–12 in
25–30 cm

short, hairy tail

9–11 in
22–28 cm

4¾–6 in
12–15 cm

4¾–6 in
12–15 cm

HOPI CHIPMUNK
Tamias rufus
Chipmunks from different parts of N. America form separate species. This species occurs in Utah, Colorado, and adjacent parts of Arizona.

EASTERN CHIPMUNK
Tamias striatus
This striped, ground-living species is a common sight in forest campgrounds of the E. US, where it has become quite tame.

BEAVERS

There are only two species of beavers in the family Castoridae. Both are exploited for their fur. One is widespread across North America, and the other in patches throughout most of Europe. Both build dams out of stones, mud, and trees, creating habitats for other species.

glossy, brown fur

BEAVER
Castor sp.
European and N. American beavers are different species, but both live a similar, semiaquatic life in rivers and lakes.

2½–4 ft
0.8–1.2 m

flat, scaly tail

DORMICE

The Gliridae occur in wooded habitats across Europe, sub-Saharan Africa, and patchily in Central Asia, with a single species in Japan. Resembling small nocturnal squirrels, all but one are soft-furred, arboreal species. Many are threatened or are declining in numbers.

2¾–6 in
7–15 cm

AFRICAN DORMOUSE
Graphiurus sp.
There are 15 species of African dormice. They are all similar in appearance and inhabit forested areas of sub-Saharan Africa.

bushy tail

GOPHERS

Members of the family Geomyidae, found in North America, live alone in shallow burrows that allow them to access roots and leaves. They carry their food in cheek pouches and store it in underground chambers.

3¼–8 in
8–20 cm

BOTTA'S POCKET GOPHER
Thomomys bottae
A small burrowing creature of soft soil and grassy areas, this species digs up mounds of earth, which can damage farm machinery.

14–20 in
35–50 cm

gingery brown fur

MARMOT
Marmota sp.
Several species of marmots
inhabit mountain grasslands and
rocky areas of N. America, while
some occur in Eurasia. Marmots
hibernate for up to eight months,
depending on the species.

12½–16 in
32–40 cm

BLACK-TAILED PRAIRIE DOG
Cynomys ludovicianus
Active during daytime,
this ground squirrel lives
in large groups in "towns"
consisting of extensive
communal burrows.

8½–10½ in
21–27 cm

HARRIS'S ANTELOPE SQUIRREL
Ammospermophilus harrisii
An agile inhabitant of the
Sonora Desert and northern
Mexico, this rodent is active
in the heat of the day, but
hibernates during winter.

POCKET MICE AND KANGAROO RATS

Mostly common, with a few rare species, the Heteromyidae are found in
varied habitats from Canada to Central America, with kangaroo rats living
mainly in deserts. Pocket mice typically run on four feet, but kangaroo
rats hop on two large hind feet.

4 in
10 cm

longer hairs of tail tuft

2¾–3½ in
7–9 cm

DESERT POCKET MOUSE
Chaetodipus penicillatus
This species is one of many small nocturnal
rodents that inhabit the open, sandy deserts
of the S.W. US and N. Mexico.

MERRIAM'S KANGAROO RAT
Dipodomys merriami
Bounding away with its tail outstretched,
this nocturnal inhabitant of the N. American
deserts looks like a miniature kangaroo.

JUMPING MICE

The families Dipodidae (jerboas), Zapodidae
(jumping mice), and Sminthidae (birch mice)
are made up of rodents that have a long tail and
powerful hind feet that enable them to jump
around like miniature kangaroos.

MEADOW JUMPING MOUSE
Zapus hudsonius
Found in the cool grasslands of
northern N. America, this species
also has isolated populations in the
mountains of Arizona and New Mexico.

2¾–4¼ in
7–11 cm

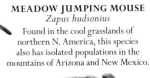

LESSER EGYPTIAN JERBOA
Jaculus jaculus
This desert species is found across
N. Africa, from Senegal to Egypt,
south to Somalia, and east to Iran.

4½–5 in
11–13 cm

EDIBLE DORMOUSE
Glis glis
5–7½ in
13–19 cm
Known in Germany as "the seven-sleeper,"
this dormouse hibernates for more than
six months. It is eaten in Mediterranean
countries, such as Slovenia.

short ears

COMMON DORMOUSE
Muscardinus avellanarius
An excellent climber and
jumper, this dormouse feeds on
flowers, fruits, and insects
at night, in shrubby woodland
across Europe.

2¼–3½ in
6–9 cm

densely furred tail

VOLES, LEMMINGS, AND MUSKRATS

About 765 chubby, short-tailed rodents, including
hamsters, voles, lemmings, and muskrats, make up
the family Cricetidae. They are found worldwide,
from western Europe to Siberia and the Pacific coast.

russet fur on back

3¼–5½ in
8–14 cm

3½–4¾ in
9–12 cm

4¾–9 in
12–23 cm

BANK VOLE
Myodes glareolus
Active mostly at dawn and
night, this typical vole inhabits
scrub, woodlands, and gardens
across most of W. Europe
and east into Russia.

COMMON VOLE
Microtus arvalis
A common burrowing inhabitant
of grasslands across most of
N. Europe, this vole also occurs
farther east and in Russia.

EURASIAN WATER VOLE
Arvicola amphibius
In Britain and parts of Europe, this vole occurs near
waterside habitats, but in Russia and Iran, it lives away
from water, digging extensive burrow systems.

4½–6 in
11–15 cm

3¼–4¾ in
8–12 cm

10–12 in
25–30 cm

NORWAY LEMMING
Lemmus lemmus
The main small mammal of the European
tundra, the lemming's fluctuating
population affects the breeding success
of many Arctic predators.

STEPPE VOLE
Lagurus lagurus
The only species in its
genus, this vole lives in the dry,
grassy plains that stretch from
Ukraine to W. Mongolia.

MUSKRAT
Ondatra zibethicus
Originally an inhabitant of rivers,
ponds, and streams in N. America,
the muskrat has been introduced to
Europe and is now widespread.

»

» VOLES, LEMMINGS, AND MUSKRATS

2¾–4¾ in
7–12 cm

relatively long tail

STRIPED DWARF HAMSTER
Cricetulus barabensis

By raiding crops to gather seeds and grain, this rodent can become a serious pest in farming areas. Farmers destroy its burrows while plowing.

2¼–3¼ in
6–8 cm

6½–12½ in
17–32 cm

ROBOROVSKI'S DESERT HAMSTER
Phodopus roborovskii

From the dry grasslands of C. Asia, this tiny animal is popular as a pet. During the breeding season, it can produce three to four litters.

COMMON HAMSTER
Cricetus cricetus

This burrowing species lives alone and hibernates underground through the winter. Its burrows may contain up to 145 lb (65 kg) of stored food.

golden orange fur

4¾–6½ in
12–17 cm

4¾–6½ in
12–17 cm

LONG-HAIRED GOLDEN HAMSTER
Mesocricetus auratus

The golden hamster has been selectively bred to create exotic forms, such as this albino variety, that would not survive in the wild.

GOLDEN HAMSTER
Mesocricetus auratus

Originally from Syria, where it is now an endangered species, this hamster has become a favorite household pet in Europe and N. America.

3½–4¼ in
9–11 cm

4¾–8 in
12–20 cm

WHITE-FOOTED MOUSE
Peromyscus leucopus

A very common and adaptable creature, this deer mouse inhabits almost every type of land habitat in central and the E. US.

HISPID COTTON RAT
Sigmodon hispidus

This short-lived, surface-dwelling herbivore occurs in grassland habitats in the S. US and Mexico.

MICE, RATS, AND RELATIVES

One fifth of all mammal species belong to this huge family. The Muridae are distributed almost worldwide, including the polar regions. Some species carry serious diseases, and others are significant agricultural pests. However, several are used in medical research and are commonly kept as pets.

4–4¾ in
10–12 cm

NORTHEAST AFRICAN SPINY MOUSE
Acomys cahirinus

Like other spiny mice, this rodent has stiffened hairs on its body for protection and very thin skin, which permits easy cooling in its hot, dry habitat.

3¼–5 in
8–13 cm

ARABIAN SPINY MOUSE
Acomys dimidiatus

This mouse was previously considered the same species as the N.E. African one, except that it occurs east of the Red Sea.

4–7 in
10–18 cm

3½–5½ in
9–14 cm

SHAW'S JIRD
Meriones shawii

Common in deserts of N. Africa and the Middle East, this rodent does not hibernate, but survives the winter by storing up to 22 lb (10 kg) of food in its burrows.

MONGOLIAN JIRD
Meriones unguiculatus

In the wild, this gerbil lives in the dry steppes of C. Asia, forming large social groups. Many are now kept as pets.

3½–5 in
9–13 cm

PALLID GERBIL
Gerbillus floweri

Like most small desert rodents, this gerbil has a pale body. It is widespread in N. Africa and the Middle East, where its coloration provides good camouflage.

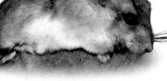

4–5 in
10–13 cm

FAT-TAILED JIRD
Pachyuromys duprasi

Like many small, desert-living mammals, this species stores fat in its tail. The tail has no fur, allowing it to radiate excess body heat.

pale fur, with grayish-brown markings

NORTHERN LUZON GIANT CLOUD RAT
Phloeomys pallidus
The two species of cloud rat inhabit high forests of the Philippines, but are rarely seen. They are the largest of the mouselike rodents.

15–17 in
38–43 cm

15½–17½ in
39–44 cm

SOUTHERN LUZON GIANT CLOUD RAT
Phloeomys cumingi
Like the smaller northern species, this rat spends the day in hollow trees or burrows, and has one young at a time.

HOUSE MOUSE
Mus musculus
A highly adaptable species, this small, slim rodent has followed humans all around the world, even living in the sub-Antarctic.

2¾–4 in
7–10 cm

ALBINO DOMESTIC MOUSE
Mus musculus
The captive-bred albino house mouse has become common as a pet, and is also widely used in medical and scientific research.

2¾–4 in
7–10 cm

3½–5 in
9–13 cm

YELLOW-NECKED FIELD MOUSE
Apodemus flavicollis
This nocturnal woodland species is difficult to distinguish from the wood mouse over much of its European range.

3½–4¼ in
9–11 cm

WOOD MOUSE
Apodemus sylvaticus
This is the most abundant wild mouse in Europe, living in every terrestrial habitat, even in the mountains.

white underparts

small ears

long whiskers on nose, eyebrows, and cheeks

HARVEST MOUSE
Micromys minutus
Europe's smallest mouse, this species lives in a variety of grassy habitats, including reed beds and cornfields.

2–3¼ in
5–8 cm

robust body

4½–10 in
11–26 cm

BROWN RAT
Rattus norvegicus
Also called the common rat, this is now a global pest, having traveled on ships and colonized even remote islands.

BLACK RAT
Rattus rattus
Often called the "ship rat" because of its ability to live aboard ships, this is also the species whose fleas transmit bubonic plague.

grayish-brown fur

3½–5½ in
9–14 cm

12–14 in
30–35 cm

5½–11½ in
14–29 cm

TYPICAL STRIPED GRASS MOUSE
Lemniscomys striatus
This distinctive, boldly marked species is a common inhabitant of grassy habitats across most of sub-Saharan Africa.

MALAGASY GIANT JUMPING RAT
Hypogeomys antimena
The sole species of giant hopping rat, this is the largest rodent in Madagascar, found only in the sandy forests of the west coast.

BAMBOO RATS AND RELATIVES

The family Spalacidae includes blind mole rats, bamboo rats, root rats, and zokors. Big, protruding incisors characterize the blind mole rats. Adapted to an underground life, they have no external eyes and ears. The bamboo rats of eastern Asia have visible eyes.

6½–14 in
17–35 cm

6–10 in
15–26 cm

GREATER MOLE RAT
Spalax microphthalmus
This blind species has sensory bristles running from its snout to its eye sockets. It is native to the steppes of Ukraine and S.E. Russia.

LESSER BAMBOO RAT
Cannomys badius
Found from Nepal to Vietnam, this single species of its genus digs deep burrows in forests, grassy areas, and sometimes gardens.

SPRINGHARES

Resembling rabbits in size and behavior, the Pedetidae differ in producing only one young at a time. They breed all year round, but living in dry, open habitats makes them vulnerable to predators.

13–18 in
33–46 cm

EAST AFRICAN SPRINGHARE
Pedetes surdaster
Springhares escape their nocturnal predators with kangaroolike leaps. Less common than *Pedetes capensis*, this species lives on the Serengeti plain in Africa.

SOUTH AFRICAN SPRINGHARE
Pedetes capensis
This species leaves its burrow at night to nibble on the grass and herbs found in the dry regions of southern Africa.

13–18 in
33–46 cm

long, tufted tail

AFRICAN MOLE-RATS

Living underground, the Heterocephalidae (naked mole-rats) and the Bathyergidae (mole-rats) burrow in sand and soft soils to feed on roots, using their protruding front teeth as shovels. Their lips close behind the teeth to keep the mouth from filling with dirt.

2¾–4½ in
7–11 cm

long, protruding incisor teeth

long, rounded tail

NAKED MOLE-RAT
Heterocephalus glaber
This highly social animal lives in a colony, where each individual performs different, specialized jobs to help the colony as a whole.

4–7½ in
10–19 cm

COMMON MOLE-RAT
Cryptomys hottentotus
Found from Tanzania to S. Africa, this common species lives in soft soils and farmland, where it feeds mainly on roots.

NAMAQUA DUNE MOLE-RAT
Bathyergus janetta
Native to Namibia and southwestern S. Africa, this species uses its forefeet rather than its teeth to dig burrows.

6½–9½ in
17–24 cm

NEW WORLD PORCUPINES

Tree-dwelling animals in American forests, the Erethizontidae have short spines, which are generally less than 4 in (10 cm) long. Most have a prehensile tail to grip branches.

2–4¼ ft
0.6–1.3 m

NORTH AMERICAN PORCUPINE
Erethizon dorsatus
This forest dweller is found throughout N. America, from Alaska to Mexico. Its spines are hidden among its shaggy fur.

17½–22 in
44–56 cm

BRAZILIAN PORCUPINE
Coendou prehensilis
This nocturnal species inhabits forested areas of S. America and Trinidad. It sleeps by day and forages for leaves and shoots at dusk.

OLD WORLD PORCUPINES

Found across most of Africa and southern Asia, the 11 species of the family Hystricidae family live in burrows. They are covered with long, stiff quills that protect them against most predators. When attacked, they often rattle their quills as a reminder of how sharp and impenetrable these are.

CRESTED PORCUPINE
Hystrix cristata
Widespread across the northern half of Africa, except the Sahara, the crested porcupine is a familiar nocturnal rodent.

18–36½ in
45–93 cm

spines are modified hair

30–39 in
75–100 cm

CAPE PORCUPINE
Hystrix africaeaustralis
Found in savanna across most of southern Africa, this species forages at night, alone or in groups, sniffing out roots and berries.

VISCACHAS AND CHINCHILLAS

All six species of the family Chinchillidae from South America have a prominent tail and large hind feet. They normally live in social groups, inhabiting burrows or rocky outcrops. Most species are now quite rare because of their exploitation for fur and their status as pests.

large ears help regulate body temperature

long whiskers aid spatial awareness

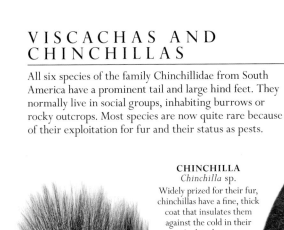

CHINCHILLA
Chinchilla sp.
Widely prized for their fur, chinchillas have a fine, thick coat that insulates them against the cold in their Andean home.

9–15 in
22–38 cm

bushy tail balances the body

12–18 in
30–45 cm

MOUNTAIN VISCACHA
Lagidium viscacia
This agile rodent has dense fur to protect it from the cold at night. It lives on steep, rocky, mountain slopes.

DASSIE RAT

This single species in the family Petromuridae is found only in southern Africa. Its peculiar, flat skull and soft ribs are adaptations to living in crevices and under stones, and distinguish it from all other rodents.

DASSIE RAT
Petromus typicus
Living on dry, rocky hillsides, this rodent emerges from crevices at dawn and dusk to forage for seeds and shoots.

5–10 in
13–25 cm

CANE RAT

The two species of the family Thryonomyidae have pale brown, coarse, flattened hair that blends with their surroundings in dry grass and cane thickets. Cane rats produce two small families of well-developed young per year.

16–30 in
41–77 cm

CANE RAT
Thryonomys sp.
There are two species of African cane rats: one lives in savanna grasslands and the other in reedbeds and marshes.

PACARANA

The single species of the family Dinomyidae is a shy, bulky, slow-moving, and almost defenseless animal. It lives alone or in pairs in montane forests, where it is preyed upon by jaguars and humans.

PACARANA
Dinomys branickii
Threatened by loss of its S. American forest habitat, and hunted for food, the pacarana is now an endangered species.

28–32 in
70–80 cm

GUINEA PIGS, MARAS, AND CAPYBARA

Including some of South America's most widespread and abundant rodents, the Caviidae occur from mountain meadows to tropical flood plains, and breed all year round. All but the maras and capybara are short-legged and dumpy animals.

BRAZILIAN GUINEA PIG
Cavia aperea
Guinea pigs mostly live in lowland habitats, but this species is also found in the Andes, from Peru to Chile.

8–16 in
20–40 cm

ROSETTE GUINEA PIG
Cavia porcellus
Several coat variations occur in pet guinea pigs. In this one, the fur forms substantial spiky whorls over the body.

8–16 in
20–40 cm

long ears

23½–32 in
60–80 cm

8–16 in
20–40 cm

LONG HAIR GUINEA PIG
Cavia porcellus
Often kept as a pet, in captivity this long-haired breed needs grooming to prevent its fur from becoming tangled.

SHORT HAIR GUINEA PIG
Cavia porcellus
First domesticated more than 500 years ago for food, guinea pigs have since become popular pets worldwide.

8–16 in
20–40 cm

PATAGONIAN MARA
Dolichotis patagonum
Extensive shared burrow systems and communal breeding habits are among the unusual features of these long-legged rodents.

CRESTED PORCUPINE
Hystrix cristata

With its long, barbed quills raised in threat, the crested porcupine is an intimidating prospect. A predator that has had the painful experience of being spiked is unlikely ever to attempt another attack. Lions, hyenas, and even people have been known to die from infected quill injuries. Despite this impressive means of defense, the porcupine is a peaceable and rather nervous animal, easily startled, and more likely to flee danger than stand its ground. Porcupines live alone or in family groups, sharing extensive burrow systems. Crested porcupines occur throughout much of northern Africa. The species was also once widespread in southern Europe—the population found in Italy may be a relic of former times or the result of a more recent introduction, perhaps by the Romans.

SIZE 18–36½ in (45–93 cm)
HABITAT Savanna grassland, light woodland, rocky terrain
DISTRIBUTION Northern Africa as far south as Tanzania, except Sahara Desert; also Italy
DIET Mainly roots, fruits, and tubers; occasionally carrion

< SPIKY HAIRDO
Raising the quills creates an illusion of size. A cornered porcupine will stand tall and try to bluff its way out of trouble. If this fails, it will turn tail and charge backward, shoving its spiky rear end in the face of its attacker.

EAR >
The small ears are largely hidden in the coarse hair. The porcupine has good hearing, which it uses to avoid danger, shuffling off into the night if it hears another animal approaching.

∨ EYE
A porcupine's eyesight is poor, but then there is often little to see in the darkness of an African night. It uses its sharp senses of hearing and smell to find its way.

∨ MOUTH AND TEETH
Porcupines have the specialized gnawing teeth typical of rodents, enabling them to chew tough roots and tubers. The muscles that control the jaws are immensely powerful.

QUILLS-A-QUIVER >
Porcupine quills are greatly enlarged hairs. They can be raised by larger versions of the tiny muscles that create goosebumps in our own skin.

like its quills, the porcupine's thin coat of coarse hair can also be raised in alarm

RATTLING TAIL ∧
The quills of the tail are swollen and hollow. The porcupine shakes them when alarmed and the soft rattling sound warns its enemies that this is a well-armed opponent.

< ∧ FEET AND CLAWS
Porcupines walk on the flat soles of their feet, with a slightly clumsy, shambling gait. The soles are hairless and padded. They have short toes and strong claws, well suited to digging.

CAPYBARA

The largest rodent, the capybara, is one of four species in the Caviidae subfamily Hydrochoerinae. Capybaras generally breed once a year; the offspring are born at the end of the wet season, when the grass is most nutritious. They may live up to six years.

coarse fur dries quickly

CAPYBARA
Hydrochoeris hydrochaeris
As big as a pig, this is the world's largest rodent. It lives a semiaquatic life in the swamps of S. America.

3¼–4¼ ft
1–1.3 m

small, rounded ears

HUTIAS, SPINY RATS, AND COYPU

A Central and South American family, the Echimyidae comprises 99 species that live in a variety of habitats, and includes tree-dwelling and burrowing spiny rats, and the semiaquatic coypu. Although they vary in size, fur type, and dentition as well as diet, phylogenetic and molecular analyses suggest they belong in the same group.

6½–12 in
16–30 cm

SPINY RAT
Proechimys sp.
These rodents have a spiny protective coat. This feature has evolved in parallel to the spiny mice of Africa.

12–17 in
30–43 cm

DESMAREST'S CUBAN HUTIA
Capromys pilorides
This hutia is common in Cuba. Other surviving hutia species face extinction because of a loss of their habitat and hunting.

18½–23 in
47–58 cm

prominent incisor teeth

COYPU
Myocastor coypus
This animal has distinctively shaggy fur and huge orange front teeth. It also has webbed hind feet for swimming and a thick, scaly tail.

long, rounded tail

PACAS

This family includes two species of nocturnal rodent from Central and South America. The Cuniculidae resemble small pigs, as they poke around on the forest floor for fruits, seeds, and roots.

20–30 in
50–75 cm

LOWLAND PACA
Cuniculus paca
Mainly a forest inhabitant, this rodent lives in the northern part of S. America, from Mexico to Paraguay.

DEGUS, ROCK RATS, AND RELATIVES

The molar teeth of these small, silky-furred rats form a figure-eight shape when worn, giving them their Latin name. The Octodontidae are widespread in the southern part of South America.

6½–9 in
16–22 cm

DEGU
Octodon degus
This species is found on the western slopes of the Andes in Chile. Its tail breaks off easily if caught by a predator.

AGOUTIS

Active during the day, these long-legged, running rodents of the family Dasyproctidae are very shy. They breed all year round but have only two young at a time, which can run within an hour of birth.

RED-RUMPED AGOUTI
Dasyprocta leporina
Found in forested areas of northeastern S. America and the Lesser Antilles, this species can be identified by its light orange rump.

19–23½ in
48–60 cm

19–23½ in
48–60 cm

17–23 in
43–58 cm

CENTRAL AMERICAN AGOUTI
Dasyprocta punctata
With a range that extends from Mexico south to Argentina, this agouti mainly feeds on fruits, but may also eat crabs. Pairs are thought to stay together for life.

AZARA'S AGOUTI
Dasyprocta azarae
An inhabitant of the forests of S. Brazil, Paraguay, and N. Argentina, this agouti barks when alarmed. It eats a variety of seeds and fruits.

TREE SHREWS

These small mammals look and behave like squirrels. They are active in the daytime and spend much of their time foraging for food on the ground.

Tree shrews are native to the tropical rainforests of Southeast Asia. They are unrelated to the true shrews and have their own order, Scandentia. Sharp claws on all their fingers and toes enable tree shrews to climb trees rapidly but, despite their name, most of them are only partly arboreal. Their mixed diet includes insects, worms, fruits, and sometimes small mammals, reptiles, and birds.

Some tree shrews are solitary, while others live in pairs or groups. They are rapid breeders, raising their young in nests in tree crevices or on branches. The female pays her brood little attention, making only the occasional short visit to suckle them.

PHYLUM	CHORDATA
CLASS	MAMMALIA
ORDER	SCANDENTIA
FAMILIES	2
SPECIES	23

PEN-TAILED TREE SHREW

The family Ptilocercidae has only one member, the Southeast Asian pen-tailed tree shrew. The species is named for its long, feathery tail, which resembles a quill pen. This aids balance when climbing.

5–6 in
13–15 cm

PEN-TAILED TREE SHREW
Ptilocercus lowii
This species has a rather spindly tail with a brushy tip, unlike most other tree shrews, which have a thick, bushy tail.

TREE SHREWS

Members of the family Tupaiidae have a long snout, which is used to locate insects and other invertebrates as well as fruits and leaves. Tree shrews also have sharp claws on all their fingers and toes, helping them to climb trees rapidly.

LARGE TREE SHREW
Tupaia tana
Tree shrews are diurnal inhabitants of S.E. Asian forests. This species is found in Borneo and Sumatra and on nearby islands.

long snout

6½–8½ in
17–21 cm

elongated claws help hold on to tree branches

COLUGOS

The two species that form the order Dermoptera are gliding, rather than flying, mammals. They are found in the rainforests of Southeast Asia.

The colugos are distinguished by a furry membrane that stretches from the neck to the tips of the fingers and tail. When the limbs are splayed this membrane spreads out, allowing the lemur to maneuver in the air as it glides from tree to tree. The distance covered in a glide may be more than 330 ft (100 m). Colugos live in the canopy, hanging upside down from branches or sheltering in cracks or hollow trees by day and emerging at night to feed on fruits and leaves. They are almost unable to move on the ground.

The teeth of colugos are unlike those of any other mammal. In the lower jaw, the teeth are arranged like combs and are believed to be used for grooming as well as for feeding.

PHYLUM	CHORDATA
CLASS	MAMMALIA
ORDER	DERMOPTERA
FAMILIES	1
SPECIES	2

FLYING LEMURS

The two members of the Cynocephalidae have forward-facing eyes that provide depth perception This helps them judge distances when gliding from tree to tree. They also have unusual comblike lower teeth, used to strain food such as fruits and flowers.

SUNDA COLUGO
Cynocephalus variegatus
This lemur lives alone or in small groups, inhabiting tree-holes or resting high on treetops in the tropical forests of S.E. Asia and the Indonesian islands.

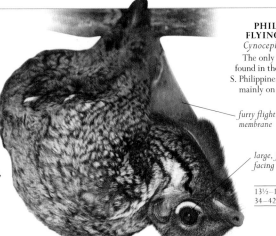

furry flight membrane

large, forward-facing eyes

13½–16½ in
34–42 cm

PHILIPPINE FLYING LEMUR
Cynocephalus volans
The only local species found in the forests of the S. Philippines, where it feeds mainly on young leaves.

13½–16½ in
34–42 cm

PRIMATES

The order to which humans belong, primates have a large brain relative to their size and forward-facing eyes which give them 3-D vision.

With the exception of a few species, including humans, primates are restricted to tropical and subtropical regions of the Americas, Africa, and Asia. They range in size from a mouse lemur, weighing 1 oz (30 g), to a gorilla weighing up to 440 lb (200 kg).

Primates rely more on vision than on smell. Many are arboreal and have characteristics such as stereoscopic vision, allowing good judgment of distance for jumping between trees; opposable thumbs and a prehensile tail for grasping branches; long legs for leaping; and long arms for swinging. Some primates have specialized diets, but many other species are omnivorous.

MAJOR GROUPS

Primates are divided into two suborders. The Strepsirrhini includes the mostly nocturnal lemurs, lorises, galagos, and their relatives. These species have a better developed sense of smell than other primates. The Haplorrhini includes the New and Old World monkeys and the apes, many of which are diurnal, and so more reliant on vision than the Strepsirrhini.

SOCIAL PATTERNS

Most primates are highly social, living in small family groups, single-male harems, or large mixed-sex troops. Many species are characterized by high levels of male competition for females. Sexual selection favoring the largest or most dominant males has led to sexual dimorphism—differences between males and females in body size or features such as canine teeth. The different sexes may also be differently colored, a form of dimorphism known as sexual dichromatism.

Most New World monkeys are monogamous, with both parents sharing responsibility for raising their young. Old World monkeys tend to live in groups dominated by related females, and the males take little or no part in parental duties. Primates are generally slow to mature fully and slow to reproduce, but they are relatively long-lived. The larger apes have a potential lifespan of up to 45 years in the wild, and longer in captivity.

PHYLUM	CHORDATA
CLASS	MAMMALIA
ORDER	PRIMATES
FAMILIES	16
SPECIES	506

DEBATE

PRIMATE HOTSPOTS

Today scientists list more species of primates than a decade ago—most the result of subspecies (geographical variants) being recognized as new species. In the Amazon basin populations of monkeys separated by rivers and mountain ranges can differ in subtle ways—such as in the structure of their chromosomes. Some scientists believe that these monkeys diverged when their forest habitats became isolated due to geological activity many thousands of years ago. Such forests may contain many species isolated in this way and are a particular focus of conservation groups aiming to protect biodiversity.

GALAGOS

Native to sub-Saharan Africa, the members of the family Galagidae inhabit a wide range of woodland and forest habitats, including scrub and wooded savanna. They have longer hind legs than forelegs, which help them to take large leaps between trees in forests. These primates often wash their hands and feet with urine, which may help to improve their grip and leave scent-trails. All are nocturnal.

large, movable ears

thick fur

huge eyes

11–18½ in
28–47 cm

SILVERY GREATER GALAGO
Otolemur monteiri
This species has the same range as the brown greater galago but prefers denser vegetation. Despite its name, melanistic (black) forms are common.

BROWN GREATER GALAGO
Otolemur crassicaudatus
One of the largest galagos, this primate lives in a variety of forests in southern Africa. This omnivore uses its comblike teeth to scrape gum from trees.

10–16 in
26–40 cm

LORISES, POTTOS, AND ANGWANTIBOS

These small, nocturnal omnivores have a short tail, and their forelimbs and hind limbs are of equal length. The Lorisidae have opposable thumbs for grasping branches and they move much more slowly and sedately through the trees than the galagos—climbing rather than leaping.

dense fur

RED SLENDER LORIS
Loris tardigradus
Native to Sri Lanka, this slender primate moves carefully through the forest canopy on its long limbs.

7–8½ in
18–21 cm

opposable thumb

dark rings around huge eyes

vicelike grip

9–10 in
22–26 cm

12–13½ in
30–34 cm

8–9 in
20–23 cm

12–16 in
30–40 cm

SUNDA SLOW LORIS
Nycticebus coucang
As its name suggests, this loris moves slowly and deliberately through the trees. It inhabits tropical forests in S.E. Asia.

PYGMY SLOW LORIS
Nycticebus pygmaeus
This species inhabits thick tropical rainforests and bamboo forests in Laos, Cambodia, Vietnam, and S. China.

WEST AFRICAN POTTO
Perodicticus potto
A shy species, this primate lives in thick rainforests across equatorial Africa. A bony shield at the nape of its neck protects it against predators.

GOLDEN ANGWANTIBO
Arctocebus aureus
Also called the golden potto, this species inhabits the understory of moist lowland forest in equatorial W. and C. Africa.

SENEGAL BUSH BABY
Galago senegalensis
Widespread across C. Africa and parts of E. Africa, the lesser, or Senegal, bush baby inhabits arid savanna woodland.

large ears

TARSIERS

The small Tarsiidae are named for their greatly elongated anklebone or tarsus. These arboreal species have long limb bones, elongated fingers, and a long, thin tail. Their round head has very large eyes, which help them to see at night while hunting for insects.

4¾–8 in
12–20 cm

4¾–6½ in
12–17 cm

silky fur

PHILIPPINE TARSIER
Tarsius syrichta
Endemic to a variety of rainforests and scrub in the Philippines, this tarsier possesses the largest eyes relative to its body size among all mammals.

MOHOLI BUSH BABY
Galago moholi
This small, shy, galago lives in small groups in southern Africa. Lithe and agile, it leaps through woodland, where it eats insects and tree gum.

4½–5½ in
11–14 cm

4½–5½ in
11–14 cm

WESTERN TARSIER
Tarsius bancanus
Adapted for clinging and leaping between trees, this species, also called the Horsfield's tarsier, lives in the tropical rainforests of Sumatra and Borneo.

DEMIDOFF'S BUSH BABY
Galagoides demidovii
Also called the dwarf bush baby, this small primate uses its long hind legs to jump through rainforest canopies of W. and C. Africa.

4¾–6½ in
12–17 cm

long tail

long, slender tail

LEMURS

Found in forests across Madagascar, the family Lemuridae is made up of the most typical lemurs: mainly arboreal and quadrupedal. Most are cathemeral, meaning they are active during the day and night. In several species, males and females are differently colored.

15–16 in
38–40 cm

thick, woolly fur

16–16½ in
40–42 cm

15½–18 in
39–46 cm

16–20 in
40–50 cm

GREATER BAMBOO LEMUR
Prolemur simus
One of the rarest lemurs, this species lives in S.E. Madagascar, where it feeds almost exclusively on giant bamboo.

RING-TAILED LEMUR
Lemur catta
Living in groups of up to 25 individuals, this lemur often spends time on the ground. Its diet consists of fruits, vegetation, sap, and bark.

BLACK-AND-WHITE RUFFED LEMUR
Varecia variegata
The largest lemur, this species eats a high proportion of fruits. Unusually for lemurs, it makes a leafy nest for its young.

BANDRO
Hapalemur alaotrensis
Also called the Lac Alaotra bamboo lemur, this critically endangered species lives only in papyrus marsh and reedbeds around Lake Alaotra, Madagascar's largest lake.

14–16½ in
35–42 cm

RED-BELLIED LEMUR
Eulemur rubriventer
A monogamous species, this lemur lives in small groups that consist of a mated pair and their dependent young.

RED COLLARED LEMUR
Eulemur collaris
This lemur has a scent gland on its wrist, with which it perfumes its long, furry tail for use in communication.

15–16½ in
38–42 cm

tail is as long as body

15½–16½ in
39–42 cm

WHITE-HEADED LEMUR
Eulemur albifrons
Only the male of this species has distinctive white fur around its black face; the female has a gray face.

red cheek patch in male

12½–14½ in
32–37 cm

MONGOOSE LEMUR
Eulemur mongoz
Primarily nocturnal during the dry season, the mongoose lemur becomes more diurnal at the start of the wet season.

♂

BLACK LEMUR
Eulemur macaco
In this sexually dichromatic species, the colors of males and females vary. Only the male is black; the female is gray-brown with white ear tufts.

grasping hands

15–18 in
38–45 cm

♀

MOUSE, DWARF, AND FORK-MARKED LEMURS

Among the smallest of all primates, the members of the family Cheirogaleidae have short limbs and large eyes. All are nocturnal, arboreal inhabitants of forests in Madagascar, and become torpid to survive the dry season.

PALE FORK-MARKED LEMUR
Phaner pallescens
Adapted to eat tree gum, this lemur has a long tongue and large premolars for chiseling bark.

9–12 in
22–30 cm

10–12 in
26–30 cm

short limbs

4¾–6 in
12–15 cm

GRAY MOUSE LEMUR
Microcebus murinus
An omnivore, this lemur's diet includes insects, flowers, and fruits. The young are carried in their mother's mouth if moved (rather than clinging to her fur).

EASTERN RUFOUS MOUSE LEMUR
Microcebus rufus
Inhabiting a range of forest habitats, this omnivorous species consumes a variety of fruits, insects, and gum.

4–6 in
10–15 cm

GREATER DWARF LEMUR
Cheirogaleus major
This solitary lemur feeds mainly on fruits and nectar. It stores fat in its tail during the wet season.

SPORTIVE LEMURS

The family Lepilemuridae consists of the sportive lemurs of Madagascar. These medium-sized lemurs have a prominent muzzle and large eyes, and are strictly arboreal and nocturnal. Their low-energy, leafy diet means that they are among the least active of all primates.

9–10 in
23–26 cm

7½–10 in
19–26 cm

WHITE-FOOTED SPORTIVE LEMUR
Lepilemur leucopus
Grayish above and white below, this lemur spends lengths of time perched vertically on tree trunks, in between bouts of foraging for food.

BLACK-STRIPED SPORTIVE LEMUR
Lepilemur dorsalis
This species lives in moist forest in N.W. Madagascar and nearby islands. It has a blunt muzzle and small ears.

SIFAKAS AND RELATIVES

The largest lemurs—the indri and sifakas—and the smaller avahis or woolly lemurs form the Indriidae of Madagascar. All members have long, powerful limbs for leaping between trees, and, with the exception of the indri, a long tail.

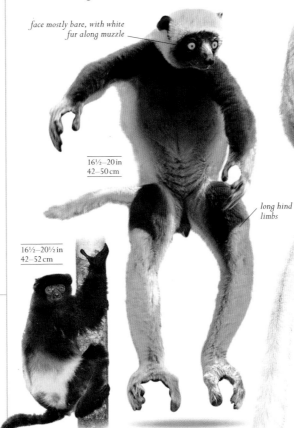

face mostly bare, with white fur along muzzle

16½–20 in
42–50 cm

long hind limbs

16½–20½ in
42–52 cm

16–20 in
40–50 cm

VERREAUX'S SIFAKA
Propithecus verreauxi
Found in S.W. Madagascar, this sifaka is capable of leaping through cactuslike vegetation on its long legs without being injured.

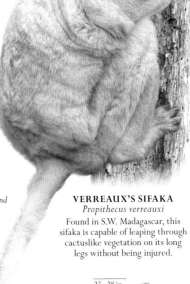

MILNE-EDWARD'S SIFAKA
Propithecus edwardsi
Occurring in small family groups in southeastern Madagascar, this sifaka has a large opposable thumb for clinging to tree trunks.

COQUEREL'S SIFAKA
Propithecus coquereli
Like other sifakas, this primate crosses open ground by skipping on its hind legs, using its forearms for balance.

25–28 in
64–72 cm

INDRI
Indri indri
The largest member of the lemur group, this is the only species to have a short, vestigial tail.

AYE-AYE

There is only one species in the Daubentoniidae—the aye-aye of Madagascar. This nocturnal primate has large, bare ears; a shaggy coat; very long fingers; and continually growing incisor teeth.

12–14½ in
30–37 cm

AYE-AYE
Daubentonia madagascariensis
The aye-aye uses its elongated middle finger for locating and extracting insect larvae from dead wood.

HOWLER, SPIDER, AND WOOLLY MONKEYS

The howler, spider, and woolly monkeys, and muriquis of the Atelidae are the largest New World monkeys. All have a prehensile tail that is used as a fifth limb when moving through the trees. Spider monkeys have longer limbs than the other members of the family.

19–25 in
48–63 cm

20–28 in
50–71 cm

GUATEMALAN BLACK HOWLER
Alouatta pigra
Found in the Yucatán Peninsula of Mexico, Belize, and Guatemala, this species forms groups of up to 11 individuals.

18–25 in
46–63 cm

MANTLED HOWLER
Alouatta palliata
Named after the "mantle" of long guard hair on its flanks, this howler monkey lives in C. America and northern S. America.

large larynx for howling

VENEZUELAN RED HOWLER
Alouatta seniculus
This howler monkey has an enlarged hyoid bone in the neck that allows it to make calls that can be heard several miles away.

12–25 in
30–64 cm

COLOMBIAN BLACK SPIDER MONKEY
Ateles fusciceps rufiventris
Like other spider monkeys, this subspecies does not have thumbs on its hands. It is found in Colombia and Panama.

12–25 in
31–63 cm

GEOFFROY'S SPIDER MONKEY
Ateles geoffroyi
Active during the day, this fruit-eating species lives in relatively large groups of up to 35 individuals in forests throughout C. America.

18–31 in
46–78 cm

SOUTHERN MURIQUI
Brachyteles arachnoides
Also known as the woolly spider monkey, this species, found in Brazilian forests, is critically endangered due to the loss of its habitat.

GRAY WOOLLY MONKEY
Lagothrix cana
A powerfully built species, this monkey lives in large troops in the primary forests of Brazil, Bolivia, and Peru.

HUMBOLDT'S WOOLLY MONKEY
Lagothrix lagotricha
One of the largest New World monkeys, this species lives in lowland primary forest in the upper Amazon basin.

hairless tail tip

18–26 in
45–65 cm

18–26 in
46–65 cm

NIGHT MONKEYS

Also called owl monkeys, the members of the Aotidae are the only nocturnal monkeys in the New World. They are small, with large eyes set in a flat, round face with dense woolly fur. They have a well-developed sense of smell.

14–16½ in
35–42 cm

BLACK-HEADED NIGHT MONKEY
Aotus nigriceps
This monogamous species inhabits the central and upper Amazon basin in primary and secondary forests of Brazil, Bolivia, and Peru.

12–15 in
30–38 cm

NORTHERN NIGHT MONKEY
Aotus trivirgatus
Also called the three-striped monkey, this species is most active on moonlit nights. It lives in forests in Venezuela and northern Brazil.

TITI, SAKI, AND UAKARI MONKEYS

This family contains a range of small- to medium-sized monkeys. The Pitheciidae are diurnal, arboreal, and social. All share a common dental makeup, including large, splayed canine teeth that help them to deal with tough seeds and fruits.

black hair tipped with white

12½–16½ in
32–42 cm

WHITE-FACED SAKI
Pithecia pithecia
Male white-faced or Guianan sakis are black with pale fur around the face, whereas females are gray-brown.

13½–16½ in
34–42 cm

BLACK-BEARDED SAKI
Chiropotes satanas
This species is known from S. Amazonia. The males possess a beard and large swellings on their forehead.

14–21 in
36–53 cm

14½–19 in
37–48 cm

MONK SAKI
Pithecia monachus
A shy primate, this species lives in the high canopy of forests in N.W. Brazil, Peru, Colombia, and Ecuador.

RIO TAPAJÓS SAKI
Pithecia irrorata
Also called the bald-faced saki, this species lives in W. Brazil, N. Bolivia, and E. Peru. It eats predominantly seeds.

hairless, red face

12–16½ in
31–42 cm

BLACK-FRONTED TITI
Callicebus nigrifrons
A fruit-eater, this titi is an inhabitant of Atlantic coastal forest around São Paulo in S.E. Brazil.

11–14 in
28–36 cm

long, shaggy coat

COLLARED TITI MONKEY
Callicebus torquatus
Also known as the yellow-handed titi, this species prefers unflooded forests on sandy soils in Brazil. It feeds mostly on fruits and seeds.

10½–13½ in
27–34 cm

COPPERY TITI
Plecturocebus cupreus
Found in rainforests of the S.W. Amazon Basin, the coppery titi is monogamous and territorial. It feeds mainly on fruits.

14–22 in
35–56 cm

SPIX'S BLACK-HEADED UAKARI
Cacajao ouakary
A highly social species, this uakari lives in groups of 30 or more individuals in the N.W. of the Amazon basin.

14–22½ in
36–57 cm

RED BALD-HEADED UAKARI
Cacajao calvus rubicundus
Several subspecies of the bald-headed uakari inhabit seasonally flooded forests of the Amazon basin. The redness of the face is thought to signal the health of the monkey.

MARMOSETS AND TAMARINS

The Callitrichidae consists of relatively small social monkeys, which inhabit a variety of forests throughout tropical and subtropical Central and South America. They are all diurnal and arboreal and have forward-facing eyes and a short muzzle. Marmosets and tamarins have a long, nonprehensile tail, lack the third molar, and have claws instead of nails.

7½–10 in
19–25 cm

8–9 in
20–23 cm

GOELDI'S MONKEY
Callimico goeldii
Inhabiting dense forest undergrowth, such as that of bamboo forest in the upper Amazon, this monkey makes forays into the tree canopy for fruits.

SILVERY MARMOSET
Mico argentatus
A specialized gum-eater (gumivore), this marmoset has huge ears, narrow jaws, and short canine teeth for cutting into tree bark.

COMMON MARMOSET
Callithrix jacchus
The female of this species generally mates with two males, both of whom assist with parental care of the usually twin offspring.

6½–8½ in
16–21 cm

long, curved claws

7–9 in
18–23 cm

GEOFFROY'S TUFTED-EAR MARMOSET
Callithrix geoffroyi
Also called the white-headed marmoset, this species uses scent-marks to deter others from using the holes it makes in tree bark to extract gum.

8–9 in
20–23 cm

BLACK-TUFTED-EAR MARMOSET
Callithrix penicillata
A monogamous species, this primate is diurnal and lives high in the rainforest canopy, where it feeds on tree sap.

4¾–6 in
12–15 cm

PYGMY MARMOSET
Cebuella pygmaea
The smallest monkey in the world, this species eats tree gum in seasonally flooded forests of the upper Amazon basin.

fine, yellow hair on back

9–10 in
23–26 cm

white moustache

EMPEROR TAMARIN
Saguinus imperator
Distinguished by its long, white moustache, this tamarin lives in the tropical forests in Peru, Brazil, and Bolivia.

8½–11 in
21–28 cm

WHITE-LIPPED TAMARIN
Saguinus labiatus
The dominant female of this species releases chemical signals, called pheromones, that suppress reproduction in other female members of the group.

9–13 in
23–33 cm

BARE-FACED TAMARIN
Saguinus bicolor
Also called the pied tamarin, this tree-dwelling primate lives in lowland forest in the central Amazon region near Manaus, Brazil.

8–10 in
20–25 cm

COTTON-TOP TAMARIN
Saguinus oedipus

Inhabiting a very restricted range in N.W. Colombia and Panama, this species feeds mainly on insects and fruits.

GOLDEN-HANDED TAMARIN
Saguinus midas

Found in northeast S. America, this tamarin has brightly colored hands and feet, with claws on all digits except the big toe, which has a nail.

8½–11 in
21–28 cm

banded tail

8–10½ in
20–27 cm

SADDLE-BACK TAMARIN
Saguinus fuscicollis

This tamarin lives in secondary forest and forest edges in the upper Amazon basin, foraging for insects, fruits, nectar, and tree sap and gum.

GOLDEN LION TAMARIN
Leontopithecus rosalia

Among the most endangered monkeys in the world, this species is found only in the Atlantic coastal forest of S.E. Brazil.

10–13 in
26–33 cm

9–10 in
22–26 cm

GOLDEN-HEADED LION TAMARIN
Leontopithecus chrysomelas

Restricted to the Atlantic forest of southern Bahia in N.E. Brazil, this primate raises its mane to look bigger when in danger.

SQUIRREL MONKEYS AND CAPUCHINS

The family Cebidae once contained several monkey groups, but now only the squirrel monkeys and capuchins remain. They have large ears, a long tail, and forelimbs that are shorter than the hind limbs. Squirrel monkeys, of which there are eight Central and South American species, belong to the genus *Saimiri*. The capuchins comprise two genera, the tufted *Sapajus* and the untufted *Cebus*.

naked face

13–18 in
33–45 cm

WHITE-HEADED CAPUCHIN
Cebus capucinus

The only capuchin found in C. America, the range of this primate extends from Honduras to the coasts of Colombia and Ecuador.

10–14½ in
25–37 cm

prehensile tail

14½–18 in
37–46 cm

WEEPER CAPUCHIN
Cebus olivaceus

Native to north-central S. America, this capuchin often uses its prehensile tail to support its body while feeding with its hands.

bright yellow limbs

nails on digits

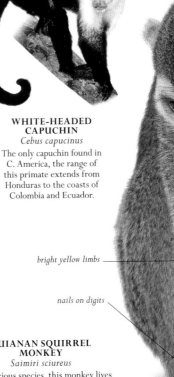

10½–12½ in
27–32 cm

long tail with a black tip

BLACK-CAPPED SQUIRREL MONKEY
Saimiri boliviensis

Males of this species become fatter around the neck and shoulders during the breeding season, when they compete for mates.

GUIANAN SQUIRREL MONKEY
Saimiri sciureus

A gregarious species, this monkey lives in large groups in a wide variety of forests in northern S. America to north-eastern S. America.

GUIANAN SQUIRREL MONKEY
Saimiri sciureus

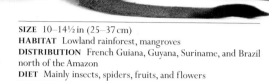

Inquisitive and intelligent, the Guianan squirrel monkey has the largest brain mass to body mass of any primate. It is also very social, living in large mixed-sex groups of about 15–50 individuals. Guianan squirrel monkeys are synchronous breeders, with all young being born over a single week during the wet season (January–February) after a gestation period of about six months. The young initially cling to their mother's abdomen but then start riding on her back at about two weeks old. Their development is rapid, and the youngsters are weaned at six months and are fully independent four months later. Squirrel monkeys can live for up to 20 years.

SIZE 10–14½ in (25–37 cm)
HABITAT Lowland rainforest, mangroves
DISTRIBUTION French Guiana, Guyana, Suriname, and Brazil north of the Amazon
DIET Mainly insects, spiders, fruits, and flowers

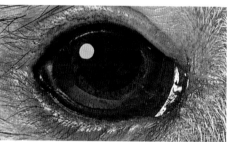

EYE >
Sharp vision and good depth perception are a result of having large, close-set, forward-facing eyes. These features are of vital importance to a monkey that spends most of its time in the trees.

< NOSTRILS
Broad, laterally facing nostrils and a short nasal region belie an excellent sense of smell. Scent is important for social communication between members of the group and for finding mates and lost infants.

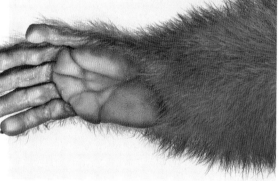

^ HAND
Although the hand is used to grip branches, hold food, self-groom, and perform other tasks, the fingers do not move independently, so the thumb and first finger cannot be opposed and used to pick things up.

^ TEETH
The Guianan squirrel monkey has 36 small, sharp teeth. The canines are longer and more prominent than the other teeth, and males have larger upper canines than females.

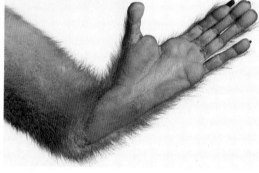

^ FOOT
Unlike the hand, the foot has a large, opposable digit—the big toe. When used in conjunction with the other four toes, it allows the foot to grasp branches.

BALANCING ACT >
The tail is at least as long as the head and body lengths combined. It is used to help with balance on tree branches, as squirrel monkeys move on all fours.

body covered by a short, thick coat

"Gothic" arches of white hair meet to form deep V-shape between the eyes

∨ **FACIAL FEATURES**
The face of the Guianan squirrel monkey has several distinctive features: the white hair above each eye forms a slightly pointed arch, the ears are hairy and have ear tufts, and there are conspicuous dark hairs (vibrissae) just above the eyes.

OLD WORLD MONKEYS

Widely distributed across Africa and Asia, the Cercopithecidae have closely spaced, downward-facing nostrils (catarrhine) and flattened nails. Most are diurnal and arboreal; however, baboons are mainly terrestrial. Guenons, baboons, and macaques are omnivorous, with strong jaws, cheek pouches, and a simple stomach. Colobus and leaf monkeys are leaf-eaters (folivorous) with a complex stomach, and lack cheek pouches.

17–21 in
43–53 cm

TOQUE MACAQUE
Macaca sinica
The smallest of the macaques, this species is endemic to the wet forests on the island of Sri Lanka.

forward-facing eyes give 3-D vision

14½–26 in
37–66 cm

21½–25 in
54–64 cm

17½–22½ in
44–57 cm

CELEBES CRESTED MACAQUE
Macaca nigra
Endemic to Indonesia's Sulawesi Island, this monkey has bare, pink swellings on its rump. These become particularly swollen in females that are ready to mate.

BARBARY MACAQUE
Macaca sylvanus
The only macaque outside Asia, the barbary macaque lives in the high cedar and oak forests of Algeria and Morocco.

12–25 in
31–63 cm

lionlike mane

16–24 in
40–61 cm

CRAB-EATING MACAQUE
Macaca fascicularis
Besides eating crabs, as its name suggests, this omnivorous S.E. Asian species also forages for insects, frogs, fruits, and seeds.

LION-TAILED MACAQUE
Macaca silenus
Endemic to the Western Ghats, mountains in S.W. India, this arboreal macaque lives mainly in wet, monsoon forests.

RHESUS MACAQUE
Macaca mulatta
This macaque lives in dry, open areas from W. Afghanistan through India to N. Thailand and China. Adults swim distances of up to ½ mile (0.8 km) between islands.

19–26 in
48–65 cm

sandy coloration

STUMP-TAILED MACAQUE
Macaca arctoides
Both arboreal and terrestrial, this species is found in tropical and subtropical wet forests in S.E. Asia.

14–23½ in
35–60 cm

thick fur

17–29 in
43–74 cm

*forearms of similar
length to hind legs*

**SOUTHERN
PIG-TAILED MACAQUE**
Macaca nemestrina
This monkey inhabits humid
areas, including rainforest
and swamps in S.E. Asia. Its
diet consists mainly of fruits.

BONNET MACAQUE
Macaca radiata
Often found close to
human habitation, this
resident of S. India is
omnivorous and may rely
on humans for food.

18–25½ in
46–65 cm

JAPANESE MACAQUE
Macaca fuscata
Inhabiting the most northerly range
of any nonhuman primate, the
Japanese macaque uses hot springs
to keep warm in winter.

*opposable
thumbs*

8–27 in
20–68 cm

13½–20½ in
34–52 cm

**WHITE-THROATED
GUENON**
Cercopithecus albogularis
Also called the Sykes'
monkey, this tree-dwelling
omnivore is distributed
in E. and S.E. Africa,
including Zanzibar and
the Mafia Islands.

RED-TAILED MONKEY
Cercopithecus ascanius
An arboreal species of moist forest habitats in
C. Africa, the red-tailed monkey has large
cheek pouches, in which it stores fruits.

18–28 in
45–70 cm

*long,
furry tail*

L'HOEST'S MONKEY
Allochrocebus lhoesti
Also known as the mountain
monkey, this arboreal guenon
lives in moist primary forests
in the highlands of C. Africa.

16½–23½ in
42–60 cm

16–23 in
40–58 cm

DIANA MONKEY
Cercopithecus diana
An inhabitant of the high canopy
of primary forests in W. Africa, the
Diana monkey rarely descends
to the ground.

**DE BRAZZA'S
MONKEY**
Cercopithecus neglectus
This is a semiterrestrial
species of C. African swamp
forests. Males are larger
than females and have a
distinctive blue scrotum.

15½–28 in
39–71 cm

15–25 in
38–63 cm

18–26 in
45–66 cm

22–34 in
55–85 cm

BLUE MONKEY
Cercopithecus mitis
Social groups of this African native
number up to 40 individuals and
consist of a dominant alpha male,
several females, and their offspring.

MONA MONKEY
Cercopithecus mona
A tree-dwelling species, this
monkey inhabits rainforest
and mangrove forests from
Ghana to Cameroon.

GOLDEN-BELLIED MANGABEY
Cercocebus chrysogaster
This mainly terrestrial mangabey lives
in rainforests and swamp forests in the
Congo basin. It forages by day in
groups of up to at least 35 individuals.

**NORTHERN BLACK-
CRESTED MANGABEY**
Lophocebus aterrimus
An arboreal species that prefers
rainforest, this mangabey is
found in the Democratic
Republic of Congo.

》

VERVET MONKEY
Chlorocebus pygerythrus
An inhabitant of savanna and open woodland, this monkey is distributed from Ethiopia through E. Africa to South Africa.

12–28 in
30–70 cm

GRIVET
Chlorocebus aethiops
A semiterrestrial species, the grivet lives in N.E. Africa. It is characterized by a green tinge on its upper face.

16–26 in
40–66 cm

19–35 in
48–88 cm

PATAS MONKEY
Erythrocebus patas
The long limbs and short digits of this primate make it well-adapted for running. It is found from W. to E. Africa.

ANGOLAN TALAPOIN
Miopithecus talapoin
The smallest of the Old World monkeys, the arboreal talapoin lives in wet and swampy forests in W. and C. Africa.

10–18 in
26–45 cm

speckled olive-gray fur

blue flanges on either side of the nose

22–43 in
55–110 cm

MANDRILL
Mandrillus sphinx
Found in rainforests in western C. Africa, this species has distinctive facial markings that are duller in females and juveniles than in males.

18–33 in
45–83 cm

DRILL
Mandrillus leucophaeus
A large terrestrial monkey of lowland mature rainforest, the drill is found only in Cameroon, Nigeria, and the Republic of Equatorial Guinea.

baboon tail typically appears broken

olive-gray body

20–34 in
51–85 cm

20–43 in
50–114 cm

CHACMA BABOON
Papio ursinus
One of the largest baboons, this species inhabits woodland, savanna, steppe, semidesert, and montane habitats across southern Africa.

YELLOW BABOON
Papio cynocephalus
An opportunistic baboon, this omnivore's diet includes seed pods, roots, insects, and other monkeys. It lives in S. and E. Africa.

20–35 in
50–90 cm

red-brown face

20–37 in
50–95 cm

14–34 in
35–86 cm

HAMADRYAS BABOON
Papio hamadryas
This species is found in E. and C. Africa, particularly Ethiopia. The male has a long silver-gray shoulder cape.

GUINEA BABOON
Papio papio
One of the smallest baboons, this species also has one of the smallest ranges, confined to western equatorial Africa.

muscular limbs help to run fast

OLIVE BABOON
Papio anubis
This species forms troops of up to 100 baboons in the savanna and grassland steppe of central sub-Saharan Africa.

20–30 in
50–75 cm

GELADA
Theropithecus gelada
A grassland grazer from the highlands of
Ethiopia, this primate has a distinctive
bare patch of skin on its chest.

**GOLDEN SNUB-NOSED
MONKEY**
Rhinopithecus roxellana
This monkey has thick fur,
which helps it to survive in
high montane forests in
W. and C. China.

18½–33 in
47–83 cm

24–30 in
61–76 cm

*mantle of
white hair*

*white "cloak"
over back*

19½–30 in
49–75 cm

PROBOSCIS MONKEY
Nasalis larvatus
A competent swimmer, this monkey
inhabits mangrove and lowland riverine
rainforest in Borneo. It is named after
the male's large nose.

18½–27 in
47–68 cm

ANGOLA COLOBUS
Colobus angolensis
A predominantly arboreal
species, this monkey inhabits
a variety of forest habitats
in Angola, Congo, and other
neighboring countries.

GUEREZA
Colobus guereza
Also called the eastern
black-and-white colobus, this
species is widespread in the
moist, tropical forests of
C. and E. Africa.

*long tail
with a white,
bushy tip*

16–31 in
41–78 cm

NORTHERN PLAINS GRAY LANGUR
Semnopithecus entellus
Also called the Hanuman langur,
this gray monkey is found in S. Asia,
including India and Pakistan.

*bright orange
coloration*

23–25 in
58–64 cm

17–26 in
43–65 cm

TUFTED GRAY LANGUR
Semnopithecus priam
Found in southeast India and Sri Lanka,
this langur lives in many different habitats.
Its diet consists mainly of tree foliage.

JAVAN LANGUR
Trachypithecus auratus
Most males and females
of this species are black
in color. However, some
individuals retain their
juvenile orange coloration
into adulthood.

»

MANDRILL
Mandrillus sphinx

The mandrill is the largest of all monkeys, and the male is particularly spectacular. Mandrills live in troops that usually contain a dominant male, several females, a gaggle of youngsters, and various non-breeding, lower-ranking males. Sometimes several groups merge to form troops of 200 or more. Mandrill society has a strict hierarchy. Individuals advertise their status with colorful patches of skin on the face and rump. A dominant male is fearsome-looking, and he has a temperament to match. The brightness of the skin pigments is controlled by hormones and color is a good indicator of strength and ferocity. A rival must be very sure of himself before challenging such a magnificent animal, and serious fights tend to occur only between well-matched individuals.

SIZE 22–43 in (55–110 cm)
HABITAT Dense rainforest
DISTRIBUTION Western Central Africa, from S. Cameroon to S.W Republic of Congo
DIET Mainly fruit

∨ **LONG, HARD GLARE**
Forward-facing eyes provide the mandrill with stereoscopic vision. They see in full color, which helps them to locate ripe fruit and to distinguish the visual signals of other individuals.

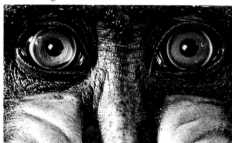

< **NOSTRILS**
In a mature male, the skin around the nostrils and the center of the snout is scarlet. Females and young mandrills have a black nose.

< **TEETH**
The long canine teeth are used principally for fighting and in display. The cheek teeth are smaller, with knobbly surfaces, which are used for grinding up plant material.

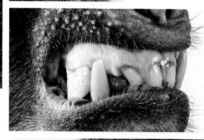

grooves on either side of the nose

∧ **GRASP**
Mandrill thumbs are short, but fully opposable—like those of the great apes—for grasping and manipulating objects. The fingers are long and very strong, with stout nails.

∧ **HIND FEET**
The hind feet resemble the hands, in that the toes are long and able to grasp. Mandrills are good climbers, and often sleep among the branches of trees.

∧ **PLAIN RUMP**
All mandrills have a hairless rump and a short tail. The rump of the subordinate male has less color than the alpha male.

short, tufted tail

relatively short hind legs

ALL FOURS >
Mandrills spend most of their time at ground level, where they move around on all fours, and typically roam 3–6 miles (5–10 km) a day.

long, powerful arms

ALPHA MALE >
The skin color displayed by an individual changes depending on breeding condition and mood. The alpha male displays the most vivid red and blue hues on its face and rump.

prominent brow ridges shade the eyes from bright sunlight

coat is made up of long, coarse hairs

∧ POWERFUL PROFILE
The mandrill's head is bulky, accommodating powerful jaw muscles needed to crush tough plant foods. Its ears are relatively small, though it has excellent hearing. Only high-ranking males have an orange beard.

GIBBONS

The gibbons, or lesser apes, of the family Hylobatidae are medium-sized fruit-eating primates. They have no tail and characteristically move by brachiation—swinging between trees using their very long forearms. Every day gibbon families sing to reinforce family and pair bonds and announce territorial ownership. Some species have an enlarged throat sac to amplify the sound.

AGILE GIBBON
Hylobates agilis
Although its coat coloration varies, all agile gibbons have white eyebrows, and males also have white cheeks. This species lives in Thailand, Indonesia, and Malaysia.

18–25 in
45–64 cm

SILVERY JAVAN GIBBON
Hylobates moloch
Endemic to western Java, Indonesia, males and females of the silvery gibbon are silver-gray with a dark cap.

18–25 in
45–64 cm

MÜLLER'S GIBBON
Hylobates muelleri
Also known as the gray gibbon, this species lives in Borneo. Monogamous pairs spend an average of 15 minutes per day singing duets.

17½–25 in
44–64 cm

black cap of female

PILEATED GIBBON
Hylobates pileatus
Female pileated gibbons are silver-gray with a black face, chest, and cap; males are black. They are found in Thailand, Cambodia, and Laos.

17½–25 in
44–64 cm

white feet and hands

32 in
81 cm

WESTERN HOOLOCK GIBBON
Hoolock hoolock
Male hoolock gibbons are black, but females are tan with dark brown cheeks. They are found in China, N.E. India, and N.W. Myanmar.

16½–23 in
42–59 cm

pronounced hair on crown of male

NORTHERN WHITE-CHEEKED GIBBON
Nomascus leucogenys
Northern white-cheeked gibbons are creamy colored when they are born. They change color by two years of age.

18–25 in
45–64 cm

♂

naked palms

18–25 in
45–64 cm

♀

LAR GIBBON
Hylobates lar
Variable in coat coloration, the white-handed or lar gibbon lives in forests in Thailand, Malaysia, Sumatra, Myanmar, and Laos.

silvery white saddle of hair extends to rump and thighs

BUFF-CHEEKED CRESTED GIBBON
Nomascus gabriellae
Male buff-cheeked gibbons are black with pale cheeks; females are buff with a black cap. They are found in Cambodia, Laos, and Vietnam.

28–35 in
71–90 cm

SIAMANG
Symphalangus syndactylus
The largest of the gibbons, the siamang lives on the island of Sumatra, Indonesia, and on the Malay Peninsula.

HUMANS AND APES

The family Hominidae contains the largest primates—the great apes and humans. Orangutans are arboreal, whereas chimpanzees, gorillas, and humans spend the majority of their time on the ground. Chimpanzees and gorillas move on all fours by "knuckle-walking." None of the great apes has a tail. Males are generally larger than females and all species have a relatively large braincase.

EASTERN GORILLA
Gorilla beringei
The largest primate, the two eastern gorilla subspecies inhabit montane cloud forest and lowland forest in the eastern Democratic Republic of Congo, Rwanda, and Uganda.

39–47 in
101–120 cm

heavily built body

domed forehead

strong hands and feet

BORNEAN ORANGUTAN
Pongo pygmaeus
A large arboreal fruit eater, the Bornean orangutan lives in the canopy of primary rainforest on the island of Borneo.

28–38 in
72–97 cm

very long arms

coarse, shaggy red-brown coat

grasping hands and feet

27–39 in
68–99 cm

SUMATRAN ORANGUTAN
Pongo abelii
The largest arboreal primate, the Sumatran orangutan is restricted to fragments of primary tropical forest in N. Sumatra.

BONOBO
Pan paniscus
Slighter than the common chimpanzee, the bonobo or pygmy chimpanzee lives in humid tropical forest in the Democratic Republic of Congo.

28–33 in
70–83 cm

WESTERN GORILLA
Gorilla gorilla
Two subspecies of western gorilla live in lowland tropical and swamp forests in western C. Africa. Adult males are called silverbacks.

40½–42 in
103–107 cm

COMMON CHIMPANZEE
Pan troglodytes
Four subspecies of the common chimpanzee are distributed in dry and moist forests and savanna woodlands in equatorial Africa.

28–38 in
70–96 cm

HUMAN
Homo sapiens
Characterized by a bipedal posture and a lack of body hair, humans permanently inhabit every terrestrial habitat, with the exception of Antarctica.

4–7 ft
1.2–2.1 m

♂ ♀

BATS

The only mammals capable of powered flight, bats are primarily nocturnal. Many species use echolocation to navigate and find food.

Bats are found worldwide in many different habitats, including tropical, subtropical, and temperate forests, savanna grasslands, deserts, and wetlands. Most fruit bats, as their name suggests, eat fruits. All other bats, once referred to as microbats, are mainly insect eaters. However, some bats drink nectar and eat pollen, a few suck blood, and some eat vertebrates, such as fishes, frogs, and bats.

Their greatly elongated arm, hand, and finger bones support an elastic wing membrane for flight. Many bats also have a tail membrane between their legs. Bats typically rest upside down, hanging from their strong toes and claws.

THE SENSE OF ECHOLOCATION

Fruit bats rely predominantly on vision and smell, whereas microbats use a specialized sense called echolocation to avoid hitting objects, and to detect prey in the dark. They emit pulses of sound through their mouth or nose and form a "sound" image of their surroundings from returning echoes. Those species that emit echolocation calls through their nose usually have elaborate facial ornamentation, called noseleaves, for focusing the sound. Bats have very sensitive hearing, often highly tuned to the frequency of their returning echoes. Some species listen to prey-generated sounds, such as the rustle that an insect makes when walking on a leaf.

HABITS AND ADAPTATIONS

Bats are very social animals, living in colonies of hundreds or thousands and, exceptionally, millions of animals. They roost in trees and caves or in buildings, bridges, and mines. Temperate species either migrate to warmer climates or hibernate during winter. They may also become torpid if food becomes short in other seasons. Many interesting reproductive adaptations have developed, including the storage of sperm, delayed fertilization, and delayed implantation to ensure that the young are born at the optimum time of year.

PHYLUM	CHORDATA
CLASS	MAMMALIA
ORDER	CHIROPTERA
FAMILIES	21
SPECIES	About 1,400

EVOLUTIONARY DEBATE

Morphological and genetic analyses of the relationships between bat families do not always agree with one another. Genetic studies suggest that all bats evolved from one common ancestor and that flight evolved only once. However, molecular studies, although still proposing two bat groups, also indicate that some echolocating bats, such as the horseshoe bats, are more closely related to non-echolocating fruit bats (Yinpterochiroptera) than to other echolocating microbats (Yangochiroptera). This is an apparent conundrum unless echolocation has evolved twice or has later been lost in the fruit bats.

FRUIT BATS

The bats of the family Pteropodidae are distributed across the tropical and subtropical regions of the Old World. They have a doglike face with simple ears and large eyes. They use vision and smell to locate their food, except for members of the genus *Rousettus*, which echolocate by producing tongue clicks. These bats feed on fruits, nectar, and pollen. They have claws on both the thumb and the second finger.

2–3 in
5–7.5 cm

BLOSSOM BAT
Syconycteris australis
This specialized nectar-feeder occurs from Papua New Guinea to eastern coastal Australia. It has a pointed muzzle and a brush-tipped tongue for probing flowers.

elastic skin membrane

4¼–7 in
11–18 cm

FRANQUET'S EPAULETTED FRUIT BAT
Epomops franqueti
Also called the singing fruit bat because of the high-pitched call of the males, this species occurs in W. and C. Africa.

1½–3¼ in
4–8 cm

LESSER LONG-TONGUED FRUIT BAT
Macroglossus minimus
Occurring in S.E. Asia, this fruit bat feeds on nectar and pollen from flowers, using its long tongue.

3¼–4½ in
8–11 cm

SHORT-NOSED FRUIT BAT
Cynopterus sphinx
This is the only fruit bat that makes tents from palm leaves. It is found across S.E. Asia and the Indian subcontinent.

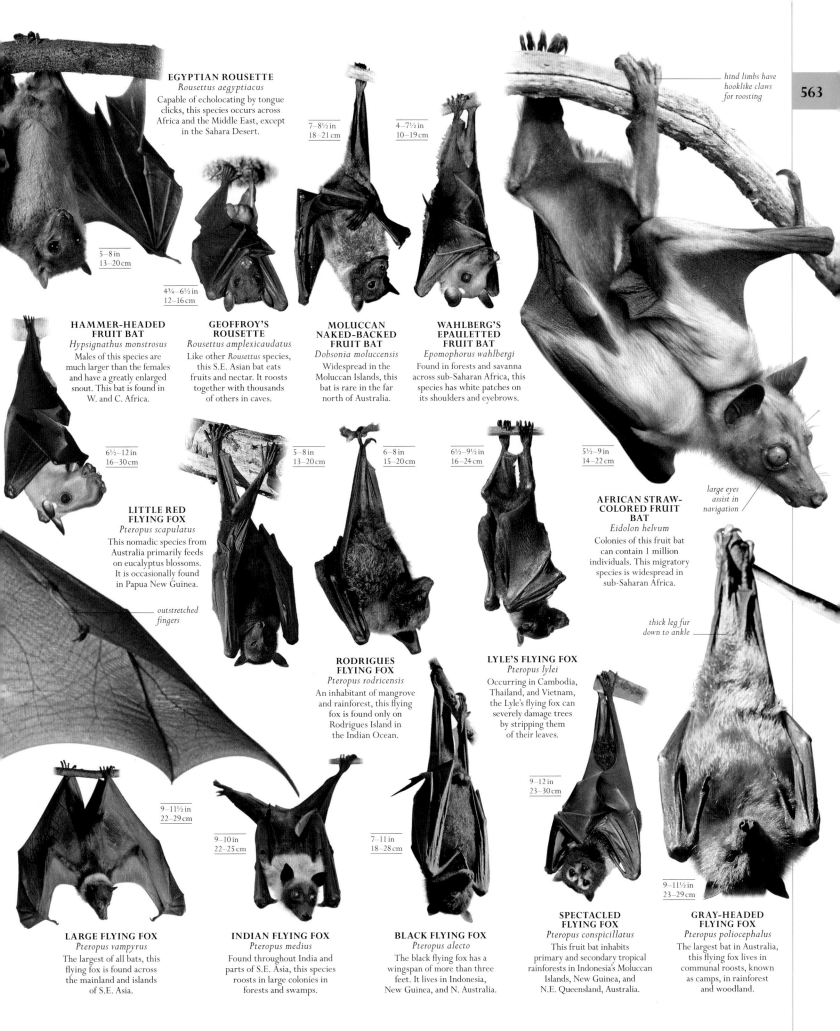

EGYPTIAN ROUSETTE
Rousettus aegyptiacus
Capable of echolocating by tongue clicks, this species occurs across Africa and the Middle East, except in the Sahara Desert.

5–8 in
13–20 cm

7–8½ in
18–21 cm

4–7½ in
10–19 cm

hind limbs have hooklike claws for roosting

4¾–6½ in
12–16 cm

HAMMER-HEADED FRUIT BAT
Hypsignathus monstrosus
Males of this species are much larger than the females and have a greatly enlarged snout. This bat is found in W. and C. Africa.

GEOFFROY'S ROUSETTE
Rousettus amplexicaudatus
Like other *Rousettus* species, this S.E. Asian bat eats fruits and nectar. It roosts together with thousands of others in caves.

MOLUCCAN NAKED-BACKED FRUIT BAT
Dobsonia moluccensis
Widespread in the Moluccan Islands, this bat is rare in the far north of Australia.

WAHLBERG'S EPAULETTED FRUIT BAT
Epomophorus wahlbergi
Found in forests and savanna across sub-Saharan Africa, this species has white patches on its shoulders and eyebrows.

6½–12 in
16–30 cm

5–8 in
13–20 cm

6–8 in
15–20 cm

6½–9½ in
16–24 cm

5½–9 in
14–22 cm

large eyes assist in navigation

AFRICAN STRAW-COLORED FRUIT BAT
Eidolon helvum
Colonies of this fruit bat can contain 1 million individuals. This migratory species is widespread in sub-Saharan Africa.

LITTLE RED FLYING FOX
Pteropus scapulatus
This nomadic species from Australia primarily feeds on eucalyptus blossoms. It is occasionally found in Papua New Guinea.

outstretched fingers

thick leg fur down to ankle

RODRIGUES FLYING FOX
Pteropus rodricensis
An inhabitant of mangrove and rainforest, this flying fox is found only on Rodrigues Island in the Indian Ocean.

LYLE'S FLYING FOX
Pteropus lylei
Occurring in Cambodia, Thailand, and Vietnam, the Lyle's flying fox can severely damage trees by stripping them of their leaves.

9–11½ in
22–29 cm

9–12 in
23–30 cm

9–10 in
22–25 cm

7–11 in
18–28 cm

9–11½ in
23–29 cm

LARGE FLYING FOX
Pteropus vampyrus
The largest of all bats, this flying fox is found across the mainland and islands of S.E. Asia.

INDIAN FLYING FOX
Pteropus medius
Found throughout India and parts of S.E. Asia, this species roosts in large colonies in forests and swamps.

BLACK FLYING FOX
Pteropus alecto
The black flying fox has a wingspan of more than three feet. It lives in Indonesia, New Guinea, and N. Australia.

SPECTACLED FLYING FOX
Pteropus conspicillatus
This fruit bat inhabits primary and secondary tropical rainforests in Indonesia's Moluccan Islands, New Guinea, and N.E. Queensland, Australia.

GRAY-HEADED FLYING FOX
Pteropus poliocephalus
The largest bat in Australia, this flying fox lives in communal roosts, known as camps, in rainforest and woodland.

LYLE'S FLYING FOX
Pteropus lylei

Lyle's flying fox is a medium-sized representative of the Old World fruit bat family, the Pteropodidae. Fruit bats are social animals; many hundreds may gather in roost trees during the day to rest and groom, dispersing at dusk to find ripe fruits. Although they may cause some damage to trees, many fruit bats are important pollinators and seed dispersers of tropical plants, including many commercial crops. Fruit bats are found in the tropical regions of Africa, Asia, and Australia, although this species is only found in Cambodia, Thailand, and Vietnam. They inhabit forested areas, including mangroves and fruit orchards.

SIZE 6½–9½ in (16–24 cm)
HABITAT Forests
DISTRIBUTION S.E. and E. Asia
DIET Fruits and leaves

all bats have a thumb claw, but only Old World fruit bats have a claw on the second finger, too

˅ CANINE FACE
Fruit bats have large eyes for navigation and a large nose for sniffing out fruits, pollen, and nectar—giving them a doglike face. Species that use echolocation have larger ears and smaller eyes.

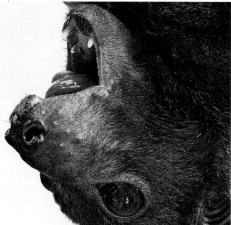

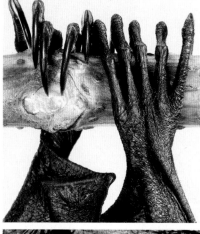

˂ CLINGING CLAWS
Bats' sharp, curved claws are perfect for clinging to tree branches. When roosting, tendons lock in place to ensure the bat's claws stay bent without the muscles needing to contract.

˄ HAND WING
A bat's wing is formed of elongated forearm and finger bones supporting a thin elastic membrane attached to the sides of the body. It has a large surface area, giving lift in flight.

˄ TAIL OR NO TAIL?
Pteropus bats lack a tail, but some have a partial membrane supported by cartilaginous spurs called calcars, which protrude from the ankle.

WALKING UPSIDE DOWN ˃
Fruit bats use their large clawed thumbs to help them move along the branches of their roost tree. The claw can also be used to manipulate fruit when feeding.

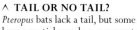

˄ WELL WRAPPED
When resting, most fruit bats hang upside down with their leathery wings folded around their bodies. Bats roosting in the open risk overheating, so they flap their wings and cover themselves in saliva to help keep cool in hot weather.

reddish brown,
foxy coloring

< HANGING AROUND
Few bats are able to move
well on the ground, let
alone take off from a flat
surface. By hanging upside
down they are able to
take flight quickly. During
the day, they hang upside
down to sleep, nestled
together for warmth.

large eyes are adapted
for excellent vision,
especially at night

upright, foxlike ears detect
sounds beyond the range of
human hearing

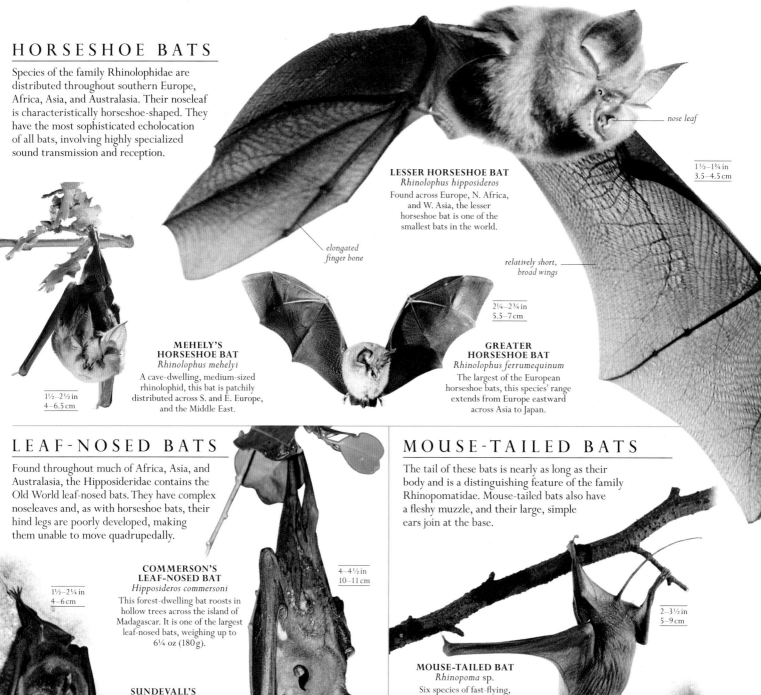

HORSESHOE BATS

Species of the family Rhinolophidae are distributed throughout southern Europe, Africa, Asia, and Australasia. Their noseleaf is characteristically horseshoe-shaped. They have the most sophisticated echolocation of all bats, involving highly specialized sound transmission and reception.

nose leaf

LESSER HORSESHOE BAT
Rhinolophus hipposideros
Found across Europe, N. Africa, and W. Asia, the lesser horseshoe bat is one of the smallest bats in the world.

1½–1¾ in
3.5–4.5 cm

elongated finger bone

relatively short, broad wings

2¼–2¾ in
5.5–7 cm

MEHELY'S HORSESHOE BAT
Rhinolophus mehelyi
A cave-dwelling, medium-sized rhinolophid, this bat is patchily distributed across S. and E. Europe, and the Middle East.

1½–2½ in
4–6.5 cm

GREATER HORSESHOE BAT
Rhinolophus ferrumequinum
The largest of the European horseshoe bats, this species' range extends from Europe eastward across Asia to Japan.

LEAF-NOSED BATS

Found throughout much of Africa, Asia, and Australasia, the Hipposideridae contains the Old World leaf-nosed bats. They have complex noseleaves and, as with horseshoe bats, their hind legs are poorly developed, making them unable to move quadrupedally.

COMMERSON'S LEAF-NOSED BAT
Hipposideros commersoni
This forest-dwelling bat roosts in hollow trees across the island of Madagascar. It is one of the largest leaf-nosed bats, weighing up to 6¼ oz (180 g).

4–4½ in
10–11 cm

1½–2¼ in
4–6 cm

SUNDEVALL'S LEAF-NOSED BAT
Hipposideros caffer
A savanna species, this leaf-nosed bat roosts in caves and buildings throughout Africa except in the Sahara and central forested regions.

MOUSE-TAILED BATS

The tail of these bats is nearly as long as their body and is a distinguishing feature of the family Rhinopomatidae. Mouse-tailed bats also have a fleshy muzzle, and their large, simple ears join at the base.

2–3½ in
5–9 cm

MOUSE-TAILED BAT
Rhinopoma sp.
Six species of fast-flying, insectivorous mouse-tailed bat live in arid and semiarid regions of N. Africa, the Middle East, and India.

first finger (thumb)

HOG-NOSED BAT

The sole member of the family Craseonycteridae, the tiny hog-nosed bat has long, broad wings that allow it to hover. It lacks a tail and calcars (cartilaginous extensions of the ankle).

1–1¼ in
3–3.5 cm

KITTI'S HOG-NOSED BAT
Craseonycteris thonglongyai
One of the world's smallest mammals, and often called the bumblebee bat, Kitti's hog-nosed bat lives in riverside caves in Thailand and Myanmar.

MEGADERMATIDAE

The family Megadermatidae comprises six fairly large echolocating bat species. Carnivorous and insectivorous, they have large ears and eyes and a broad tail membrane but little or no tail.

4–5 in
10–13 cm

GHOST BAT
Macroderma gigas
Endemic to N. Australia, the ghost bat is one of the largest microbats. It preys on vertebrates, such as frogs and lizards.

SHEATH-TAILED BATS

The Emballonuridae are more commonly known as sheath-tailed bats because just the tip of their tail protrudes through the tail membrane, giving it the appearance of being sheathed. Many have glandular odor-storing sacs on their wings.

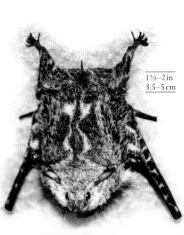

3–3¼ in
7.5–8.5 cm

1½–2 in
4–5 cm

1½–2 in
3.5–5 cm

2–2¼ in
4.5–6 cm

HILDEGARDE'S TOMB BAT
Taphozous hildegardeae
Dependent on caves for roosting, Hildegarde's tomb bat feeds on insects in the coastal forests of Kenya and Tanzania.

LESSER SHEATH-TAILED BAT
Emballonura monticola
The short tail of this bat appears to retract into a sheath when it stretches its legs. It lives in Indonesia, Malaysia, Myanmar, and Thailand.

PROBOSCIS BAT
Rhynchonycteris naso
These bats spend the day roosting in groups on the underside of branches. They live in tropical forests in C. and S. America.

GREATER SAC-WINGED BAT
Saccopteryx bilineata
Male greater sac-winged bats attract females by a pungent secretion in their wing sacs. The species is found in C. and S. America.

NEW WORLD LEAF-NOSED BATS

The Phyllostomidae contains species that are distributed from southwestern US to northern Argentina. Most have large ears and a noseleaf, shaped like a spearhead, to enhance echolocation.

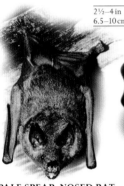

2½–4 in
6.5–10 cm

2¼–3 in
5.5–7.5 cm

2–2¼ in
4.5–6 cm

2–2¾ in
4.5–7 cm

2–2¼ in
4.5–6 cm

GERVAIS' FRUIT-EATING BAT
Artibeus cinereus
Preferring to roost in palm trees, Gervais' fruit-eating bat lives in S. America, including Venezuela, Brazil, and the Guianas.

SEBA'S SHORT-TAILED BAT
Carollia perspicillata
Inhabiting moist evergreen and dry deciduous forests across much of C. and S. America, this species is a generalist fruit eater.

PALE SPEAR-NOSED BAT
Phyllostomus discolor
This species emits echolocation calls through its nose. It is found in C. and northern S. America.

COMMON TENT-MAKING BAT
Uroderma bilobatum
Found in lowland forests from Mexico to central S. America, this bat creates shelters by biting palm and banana leaves.

3¼–4¼ in
8–10.5 cm

2¼–3 in
6–7.5 cm

GEOFFROY'S TAILLESS BAT
Anoura geoffroyi
A specialized nectar feeder, this species has a long muzzle, elongate cheek teeth, and a brush-tipped tongue. It lives in C. and S. America.

SILKY SHORT-TAILED BAT
Carollia brevicaudum
This leaf-nosed bat is widespread in C. America and the Amazon region. It helps to restore degraded forests by dispersing the seeds of fruit trees.

2¾–3¾ in
7–9.5 cm

2¼–3¼ in
6–8 cm

2¾–3¾ in
7–9.5 cm

FRINGE-LIPPED BAT
Trachops cirrhosus
The fringe-lipped bat catches frogs by listening for their calls. It lives in tropical forests in C. and northern S. America.

WHITE-LINED BROAD-NOSED BAT
Platyrrhinus lineatus
Named for the white lines on its face and back, this broad-nosed bat roosts in damp forests of central S. America.

CALIFORNIAN LEAF-NOSED BAT
Macrotus californicus
Primarily hunting for moths by sight rather than echolocation, the Californian leaf-nosed bat lives in N. Mexico and the S.E. US.

COMMON VAMPIRE BAT
Desmodus rotundus
Famous for feeding on other mammals' blood, the common vampire bat occurs in a variety of habitats in Mexico and C. and S. America.

MORMOOPIDAE

The family Mormoopidae contains the naked-backed and moustache bats. Several species have wings that join across their back and they have a fringe of stiff hairs around the muzzle.

2–2¼ in
4.5–6 cm

DAVY'S NAKED-BACKED BAT
Pteronotus davyi
Unlike most other bats, this bat's wing membranes meet across its back. It is distributed from Mexico to S. America.

NOCTILIONIDAE

The two species of bulldog bats from the family Noctilionidae have long legs and large feet and claws. They have full lips and cheek pouches for storing food while flying.

3¼–4 in
8–10 cm

GREATER BULLDOG BAT
Noctilio leporinus
Specialized for snatching fishes from the water's surface with its claws, the greater bulldog or fishing bat lives in tropical C. and S. America.

NATALIDAE

The funnel-eared bats of the family Natalidae are small and slender with large ears. Adult males have a sensory structure, called a natalid organ, on their forehead.

1½–1¾ in
3.8–4.3 cm

MEXICAN FUNNEL-EARED BAT
Natalus mexicanus
An insectivorous species, the Mexican funnel-eared bat lives in C. and S. America. It roosts in caves and, sometimes, in abandoned mines.

SLIT-FACED BATS

The slit-faced bats of the family Nycteridae have a furrow from their nostrils to a pit between their eyes. The cartilage at the end of their tail ends in a "Y" shape.

2½–3 in
6.5–7.5 cm

MALAYAN SLIT-FACED BAT
Nycteris tragata
The facial furrow of this slit-faced bat helps to direct its echolocation calls. It is found in tropical forest in Myanmar, Malaysia, Sumatra, and Borneo.

MYSTACINIDAE

The one extant member of the Mystacinidae has extra talons on its thumb and toe claws. Its tough, leathery wings can be folded against the body when moving on the ground.

2¼–3¼ in
6–8 cm

LESSER NEW ZEALAND SHORT-TAILED BAT
Mystacina tuberculata
An agile mover on the ground, the New Zealand short-tailed bat sniffs out prey in leaf litter on the forest floor.

THYROPTERIDAE

The family Thyropteridae contains five species of disk-winged bats. Suction cups on the wrists and ankles allow them to attach to smooth-surfaced tropical leaves when roosting.

1½–2 in
4–5 cm

SPIX'S DISK-WINGED BAT
Thyroptera tricolor
An insectivorous inhabitant of lowland forests from S. Mexico to S.E. Brazil, Spix's disk-winged bat roosts head upward inside furled leaves.

MOLOSSIDAE

Free-tailed bats of the family Molossidae have a distinctive tail that extends beyond the edge of the tail membrane. They are robust, stocky bats with long, narrow wings for fast flight. Their wing and tail membranes are particularly leathery.

3¼–3½ in
8–9 cm

4½–5½ in
11–14 cm

2–2½ in
4.5–6.5 cm

LARGE-EARED FREE-TAILED BAT
Otomops martiensseni
This African species, like the European free-tailed bat, has exceptionally low-frequency echolocation calls, which are clearly audible to humans.

BRAZILIAN FREE-TAILED BAT
Tadarida brasiliensis
Colonies of the Mexican free-tailed bat roosting in caves and beneath bridges in Texas and Mexico can number in the millions.

EUROPEAN FREE-TAILED BAT
Tadarida teniotis
The only free-tailed bat to live in Europe, this bat's range extends from the Mediterranean to S. and S.E. Asia.

2¾–3¼ in
7–8 cm

MILLER'S MASTIFF BAT
Molossus pretiosus
An insectivorous bat of lowland dry forests, open savanna, and cactus scrub, Miller's mastiff bat is distributed from Mexico to Brazil.

VESPERTILIONIDAE

Consisting of 496 species, the Vespertilionidae is the largest family of bats. Known as evening or common bats, they are distributed worldwide, excluding the polar regions, and are mostly insectivorous. They typically have a plain nose and small eyes.

2–2¼ in
5–6 cm

SCHREIBER'S BENT-WINGED BAT
Miniopterus schreibersii
Possessing long finger bones and broad wings, Schreiber's bat is patchily distributed in southwestern Europe and N. and W. Africa.

ears joined at base

2¼–3½ in
6–9 cm

1½–2¼ in
4–6 cm

2–3¼ in
5–8.5 cm

GRAY LONG-EARED BAT
Plecotus austriacus
The gray long-eared bat's ears are almost the same length as its body. It lives in southern and C. Europe and N. Africa.

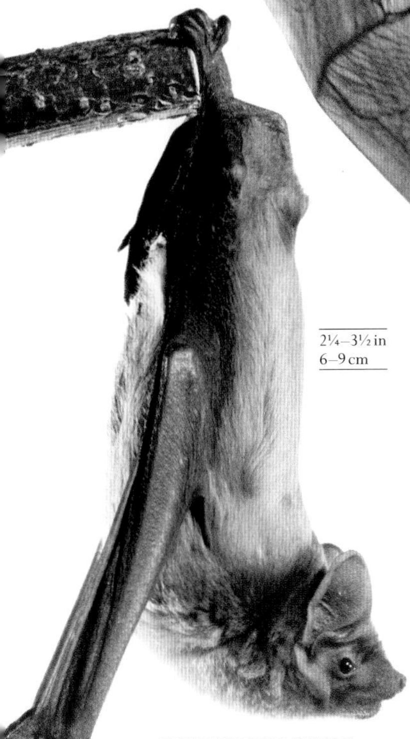

BIG BROWN BAT
Eptesicus fuscus
Often found roosting in buildings, the insectivorous big brown bat is distributed from S. Canada to N. Brazil and some Caribbean islands.

2–2½ in
4.5–6.5 cm

COMMON NOCTULE
Nyctalus noctula
A fast and powerful flier with narrow, blackish-brown wings, the noctule is found across N.E. Europe and parts of Asia.

EUROPEAN PARTICOLORED BAT
Vespertilio murinus
This pale-bellied, dark-backed bat is native to mountain, steppe, and forest habitats from E. and C. Europe to Asia. It is also found in urban areas.

1½–2 in
4–5 cm

NATTERER'S BAT
Myotis nattereri
From N.W. Africa, across Europe to S.W. Asia, Natterer's bat catches insects in its fringed tail membrane during a slow, hovering flight.

1½–2 in
4–5 cm

1½–2¼ in
3.5–5.5 cm

1¾–2¼ in
4.5–5.5 cm

2–2¼ in
4.5–6 cm

FRINGED MYOTIS
Myotis thysanodes
Named after a fringe of hairs along the edge of its tail membrane, this myotis bat lives in western N. America.

2–2¼ in
4.5–5.5 cm

EASTERN PIPISTRELLE
Perimyotis subflavus
The eastern pipistrelle hibernates in rock crevices, mines, and caves during winter. Its range covers eastern N. America, from S. Canada to N. Honduras.

COMMON PIPISTRELLE
Pipistrellus pipistrellus
The most widespread pipistrelle species, the common pipistrelle's range extends from W. Europe to the Far East and N. Africa.

NATHUSIUS' PIPISTRELLE
Pipstrellus nathusii
A long-distance migrant capable of journeys of over 1,200 miles (1,900 km) in spring and fall, this species mainly occurs in E. and C. Europe.

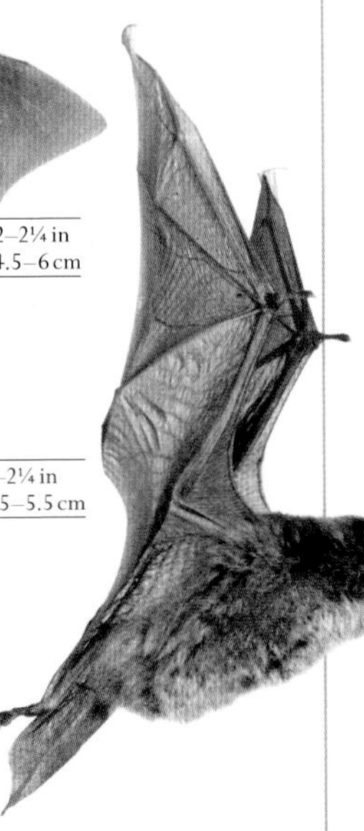

DAUBENTON'S BAT
Myotis daubentonii
This Eurasian bat has relatively large feet, which help it to catch flying insects emerging from the surface of bodies of water.

HEDGEHOGS, MOLES, AND RELATIVES

Four families in two orders have recently been grouped together in a new order Eulipotyphla, which means "truly fat and blind." All of its members are small, long-snouted, primarily insectivorous mammals.

The four families in the Eulipotyphla are the Erinaceidae (hedgehogs and gymnures), Solenodontidae (solenodons), Talpidae (moles and desmans), and Soricidae (shrews). Most species of Erinaceidae have spines—only the gymnures have fur—and are predominantly terrestrial and nocturnal. The Solenodontidae is the smallest family but contains the two largest species, both of which live on Caribbean islands. The Talpidae is the most diverse of the four families, comprising 54 species

and including several types of burrowing mole and the subaquatic desmans. The Soricidae is the largest family, containing about 450 species, including one of the smallest of all mammals, the Etruscan shrew.

VENOMOUS SALIVA

Well adapted to their insectivorous way of life, the long, mobile, cartilaginous snout of shrews, moles, desmans, and solenodons contains numerous simple, sharp-pointed teeth, which they use to catch and kill prey, such as earthworms, other invertebrates, and small vertebrates. Some species produce venomous saliva, which in solenodons runs along a groove in one of their lower incisor teeth and helps to subdue larger prey before it is killed. By contrast, hedgehogs do not produce toxic saliva but some are immune to snake venom.

PHYLUM	CHORDATA
CLASS	MAMMALIA
ORDER	EULIPOTYPHLA
FAMILIES	4
SPECIES	530

Immunity to snake venom allows a hedgehog to take advantage of any snake it comes across as a potential food source.

HEDGEHOGS AND GYMNURES

With a long, sensitive snout and a short, hairy tail, members of the family Erinaceidae eat almost anything, from invertebrates and fruit to birds' eggs and carrion. The hedgehogs, from Eurasia and Africa, are covered in sharp protective spines, while the southeast Asian gymnures have normal hair and look more like rats or opossums. Gymnures (also called moonrats) have especially well-developed scent glands that exude a strong garliclike odor for marking territories.

WEST EUROPEAN HEDGEHOG
Erinaceus europaeus
Found throughout W. Europe, this species inhabits woodland, farmland, and gardens. In cooler areas, it hibernates in a nest of leaves and grass.

8–10 in
20–25 cm

pale spines with darker bands

6½–7½ in
17–19 cm

SOUTHERN AFRICAN HEDGEHOG
Atelerix frontalis
Inhabiting grassland, scrub, and gardens in southern Africa, this species has a white band across its forehead, contrasting with its dark face.

8–10½ in
20–27 cm

NORTH AFRICAN HEDGEHOG
Atelerix algirus
Found in varied habitats in the Mediterranean region, this long-legged hedgehog with a pale face and underparts has a spineless "parting" on its crown.

5½–10 in
14–26cm

AFRICAN PYGMY HEDGEHOG
Atelerix albiventris
Because the innermost digit of the hind feet is reduced or absent, this species is also known as the four-toed hedgehog.

5–9½ in
13–24cm

DESERT HEDGEHOG
Paraechinus aethiopicus
This small African and Middle Eastern hedgehog is immune to the venom of snakes and scorpions, which form a large part of its diet.

FAMILY RELATIONS

Molecular analyses of the Eulipotyphla families suggest that they share a common ancestor and so form a natural group. The results also indicate that the shrews are more closely related to the hedgehogs and gymnures than to the talpids and solenodons as was previously thought. However, the relationship of the talpids to the shrews and hedgehogs and gymnures has yet to be resolved.

6½–11 in
16–28 cm

**COMMON
LONG-EARED HEDGEHOG**
Hemiechinus auritus
Long ears help radiate heat and keep this nocturnal hedgehog cool in the deserts of N. Africa and C. Asia.

10–18 in
25–46 cm

MOONRAT
Echinosorex gymnurus
Resembling a large rat, the nocturnal, white-coated moonrat lives in swamps and other wet habitats in Malaysia.

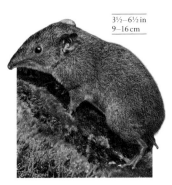

3½–6½ in
9–16 cm

SHORT-TAILED GYMNURE
Hylomys suillus
Usually solitary, this S.E. Asian gymnure lives in forested areas with dense undergrowth. It uses its long, mobile snout to find invertebrate prey but will also eat fruits.

SOLENODONS

The Solenodontidae are the oldest extant lineage of the group and are thought to have separated from other Eulipotyphla about 75 million years ago. They have an elongated, flexible, cartilaginous snout; a long, naked, scaly tail; small eyes; and coarse, dark fur. Unusually among mammals, they also have venomous saliva, which is used to subdue prey ranging from invertebrates to small reptiles.

10½–19½ in
27–49 cm

8–14 in
20–36 cm

shaggy, brown coat

HISPANIOLAN SOLENODON
Solenodon paradoxus
One of two extant species in its family, this solenodon is known only on Hispaniola, an island to the east of Cuba in the Caribbean.

CUBAN SOLENODON
Solenodon cubanus
With longer, finer fur than its Hispaniolian relative, this secretive, nocturnal burrower was erroneously believed to be extinct during the 20th century.

MOLES AND DESMANS

Small, dark insectivores with a cylindrical body, short, dense fur, and a very sensitive, hairless, tubular snout, moles of the family Talpidae are well adapted for a life of burrowing—the forelegs have powerful claws, and the hands are turned permanently outward in the form of a shovel. In contrast, the aquatic desmans have webbed paws fringed with stiff hairs and a long, flattened tail to aid in swimming.

STAR-NOSED MOLE
Condylura cristata
A semiaquatic N. American mole, this species has 11 pairs of pink, fleshy appendages around its snout, which detect prey by touch.

3¾–5 in
9.5–13 cm

4¼–6½ in
11–16 cm

SMALL JAPANESE MOLE
Mogera imaizumii
This small mole from Japan is found in soft, deep soils, and is distinguished from close relatives by dental features.

EUROPEAN MOLE
Talpa europaea
Secretive due to its burrowing lifestyle, this mole creates extensive networks of permanent tunnels, often marked by distinctive surface mole-hills.

4–6¼ in
10–15.5 cm

dense, waterproof coat

RUSSIAN DESMAN
Desmana moschata
The largest member of its family, this desman's webbed hind feet and long, flattened tail are adaptations for swimming in search of food.

7½–9½ in
19–24 cm

4¼–6½ in
11–16 cm

5–6¼ in
13–15.5 cm

EASTERN MOLE
Scalopus aquaticus
A burrowing mole from N. America, typically in moist sandy soils, this species has ears and eyes covered by skin and fur respectively.

PYRENEAN DESMAN
Galemys pyrenaicus
Feeding in Pyrenean mountain streams, this desman rarely digs burrows; it shelters in rock crevices or the holes of water voles.

SHREWS

With a pointed snout, velvety fur, a long tail, and sharp, simple teeth, members of the Soricidae are largely insectivorous, but also eat seeds, fruits, and carrion. Mostly terrestrial, they are highly active, needing to eat at least 80 percent of their own body weight daily. They have poor vision, but excellent senses of hearing and smell, and use echolocation to find their way around.

LESSER WHITE-TOOTHED SHREW
Crocidura suaveolens
Like others in its genus, this European species lacks the iron deposits that produce red tips on the teeth of many shrews.

2–3¼ in
4.5–8 cm

3½–6½ in
9–16 cm

HOUSE SHREW
Suncus murinus
Native to S. Asia but introduced elsewhere in Asia and Africa, this adaptable, uniformly gray-brown shrew is often associated with human habitation.

2½–3¾ in
6.5–9.5 cm

REDDISH-GRAY SHREW
Crocidura cyanea
Males of this southern African forest shrew use a strong, musky scent to mark their territories.

pale feet

3½–4¾ in
9–11.5 cm

2–2¾ in
5–7 cm

1½–2 in
3.5–5 cm

CRAWFORD'S GRAY SHREW
Notiosorex crawfordi
From arid parts of N. America, this shrew can survive without water. To conserve fluid, it produces highly concentrated urine.

NORTHERN SHORT-TAILED SHREW
Blarina brevicaudus
This large and venomous N. American species usually forages in tunnels or under leaf litter or snow, rather than above ground.

ETRUSCAN SHREW
Suncus etruscus
One of the smallest living mammals, weighing only ¹⁄₁₆ oz (2 g), this shrew occurs in S. Europe and the Middle East, with closely related forms in Asia.

2¼–3¼ in
5.5–8 cm

COMMON SHREW
Sorex araneus
The most common shrew in N. Europe, it is active day and night throughout the year, searching for food.

2¼–3 in
6–7.5 cm

PYGMY SHREW
Sorex minutus
Smaller than the common shrew, with which it often occurs, this species is distinguished by its relatively longer and hairier tail.

1½–2½ in
4–6.5 cm

short, dense, velvety fur

ALPINE SHREW
Sorex alpinus
The tail of this dark C. European shrew is as long as its head and body. It aids balance when climbing trees.

2–2¾ in
5–7 cm

EURASIAN WATER SHREW
Neomys fodiens
The stiff hairs on its feet and tail improve the swimming efficiency of this large, sharply bicolored shrew, which hunts mainly in water.

3–4¼ in
7.5–10.5 cm

NORTH AMERICAN LEAST SHREW
Cryptotis parvus
A ferocious hunter, this species tackles a variety of prey. It also bites the tail of lizards which, when shed, provides an easy meal.

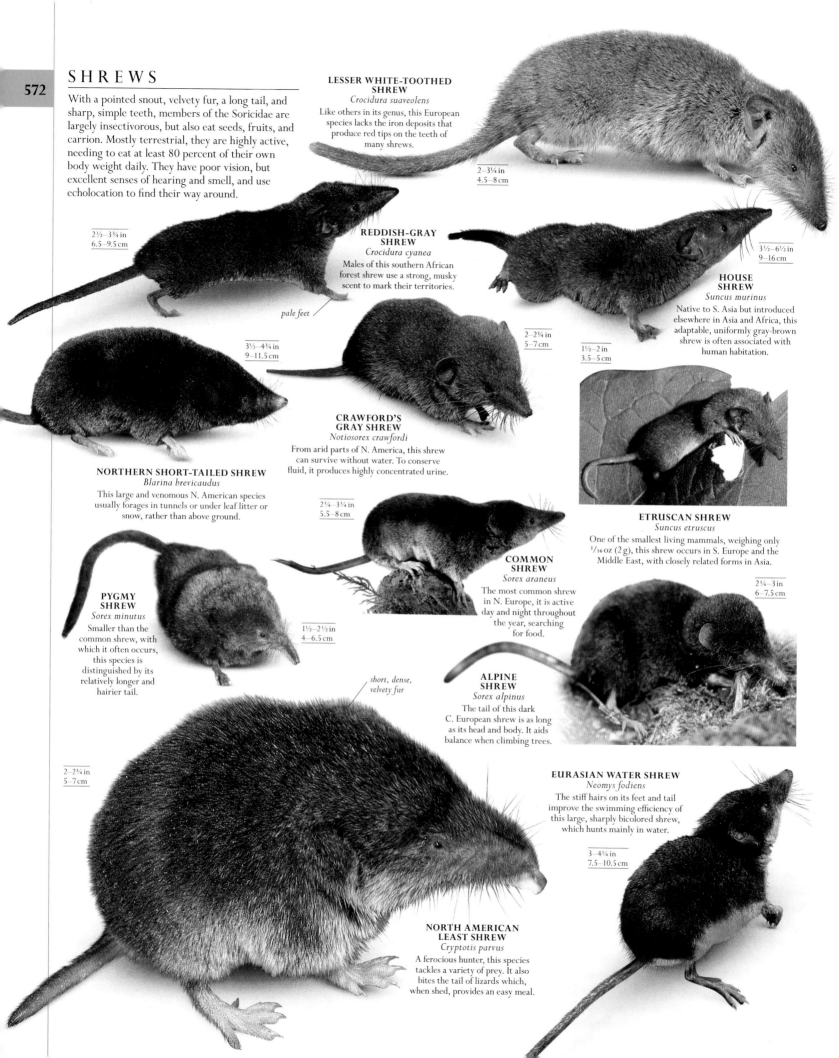

PANGOLINS

With a body covered in large keratin scales, pangolins are also known as scaly anteaters, reflecting their appearance and diet.

Mainly nocturnal and with small eyes, pangolins use their excellent sense of smell to find food. Although somewhat similar in appearance to the American armadillos, and sharing a similar diet, pangolins belong to a different order, Pholidota, and are most closely related to the carnivores.

The horny scales, which cover all exposed parts of the body, can make up one fifth of a pangolin's body weight: despite this, they are excellent swimmers. There are both terrestrial and arboreal species: ground-dwellers live in deep burrows, while tree-dwellers live in holes in trees.

Pangolins have powerful claws that are used to excavate insect nests. Food is captured on the sticky tongue, which extends up to 16 in (40 cm) from the toothless mouth. Because of the size of their front claws, pangolins walk on their wrists, the forepaws curled over to protect the claws.

SAFE FROM ATTACK

The role of the scaly coat is as protection against predators, enhanced by the tendency of pangolins to roll into a ball when threatened or sleeping. Further defense is provided by their ability to emit a foul-smelling chemical from their anal glands, which is also used to mark territory. Nevertheless, pangolins are heavily exploited for their meat, scales, and perceived value in traditional Chinese medicine.

PHYLUM	CHORDATA
CLASS	MAMMALIA
ORDER	PHOLIDOTA
FAMILIES	1
SPECIES	8

The long, sticky tongue of this African ground-dwelling pangolin is used for drinking as well as capturing insects.

MANIDAE

The eight pangolin species of the family Manidae live in tropical Africa and Asia. They are unique among mammals in having their skin protected by large, horny scales. When threatened, they can curl up into a ball, the sharp plate edges providing additional defense. Large, powerful front claws are used to excavate ant hills and termite mounds in search of food, captured by sticky saliva on their long, probing tongue.

long fleshy nose

30 scales on tail

16–26 in
40–65 cm

SUNDA PANGOLIN
Manis javanica
Females of this semiarboreal Asian species usually have one offspring, which is carried on the tail when a few days old.

AFRICAN TREE PANGOLIN
Manis tricuspis
This arboreal pangolin from equatorial Africa has pale fur and distinctive three-pointed scales, although the points can become worn with age.

10–17 in
25–43 cm

LONG-TAILED PANGOLIN
Manis tetradactyla
This small pangolin from W. Africa lives high in the forest canopy. It has a prehensile tail two-thirds of its total body length.

12–16 in
30–40 cm

20–30 in
51–75 cm

INDIAN PANGOLIN
Manis crassicaudata
Its overlapping scales and ability to emit a noxious defensive fluid can protect this pangolin even from the attentions of tigers.

broad, rounded scales

18–22 in
45–55 cm

GROUND PANGOLIN
Manis temminckii
The only pangolin in southern and eastern Africa, this species is very secretive. It is nocturnal and walks on four feet, or two if foraging for food.

CARNIVORES

Predominantly meat-eaters, carnivores have a body that is adapted for hunting and their teeth are specialized for grasping and killing prey.

The first known carnivore fossil dates to 55 million years ago. They were small, tree-dwelling, rather catlike mammals, but their descendants, which include some of the largest predators on earth, display a range of forms and lifestyles. The order Carnivora ranges in size from the least weasel, 10 in (26 cm) long, to the southern elephant seal, which measures up to 16 ft (5 m) from nose to tail. The order includes the world's fastest land animal, the cheetah, and the famously idle giant panda. Carnivores are naturally present on every continent except Australia, where they have been introduced by man. They are not restricted to dry land—the 34 species of seal, sea lion, and walrus are more at home in the oceans.

CHARACTERISTICS

With such diversity, it can be difficult to ascertain what the animals of this order have in common. The most important shared characteristic of the group is their teeth. All members of Carnivora have four long canine teeth and a distinctive set of cheek teeth known as carnassials, which are modified for cutting meat. The sharp edges of the carnassials work like a pair of scissor blades when the animal opens and shuts its jaws.

Most carnivores eat at least some meat, but few are exclusively carnivorous. Several species, such as the foxes and raccoons, are omnivores, eating a wide range of plant and animal foods. One species, the giant panda, is almost completely herbivorous, with a diet consisting of mainly bamboo.

SOLITARY OR SOCIAL

Carnivores may lead solitary lives, like most weasels and bears, or be highly social, like wolves, lions, and meerkats. Animals in the latter category live in highly organized co-operative groups, sharing responsibility for hunting, rearing young, and protecting their territory. Seals and sea lions are generally colonial during the breeding season, when they are obliged to return to dry land to mate and give birth; some species gather in hundreds or even thousands on favoured beaches.

PHYLUM	CHORDATA
CLASS	MAMMALIA
ORDER	CARNIVORA
FAMILIES	16
SPECIES	288

DEBATE

PINNIPED OR CARNIVORE?

Taken at face value, it seems unlikely that the aquatic seals, sea lions, and walruses, known as pinnipeds from the Latin for "web-footed," could belong to the same group as weasels and wildcats. But the structure of their skull and teeth, and the information coded in their DNA, tells a different story. Pinnipeds have limbs modified for swimming, but—unlike whales—they are not wholly aquatic and must return to land in order to breed. Fossil and molecular evidence suggests that seals, sea lions, and the walrus share a common bear- or weasellike ancestor, which diverged from the other carnivores about 23 million years ago.

DOGS, FOXES, AND RELATIVES

The Canidae are medium-sized, long-legged mammals, most of which have a bushy tail and erect ears. They are swift, intelligent predators, though most also eat plant foods. The highly social gray wolf is the ancestor of the domestic dog, a species domesticated by humans more than 14,000 years ago, and now hugely varied in size and form.

20–30 in
50–75 cm

ARCTIC FOX
Alopex lagopus
This sturdy fox lives in the world's northernmost regions. Its variable color forms include snowy white.

short, pointed face

large ears

15–32 in
38–80 cm

BLANFORD'S FOX
Vulpes cana
A strictly nocturnal species, this fox inhabits steppe country of the Arabian Peninsula and the Middle East. It feeds upon invertebrates and fruits.

yellowish-white underside

18–23½ in
45–60 cm

BENGAL FOX
Vulpes bengalensis
This agile, omnivorous fox inhabits open country in Nepal and India. Pairs remain together from year to year and rear several litters together.

15½–22½ in
39–57 cm

CORSAC FOX
Vulpes corsac
A social, pack-dwelling animal of the Asian steppes, the Corsac fox is an opportunistic hunter of small animals. It also eats vegetable matter.

19–20½ in
48–52 cm

KIT FOX
Vulpes macrotis
Found in the S.W. US, the kit fox is an efficient digger. Families live in burrows with up to 20 entrances.

SWIFT FOX
Vulpes velox
This close relative of the kit
fox occurs in the C. US and
has been reintroduced in
Canada, where it became
extinct in 1938.

18½–22 in
47–55 cm

FENNEC FOX
Vulpes zerda
The distinctive ears of this
tiny, nocturnal N. African
fox serve a dual purpose:
providing keen hearing and
shedding excess body heat.

13–16 in
33–41 cm

pointed ears,
black on top

23–35 in
59–90 cm

14–22 in
35–55 cm

18–23½ in
46–60 cm

RÜPPELL'S FOX
Vulpes rueppellii
Ranging from N. Africa to
Pakistan, this small, social fox
survives desert conditions
by eating a wide range of
plants and animals.

RED FOX
Vulpes vulpes
This highly adaptable hunter
and scavenger is the world's
most widespread carnivore,
occurring throughout most
of the northern hemisphere.

black
lower limb

large, bushy
tail, with a
white tip

BAT-EARED FOX
Otocyon megalotis
Inhabiting open grassland
and scrubby savanna in
S. and E. Africa, this social
fox feeds chiefly on termites
and beetles.

brown-black
body fur

PAMPAS FOX
Lycalopex gymnocercus
A native of temperate
grassland in S. America, this
solitary fox hunts a variety
of small animals and
occasionally sheep.

23½–29 in
60–74 cm

21½–26 in
54–66 cm

19½–28 in
49–70 cm

NORTHERN GRAY FOX
Urocyon cinereoargenteus
This fox is a relatively common inhabitant of forested
areas throughout the Americas, though it avoids
areas where coyotes and bobcats live.

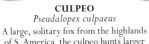

18–36 in
45–92 cm

22½–30 in
57–77 cm

CULPEO
Pseudalopex culpaeus
A large, solitary fox from the highlands
of S. America, the culpeo hunts larger
prey than most other fox species.

CRAB-EATING FOX
Cerdocyon thous
An adaptable omnivore from the grasslands and
temperate and tropical forests of S. America, this
species feeds on fruits, carrion, and small animals.

RACCOON DOG
Nyctereutes procyonoides
This unusual canid inhabits wetlands
in E. Asia. It climbs well and is the
only species of dog to hibernate.

ginger ears

26—41 in
65—105 cm

26—35 in
65—90 cm

26—30¾ in
65—78 cm

GOLDEN JACKAL
Canis aureus
Sometimes described as the archetypal canid, this
swift-footed, social, and opportunistic hunter and
forager is widespread across Africa and Asia.

BLACK-BACKED JACKAL
Lupulella mesomelas
The largest of the African jackals, this
adaptable, omnivorous canid lives in
family groups. It is active day and night.

SIDE-STRIPED JACKAL
Lupulella adusta
A widespread African jackal, this species
is a nocturnal scavenger and hunter. It is
often persecuted by farmers.

33—39 in
84—101 cm

COYOTE
Canis latrans
Native to N. and C. America, this
widespread and common species
dwells in packs and occasionally
mates with the gray wolf.

29—37 in
74—94 cm

ETHIOPIAN WOLF
Canis simensis
Currently listed as the
world's rarest canid, wild
packs of this threatened wolf
survive only in the remote
highlands of Ethiopia.

34—51 in
87—130 cm

3¼—4 ft
1—1.2 m

RED WOLF
Canis lupus rufus
This highly endangered
subspecies from the S.E. US
survives on special reserves,
thanks to captive breeding
in the 1980s.

ARCTIC WOLF
Canis lupus arctcos
Distinguished by a pale coat, this
subspecies of the gray wolf lives in parts
of Canada, Alaska, and Greenland.

thick fur
traps heat

3—5¼ ft
0.9—1.6 m

sharp teeth

GRAY WOLF
Canis lupus
The ancestor of the domestic dog, this
adaptable wolf has a vast distribution,
covering most of the northern hemisphere.

large feet
and claws

44—46 in
112—117 cm

DINGO
Canis familiaris
This feral form of the domestic dog was introduced
to Australia about 4,000 years ago, where it rapidly
became the continent's top predator.

GOLDEN RETRIEVER
Canis familiaris
Bred in Scotland to retrieve game, this dog is loyal and intelligent, and makes an excellent pet. It loves being in water.

2¾–3¼ ft
85–100 cm

25–31 in
63–79 cm

DALMATIAN
Canis familiaris
Initially bred as guard dogs and hunting companions, dalmatians subsequently became coach dogs, escorting horsedrawn carriages. Today they are kept as pets.

plumelike tail

dense coat

31–38 in
79–96 cm

ALASKAN MALAMUTE
Canis familiaris
Bred in Alaska as a sled dog, the malamute resembles the ancestral wolf, and may be among the earliest domestic breeds.

23½ in
60 cm

BASSET HOUND
Canis familiaris
This dog was bred as a tracking animal. Its short legs allow it to move easily through dense vegetation.

MANED WOLF
Chrysocyon brachyurus
An opportunistic omnivore, the maned wolf inhabits S. American savanna, where its long legs help it to see over tall grasses.

large, erect, mobile ears

dark muzzle

19½ in
49 cm

SMOOTH FOX TERRIER
Canis familiaris
An enthusiastic digger, small enough to enter fox burrows, this lively dog was bred to keep farms free of vermin.

rough outer fur with downy undercoat

37–45 in
95–115 cm

long legs have earned the nickname "fox on stilts"

8–12 in
20–30 cm

CHIHUAHUA
Canis familiaris
The smallest of domestic dog breeds, but bold-natured nonetheless, the chihuahua orginated in Mexico. Its coat can be of almost any color.

26–34 in
67–85 cm

ROUGH COLLIE
Canis familiaris
Originally bred as a herding dog in the Scottish Highlands, this dog has an athletic build disguised by a thick coat.

2½–4½ ft
0.8–1.4 m

22½–30 in
57–75 cm

35–55 in
0.9–1.4 m

DHOLE
Cuon alpinus
Known throughout its Asian range as a fierce predator, this wild canid lives and hunts in packs, targeting large mammals, such as deer and goats.

AFRICAN WILD DOG
Lycaon pictus
This endangered canid lives in highly organized packs, hunting and rearing young cooperatively and supporting sick and injured relatives.

BUSH DOG
Speothos venaticus
This distinctive, short-legged species is a predator of the Amazon region. It hunts mainly rodents, working alone or in a pack.

POLAR BEAR
Ursus maritimus

unlike other bears, the polar bear has a slightly convex "Roman" nose

This magnificent animal is as at home at sea as on land. It is the world's largest terrestrial predator, and its huge bulk can make it seem lumbering. In the water, however, it is transformed into a creature of effortless grace. Polar bears are true nomads and spend much of the year far from land, roaming the frozen expanse of the Arctic Ocean. In summer the melting ice forces them to retreat to land, where they sometimes come into contact with humans. Young polar bears are born in the middle of winter in a den dug by their mother. She scarcely wakes when they are born and nurses them in her sleep for three months, breaking down her own body reserves to create rich, fatty milk. In the spring, the cubs have increased their birth weight dramatically, while the mother is close to starvation. She spends the next two years teaching them to swim, hunt seals, defend themselves, and build their own snow dens. Climate change is threatening the habitat and feeding habits of polar bears, which could result in this species' extinction.

SIZE 6–9¼ ft (1.8–2.8 m)
HABITAT Arctic ice fields
DISTRIBUTION Arctic Ocean; polar regions of Russia, Alaska, Canada, Norway, and Greenland
DIET Mainly seals

FURRY EARS >
The small ears are completely covered with fur to prevent frostbite. Polar bears have good hearing but rely mainly on their sense of smell to locate their prey.

< DARK EYE
The dark eyes and nose are the most conspicuous parts of the bear's body. Polar bears have good eyesight, comparable to that of humans.

∧ CRUSHING BLOW
The front paws are the principal means of dispatching prey. Polar bears can smell seals below the ice and break into their lairs by rising up on their hind legs and crashing down with their forepaws. Then, they grab the seal and pull it out.

∧ INSULATED PAWS
Polar bears have fur on the soles of their feet, which insulates them against the cold. The furry soles also keep them from slipping when walking on ice.

∧ STUBBY TAIL
The polar bear has little use for a long tail, so the appendage is reduced to a stub—all but hidden in the dense fur.

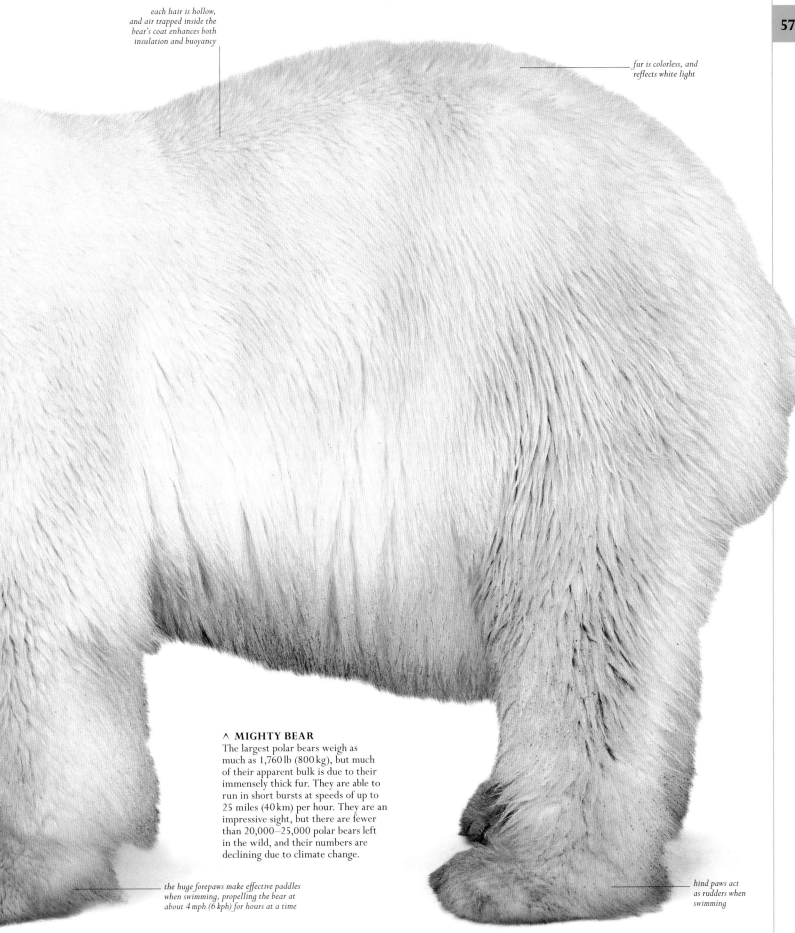

each hair is hollow, and air trapped inside the bear's coat enhances both insulation and buoyancy

fur is colorless, and reflects white light

∧ **MIGHTY BEAR**

The largest polar bears weigh as much as 1,760 lb (800 kg), but much of their apparent bulk is due to their immensely thick fur. They are able to run in short bursts at speeds of up to 25 miles (40 km) per hour. They are an impressive sight, but there are fewer than 20,000–25,000 polar bears left in the wild, and their numbers are declining due to climate change.

the huge forepaws make effective paddles when swimming, propelling the bear at about 4 mph (6 kph) for hours at a time

hind paws act as rudders when swimming

BEARS

The Ursidae are large, stocky-yet-agile animals native to Europe, Asia, and the Americas. Most bears are omnivorous, with plant material making up most of their diet. However, the polar bear is a specialized meat-eater and the giant panda is almost entirely herbivorous. Bears live alone except for mothers caring for their young.

BROWN BEAR
Ursus arctos
This bear's vast range, including Asia, N. Europe, and N. America, testifies to its dietary flexibility. It eats seasonal foods, including berries and spawning salmon.

4–6 ft
1.2–1.8 m

5–9¼ ft
1.5–2.8 m

white face with black eyes and ears

five claws grow to 4 in (10 cm)

GIANT PANDA
Ailuropoda melanoleuca
This endangered bear lives in central Chinese forests. Despite being a carnivore, it eats mostly bamboo; this lack of adequate nutrition results in a low-energy lifestyle.

EARED SEALS

Members of this family are distinguished from true seals by their small external ears and by their limbs, which can be used for moving on land—they are able to walk on their flippers, albeit in an ungainly manner. The Otariidae are excellent swimmers, but their dives are short and shallow compared with those of some true seals. They occur in most oceans except the North Atlantic.

STELLER'S SEA LION
Eumetopias jubatus
Also known as the northern sea lion, this North Pacific species is the largest of the eared seals. It eats mainly fishes, but may take smaller seals.

thick neck

black flippers

6½–11 ft
2–3.3 m

JUVENILE

ADULT

dark brown or black pup

NEW ZEALAND SEA LION
Phocarctos hookeri
Found only in the waters around New Zealand, this rare species breeds only on a few offshore islands.

6–8¾ ft
1.8–2.7 m

4¼–8¼ ft
1.3–2.5 m

AUSTRALIAN SEA LION
Neophoca cinerea
This relatively rare species has breeding colonies restricted to Western and South Australia. Small groups stay together even outside the breeding season.

6–8½ ft
1.8–2.6 m

SOUTHERN SEA LION
Otaria bryonia
Thickset with a blunt face, this resident of S. America and the Falkland Islands in the South Atlantic occasionally hunts cooperatively and may enter rivers in search of fishes.

streamlined body tapers from shoulder to tail

6½–7¾ ft
2–2.4 m

dog-like muzzle with whiskers

CALIFORNIA SEA LION
Zalophus californianus
Widely recognized as the "performing seal," this species is agile on land and in water, often leaping partly clear of the surface.

NORTHERN FUR SEAL
Callorhinus ursinus
Except when breeding, this species lives far offshore in the North Pacific. Males can be over five times heavier than females.

5–7 ft
1.5–2.1 m

AMERICAN BLACK BEAR
Ursus americanus

With about 850,000–950,000 individuals across varied habitats in N. America, this is the world's most common bear. Some have brown or blond fur.

4–6¼ ft
1.2–1.9 m

ASIATIC BLACK BEAR
Ursus thibetanus

A widespread forest-dweller, this bear is variable in appearance, habitat, and behavior. In the tropics, only pregnant females hibernate.

3½–6¼ ft
1.1–1.9 m

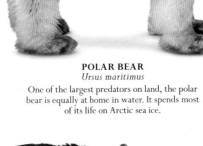

all-white body

relatively long neck

partially furred paw-pads provide extra grip on ice

6–9¼ ft
1.8–2.8 m

POLAR BEAR
Ursus maritimus

One of the largest predators on land, the polar bear is equally at home in water. It spends most of its life on Arctic sea ice.

4½–6¼ ft
1.4–1.9 m

SLOTH BEAR
Melursus ursinus

This shaggy Indian bear can live in a range of habitats. It uses its huge claws to open termite mounds, sucking up the panicking insects.

SUN BEAR
Helarctos malayanus

This shy S.E. Asian forest bear eats insects, honey, fruits, and plant shoots. It is active by day, becoming nocturnal if disturbed.

3¼–5 ft
1–1.5 m

4¼–6¼ ft
1.3–1.9 m

SPECTACLED BEAR
Tremarctos ornatus

A vulnerable species of Andean cloud forests, this bear is an excellent climber. Its varied diet includes fruits, shoots, and meat.

BROWN FUR SEAL
Arctocephalus pusillus

This species has two distinct populations—one South African, the other Australian. Both have suffered from excessive hunting.

velvety fur traps warm air next to the skin

4–7½ ft
1.2–2.3 m

3½–5¼ ft
1.1–1.6 m

GALAPAGOS FUR SEAL
Arctocephalus galapagoensis

The smallest of the eared seals, this species is also the least variable, with males only slightly larger than females.

pointed muzzle

NEW ZEALAND FUR SEAL
Arctocephalus forsteri

This seal breeds on the rocky coasts of New Zealand and Australia. Now protected by law, it is no longer hunted and its numbers are increasing.

5–6¼ ft
1.5–1.9 m

massive neck with a coarse mane

GUADALUPE FUR SEAL
Arctocephalus townsendi

Distinguished by its long, tapering nose, this fur seal breeds on rocky beaches and caves accessible only from the sea.

3½–6½ ft
1.1–2 m

SOUTH AMERICAN FUR SEAL
Arctocephalus australis

This voracious predator of fishes, squid, and crustaceans breeds on the rocky beaches of S. America and the Falkland Islands.

5–6¼ ft
1.5–1.9 m

4–6½ ft
1.2–2 m

ANTARCTIC FUR SEAL
Arctocephalus gazella

This species breeds on the scattered islands of the Southern Ocean. It is recovering from excessive hunting in the past.

fore flippers

CALIFORNIA SEA LION
Zalophus californianus

The California sea lion lives in polar, temperate, and subtropical waters and on land, and therefore needs to cope with a wide range of environmental conditions. The waters in which it feeds are cold but the beaches on which this sea lion breeds can be hot, so it needs an efficient way of regulating its internal temperature. Its body has an insulating layer of blubber 1in (2.5cm) thick to help keep the sea lion warm when hunting in water, and if it becomes too hot on land, it can cool down by returning to the sea. The flippers, which have little fur and lack blubber, have a heat exchange system that can retain or expel body heat by transferring it between arteries and veins at the flipper bases and by controlling blood flow to the skin. The fore flippers may also be waved in the air to help with cooling—a behavior seen both in water and on land. California sea lions are hunted by sharks and orcas.

flexible back gives the sea lion agility and maneuverability when hunting fishes

streamlined body reduces drag

EAR >
Unlike seals, sea lions have small, pointed, external ear flaps that fold down and back against the head. They can hear well both on land and under water.

∧ DIVING
Dives last for up to ten minutes during which the sea lion's heart rate falls to about 20 beats per minute.

gently sloping forehead

black, leathery nose pad

∧ MOUTH
The mouth contains 34 teeth. Initially white, they turn black with increasing age. The four large, recurved canines are used to grasp and hold onto prey.

∧ NOSE
A sea lion's nostrils are only opened when inhaling or exhaling. The sensory whiskers help in working out the size and shape of objects when under water.

∨ FORE FLIPPER
Swimming is powered by the long fore flippers, which enable the sea lion to achieve speeds of over 13 mph (21 kph).

HIND FLIPPER >
In water, the hind flippers trail behind the body and are used for steering, but for moving on land they are rotated forward and used with the front flippers.

> FEMALE SEA LION
California sea lions are sexually dimorphic. Females are smaller and less muscular, particularly around the neck and shoulders, and their fur is lighter. They also have a gently sloping forehead compared to the more upright one of males.

SIZE 6½–7¾ ft (2–2.4 m)
HABITAT Coasts and offshore waters
DISTRIBUTION Eastern North Pacific
DIET Mainly fishes

sea lions can see well on land and even better
under water, helping them to locate prey

short, coarse fur

sensory whiskers
(vibrissae) grow up
to 8 in (20 cm) long

long, ivory tusks

WALRUS

The walrus is the only member of the family Odobenidae. This enormous Arctic pinniped has a huge, blubbery body, long tusks in both sexes, and a mustache of sensitive whiskers for detecting food. Thousands of walruses often haul out of the water in large groups on beaches and ice floes.

8–11 ft
2.5–3.5 m

thick, creased skin

WALRUS
Odobenus rosmarus
The walrus frequents shallow waters in the Arctic. The male walrus is twice the size of the female, which it attracts using underwater bellows.

paddlelike front flippers

blunt body tapering at tail

EARLESS SEALS

Also called true seals, the Phocidae are more adapted to life in water than their cousins, walruses and sea lions. In place of ear flaps, they have tiny holes on the sides of their head. Their flippers are useless on land, but propel them with speed and agility in water. Mostly inhabiting cool temperate and polar waters, seals prey on fishes and invertebrates, while leopard seals hunt penguins.

5½–8¼ ft
1.7–2.5 m

ROSS SEAL
Ommatophoca rossii
This rare, fast-swimming seal spends most of its life hunting squid beneath Antarctic pack ice. Males are smaller than females.

small head

short flippers

few, short whiskers

9¼–10 ft
2.8–3 m

LEOPARD SEAL
Hydrurga leptonyx
This fearsome predator mostly ambushes smaller seals, fishes, and penguins at the edge of Antarctic pack ice, though it also eats krill.

8¼–11 ft
2.5–3.4 m

WEDDELL SEAL
Leptonychotes weddellii
The most southerly dwelling wild mammal, this expert diver specializes in long, deep dives under Antarctic ice shelves. Females may be larger than males.

6½–8½ ft
2–2.6 m

GRAY SEAL
Halichoerus grypus
The gray seal lives in the North Atlantic, with large numbers breeding around Britain. Males are up to three times heavier than females.

5¼–7½ ft
1.6–2.3 m

BEARDED SEAL
Erignathus barbatus
This large, Arctic seal feeds on bottom-dwelling fishes and invertebrates, which it locates partly by touch, using its long, stiff whiskers.

6–7¾ ft
1.8–2.4 m

5¼–5½ ft
1.6–1.7 m

short flippers

HAWAIIAN MONK SEAL
Monachus schauinslandi
Along with its Mediterranean relative, this is one of only two surviving, highly endangered and protected species of monk seals. Fewer than 1,000 of this species remain.

HARP SEAL
Pagophilus groenlandicus
This small seal migrates south in winter and north in summer in noisy groups, following the Arctic pack ice, on which it hauls out and rests.

6½–7¾ ft
2–2.4 m

CRABEATER SEAL
Lobodon carcinophaga
Despite its name, this agile Antarctic seal eats
mainly krill, which it filters from the water,
using specially modified teeth.

6½–8¾ ft
2–2.7 m

**HOODED
SEAL**
Cystophora cristata
This solitary, Arctic seal has
an unusual, inflatable nose that
droops over its mouth. The
pups become independent
at just five days old.

inflatable muzzle of male

7–13 ft
2.1–4 m

10–16 ft
3–5 m

**SOUTHERN
ELEPHANT SEAL**
Mirounga leonina
The huge male of this
Southern Ocean species has
a trunklike nose and is the
largest carnivore, weighing
up to 6,610 lb (3,000 kg).

♀

NORTHERN ELEPHANT SEAL ♂
Mirounga angustirostris
This is a large seal of the North Pacific. The male has a
long, trunklike nose. Like its southern relative, it was
almost hunted to extinction, but is now recovering.

5–6 ft
1.5–1.8 m

SPOTTED SEAL
Phoca largha
This small seal is mainly found on
ice floes off the northern coasts of
Siberia and Yukon, Canada. Adults
form stable pairs in order to breed.

*large eyes set
back on head*

4–6½ ft
1.2–2 m

3½–5¼ ft
1.1–1.6 m

*spots, rings, and
blotches on body*

**COMMON
SEAL**
Phoca vitulina
Also known as the
harbor seal, this dainty seal
is widespread along temperate
coasts. It hauls on to sandy
beaches and sheltered reefs.

RINGED SEAL
Pusa hispida
This small seal mainly inhabits
Arctic ice shelves. Pups are born
in lairs beneath the ice to protect
them from predators.

4–5 ft
1.2–1.5 m

BAIKAL SEAL
Pusa sibirica
This small, freshwater
seal is found in Lake
Baikal, Siberia. In winter,
it uses its teeth and claws
to maintain breathing
holes in the ice.

5 ft
1.5 m

CASPIAN SEAL
Pusa caspica
About 100,000 of this small species live in the
Caspian Sea. Unusually, the male is believed to take
only one mate, without fighting off other males.

8–12½ in
20–32 cm

SKUNKS AND RELATIVES

This small group of cat-sized mammals from the Americas is named for its most distinctive behavior—that of squirting foul-smelling musk at an aggressor. The family name, Mephitidae, comes from the Latin word for "bad smell."

RACCOONS AND RELATIVES

Raccoons, kinkajous, and olingos of the family Procyonidae are nimble New World carnivores. Most species are omnivorous and eat mainly plant material—especially fruits—and also insects, snails, small birds, and mammals. The raccoon is the largest of the group.

PALAWAN STINK BADGER
Mydaus marchei

This bumbling cousin of the American skunks is found only on the Philippine islands of Palawan and Calamian. It eats mainly invertebrates.

12½–19½ in
32–49 cm

HUMBOLDT'S HOG-NOSED SKUNK
Conepatus humboldtii

A native of S. Chile and Argentina, this small skunk detects underground invertebrate prey by scent and digs it up.

SOUTH AMERICAN COATI
Nasua nasua

This species lives in loose, female-dominated colonies. An agile climber, it mainly eats fruits, supplemented out of season with animal prey.

EASTERN SPOTTED SKUNK
Spilogale putorius

A relatively small, weasellike skunk from the E. US, this species is more agile than other skunks and climbs well.

11–12 in
28–31 cm

HOODED SKUNK
Mephitis macroura

This widespread C. American skunk occupies a range of habitats, feeding opportunistically on fruits, eggs, and small animals.

17–27 in
43–68 cm

17–23 in
43–58 cm

7½–13 in
19–33 cm

white stripe warns predators of foul secretions

WHITE-NOSED COATI
Nasua narica

This social C. American omnivore spends the day foraging at ground level. It climbs well, however, and often sleeps in trees.

long hairs on tail and back rise when animal is alarmed

STRIPED SKUNK
Mephitis mephitis

This nocturnal omnivore ranges from Canada to Mexico. It does not hibernate but can experience periods of torpor in winter.

6½–16 in
17–40 cm

tail has faint dark rings

RED PANDA

This tree-dwelling, herbivorous mammal is now placed in its own family, the Ailuridae. Previously, zoologists had classified it with the bears but now it is thought to be most closely related to the group that includes skunks, raccoons, and weasels.

WEASELS AND RELATIVES

Members of the family Mustelidae occur throughout Eurasia, Africa, and the Americas. They are characterized by a sinuous body and short legs, though the badgers and wolverines are more robust. Most are active predators and good swimmers.

8–14 in
20–36 cm

EUROPEAN MINK
Mustela lutreola

A semiaquatic predator, this species was once widespread in C. and W. Europe, but is now much less common than the introduced American mink.

12–17 in
30–43 cm

AMERICAN MINK
Neovison vison

A fierce and stealthy predator, this mink is also a skilled swimmer. It has been introduced worldwide by fur farmers.

RED PANDA
Ailurus fulgens

This raccoonlike mammal inhabits temperate forest in the Himalayas. It eats bamboo, fruits, small animals, and birds' eggs.

20–30¾ in
51–78 cm

tail has conspicuous black tip

12–16 in
30–40 cm

KINKAJOU
Potos flavus
A nocturnal tree-dweller from
C. and S. America, the kinkajou
uses its long tongue to pluck fruits
and collect honey from bee nests.

prehensile tail

16–30 in
41–76 cm

OLINGUITO
Bassaricyon neblina
Shy, nocturnal fruit-eaters, the olinguito is one
of four species of olingo. It lives in the Andean
cloud forests of Ecuador and Colombia.

*long fur is
grizzled*

NORTHERN RACCOON
Procyon lotor
This adaptable opportunist
favors wooded and
scrubby habitats across
N. America, but often
thrives in towns by
scavenging human refuse.

*black "bandit
mask" around eyes*

12–14½ in
30–37 cm

17½–24 in
44–62 cm

RINGTAIL
Bassariscus astutus
An agile omnivore from
C. America, this species
forages by night for fruits and
small animal prey, pausing
often to scent-mark territory.

8–18 in
20–46 cm

**EUROPEAN
POLECAT**
Mustela putorius
This lively nocturnal species,
found in forests and meadows of
C. and W. Europe, is the ancestor
of the domestic ferret.

8–10 in
20–26 cm

*slender,
elongated
neck*

4¼–10 in
11–26 cm

LEAST WEASEL
Mustela nivalis
The smallest carnivore, yet
a fierce and highly successful
predator, this weasel specializes
in hunting mice.

pointed muzzle

7½–13½ in
19–34 cm

STOAT
Mustela erminea
This lithe and ferocious little
predator is found throughout
much of the northern hemisphere.
Its most northerly populations
turn white in winter.

16–20 in
40–50 cm

BLACK-FOOTED FERRET
Mustela nigripes
Extinct in the wild in the late 20th
century, this slender burrowing ferret
has been reintroduced to reserves
in the midwestern US.

**LONG-TAILED
WEASEL**
Mustela frenata
A widespread species
of the Americas, this
weasel hunts mice and
voles. Northern individuals
turn white in winter.

» WEASELS AND RELATIVES

four white stripes from head to tail

22–28 in
55–70 cm

HOG BADGER
Arctonyx collaris
Native to S.E. Asia, this badger uses its elongated snout to root around for edible morsels on forest floors.

22–35 in
56–90 cm

29–38 in
74–96 cm

AMERICAN BADGER
Taxidea taxus
This stout, burrowing animal inhabits grassland and woodland throughout central N. America. It eats a huge variety of plants and animals.

16½–28 in
42–72 cm

white stripe from nose, down the back to the rump

EUROPEAN BADGER
Meles meles
This stocky badger inhabits wooded terrain across much of Europe and Asia. It lives communally in extensive burrow systems called setts.

HONEY BADGER
Mellivora capensis
An exceptionally feisty animal, this species lives in W. and S. Asia, and Africa. It raids beehives for honey, but also eats termites, scorpions, and porcupines.

18½–22 in
47–55 cm

GREATER GRISON
Galictis vittata
This adaptable omnivorous weasel with badgerlike markings is found in tropical forest and grassland in C. and S. America.

26–41 in
65–105 cm

SABLE
Martes zibellina
This fierce predator from the forests of Siberia, China, and Japan is hunted in turn by humans for its luxuriantly soft, silky fur.

14–22 in
35–56 cm

WOLVERINE
Gulo gulo
This large mustelid is widespread in N. America and Eurasia. Its voracious feeding habits have earned it its other name, "glutton."

dense, dark brown coat

flattened, tapering tail

18–23 in
45–58 cm

16–21½ in
40–54 cm

18–26 in
45–65 cm

FISHER
Martes pennanti
Found in dense N. American forests, this large marten rarely eats fishes, and is one of the few predators to tackle porcupines.

EUROPEAN PINE MARTEN
Martes martes
Present in forests throughout much of Europe, this lively hunter is seldom seen, as it is nocturnal and wary of humans.

BEECH MARTEN
Martes foina
Widespread in Eurasia, this marten emerges from a rock crevice or hollow log at dusk in search of small mammals, birds, and seasonal fruits.

AFRICAN ZORILLA
Ictonyx striatus
A striped polecat, this African species hunts a range of prey by night. It rests by day in hollow logs or burrows.

11–15 in
28–38 cm

dorsal stripes meet at the tail

AFRICAN STRIPED WEASEL
Poecilogale albinucha
Found in C. and southern Africa, this species lives in self-dug burrows. It emerges at night to hunt for smaller animals, especially rodents, which it tracks by scent.

9½–13 in
24–33 cm

29–35 in
73–88 cm

14–18½ in
36–47 cm

long front claws to dig up buried insects

3¼–4¼ ft
1–1.3 m

GIANT OTTER
Pteronura brasiliensis
This S. American predator needs 2¼ lb (3 kg) of fish a day to sustain itself. About 1,000–5,000 individuals of this endangered species remain.

AFRICAN CLAWLESS OTTER
Aonyx capensis
Found alongside water in forests and wetland throughout much of sub-Saharan Africa, this large otter eats mainly crabs, frogs, and fishes.

grayish-white flecks on face and throat

ASIAN SMALL-CLAWED OTTER
Aonyx cinereus
The world's smallest otter, this species lives in wetlands in India and S.E. Asia. Its survival is threatened by habitat loss and pollution.

short, blunt claws

20–32 in
50–82 cm

23–29 in
58–73 cm

3¼–4 ft
1–1.2 m

EURASIAN OTTER
Lutra lutra
Eurasian otters thrive in both river and coastal habitats, as long as they have access to fresh water for drinking and washing.

NORTH AMERICAN RIVER OTTER
Lontra canadensis
Widespread in N. America, this species inhabits well-vegetated rivers and lake shores. It eats mainly fishes and crayfish, but also hunts small land animals.

SEA OTTER
Enhydra lutris
The sea otter hunts fishes and shellfish in the cool North Pacific, relying on its incredibly dense fur to keep warm.

CATS

Members of the cat family, or Felidae, are among the most specialized of meat-eaters, and many species consume no vegetable material at all. As a group, cats are supremely athletic, having a supple, muscular body that is well adapted to running, climbing, leaping, and swimming. Their short jaws contain sharp teeth adapted for stabbing (canines) and slicing (carnassials). They have retractile claws.

LEOPARD
Panthera pardus

This most adaptable of the big cats occurs widely throughout Africa and southern Asia. It often hides its prey in trees, away from other hunters.

3–6¼ ft
0.9–1.9 m

INDOCHINESE CLOUDED LEOPARD
Neofelis nebulosa

A large, nocturnal, S.E. Asian forest cat with cloud-shaped coat markings, this species is declining due to hunting and habitat loss.

26–42 in
67–107 cm

BLACK LEOPARD
Panthera pardus

Melanistic black coloration is not uncommon among leopards. Also known as black panthers, they are found mainly in dense, moist forests in S.E. Asia.

3–6¼ ft
0.9–1.9 m

JAGUAR
Panthera onca

The America's only big cat, the jaguar is an excellent climber and swimmer. Its broad prey base includes deer, turtles, and fishes.

4–5½ ft
1.2–1.7 m

TIGER
Panthera tigris

The world's largest cat hunts using stealth and raw power to take prey as large as oxen. Fewer than 3,900 remain in the wild in Asia.

4½–9½ ft
1.4–2.9 m

LION
Panthera leo

Africa's top predator lives in family groups known as prides. Females work together to bring down prey including zebra and antelope.

♂

♀

thick mane

5¼–8¼ ft
1.6–2.5 m

3–4 ft
0.9–1.2 m

SNOW LEOPARD
Uncia uncia
At home in the remote high mountains of C. Asia,
the snow leopard lives alone and hunts wild sheep
and goats, deer, and marmots.

32–43 in
80–110 cm

**EURASIAN
LYNX**
Lynx lynx
This large lynx is big
enough to tackle small deer.
One kill will feed an
individual for about a week.

27–32 in
68–82 cm

IBERIAN LYNX
Lynx pardinus
Now being bred in
captivity, this lynx is
probably the world's
most endangered cat,
with fewer than 400
remaining wild
in Spain.

large, tufted ears

short tail

24–42 in
61–106 cm

CARACAL
Caracal sp.
Nocturnal hunters of
medium-sized prey, such
as hyraxes and small
antelope, caracals occupy
dry scrublands in Africa
and S.W. Asia.

26–41 in
65–105 cm

BOBCAT
Lynx rufus
This adaptable stalk-and-pounce predator is named
for its short "bob" tail. It occurs throughout
N. America and hunts mainly rabbits.

small head with
high-set eyes

21–26 in
53–67 cm

30–42 in
76–107 cm

BAY CAT
Catopuma badia
This rare cat has two color morphs, gray
and the more common red. It is found
only on the island of Borneo.

**ASIAN
GOLDEN CAT**
*Catopuma
temminckii*
This large, golden
brown, occasionally
spotty, cat inhabits
forested parts of
S.E. Asia. Pairs
cooperate to hunt
and rear young.

26–41¼ in
66–105 cm

CANADIAN LYNX
Lynx canadensis
The Canadian lynx inhabits
dense forests and tundra. Its
numbers fluctuate with the
availability of its preferred
prey, the snowshoe hare.

CHEETAH
Acinonyx jubatus
The fastest animal on four
legs, reaching 63 mph
(102 kph), the cheetah uses
speed to catch antelope on
the African savanna.

4–5 ft
1.2–1.5 m

long tail
aids balance

furry ears can swivel independently to scan surroundings for sounds of prey or danger

TIGER
Panthera tigris

The largest and most striking of the big cats, the tiger is a powerful predator with almost supernatural grace and agility. Its natural range extends from the tropical jungles of Indonesia to the snowy expanses of Siberia, where the largest individuals are found. A full-grown male may weigh anything up to 660 lb (300 kg), but despite this bulk it can leap up to 33 ft (10 m) in a single bound. Adult tigers live alone, except for females with cubs—mothers watch over their young for two years or more, while teaching them vital survival skills.

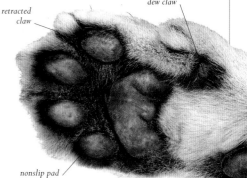

SIZE 4½–9½ ft (1.4–2.9 m)
HABITAT Forests, swamps, scrub thickets, savanna, and rocky landscapes
DISTRIBUTION India to China, Siberia, the Malay peninsula, and Sumatra
DIET Mainly hoofed animals, such as deer and pigs; may also catch smaller mammals and birds

sense of smell is surprisingly poor, although tigers do use scent marks to define their territory

long whiskers allow the tiger to feel its way through dense undergrowth in almost total darkness

ROUND PUPIL >
Unlike small cats, whose pupils contract to vertical slits, tiger pupils are always round. They expand to provide excellent night vision and contract to pinpricks in bright light.

WHITE EAR FLASH ˅
The prominent white spot on the back of each ear is thought to aid communication. Cubs following their mother will notice ear movements that may signal danger.

˄ STAB AND SLICE
Four long canine teeth deliver the killing bite to a tiger's prey. Blade-edged cheek teeth called carnassials slice through the meat with ease.

retracted claw

dew claw

FORELIMBS >
The tiger has long legs and large feet, which allow it to run fast, leap large distances, and knock prey as large as oxen to the ground with a single, deadly swipe.

nonslip pad

˂ KILLER IN STRIPES
A fiery orange coat marked with bold black stripes provides superb camouflage as the tiger moves through sun-dappled vegetation. The white tigers seen in zoos are usually bred in captivity and are extremely rare in the wild. Indeed, tigers have been hunted almost to extinction; globally, there are fewer than 3,900 left in the wild.

˄ PADDED PAW
Tigers have five toes on their front paws—four of which are weight-bearing and the fifth is the dew claw. The claws are withdrawn fully into the paw when not in use.

˂ TAIL END
The long tail is typically held curved just above the ground. The tiger uses it to aid balance when chasing prey or climbing.

≫ CATS

14–20 in
35–50 cm

SIAMESE CAT
Felis catus
This elegant and sociable
breed originated in Thailand.
Siamese kittens are born
cream and develop their dark
extremities as they grow.

14–20 in
35–50 cm

TABBY CAT
Felis catus
The tabby is not a breed but a coat pattern
found in many breeds. It resembles the
pattern seen in the ancestral wild cat.

14–20 in
35–50 cm

SPHYNX CAT
Felis catus
Hairless except for a peachy fuzz,
sphynx cats were developed in
Canada. As they feel the cold, they
are most often kept as house cats.

14–20 in
35–50 cm

CORNISH REX
Felis catus
The Cornish rex
has a coat made up
only of down hairs.
The curl of the fur
is caused by a
genetic mutation.

18½–26 in
47–66 cm

**EUROPEAN
WILD CAT**
Felis silvestris silvestris
This elusive but ferocious
predator is declining due
to persecution, habitat
loss, and interbreeding
with feral domestic cats.

PERSIAN CAT
Felis catus
A long-established and popular
breed, this domestic cat is
characterized by long hair
and a short muzzle.

14–20 in
35–50 cm

14–20 in
35–50 cm

MANX CAT
Felis catus
Short-tailed cats appeared
naturally on the Isle of Man
over 300 years ago. The trait
spread rapidly in the small
island population.

*yellowish-gray
to reddish-
brown fur*

18½–29 in
47–74 cm

INDIAN DESERT CAT
Felis lybica ornata
This small felid, also known
as the Asian steppe wild cat,
can be distinguished from
the two African subspecies
by its brown-spotted,
sandy-gray coat.

JUNGLE CAT
Felis chaus
A large and relatively common wild cat occuring
from Egypt to Indonesia, the jungle cat prefers
grassland and swampy habitats.

24–34 in
61–85 cm

SAND CAT
Felis margarita
A small desert specialist found in N. Africa, Arabia, and Kazakhstan, the sand cat hunts gerbils and other nocturnal rodents.

15½–20½ in
39–52 cm

BLACK-FOOTED CAT
Felis nigripes
An opportunistic solitary hunter, the black-footed cat is threatened by persecution and habitat loss in its native southern Africa.

14–20½ in
36–52 cm

18–26 in
46–65 cm

SERVAL
Leptailurus serval
Found in grassland habitats throughout much of Africa, this distinctive-looking cat is an agile predator of small mammals.

23–36 in
59–92 cm

PALLAS'S CAT
Otocolobus manul
This short-legged cat lives in the stony deserts of C. Asia. Its coat provides useful camouflage when stalking pikas, gerbils, and grouse.

18–24 in
45–62 cm

MARBLED CAT
Pardofelis marmorata
A rare forest species adapted to life in the trees, the S.E. Asian marbled cat is an expert climber and preys mainly on birds.

14–19 in
35–48 cm

pointed ears

RUSTY-SPOTTED CAT
Prionailurus rubiginosus
This lively cat from India and Sri Lanka stalks most of its prey on the ground, but is also an excellent climber.

retractile claws

22½–45 in
57–115 cm

FISHING CAT
Prionailurus viverrinus
A large cat with a patchy distribution in S. and S.E. Asia, the endangered fishing cat also eats waterfowl and land animals.

18–20½ in
45–52 cm

FLAT-HEADED CAT
Prionailurus planiceps
This unusual water-loving cat from S.E. Asia hunts mainly fishes and crustaceans, which it finds by dipping in its head or groping with its paws.

»

sandy colored coat

round head with erect ears

large canine teeth for killing prey

19½–33 in
49–83 cm

2¾–5¼ ft
0.9–1.6 m

long hind legs for exceptional sprinting and leaping power

PUMA
Puma concolor
Also known as the cougar, panther, and mountain lion, the puma occupies rugged terrain across a vast range from Canada to Argentina.

JAGUARUNDI
Herpailurus yagouaroundi
One of the largest and most widespread S. American cats, the jaguarundi hunts small mammals by day in diverse habitats.

HYENAS AND AARDWOLF

The small family of Hyaenidae contains three species of scavenging hyena and the aardwolf, which specializes in eating insects. Hyenas are characterized by a stocky, doglike body with short hind legs and powerful bone-crushing jaws. They are highly intelligent and live in family groups, called clans.

AARDWOLF
Proteles cristata
This dainty, weak-jawed cousin of the hyenas eats only insects. It lives in southern and E. Africa in the dry grassland favored by termites.

22–31 in
55–80 cm

powerful neck and forequarters

SPOTTED HYENA
Crocuta crocuta
This efficient scavenger is also a proficient hunter of ungulates. It lives in unforested areas of sub-Saharan Africa.

4¼–5¼ ft
1.3–1.6 m

3¼–4 ft
1–1.2 m

STRIPED HYENA
Hyaena hyaena
The small striped hyena is found in open country from N. Africa to India. It eats a mixed diet of scavenged meat, small prey animals, and fruits.

3½–4½ ft
1.1–1.4 m

BROWN HYENA
Hyaena brunnea
This social southern African scavenger survives desert conditions by foraging at night and supplementing its diet with watery fruits.

COLOCOLO
Leopardus colocolo
Highly adaptable, the mainly nocturnal colocolo is found in a wide variety of habitats in S. America, from forest to grassland and swamp.

16½–31 in
42–79 cm

23–25 in
58–64 cm

17–35 in
43–88 cm

spotted coat varies from gray to golden

15–22 in
38–56 cm

ANDEAN CAT
Leopardus jacobita
This extremely rare species is restricted to remote uplands, where it is seldom seen by humans. It eats chinchillalike rodents called viscachas.

GEOFFROY'S CAT
Leopardus geoffroyi
An adaptable hunter of small mammals, fishes, and birds, Geoffroy's cat stalks grasslands, forests, and wetlands from Bolivia to S. Argentina.

17–31 in
43–79 cm

large eyes adapted to nocturnal lifestyle

ONCILLA
Leopardus tigrinus
This spotted forest-dweller is widespread from Costa Rica to Argentina, where it hunts rodents, opossums, and birds. It is solitary and nocturnal.

lithe body adapted for climbing

MARGAY
Leopardus wiedii
Seldom seen because of its rarity and preference for dense forest cover, the margay is found from Mexico to northern S. America.

28–39 in
72–100 cm

OCELOT
Leopardus pardalis
Ocelots occupy forested parts of C. and S. America. They hunt nocturnally, targeting rodents and other prey on land and in water.

strong claws are hidden within fleshy sheaths

MALAGASY CARNIVORES

The large island of Madagascar has been separate from other land masses for the last 88 million years, allowing its resident mammals to evolve independently. Its native carnivores are now classified in their own family, the Eupleridae. They are very varied in appearance, having diversified to fill the niches occupied elsewhere by predators such as cats, weasels, and mongooses.

FOSSA
Cryptoprocta ferox
The catlike fossa is the largest Madagascan carnivore. It hunts mainly lemurs, but it will eat almost any small animal it can catch.

12–15 in
30–38 cm

stocky body

23½–32 in
60–80 cm

16–18 in
40–45 cm

18–20 in
45–50 cm

foxlike pointed muzzle

SPOTTED FANALOKA
Fossa fossana
A small civetlike animal, the spotted fanaloka inhabits Madagascar's wet forests. It hunts invertebrate prey on land and in water.

FALANOUC
Eupleres goudotii
A ground-dwelling, nocturnal forest animal, the falanouc is an adept digger. It uses its large feet to unearth invertebrate prey.

RING-TAILED VONTSIRA
Galidia elegans
The Madagascan equivalent of mongooses, vontsiras are active forest animals. They feed opportunistically on most plant and animal foods.

∨ DEMOLITION EXPERT

This tough scavenger can reduce a carcass to nothing in minutes. It will attempt to devour huge chunks of meat before other scavengers arrive. Then, if time allows, it dismembers the remains and stashes them nearby. Often, all that is left of a herbivore, such as an antelope, is the green stomach contents.

crest of long hairs along the back lies flat when the animal is relaxed

shorter hind legs add to the front-heavy, skulking appearance

SIZE 3¼–4 ft (1–1.2 m)
HABITAT Open country, savanna grassland, and scrub to semidesert
DISTRIBUTION North and East Africa, Middle East to East India
DIET Mainly carrion

muscular shoulders and neck allow the hyena to carry or drag its own bodyweight in carrion

large ears follow sound from any direction

STRIPED HYENA
Hyaena hyaena

Maligned in folklore as skulking, cowardly scavengers, most hyenas are also adept hunters in their own right. The striped hyena is less bold and sociable than its spotted cousin, and, in Africa at least, individuals tend to live a solitary life except when breeding. Elsewhere, such as in Israel and India, striped hyenas more often live in family groups, which are dominated by a single adult female. Carrion forms the bulk of the striped hyena's diet, but it is also partial to fruits, especially melons, which provide valuable water. Striped hyenas are unpopular with farmers, who fear attacks on livestock and damage to crops, but these animals increasingly need conservation, with perhaps fewer than 10,000 remaining in the wild.

∨ THROAT
The black throat patch probably contributes to the animal's camouflage. The thicker hair in this area may provide some protection against serious injury when fighting.

∧ NOSE FOR TROUBLE
Scent is an important sense to hyenas. Individuals often pause in their daily routine to daub scent from a gland under the tail on rocks and tussocks around their home range.

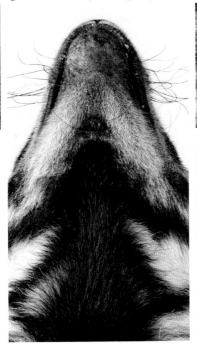

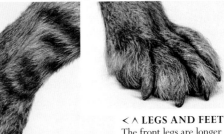

∧ MOUTH AND TEETH
The hyena's stout jaws are operated by immensely powerful muscles. The large cheek teeth (molars and premolars) are easily capable of crushing bones.

< FUR
The striped hyena is smaller than the spotted or brown hyena, and is distinguished by black stripes on its legs and flanks. The pattern aids camouflage in its dusty habitat.

< ∧ LEGS AND FEET
The front legs are longer than the back and terminate in compact paws with four digits on each. The doglike claws are short, blunt, and nonretractile.

MONGOOSES

Mongooses are small, slender-bodied, mostly ground-dwelling carnivores in the family Herpestidae. They live in warm temperate to tropical parts of Africa and Eurasia. Some species live in complex social groups. Together with the Malagasy carnivores, mongooses are thought to be closely related to hyenas.

wedge-shaped head

10–18 in
26–46 cm

YELLOW MONGOOSE
Cynictis penicillata
Inhabiting dry savanna in southern Africa, this mongoose lives in groups led by a dominant male, but forages independently.

WHITE-TAILED MONGOOSE
Ichneumia albicauda
Found in dry habitats in Africa and the southern part of the Arabian Peninsula, this large insect-eater also feeds on vertebrates and ripe berries.

18½–27 in
47–69 cm

sharp, nonretractile claws

6½–9 in
16–23 cm

COMMON DWARF MONGOOSE
Helogale parvula
This small, energetic carnivore forages in packs in African grassland, woodland, and brush country. It eats large invertebrates, such as crickets and scorpions.

12–14½ in
30–37 cm

COMMON CUSIMANSE
Crossarchus obscurus
This forest-dwelling animal from W. Africa lives and hunts in nomadic troops. It is said to make an excellent pet.

SLENDER MONGOOSE
Galerella sanguinea
A widespread species in Africa, the slender mongoose generally lives alone. It is active by day and busiest just before dusk.

12½–13½ in
32–34 cm

MEERKAT
Suricata suricatta
This species lives in groups in semidesert habitats. All members help babysit, maintain the burrow, and take turns standing sentry while others forage.

pointed nose

9½–11½ in
24–29 cm

12–16 in
30–40 cm

BANDED MONGOOSE
Mungos mungo
Found in troops in sub-Saharan woodland, this mongoose lives in dens, often excavated in termite mounds. Males and females can lead their troop deep into a neighboring troop's home range in pursuit of copulations.

18–21 in
45–53 cm

22–24 in
56–61 cm

EGYPTIAN MONGOOSE
Herpestes ichneumon
Not restricted to Egypt, this grizzled gray mongoose occurs in open grassy habitats from Spain to South Africa.

coat marked with broken bands

INDIAN GRAY MONGOOSE
Urva edwardsii
Frequenting forests and plantations, this species often hunts close to human habitation, where it makes itself useful by killing mice and rats.

13–19 in
33–48 cm

INDIAN BROWN MONGOOSE
Urva fusca
This uncommon species occupies the jungles of S. India and Sri Lanka. Like other mongooses, it can kill snakes, but prefers easier prey.

15½–18½ in
39–47 cm

RUDDY MONGOOSE
Urva smithii
This little-known Indian forest-dweller hunts birds, reptiles, and smaller mammals. Its tail is sometimes longer than its body.

slender tail

AFRICAN PALM CIVET

The secretive and nocturnal African palm civet is the sole member of the Nandiniidae, which is thought to have split from civet and catlike ancestors 44.5 million years ago.

16½–28 in
42–71 cm

AFRICAN PALM CIVET
Nandinia binotata
This is a very common but shy tree-dwelling species from C. Africa. Though an omnivore, it eats mostly fruits.

CIVETS, GENETS, AND LINSANGS

With most species possessing a boldly patterned coat, the civets and genets (family Viverridae) and the lingsangs (family Prionodontidae) resemble long-tailed cats, but are less specialized meat-eaters. When threatened, these shy and nocturnal animals squirt a jet of foul-smelling fluid from stink glands near the base of the tail.

24–38 in
61–97 cm

BINTURONG
Arctictis binturong
Possessing a prehensile tail, this S.E. Asian species moves methodically through the forest canopy in search of fruits and small animals.

26–33 in
67–84 cm

AFRICAN CIVET
Civettictis civetta
An opportunistic animal, this large, ground-dwelling omnivore lives alone and marks its territory with a strong, musky scent.

20–34 in
51–87 cm

MASKED PALM CIVET
Paguma larvata
This agile, solitary, tree-dwelling native of Indochina eats fruits, insects, and small vertebrates.

16½–28 in
42–71 cm

ASIAN PALM CIVET
Paradoxurus hermaphroditus
With a natural range extending from Pakistan to Indonesia, this fruit-loving civet is considered a pest in palm and banana plantations.

long tail for balance when climbing trees

rows of black spots

19½–27 in
49–68 cm

SMALL INDIAN CIVET
Viverricula indica
This small, ground-dwelling civet inhabits forests, grasslands, and bamboo thickets from Pakistan to China and Indonesia.

18–20½ in
46–52 cm

COMMON GENET
Genetta genetta
This widespread predator of small mammals and birds occupies scrub and forests of Africa and S. Europe.

17–23 in
43–58 cm

CAPE GENET
Genetta tigrina
This genet is found in eastern South Africa and Lesotho. It eats mostly invertebrates, but will tackle prey as large as geese.

large eyes to see in the dark

soft, velvety fur

21½–30 in
54–77 cm

MALAY CIVET
Viverra tangalunga
Restricted to the tropical forests of Malaysia, Indonesia, and the Philippines, this nocturnal forager mostly hunts for prey on the ground.

BANDED LINSANG
Prionodon linsang
Also called the tiger civet, this shy species lives in tree holes in the S.E. Asian jungle, hunting rats, squirrels, lizards, and birds.

15–18 in
38–45 cm

long, thick tail

ODD-TOED UNGULATES

All members of the Perissodactyla are browsers and grazers of plants. They appear to have little in common, but are linked through a series of extinct intermediate forms.

Modern odd-toed ungulate families range from graceful horses to piglike tapirs and massive rhinoceroses. Unlike the Artiodactyla, they have a relatively simple stomach, and rely on bacteria in the caecum, a pouchlike extension of the large intestine, and colon to digest plant cellulose.

In prehistoric times, odd-toed ungulates were among the most important herbivorous mammals, sometimes becoming the dominant herbivores in grassland and forest ecosystems. For a variety of reasons, including competition with even-toed ungulates, most of these species are now known only from the fossil record.

WEIGHT-BEARING TOES

Odd-toed ungulates rest their weight primarily on the third toe of each foot. Indeed, horses have lost the rest of their toes, and the single remaining digit is protected with a well-developed horny hoof. The other two families have retained more toes—three on all four feet in the case of rhinoceroses, and three on the hind legs and four on the front in the case of tapirs.

HORSE POWER

Once distributed across the world, apart from Antarctica and Australasia, extant species in this order are mainly native to Africa and Asia. Only some of the tapir species are found in the Americas, for although the horse family evolved in that region, it had died out by the end of the Pleistocene around 10,000 years ago. The Spanish conquistadors reintroduced the modern domestic horse to the Americas in the 15th century.

Horses have a long history of domestication, especially as transport and pack animals and to provide power for agriculture and forestry. The first equid to be domesticated appears to have been the donkey, some 5,000 years ago, with horses following 4,000 years ago. Today there are more than 250 horse breeds worldwide.

PHYLUM	CHORDATA
CLASS	MAMMALIA
ORDER	PERISSODACTYLA
FAMILIES	3
SPECIES	18

The black rhinoceros is a browser and has a prehensile upper lip, which it uses to grasp twigs and leaves.

RHINOCEROSES

A huge, barrel-shaped body and a large head bearing one or two horns make the five species of the family Rhinocerotidae unmistakable. While their bulk, horns, and protective skin ensure they have few natural enemies, these mostly solitary animals are threatened by hunting and the destruction of their habitat. These herbivores can ferment food in their hindgut, which enables them to eat woody as well as leafy matter.

3¼–5 ft
1–1.5 m

rough skin with few hairs

SUMATRAN RHINOCEROS
Dicerorhinus sumatrensis
This critically endangered species from S.E. Asian forests is the smallest rhinoceros. The smaller of its two horns is usually a mere stub.

long front horn

4½–5½ ft
1.4–1.7 m

prehensile upper lip

BLACK RHINOCEROS
Diceros bicornis
Smaller but more aggressive than the white rhinoceros, this critically endangered animal of sub-Saharan Africa has a prehensile upper lip to draw twigs and leaves into its mouth.

TAPIRS

Inhabiting tropical forests, the Tapiridae are found in Southeast Asia, and South and Central America. These large herbivores have a short prehensile trunk that can grasp foliage overhead and be used as a snorkel under water. Another distinctive feature is the protruding rump with a short tail. Young tapirs have a striped and spotted coat. Splayed hoofs, with four toes on the front feet and three on the rear, help tapirs walk on soft ground.

dual coloration adds camouflage to body

MALAYAN TAPIR
Tapirus indicus
This striking, two-toned tapir is the largest and only Asian species. Males and females create overlapping trails around their rainforest territory in S.E. Asia.

3¼–4¼ ft
1–1.3 m

short, flexible trunk

white, saddle-shaped marking

2½–4 ft
0.8–1.2 m

2½–4 ft
0.8–1.2 m

32–35 in
80–90 cm

SOUTH AMERICAN TAPIR
Tapirus terrestris
Despite its size, this shy tapir is unobtrusive. It moves easily through jungle undergrowth. It is a favored prey of crocodiles.

BAIRD'S TAPIR
Tapirus bairdii
The largest land mammal in S. America, this tapir favors waterside habitats in dense jungles and swamps for swimming and wallowing.

MOUNTAIN TAPIR
Tapirus pinchaque
The smallest of all tapirs, this species inhabits montane cloud forest in the northern Andes. It has a thick, woolly coat and a white lower lip.

characteristic hump in front of shoulder

5–6 ft
1.5–1.8 m

elongated head

5½–6½ ft
1.7–2 m

INDIAN RHINOCEROS
Rhinoceros unicornis
Living in grasslands, forests, and wetlands of India and Nepal, this solitary, single-horned rhinoceros has thick folds of skin around its neck.

WHITE RHINOCEROS
Ceratotherium simum
Found in the African savanna, this social species is also the heaviest rhinoceros. "White" is a corruption of "wide"—for the animal's square mouth, adapted for grazing.

square mouth

three toes

5–5½ ft
1.5–1.7 m

JAVAN RHINOCEROS
Rhinoceros sondaicus
Once widespread in S.E. Asia, this solitary, nocturnal browser is one of the world's rarest animals. Its small horn does not exceed 8 in (20 cm) in length.

»

WHITE RHINOCEROS
Ceratotherium simum

Despite its unnerving appearance, this giant of the African plains is a mild-tempered vegetarian. The huge horns are used almost entirely in self-defense or for protecting the young. Adult rhinoceroses usually live alone, although they sometimes form loose groups to share feeding areas. Males are territorial and mark their territory with pungent urine and droppings. They will compete for the right to mate with a certain female, but most disputes are solved with a bout of showing off, after which the weaker male backs down. This species has suffered a drastic decline in numbers and range as a result of habitat loss and hunting.

SIZE 5–6 ft (1.5–1.8 m)
HABITAT Savanna grassland
DISTRIBUTION C. and S. Africa
DIET Grass

∨ SHORTSIGHTED
Rhinoceroses are somewhat shortsighted. Having eyes located on the sides of the head gives them a wide field of vision, but impairs their ability to see straight ahead.

∨ HAIRY EARS
The ears are the hairiest part of a rhinoceros's body. They provide excellent hearing, and swivel to detect sounds from any direction.

WIDE MOUTH >
The white rhinoceros is the largest animal to survive on an exclusive diet of grass. Its wide, straight mouth is shaped for maximum efficiency when cropping short grass.

< HORN OF HAIR
Rhinoceros horn is made of a protein called keratin. The same substance is found in hair and nails—in fact, the rhinoceros's horn is little more than a mass of compacted hair.

front horn of a mature rhino can grow up to 5 ft (1.5 m) long

< RHINO HIDE
Despite the species' name, the tough, wrinkled skin of the white rhinoceros is gray, not white. Up to ¾ in (2 cm) thick, the skin is made up of layers of collagen, arranged in a crisscross pattern.

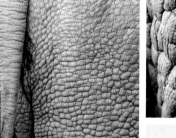

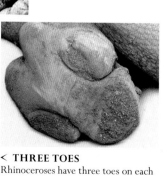

< UNDERSIDE OF FOOT
The unusual shape of a rhinoceros's foot makes the animal relatively easy to track. Some experienced trackers can even recognize individual animals from their footprint.

> GENTLE GIANT
White rhinoceroses suffer from an undeserved reputation for bad temper. In reality they are peace-loving, even timid, animals. Under normal circumstances they charge only when provoked or confused. The black rhinoceros is much more aggressive.

∧ TELLING TAIL
The short tail hangs down when the rhinoceros is relaxed and curls up like that of a pig at times of excitement, such as when mating.

< THREE TOES
Rhinoceroses have three toes on each foot. Most of the weight is borne on the middle toe, with the smaller side toes aiding balance and grip.

large nostrils and an
excellent sense of smell
make up for weak eyesight

HORSES AND RELATIVES

Although the Equidae are a large family in the fossil record, just nine species of horse, donkey, asses, and zebra remain. They live in herds or transient groups, inhabiting open grassland and desert. They have all-around vision and their movable, sensitive ears alert them to predators. These swift animals have slender legs with a single hoofed toe, and their coat is short except for longer hair on the mane and tail.

stiff, striped mane

distinctive dewlap

4–5 ft
1.2–1.5 m

MOUNTAIN ZEBRA
Equus zebra
Found in S.W. Africa, this species lives in dry, rocky, mountainous habitats. It has broad stripes on the hindquarters that contrast with its narrow stripes elsewhere.

narrow, intermediate stripes

4¼–4½ ft
1.3–1.4 m

CHAPMAN'S ZEBRA
Equus quagga antiquorum
This zebra from southern Africa is a subspecies of the plains zebra. It has characteristic shadow stripes on its body that alternate with the black ones.

4¼–4½ ft
1.3–1.4 m

GRANT'S ZEBRA
Equus quagga boehmi
The smallest of the six subspecies of the plains zebra, this mammal has broad, defined stripes. It is a native of the savanna in E. Africa.

5–5¼ ft
1.5–1.6 m

GREVY'S ZEBRA
Equus grevyi
The largest wild species in its family, this big-eared zebra from E. Africa has narrow, variable stripes on its body and a white belly.

4–4¼ ft
1.2–1.3 m

KHUR
Equus hemionus khur
A fast runner from arid Asian grasslands, this subspecies is now found in the wild only in a sanctuary in Gujarat, India.

dorsal stripe

4–4¼ ft
1.2–1.3 m

KULAN
Equus hemionus kulan
A little larger than most donkeys, this subspecies of the Asiatic wild ass has a black dorsal stripe edged with white, and a short, upright mane.

4–5 ft
1.2–1.5 m

PERSIAN ONAGER
Equus hemionus onager
Extinct in much of its former Asian range and now found only in parts of Iran, this subspecies has a brown dorsal stripe on its back.

4¼–4½ ft
1.3–1.4 m

KIANG
Equus kiang
The largest wild ass, the kiang inhabits the Tibetan Plateau. Its chestnut-colored coat is the darkest of all Asian wild asses, and is woolly in winter.

gray-brown coat

striped legs

4–4½ ft
1.2–1.4 m

SOMALI WILD ASS
Equus africanus somalicus
The ancestor of the donkey, this wild ass from N.E. Africa has a short, gray coat, and striped, zebralike legs.

3–5½ ft
0.9–1.7 m

DONKEY
Equus asinus
The domesticated form of the African wild ass, this subspecies has a worldwide distribution as a transport and pack animal.

PRZEWALSKI'S HORSE
Equus przewalskii
The last true wild horse, this species often bears faint leg stripes. It once survived only in captivity, but is now being reintroduced into the wild in Mongolia and China.

pale muzzle

4–5 ft
1.2–1.5 m

pale, yellowish-brown flanks

brown legs, often with faint stripes

single toe encased in a hoof

4–4¼ ft
1.27–1.3 m

EXMOOR PONY
Equus caballus
This rare, primitive, and hardy pony survives semi-wild in Exmoor, England. It is always dun, bay, or brown with black points.

SHIRE HORSE
Equus caballus
A large, powerful horse, bred in England from Dutch stock, this breed is still used in agriculture and forestry for pulling heavy loads.

5½–6¼ ft
1.7–1.9 m

distinctive "dished" face

ARABIAN HORSE
Equus caballus
A swift desert horse, the Arab has had a huge influence on the development of the modern racehorse.

5 ft
1.5 m

PAINT HORSE
Equus caballus
This American horse combines a dark coat, ranging from chestnut to black, with variable white areas.

5–5¼ ft
1.5–1.6 m

silky mane

MULE
Equus asinus x *E. caballus*
Built like a horse but with the paternal donkey's head and long ears, this generally sterile hybrid is a strong pack animal.

4–6 ft
1.25–1.8 m

HINNY
Equus caballus x
E. asinus
The offspring of a male horse and a female donkey, the hinny has a donkeylike body and the head, ears, and mane of a horse.

variable coat color

3¼–4¼ ft
1–1.3 m

EVEN-TOED UNGULATES

One of two orders of hoofed mammals, the Artiodactyla stand on two or four toes. Most are herbivorous, and have a multi-chambered, fermenting stomach.

Even-toed ungulates are often described as cloven hoofed. This is because the tip of each toe has its own hoof—a hard, rubbery, sole surrounded by a thick nail. The hoof wears down through use, but grows continually. Only the camelids lack hoofs—although they walk on two toes, the hoof is reduced to a small nail.

CHEWING THE CUD

With long legs and their feet protected with hooves, even-toed ungulates often range extensively across grassland and forested habitats in search of food. Most families are grazers of grass and herbs or browsers of shoots and leaves. Their digestive system is well adapted to a diet of tough vegetable matter: the stomach has three or four chambers and contains bacteria that digest the cellulose in plant cell walls, releasing nutrients inside. To help with this, partly digested food (cud) is regurgitated for further

chewing, a process known as rumination. Artiodactyls also have large, broad cheek teeth to grind up their tough food, and long intestines. Pigs and peccaries are somewhat different: they do not ruminate, have more generalized teeth, and a more omnivorous diet. Sometimes the canine teeth are enlarged into tusks for defense, fighting, and rooting for food.

DOMESTICATION

Introduced into Australasia, members of this order are found on every continent except Antarctica. They vary greatly in shape and in size, with shoulder height ranging from 8 in (20 cm) in the mouse deer to almost 13 ft (4 m) in the giraffe. Many species are hunted in the wild as food by humans, but others have been domesticated and are economically important—cattle, llamas, sheep, and pigs are used for meat, leather, wool, dairy products, and transport. Domestication has given rise to animal breeds with a distinctive appearance that matches their specific purpose.

PHYLUM	CHORDATA
CLASS	MAMMALIA
ORDER	ARTIODACTYLA
FAMILIES	10
SPECIES	384

DEBATE

ONE SPECIES OR MORE?

During the last ten years, over 100 new bovid species have been described, often because taxonomists, scientists who classify things, have categorized different races (subspecies) of certain bovids as species in their own right. Genetic evidence is used to substantiate (or sometimes refute) these divisions. Giraffes, for example, were previously described as a single species containing a number of subspecies. However, recent DNA evidence supports their division into four species—the Masai (*G. tippelskirchi*), reticulated (*G. reticulata*), northern (*G. camelopardalis*; three subspecies), and southern giraffes (*G. giraffa* ; two subspecies).

PECCARIES

Found in the Americas, the Tayassuidae share many features with pigs, such as small eyes and a snout that ends in a cartilaginous disk. But they have a more complex, chambered stomach, and short, straight tusks.

white neck collar

long snout

20½–27 in
52–69 cm

CHACOAN PECCARY
Catagonus wagneri
Found in the Chaco ecoregion of central S. America, this large peccary was first described from fossils, and was discovered to be a living species only in 1975.

16–23½ in
40–60 cm

12–20 in
30–50 cm

WHITE-LIPPED PECCARY
Tayassu pecari
Living in large herds in C. and S. America, this species will group together and face off a predator, but they are fast and will flee if necessary.

COLLARED PECCARY
Pecari tajacu
Widespread in the tropical and subtropical Americas, this diurnal pig is extremely social in nature. It is considered destructive in agricultural areas, as it often raids crops.

PIGS

Uniquely in their order, the Old World Suidae have four toes, although they walk only on the middle two. They have a simple stomach, unlike the chambered stomach of most of their relatives. Largely omnivorous, they dig with their snout and tusks for food. They have bristly coats and a short tail that ends in a long tassel.

shaggy mane

26–32 in
65–80 cm

MOLUCCAN BABIRUSA
Babyrousa babyrussa
Males of this species have remarkable tusks: the upper canines grow upward, pierce the snout, and curve backward. It is native to some Indonesian islands.

22–34 in
55–85 cm

2½–3½ ft
75–110 cm

GIANT FOREST HOG
Hylochoerus meinertzhageni
Unlike most pigs, this large, nocturnal wild species from Africa has a thick covering of black and reddish fur.

23½–34 in
60–85 cm

COMMON WARTHOG
Phacochoerus africanus
This pig has a warty face and two pairs of tusks. It often feeds on its front knees when grazing, and runs with its tail erect.

BUSHPIG
Potamochoerus larvatus
Found in African forests and reedbeds, this pig has brown fur, and a pale mane, which stands upright when the animal is alarmed.

rounded back

22–32 in
55–80 cm

RED RIVER HOG
Potamochoerus porcus
This richly colored pig from C. Africa has a pair of lumps on its snout, and distinctive white facial markings.

8–10 in
20–25 cm

PYGMY HOG
Porcula salvania
Once spread across India and Nepal, this tiny, dark brown pig with a sharply tapered snout is now critically endangered.

35 in
90 cm

VISAYAN WARTY PIG
Sus cebifrons
Three pairs of fleshy warts on the face protect this boar against the tusks of rival males while fighting.

35 in
90 cm

BEARDED PIG
Sus barbatus
Found in the forests of S.E. Asia, this pig often moves in large herds. It sports a whitish "beard" and distinct tail tassels.

narrow mane of longer hair

28–32 in
70–80 cm

23½–32 in
60–80 cm

22–43 in
55–110 cm

long snout ends in a large disk of cartilage

WILD BOAR
Sus scrofa
Widespread in Eurasia, the bristly wild boar is the main ancestor of domestic pigs. The piglets have stripes along the body for camouflage in dense thickets.

PIÉTRAIN PIG
Sus scrofa domesticus
A Belgian domestic breed that produces high-quality, lean meat, this pig has dark spots set within paler blotches on its body.

MIDDLE WHITE PIG
Sus scrofa domesticus
Reared in England for pork, this pig is an unpigmented domestic breed, characterized by its round form and short, upturned nose.

»

MUSK DEER

Found mainly in Asian montane forests, the solitary, nocturnal members of the family Moschidae are named for the musk gland of adult males. These small, stocky deer have enlarged upper canines and longer hind limbs that help them climb rough terrain.

brittle, sandy brown coat

elongated, tusklike canines used for fighting

20–23½ in
50–60 cm

ALPINE MUSK DEER
Moschus chrysogaster
Inhabiting highland forests from S.E. China to the north of India, this is one of the larger species of musk deer and has long, harelike ears. The male's musk sac produces musk, which is used to mark territory.

CHEVROTAINS

Found in African and Asian tropical forests, the Tragulidae are somewhat similar to small deer, but lack horns or antlers. In males, the enlarged upper canines project from either side of the lower jaw. Like pigs, chevrotains have four toes on each foot and somewhat short legs. The stomach is chambered, to ferment and digest tough plant food.

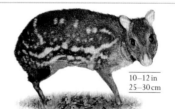

10–12 in
25–30 cm

WHITE-SPOTTED CHEVROTAIN
Moschiola meminna
Nocturnal and secretive, this tiny chevrotain from India and Sri Lanka has spots and white stripes on its body.

distinctive markings

12–14 in
30–36 cm

WATER CHEVROTAIN
Hyemoschus aquaticus
Large and boldly marked with stripes and spots, this chevrotain of W. and C. African forests is a good swimmer and diver.

12–14 in
30–35 cm

8–14 in
20–35 cm

GREATER INDO-MALAYAN CHEVROTAIN
Tragulus napu
Although very small, this is the largest Asian chevrotain. Its pointed head has dark stripes running from the large eyes to the black nose.

JAVA CHEVROTAIN
Tragulus javanicus
This is the world's smallest hoofed mammal. Its relationship with other chevrotains in Southeast Asian forests is poorly understood.

DEER

Distributed almost globally, the Cervidae are largely absent from Africa and introduced in Australia. They occupy forested and open habitats, but many favor the transitional zones in between. The size and form of the antlers of each species is unique. Unlike permanently horned animals, such as antelopes, male deer shed and regrow their antlers each year; females, with one exception, lack antlers but may have only short stubs.

3½–5¼ ft
1.1–1.6 m

37–43 in
95–110 cm

RUSA DEER
Rusa timorensis
This Indonesian forest deer was introduced into arid Australian bushlands, where it thrives. It has large ears and antlers for its size.

SAMBAR
Rusa unicolor
A large, dark brown deer with a prominent mane, the sambar browses in woodlands throughout S. Asia, up to the Himalayan foothills.

26–30 in
65–75 cm

VISAYAN SPOTTED DEER
Rusa alfredi
Endemic to the Philippines, this nocturnal deer has short legs, a distincitve crouching posture, and a dense pattern of pale spots.

18–20 in
45–50 cm

short tail

REEVES' MUNTJAC
Muntiacus reevesi
Native to eastern Asia, this species was introduced in western Europe. Despite its small size, it damages woodlands by grazing and browsing.

20–28 in
50–70 cm

long, slender legs

RED MUNTJAC
Muntiacus muntjak
This S. Asian deer uses its short, one-branched antlers and elongated upper canines to defend itself against predators and its territory against rivals.

PÈRE DAVID'S DEER
Elaphurus davidianus
Known only from captive populations, this deer was described in 1865 by a French missionary, Father Armand David, and has been reintroduced to China.

3½–4½ ft
1.1–1.4 m

28–37 in
70–95 cm

AXIS DEER
Axis axis
Introduced in Australia and N. America, this common Indian forest deer has lyre-shaped antlers. It is a favorite prey of tigers.

HOG DEER
Axis porcinus
Found in Asian forests, this deer is named after its hoglike habit of running head down, going under rather than leaping over obstacles.

22–30 in
55–75 cm

palmate antlers

tail white below

30–37 in
75–95 cm

4–4½ ft
1.2–1.4 m

long, pointed, branched antlers

20–28 in
50–70 cm

TUFTED DEER
Elaphodus cephalophus
This small deer lives in Asian montane forests. Males of this species have small antlers and short tusks, with a tuft of black hair on the forehead.

COMMON FALLOW DEER
Dama dama
Often domesticated as a source of venison, this species is distinguished by its spotted coat and flattened, palmate antlers.

BARASINGHA
Rucervus duvaucelii
This wetland deer from India was introduced in the US for sport; the multi-pointed, branched antlers of males are highly prized.

4¼–5½ ft
1.3–1.7 m

3–4¼ ft
0.95–1.3 m

WESTERN RED DEER
Cervus elaphus
From Europe, Turkey, and N. Africa, this deer varies across its range in bulk, antler size, and the prominence of its mane.

23½–45 in
60–115 cm

WAPITI
Cervus canadensis
Although this N. American and Asian deer is similar in appearance to the red deer, genetic analysis has confirmed it is a species in its own right.

shaggy fur on neck

reddish-brown coat

SIKA DEER
Cervus nippon
Distinguished by its stout, upright antlers, the sika breeds with red deer, especially when introduced outside its native E. Asia.

30–41¼ in
75–105 cm

22–41¼ in
55–105 cm

grayish-brown coat

MULE DEER
Odocoileus hemionus
This species has a black-tipped tail and forked antlers, which distinguish it from the white-tailed deer that shares its range in western N. America.

WHITE-TAILED DEER
Odocoileus virginianus
Found from Canada to Peru, and introduced in Europe and New Zealand, this deer flashes the white underside of its tail when alarmed.

antlers vary
greatly in
shape

paddle used for
scraping away snow

thick neck

hairy
nose pad

26–33 in
65–84 cm

EUROPEAN ROE DEER
Capreolus capreolus
A small European scrub and forest
deer, its fur is reddish-brown in
summer, becoming darker, sometimes
almost black, in winter.

CARIBOU
Rangifer tarandus
The soles of this deer's feet shrink
in winter. The exposed rims
of the hoofs provide traction
when walking on ice.

2¼–4½ ft
0.7–1.4 m

3½–4¼ ft
1.1–1.3 m

MARSH DEER
Blastocerus dichotomus
Adapted to wetland habitats, the marsh
deer is the largest S. American deer. It swims
well and has a membrane between the
two toes of each foot that aids walking
on soft ground.

6–7 ft
1.8–2.1 m

MOOSE
Alces americanus
The world's largest deer, inhabiting
the forests of N. America, the male's
palmate antlers may span 6½ ft (2 m)
and each has up to 20 points.

20–26 in
50–65 cm

**COMMON BROWN
BROCKET**
Mazama gouazoubira
A solitary deer of
scrubland and forest
thickets in C. and
S. America, it feeds
mainly on fruits
and cacti during
the dry season.

23½–32 in
60–80 cm

**COMMON
RED BROCKET**
Mazama americana
From the jungles of S. America, this small,
solitary deer prefers fruits rather than leaves.
Males have short, unbranched antlers.

SOUTHERN PUDU
Pudu puda
One of the world's smallest deer, this
stocky animal lives in the temperate
rainforests of Argentina and Chile.

23½–28 in
60–70 cm

20–22 in
50–55 cm

12–16 in
30–40 cm

PAMPAS DEER
Ozotoceros bezoarticus
A slender deer of
S. American grasslands
and wetlands, the
pampas deer stands
on its hind legs
to browse from
tree branches.

**CHINESE
WATER DEER**
Hydropotes inermis
Uniquely among deer,
neither sex grows antlers.
Their long upper canines
protrude as tusks, up to
3¼ in (8 cm) long in males.

PRONGHORN

Widely represented in the North
American fossil record, the
Antilocapridae included species with
bizarrely shaped or multiple horns.
Today only the pronghorn survives.
With a similar body shape and cloven
hoofs, the pronghorn resembles
antelopes (Bovidae), but unlike them
it lacks lateral toes and sheds its
horns outside the breeding season.

white neck
band

34–35 in
86–88 cm

PRONGHORN
Antilocapra americana
The fastest mammal in the
New World, pronghorns
form large herds in open
grassland and are the
ecological counterpart
of Old World antelopes.

BOVIDS

A large and varied family, the Bovidae are found on all continents except Antarctica. Despite their diversity, they share common features: males (and in some species, females) have permanent unbranched horns, often twisted or fluted, and a complex, four-chambered, ruminant's stomach.

COMMON ELAND
Taurotragus oryx
With spiral horns, the eland is the largest antelope, found in open grassland from Ethiopia to South Africa. Males sometimes develop white stripes down their flanks.

4¼–6 ft
1.3–1.8 m

prominent dewlap

NILGAI
Boselaphus tragocamelus
The largest Asian antelope, with a robust body sloping down from the shoulder, male nilgai develop a blue-gray coat; females are yellow-brown in color.

4–4½ ft
1.2–1.4 m

loosely spiraled horns

large ears

22–26 in
55–66 cm

FOUR-HORNED ANTELOPE
Tetracerus quadricornis
This solitary Asian forest species usually develops two pairs of horns: one between the ears and the other on the forehead.

32–47½ in
82–121 cm

NYALA
Tragelaphus angasii
This southern African forest antelope has spiral horns and a dark brown coat, with vertical white stripes on the sides.

white chest patch

23¾–39 in
61–100 cm

CAPE BUSHBUCK
Tragelaphus sylvaticus
Widespread in sub-Saharan forests, this species has varied stripes and spots, especially on the face, ears, and tail.

3–4¼ ft
0.9–1.3 m

ZAMBEZI SITATUNGA
Tragelaphus selousi
Inhabiting southern African swamps, this excellent swimmer will often take refuge in pools of water when threatened by a predator.

SOUTHERN LESSER KUDU
Ammelaphus australis
Males and females of this antelope of arid scrubland in N.E. Africa have 7–14 vertical white stripes on their body.

4–4¼ ft
1.2–1.3 m

chestnut coat with white stripes

BONGO
Tragelaphus eurycerus
Well camouflaged in the dense forests of W. and C. Africa, female bongos are usually more brightly colored than the males. Both sexes have spiral horns.

4–5¼ ft
1.2–1.6 m

ZAMBEZI KUDU
Strepsiceros zambesiensis
Male Zambezi kudu have some of the most magnificent horns of any antelope, with two and a half twists when fully grown.

3¼–3½ ft
1–1.1 m

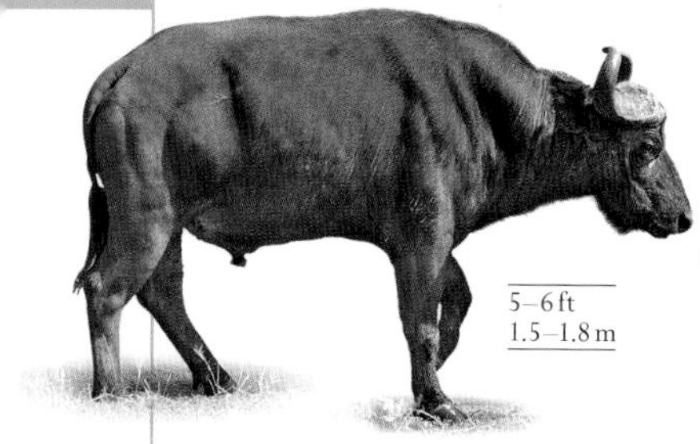

AFRICAN BUFFALO
Syncerus caffer
Unpredictable and dangerous, this buffalo cannot be domesticated. Grassland forms have more curved horns than the smaller forest-dwellers.

5–6 ft
1.5–1.8 m

5–6¼ ft
1.5–1.9 m

ASIAN WATER BUFFALO
Bubalus bubalis
Predominantly domesticated, especially for their power and milk, a few wild water buffalo, *B. arnee*, remain in S. Asia. Coloring and horns are variable.

23½–39 in
60–100 cm

LOWLAND ANOA
Bubalus depressicornis
The smallest of all wild cattle, native to the rainforests of Sulawesi, the horns are straight and upright compared to other buffalo.

distinctive hump over the shoulders

short, curved horns

shorter hair on rear

large head

AMERICAN BISON
Bison bison
Once roaming N. America in huge herds, the few remaining wild bison are dwarfed by captive populations raised for meat and hides.

5–6½ ft
1.5–2 m

shaggy, dark brown coat

WISENT
Bison bonasus
With shorter hair but longer horns than the American bison, this species is now restricted to E. European and Russian primary forests.

5–6½ ft
1.5–2 m

BANTENG
Bos javanicus
Native to S.E. Asia and domesticated locally as draft animals, this bovid is brown, with white lower legs, muzzle, rump, and eye spots.

5¼ ft
1.6 m

YAK
Bos mutus
Inhabiting montane parts of C. Asia, insulated with long, shaggy fur, wild yaks have a dark coat, but domesticated ones are more varied and often have white markings.

4½–6½ ft
1.4–2 m

GAUR
Bos gaurus
The largest species of wild cattle, this muscular Asian forest bovine is dark brown, but paler on the muzzle and lower legs.

5½–7¼ ft
1.7–2.2 m

long horns from which this breed gets its name

TEXAN LONGHORN
Bos taurus taurus
With a wide variety of colors, and impressive spreading horns, this breed is hardy and well suited to an extensive ranching system.

4–5 ft
1.2–1.5 m

HEREFORD
Bos taurus taurus
Originating in England, the Hereford is a beef breed that has deep, muscular forequarters and a docile temperament.

4½–5 ft
1.4–1.5 m

ANKOLE
Bos taurus taurus
Originating in Africa, the ankole's huge horns, up to 3½ ft (1.8 m) long and very thick, help to keep it cool in hot conditions.

4–5 ft
1.2–1.5 m

14–16½ in
35–42 cm

MAXWELL'S DUIKER
Philantomba maxwellii
With a gray-brown coat, this small W. African rainforest species has few distinctive features, apart from its pale facial markings.

12–16 in
30–40 cm

ZIMBABWE BLUE DUIKER
Philantomba bicolor
This small forest antelope from S.E. Africa has simple conical horns. It eats mainly newly fallen leaves, fruits, and seeds, but also insects and fungi when available.

16–20 in
40–50 cm

ZEBRA DUIKER
Cephalophus zebra
This is the only duiker without a predominantly plain coat. The stripes provide camouflage in its W. African forest-edge habitat.

4–4¼ ft
1.2–1.3 m

JERSEY
Bos taurus taurus
Famed for its rich, creamy milk, this breed was developed on the island of Jersey from stock imported from France.

26–34 in
65–85 cm

WESTERN YELLOW-BACKED DUIKER
Cephalophus silvicultor
Found from west to central Africa, this large duiker has a dark gray coat with a conspicuous white or yellow patch on the back.

21½–23 in
54–58 cm

BLACK-FRONTED DUIKER
Cephalophus nigrifrons
A dark forehead and eye glands, contrasting with paler brows, give this forest dweller from C. Africa a distinctive facial appearance.

4–4½ ft
1.2–1.4 m

BRAHMAN
Bos taurus indicus
Originating in Asia, the brahman is now farmed throughout the tropics. Also known as the zebu, it has a characteristic hump on the back.

21½–22 in
55–56 cm

15½–27 in
39–68 cm

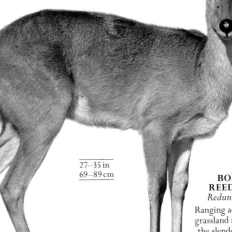

26–59 in
65–150 cm

27–35 in
69–89 cm

OGILBY'S DUIKER
Cephalophus ogilbyi
From the rainforests of W. Africa, this duiker has well-developed hindquarters, with a red-brown rump.

COMMON DUIKER
Sylvicapra grimmia
A widespread sub-Saharan antelope with tiny horns, this species occupies a wide range of habitats, often scavenging on fallen fruit.

BOHOR REEDBUCK
Redunca bohor
Ranging across rolling grassland in C. Africa, the slender female of this species contrasts with the thick-necked, horned male.

SOUTHERN REEDBUCK
Redunca arundinum
A robust antelope in grasslands in southern C. Africa, this species has conspicuous black foreleg markings. Only the males have horns.

RED LECHWE
Kobus leche
A highly gregarious antelope, favoring southern C. African marshlands, its long legs enable the lechwe to run well through shallow water.

32–39 in
82–100 cm

34–44 in
87–112 cm

UGANDA KOB
Kobus thomasi
A sociable antelope from E. Africa, the Uganda kob has a distinct white throat patch. Males have ridged, lyre-shaped horns.

30–37 in
77–94 cm

4–4½ ft
1.2–1.4 m

4–4½ ft
1.2–1.4 m

COMMON WATERBUCK
Kobus ellipsiprymnus ellipsiprymnus
Despite its name, this African waterbuck is a savanna and woodland antelope, although it will seek refuge from predators in water.

DEFASSA WATERBUCK
Kobus ellipsiprymnus defassa
From W. and C. Africa, this subspecies of waterbuck has a solid white rump, rather than a white crescent surrounding the tail.

PUKU
Kobus vardonii
Very similar to the kob, but from southern C. Africa, the puku is slightly smaller and more stockily proportioned.

ROAN ANTELOPE
Hippotragus equinus
The ridged horns of this sub-Saharan savanna antelope grow in a smooth curve. The animal has a distinctive black and white face.

49½–57 in
126–145 cm

37–45 in
95–115 cm

ADDAX
Addax nasomaculatus
This endangered antelope from the Sahara, has long horns with two or three spiral twists, and a pale sandy or whitish coat.

long, thin, curved horns

SCIMITAR-HORNED ORYX
Oryx dammah
Formerly ranging across the Sahara, this whitish oryx was almost hunted to extinction in the 20th century, but has been reintroduced to some areas.

53–55 in
135–140 cm

SOUTHERN SABLE ANTELOPE
Hippotragus niger
A powerful antelope with a dark coat and striking white face markings, its magnificent horns can reach more than 3¼ ft (1 m) in length.

2½–3¼ ft
0.8–1 m

ARABIAN ORYX
Oryx leucoryx
White-coated, with long, straight horns, this antelope has recently been reintroduced to parts of its former Middle Eastern range.

GALLA ORYX
Oryx gallarum
Gray-brown, marked with black on the face, tail, flanks, and front legs, this E. African desert oryx has very long, barely curved horns.

rusty colored chest and neck

3¼–4¼ ft
1–1.3 m

relatively short, sturdy legs

3¼–4 ft
1–1.25 m

44–49 in
112–125 cm

GEMSBOK
Oryx gazella
The largest oryx, the gemsbok frequents arid habitats in southern Africa, but has also been successfully established in N. America.

WESTERN HARTEBEEST
Alcelaphus major
A large, long-faced antelope
that lives on open grassland
in E. Africa. Hartebeest
species differ in their
coloration and horn shape.

56 in
143 cm

4–4¼ ft
1.2–1.3 m

RED HARTEBEEST
Alcelaphus caama
Chestnut brown in color,
with a darker face and tail, this
antelope is most active at
dawn and dusk.

KLIPSPRINGER
Oreotragus sp.
Meaning "rock jumper" in
Afrikaans, the klipspringer inhabits
rocky outcrops. It can leap more
than ten times its own height.

16½–22½ in
42–57 cm

20–25 in
51–64 cm

**LICHTENSTEIN'S
HARTEBEEST**
Alcelaphus lichtensteinii
Distinguished by its strongly
curved horns that turn inward
at their tips, this hartebeest is
found in savanna and floodplain-
grassland in C. Africa.

4–4½ ft
1.2–1.4 m

sloping back

heavy, recurved horns

CENTRAL ORIBI
Ourebia hastata
This graceful, long-
necked antelope from
E. Africa west to Angola
has a distinctive white
brow line and a large,
dark facial gland.

black face

32–39 in
80–100 cm

long, black
throat hair

BONTEBOK
Damaliscus pygargus dorcas
With a striking white face patch,
this South African antelope was hunted
almost to extinction. It is now found
only in protected areas.

dark gray coat

SERENGETI TOPI
Damaliscus jimela
The Serengeti topi inhabits the grasslands of
E. Africa. Males often stand on termite mounds to
defend their territory and look for predators.

4¼–5¼ ft
1.3–1.6 m

BLUE WILDEBEEST
Connochaetes taurinus
A highly gregarious African
antelope, also known as
a brindled gnu, this species
is found in savanna regions
of southern Africa.

3½–4¼ ft
1.1–1.3 m

KALAHARI SPRINGBOK
Antidorcas hofmeyi
With its numbers much reduced
by hunting, this agile antelope,
with lyre-shaped horns, browses
the arid lands of southern Africa.

30–34 in
77–87 cm

SERENGETI THOMSON'S GAZELLE
Eudorcas nasalis
The most numerous gazelle on the
plains of E. Africa, this species has
a broad black stripe on its sides.

21–26 in
53–67 cm

BLACKBUCK
Antilope cervicapra
With the ability to reach speeds of
50 mph (80 kph), the blackbuck
inhabits grassland and open
woodland in India and Pakistan.

23½–34 in
60–85 cm

*strongly
ridged horns*

DORCAS GAZELLE
Gazella dorcas
Inhabiting deserts from
N. Africa to the Middle East,
this small gazelle can survive
without drinking, taking
moisture from its food.

22–26 in
55–65 cm

MOUNTAIN GAZELLE
Gazella gazella
From the mountains and plains
of the Middle East, this gazelle
has several isolated subspecies,
some extremely rare and
threatened by poaching.

23½–26 in
60–65 cm

22–23½ in
55–60 cm

SAND GAZELLE
Gazella marica
Like its close relative the
goitered gazelle (*G. subguttorosa*), this
C. Asian gazelle is unusual in that
only the male has horns.

14–18 in
35–45 cm

15–17 in
38–43 cm

GÜNTHER'S DIK-DIK
Madoqua guentheri
A dik-dik of the E. African
semideserts, this species has
a long, elastic muzzle that
can be inflated in order to
regulate its temperature.

DAMARA DIK-DIK
Madoqua damarensis
A tiny antelope, named
for its sharp alarm note,
the Damara dik-dik has
an elongated, mobile
snout. It forms
territorial pairs.

22½–31 in
57–79 cm

WESTERN SAIGA
Saiga tatarica
Now highly endangered and
found only in western Asia, this
saiga's enlarged, flexible nose
warms cold winter air and filters
out summer dust.

32–41 in
80–105 cm

SOUTHERN GERENUK
Litocranius walleri
Long-necked and standing
on its hind legs, the
E. African gerenuk browses
leaves growing out of the
reach of other antelopes.

COASTAL SUNI
Neotragus moschatus
A tiny, reddish antelope
from S.E. Africa, the suni
is nocturnal and spends
much of its time hidden
in dense scrub.

13–14 in
33–36 cm

*lower leg
lacks muscle*

DAMA GAZELLE
Nanger dama
This rare Saharan gazelle is strikingly bicolored. The amount of white varies between subspecies, although all have a white throat patch.

3¼–4 ft
1–1.2 m

GRANT'S GAZELLE
Nanger granti
Frequenting the E. African plains, this gazelle can survive without drinking water, and so does not follow the migrations of many of its relatives.

30–37 in
75–94 cm

SOEMMERRING'S GAZELLE
Nanger soemmerringii
Similar to, but much rarer than Grant's gazelle, this E. African species is distinguished by its stronger face pattern and larger white rump patch.

24–35 in
60–90 cm

34–39 in
86–98 cm

STEENBOK
Raphicerus campestris
A small E. African bush antelope, the steenbok has particularly large, white-lined ears, with black margins and internal markings.

18–23½ in
45–60 cm

18–24 in
45–60 cm

SHARPE'S GRYSBOK
Raphicerus sharpei
Shy and solitary, with stubby horns, this nocturnal antelope from eastern Africa takes refuge from predators in the burrows of aardvarks.

COMMON IMPALA
Aepyceros melampus
An African plains antelope, with lyre-shaped horns in the male, the impala is an important source of food for the big cats of the region.

ALPINE CHAMOIS
Rupicapra rupicapra
Living high among mountain rocks in the upland blocks of S. Europe and Asia Minor, this species occurs in isolated populations. Each has a subtly different appearance.

28–34 in
70–85 cm

HIMALAYAN BROWN GORAL
Naemorhedus goral
A rough-haired, goatlike browser with curved horns, the goral lives in small herds in the forests of the Himalayas.

22½–31 in
57–78 cm

INDOCHINESE SEROW
Capricornis maritimus
A goat-antelope, with coarse fur and a distinct mane, the Indochinese serow is a browser of leaves and shoots and sometimes a little grass.

34–37 in
85–94 cm

white coat has dense underfur for warmth

3–3½ ft
0.9–1.1 m

MOUNTAIN GOAT
Oreamnos americanus
A sure-footed climber from the northern Rocky Mountains, this goat has a dense, woolly white coat to protect it from low temperatures and high winds.

» BOVIDS

BARBARY SHEEP
Ammotragus lervia
Also known as the aoudad, this native of arid mountain regions in N. Africa stands motionless when threatened, making it difficult to see.

2½–3½ ft
0.8–1.1 m

curved horns

long hair on throat and front legs

MUSKOX
Ovibos moschatus
An inhabitant of Arctic tundra, the muskox has a shaggy coat, with dense insulating underwool, which provides protection against the elements.

4–5 ft
1.2–1.5 m

BHUTAN TAKIN
Budorcas whitei
Living in small herds in the mountain forests of N.E. India, China, and Bhutan, the Bhutan takin has shaggy fur and a broad, arched muzzle.

3½–4½ ft
1.1–1.4 m

HIMALAYAN TAHR
Hemitragus jemlahicus
Living on rocky Himalayan mountain slopes, the tahr has hoofs with rubbery soles, which give additional grip on precipitous or unstable ground.

26–39 in
65–100 cm

BHARAL
Pseudois nayaur
From rocky deserts to mountain slopes on the Tibetan Plateau, this species stays close to cliffs, its refuge from predators.

31–36 in
78–92 cm

WALIA IBEX
Capra walie
Due to the the lack of seasonal changes in the Ethiopian mountains, this rare species of ibex, unlike other ibexes, breeds all year round.

26–39 in
65–100 cm

26–37 in
65–95 cm

26–43 in
65–110 cm

ALPINE IBEX
Capra ibex
Living above the tree line in the Alps, this ibex has recurved horns up to 3¼ ft (1 m) long that are especially impressive on males.

NUBIAN IBEX
Capra nubiana
The Nubian ibex inhabits the desert mountains of the Middle East. Mature males of this species can weigh more than twice as much as females.

MARKHOR
Capra falconeri
The largest wild goat, from the mountains of C. Asia, is endangered. It is hunted for its impressive corkscrew horns and meat.

26–41 in
65–104 cm

3–3½ ft
0.9–1.1 m

28–39 in
70–100 cm

28–35 in
70–90 cm

ANGORA GOAT
Capra hircus
This breed of goat, originating in Turkey, has a fleece that is highly valued as the source of mohair, a durable silky fiber.

BAGOT GOAT
Capra hircus
One of more than 300 goat breeds, the Bagot was bred in England from stock brought home in the 13th century by the returning Crusaders.

GOLDEN GUERNSEY GOAT
Capra hircus
A small goat, often with long hair, this rare breed is kept for milk and showing and originated on Guernsey in the Channel Islands.

MOUFLON
Ovis aries orientalis
With a reddish coat and pale saddle, this native of Asia Minor has been established on several Mediterranean islands since Neolithic times.

35–39 in
90–100 cm

MANX LOAGHTAN SHEEP
Ovis aries
Originating on the Isle of Man, this primitive, hardy breed, kept for its meat, has brown wool and usually four horns.

26–32 in
65–80 cm

COTSWOLD SHEEP
Ovis aries
Originating in England, this white-faced breed is hardy and dual-purpose, kept for its long wool and its meat.

26–39 in
65–100 cm

26–32 in
65–80 cm

JACOB SHEEP
Ovis aries
This ancient, hardy breed, with a patterned coat, is said to have originated in Palestine. It has as many as three pairs of horns.

♀

3–4 ft
0.9–1.2 m

MARCO POLO ARGALI
Ovis polii
This mountain sheep is named after Marco Polo (*c.*1254–1324), the first person to describe a wild sheep from C. Asia. The males have the largest horns of any of the argali species.

FAT-TAILED SHEEP
Ovis aries
This breed, found mainly in Africa and Asia, tolerates dry conditions, relying on fat stored in its swollen tail and hindquarters.

26–43 in
65–110 cm

massive, curving horns

DALL SHEEP
Ovis dalli
Creamy white or brown, with curved yellowish horns, dall sheep live in the subarctic mountains of Canada and Alaska.

31–43 in
79–109 cm

short legs

BIGHORN SHEEP
Ovis canadensis
With mountain and desert forms in N. America, males use their impressive horns to establish a dominance hierarchy. Only the high-ranking males secure access to females.

30–44 in
76–112 cm

35–42 in
90–107 cm

SNOW SHEEP
Ovis nivicola
With pale wool and dark legs, this Siberian sheep is extremely agile and can move rapidly over rough, mountainous terrain.

ROTHSCHILD'S GIRAFFE
Giraffa camelopardalis rothschildi

A subspecies of the northern giraffe, Rothschild's giraffes live in small, single-sex groups. Females often move from one group to another, whereas adult males may be solitary. The sexes interact only to mate, the female producing a single calf after a gestation period of about 450 days. Calves are usually weaned at about 12 months. A giraffe's movement is restricted to walking and galloping—giraffes cannot trot or swim. When walking, the front and back legs on the same side move forward at the same time, and an adult can cover about 15 ft (4.5 m) in just one stride. When galloping away from predators, such as lions and African wild dogs, giraffes can reach speeds of up to 34 mph (55 kph).

SIZE 5–5½ ft (1.5–1.7 m)
HABITAT Open woodland, savanna grassland
DISTRIBUTION South Sudan, Kenya, and Uganda
DIET Leaves, shoots, seeds, and fruits of trees

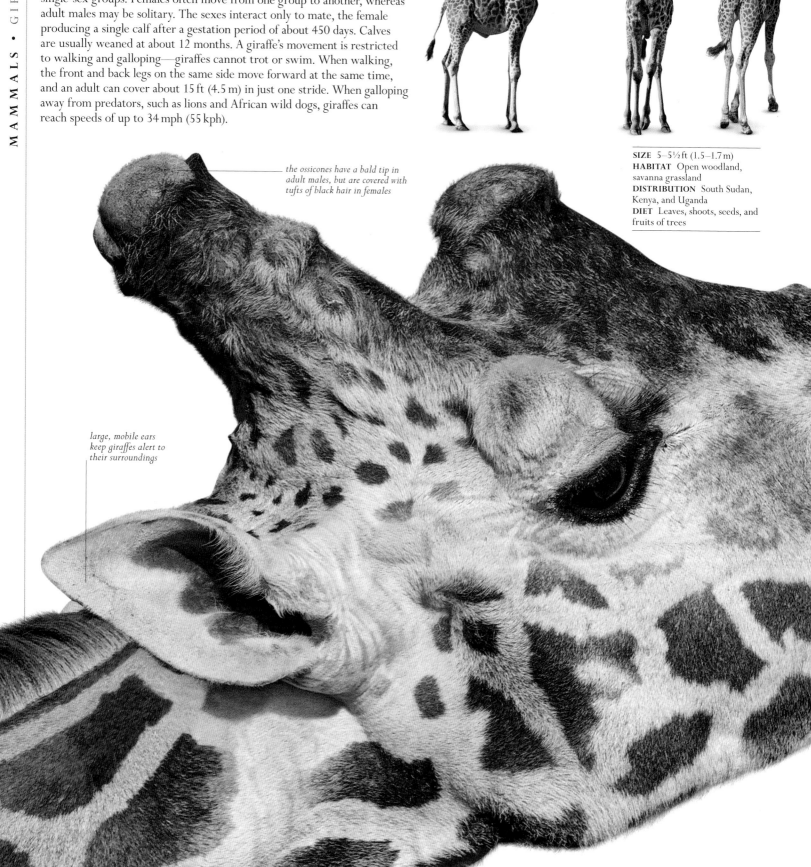

the ossicones have a bald tip in adult males, but are covered with tufts of black hair in females

large, mobile ears keep giraffes alert to their surroundings

∧ EYE
The large eyes on the side of the head have excellent vision. This, combined with the giraffe's height, allows them to spot predators while they are still some distance away.

∨ NOSTRIL
The nostrils can close to prevent debris from entering the nasal passages during sand and dust storms, and act as a barrier to ants while feeding.

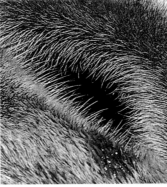

TAIL >
The tail can be up to 8 ft (2.5 m) long, and the tuft of black hair at its tip is used to deter insects.

< CURLED TONGUE
The prehensile tongue and upper lip can pluck buds from between the thorns of a tree as well as strip a branch of its leaves.

∧ SKIN
Dark skin patches have large sweat glands that help maintain a constant body temperature by dissipating heat.

FOOT >
Each elongated leg ends in a foot with a cloven hoof that has two weight-bearing digits.

giraffes have a good sense of smell

colored skin patches vary in size and shape, and tend to darken with age in males

the longer, coarser hairs on the chin and nose have specialized sensory nerve receptors

dark coloration is thought to prevent the tongue from being sunburned while feeding

ossicones are skin-covered, hornlike bony outgrowths from the skull

short, upright mane extends down the back of the neck from the ears to the shoulders

∧ AT FULL STRETCH
A long neck, head, and tongue allow a giraffe to browse from trees that few other mammals can reach, thereby reducing competition for food. Having a long face also helps to protect the eyes from thorns while feeding.

> HEAD
All giraffes have two ossicones that are soft at birth but gradually harden as calcium builds up. Male Rothschild's giraffes develop a third smaller ossicone on their forehead.

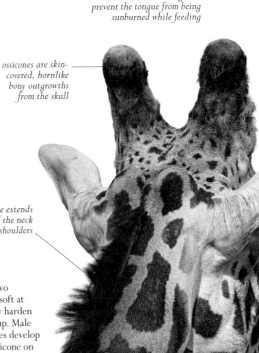

GIRAFFES AND OKAPI

Diverse in the fossil record, the Giraffidae is represented now by just five species from sub-Saharan Africa. Although giraffes live in a different habitat than the okapi, they share some common features, including a long, dark tongue, ossicones (horns) covered in skin, and lobed canine teeth. Otherwise they resemble the bovids, with cloven hoofs, a four-chambered stomach, and the incisors (upper front teeth) replaced by a horny pad.

5–5½ ft
1.5–1.7 m

OKAPI
Okapia johnstoni
Restricted to the rainforests of C. Africa, the okapi's long neck and flexible blue tongue show clear similarities to those of giraffes.

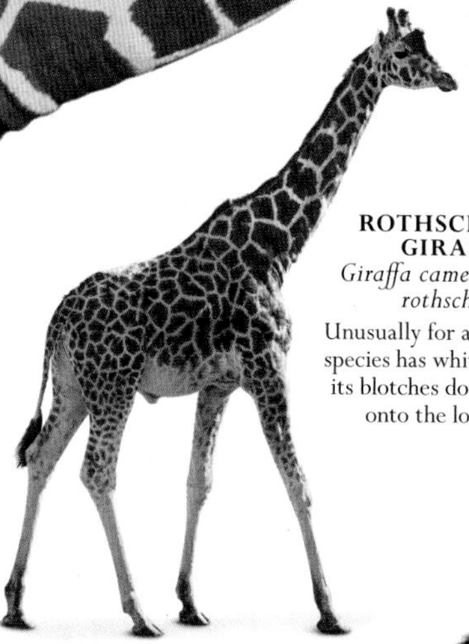

short horns covered in skin

large ears

short upright mane

short body with sloping back

5–5½ ft
1.5–1.7 m

ROTHSCHILD'S GIRAFFE
Giraffa camelopardalis rothschildi
Unusually for a giraffe, this species has white "socks"— its blotches do not extend onto the lower leg.

8¾–10 ft
2.7–3 m

MASAI GIRAFFE
Giraffa tippelskirchi
With a neck up to 7¾ ft (2.4 m) long on top of the shoulder height, giraffes are the world's tallest mammals.

8¾–10 ft
2.7–3 m

8¾–10 ft
2.7–3 m

irregular blotches

RETICULATED GIRAFFE
Giraffa reticulata
Ranging from northern Kenya to Ethiopia, this species of giraffe has large polygonal blotches, often with pale centers, on a pale background.

ANGOLAN GIRAFFE
Giraffa giraffa angolensis
One of two subspecies of southern giraffe, the Angolan giraffe is found in Namibia, Zambia, Botswana, and Zimbabwe.

CAMELS AND RELATIVES

Unique within their order, members of the family Camelidae have just two toes, but no hoofs. Each toe has a small nail at its tip and a soft footpad, which can shift slightly to help maintain grip in mountainous terrian and prevent sinking in soft sand. They also have distinctive teeth, oval red blood cells, a three-chambered stomach, and a leg musculature that means they lie down resting on their knees.

DROMEDARY
Camelus dromedarius
Used for transport in arid regions, this Arabian species is supremely adapted to desert life. Only the feral Australian populations show wild characteristics.

6–6½ ft
1.8–2 m

34–35 in
85–90 cm

BACTRIAN CAMEL
Camelus bactrianus
This two-humped camel has been widely domesticated. The wild species, *Camelus ferus*, is restricted to critically small populations in the Asian deserts.

6–6½ ft
1.8–2 m

LLAMA
Lama glama
Derived from the guanaco, the llama is a valuable pack and meat animal. Originally from the Andes, this domestic species is now more widely distributed in Europe and N. America.

VICUÑA
Vicugna vicugna
The smaller of the two wild Andean camelids, the vicuña produces a fine wool, which led to its domestication and development of the alpaca.

34–35 in
85–90 cm

long, woolly coat

35–51 in
90–130 cm

GUANACO
Lama guanicoe
Native to the arid mountains of S. America, the guanaco has high levels of oxygen-carrying hemoglobin in the blood, enabling it to thrive at extreme altitudes.

40–42 in
102–106 cm

nail on tip of toe

ALPACA
Vicugna pacos
Kept in herds that graze high in the Andes, and now elsewhere in the world, this domestic species is an important source of wool.

HIPPOPOTAMUSES

Among the Artiodactyla, the family Hippopotamidae is unique in that its members walk on four toes on each foot. They also have a huge, barrel-shaped body, short but stout legs, and a large head. Their wide mouth, with tusklike canine teeth, is used in feeding, fighting, and defense. Reflecting an amphibious lifestyle, the nostrils and eyes are on the top of the snout and the skin is smooth, lacking sweat glands.

PYGMY HIPPOPOTAMUS
Choeropsis liberiensis
An inhabitant of the forested swamps of W. Africa, this species is similar in shape to its larger relative, but its muzzle is proportionally smaller.

30–39 in
75–100 cm

5–5½ ft
1.5–1.65 m

COMMON HIPPOPOTAMUS
Hippopotamus amphibius
A solitary nocturnal grazer and sociable daytime wallower, this species is now found mainly in eastern and southern Africa.

BACTRIAN CAMEL
Camelus bactrianus

The Bactrian camel is an exceptionally hardy animal, built to survive in the harsh desert landscapes of southern Asia, where temperatures range between 104°F (40°C) in summer and -20°F (-29°C) in winter. Camels are adapted to travel long distances over difficult terrain in search of food—such as grasses, leaves, and shrubs—which is scarce. When water is available, the Bactrian camel is able to gulp more than 26 gallons (100 liters) in 10 minutes. It can also survive by drinking salty water if necessary. Almost all of the world's Bactrian camels are domesticated; fewer than 1,000 animals remain in the wild, in remote and inhospitable parts of China and Mongolia. While the genetic status of the two domestic species of camel has long been established, the wild two-humped camel has only recently been recognized as a separate species, *Camelus ferus*, based on molecular genetic data.

closable nostrils keep sand out

small, furry ears

thick fur keeps the camel warm, and also protects it from sunburn

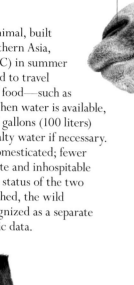

SIZE 6–6½ ft (1.8–2 m)
HABITAT Stony desert, steppe, and rocky plains
DISTRIBUTION Asia
DIET Herbivorous

∨ **EYELASHES**
Two rows of extra-thick eyelashes protect the eyes from strong sunlight, and windblown sand and grit, saving precious water from being blinked away as tears.

∨ **KNEE PADS**
Camels rest on their knees, with their legs folded beneath them. The knees are protected by thick pads of skin.

shaggy mane

∧ **TEETH**
Camels swallow their food whole, then regurgitate and rechew it to aid digestion. Starving individuals have been known to eat rope and leather, which are hard to break down.

∧ **MOBILE LIPS**
Each half of the split upper lip can be controlled independently, which probably helps them browse thorny vegetation effectively without using their tongues and, thereby, losing moisture.

< ∧ **FEET**
The feet each have two toes and a tough padded sole that allows the camel to cope equally well with sharp stony ground, hot sand, or compacted snow.

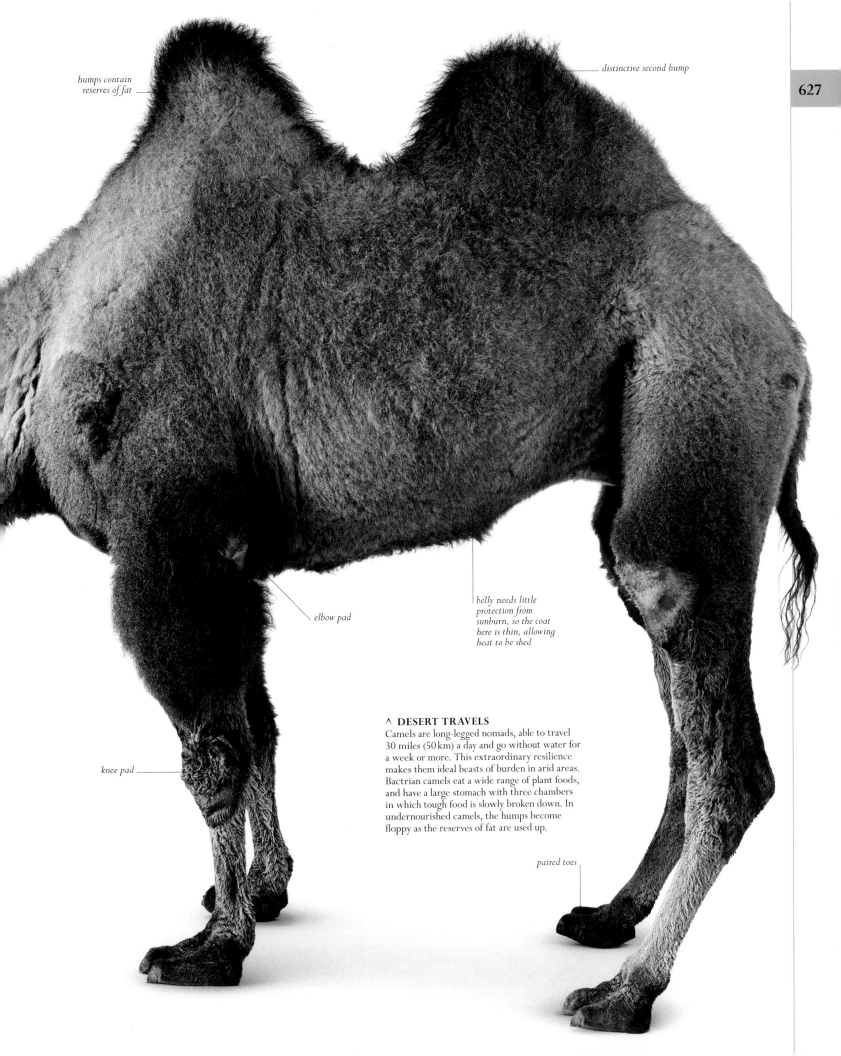

humps contain
reserves of fat

distinctive second hump

elbow pad

belly needs little
protection from
sunburn, so the coat
here is thin, allowing
heat to be shed

knee pad

∧ **DESERT TRAVELS**

Camels are long-legged nomads, able to travel
30 miles (50 km) a day and go without water for
a week or more. This extraordinary resilience
makes them ideal beasts of burden in arid areas.
Bactrian camels eat a wide range of plant foods,
and have a large stomach with three chambers
in which tough food is slowly broken down. In
undernourished camels, the humps become
floppy as the reserves of fat are used up.

paired toes

WHALES, PORPOISES, AND DOLPHINS

Collectively known as cetaceans, this group of mammals is wholly aquatic and all but six species in the order are found in coastal and marine waters.

Supremely adapted to life in water, the cetacean body is tapered and streamlined, with the forelimbs modified into flippers. There are no visible hind limbs, but the tail has horizontal flukes for added propulsion. Many species also have a dorsal fin. The skin is almost hairless and the body is insulated by a layer of blubber, which is especially thick in species living in cold water.

BREATHING AND COMMUNICATING

Capable of diving to great depths and for long periods because of their ability to store oxygen in their muscle tissues, cetaceans must come to the surface to breathe. Breathing is through nostrils, known as blowholes, which are located on the top of the head. When stale air is expelled, it may create a spout of condensation—the spout's size, angle, and shape can be used to distinguish some species, even when their body is almost fully submerged.

Most species produce sound. Some use a series of clicks for echolocation. The clicks bounce off nearby objects, alerting them to any obstacles in their path. Others communicate with each other using vocalizations that range from whistles and groans to the complex songs of many great whales. While their hearing is good, the ears are reduced to simple openings behind the eyes. External ears are both undesirable—they would affect streamlining—and unnecessary, since water is an effective medium for sound transmission.

HUNTER OR FILTER-FEEDER

Cetaceans are divided into two main groups based on their feeding habits. Toothed species, predators of fishes, large invertebrates, seabirds, seals, and sometimes smaller cetaceans, catch prey with their sharp teeth and usually swallow it whole, without chewing. In contrast, filter-feeding baleen whales have fibrous baleen plates suspended from their upper jaw that act like sieves. Water, containing invertebrates and small fishes, is taken in and then forced out through the plates by the tongue, leaving behind the organisms.

PHYLUM	CHORDATA
CLASS	MAMMALIA
ORDER	CETACEA
FAMILIES	14
SPECIES	About 90

LAND-LIVING ANCESTORS

The taxonomic position of cetaceans has long been debated; their extreme adaptation to a wholly aquatic lifestyle masks anatomical features shared with other orders. Today it is generally accepted that Cetacea is most closely related to Artiodactyla—specifically the hippopotamus family—and so they are grouped together as Cetartiodactyla, at order or superorder level. Evidence comes mainly from genetic and molecular studies, but there is increasing anatomical support, including strong similarities between the ankle bones of artiodactyls and some fossil ancestors of whales.

RIGHT WHALES

Found in cool temperate and polar waters, the Balaenidae were considered the "right" ones to hunt, as they are easily approached, often close to shore, and have a thick layer of oily blubber—an adaptation to their cool water habitats. Right whales and the bowhead whale lack dorsal fins and throat grooves and their strongly curved jaws support the longest baleen plates of any whale.

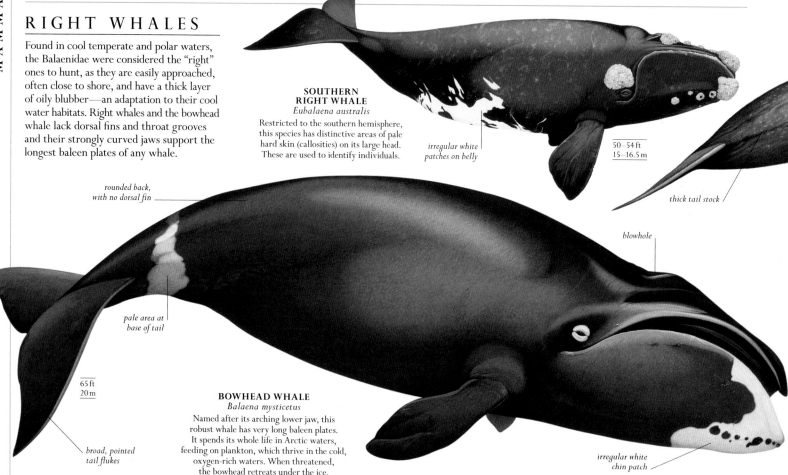

SOUTHERN RIGHT WHALE
Eubalaena australis
Restricted to the southern hemisphere, this species has distinctive areas of pale hard skin (callosities) on its large head. These are used to identify individuals.

irregular white patches on belly

50–54 ft
15–16.5 m

thick tail stock

blowhole

rounded back, with no dorsal fin

pale area at base of tail

65 ft
20 m

broad, pointed tail flukes

BOWHEAD WHALE
Balaena mysticetus
Named after its arching lower jaw, this robust whale has very long baleen plates. It spends its whole life in Arctic waters, feeding on plankton, which thrive in the cold, oxygen-rich waters. When threatened, the bowhead retreats under the ice.

irregular white chin patch

PYGMY RIGHT WHALE

Consisting of one species, the family Neobalaenidae is confined to waters in the southern hemisphere. Unlike true right whales, the pygmy right whale has a small, prominent dorsal fin, but no callosities on the head. Its baleen plates are ivory in color.

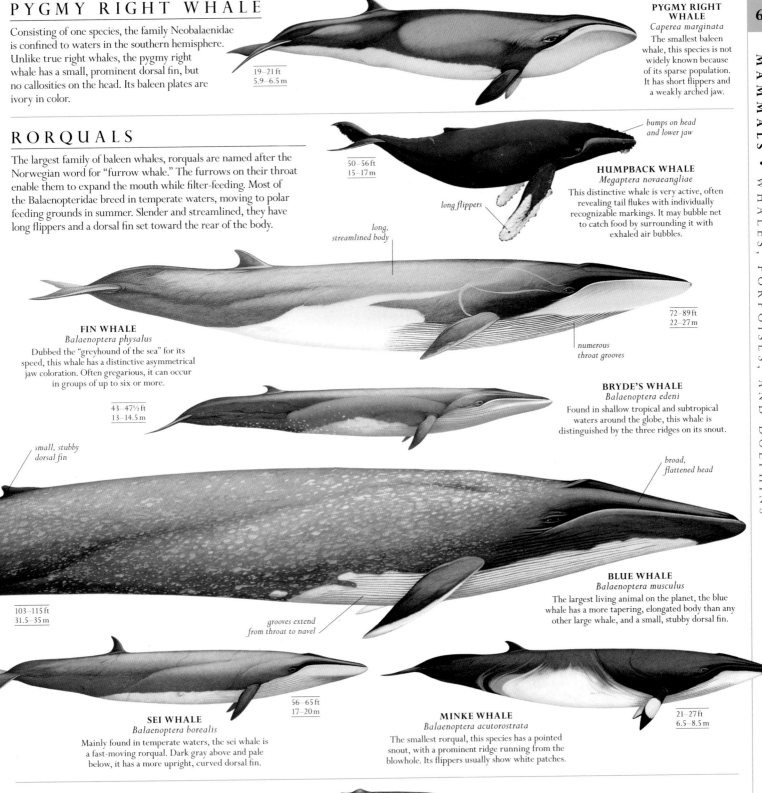

19–21 ft
5.9–6.5 m

PYGMY RIGHT WHALE
Caperea marginata
The smallest baleen whale, this species is not widely known because of its sparse population. It has short flippers and a weakly arched jaw.

RORQUALS

The largest family of baleen whales, rorquals are named after the Norwegian word for "furrow whale." The furrows on their throat enable them to expand the mouth while filter-feeding. Most of the Balaenopteridae breed in temperate waters, moving to polar feeding grounds in summer. Slender and streamlined, they have long flippers and a dorsal fin set toward the rear of the body.

bumps on head and lower jaw

50–56 ft
15–17 m

long flippers

HUMPBACK WHALE
Megaptera novaeangliae
This distinctive whale is very active, often revealing tail flukes with individually recognizable markings. It may bubble net to catch food by surrounding it with exhaled air bubbles.

long, streamlined body

FIN WHALE
Balaenoptera physalus
Dubbed the "greyhound of the sea" for its speed, this whale has a distinctive asymmetrical jaw coloration. Often gregarious, it can occur in groups of up to six or more.

43–47½ ft
13–14.5 m

72–89 ft
22–27 m

numerous throat grooves

BRYDE'S WHALE
Balaenoptera edeni
Found in shallow tropical and subtropical waters around the globe, this whale is distinguished by the three ridges on its snout.

small, stubby dorsal fin

broad, flattened head

103–115 ft
31.5–35 m

grooves extend from throat to navel

BLUE WHALE
Balaenoptera musculus
The largest living animal on the planet, the blue whale has a more tapering, elongated body than any other large whale, and a small, stubby dorsal fin.

SEI WHALE
Balaenoptera borealis
Mainly found in temperate waters, the sei whale is a fast-moving rorqual. Dark gray above and pale below, it has a more upright, curved dorsal fin.

56–65 ft
17–20 m

MINKE WHALE
Balaenoptera acutorostrata
The smallest rorqual, this species has a pointed snout, with a prominent ridge running from the blowhole. Its flippers usually show white patches.

21–27 ft
6.5–8.5 m

GRAY WHALE

The sole species in the family Eschrichtiidae is now restricted to the eastern North Pacific, having been hunted to extinction in the Atlantic. It undertakes vast annual migrations—the longest of any mammal—from the Bering Sea to breed in subtropical waters, especially off Baja California, Mexico.

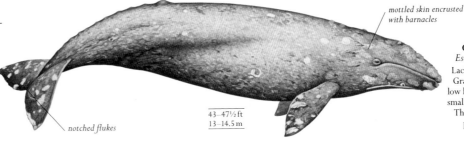

mottled skin encrusted with barnacles

notched flukes

43–47½ ft
13–14.5 m

GRAY WHALE
Eschrichtius robustus
Lacking a dorsal fin, the Gray whale's back has a low hump, with a series of smaller "knuckles" behind. The throat grooves are poorly developed.

BEAKED WHALES

The Ziphiidae are found in the open ocean, typically in small groups that congregate around underwater canyons. They feed near the sea floor, diving for an hour or more. Their facial beak usually has one or two pairs of teeth, which are for display only—food is simply sucked in. Because of their remote habitat and ability to dive deep, few of the 22 species in the family are familiar; some have never been sighted alive.

15–16 ft
4.5–5 m

14–15¾ ft
4.3–4.8 m

GRAY'S BEAKED WHALE
Mesoplodon grayi
Found throughout the southern hemisphere, this is one of the few beaked whales with numerous teeth. Its long, slender beak is often whitish.

BLAINVILLE'S BEAKED WHALE
Mesoplodon densirostris
An arched lower jaw and flattened forehead make this widespread species one of the more easily recognized beaked whales.

black face mask

15–17¼ ft
4.7–5.3 m

STRAP-TOOTHED WHALE
Mesoplodon layardii
This is a boldly marked species from the southern hemisphere. The male has long, curved teeth, which can meet over the upper jaw.

white oval patch with forward-pointing extensions

HUBB'S BEAKED WHALE
Mesoplodon carlhubbsi
Presumed to inhabit the North Pacific, but rarely seen, this species has white patches on its beak and head. The male has two prominent teeth.

18–20½ ft
5.5–6.3 m

13¾–16 ft
4.2–5 m

15–17¼ ft
4.7–5.3 m

GINKGO-TOOTHED BEAKED WHALE
Mesoplodon ginkgodens
This species is known mainly from strandings (washing ashore on beaches) in the Pacific and Indian oceans. The males have distinctively broad, triangular teeth.

GERVAIS' BEAKED WHALE
Mesoplodon europaeus
This slender whale is seen most often around the Canary Islands in the Atlantic Ocean. The male's teeth are small and at the tip of the beak.

SPERM WHALE

With a small lower jaw and rounded forehead, the sole member of the Physeteridae has a front-heavy appearance. The large head includes an organ filled with a waxy substance called spermaceti oil that acts as a ballast while diving, and is used to detect prey by echolocation. Sperm whales feed at great depths, especially on squid. The nasal bones are asymmetrical, and the blowhole is on the left side of the head.

36–50 ft
11–15 m

broad, triangular tail flukes

SPERM WHALE
Physeter catodon
The sperm whale is the largest living toothed animal, with teeth only in its lower jaw. Found worldwide, it displays its large, triangular tail flukes as it embarks on a deep dive.

unique wrinkly skin

PORPOISES

Generally smaller but stouter than dolphins, the Phocoenidae have a small, rounded head and blunt jaws. The dorsal fin is triangular rather than curved. Most distinctively, porpoises have flattened, spade-shaped teeth. Active predators of fishes, squid, and crustaceans, porpoises use sound to find prey and communicate. The seven species occur mainly in shallow coastal waters.

5½–8 ft
1.7–2.5 m

DALL'S PORPOISE
Phocoenoides dalli
This highly active porpoise from the North Pacific has a stout, bicolored body and a small head. It prefers the open ocean.

warty ridge

large, rounded dorsal fin

flukes

white area may increase with age

5–8 ft
1.5–2.5 m

SPECTACLED PORPOISE
Phocoena dioptrica
Although rarely seen in its subantarctic range, this species is readily recognized by its sharply demarcated colors. The upper blue-black skin makes it difficult to see from above.

INDIAN FINLESS PORPOISE
Neophocaena phocaenoides
Found throughout Asian coastal waters, including a freshwater population in China, this porpoise has a low warty ridge instead of a dorsal fin.

4½–5½ ft
1.4–1.7 m

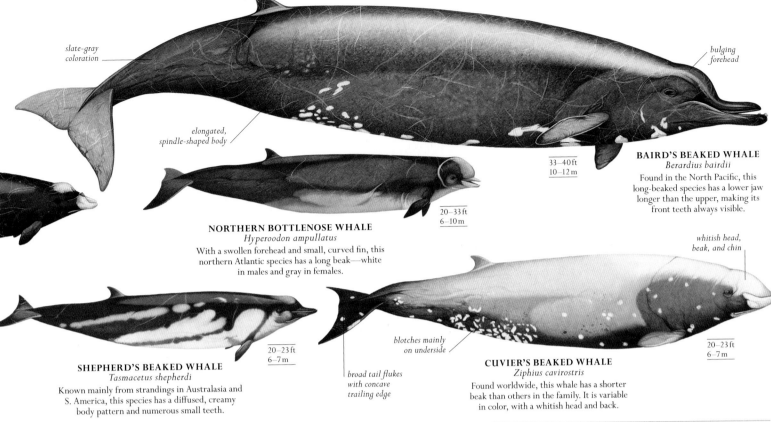

slate-gray coloration

bulging forehead

elongated, spindle-shaped body

33–40 ft
10–12 m

BAIRD'S BEAKED WHALE
Berardius bairdii
Found in the North Pacific, this long-beaked species has a lower jaw longer than the upper, making its front teeth always visible.

20–33 ft
6–10 m

NORTHERN BOTTLENOSE WHALE
Hyperoodon ampullatus
With a swollen forehead and small, curved fin, this northern Atlantic species has a long beak—white in males and gray in females.

whitish head, beak, and chin

20–23 ft
6–7 m

SHEPHERD'S BEAKED WHALE
Tasmacetus shepherdi
Known mainly from strandings in Australasia and S. America, this species has a diffused, creamy body pattern and numerous small teeth.

blotches mainly on underside

broad tail flukes with concave trailing edge

CUVIER'S BEAKED WHALE
Ziphius cavirostris
Found worldwide, this whale has a shorter beak than others in the family. It is variable in color, with a whitish head and back.

20–23 ft
6–7 m

PYGMY AND DWARF SPERM WHALES

The Kogiidae are small whales with a strongly recurved dorsal fin. They have a characteristic pale, crescent-shaped marking behind each eye that looks somewhat like the gill opening of a shark.

huge square head

small lower jaw

PYGMY SPERM WHALE
Kogia breviceps
One of the smallest whales, this species is an inhabitant of deep temperate and tropical waters. It is known mainly from strandings.

8¾–13¾ ft
2.7–4.2 m

NARWHAL AND BELUGA

The Monodontidae form a small family of two medium-sized Arctic whales that are rather unusual in appearance. They are highly gregarious species and are found in bays, estuaries, and fjords, and along the edges of pack ice, sometimes in herds of several hundred. Neither species has a true dorsal fin; both have a swollen, rounded forehead, which can change shape when they produce sounds, of which they have a wide range.

tusk protrudes from upper jaw of males

12–16 ft
3.7–5 m

NARWHAL
Monodon monoceros
Mottled gray and brown in color, the narwhal has only two teeth, one of which grows into a twisted tusk up to 10 ft (3 m) long in adult males.

body scarred by polar bear

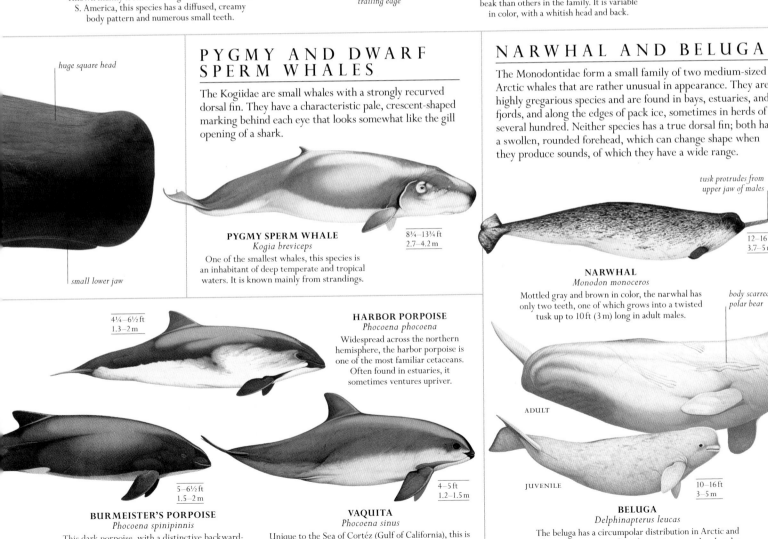

4¼–6½ ft
1.3–2 m

HARBOR PORPOISE
Phocoena phocoena
Widespread across the northern hemisphere, the harbor porpoise is one of the most familiar cetaceans. Often found in estuaries, it sometimes ventures upriver.

ADULT

5–6½ ft
1.5–2 m

BURMEISTER'S PORPOISE
Phocoena spinipinnis
This dark porpoise, with a distinctive backward-pointing dorsal fin, is one of the most abundant cetaceans around the coasts of S. America.

4–5 ft
1.2–1.5 m

VAQUITA
Phocoena sinus
Unique to the Sea of Cortéz (Gulf of California), this is the smallest and rarest porpoise, often inhabiting lagoons so shallow that its back protrudes above the surface.

JUVENILE

10–16 ft
3–5 m

BELUGA
Delphinapterus leucas
The beluga has a circumpolar distribution in Arctic and subarctic waters, spending the winter around and under pack ice. Uniquely, it is entirely white as an adult.

OCEANIC DOLPHINS

Found worldwide, often in shallow seas over continental shelves, the Delphinidae typically have a curved dorsal fin, protruding beak, and swollen forehead. Small- to medium-sized, their colors and patterns vary widely. Most mainly eat fishes and travel in herds or "pods." Larger species are collectively known as blackfishes.

HOURGLASS DOLPHIN
Lagenorhynchus cruciger

5¼–6¼ ft
1.6–1.9 m

Rarely seen, this subantarctic dolphin has two white flank lobes. It frequently associates with fin whales, and was valued by whalers, who used the dolphins to locate the whales.

WHITE-BEAKED DOLPHIN
Lagenorhynchus albirostris

7¼–10 ft
2.4–3.1 m

Highly acrobatic, this North Atlantic species frequently rides the bow wave of boats. It has a deeply curved dorsal fin and a thick, whitish beak.

8–9¼ ft
2.5–2.8 m

ATLANTIC WHITE-SIDED DOLPHIN
Lagenorhynchus acutus

Restricted to cool, North Atlantic waters, this dolphin has well-defined black, gray, and white markings, with a yellow flank stripe.

sickle-shaped dorsal fin

5¼–7 ft
1.6–2.1 m

DUSKY DOLPHIN
Lagenorhynchus obscurus

Widely distributed in coastal waters across the southern hemisphere, this is a gregarious and highly acrobatic dolphin, frequently leaping as it feeds.

RISSO'S DOLPHIN
Grampus griseus

This species has a distinctive rounded head and mostly gray body that becomes lighter and scarred with age. It is beakless and has a sickle-shaped dorsal fin.

broad, dark flukes

scarred body

12–13 ft
3.8–4.1 m

6½–7¼ ft
2–2.2 m

PEALE'S DOLPHIN
Lagenorhynchus australis

Found inshore around southern S. America, this dolphin shares its white "armpit" patch with *Cephalorhynchus* species, and may be closely related to that genus.

5 ft
1.5 m

COMMERSON'S DOLPHIN
Cephalorhynchus commersonii

A small, piebald, beakless dolphin from the waters around southern S. America and the Indian Ocean, this excellent swimmer and jumper has broad, blunt tail flukes.

7¾–8¾ ft
2.4–2.7 m

FRASER'S DOLPHIN
Lagenodelphis hosei

This highly gregarious species is found in deep water in the southern hemisphere. Its flippers, fin, and beak are small in proportion to its stocky body.

4½–5 ft
1.4–1.5 m

HECTOR'S DOLPHIN
Cephalorhynchus hectori

Found only in New Zealand waters, Hector's dolphin is the smallest member of its family. It has a rounded dorsal fin with an undercut rear margin.

5 ft
1.5 m

TUCUXI
Sotalia fluviatilis

Although inhabiting rivers in the Amazon basin, the tucuxi is related to oceanic species, not true river dolphins. It resembles a small bottlenose dolphin.

5¼–7½ ft
1.6–2.3 m

SHORT-BEAKED COMMON DOLPHIN
Delphinus delphis

Often forming large groups, this dolphin has a characteristic hourglass flank pattern, which can be seen easily when it leaps out of the water.

RIVER AND ESTUARINE DOLPHINS

The family Iniidae comprises three species of Amazon river dolphin, which can be recognized by their small eyes, bulging forehead, and long beak, as can the Franciscana of the family Pontopriidae. They share their long, slender beak with the South Asian river dolphin, although their teeth are not visible when the mouth is closed.

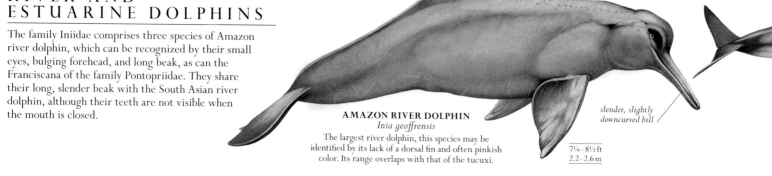

AMAZON RIVER DOLPHIN
Inia geoffrensis

The largest river dolphin, this species may be identified by its lack of a dorsal fin and often pinkish color. Its range overlaps with that of the tucuxi.

slender, slightly downcurved bill

7¼–8½ ft
2.2–2.6 m

ROUGH-TOOTHED DOLPHIN
Steno bredanensis

8½–9¼ ft
2.6–2.8 m

Occurring in most warm waters, this species has a conical head that extends into a long, slender beak, and a broad-based, pointed dorsal fin.

BOTTLENOSE DOLPHIN
Tursiops truncatus

6¼–13 ft
1.9–3.9 m

This widespread dolphin frequently interacts with humans. Offshore populations are larger and darker than those inshore, and have a shorter fin and beak.

short, pronounced beak

long, slender flippers

ATLANTIC SPOTTED DOLPHIN
Stenella frontalis

6¼–7½ ft
1.9–2.3 m

From the tropical and subtropical Atlantic, this dolphin has dark spots below and pale spots above. These increase in density as it matures.

STRIPED DOLPHIN
Stenella coeruleoalba

7¼–8½ ft
2.2–2.6 m

Found in temperate and tropical parts of all oceans, this acrobatic species has a distinctive pattern of blue stripes and wedges.

SOUTHERN RIGHT WHALE DOLPHIN
Lissodelphis peronii

10 ft
3 m

Within its range, the only dolphin to lack a dorsal fin, this distinctive black and white species occurs throughout the cool oceans of the southern hemisphere.

FALSE KILLER WHALE
Pseudorca crassidens

16½–20 ft
5.1–6.1 m

This uniformly dark species is widespread in shallow temperate and tropical waters. It feeds mainly on large fishes and squid, but will also tackle dolphins.

MELON-HEADED WHALE
Peponocephala electra

9¼ ft
2.8 m

An offshore tropical species, this whale has a rounded head and a tall, pointed dorsal fin. It is uniformly gray, with a darker face mask.

PYGMY KILLER WHALE
Feresa attenuata

7–8½ ft
2.1–2.6 m

Lacking a beak, this small, dark, robust species can be very aggressive. It kills and eats other dolphins within its circumtropical range.

tall dorsal fin

ORCA
Orcinus orca

23–32 ft
7–9.8 m

Boldly marked, with a tall dorsal fin, this species is at the top of marine food chain. It feeds on fishes, seals, sharks, and other cetaceans.

white underparts

large, broad flippers

LONG-FINNED PILOT WHALE
Globicephala melas

18½–22 ft
5.7–6.7 m

Prone to mass stranding, this sociable species is widespread in temperate waters. Its bulbous forehead can be seen as it lifts its head out of the water to look around.

SOUTH ASIAN RIVER DOLPHIN

The sole species in the family Platanistidae has two virtually identical subspecies in the rivers Indus and Ganges. Their long teeth are visible even when the mouth is closed. Their tiny eyes lack lenses and they are effectively blind.

SOUTH ASIAN RIVER DOLPHIN
Platanista gangetica

From gray to pale blue to brown, this dolphin has a distinctive long beak, large flippers, and triangular dorsal hump. It navigates and hunts using echolocation.

robust, uniformly colored body

sharp teeth

5½–8½ ft
1.7–2.6 m

FRANCISCANA
Pontoporia blainvillei

4–4½ ft
1.2–1.4 m

Inhabiting estuaries and coastal waters of eastern S. America, this species has, proportionally, the longest beak of any cetacean at 15 percent of its body length.

proportionally broad flukes

GLOSSARY

ABDOMEN
The hind part of the body. The abdomen lies below the ribcage in mammals, and behind the thorax in arthropods.

ACCESSORY FRUIT
A fruit that is formed from the flower's ovary and another structure, such as the swollen base of the flower. Examples of accessory fruits are apples and figs.

ADIPOSE FIN
A small fin behind the dorsal fin of a fish. It is mostly fatty tissue covered with skin.

AGARIC
Mushroomlike fruit body of a fungus, consisting of stem and cap.

ALKALOID
Bitter, sometimes poisonous, chemical produced by certain plants or fungi.

ALLUVIAL (DEPOSIT)
Concentrations of material that have been separated by weathering from the host rock, then deposited in rivers or streams.

AMPHIBOLE
Group of common rock-forming minerals, often with complex composition. Most are ferro-magnesian silicates.

ANAL FIN
The unpaired fin of a fish on the lower side of its body, behind the anus.

ANGIOSPERM
A seed-producing plant that encloses its seeds in a fruit and produces flowers. *See also* gymnosperm.

ANNUAL
A plant that completes its life cycle, from germination to death, in a single season of growth.

ANTENNA (PL. ANTENNAE)
A sensory feeler on the head of arthropods and some other invertebrates, such as mollusks. Antennae are always present in pairs, and can be sensitive to touch, sound, heat, and taste. Their size and shape varies widely according to the way in which they are used.

ANTHER
In flowering plants, the podlike structure on the stamen that produces pollen.

ANTLER
A bony growth on the head of deer. Unlike horns, antlers often branch, and in most cases they are grown and shed every year in a cycle linked with the breeding season.

ARBOREAL
Living fully or partly—known as semiarboreal—in trees.

ASCUS (PL. ASCI)
Microscopic saclike structure of sac fungi (Ascomycota) that produces spores.

ASEXUAL REPRODUCTION
A form of reproduction that involves just one parent, producing offspring that are genetically identical to each other (clones). It is most common in microbes, plants, and invertebrates.

BALEEN
A fibrous substance used by some whales for filtering food from water. Baleen grows in the form of plates with frayed edges, which hang from a whale's upper jaw. The baleen plates trap food, which the whale then swallows.

BASAL ANGIOSPERMS
Five orders of flowering plants with certain primitive features that diverged away from the main evolutionary line of flowering plants before others. Includes water lilies.

BASIDIUM (PL. BASIDIA)
Microscopic clublike structure of mushrooms and relatives (Basidiomycota) that produces spores.

BEAK
A set of narrow protruding jaws, usually without teeth. Beaks have evolved separately in many groups of vertebrates, including birds, tortoises, and some whales. The beak of a bird is also called a bill.

BERRY
A fleshy, many-seeded fruit of a plant that develops from a single ovary. Many fruits popularly called berries are not true berries but compound fruits—an example of which is raspberry.

BIENNIAL
A plant that completes its life cycle, from germination to death, in two years. It typically stores up food in the first season that can be used for reproduction in the second season.

BIOLUMINESCENCE
The production of light by living organisms.

BIPEDAL
An animal that moves on two legs.

BRACKET
Shelflike fruit body of a fungus.

BRACT
Modified leaf, often brightly colored, beneath a flower or flower cluster.

BROMELIAD
Flowering plant of the family Bromeliaceae. Almost exclusively tropical American, most live as rainforest epiphytes: perched in the rainforest branches without getting nourishment from the trees. Many form rosettes of leaves that collect rainwater, forming treetop pools that are important nurseries of insect larvae and frog tadpoles.

BROWSER
An animal that feeds on the leaves of trees and shrubs rather than on grasses.

BULB
The underground shoot of a plant, consisting of modified leaves and used for food storage during periods of dormancy or for asexual reproduction.

CALYX
Outer cuplike whorl of a flower, made up of sepals.

CAMOUFLAGE
Colors or patterns that enable an animal to merge with its background. It is often used for protection against predators and for concealment when approaching prey.

CANINE (TOOTH)
In mammals, a tooth with a single sharp point that is shaped for piercing and gripping prey. Canine teeth are located toward the front of the jaws, and are highly developed in carnivores.

CARAPACE
A hard shield on the back of some animals, including crustaceans, arachnids, and some reptiles. In turtles and tortoises, the carapace is the upper part of the shell.

CARBOHYDRATE
Foodstuff, such as sugar or starch, that provides energy.

CARNASSIAL (TOOTH)
In mammalian carnivores, a bladelike upper premolar and lower molar tooth that have evolved for slicing through flesh.

CARNIVORE
Any animal that eats meat. The word carnivore can also be used in a more restricted sense to mean mammals of the order Carnivora.

CARPEL
Female reproductive part of a flower, divided into ovary, style, and stigma. Also known as the pistil.

CARTILAGE
A rubbery substance forming part of vertebrate skeletons. In most vertebrates, it lines the joints, but in cartilaginous fishes it forms the whole skeleton.

CATKIN
Hanging cluster of flowers, usually with simple flowers all of the same sex.

CECUM
Pouch of the digestive tract, often used in the digestion of plant food.

CELL
The smallest unit of an organism that can exist on its own.

CELLULOSE
A complex carbohydrate found in plants. It is used by plants as building material, and has a resilient chemical structure that animals find hard to digest. Plant-eating animals break it down in their stomach with the aid of microorganisms.

CHITIN
A tough carbohydrate making up cell walls of fungi and certain animal exoskeletons.

CHLOROPHYLL
Green pigment found in chloroplasts used to trap light energy in photosynthesis.

CHLOROPLAST
Granule inside the cell of a photosynthesizing eukaryote used for photosynthesis. *See* photosynthesis.

CHROMOSOME
Microscopic filament inside a cell that carries genetic information (DNA).

CLIMBER
A plant that grows up a vertical surface, such as a rock or a tree, using it for support. Climbers do not gain nourishment from other plants, but may weaken them by blocking out light.

CLOACA
An opening toward the rear of the body that is shared by several body systems. In some vertebrates—such as bony fishes and amphibians—the gut, kidneys, and reproductive systems all use this opening, mainly for waste.

CLONES
Genetically identical individuals. Two or more identical organisms that share exactly the same genes.

CLOVEN-HOOFED
Having hoofs that look as though they are split in half. Most cloven-hoofed mammals, such as deer and antelope, actually have two hoofs, arranged on either side of a line that divides the foot in two.

COCOON
A case made of open, woven silk. Many insects spin a cocoon before they begin pupation, and many spiders spin one to hold their eggs.

COLONY
A group of animals belonging to the same species that spend their lives together, often dividing up the tasks involved in survival. In some colonial species, particularly aquatic invertebrates, the colony members are permanently fastened together. In others, such as ants, bees, and wasps, members forage for food independently but live in the same nest.

COMMENSAL
Living in close relationship with another species without either helping it or damaging it.

COMPOUND
A substance made up of two or more elements chemically reacted together.

COMPOUND EYE
An eye that is divided into separate compartments, each with its own set of lenses. The number of compartments the eye can contain varies from a few dozen to thousands. Compound eyes are a common feature of arthropods.

COMPOUND LEAF
A leaf with a blade divided into smaller leaflets. *See also* simple leaf.

CONE
The reproductive structure of a plant consisting of a cluster of scales or bracts used for producing spores, ovules, or pollen. Cones are found on various trees, particularly pine trees.

CORM
The underground storage organ of a plant, formed from a swollen stem base.

COROLLA
Inner whorl of a flower made up of petals.

CRYPTIC (COLORATION)
Coloration and markings that make an animal hard to see against its background.

CRYSTAL
A solid that has a definite internal atomic structure, producing a characteristic external shape and certain physical and optical properties.

CYTOPLASM
The jellylike interior of a cell. In the cells of eukaryotes it is restricted to the region around the nucleus.

DECIDUOUS
A plant that seasonally loses its leaves. For instance, many temperate plants lose their leaves in winter.

DELAYED IMPLANTATION
In mammals, a delay between the fertilization of an egg and the subsequent development of an embryo. This delay allows birth to occur when conditions, such as food availability, are favorable for raising young.

DNA (DEOXYRIBONUCLEIC ACID)
The chemical substance found in the cells of all living organisms that determines their inherited characteristics.

DICOTS (DICOTYLEDONS)
Group of flowering plants with two cotyledons (seed leaves).

DIOECIOUS
A plant that has male and female parts on different individuals.

DOMESTICATED
An animal that lives fully or partly under human control. Some such animals look identical to their counterparts in the wild, but many have been bred to produce artificial varieties not found in nature.

DORSAL FIN
The unpaired fin of a fish on the back of its body.

ECHOLOCATION
A method of sensing nearby objects by using pulses of high-frequency sound. Echoes bounce back from obstacles and other animals, allowing the sender to build up a picture of its surroundings. It is used by some bats, cave-dwelling birds, and toothed whales.

ECLIPSE PLUMAGE
In some birds, particularly waterfowl, an unobtrusive plumage adopted by the males once the breeding season is over.

ECOSYSTEM
A collection of species living in the same habitat, together with their physical surroundings.

ECTOPARASITE
An organism that lives parasitically on the surface of another organism's body. Some animal ectoparasites spend all their lives on their hosts, but many—including fleas and ticks—develop elsewhere and climb onto the host in order to feed.

ECTOTHERMIC
Having a body temperature that is dictated principally by the temperature of the surroundings. Ectothermic is also known as cold-blooded.

ELEMENT
Chemical substance that cannot be broken down into a simpler chemical form.

EMBRYO
A young animal or plant at a rudimentary stage of development.

ENDEMIC
A species native to a particular geographic area, such as an island, forest, mountain, state, or country, and which is found nowhere else.

ENDOPARASITE
An organism that lives parasitically inside the body of another organism, either feeding directly on its tissues or stealing some of its food. Endoparasites frequently have complex life cycles involving more than one host.

ENDOSKELETON
An internal skeleton, typically made of bone. Unlike an exoskeleton, this kind of skeleton can grow in step with the rest of the body.

ENDOTHERMIC
Able to maintain a constant, warm body temperature, regardless of external conditions. Also known as warm-blooded.

ENZYME
A group of substances produced by all living things that promotes a chemical process, such as photosynthesis or digestion.

EPIPHYTE
A plant or plantlike organism (such as an alga or lichen) that lives on the body of another plant without getting any nourishment from it.

EUDICOTS (EUDICOTYLEDONS)
Any one of the higher dicot flowering plant orders that together contain the majority of flowering plants.

EUKARYOTE
An organism whose cells have a nucleus. Protists, fungi, plants, and animals are eukaryotes.

EVAPORITE (DEPOSIT)
Sedimentary rock or mineral resulting from the evaporation of water from mineral-bearing fluids, usually sea water.

EVERGREEN
A plant that does not lose its leaves seasonally, such as a conifer.

EXTERNAL FERTILIZATION
In reproduction, a form of fertilization that takes place outside the female's body, usually in water, for example in most fishes.

EXOSKELETON
An external skeleton that supports and protects an animal's body. The most complex exoskeletons, formed by arthropods, consist of rigid plates that meet at flexible joints. This kind of skeleton cannot grow, and has to be shed and replaced at periodic intervals.

FERAL
An animal that comes from domesticated stock but which has subsequently taken up life in the wild. Examples include city pigeons, cats, and horses.

FERTILIZATION
The union of an egg cell and sperm, which creates a cell capable of developing into a new organism.

FLAGELLUM (PL. FLAGELLA)
A whiplike structure of a cell used for propulsion. It is the main structure of locomotion in flagellate protists.

FLOWER
Reproductive structure of the largest group of seed plants, typically consisting of sepals, petals, stamens, and carpels.

FOOD CHAIN
A series of organisms, each of which is eaten by the next.

FOSSIL
Any record of past life preserved in Earth's crust. Fossils include bones, shells, footprints, excrement, and burrows.

FRUIT
A fleshy structure of a plant that develops from the ovary of the flower and contains one or more seeds. Fruits can be simple, like berries, or compound, where the fruits of separate flowers are merged. *See also* accessory fruit.

FRUIT BODY
Fleshy spore-producing structure of a fungus, typically in the shape of a mushroom (agaric) or bracket.

GALL
A tumorlike growth in a plant that is induced by another organism (such as a fungus or an insect). By triggering the formation of galls, animals provide themselves with a safe hiding place and a convenient source of food.

GAMETE
A sex cell. In animals, this is either a sperm cell or an unfertilized egg cell.

GENE
The basic unit of heredity in all living things, typically a segment of DNA that provides the coded instructions for a particular protein.

GERMINATION
Developmental stage where a seed or a spore begins to grow.

GESTATION (PERIOD)
Length of time from fertilization to birth in a live-bearing animal.

GILL
In fishes, amphibians, crustaceans, and mollusks, an organ used for extracting oxygen from water. Unlike lungs, gills are out-growths of the body. In fungi, gills are the bladelike, spore-producing structures under the cap of agaric fungi.

GRAZER
An animal that feeds on grass or algae.

GYMNOSPERM
A seed plant that doesn't enclose its seeds in a fruit. Many gymnosperms carry their seeds on cones. *See also* angiosperm.

HERB (HERBACEOUS PLANT)
A nonwoody plant, usually much shorter than shrubs or trees.

HERBIVORE
An animal that feeds on plants or algae.

HERMAPHRODITE
An organism that has both male and female sex organs.

HIBERNATION
A period of dormancy in winter. During hibernation, an animal's body processes drop to a low level to conserve energy.

HORMONE
A chemical signal produced by one part of the body that changes the behavior of another part of the body.

HORN
In mammals, a pointed growth on the head. True horns are hollow, and are often curved.

HOST
An organism on or in which a parasite, or symbiont, feeds.

HYDROTHERMAL VEIN
Sheet-shaped mass of material altered or deposited by water heated by igneous activity in rock.

HYPHA (PL. HYPHAE)
Microscopic threadlike structure making up the body of a fungus. Many such hyphae make up a mass that is known as a mycelium.

IGNEOUS ROCK
Rock formed from erupted volcanic lava or solidified magma.

INCISOR (TOOTH)
In mammals, a flat tooth at the front of the jaw that is shaped for slicing or gnawing.

INCUBATION
In birds, the period when a parent sits on the eggs and warms them, allowing them to develop. Incubation periods range from under 14 days to several months.

INFLORESCENCE
A flower cluster or a single flower.

INORGANIC
Chemical substance that is not based on the element carbon.

INSECTIVORE
An animal that feeds on insects.

INTERNAL FERTILIZATION
In reproduction, a form of fertilization that takes place inside the female's body. Internal fertilization is a characteristic of many land animals, particularly insects and vertebrates.

KEEL
In birds, an enlargement of the breastbone that is anchored to the muscles that are used in flight.

KERATIN
A tough structural protein found in hair, claws, and horns.

LARVA (PL. LARVAE)
An immature but independent animal that looks completely different from an adult of the species. A larva develops the adult shape by metamorphosis; in many insects, the change takes place in a resting stage that is called a pupa.

LAVA
Molten rock that has erupted from a volcano, and then hardens.

LEGUME
Plants of the pea family of flowering plants, the Fabaceae. They are important for having root nodules that contain nitrogen-fixing bacteria.

LEK
A communal display area used by male animals (particularly birds) during courtship. The same location is often revisited for many years.

LICHEN
Mutualistic association between a fungus and a photosynthetic alga. From this relationship, the fungus obtains sugars, and the alga obtains minerals.

LIFE CYCLE
The developmental sequence of an organism from gametes (sex cells) to death.

LOCOMOTION
Movement from place to place.

LUSTER
The shine or look of a mineral due to the reflection of light off its surface.

MAGMA
Rock in a molten state and lying below Earth's surface.

MAGNOLIIDAE
Group of flowering plants consisting of four orders with certain primitive characteristics, such as undifferentiated tepals instead of sepals and petals.

MANDIBLE
The paired jaws of an arthropod, or, in vertebrates, the bone that makes up all or part of the lower jaw.

METABOLISM
The complete array of chemical processes that takes place inside an animal's body. Some of these processes release energy by breaking down food, while others use energy, for example, by making the body's muscles contract.

METAMORPHIC ROCK
Rock that has been changed by heat, pressure, or both to form new rock consisting of new minerals.

METAMORPHOSIS
A change in body shape undergone by many animals—particularly invertebrates—as they grow from being a juvenile to an adult. In insects, metamorphosis can be complete or incomplete. Complete metamorphosis involves a total shape change during a resting stage, which is called a pupa. Incomplete metamorphosis involves a series of less drastic changes, and these occur each time the young animal molts.

MIGRATION
A journey undertaken to a different region, following a well-defined route. Most migratory animals move with the seasons to take advantage of good breeding conditions in one place, and a suitable wintering climate in another.

MIMICRY
A form of camouflage in which an animal resembles another animal or an inanimate object, such as a twig or a leaf. Mimicry is very common in insects, with many harmless species imitating ones that have dangerous bites or stings.

MINERAL
An inorganic, naturally occurring material with a constant chemical composition and regular internal atomic structure.

MITOCHONDRION (PL. MITOCHONDRIA)
Granule inside the cell of a eukaryote used for respiration. Mitochondria use up oxygen to release energy.

MOLAR (TOOTH)
In mammals, a tooth at the rear of the jaw. Molar teeth often have a flattened or ridged surface, and deep roots. They are normally used for chewing.

MONOCOTS (MONOCOTYLEDONS)
Group of flowering plants with a single seed leaf (cotyledon).

MONOECIOUS
A plant that has separate male and female parts on the same individual.

MONOGAMOUS
Mating with a single partner, either during the course of one breeding season or throughout life. Monogamous partnerships are common in animals that care for the development of their young.

MOLT
The shedding of fur, feathers, or skin so that it can be replaced. Mammals and birds molt to keep their fur and feathers in good condition, to adjust their insulation, or so they can be ready to breed. Arthropods, such as insects, molt their exoskeleton in order to grow.

MUTUALISM
A relationship between two different species within an ecological community where both species benefit. For instance, a flowering plant and a pollinating insect share a mutualistic relationship.

MYCELIUM
Mass of threadlike hyphae making up the body of a fungus.

MYCORRHIZA (PL. MYCORRHIZAE)
The mutualistic association between a fungus and the roots of a plant. The fungus obtains sugars, while the plant increases its mineral uptake by absorbing minerals via the fungus's extensive mycelium.

NICHE
An organism's place and role in its habitat. Although two species may share the same habitat, they never share the same niche.

NITROGEN FIXATION
A chemical process whereby atmospheric nitrogen is converted into more complex nitrogen-containing substances, such as protein. Nitrogen fixation is undertaken by certain types of microorganisms.

NODE
The point where two sections of a plant's stem join, from which one or more leaves, shoots, branches, or flowers arise.

NUCLEUS (PL. NUCLEI)
The structure inside the cell of a eukaryote that contains chromosomes.

NUT
Dry, hard-shelled fruit of some plants, usually a single seed.

NYMPH
An immature insect that looks similar to its parents but which does not have functioning wings or reproductive organs. A nymph develops its adult form by metamorphosis, changing slightly each time it molts.

OMNIVORE
An animal that eats both plants and other animals as its primary food source.

OPERCULUM
A cover or lid. In some gastropod mollusks, an operculum is used to seal the shell when the animal has withdrawn inside. In bony fishes, an operculum on each side of the body protects the chamber containing the gills.

OPPOSABLE
Able to be pressed together from opposite directions. For example, many primates have opposable thumbs, which can be pressed against the fingers so objects can be grasped.

ORGAN
The structure of a body that carries out a particular function. Examples are the heart, skin, or a leaf.

ORGANELLE
A specialized structure that forms part of a plant or animal cell.

ORGANIC
Chemical substance based on carbon.

OVIPAROUS
Reproducing by laying eggs.

OVIPOSITOR
An egg-laying tube extending out of the body of some female animals, especially in insects.

OVULE
Structure that contains the egg of a seed plant. These are encased in an ovary in flowering plants, but naked in gymnosperms. After fertilization the ovule becomes the seed.

PARASITE
An organism that lives on or in another host organism, gaining advantage (such as nourishment) while causing the host harm. Most parasites are much smaller than their host and have complex life cycles involving the production of huge numbers of young. Parasites often weaken their host but generally do not kill them.

PARTHENOGENESIS
A form of reproduction in which an egg cell develops into a young animal without having to be fertilized, producing offspring that are genetically identical to the parent. In animals that have separate sexes, young produced by parthenogenesis are always female. Parthenogenesis is common in invertebrates.

PECTORAL FIN
One of the two paired fins positioned toward the front of a fish's body, often just behind its head. Pectoral fins are usually highly mobile and are normally used for maneuvering.

PELVIC FIN
The rear paired fins in fishes, which are normally positioned close to the underside, sometimes near the head but more often towards the tail. Pelvic fins are generally used as stabilizers.

PERENNIAL
A plant that normally lives for more than one season.

PERIANTH
The outer two whorls of a flower (the calyx, made up of sepals, and the corolla, made up of petals); especially where the two are undifferentiated.

PETAL
One of the parts of the corolla of a flower. Petals are often brightly colored to attract pollinating animals.

PETIOLE
The stalk of a leaf.

PHEROMONE
A chemical produced by one animal that has an effect on other members of its species. Pheromones are often volatile substances

that spread through the air, triggering a response from animals some distance away.

PHOTOSYNTHESIS
A process whereby organisms use light energy to make food and oxygen; photosynthesis occurs in plants, algae, and many microorganisms.

PLACENTA
An organ developed by an embryo mammal that allows it to absorb nutrients and oxygen from its mother's bloodstream before it is born.

PLANKTON
Floating organisms, many of them microscopic, that drift in open waters, particularly near the surface of the sea. Planktonic organisms can often move, but most are too small to make any headway against strong currents. Planktonic animals are known as zooplankton. Planktonic algae are called phytoplankton.

PLASTRON
The lower part of the shell structure of tortoises and turtles.

POLLEN
Tiny grains produced by seed plants that contain male gametes for fertilizing the female egg of either a flowering plant or a coniferous plant.

POLYGAMOUS
A reproductive system in which individuals mate with more than one partner during the course of a single breeding season.

PREDATOR
An animal that catches and kills other animals, referred to as its prey. Some predators attack and catch their prey by lying in wait, but most actively pursue and attack other animals.

PREHENSILE
Able to curl around objects and grip them.

PREMOLAR (TOOTH)
In mammals, a tooth positioned midway along the jaw between the canine and the molar teeth. In carnivores, specialized premolar teeth act like shears, slicing through flesh.

PROBOSCIS
An animal's nose, or set of mouthparts with a noselike shape. In insects that feed on fluids, the proboscis is often long and slender, and can usually be stowed away when not in use.

PROKARYOTE
An organism whose cells do not have a nucleus. The Archaea and Bacteria used to be classified as Prokaryota.

PRONOTUM
The part of an insect's cuticle that covers the first segment of its thorax, often hardening into a shell.

PROTEIN
Substance found in food, such as meat, fish, cheese, and beans, that is used for growth and for carrying out a variety of essential biological functions.

PSEUDOPOD
Temporary projection of a cell, such as an amoeba or a white blood cell, which is used for creeping forward or for catching food.

PUPA (PL. PUPAE)
In insects, a stage during which the larval body is broken down and rebuilt as an adult. During the pupal stage the insect does not feed and usually cannot move, although some pupae may wriggle if they are touched. The pupa is protected by a hard case, which itself is sometimes wrapped in silk.

QUADRUPEDAL
An animal that walks on four legs.

RAPTOR
A bird of prey.

RHIZOME
A creeping or underground stem that can send out new shoots.

ROCK
Material made up of one or more minerals.

ROOT NODULE
Spherical swelling on the root of a legume that contains nitrogen-fixing bacteria.

RUMINANT
A hoofed mammal that has a specialized digestive system with several stomach chambers. One of these—the rumen—contains large numbers of microorganisms which help break down plant food. To speed this process, a ruminant usually regurgitates its food and rechews it, a process called chewing the cud.

SALLY
A bird's short flight from a perch to catch an invertebrate, often in mid-air.

SCUTE
A shieldlike plate or scale that forms a bony covering on some animals.

SEDIMENTARY ROCK
Rock formed by the consolidation and hardening of rock fragments, organic remains, or other material.

SEED
Developmental stage of a seed plant, consisting of an encapsulated embryo.

SEPAL
One of the parts of the calyx of a flower, usually small and leaflike and enveloping the unopened flower bud.

SEXUAL DIMORPHISM
Showing physical differences between males and females. In animals that have separate sexes, males and females always differ, but in highly dimorphic species, such as elephant seals, the two sexes look very different and are often unequal in size.

SHOOT
The aerial part of plant. It is a new growth, usually growing upward.

SHRUB
Woody perennial plant with multiple stems.

SIMPLE LEAF
A leaf with an undivided blade.

SPERMATOPHORE
A packet of sperm that is transferred either directly from male to female, or indirectly—for example, by being left on the ground. Spermatophores are produced by a range of animals, including salamanders, squid, and some arthropods.

SPIRACLE
In some fishes, an opening behind the eye that lets water flow into the gills. In insects and myriapods, the spiracle is an opening on the body wall that lets air into the tracheal system.

SPORE
A single cell containing half the quantity of genetic material of typical body cells. Unlike gametes, spores can divide and grow without being fertilized. Spores are produced by fungi, algae, and plants.

SPOROPHYTE
The spore-producing stage of a plant. It is the dominant (visible) stage of ferns and seed plants.

STAMEN
Male reproductive part of a flower. It has an anther borne on a long filament.

STEREOSCOPIC (VISION)
Vision in which the two eyes face forward, giving overlapping fields of view and allowing the animal to judge distance. Also called binocular vision.

STOMA (PL. STOMATA)
Tiny adjustable pore on the surface of a plant that allows the exchange of gases for photosynthesis and respiration.

SWIM BLADDER
A gas-filled bladder that most bony fishes use to regulate their buoyancy. By adjusting the gas pressure inside the bladder, a fish can become neutrally buoyant, meaning that it neither rises nor sinks.

SYMBIOSIS
Any relationship between two different species within an ecological community. Examples of symbiotic relationships are predator–prey, parasite–host, and mutualism.

TEPAL
The outer part of a flower undifferentiated into sepals and petals. Tepals collectively make up the perianth.

TERRESTRIAL
Living wholly or mainly on the ground.

TERRITORY
An area defended by an animal, or group of animals, against other members of the same species. Territories often include resources, such as food supply, that help the male attract a mate.

THORAX
The middle region of an arthropod's body. The thorax contains powerful muscles and, if the animal has any, bears legs and wings.

In vertebrates with four limbs, the thorax is the chest.

TORPOR
A sleeplike state in which body processes slow to a fraction of their normal rate. Animals usually become torpid to survive difficult conditions, such as extreme cold or lack of food.

TRACHEAL SYSTEM
A system of minute tubes that arthropods (for example, insects) use to carry oxygen into their bodies. Air enters the tubes through openings called spiracles, and then flows through the tracheae to reach individual cells.

TREE
A woody perennial plant usually with a well-defined stem, or trunk, and a crown of branches above.

UNGULATE
A hoofed mammal.

UTERUS
In female mammals, the part of the body that contains and usually nourishes developing young. In placental mammals, the young are connected to the wall of the uterus via a placenta.

VECTOR
An organism that transmits a disease-causing parasite from one host to another.

VEIL
Thin skin- or weblike tissue that protects the fruit body of a fungus.

VESTIGIAL
Relating to an organ that is atrophied or nonfunctional.

VIVIPAROUS
Reproducing by giving birth to live young.

VOLVA
A saclike remnant of a veil at the base of a fungus's fruit body stem.

WOODY PLANT
A plant that has wood—a type of strengthening tissue found in plants consisting of thick-walled water-transporting vessels.

ZYGODACTYL FEET
A specialized arrangement of the feet in which the toes are arranged in pairs, with the second and third toes facing forward and the first and fourth toes facing backward. This adaptation helps birds to climb and perch on tree trunks and other vertical surfaces. Several groups of birds have zygodactyl feet, including parrots, cuckoos, turacos, owls, and toucans, woodpeckers, and their relatives.

INDEX

Page numbers in **bold** type refer to feature pages or introductions to animal groups.

643

ACKNOWLEDGMENTS

Consultants at the Smithsonian Institution:

Dr. Don E. Wilson, Senior Scientist/Chair of the Department of Vertebrate Zoology;
Dr. George Zug, Emeritus Research Zoologist, Department of Vertebrate Zoology, Division of Amphibians and Reptiles;
Dr. Jeffrey T. Williams: Collections Manager, Department of Vertebrate Zoology

Dr. Hans-Dieter Sues, Curator of Vertebrate Paleontology/Senior Research Geologist, Department of Paleobiology

Paul Pohwat, Mineral Collection Manager, Department of Mineral Sciences;
Leslie Hale, Rock and Ore Collections Manager, Department of Mineral Sciences;
Dr. Jeffrey E. Post, Geologist/Curator, National Gem and Mineral Collection, Department of Mineral Sciences

Dr. Carla Dove, Program Manager, Feather Identification Lab, Division of Birds, Department of Vertebrate Zoology

Dr. Warren Wagner, Research Botanist/ Curator, Chair of Botany, and Staff of the Department of Botany

Gary Hevel, Museum Specialist/Public Information Officer, Department of Entomology;
Dana M. De Roche, Department of Entomology

Department of Invertebrate Zoology:
Dr. Rafael Lemaitre: Research Zoologist/ Curator of Crustacea;
Dr. M. G. (Jerry) Harasewych, Research Zoologist;
Dr. Michael Vecchione, Adjunct Scientist, National Systemics Laboratory, National Marine Fisheries Service, NOAA;
Dr. Chris Meyer, Research Zoologist;
Dr. Jon Norenburg, Research Zoologist;
Dr. Allen Collins, Zoologist, National Systemics Laboratory, National Marine Fisheries Service, NOAA;
Dr. David L. Pawson, Senior Research Scientist;
Dr. Klaus Rutzler, Research Zoologist;
Dr. Stephen Cairns, Research Scientist / Chair

Additional consultants:

Dr. Diana Lipscomb, Chair and Professor Biological Sciences, George Washington University

Dr. James D. Lawrey, Department of Environmental Science and Policy, George Mason University

Dr. Robert Lücking, Research Collections Manager/Adjunct Curator, Department of Botany, The Field Museum

Dr. Thorsten Lumbsch, Associate Curator and Chair, Department of Botany, The Field Museum

Dr. Ashleigh Smythe, Visiting Assistant Professor of Biology, Hamilton College

Dr. Matthew D. Kane, Program Director, Ecosystem Science, Division of Environmental Biology, National Science Foundation

Dr. William B. Whitman, Department of Microbiology, University of Georgia

Andrew M. Minnis: Systematic Mycology and Microbiology Laboratory, USDA

Dorling Kindersley would like to thank the following people for their assistance with this book:
David Burnie, Kim Dennis-Bryan, Sarah Larter, and Alison Sturgeon for structural development; Hannah Bowen, Sudeshna Dasgupta, Jemima Dunne, Angeles Gavira Guerrero, Cathy Meeus, Andrea Mills, Manas Ranjan Debata, Paula Regan, Alison Sturgeon, Andy Szudek, and Miezan van Zyl for additional editing; Avanika, Helen Abramson, Niamh Connaughton, Sonali Jindal, Anita Kakkar, Nayan Keshan, Chhavi Nagpal, Manisha Majithia, and Claire Rugg for editorial assistance; Sudakshina Basu, Steve Crozier, Clare Joyce, Edward Kinsey, Amit Malhotra, Pooja Pipil, Aparajita Sen, Neha Sharma, Nitu Singh, Sonakshi Sinha, and George Thomas for additional design; Amy Orsborne for jacket design; Richard Gilbert, Ann Kay, Anna Kruger, Constance Novis, Nikky Twyman, and Fiona Wild for proofreading; Sue Butterworth for the index; Claire Cordier, Laura Evans, Rose Horridge, and Emma Shepherd from the DK picture library; Syed Mohammad Farhan, Vijay Kandwal, Ashok Kumar, Nityanand Kumar, Pawan Kumar, Mrinmoy Mazumdar, Shanker Prasad, Mohd Rizwan, Vikram Singh, Bimlesh Tiwary, Anita Yadav, and Tanveer Zaidi for technical support; Mohammad Usman for production; Stephen Harris for reviewing the plants chapter; Dr. Gregory Kenicer for his taxonomic advice on plants; and Derek Harvey, for his tremendous knowledge and unstinting enthusiasm for this book.

The publisher would also like to thank the following companies for their generosity in allowing Dorling Kindersley access to their collections for photography:
Anglo Aquatic Plant Co Ltd, Strayfield Road, Enfield, Middlesex EN2 9JE, http://angloaquatic.co.uk; **Cactusland,** Southfield Nurseries, Bourne Road, Morton, Bourne, Lincolnshire PE10 0RH, www.cactusland.co.uk; **Burnham Nurseries Orchids,** Burnham Nurseries Ltd, Forches Cross, Newton Abbot, Devon TQ12 6PZ, www.orchids.uk.com; **Triffid Nurseries,** Great Hallows, Church Lane, Stoke Ash, Suffolk IP23 7ET, www.triffidnurseries.co.uk; **Amazing Animals,** Heythrop, Green Lane, Chipping Norton, Oxfordshire OX7 5TU, www.amazinganimals.co.uk; **Birdland Park and Gardens,** Rissington Rd, Bourton-on-the-Water, Gloucestershire GL54 2BN, www.birdland.co.uk; **Virginia Cheeseman F.R.E.S.,** 21 Willow Close, Flackwell Heath, High Wycombe, Buckinghamshire HP10 9LH, www.virginiacheeseman.co.uk; **Colchester Zoo,** www.colchester-zoo.com; **Cotswold Falconry Centre,** Batsford Park, Batsford, Moreton in Marsh, Gloucestershire GL56 9AB, www.cotswold-falconry.co.uk; **Cotswold Wildlife Park,** Burford, Oxfordshire OX18 4JP, www.cotswoldwildlifepark.co.uk; **Emerald Exotics; Shaun Foggett,** www.crocodilesoftheworld.co.uk.

Picture credits
Alamy Images: agefotostock / Marevision 111fbl, 343cra, The Africa Image Library 557, Amazon Images 549, Arco Images GmbH / Huetter C 601, Art Directors & TRIP 143, blickwinkel 144, 146, 188, 307, 325, 326, 569, 615, blickwinkel / Hartl 336clb, Christian Hütter 379bc, Design Pics Inc / Milo Burcham 5ca, 249br, 323br, 506–507, Steffen Hauser / botanikfoto 142, 159tc, Penny Boyd 600, Brandon Cole Marine Photography 521, BSIP SA 93, James Caldwell 267, Rosemary Calvert 20, CuboImages srl 147, Andrew Darrington 291, Danita Delimont 151, Garry DeLong 105, Paul Dymond 458, Emilio Ereza 348, David Fleetham 324, Florapix 148, Florida Images 148, FLPA 547cl, Frank Hecker 328tl, Minden Pictures 621cla, Paulo Oliveira 328cla, 339tr, Poelzer Wolfgang 343cr; Jane Gould 22–23b, Martin Fowler 182, Les Gibbon 305, Rupert Hansen 29, Chris Hellier 143, Imagebroker / Arco / G. Lacz 596cr, Imagebroker / Florian Kopp 580, Indiapicture / P S Lehri 611, Interphoto 29, T. Kitchin & V. Hurst 23, Chris Knapton 27, S & D & K Maslowski / FLPA 28, Carver Mostardi 567, Tsuneo Nakamura / Volvox Inc 585, The Natural History Museum, London 258, Nature Picture Library / Sue Daly 252cb, National Geographic Image Collection / Joel Sartore 571br, Nic Hamilton Photographic 29, Pictorial Press Ltd, 28, Matt Smith 166, Stefan Sollfors 266, Sylvia Cordaiy Photo Library Ltd 17, Natural Visions 149, Joe Vogan 576, Wildlife GmbH 28, 130, 151, WoodyStock 155; **Maria Elisabeth Albinsson:** CSIRO 95cr, 100bc; **Algaebase.org:** Robert Anderson 104bc, Colin Bates 104tr, Mirella Coppola di Canzano (c) University of Trieste 104cla, Prof MD Guiry 103clb, 104, Razy Hoffman 103bc, E.M.Tronchin & O.De Clerck 104crb; **Ardea:** Ian Beames 545, John Cancalosi 394, John Clegg 259, 272, Steve Downer 316, 566, Jean-Paul Ferrero 274, 511, 601, Kenneth W Fink 563, 612, Francois Gohier 612, Joanna Van Gruisen 610, Steve Hopkin 259, 263, 303, Tom & Pat Leeson 589, Ken Lucas 34, 273, 304, 317, Ken Lucas 595, Thomas Marent 596, John Mason 291, Pat Morris 35, 519, 567, 572, Pat Morris 507, 595, Gavin Parsons 270, David Spears (Last Refuge) 265, David Spears / Last Refuge 271, Peter Steyn 519, Andy Teare 589, Duncan Usher 267, M Watson 374, 610, 624; **Australian National Botanic Gardens:** © M.Fagg 169, B. Fuhrer 111cla; **Nick Baker, ecologyasia:** 566; **Jón Baldur Hlíðberg (www.fauna.is):** 327cr, 338br, 339, 341br, 344c, 509; **Bar Aviad:** Bar Aviad 574; **Michael J Barritt:** 511; **Dr. Philippe Béarez / Muséum national d'histoire naturelle, Paris:** 336tr; **Photo Biopix.dk:** N. Sloth 105, 105, 113, 115, 261, 265, 267, 271, 273, 275, 279, 285, 286, 288, 294, 302; **Biosphoto:** Jany Sauvanet 539; **Ashley M. Bradford:** 293cl; **(c) Brent Huffman / Ultimate Ungulate Images:** Brent Huffman 610, 621; **David Bygott:** 35, 521; **Ramon Campos:** 510; **David Cappaert:** 284tc; **CDC:** Courtesy of Larry Stauffer, Oregon State Public Health Laboratory 93bc, Dr. Richard Facklam 93cla, Janice Haney Carr 32cr, 93tc, Segrid McAllister 93cra; **Tyler Christensen:** 302; **Josep Clotas:** 328; **Patrick Coin:** Patrick Coin 263; **Niall Corbet:** 538; **caronsteelephotography.com:** © 6–7; **Corbis:** 13, 22, Theo Allofs 19, 122, 408, Alloy 12, Steve Austin 122, Hinrich Baesemann 424, Barrett & MacKay / All Canada Photos 29, 31, E. & P. Bauer 471, Tom Bean 14, Annie Griffiths Belt 432, Biodisc 33, 238, Biodisc / Visuals Unlimited 286, Jonathan Blair 38, Tom Brakefield 21, 24, 29, Frank Burek 19, Janice Carr 90, W. Cody 19, 107, 118, Brandon D. Cole 321, Richard Cummins 20, Tim Davis 31, Renee DeMartin 24, Dennis Kunkel Microscopy, Inc / Visuals Unlimited 33, 100, Dennis Kunkel Microscopy, Inc. 33, 93, DLILLC 24, 31, 416, 442, Pat Doyle 26, Wim van Egmond 98, Ric Ergenbright 13, Ron Erwin 24, Eurasia Press / Steven Vidler 411, Neil Farrin / JAI 27, Andre Fatras 26, Natalie Fobes 211, Patricia Fogden 354, Christopher Talbot Frank 16, Stephen Frink 350, Jack Goldfarb / Design Pics 19, C. Goldsmith / BSIP 22, Mike Grandmaison 118, Franck Guiziou / Hemis 19, Don Hammond / Design Pics 19, Martin Harvey / Gallo Images 19, 31, Helmut Heintges 24, Pierre Jacques / Hemis 19, Peter Johnson 18, 456, Don Johnston / All Canada Photos 264, Mike Jones 408, Wolfgang Kaehler 18, 26, 238, Karen Kasmauski 27, Steven Kazlowski / Science Faction 16, Layne Kennedy 38, Antonio Lacerda / EPA 15, Frans Lanting 14, 18, 19, 23, 31, 249, 250, 376, 425, 460, 507, Frederic Larson / San Francisco Chronicle 20, Lester Lefkowitz 18, Charles & Josette Lenars 21, Library of Congress - digital ve / Science Faction 28, Wayne Lynch / All Canada Photos 528, Bob Marsh / Papilio 212, Chris Mattison 374, Joe McDonald 354, 533, Momatiuk / Eastcott 31, moodboard 16, 323, 375, Sally A. Morgan 25, Werner H. Mueller 19, David Muench 108, NASA 13, David A. Northcott 389, Owaki - Kulla 15, 19, William Perlman 32, Photolibrary 30, Patrick Pleau / EPA 19, Louie Psihoyos / Science Faction 16, Ivan Quintero / EPA 20, Radius Images 107, 112, Lew Robertson 19, Jeffrey Rotman 19, 332, Kevin Schafer 27, David Scharf / Science Faction 28, Dr. Peter Siver 89, 91, Paul Souders 13, 19, 24, Keren Su 18, Glyn Thomas / moodboard 19, Steve & Ann Toon / Robert Harding World Imagery 415, Craig Tuttle 323, 409, Jeff Vanuga 22, Visuals Unlimited 14, 33, 92, 98, 100, Kennan Ward 13, Michele Westmorland 18, Stuart Westmorland 507, Ralph White 90, Norbert Wu 325, 332, Norbert Wu / Science Faction 320, 323, Yu Xiangquan / Xinhua Press 21, Robert Yin 274, Robert Yinn 322, Frank Young 239, Frank Young / Papilio 211; **Alan Couch:** 511; **David Cowles:** David Cowles at http: / / rosario.wallawalla.edu / inverts 258; **Dick Haaksma:** 111ca; **Greg Kenicer:** 111br; **Whitney Cranshaw:** 290cl; **Alan Cressler:** 262; **Craig Jackson, PhD:** 519cr; **CSIRO:** 336cra; **Matt Carter:** 481bl; **Michael J Cuomo:** www.phsource.us 258; **Ignacio De la Riva:** 365c; **Frank Steinmann:** 362tr; **Dr. Frances Dipper:** 252, 253, 253cla; **Jane K. Dolven:** 98bc; **Stepan Koval:** 111cra; **Dorling Kindersley:** Andy and Gill Swash 409cb/Hoatzin, Blackpool Zoo, Lancashire, UK 582–583 (all images), Centre for Wildlife Gardening / London Wildlife Trust 172bc, 174–175 (all images), Colchester Zoo 522–523 (all images), Demetrio Carrasco / Courtesy of Huascaran National Park 145, Frank Greenaway / Natural History Museum, London 291cr, George Lin 409bl, Greg Dean / Yvonne Dean 409cb, 409cb/Cuckoo, Hanne Eriksen / Jens Eriksen 409crb, Jan-Michael Breider 409ca, Natural History Museum, London 284, Roger Charlwood / Liz Charlwood 409c, Roger Tidman 409clb; **Dreamstime.com:** 614, Allnaturalbeth 350cr, Amskad 303, Anolis01 549bl, Amwu 393tr, Anton Zhuravkov 618cb, Antos777 343crb, John Anderson 348, Argestes 162, Michael Blajenov 611, Mikhail Blajenov 619, Steve Byland 534, Bibek Basumatary 149cr, Bluehand 349cb, Bonita Chessier 548, Musat Christian 556, Mzedig 618clb, Clickit 613, Colette6 611, Chonticha Wat 538cla, Ambrogio Corralloni 532, Cosmln 302, David Havel 474–475c, Davthy 547, Dbmz 301, Destinyvispro 621, Docbombay 534, Edurivero 612, Henketv 26–27t, Katharina Notarianni 532tl, Stefan Ekernas 557, Stefan Ekernas 507, Michael Flippo 615, Joao Estevao Freitas 263, Geddy 272, Eric Geveart 560, Daniel Gilbey 618, Katerynakon 96cb, Maksum Gorpenyuk 611, Jeff Grabert 597, Morten Hilmer 584, Iorboaz 610, Eric Isselee 35, 529, 533, 535, 542, 614, 615, 617, Industryandtravel 385br, Isselee 536, 577br, Jontimmer 612, Jemini Joseph 611, Juliakedo 618, Valery Kraynov 33, 162, Adam Larsen 509, Sonya Lunsford 611, Stephen Meese 609, Milosluz 263, Jason Mintzer 532, Mlane 170, Nina Morozova 133, 148, Derrick Neill 532, Duncan Noakes 619, outdoorsman 394cb, 591, Pancaketom 569, Pipa100 339clb, Planetfelicity 328cl, 339crb, Natalia Pavlova 581, Shane Myers 323bl, 374–375, Susan Pettitt 603, Xiaobin Qiu 580, Rajahs 584, Laurent Renault 548, Derek Rogers 584, Dmitry Rukhlenko 614, Sdecoret 22bl, Steven Russell Smith Photos 569, Ryszard 303, Benjamin Schalkwijk 556, Olga Sharan 609, Paul Shneider 611, Sloth92 555, 556, Smellme 547, 620, Tatiana Belova 345ca, Nico Smit 596, 617, Nickolay Stanev 617, Tampatra1 12–13c, Vladimirdavydov 294, Oleg Vusovich 585, Leigh Warner 595, Worldfoto 563, Judy Worley 602, Zaznoba 555; **Shane Farrell:** 264cr, 284br; **Carol Fenwick (www.carolscornwall.com):** 104br; **Hernan Fernandez:** 510; **Flickr.com:** Ana Cotta 549, Pat Gaines 586, Sonnia Hill 152, Barry Hodges 131, Emilio Esteban Infantes 131, Marj Kibby 136, Kate Knight 170, Ron Kube, Calgary, Alberta, Canada 587, John Leverton 603, John Merriman 170, Moonmoths 297br, Marcio Motta MSc. Biologist of Maracaja Institute for Mammalian Conservation 597, Jerry R. Oldenettel 161, Jennifer Richmond 159; **Florida Museum of Natural History:** Dr. Arthur Anker 518; **Fotolia:** poco_bw 337c; **FLPA:** 30, Nicholas and Sherry Lu Aldridge 105, Ingo Arndt / Minden Pictures 211, 245, 320, Fred Bavendam 313, 328, Fred Bavendam / Minden Pictures 272, 275, 313, 318, 319, Stephen Belcher / Minden